DIGITAL CIRCUITS AND LOGIC DESIGN

Dr. SANJAY SHARMA
B.E., M.Tech, Ph.D, MIETE
Professor and Head
Department of Electronics & Communication Engg.
Krishna Institute of Engg. & Technology,
Ghaziabad, India

Edited By

DEEKSHA SHARMA

&

V. SHARMA

S.K. KATARIA & SONS®
Publisher of Engineering Books
4885/109, Prakash Mahal, Dr. Subhash Bhargav Lane,
Opposite Delhi Medical Association, Daryaganj, New Delhi–110002 (INDIA)
Phone: +91-11-43551243; *Mobile:* +91-9871775858, 7678425209
e-mail: katariabook@gmail.com, katariabooks@yahoo.com
Website: www.skkatariaandsons.com
(e-books, subscriptions and bulk buying also available on our website)

Published by:

S.K. Kataria & Sons®
4885/109, Prakash Mahal, Dr. Subhash Bhargav Lane,
Opposite Delhi Medical Association, Daryaganj, New Delhi-110002 (INDIA)
Phone: +91-11-43551243, *Mobile:* +91-9871775858, 7678425209
e-mail: katariabook@gmail.com, katariabooks@yahoo.com
Website: www.skkatariaandsons.com

First Edition : August, 2009; Fourth Edition: July, 2012; Reprint: 2025

Price: ₹ 695/-

Laser Typesetting by : Plus Computers, Delhi *Printed at :* New A.S. Printing Press, Delhi

Preface to the Fourth Edition...

I feel happy in presenting the Fourth Revised edition of this book to the students. In fact, the favourable and warm reception which the first, second and third editions have earned, has been a matter of great satisfaction for me. The author wishes to express his deep gratitude to the large number of readers who have used it and a particular to those of them, who sent valuable suggestions for the improvements of the book.

In this addition, the author has made sincere efforts to make the book up-to-date.

The present edition has been thoroughly revised and a lot of useful material has been added. To make it more useful, several solved examples and problems have been added from the point of view of Examination. The mistakes and misprints, which had crept in last edition, have been eliminated in this edition. The author hopes that all these features will make the book even more useful and attractive to the readers

Any errors, misprints and suggestions for the improvements of the book, brought to my notice would be gratefully and thankfully acknowledged and incorporated in the next edition.

Dr. Sanjay Sharma

July, 2012

Preface to the First Edition...

Digital Circuits and Logic Design is a compulsory subject for the engineering students from all the disciplines. This book is primarily intended to serve as a textbook in accordance with the syllabus of Digital Circuits and Logic Design offered by PTU, Jalandhar.

The objective of the book is to provide a clear explanation of the operation of all important concepts of Digital Circuits and Logic Design in general use today.

This book has been written in a very simple and lucid manner. Every effort has been made to make the treatments simple and comprehensive. Throughout the book, the stress has been given on fundamental concepts through illustrative examples. Neat and Clear diagrams have been used for explanation.

Summary of each chapter is provided to help the students to grasp the subject clearly. A large number of self explanatory accurate diagrams and tables have been used to supplement the text. Exercises and problems are also provided. These have been selected from examination papers of various Indian Universities. This will help the students to prepare accordingly.

This book is also intended as a textbook for the use in Electronic and Communication courses in polytechnics, IETE, AMIE and for various Engg. examinations. Thus, primarily designated as an undergraduate textbook for students of science and technology, this study will provide balanced coverage of digital circuits. It also serves as a quick reference for practising professionals.

The author hopes that with these unique features the book will fulfill the genuine requirements of the student community. Lastly, much originality cannot be claimed in a book of this kind. The author takes this opportunity to place on record his indebtedness to the large number of books and journals that he has freely consulted in the preparation of this book.

Dr. Sanjay Sharma
August, 2009

Acknowledgement...

I take this opportunity to express my gratitude and thanks to the respected Director General of KIET, **REAR ADMIRAL Dr. AJAY SHARMA** for his valuable technical suggestions and constant encouragement, without which this book would not have come into existence.

I am specially grateful to **Dr. S. NASEEM AHMAD**, Professor and Head, Department of Electronics and Communication Engg., Jamia Millia Islamia, New Delhi for his time-to-time, much needed valuable guidance.

I would like to express my sincere thanks to **Mr. Vibhav Sachan** (Associate Professor, E&C, KIET), and **Mr. Manish Gupta** (Lecturer, E&C, MIT, Modinagar) for constant inspiration and encouragement.

Special thanks to **Dr. B.P. Singh** (Former Professor, Electronics & Comm. Engg., MMMEC, Gorakhpur), **Dr. R.K. Chauhan** (Professor, Electronics & Comm. Engg., MMMEC, Gorakhpur), **Dr. G.S. Tripathi** (Electronics & Comm. Engg., MMMEC, Gorakhpur), **Dr. Arun Kumar** (Former Professor, Electronics & Comm. Engg., MMMEC, Gorakhpur) and **Dr. C.B. Tripathi** (Professor Electronics & Comm. Engg., MMMEC Gorakhpur) and **Dr. Vipin Kumar** (H.O.D. Physics, KIET), and **Mr. Manish Gupta** (Lecturer, MIT, Modinagar) for their guidance and valuable suggestions.

I am grateful to my younger brothers **Rajeev** and **Rakesh,** daughter **Ritika**, for inspiring me for this project.

I deeply express my heartful thanks to the publishers **S.K. Kataria & Sons** for publishing this book in such a beautiful get-up and well in time.

Lastly, I am also grateful to **Mr. Manoj Tyagi** and his staff from **Plus Computer's** for their hard work.

Dr. Sanjay Sharma

DEDICATED
TO
MY LOVING SON
MRIDUL

Contents...

9. MEMORY DEVICES 474–504

PART—A

1. **Number Systems:** Binary, Octal, Decimal, Hexadecimal. Number base conversions, 1's, 2's, rth's complements, signed Binary numbers, Binary Airthmetic, Binary codes; Weighted BCD, Gray code, Excess-3 code, ASCII–conversion from one code to another.
2. **Boolean Algebra:** Boolean postulates and laws – De-Morgan's Theorem, Principle of Duality, Boolean expression – Boolean function, Minimization of Boolean expressions – Sum of Products (SOP), Product of Sums (POS), Minterm, Maxterm, Canonical forms, Conversion between canonical forms, Karnaugh map Minimization, Quine-McCluskey method – Don't care conditions.
3. **Logic GATES:** AND, OR, NOT, NAND, NOR, Exclusive-OR and Exclusive-NOR, Implementations of Logic functions using gates, NAND-NOR implementations, study of logic families like RTL, DTL, DCTL, TTL, MOS, CMOS, ECL and their characteristics.
4. **Combinational Circuits:** Design procedure – Adders, Subtractors, Serial adder/ Subtractor, Parallel adder/encoder/decoder, parity checker, code converters, Implementaion of combinational logic using MUX.

Above syllabus has been covered in following chapters:

1

NUMBER SYSTEMS

Syllabus

Number Systems: Binary, Octal, Decimal, Hexadecimal. Number base conversions, 1's, 2's, rth's complements, signed Binary numbers, Binary Airthmetic, Binary codes; Weighted BCD, Gray code, Excess-3 code, ASCII–conversion from one code to another.

1.1 INTRODUCTION TO DIGITAL SYSTEMS

In modern world of Electronics, the term 'digital' is usually associated with a computer. This is due to the fact that the term 'digital' is derived from the way the computers perform operations, by counting digits. Nowadays, digital concepts are usually applied to various problems that could only be solved by analog methods several years ago. Fast and reliable solutions using digital techniques proved the tremendous power and usefulness of digital systems. In modern times, digital systems find applications in a wide variety of industrial and consumer products such as automated industrial machinery, pocket calculators, microprocessors, digital computers, digital watches, TV games, signal processing and so on. Infact, the areas of application of digital systems have been increasing every day.

Also, the digital computer can be programmed to perform different tasks. The users are allowed to specify and change the program or the data according to the specific requirements. Because of this flexibility, general purpose digital computers can perform a variety of information processing tasks that range over a very wide spectrum of applications.

For many years, the main application of digital electronics was only in the computer systems. But today, the digital electronics is used in many other application such as:

Chapter Outline

We make a living by what we get, we make a life by what we give. – **Sir Winston Churchill.**

1. Television
2. Communication systems
3. Radar, navigation and guidance systems
4. Military systems
5. Medical equipments
6. Industrial process control
7. Consumer electronics

1.2. INTRODUCTION TO ANALOG AND DIGITAL SIGNALS AND SYSTEMS*

In day-to-day life, we come across various types of signals and systems.

The signals may be broadly classified into following two categories :

(i) Analog signals
(ii) Digital signals

(i) Analog Signals

Analog signals are the signals which may have infinite number of different magnitudes or values. They vary continuously with time. Sine wave, triangular wave etc. are few examples of analog signals.

(ii) Digital Signals

A signal is known as a digital signal if it has only a finite number of predetermined distinct magnitudes. Depending upon the number of distinct magnitudes, the digital signals may be classified as under :

S.No.	Number of distinct magnitudes	Type of digital signal
1.	2	Binary
2.	8	Octal
3.	16	Hexadecimal

Actually, the digital signals are discrete time signals, i.e., they are not continuous with time as shown in figure 1.1.

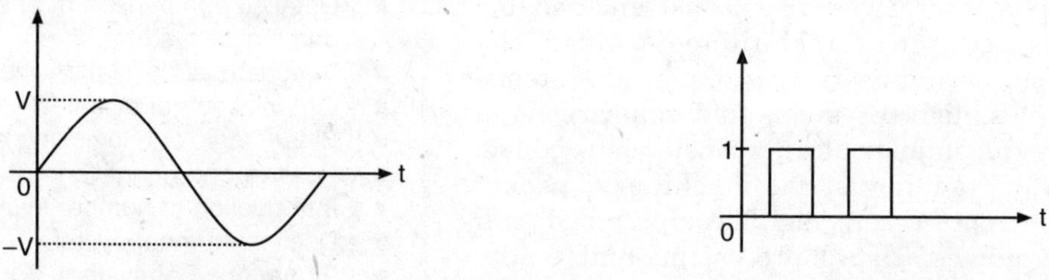

(a) Analog signal (b) Digital signal

Fig. 1.1.

1.2.1 Performance Comparison of Digital and Analog Signals

S.No.	Parameter of comparison	Analog signals	Digital signals
1.	Number of values	Infinite	Finite (2, 8, 16 etc.)
2.	Nature of signals	Continuous-time	Discrete-time
3.	Sources of signals	Signals generators, transducers etc.	Computers, A to D converters etc.
4.	Examples	Sine wave, triangular wave	Binary signal

* Analog signals vary gradually and continuously, while digital signals produce discrete voltage levels commonly referred to as HIGH and LOW.

1.3. ANALOG SYSTEMS

The systems which process the analog signals are known as the analog systems. The analog systems, and analog measurement systems have been used for a long time. There are some certain advantages of these systems, but more than the advantages, there are some drawbacks.

DO YOU KNOW?

The two basic ways of representing the numerical value of physical quantities are analog (continuous) and digital (discrete).

Examples

Few examples of analog systems may be listed as under:

(i) Filters circuits
(ii) Amplifier circuits
(iii) Signal generators
(iv) Motor speed controllers

It is important to note that the displays used for the analog systems are the analog meters (having moving pointers).

1.3.1. Drawbacks of Analog Systems

(i) Analog systems are less accurate.
(ii) Analysis of analog systems require time-domain analysis which requires very sound mathematical background.
(iii) Analog systems are usually affected more by the undesired electrical disturbance called noise.
(iv) Once the noise contaminates an analog signal, it is impossible to separate the signal from noise. The signal contaminated by noise leads to degradation of quality of output.
(v) The performance of analog systems degrades because of component ageing.
(vi) Their performance changes with variation in temperature.
(vii) Analog systems have less reliability.
(viii) Analog systems have less versatility since no software is included.
(ix) Because of the use of analog meters, it is quite difficult to take accurate reading.
(x) Storage and processing of data is not possible since they do not have memory.
(xi) In analog systems, computerized control cannot be exercised.

1.3.2 Performance Comparison of Analog and Digital Systems

S. No.	Parameter of comparison	Analog systems	Digital systems
1.	Type of signals processed	Analog signals	Digital signals
2.	Type of display	Analog meters	Digital displays using LED and LCD
3.	Accuracy of system	Small	High
4.	Design complexity	Difficult to design	Easier to design
5.	Memory attached	No memory	They have memory
6.	Storage of information	Not possible	Possible
7.	Noise effect	High	Small
8.	Versatility	Small	High
9.	Distortion	High	Small
10.	Effect of temperature and ageing on performance	High	Small
11.	Communication between systems	Not easy	Easy
12.	Few Examples	Filters, amplifiers, power supplies, signal generators	Counters, registers, micro-processors, computers

1.4 SALIENT FEATURES OF DIGITAL SYSTEMS

As a matter of fact, digital systems have following salient features (advantages) over analog systems :

(i) Digital systems manipulate discrete elements of information.

(ii) Discrete elements are nothing but the *digits* such as 10 decimal digits, 26 letters of alphabets and so on.

(iii) Digital systems make use of physical quantities called *signals* to represent discrete elements.

(iv) In digital systems, the signals have two discrete values and are therefore said to be binary.

(v) A signal in digital system represents one binary digit, called a *bit*. The bit has a value either 0 or 1.

(vi) Discrete elements of information are represented with group of bits called *binary codes*. As an example, the decimal digits 0 through 9 are represented in a digital system with 4-bit code called *BCD (Binary Coded Decimal) code*.

(vii) The digital systems like a digital computer is a programmable device which may be programmed to perform a variety of tasks.

(viii) The digital systems like a digital computer is an interconnection of digital modules. The major units/modules of digital computer are a central processing unit, memory unit and input-output unit. The memory unit stores programs and data. The central processing unit (CPU) performs arithmetic and logical operations on data stored in the memory specified by the program. Further, input device such as keyboards is used to store program and data into the memory and display device such as printer, monitor is used to print/display the result of operation.

KARL FRIEDRICH GAUSS

In 1799 the *German mathematician Karl Friedrich Gauss*, at the ripe age of 22, proved the fundamental theorem of algebra, namely that every algebraic equation in one unknown has a root in the form of a complex number. He showed that every equation of the *n*th order has exactly n solutions (roots), no more and no less. Gauss was also one of the first to give a coherent account of complex numbers and to interpret them as points in a complex plane.

Thus, based upon discussion, we may define digital system as under :

Definition : A digital system may be defined as an interconnection of digital modules and it is a system that manipulates discrete elements of information that is represented internally in the binary form.

1.5. DETAILED COMPARISON OF ANALOG AND DIGITAL SYSTEMS

Analog systems process information that varies continuously, i.e., they process time-varying signals that may take on any value across a continuous range of voltage, current or any physical parameter. On the other hand, the digital systems use discrete quantities to represent information. Discrete means distinct or separated as opposed to continuous or connected. Figure 1.2 shows the difference between two signals : continuous and discrete.

The digital systems represent levels of discrete signals, usually, in a binary number system. The binary number system has only two characters : 0 and 1. The numbers formed by these two characters are used to represent the discrete levels of digital signal.

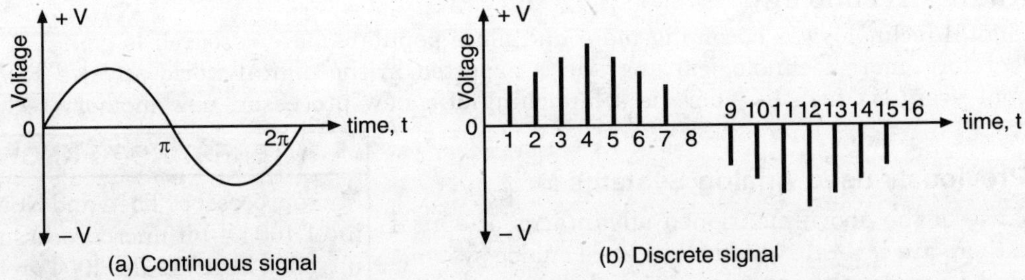

Fig. 1.2. Illustration of continuous-time and discrete-time signals.

Nowadays, digital systems have justified their importance and advantages over the analog systems.

1.5.1. Some Special Advantages Offered by Digital Systems

1. Easy to Design

Digital systems are quite easy to design than analog systems, because digital design involves logic design which does not require special maths skills and its behaviour may be visualized part-by-part.

To visualize the behaviour of analog systems is quite difficult. Actually, it requires the special insights about the analog components such as capacitors, inductors and transistors used in the analog systems.

2. Reliability and Reproducibility of Results

The output of analog systems varies with the variation in temperature, power-supply voltage, component-aging, and several other factors. Therefore, it is difficult to reproduce the same result each time with the same sets of inputs and circuit components. However, this is not the case with digital systems. Digital systems always reproduce exactly same results with same set of inputs and circuit components.

3. Flexibility

Digital systems are more flexible to design since their design involves a set of logical steps and have various discrete and integrated options available to the user.

4. Functionality

It is quite easy to provide additional functionality in digital systems than analog systems. As an example, in digital systems, it is quite easy to add password facility. However, in analog systems, it is a difficult job.

5. Programmability

Nowadays, digital design is carried out by writing programs, in *hardware description language* (HDL). These languages allow us to stimulate and test the performance of digital circuits. This feature is quite useful in designing critical digital systems for various applications.

6. Device Speed

Today's digital devices are very fast and they can produce 500 millions or more results since their operating speed is smaller than 2 nanoseconds.

7. Economy in Manufacturing

Digital circuits can easily be integrated into a single chip and can be produced in large quantity to have smaller cost compared to discrete circuits.

DO YOU KNOW?

Most quantities in the real world are analog, but digital techniques are generally superior to analog techniques, and most of the predicted advances will be in the digital realm.

It is always the simple that produces the marvelous. — **Anonymous**

8. Upgrading Technology

As digital technology is becoming more and more popular, more research is going on in this field. Therefore, more technological upgrade is expected in the digital world and we know that after about every six months, there is a launching of a new processor, new memory technology etc.

1.5.2. Previously used Analog Systems

Because of the above mentioned advantages of digital systems, there are several systems that were analog systems earlier, but in modern times, they are digital systems. Let us consider few of them.

> **INTERESTING TO KNOW?**
>
> Microprocessors Then and Now. The intel 4004 4-bit microprocessor was unveiled in 1971. The 4004 processor contained about 2300 transistors. A recent chip, the mobile Intel Pentium 4 Processor, includes 55 million transistors and is available at speeds to 2 GHz (gigahertz).

1. Cameras

The majority cameras still make use of silver-halide film to record various images. However, the continuous increase in density of digital memory chips has encouraged the development of digital cameras that record a picture in a graphical format. The digital cameras provide additional facility that they can store thousands of pictures and has an ability to process them as well.

2. Audio Recording

The audio recording upon vinyl or magnetic tape would become no where in very few years. Since, digital compact disc (CDs) are now commonly used for audio recording. The storing capacity of one CD is as huge as six billion bits and its sound reproducing capacity is far better than magnetic tapes.

3. Video Recording

Like audio recording, nowadays, the digital recording is achieved using digital versatile disc (DVD). Actually, a DVD can store video information in a highly compressed digital format known as MPEG-2. Infact, the capacity of a single layer, single-side DVD is about 35 billion bits, which is sufficient for about 2 hours of high-quality video, and a two-layer, double-sided disc has capacity four times that of a single-side.

4. Telephone System

Most of the telephone exchanges now make use of digital signals in place of analog signals.

5. Special Effects in Advertisement and Movies

The digital processing and creation have allowed to add special effects in advertisement and movies for various applications.

1.5.3. Examples of Digital Systems*

Some of the examples of digital systems used in day-to-day life may be listed as under :

1. In communication systems
2. Business transactions
3. Traffic control
4. Medical treatment
5. Industrial applications
6. Digital telephones
7. Digital TV
8. Digital cameras
9. Weather monitoring
10. Digital computers

1.6. VARIOUS BASIC DIGITAL DEVICES

1. Logic Gates

Logic gates are the basic elements that make up a digital system. An electronic gate is a

* Most modern electronics equipment contains both analog and digital circuitry.

circuit which is able to operate on a number of binary inputs in order to perform a particular logical function. The types of gates available are the NOT, AND, OR, NAND, NOR, exclusive-OR and the exclusive-NOR. Except for the exclusive-NOR gate, they are available in monolithic integrated circuit form. The gate is a digital circuit with one or more inputs but only one output. By connecting the different gates in different ways, we can build various circuits that can perform arithmetic and other functions associated with the human brain since they simulate mental processes.

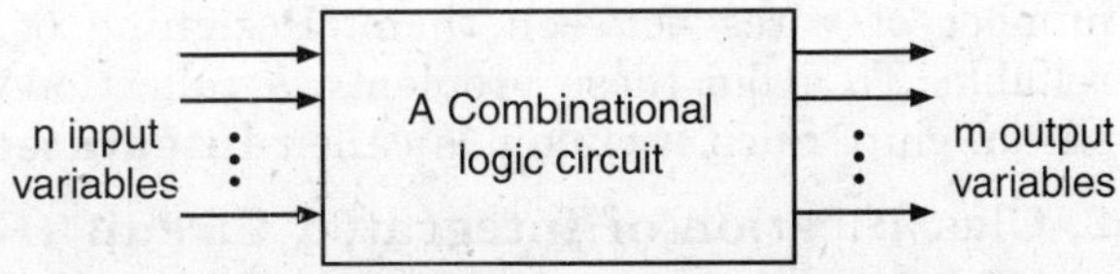

Fig. 1.3. A Block diagram of a combinational circuit.

2. Combinational Circuits

When logic gates are connected together to produce a specified output for certain specified combinations of input variables, with no storage involved, the resulting circuit is known as **combinational logic.** In combinational logic, the output variables are at all times dependent upon the combination of input variables.

Basically, a combinational circuit consists of input variables, logic gates, and output variables. The logic gates accept signals from the input variables and generate output signals. In fact, this process transforms binary information from the given input data to the required output data. Figure 1.3 depicts a block diagram of a combinational circuit. As illustrated in figure 1.3, a combinational circuit accepts *n*-input, binary variables and generates m output variables depending upon the logical combination of gates.

3. Sequential Circuits

There are several applications in which digital outputs are required to be generated in accordance with the sequence in which the input signals are received. This requirement cannot be satisfied using a combinational logic system. These applications require outputs to be generated that are not only dependent on the present input conditions but they also depend upon the past history of these inputs. The past history is provided by feedback from the output back to the input.

Figure 1.4 shows a block diagram of sequential circuit. As illustrated in figure 1.4, memory elements are connected to the combinational circuit as a feedback path.

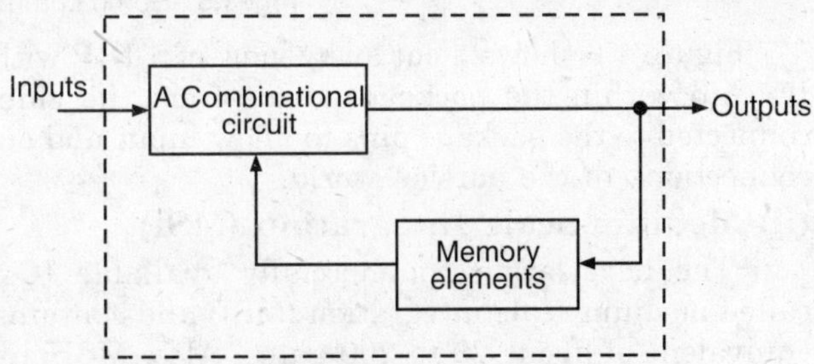

Fig. 1.4. Block diagram of sequential circuit.

To have a sequential system, we must know what has happened in the past. Therefore, we must have a storage device to retain this information until we are ready to use it. The basic unit for this storage is the flip-flop (abbreviated as FF). It is a device with two stable states, *i.e.,* its output is either 0 (logic 0) or +5 V dc (logic 1). A flip-flop maintains its output state (either 0 or 1) until directed by an input signal to change its state. This means that it can store 1-bit information. Therefore, it is also called as 1-bit memory unit.

1.7. ORGANIZATION OF DIGITAL ICs

1. Need of Integrated Circuit (IC)

The traditional method of combinational circuit design involves simplification and realization of logic functions using logic gates. This method of designing combinational circuit is effective for small circuit. When circuit complexity is more, more number of gates are required with more number of wires between them. Designing of such circuits may be time-consuming and less reliable. To avoid these problems, a collection of one or more gates are fabricated on a single silicon chip. Such a circuit is called **Integrated Circuit** (IC).

2. Classification of Integrated Circuit (IC)

The integrated circuits may be classified using following four levels of integration :

(i) Small-Scale Integration (SSI)
(ii) Medium-Scale Integration (MSI)
(iii) Large-Scale Integration (LSI), and
(iv) Very-Large-Scale Integration (VLSI)

(i) Small-Scale Integration (SSI)

The simplest type of digital ICs are placed in the small-scale integration (SSI) category. These ICs have up to 20 equivalent gate circuits on a single chip. These include basic gate functions and flip-flops. SSI ICs are fabricated in one of the two main configurations :

(a) Dual-in-line package (DIP)
(b) Flat pack

Figure 1.5 shows these packages for 14-pin and 16-pin ICs.

(a) Dual-in-line package (DIP) (b) Flat pack

Fig. 1.5. Some common IC packages.

Figure 1.6 shows a cut away view of a DIP with the IC chip within the package. Leads from the chip are connected to the package pins to allow input and output connections to the outside world.

Fig. 1.6. A Cut away view of a dual-in-line 14-pin package.

(ii) Medium-Scale Integration (MSI)

The next larger commercially available ICs are called medium-scale integration (MSI) and contains the equivalent of about 20 to 200 gates. Also, MSI circuits include the more complex logic functions such as encoders, decoders, counters, multiplexers, demultiplexers, registers, arithmetic circuits, small memories and others. Figure 1.7 depicts a 24-pin DIP and flat pack and a 28-pin chip carrier in which MSI functions are packaged.

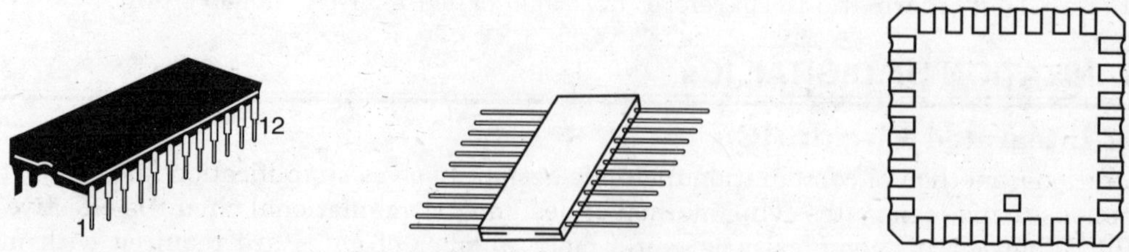

(a) 24-pin Dual-in-Line (b) 24-pin Flat Pack (c) 28-pin ceramic chip carrier package

Fig. 1.7.

(iii) Large-Scale Integration (LSI)

Large-scale integration (LSI) ICs are still bigger and contain the equivalent of 200 to 200,000 gates or more. LSI ICs include memories, microprocessors, programmable logic devices and customized devices. As a matter of fact, most of the LSI ICs are usually fabricated in dual-in-line packages.

(iv) Very Large-Scale Integration (VLSI)

Beyond large-scale integration, there is also a category which is known as very large-scale integration (VLSI). The complexity in VLSI is usually stated in terms of transistor count than gate count. Any IC with over 1,000,000 transistor is VLSI IC. It includes advance microprocessors, large memories, larger programmable logic devices and customized devices. The digital ICs reduce system cost, size, power and the number of external wire connections improving reliability of the system.*

> **DO YOU KNOW?**
>
> Almost all of the digital circuits used in modern digital systems are integrated circuits (ICs). The wide variety of available logic ICs has made it possible to construct complex digital systems that are smaller and more reliable than their discrete-component counterparts.

1.8. BINARY LOGIC AND LOGIC LEVELS

As a matter of fact, logic is defined as a statement which is true if some condition is satisfied and false if that particular condition is not satisfied. Here, let us briefly discuss few examples of some logical statements.

(i) A bulb turns ON, if we close the switch, otherwise it is OFF.

(ii) You will pass with good marks, if you study well.

(iii) I will miss the bus, if I am late.

Thus, each logical statement has two possible outcomes. Such a logic which has only two possible outcomes is called as the **Binary Logic**. The two possible outcomes of a binary logic are represented by **two voltage levels.**

The true outcome is denoted by a **HIGH** level and called as **Logic 1 level** and the false outcome is denoted by a **LOW** level and it is called as **Logic 0 level.**

Table 1.1 obviously illustrates the concept of logic levels.

Table 1.1 Binary logic and logic levels

S.No.	Statement	Two possible outcomes	Logic levels assigned
1.	A lamp may be	OFF	Low (Logic 0)
		ON	High (Logic 1)
2.	A gate may be	Closed	Low (Logic 0)
		Open	High (Logic 1)
3.	Your examination result may be	Fail	Low (Logic 0)
		Pass	High (Logic 1)

Important Point : The digital circuits are also known as the logic circuits because they work on the basis of some logic.

* Digital circuits have become very popular because of the availablity of low-cost digital ICs. Other advantages of digital circuitry are computer compatibility, memory, ease of use, simplicity of design, accuracy, and stability.

1.9. POSITIVE AND NEGATIVE LOGIC

The digital circuits will have only two distinct output voltage levels.

These two voltage levels will be used to represent the logic 0 and logic 1 outputs of the logic circuit as shown in figure 1.8.

The high and low voltage levels of the output voltage can be assigned to represent logic 0 or logic 1 levels by using following two different types of logic :

(i) Positive logic

(ii) Negative logic

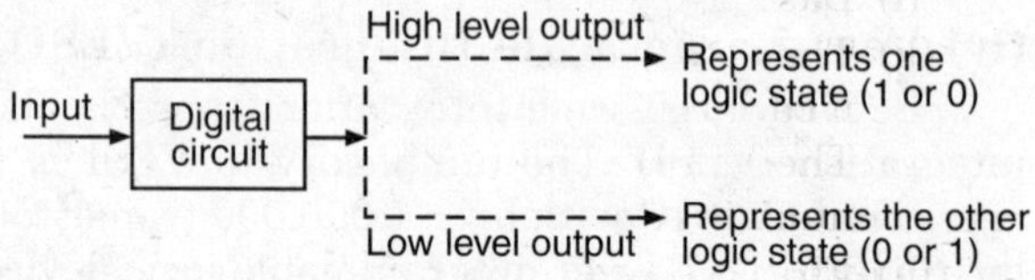

Fig. 1.8.

(i) Positive Logic*

A LOW voltage level represents logic 0 state and a comparatively HIGH output voltage level represents logic 1 state as shown in figure 1.9(a).

As an example, 0 volts represent a logic 0 state and +5 V represents logic 1 state. This is known as positive logic.

Positive Logic :	Logic 0 (LOW) = 0 V Logic 1 (HIGH) = +5 V

Negative Logic

A LOW voltage level represents logic 1 state and a HIGH output voltage level represents logic 0 state as shown in figure 1.9(*b*).

As an example, 0 volts represent a logic 1 state and +5 V represent logic 0 state. This is known as negative logic.

Negative Logic :	Logic 0 (LOW) = +5 V Logic 1 (HIGH) = 0 V

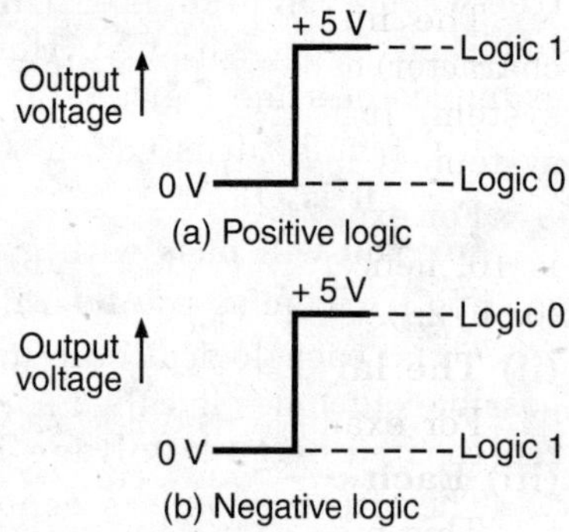

Fig. 1.9.

Important Note : In this chapter, we shall consider only positive logic. Also, we shall assume that the logic 0 level corresponds to 0 Volts and logic 1 level corresponds to +5 V.

1.10. INTRODUCTION TO NUMBER SYSTEMS

1. Definition and Importance

Number system is a basis for counting various items. On hearing the word 'number', all of us immediately think of the familiar decimal number system with its 10 digits : 0, 1, 2, 3, 4, 5, 6, 7, 8 and 9.

As a matter of fact, modern computers communicate and operate with binary numbers which use only the digits 0 and 1. Let us consider decimal number 18. This number is represented in binary as 10010. In the example, if decimal number is considered, we require only two digit to represent the number, whereas if binary number is considered, we require five digits. Therefore, we can say that, when decimal quantities are represented in the binary form, they take more digits. For large decimal numbers people have to deal with very large binary strings and therefore, they do not like working with binary numbers. This fact gave rise to three new number systems : Octal, Hexadecimal and Binary Coded Decimal (BCD). These number systems represent binary

* Logic level indicators may take the form of simple LED and resistor circuits, voltmeters, or logic probes. Light-emitting diode logic level indicators will probably be found on your laboratory equipment.

number in a compressed form. Therefore, these number systems are now widely used to compress long strings of binary numbers.

2. Few Important Points to Define a Number System

To define any number system, we have to specify following aspects :

(i) Base of the number system such as 2, 8, 10 or 16.

(ii) The base decides the total number of digits available in that number system. For example, if base is 2, we have 2-digits (0 and 1), if base is 10, we have 10-digits (0 through 9) and so on. Therefore, total number of digits available in the number system is equal to the base of the number system.

(iii) First digit in the number system is always zero (0) and last digit in the number system is always, base –1. As an example, in decimal system, last digit is 9 (10 – 1).

1.10.1. Few Common Aspects to All Numbering Systems

It will be interesting to know that all the numbering systems have a few common aspects as under :

(i) Base or Radix

The number or values that a digit (one character) can assume is equal to the base of the system. It is also called as the Radix of the system. It is denoted by r.

For example, for a decimal system, the base is 10, hence, every digit can assume 10 values (0, 1, 2, ..., 9).

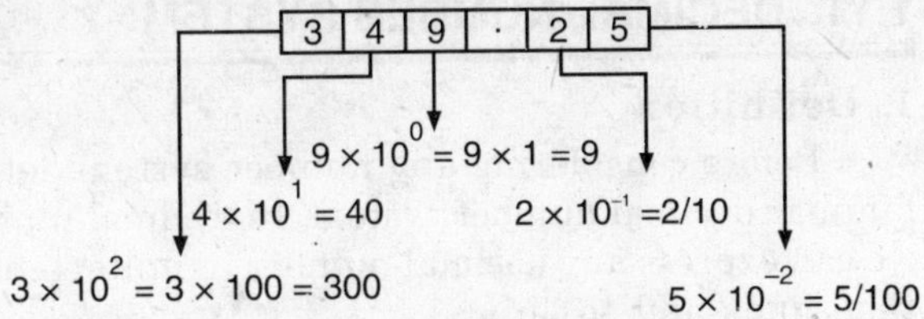

Fig. 1.10. Numbers have positional importance.

(ii) The largest value of a digit is always one less than the base

For example, the largest digit in a decimal system is 9. (One less than the base 10).

(iii) Each digit position (*i.e.,* place) represents a different multiple of base

This means that the numbers have positional importance. For example, let us consider the decimal number $(349.25)_{10}$.

Hence, we can implement a common rule for all the numbering systems as under :

For a general number, we have to multiply each of digit by some power of base (B) or radix as shown in figure 1.11.

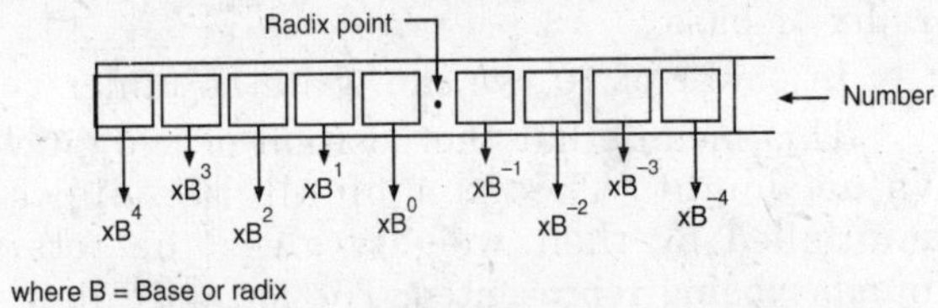

Fig. 1.11. Positional weights of a number

4. Column Numbers

The column number is the number assigned to the digits placed in relation with the radix point. Column numbers to the left of the radix point start with 0 and go up (0, 1, 2, ...) as shown in figure 1.11. Column numbers to the right of the radix point start from –1 and become more and more negative (–1, –2, –3, –4, ...) as shown in figure 1.11.

1.10.2. Various Numbering Systems

Various numbering systems used in practice and their bases are as shown in Table 1.2.

Table 1.2 Various number systems and their bases

S.No.	Name of number system	Base/Radix
1.	Binary	2
2.	Octal	8
3.	Decimal	10
4.	Duodecimal	12
5.	Hexadecimal	16

Table 1.3 illustrates base, 1st digit, last digit and available digits in various number systems.

Table 1.3.

S.No.	Number system	Base	First digit	Last digit	All digits/characters
1.	Binary	2	0	1	0, 1
2.	Octal	8	0	7	0, 1, 2, 3, 4, 5, 6, 7
3.	Decimal	10	0	9	0, 1, 2, 3, 4, 5, 6, 7, 8, 9
4.	Hexadecimal	16	0	F	0, 1, 2, 3, 4, 5, 6, 7, 8, 9, A, B, C, D, E, F

Important Point : In hexadecimal number system, meaning of
A ≡ 10, B ≡ 11, C ≡ 12, D ≡ 13, E ≡ 14 and F ≡ 15.

1.11. DECIMAL NUMBER SYSTEM

(PTU, Jalandhar, Sem. Exam., 2006-07)

1. Definition

Before considering any number system, let us consider familiar decimal number system. In decimal number system, we can express any decimal number in units, tens, hundreds, thousands and so on.

DO YOU KNOW?

Digital or logic circuits operate on voltages that fall in prescribed ranges that represent either a binary 0 or a binary 1.

2. Representation

When we write a decimal number, say, 5678.9, we know that it can be represented as under :

$$5000 + 600 + 70 + 8 + 0.9 = 5678.9$$

The decimal number 5678.9 can also be written as $(5678.9)_{10}$, where the 10 subscript indicates the radix or base.

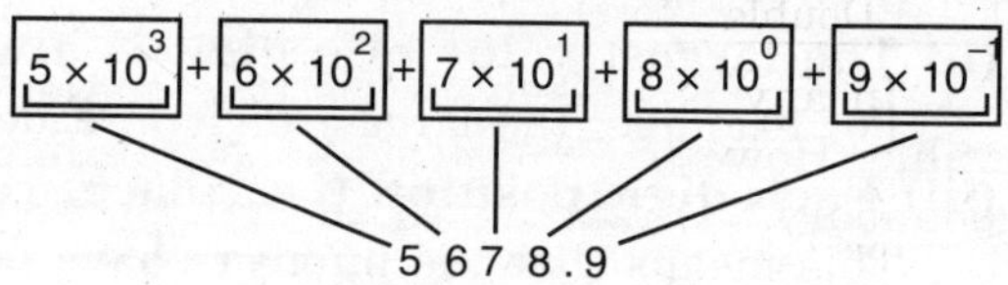

Fig. 1.12

In power of 10, we can write as under :

This means that the position of a digit with reference to the decimal point determines its value/weight. The sum of all the digits multiplied by their weights gives the total number being represented. The leftmost digit, which has the greatest weight is called the **most significant digit** and the rightmost digit, which has the least weight, is called the **least significant digit.** Figure 1.13 shows decimal digit and its weights expressed as a power of 10.

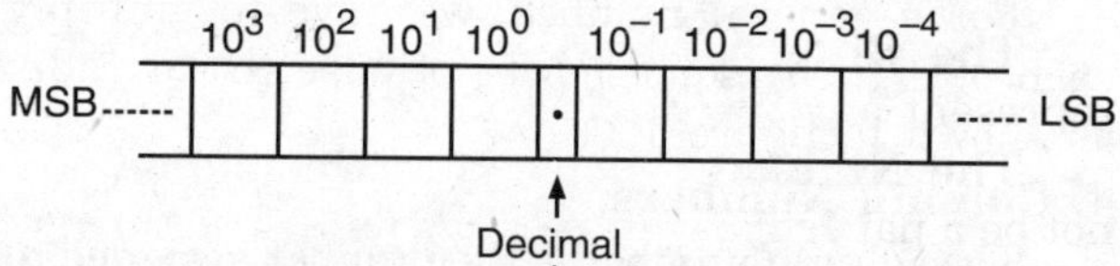

Fig. 1.13. Decimal position values as powers of 10.

EXAMPLE 1.1. Represent binary number 98.72 in power of 10.

Solution: We have $N = 9 \times 10^1 + 8 \times 10^0 + 7 \times 10^{-1} + 2 \times 10^{-2}$.

The digit 9 has a weight of 10, the digit 8 has a weight of 1, the digit 7 has a weight of 1/10 and the digit 2 has a weight of 1/100.

1.12. BINARY NUMBER SYSTEM

1. Definition

We know that decimal system with its ten digits is a base-ten system. Similarly, binary system with its two digits is a base-two system. The two binary digits (bits) are 1 and 0. Like

Everything that is done in the world is done by hope. — **Anonymous**

digital system, in binary system, each binary digit commonly known as bit, has its own value or weight. However, in binary system, weight is expressed as a power of 2, as shown in figure 1.14.

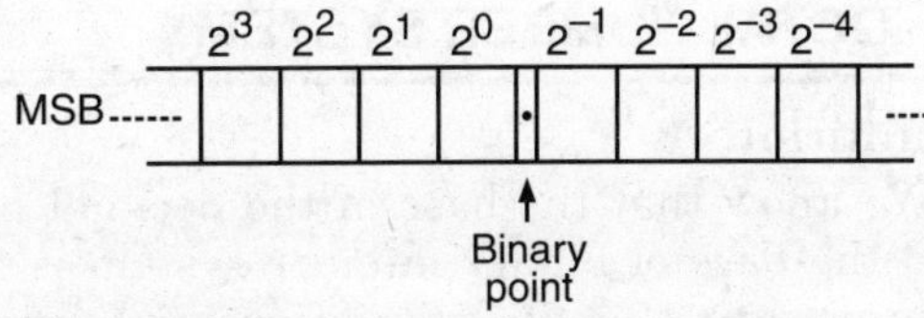

Fig. 1.14. Binary position values as a power of 2.

EXAMPLE 1.2. Represent binary number 1101.101 in power of 2 and find its decimal equivalent.

Solution: Representing given binary number in power of 2, we get,

$$N = 1 \times 2^3 + 1 \times 2^2 + 0 \times 2^1 + 1 \times 2^0 + 1 \times 2^{-1} + 0 \times 2^{-2} + 1 \times 2^{-3}$$
$$= 8 + 4 + 0 + 1 + 0.5 + 0 + 0.125$$

or $N = (13.625)_{10}$ **Ans.**

1.12.1. Binary Number Formats

We typically write binary numbers as a sequence of bits (bits is short for binary digits). We have defined boundaries for these bits. These boundaries are :

Table 1.4 Binary number formats

Name	Size (bits)	Example
Bit	1	1
Nibble	4	0101
Byte	8	0000 0101
Word	16	0000 0000 0000 0101
Double Word	32	0000 0000 0000 0000 0000 0000 0000 0101

In any number base, we may add as many leading zeros as we wish without changing its value. However, we normally add leading zeros to adjust the binary number to a desired size boundary.* For example, we can represent the number five as:

Nibble	0101
Byte	0000 0101
Word	0000 0000 0000 0101

The Bit: The smallest *unit* of data is defined as a single **bit**. With a single bit, we can represent any two distinct items like true or false, on or off, male or female, right or wrong etc.

The Nibble: A nibble is a combination of four bits. It would not be a particularly interesting data structure except for two items: BCD (binary coded decimal) numbers and hexadecimal (base 16) numbers because it takes four bits to represent a single BCD or hexadecimal digit. With a nibble, we can represent up to 16 distinct values. The structure of nibble is as shown in figure 1.15 with b3 acting as MSB and b0 as LSB.

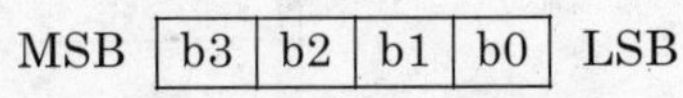

Fig. 1.15. Structure of a nibble

The Byte: A byte is a combination of 8-binary bits. The number of distinct values represented by a byte is 256, ranging from 00000000 to 11111111. This means that

Number of distinct values = $2^N = 2^4 = 16$

Where, N = Number of bits used to represent a number.

The Word: A word is a combination of 16 bits. Hence, it consists of two bytes.

The Double Word: A double word is exactly what its name implies, two words. Therefore, a double word quantity is 32 bits. Naturally, this double word can be divided into a higher order word and a low order word, four bytes, or eight nibbles.

DO YOU KNOW?

The binary number system (0 and 1) is the basic system used in digital technology.

* The place values left on the binary point in binary are 64, 32, 16, 8, 4, 2 and 1.

1.13. OCTAL NUMBER SYSTEM

1. Definition

We know that the base of the decimal number system is 10 because it uses the digits 0 to 9, and the base of binary number system is 2 because it uses digits 0 and 1. The octal number system uses first eight digits of decimal number system : 0, 1, 2, 3, 4, 5, 6 and 7. As it uses 8 digits, its base is 8.

EXAMPLE 1.3. Represent octal number 567 in power of 8 and find its decimal equivalent.

Solution: The given octal number 567 can be represented in power of 8 as under :

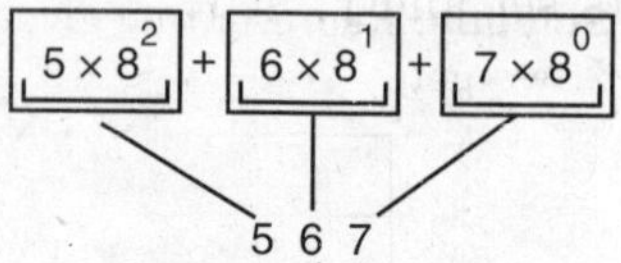

Fig. 1.16.

$$N = 5 \times 64 + 6 \times 8 + 7 \times 1 = 320 + 48 + 7$$

or $$N = (375)_{10} \quad \textbf{Ans.}$$

1.14. HEXADECIMAL NUMBER SYSTEM

(PTU, Jalandhar, Sem. Exam., 2004-05)

1. Definition

The hexadecimal number system has a base of 16 having 16 digits : 0, 1, 2, 3, 4, 5, 6, 7, 8, 9, A, B, C, D, E and F. It is another number system that is particularly useful for human communications with a computer. Although it is somewhat more difficult to interpret than the octal number system, it has become the most popular. Since its base is a power of 2 (2^4), it is easy to convert hexadecimal numbers to binary and vice-versa.

2. Relationship between Decimal, Binary and Hexadecimal Numbers

Table 1.5 shows the relationship between decimal, binary and hexadecimal. Note that each hexadecimal digit represents a group of four binary digits, called nibbles, that are fundamental parts of larger binary words.

Table 1.5. Relationship between decimal, binary and hexadecimal numbers

S. No.	Decimal	Binary	Hexadecimal
1.	0	0000	0
2.	1	0001	1
3.	2	0010	2
4.	3	0011	3
5.	4	0100	4
6.	5	0101	5
7.	6	0110	6
8.	7	0111	7
9.	8	1000	8
10.	9	1001	9
11.	10	1010	*A*
12.	11	1011	*B*
13.	12	1100	*C*
14.	13	1101	*D*
15.	14	1110	*E*
16.	15	1111	*F*

EXAMPLE 1.4. Represent hexadecimal number 3FD in power of 16 and find its decimal equivalent.

Solution: The given hexadecimal number $3FD_{16}$ can be represented in power of 16 as under :

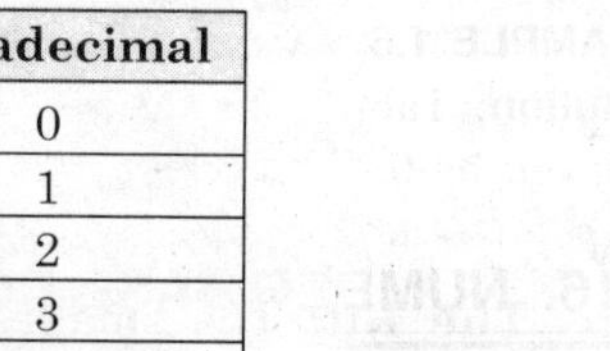

Fig. 1.17.

$$N = 3 \times 256 + F(15) \times 16 + D(13) \times 1 = 768 + 240 + 13$$

or $$N = (1021)_{10} \quad \textbf{Ans.}$$

1.15. COUNTING IN RADIX (BASE) r

In previous articles, we have discussed number systems with radix (base) r equal to 10, 2, 8 and 16. Each number system has r set of characters. For example, in decimal number system, r equals to 10 has 10 characters from 0 to 9, in binary number system r equal to 2 has 2 characters 0 and 1 and so on. In general we can say that, a number represented in radix, r, has r characters in its set and r can be any value. This has been illustrated in Table 1.6.

Table 1.6. Radix and character set

Radix (Base) r	Characters in set
2 (Binary)	0, 1
3	0, 1, 2
4	0, 1, 2, 3
:	
:	
7	0, 1, 2, 3, 4, 5, 6
8 (Octal)	0, 1, 2, 3, 4, 5, 6, 7
:	
:	
10 (Decimal)	0, 1, 2, 3, 4, 5, 6, 7, 8, 9
:	
:	
16 (Hexadecimal)	0, 1, 2, 3, 4, 5, 6, 7, 8, 9, A, B, C, D, E, F

EXAMPLE 1.5. Find the decimal equivalent of $(231.23)_4$.

Solution: We have $N = 2 \times 4^2 + 3 \times 4^1 + 1 \times 4^0 + 2 \times 4^{-1} + 3 \times 4^{-2}$

or $N = 32 + 12 + 1 + 0.5 + 0.1875 = (45.6875)_{10}$. **Ans.**

EXAMPLE 1.6. Count from 0 to 9 in radix 5.

Solution: Table 1.6 indicates that radix 5 has 5 characters. A count sequence from 0 decimal to 9 decimal is given by 0, 1, 2, 3, 4,

1.16. NUMBER BASE CONVERSIONS

1. Definition and Importance

The human beings use decimal number system while computer uses binary number system. Therefore, it is essential to convert decimal number into its equivalent binary while feeding number into the computer and to convert binary number into its decimal equivalent while displaying result of operation to the human beings. However, dealing with a large quantity of binary numbers of many bits is inconvenient for human beings. Therefore, octal and hexadecimal numbers are used as a shorthand means of expressing large binary numbers. But, it is essential to keep in mind that the digital circuits and systems work strictly in binary. Actually, we use octal and hexadecimal only as a convenience for the operators of the system.

> **DO YOU KNOW?**
>
> The two basic ways to transfer digital information are parallel– all bits simultaneously and serial– one bit at a time.

2. Conversions between Decimal, Binary, Octal and Hexadecimal Numbers

Before we discuss conversions between binary, octal and hexadecimal numbers, let us observe the number of digits in several number systems. Table 1.7 shows the decimal, binary, octal and hexadecimal numbers.

Table 1.7. Decimal, binary, octal and hexadecimal numbers

S.No.	Decimal	Binary	Octal	Hexadecimal
1.	0	0000	0	0
2.	1	0001	1	1
3.	2	0010	2	2
4.	3	0011	3	3
5.	4	0100	4	4
6.	5	0101	5	5
7.	6	0110	6	6
8.	7	0111	7	7
9.	8	1000	10	8
10.	9	1001	11	9
11.	10	1010	12	A
12.	11	1011	13	B
13.	12	1100	14	C
14.	13	1101	15	D
15.	14	1110	16	E
16.	15	1111	17	F

1.16.1. Binary to Octal Conversion

We know that the base for octal numbers is 8 and the base for binary numbers is 2. The base for octal number is the third power of the base for binary numbers. Therefore, by grouping 3 digits of binary numbers and then converting each group digit to its octal equivalent, we can convert binary number to its octal equivalent.

EXAMPLE 1.7. Convert $(1\ 1\ 1\ 1\ 0\ 1\ 1\ 0\ 0)_2$ to octal equivalent.

Solution: We have,

111	101	100
7	5	4

Therefore, Octal number = $(754)_8$. **Ans.**

1.16.2. Octal to Binary Conversion

Conversion from octal to binary is a reversal of the process explained in the previous sub-section. Each digit of the octal number is individually converted to its binary equivalent to get octal to binary conversion of the number.

EXAMPLE 1.8. Convert $(634)_8$ to binary.

Solution: We have

6	3	4
110	011	100

Therefore, Binary number = 110 011 100. **Ans.**

EXAMPLE 1.9. Convert $(725.63)_8$ to binary.

Solution: We have,

7	2	5	·	6	3
111	010	101	·	110	011

Therefore, Binary number = 111010101 . 110 011. **Ans.**

1.16.3. Binary to Hexadecimal Conversion

We know that base for hexadecimal numbers is 16 and the base for binary numbers is 2. The base for hexadecimal number is the fourth power of the base for binary numbers. Therefore, by grouping 4 digits of binary numbers and then converting each group digit to its hexadecimal equivalent, we can convert binary number to its hexadecimal equivalent.

EXAMPLE 1.10. Convert $(1101100010011011)_2$ to hexadecimal equivalent.

Solution: We have,

1 1 0 1	1 0 0 0	1 0 0 1	1 0 1 1
D	8	9	B

Therefore, Hexadecimal number = $(D89B)_H$. **Ans.**

1.16.4. Hexadecimal to Binary Conversion

Conversion from hexadecimal to binary is a reversal of the process explained in the previous sub-section. Each digit of the hexadecimal number is individually converted to its binary equivalent to get hexadecimal to binary conversion of the number.

EXAMPLE 1.11. Convert $(3FD)_H$ to binary.

Solution: We have,

3	F	D
0011	1111	1101

Therefore, Binary number = 0011 1111 1101. **Ans.**

EXAMPLE 1.12. Convert $(5A9.B4)_H$ to binary.

Solution: We have,

5	A	9	.	B	4
0101	1010	1001	.	1011	0100

Therefore, Binary number = 0101 1010 1001 . 1011 0100. **Ans.**

1.16.5. Octal to Hexadecimal Conversion (PTU, Jalandhar, Sem. Exam., 2004-05)

The easiest way to convert octal number to hexadecimal number is given below :

(i) We convert octal number to its binary equivalent.

(ii) We convert binary number to its hexadecimal equivalent.

EXAMPLE 1.13. Convert $(615)_8$ to its hexadecimal equivalent.

Solution: (i) Octal to binary

6	1	5
110	001	101

Therefore, Binary number = 110001101

(ii) Binary to hexadecimal

0 0 0 1	1 0 0 0	1 1 0 1
1	8	D

Therefore, Hexadecimal number = $18D_H$. **Ans.**

1.16.6. Hexadecimal to Octal Conversion

The easiest way to convert hexadecimal number to octal number is given below.

(i) Convert hexadecimal number to its binary equivalent.

(ii) Convert binary number to its octal equivalent.

EXAMPLE 1.14. Convert $(25B)_H$ to its octal equivalent.

Solution:

(i) Hexadecimal to binary

2	5	B
0010	0101	1011

Therefore, Binary number = 0010 0101 1011

(*ii*) Binary to octal

0 0 1	0 0 1	0 1 1	0 1 1
1	1	3	3

Therefore, Octal number $= (1133)_8$. **Ans.**

1.16.7. Converting any Radix to Decimal

In general, numbers can be represented as under :

$$N = A_{n-1}\, r^{n-1} + A_{n-2}\, r^{n-2} + ... + A_1\, r^1 + A_0\, r^0 + A_{-1}\, r^{-1} + A_{-2}\, r^{-2} + ... C_{-m}\, r^{-m}$$

where N = Number in decimal

A = Digit

r = Radix or base of a number system

n = The number of digits in the integer portion of number

m = The number of digits in the fractional portion of number

From the above general expression, we can convert number with any radix into its decimal equivalent. This has been illustrated using next few examples.

EXAMPLE 1.15. Convert binary number 1 1 0 1 . 1 to its decimal equivalent.

Solution: We have, $N = 1 \times 2^3 + 1 \times 2^2 + 0 \times 2^1 + 1 \times 2^0 + 1 \times 2^{-1}$

$= 8 + 4 + 0 + 1 + 0.5$

$= (13.5)_{10}$ **Ans.**

or

(i)	1	1	0	1	1
(ii)	8	4	2	1	0.5
(iii)	8 +	4 +	0 +	1 +	0.5 $=(13.5)_{10}$. **Ans.**

EXAMPLE 1.16. Convert $(475.25)_8$ to its decimal equivalent.

Solution: We have, $N = 4 \times 8^2 + 7 \times 8^1 + 5 \times 8^0 + 2 \times 8^{-1} + 5 \times 8^{-2}$

$= 256 + 56 + 5 + 0.25 + 0.078125$

or $N = (317.32813)_{10}$. **Ans.**

EXAMPLE 1.17. Convert $(9\ B2.1A)_H$ to its decimal equivalent.

Solution: We have, $N = 9 \times 16^2 + B(11) \times 16^1 + 2 \times 16^0 + 1 \times 16^{-1} + A(10) \times 16^{-2}$

$= 2304 + 176 + 2 + 0.0625 + 0.039$

or $N = (2482.1)_{10}$. **Ans.**

EXAMPLE 1.18. Convert $(3102.12)_4$ to its decimal equivalent.

Solution: We have, $N = 3 \times 4^3 + 1 \times 4^2 + 0 \times 4^1 + 2 \times 4^0 + 1 \times 4^{-1} + 2 \times 4^{-2}$

or $N = 192 + 16 + 0 + 2 + 0.25 + 0.125 = (210.375)_{10}$. **Ans.**

EXAMPLE 1.19. Convert $(614.15)_7$ to its decimal equivalent.

Solution: $N = 6 \times 7^2 + 1 \times 7^1 + 4 \times 7^0 + 1 \times 7^{-1} + 5 \times 7^{-2}$

or $N = 294 + 7 + 4 + 0.142857 + 0.102$

or $N = (305.24486)_{10}$. **Ans.**

1.16.8. Conversion of Decimal Numbers to any Radix Number

We have to carry out the conversion of decimal number to any radix number in two steps. In step 1, we have to convert integer part and in step 2, we have to convert fractional part. The conversion of integer part is accomplished by successive division method, and the conversion of fractional part is accomplished by successive multiplication method. Let us discuss these two methods.*

* All men and women in the field of digital electronics must be able to convert binary to decimal numbers and decimal to binary numbers.

1.16.8.1. Successive Division for Integer Part Conversion

In this method we repeatedly divide the integer part of the decimal number by r (the new radix) until quotient is zero. The remainder of each division becomes the numeral in the new radix. The remainders are taken in the reverse order to form a new radix number. This means that the first remainder is the least significant digit (LSD) and the last remainder is the most significant digit (MSD) in the new radix number. This procedure is illustrated in following examples.

EXAMPLE 1.20. Convert decimal number 37 to its binary equivalent.

Solution: Here, r is 2

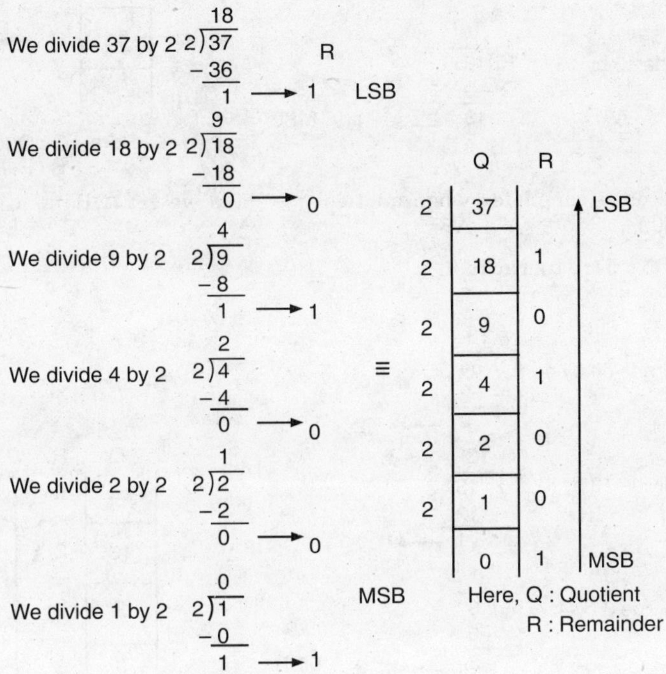

Therefore, Binary equivalent = $(100101)_2$. **Ans.**

EXAMPLE 1.21. Convert decimal number 214 to its octal equivalent.

Solution: Here, r is 8

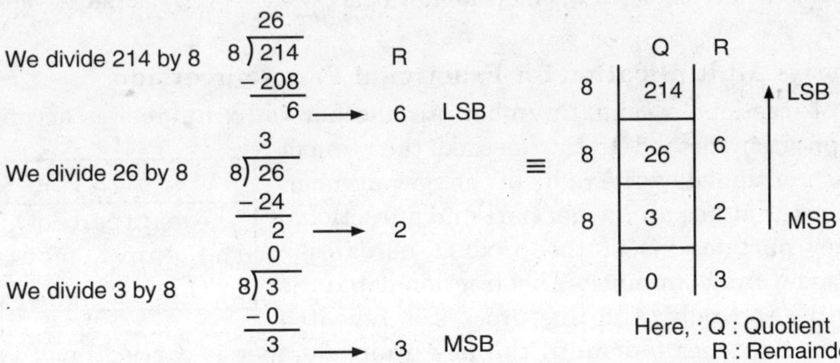

Therefore, Octal equivalent = $(326)_8$. **Ans.**

EXAMPLE 1.22. Convert decimal number 3509 to its hexadecimal equivalent.

Solution: Here, r is 16

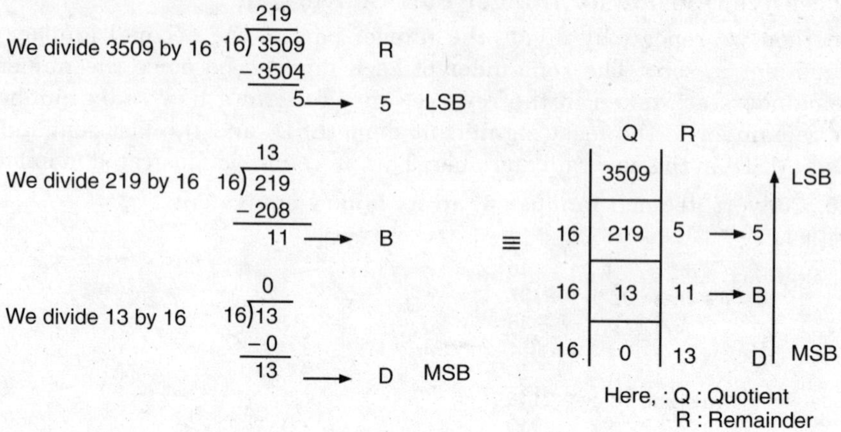

The conversion becomes completed when quotient is 0, and we get $(DB5)_{16}$ as hexadecimal equivalent to decimal number 3509. **Ans.**

EXAMPLE 1.23. Convert 54_{10} to radix 4.

Solution: Here, r is 4

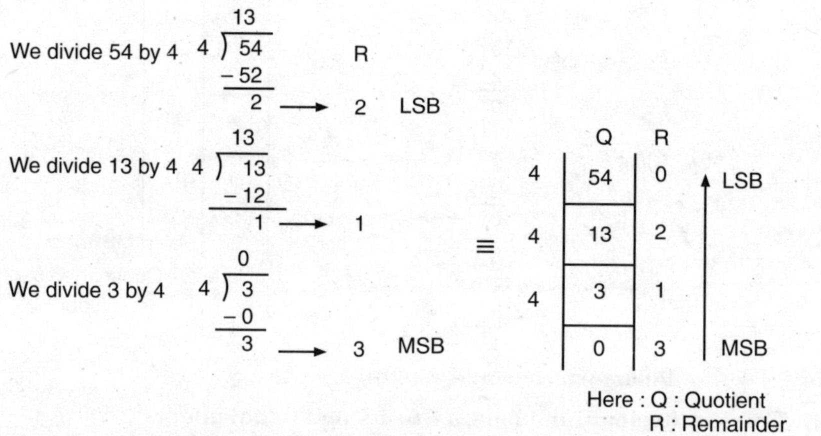

The conversion becomes completed when quotient is 0, and we get $(312)_4$ as radix 4 equivalent to decimal number 54. **Ans.**

1.16.8.2. Successive Multiplication for Fractional Part Conversion

Conversion of fractional decimal numbers to another radix number is accomplished using a successive multiplication method. In this method, the number to be converted is multiplied by the radix of the new number, producing a product that has an integer part and a fractional part. The integer part (carry) of the product becomes a numeral in the new radix number. The fractional part is again multiplied by the radix and this process is repeated until fractional part reaches 0 or until the new radix number is carried out to sufficient digits. The integer part (carry) of each product is read downward to represent the new radix number. This has been illustrated in following examples.

DO YOU KNOW?
The main parts of all computers are the input, control, memory, arithmetic/logic, and output units.

EXAMPLE 1.24. Convert 0.8125 decimal number to its binary equivalent.

Solution: We have,

Fraction		Radix		Result		Recorded Carries	
0.8125	×	2	=	1.625	=	0.625 with a carry of 1	MSB
0.625	×	2	=	1.25	=	0.25 with a carry of 1	|
0.25	×	2	=	0.5	=	0.5 with a carry of 0	↓
0.5	×	2	=	1.0	=	0.0 with a carry of 1	LSB

Reading carries downward, we obtain,

binary fraction = 0.1101, which is equivalent to 0.8125 decimal. **Ans.**

EXAMPLE 1.25. Convert 0.95 decimal number to its binary equivalent.

Solution: We have,

Fraction		Radix		Result		Recorded Carries	
0.95	×	2	=	1.9	=	0.9 with a carry of 1	MSB
0.9	×	2	=	1.8	=	0.8 with a carry of 1	|
0.8	×	2	=	1.6	=	0.6 with a carry of 1	|
0.6	×	2	=	1.2	=	0.2 with a carry of 1	|
0.2	×	2	=	0.4	=	0.4 with a carry of 0	|
0.4	×	2	=	0.8	=	0.8 with a carry of 0	↓
0.8	×	2	=	1.6	=	0.6 with a carry of 1	LSB

In this particular case, 0.8 is repeated and if we multiply further, we will get repeated sequence. If we stop here, we get 7 binary digits, 0.1111001. This answer is an approximate answer. To get more accurate answer we have to continue multiplying by 2 until we have as many digits as necessary for out application. **Ans.**

EXAMPLE 1.26. Convert 0.640625 decimal number to its octal equivalent.

Solution: We have,

Fraction		Radix		Result		Recorded Carries	
0.640625	×	8	=	5.125	=	0.125 with a carry of 5	MSB
0.125	×	8	=	1.0	=	0 with a carry of 1	↓ LSB

Reading carries downward, we obtain octal fraction = .51, which is equivalent to 0.640625 decimal. **Ans.**

EXAMPLE 1.27. Convert 0.1289062 decimal number to its hex equivalent.

Solution: We have,

Fraction		Radix		Result		Recorded Carries	
0.1289062	×	16	=	2.0625	=	2	MSB
0.0625	×	16	=	1.0	=	1	↓ LSB

Reading carries downward, we obtain hexadecimal fraction $= (0.21)_{16}$, which is equivalent to 0.1289062 decimal. **Ans.**

Now, let us carry out the conversion examples of numbers having both integer part and fractional part.

EXAMPLE 1.28. Convert decimal number 24.6 to a binary number.

Solution: We have,

(i) First, we separate out integer part and fraction part as under :

Integer part : 24, Fractional part : 0.6

> **REMEMBER**
>
> It is important to remember that in decimal counting, the units position (LSD) changes upward with each step in the count, the tens position changes upward every 10 steps in the count, the hundreds position changes upward every 100 steps in the count, and so on.

(ii) Then, we find equivalent binary number for integer part as under :

We divide 24 by 2: $2\overline{)24}$ = 12, -24, remainder 0 → 0 LSB

We divide 12 by 2: $2\overline{)12}$ = 6, -12, remainder 0 → 0

We divide 6 by 2: $2\overline{)6}$ = 3, -6, remainder 0 → 0

We divide 3 by 2: $2\overline{)3}$ = 1, -2, remainder 1 → 1

We divide 1 by 2: $2\overline{)1}$ = 0, -0, remainder 1 → 1 MSB

≡

	Q	R	
2	24		↑ LSB
2	12	0	
2	6	0	
2	3	0	
2	1	1	
	0	1	MSB

Here : Q : Quotient
R : Remainder

Then, binary equivalent of integer part = $(11000)_2$

(iii) Lastly, let us find equivalent binary number for fractional part as under :

Fraction		Radix		Result		Recorded Carries	
0.6	×	2	=	1.2	=	0.2 with a carry of 1	MSB
0.2	×	2	=	0.4	=	0.4 with a carry of 0	|
0.4	×	2	=	0.8	=	0.8 with a carry of 0	|
0.8	×	2	=	1.6	=	0.6 with a carry of 1	↓
0.6	×	2	=	1.2	=	0.2 with a carry of 1	LSB

The binary equivalent number is 11000.10011. This number is an approximation of decimal 24.6, because we have terminated the conversion of fractional part after 5 digits.

EXAMPLE 1.29. Convert decimal number 35.45 to octal number.

Solution: (i) First, we separate the integer part and the fraction part as under :

Integer part : 35, Fractional part : 0.45

(ii) Then, we find equivalent octal number for integer part as under :

$8\overline{)35}$ = 4, -32, remainder 03 → 3 LSB

$8\overline{)4}$ = 0, -0, remainder 4 → 4 MSB

≡

	Q	R	
8	35		↑ LSB
8	4	3	
	0	4	MSB

Therefore, octal equivalent of integer part = $(43)_8$

(iii) Finally, let us find equivalent octal number for fractional part as under :

Fraction		Radix		Result		Recorded Carries	
0.45	×	8	=	3.6	=	0.6 with a carry of 3	MSB
0.6	×	8	=	4.8	=	0.8 with a carry of 4	|
0.8	×	8	=	6.4	=	0.4 with a carry of 6	|
0.4	×	8	=	3.2	=	0.2 with a carry of 3	↓
0.2	×	8	=	1.6	=	0.6 with a carry of 1	LSB

The octal equivalent number is $(43.34631)_8$. This number is an approximation of decimal 35.45, because we have terminated the conversion of fractional part after 5 digits.

EXAMPLE 1.30. Convert decimal number 22.64 to hexadecimal number.*

Solution: (i) First, let us separate the integer part and fractional part as under :

Integer part : 22, Fractional part : 0.64

(ii) Then, we find equivalent hexadecimal number for integer part as under :

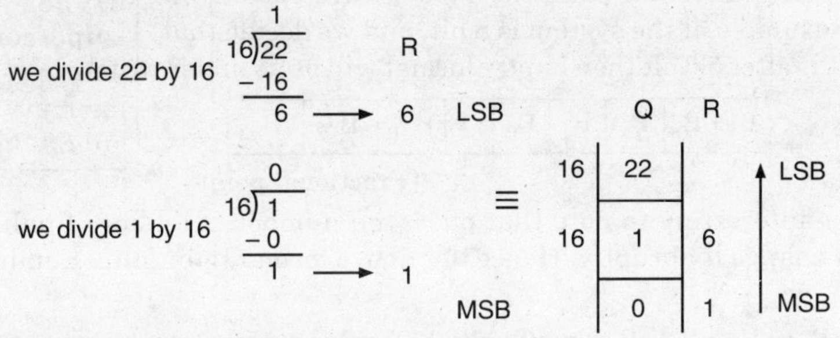

Therefore, Hexadecimal equivalent of integer part = $(16)_{16}$

(iii) Finally, let us find equivalent hexadecimal number for fractional part as under :

Fraction		Radix		Result		Recorded Carries	
0.64	×	16	=	10.24	=	0.24 with a carry of 10 = A	MSB
0.24	×	16	=	3.84	=	0.84 with a carry of 3 = 3	|
0.84	×	16	=	13.44	=	0.44 with a carry of 13 = D	↓
0.44	×	16	=	7.04	=	0.04 with a carry of 7 = 7	LSB

The hexadecimal equivalent number is $(16.A3D7)_{16}$. This number is an approximation of decimal 22.64 because we have terminated the conversion of fractional part after 4 digits.

EXAMPLE 1.31. Determine the values of base x if:

(i) $(211)_x = (152)_8$ (ii) $(193)_x = (623)_8$

Solution:

(i) We have, $(211)_x = (152)_8$

$$2 \times x^2 + 1 \times x^1 + 1 \times x^0 = 1 \times 8^2 + 5 \times 8^1 + 2 \times 8^0$$

or $$2x^2 + x + 1 = 64 + 40 + 2$$

or $$2x^2 + x + 1 = 106$$

or $$2x^2 + x - 105 = 0 \qquad \text{...(i)}$$

Solving equation (i), we obtain,

$$x = 7$$

Therefore, $(211)_7 = (152)_8$ **...Ans.**

(ii) We have, $(193)_x = (623)_8$

$$1 \times x^2 + 9 \times x^1 + 3 \times x^0 = 6 \times 8^2 + 2 \times 8^1 + 3 \times 8^0$$

or $$x^2 + 9x + 3 = 384 + 16 + 3$$

or $$x^2 + 9x + 3 = 403$$

or $$x^2 + 9x - 400 = 0 \qquad \text{...(ii)}$$

* Hexadecimal digits are widely used to represent binary numbers in the computer field.

Solving equation (ii), we get

$$x = 16$$

Therefore, $(193)_{16} = (623)_8$ ...**Ans.**

1.17. FLOATING - POINT NUMBERS

1. Basic Concepts

(PTU, Jalandhar, Sem. Exam, 2007-08)

All the numbering systems which have been discussed so far fall in the category of **base *r*** fixed point numbers. Fixed point number means that **the fractional point** (*i.e.*, decimal point) is once finalized will remain fixed. The fractional point, is fixed within each *n* bit number word. For example, if the system is 8 bit, and we decide that the fractional point is after 5th bit then binary format will be as under,

B_4	B_3	B_2	B_1	B_0	.	B_{-1}	B_{-2}	B_{-3}

Fractional point

H.K. ONNES

Superconductor Sleuth

A Dutch physicist, H.K. Onnes found the first superconductor. In 1911, Onnes realized that mercury had no resistance at 4.3° above absolute zero.

Here, it will be interesting to note that once, the number of bits are finalized for a particular system, we cannot change it abruptly. Hence, for above format, minimum number we can represent is $(0)_2$, i.e.,

$B_4\ B_3\ B_2\ B_1\ B_0\ .\ B_{-1}\ B_{-2}\ B_{-3} = (00000.000)_2$. Maximum number that we can represent is $B_4\ B_3\ B_2\ B_1\ B_0\ .\ B_{-1}\ B_{-2}\ B_{-3} = (11111.111)_2$, i.e., $(31.875)_{10}$.

To increase the range of the numbers to be represented, an obvious solution is to increase number of bits. Let us take an example for a 32 bit system. We take that integer part is of 32 bit. Then maximum number we can represent will be

$$2^{32} - 1 = 4,\ 294,\ 967,\ 295 = 4.29 \times 10^9.$$

But, if we want to represent Angstroms number, i.e., 4.07×10^{26} A°, then it is not possible in 32 bit. Secondly minimum number that we can represent in above case is $(0)_{10}$ = zero. But, if we want to represent suppose mass of electron in kilogram, i.e., 1.67×10^{-27} kg, it is not possible. Here, we may argue to increase number of bits, but just increasing number of bits is not the solution because our system must support that many number of bits. Hence, finally the conclusion is, for a fixed point number representation, we cannot represent very large number and very small number.

2. Remedy

The remedy to this problem is to switch over to **Scientific Notation.** In scientific notation, number is in the form of $N = M \times 10^E$, where M and E are fixed points numbers called, **Mantissa and Exponent** respectively. E is an integer that specifies the number of zeros to be appended to M, to obtain Floating point number N. Here, we may ask one question that why the **Version of Scientific Number (notation)** is called **FLOATING-POINT.** To understand this, we shall take simple example.

Let us consider the number $(1000.234)_{10}$

This number can be represented as under :

1000234×10^{-3}	10.00234×10^2
1000234×10^{-2}	1.000234×10^3
1000234×10^{-1}	0.100234×10^4
1000234×10^{0}	
1000234×10^{1}	

In representation of $(1000.234)_{10}$, we must have observed that decimal point is not fixed as such. It can FLOAT from number to number depending upon power of 10. Normally, in scientific notation, we represent number as $M \times B^E$. Therefore, depending upon our requirement, we set E and in M will allow decimal point to float. Therefore, it is called as floating point numbering system. In $N = \pm M \times B^E$, B = base of floating-point number system. We have taken base = 10 for decimal system. But, for binary number, it will be 2, B = 2. Let us see the format of floating-point number system.

3. Format of Floating-Point Number

Figure 1.18 shows 32 bit format for floating-point. The sign S, indicates sign of mantissa. This bit is assigned left most position to maintain compatibility with the sign conversation for fixed point binary numbers. Eight bits are used for exponent E, with remaining 23 bits for mantissa. E represents signed binary integer, so that exponent value can range from – 128 to + 127. Floating-point numbers are standardized by IEEE. According to it, we have to represent floating-point number in the form

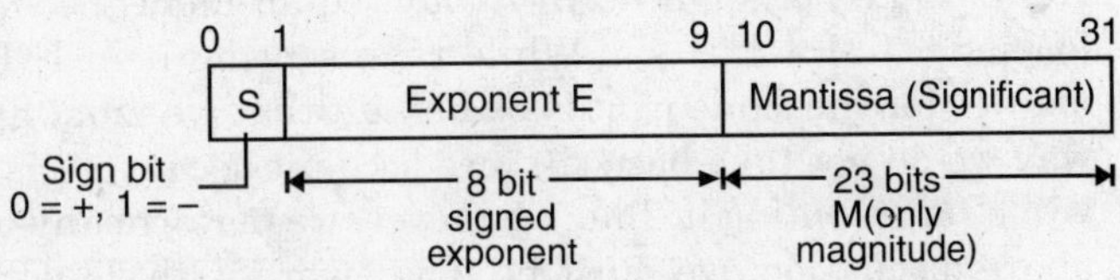

Fig. 1.18. Format of a 32 bit floating-point number

$\pm 1\,.\,M \times 2^{E' - 127}$ (Single Precision – 32 bit)

$\pm 1\,.\,M \times 2^{E' - 1023}$ (Double Precision – 64 bit).

Here M = Mantissa, E = E′ – 127. The format is shown in figure 1.18.

Fig. 1.19. IEEE 754 format

We have, $E = E' - 127$

Therefore, $E' = E + 127$

E = Exponent of number.

The range for E′ is 0 to 255.

Before we discuss exponent, first, let us understand that is 1. M. Whenever any number is given to us, to represent it in floating-point number system, we have to bring down to the format 1. M. Say for example, where the number given is 0.0110×2^6. Presently, the number has form 0. M M = 0110. But, we want 1. M. Hence, now, we shift number to left side till we get 1. M. While shifting the number we have to adjust exponent also.

0.0110×2^6 → Actual number
0.110×2^5 → After 1st shift
1.10×2^4 → after 2nd shift
} Shift operation

Now, we stop here as we have 1. M. Now, M = 10.

Therefore, the floating-point number has the following parameters.

M = 10

S = plus (positive)

$E' = E + 127 = 4 + 127 = 131 = (1\ 0\ 0\ 0\ 0\ 1\ 1)_2$

Therefore, the floating-point number is as shown in figure 1.19.

Figure 1.20 shows floating point representation of 1.10×2^4.

Hence, finally number is $(0\ 1\ 0\ 0\ 0\ 0\ 1\ 1\ 1\ 0\ ...)_2$

In hex, this number is $(4\ 3\ 8\ 0\ 0\ 0\ 0\ 0)_{16}$.

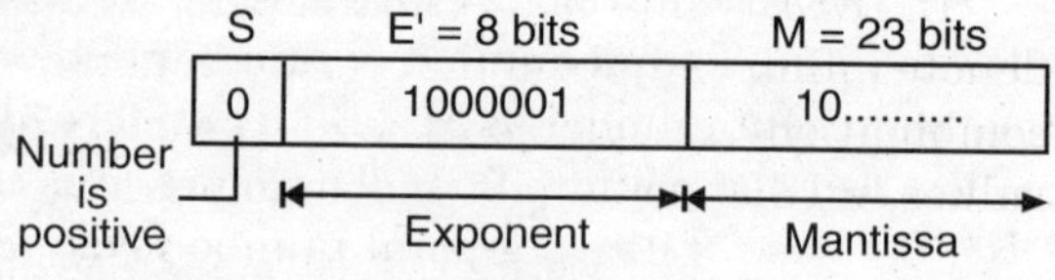

Fig. 1.20.

4. De-normalized/Normalized Floating Point Numbers

While representing 0.0110×2^6 number in floating-point, we have shifted the bits and adjusted exponent, till we get 1.M. The number we get after performing this operation is called Normalized Number, i.e., brought down to our requirement. Secondly, if we see floating-point number, it has format $\pm$ 1.M $\times$ $2^{E' - 127}$. While representing it in IEEE format, we simply forget 1 in. 1.M. The answer for the same is, it is by default that we must have 1 before 'fractional point'. Therefore, the way we do not take base (B) into account because it is default, we do not take 1 into account, only while representing it. But, whenever, we perform any operation, it will be taken into account. From above discussion, we must be now aware with de-normalized number. De-normalized number is 'RAW' number which does not follow 1.M format.

5. Range

Each floating-point format (Single precision/double precision) has capability to represent the number within *Certain* range. Generalised graphical representation is shown in figure 1.21.

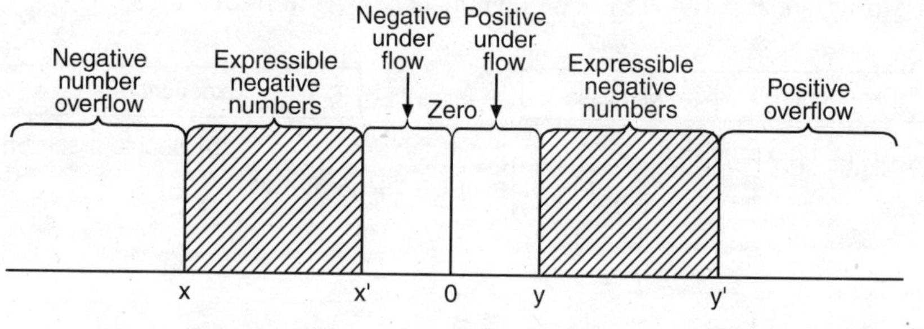

Fig. 1.21. Expressible number in floating-point representation

Figure 1.21 shows the numbers that can be expressed (+ve or –ve), by shaded part. Below and above the shaded parts, numbers are not representable. Representable range of numbers are x – x and y – y.

Five regions are not included in the group.

(i) Negative numbers, those are under negative overflow.

(ii) Negative numbers, those are under negative underflow.

(iii) Zero

(iv) Positive numbers those are under positive underflow.

(v) Positive numbers those are under positive overflow.

First, let us try to understand overflow/underflow.

Overflow occurs when an arithmetic operation results in magnitude greater than can be expressed with an exponent.

Underflow occurs when fractional part is too small. Under flow is less serious problem because the result can be approximated to zero. To represent 0, special bit pattern is designated. In IEEE format, when E = Exponent = 0 and significand (Mantissa) = 0, the number is zero.

In IEEE format, largest exponent field E = $(1\ 1\ 1\ 1\ 1\ 1\ 1\ 1)_2$, corresponds to 255, is reversed to denote infinite numbers that result from some types of overflow. In this standard, certain invalid bit combinations considered to be in a special *not a number* (NAN category). The following table shows values in IEEE for single and double precision number.

Table 1.8. Single Precision (32 bits)

Exponent (E′)	Significand (Mantissa)	Value
255	≠ 0	NAN
255	0	$(-1)^s\ \infty$
0 < e < 255	–	$(-1)^s\ 2^{E'-127}\ (1.M)$
0	≠ 0	$(-1)^s\ 2^{-126}\ (0.M)$
0	0	$(-1)^s\ 0$

Table 1.9. Double Precision (64 bits)

Exponent (E′)	Significand (Mantissa)	Value
2047	≠ 0	NAN
2047	0	$(-1)^s\ \infty$
0 < e < 2047	–	$(-1)^s\ 2^{E'-1023}\ (1.M)$
0	≠ 0	$(-1)^s\ 2^{-1022}\ (0.M)$
0	0	$(-1)^s\ 0$

EXAMPLE 1.32. Represent the following number in single precision format: (i) 0 (ii) –1 (iii) 0.5 (iv) 1 (v) 3.92×10^2

Solution:

(i) For '0', Sign, bit = 0, M = 0 and E = – 127

$(-1)^s\ 1.M \times 2^{E-127}$

Therefore, $(-1)^0\ 1.0 \times 2^0$

Hence, E′ = 0.

S	Exponent (E′)	Mantissa
0	0000 0000	000 ... 0

Therefore, floating-point number is 0 0 0 0 0 0 0 0 H (hex)

(ii) For – 1

Sign bit = 1 → Number is – ve

$(1)_{10} = 0.0\ .\ .\ .\ 1 \rightarrow$ De-normalized

$(1)_{10} = 1.0 \times 2^0 \rightarrow$ Normalized.

M = 0

E = E′ – 127

0 = E′ – 127

Therefore, $E' = 127 = (0\,1\,1\,1\,1\,1\,1\,1)_2$.

S	(E′)	Mantissa
1	0 1 1 1 1 1 1 1	0 0 0
1 bit	8 bits	23 bits

Hence, floating-point number is $(1\,0\,1\,1\,1\,1\,1\,1\,1\,1\,0\,0......0)_2$ and $(B\,F\,8\,0\,0\,0\,0\,0)_{16}$, i.e., hex.

(iii) $(0.5)_{10}$

$(0.5)_{10} = (0.1)_2 = 1.0 \times 2 - 2^{-1} \rightarrow$ Normalized

Hence, as number is +ve, S = 0

$$E = E' - 127$$

Hence, $E' = 126 = (0\,1\,1\,1\,1\,1\,1\,0)_2$

$M = 0.$

Hence,

S	E′	
1	0 1 1 1 1 1 1 0	0 0 0

Hence, floating point number is $(0\,0\,1\,1\,1\,1\,1\,1\,0\,0\,0\,.....0)_2$

and $(3\,F\,0\,0\,0\,0\,0\,0)_{16}$ i.e., hex.

(iv) $(10)_{10}$

$$(10)_{10} = (1\,0\,1\,0)_2 \rightarrow \text{De-normalized}$$

$1.0\,10 \times 2^3 \rightarrow$ Normalized

$S = 0 \rightarrow$ As number is +ve

$M = 0\,1\,0$

$E = E' - 127$

$3 = E' - 127,$

Therefore, $E' = 127 + 3 = (130)_{10} = (1\,0\,0\,0\,0\,0\,1\,0)_2$

S	E′	Mantissa - M
0	1 0 0 0 0 0 1 0	0 1 0 0....... 0

Hence, floating-point number is $(0\,1\,0\,0\,0\,0\,0\,1\,0\,0\,1\,0\,0\,....\,0)_2$ and in hex 4 1 2 0 0 0 0 0 H.

(v) $(3.92 \times 10^2)_{10}$

$$= (3920)_{10} = (1\,1\,0\,0\,0\,1\,0\,0\,0)_2 \rightarrow \text{De-normalized}$$

Hence, normalized number $\rightarrow 1.1\,0\,0\,0\,1\,0\,0\,0 \times 2^8$

Here, 5 bit = 0 $\rightarrow$ because number is + ve

$M = 1\,0\,0\,0\,0\,0\,0$

$E = E' - 127 = (8)_{10}$

Therefore, $E' = (135)_{10} = (1\,0\,0\,0\,0\,1\,1\,1)_2$.

Hence,

S	E′	
0	1 0 0 0 0 1 1 1	1 0 0 0 1 0 0 0....... 0

Hence, floating-point number in binary is $(0\,1\,0\,0\,0\,0\,1\,1\,1\,1\,0\,0\,0\,1\,0\,0\,0\,....0)_2$ and in hex 4 3 C 4 0 0 0 0 H.

1.18 COMPLEMENTS

(PTU, Jalandhar, Sem. Exam., 2008-09)

As a matter of fact, complements are used in the digital computers in order to simplify the subtraction operation and for the logical manipulations.

The man who can make hard things easy is the educator. —**R.W. Emerson**

1.18.1. Classification of Complements (PTU, Jalandhar, Sem. Exam., 2007-08)

For each radix-r system, there are following two types of complements:

(i) The radix complement

(ii) The diminished radix complement.

The radix complement is referred to as the r's complement and the diminished radix complement is referred to as (r – 1)'s complement. Let us consider the binary system with a base r = 2. Hence, the two types of complements for the binary system are 2's complement and 1's complement. Similarly, for an octal system, we have 8's complement and 7's complement. For a decimal system, we have r = 10 and the types of complements are 10's complements and 9's complement.

1.18.2. 1's Complement

The 1's complement of a binary number is the number that results when we complement each bit. The process of obtaining 1's complement has been illustrated in figure 1.22(a). This complemented value represents the negative of the original number. The 1's complement system is very easy to implement merely using inverters. Figure 1.22 (a) shows that the given 8-bit binary number is $(10010\ 100)_2$ and its 1's complement is $(0110\ 1011)_2$. The same principle of complementing bits to obtain 1's complement is applicable to binary numbers of any length. Figure 1.22 (b) shows how to obtain the 1's complement of a binary number using inverters. The binary number $b_3\ b_2\ b_1\ b_0$ is applied to inverter input. We get the 1's complement of this number the output of inverters $(\bar{b}_3, \bar{b}_2, \bar{b}_1, \bar{b}_0)$.

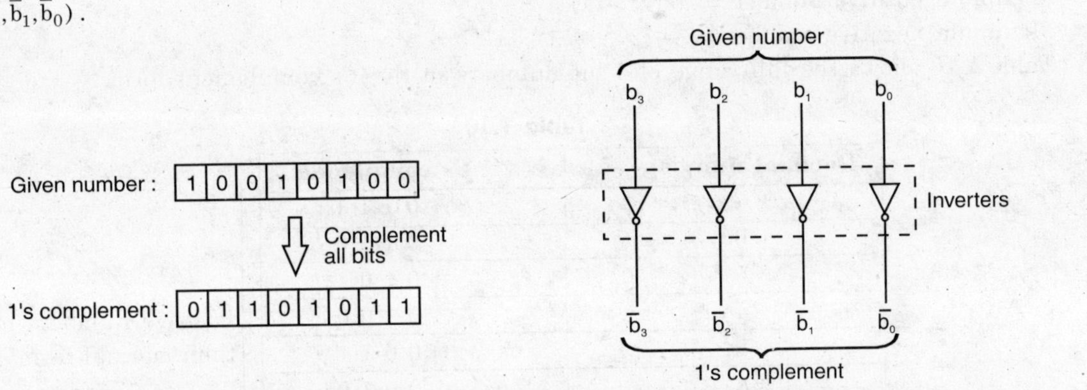

(a) 1's complement of a number (b) 1's complement can be obtained by using inverters

Fig. 1.22.

EXAMPLE 1.33. Obtain the 1's complement of the following numbers : $(1010)_2$, $(11010101)_2$.

Solution:

Given number		1's complement
1010	→	0101
11010101	→	0010 1010

Why is it called 1's complement?

This system is called as 1's complement because the number can be subtracted from 1111 1111 to obtain the result. The answer will be same as if all the 1's were changed to 0's and all the 0's to 1's. This alternative method is illustrated in the following example:

EXAMPLE 1.34. Find 1's complement of $(1001)_2$.

Solution: Let us subtract $(1001)_2$ from $(1111)_2$ as under:

$$\begin{array}{r} 1\ 1\ 1\ 1 \\ -\ 1\ 0\ 0\ 1 \\ \hline \end{array}$$

1's complement → 0 1 1 0 **Ans.**

1.18.3. Representation of Positive and Negative Numbers Using 1's Complement

Positive number in 1's complement form are represented in the same way as the positive sign magnitude numbers. This means that $(+5)_{10}$ can be represented as 0101. But, the negative numbers are the 1's complement of the corresponding positive numbers. For example, $(-5)_{10}$ is represented as under :

$N = (-5)_{10}$ ... Given number

$(+5)_{10} = 0\ 1\ 0\ 1$ ... In sign magnitude form

(first bit 0: Sign; bits 1 0 1: Magnitude)

Therefore, $N = 1\ 0\ 1\ 0$ ...1's complement form

Thus, $(-5)_{10}$ is represented as 1010 in the 1's complement form.

Range of 1's Complement Number System

1. From the discussion done till now, the maximum positive number in the 1's complement form for a 4 bit number is 0111.

Therefore, *for* n = *4, the maximum positive number is* 7, i.e., $[2^{n-1} - 1]$

2. The maximum negative number is represented as 1 1 1 1.

Hence, *For* n = *4, the maximum negative number is* $-[2^{n-1} - 1]$.

Hence, range of 1's complement number system is given by,

Maximum positive number $\Rightarrow (2^{n-1} - 1)$.

Maximum negative number $\Rightarrow -(2^{n-1} - 1)$

Table 1.10 shows the full range of 4 bit numbers in the 1's complement form.

Table 1.10

Decimal number	1's complement	
+ 7	0 1 1 1	
+ 6	0 1 1 0	
+ 5	0 1 0 1	
+ 4	0 1 0 0	Positive numbers in their normal form.
+ 3	0 0 1 1	
+ 2	0 0 1 0	
+ 1	0 0 0 1	
+ 0	0 0 0 0	
− 0	1 1 1 1	
− 1	1 1 1 0	
− 2	1 1 0 1	
− 3	1 1 0 0	
− 4	1 0 1 1	Negative numbers in their 1's complement form.
− 5	1 0 1 0	
− 6	1 0 0 1	
− 7	1 0 0 0	

One of the difficulties in using 1's complement is its representation of zero. As seen from Table 1.10, both 0 0 0 0 and 1 1 1 1 represent zero. The [0 0 0 0] is called as positive zero and [1 1 1 1] is called as the negative zero.

1.18.4. 2's Complement (PTU, Jalandhar, Sem. Exam., 2006-07)

The 2's complement of a binary number is obtained by adding 1 to the 1's complement of that number.

Therefore, 2's complement = 1's complement + 1.

EXAMPLE 1.35. Obtain the 2's complement of $(1011)_2$.

Solution:

(i) First, let us obtain 1's complement of the given number.

1's complement of 1011 is 0100.

(ii) Then, we add 1 to 1's complement.

1's complement:		0	1	0	0
we add 1 :	+				1
2's complement:		0	1	0	1

Ans.

Hence the 2's complement of 1011 is 0101.

EXAMPLE 1.36. Obtain the 2's complement of $(10110010)_2$.

Solution:									
Given number :		1	0	1	1	0	0	1	0
1's complement:		0	1	0	0	1	1	0	1
We add 1:	+								1
2's complement:		0	1	0	0	1	1	1	0

1.18.5. Representation of Positive and Negative Numbers using 2's Complement

(PTU, Jalandhar, Sem., Exam., 2008-09)

Positive numbers in 2's complement form are represented the same way as in the sign-magnitude and 1's complement form. For example, $(+6)_{10}$ is represented as 0110 in the 2's complement form. Negative numbers are the 2's complements of the corresponding positive numbers. For example, $(-6)_{10}$ is represented in the 2's complement form as under :

$$N = (-6)_{10} \text{ ... given}$$

$$(+6)_{10} = \underbrace{0110}$$

↓ invert all bits and add 1.

1010 ... 2's complement of $(-6)_{10}$

In fact, 2's complement is another method of storing negative values. A typical microcomputer represents positive and negative numbers with the help of 2's complement. The representation of four bit (+) and (–) numbers using 2's complement is shown in Table 1.11. The range of a 4-bit sign-magnitude number extends from decimal (– 7) to (+ 7). The $(0)_{10}$ and positive numbers are represented by normal binary numbers in Table 1.11. But the negative numbers are represented in the 2's complement form. However, note that the MSB (sign bit) is not changed while obtaining the 2's complement. For example, 2's complement of $(-6)_{10}$ is obtained as under :

$$(-6)_{10} = (1\,1\,1\,0)_2$$

Its 2's complement = [1] 010
→No change

Similarly, the 2's complement of other negative numbers may be obtained.

Table 1.11. Two's complement numbers

Decimal	Signed binary	2's complement	
+ 7	0 1 1 1	0 1 1 1	Positive numbers with normal reresentation
+ 6	0 1 1 0	0 1 1 0	
+ 5	0 1 0 1	0 1 0 1	
+ 4	0 1 0 0	0 1 0 0	
+ 3	0 0 1 1	0 0 1 1	
+ 2	0 0 1 0	0 0 1 0	
+ 1	0 0 0 1	0 0 0 1	
0	0 0 0 0	0 0 0 0	Unique zero
– 1	1 0 0 1	1 1 1 1	Negative numbers in the 2's complement notation
– 2	1 0 1 0	1 1 1 0	
– 3	1 0 1 1	1 1 0 1	
– 4	1 1 0 0	1 1 0 0	
– 5	1 1 0 1	1 0 1 1	
– 6	1 1 1 0	1 0 1 0	
– 7	1 1 1 1	1 0 0 1	
– 8	1 0 0 0	1 0 0 0	

Important observations from Table 1.11

Two's complement numbers have some interesting characteristics. They may be listed as under:

1. There is one unique 0.
2. The two's complement of 0 is 0.
3. The MSB of a sign-magnitude number cannot be used to express quantity. It can be used as a sign bit.
4. There are eight negative integers, seven positive integers and one zero, making a total of 16 unique states.

> Therefore, in an n bit 2's complement system, there will always be,
> (i) $2^{n-1} - 1$ positive integers
> (ii) 2^{n-1} negative integers
> (iii) and one 0
> } Total 2^n unique states

5. To convert a negative number to a positive number, we find its 2's complement.
6. 2's complement of a 2's complement of a number results in the original number itself. For example, if the given number is $(4)_{10} = (0\ 1\ 0\ 0)_2$, then its 2's complement is (1 1 0 0) and the 2's complement of (1 1 0 0) is (0 1 0 0), i.e., $(4)_{10}$.

EXAMPLE 1.37. Express the decimal numbers +25 and – 25 in the 8 bit sign magnitude, 1's complement and 2's complement forms.

Solution: 1. Representation of + 25

Here, + 25 is represented in the sign magnitude, 1's complement and 2's complement forms as under:

+ 25 = | 0 | 0 0 1 1 0 0 1 |

↓ Sign ↓ Magnitude

2. Representation of – 25 in sign-magnitude form :

	Sign	Magnitude
– 25 =	1	0 0 1 1 0 0 1

3. Representation of – 25 in 1's complements form :

+ 25 =	0	0 0 1 1 0 0 1	... Sign magnitude form
		↓ Invert all bits	
– 25 =	1	1 1 0 0 1 1 0	... 1's Complement form

4. Representation of – 25 in 2's complements form :

+ 25 =	0	0 0 1 1 0 0 1	... Sign magnitude form
		↓ Invert all bits and add 1	
–25 =	1	1 1 0 0 1 1 1	...2's Complement form

1.19. BINARY SUBTRACTION USING 1's AND 2's COMPLEMENTS

As a matter of fact, the direct binary subtraction becomes complicated as the number size increases. Therefore, we can represent the subtraction of A – B in the form of addition as : A + (– B). We can represent number B (which is to be subtracted) in its 1's complement or 2's complement form and use addition instead of subtraction to get the result of A – B.

1.19.1. Subtraction Using 1's Complement

The steps to be followed for subtraction $(A)_2 - (B)_2$ using 1's complement may be described as under :

(i) First we convert the number to be subtracted $(B)_2$ to its 1's complement.

(ii) Then, we add $(A)_2$ and 1's complement of $(B)_2$ using the rules of binary addition.

(iii) If final carry is 1, then we add it to the result of addition obtained in step 2 to get the final result of $(A)_2 - (B)_2$. Here, it may be noted that if the final carry is 1 then the subtraction is positive and is in its true form.

(iv) If the final carry produced in step 2 is 0, then the result obtained in step 2 is negative and is in the 1's complement form and requires to be inverted.

The following examples will make the concept of subtraction using 1's complement obvious. There are two possible cases depending on the magnitude and sign of the numbers involved.

Case 1. Number A and B, both positive and A > B.

Case 2. A and B both positive and A < B.

Let us discuss them one-by-one.

Case 1. A and B both positive with A > B.

EXAMPLE 1.38. Perform $(1011)_2 - (0100)_2$ using 1's complement method.

Solution: $A = (1011)_2$ and $B = (0100)_2$

(i) First, we obtain 1's complement of B as 1011.

(ii) Then, we add A and 1's complement of B.

Thus,						
A			1	0	1	1
1's complement of B	+		1	0	1	1
Final carry →	1		0	1	1	0
	+ └				→	1 (we add the final carry)
Final answer →			0	1	1	1

Here, it may be noted that since final carry was produced, the answer is positive and is in the true form.

Case 2. A and B both positive with A < B.

EXAMPLE 1.39. Subtract $(9)_{10}$ from $(4)_{10}$ using 1's complement method.

Solution: Given that, $A = (4)_{10} = (0100)_2$
$B = (9)_{10} = (1001)_2$

(i) First, we obtain 1's complement of $(9)_{10}$.
1's complement of $(9)_{10}$ or $(1001)_2$ is $(0110)_2$.

(ii) Then, we add $(4)_{10}$ and 1's complement of $(9)_{10}$.

A		0	1	0	0
1's complement of B	+	0	1	1	0
Zero final carry →	0	1	0	1	0 ← Final answer

Here, it may be noted that as the final carry is zero, the answer is negative and is in its 1's complements form.

Then, we invert the final answer.

Since the final answer is in 1's complement form, we have to invert it to obtain the true answer.

Thus, we have
$$1010 \xrightarrow{\text{Invert}} (0101)_2 = (5)_{10}.$$

Therefore, the answer is $(5)_{10}$. **Ans.**

1.19.2. Binary Subtraction using 2's Complement

If the subtraction of two binary numbers A and B is to be performed using the 2's complement, then the following steps must be followed :

(i) First, we add $(A)_2$ to the 2's complement of $(B)_2$.
(ii) If the carry is generated, then the result is positive and is in its true form.
(iii) If the carry is not produced, then the result is negative and is in its 2's complement form.

Important Point. Carry is always to be discarded.

Depending on the magnitude of numbers A and B, we shall deal with different possibilities as under :

Case 1. Both A and B are positive with A >B.

Case 2. Both A and B are positive with A < B.

Case 1. Both numbers are positive with A > B.

The subtraction (A – B) for A > B is illustrated in Example 1.39.

EXAMPLE 1.40. Perform $(9)_{10} - (5)_{10}$ using 2's complement method.

Solution:

(i) First, we obtain 2's complement of $(5)_{10}$.

Decimal	Binary	2's complement
$(5)_{10}$	$(0101)_2$	1011

(ii) Then, we add $(9)_{10}$ to 2's complement of $(5)_{10}$.

$(9)_{10}$		1	0	0	1
2's complement of $(5)_{10}$		1	0	1	1
Carry			1	1	
We discard carry ⟶	1	0	1	0	0 ⇒ $(4)_{10}$

Final carry indicates that the answer is positive and is in its true form.

Answer (0100)

Important Point. The final carry bit acts as a sign bit for the answer. If it is 1, then the answer is positive, and if it is 0, then the answer is negative.

Case 2. A and B both positive with A < B.

EXAMPLE 1.41. Perform $(4)_{10} - (9)_{10}$ using the 2's complement method.

Solution: Let us convert both the number to binary.

$$(4)_{10} = (0100)_2 \text{ and } (9)_{10} = (1001)_2$$

(i) First, we obtain 2's complement of $(9)_{10}$.

Decimal	Binary	2's complement
$(9)_{10}$	$(1001)_2$	(0111)

(ii) Then, we add $(4)_{10}$ to 2's complement of $(9)_{10}$.

We have, $(4)_{10}$		0	1	0	0
+					
2's complement of $(9)_{10}$.		0	1	1	1
Carry		1			
Final carry →	0	1	0	1	1

→ Answer is negative and is in the 2's complement form.

→ 0 carry indicates that the result is negative and is in its 2's complement form.

(iii) Finally, we convert the answer into its true form.

Answer	1	0	1	1	In 2's complement form
we subtract 1	–			1	
	1	0	1	0	
Now, we invert all bits	0	1	0	1	Answer is in true form

Therefore, thus the answer is : $(0101)_2$, i.e., $(-5)_{10}$. **Ans.**

1.20 SIGNED AND UNSIGNED BINARY NUMBERS (PTU, Jalandhar, Sem. Exam., 2008-09)

1.20.1. Unsigned Binary Numbers

1. Basic Concepts

In some applications, all the data is either positive or negative. Then we can just forget about the (+) or (–) signs, and concentrate only on the magnitude (absolute value) of numbers. For example, the smallest 8 bit binary number is 0000 0000, i.e., all zeros, and the largest 8 bit binary number is 1111 1111. Therefore, the total range of unsigned 8 bit binary numbers is from $(00)_H$ to $(FF)_H$ or from $(0)_{10}$ to $(255)_{10}$.

Important Point. It is important to note that we have not included (+) or (–) signs with these numbers.

With 16 bit numbers, the total range is given by,

Smallest:	0000	0000	0000	0000	$(0000)_H$
Largest:	1111	1111	1111	1111	$(FFFF)_H$

This represents the magnitude of 0 to 65, 535 decimal.

1	0	1	1	0	1	0	0	← Unsigned 8-bit binary number

All the bits are used for representing only the magnitude

Fig. 1.23. Unsigned 8 bit binary number

A successful man is one who can lay a firm foundation with the bricks that others throw at him.

—David Brinkley

Data of the foregoing type is called as *unsigned binary* numbers because all of the bits in a binary number are used to represent the magnitude of the corresponding decimal number.

2. Few Important features of unsigned numbers

(i) We can add or subtract the unsigned binary numbers if certain conditions are satisfied.

(ii) The microcomputers of first generation were able to process only 8 bits at a time. Therefore, with 8 bit unsigned arithmetic, all the magnitudes must be restricted between 0 and $(255)_{10}$ or $(00)_H$ to $(FF)_H$.

(iii) This means that all the numbers being added or subtracted must fall in the range 0 to 255. More important is that the answer also must fall in the range 0 to 255.

(iv) For the magnitudes greater than 255, we have to use 16 bit arithmetic.

3. Overflow

If the addition or multiplication of two 8 bit numbers results in generation of a number, greater than $(255)_{10}$, then it is said that *overflow* has taken place.

1.20.2. Sign-magnitude Numbers

If the data has positive as well as negative numbers, then the signed binary numbers must be used. The + or – signs are also represented in the binary form, i.e., by using 0 or 1. Hence, a 0 is used to represent the (+) sign and 1 is used to represent the (–) sign. The MSB of a binary number is used to represent the sign and the remaining bits are used for representing the magnitude. Also, 8-bit signed binary numbers have been shown in figure 1.24.

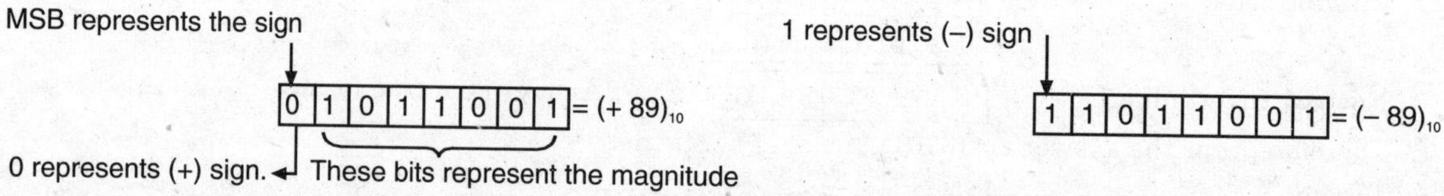

(a) A positive binary number (b) A negative binary number

Fig. 1.24. 8 bit signed binary numbers

Definition of signed numbers

The numbers shown in figure 1.24 contain a sign bit followed by magnitude bits. Numbers in this form are called as *sign-magnitude numbers* or only sign-numbers.

1.20.3. Range of Sign-magnitude Numbers

The unsigned 8 bit numbers cover the decimal range of 0 to 255. But in the sign-magnitude numbers, the MSB is utilized for representing the sign. Hence, the range gets modified. For an 8 bit sign-magnitude number, the largest negative number is (– 127) given by,

1	1	1	1	1	1	1	1	$= (-127)_{10}$

Also, the largest positive number is (+ 127) given by,

0	1	1	1	1	1	1	1	$= (+127)_{10}$

Important Point. Therefore, the range of sign-magnitude, 8 bit binary numbers is modified to decimal (– 127) to (+ 127) from 0 to 255.

The largest magnitude is 127, which is approximately half of what is for unsigned binary numbers. With the sign-magnitude numbers, we can use the 8 bit arithmetic as long as the input data range falls in decimal – 127 to + 127. It is still essential to check all the sums for overflow. If the magnitude of data is greater than 127, then 16 bit arithmetic must be used. With 16 bit numbers, the range of sign-magnitude numbers extends from decimal (– 32, 767) to (+ 32, 767).

Salient Features of sign-magnitude numbers

(i) The main advantage is their simplicity.

(ii) We can easily find the magnitude by deleting the sign bit.

Drawback

The sign-magnitude numbers have a limited use because they require complicated circuits. These numbers are often used in analog-to-digital (A to D) converters.

1.20.4. Signed Complement Numbers

(PTU, Jalandhar, Sem. Exam., 2008-09)

When arithmetic operations are implemented in a computer, it is more convenient to use a different system for representing the negative numbers. This system is called as the *signed-complement system*. In this system, a negative number is indicated by its complement. If the signed-magnitude system negates a number by changing its sign, the signed complement system will negative the number by taking its complement. Since the positive number always starts with a zero (plus) in its MSB position, the complement will always start with a 1, indicating a negative number. The signed – complement system can use either 1's complement or 2's complement, but the 2's complement is most commonly used. As an example, let us observe how to represent – $(7)_{10}$ using different techniques using 8 binary bits.

1. Signed-Magnitude representation : (Sign) | 1 | 0 | 0 | 0 | 0 | 1 | 1 | 1 |

2. Signed-1's complement representation : | 1 | 1 | 1 | 1 | 1 | 0 | 0 | 0 | (1'complement)

3. Signed-2's complement representation : | 1 | 1 | 1 | 1 | 1 | 0 | 0 | 1 | (2'complement)

In the signed-magnitude representation, the left most 1 represents (–) sign and the remaining seven bits represent the magnitude 7. In the signed 1's complement representation, – 7 is obtained by complementing all the bits of + 7 including the sign bit. In the signed 2's complement, – 7 is obtained by obtaining the 2's complement of + 7, including the sign bit. Table 1.12 shows all the possible 4 bit signed binary numbers in the three representations discussed now.

Observations

(i) Table 1.12 shows that the positive numbers in all the three representations are identical and they have a 0 in their left most position.

(ii) The signed 2's complement system has only one representation for 0 which is always positive but the other two systems have either a positive zero or a negative zero.

(iii) All the negative numbers have a 1 in the left most bit position. This is how we can distinguish the negative numbers from the positive ones.

(iv) With four bits, we can represent 16 binary numbers. But, in the signed magnitude and 1's complement systems, there are eight positive numbers including one zero and eight negative numbers.

Table 1.12. Signed binary numbers

S.No.	Decimal complement	Signed 2's complement	Signed 1's complement	Signed
1.	+ 7	0 1 1 1	0 1 1 1	0 1 1 1
2.	+ 6	0 1 1 0	0 1 1 0	0 1 1 0
3.	+ 5	0 1 0 1	0 1 0 1	0 1 0 1
4.	+ 4	0 1 0 0	0 1 0 0	0 1 0 0

S.No.	Decimal complement	Signed 2's complement	Signed 1's complement	Signed
5.	+ 3	0 0 1 1	0 0 1 1	0 0 1 1
6.	+ 2	0 0 1 0	0 0 1 0	0 0 1 0
7.	+1	0 0 0 1	0 0 0 1	0 0 0 1
8.	+ 0	0 0 0 0	0 0 0 0	0 0 0 0
9.	– 0	–	1 1 1 1	1 0 0 0
10.	– 1	1 1 1 1	1 1 1 0	1 0 0 1
11.	– 2	1 1 1 0	1 1 0 1	1 0 1 0
12.	– 3	1 1 0 1	1 1 0 0	1 0 1 1
13.	– 4	1 1 0 0	1 0 1 1	1 1 0 0
14.	– 5	1 0 1 1	1 0 1 0	1 1 0 1
15.	– 6	1 0 1 0	1 0 0 1	1 1 1 0
16.	– 7	1 0 0 1	1 0 0 0	1 1 1 1
17.	– 8	1 0 0 0	–	–

The signed magnitude system is awkward when used in computer arithmetic because the sign and magnitude need to handled separately. Hence, the signed complement system is preferred in computer arithmetic. The 1's complement imposes some difficulties and is generally not used for the arithmetic operations. However, it is useful as a logical operation because the change of 0 to 1 and 1 to 0 is equivalent to logical complement operation. The signed 2's complement system is commonly used in the computer arithmetic.

EXAMPLE 1.43. Find 2's complement of $(1\ 0\ 1\ 0\ \ 0\ 0\ 1\ 1)_2$.

Solution: We have,

```
  1 0 1 0  0 0 1 1  ← number
  0 1 0 1  1 1 0 0  ← 1's complement
+                1
  ----------------
  0 1 0 1  1 1 0 1  ← 2's complement   Ans.
```

1.20.5. Salient Features of 2's Complement Representation

(*i*) From table 1.12, we can note that there are distinct + 0 and – 0 representations in both the sign-magnitude and 1's complement systems, but the 2's-complement system has only *a* + 0 representation.

(*ii*) For 4-bit numbers, the value – 8 is representable only in the 2's complement system and not in other systems.

(*iii*) It is more efficient for logic circuit implementation, and one often used in computers, for addition and subtraction operations.

1.20.6. Few Important Points

Table 1.12 lists all possible 4 bit signed binary numbers in the three representations. Looking at the table 1.12, we can note the following points :

(*i*) Positive numbers in all three representations are identical and have 0 in the leftmost position.

(*ii*) All negative numbers have a 1 in the left most bit position.

(*iii*) The signed-2's complement system has only one representation for 0, which is always positive.

DO YOU KNOW?

Unfortunately, the decimal number system does not lend itself to convenient implementation in digital systems. For example, it is very difficult to design electronic equipment so that it can work with 10 different voltage levels (each one representing one decimal character, 0 through 9).

1.21. BINARY ARITHMETIC

1. Definition

Computer circuits do not process decimal numbers; they process binary numbers. In this article, let us discuss binary arithmetic. Binary addition is the key to binary subtraction, multiplication, and division. Hence, first let us discuss rules for binary addition.

2. Rules for Binary Addition

A		B	SUM	CARRY
0	+	0	0	0
0	+	1	1	0
1	+	0	1	0
1	+	1	0	1

EXAMPLE 1.44. Add $(1\ 0\ 1\ 0)_2$ and $(0\ 0\ 1\ 1)_2$.

Solution: We have,

	Binary		Decimal
	1	← carry	
	1 0 1 0	Decimal	10
+	0 0 1 1	+ Decimal	3
	1 1 0 1	Decimal	13

EXAMPLE 1.45. Add 28 and 15 in binary.

Solution: First let us find the binary equivalent of 28 and 15 as under :

Divisor	Number	Remainder	
2	28		↑ LSB
2	14	0	
2	7	0	
2	3	1	
2	1	1	
	0	1	MSB

Divisor	Number	Remainder	
2	15		↑ LSB
2	7	1	
2	3	1	
2	1	1	
	0	1	MSB

Thus, $(11100)_2$ is a binary equivalent of 28 and $(1111)_2$ is a binary equivalent of 15.

Addition of 28 and 15

Binary		Decimal
1 1	← carry	
1 1 1 0 0	Decimal	28
0 1 1 1 1	Decimal	15
1 0 1 0 1 1	Decimal	43

EXAMPLE 1.46. Add $(10111)_2$ and $(11001)_2$

Solution: The binary addition will take place column by column. The carry generated in the addition of bits in the previous column is to be transferred to the next column. Then, in that column, the addition of carry and the bits corresponding to the two given numbers is to be performed.

Carry		1	1	1	1	
A		1	0	1	1	1
B	+	1	1	0	0	1
Answer →	1	1	0	0	0	0

Hence, the result of addition is $(110000)_2$.

EXAMPLE 1.47. Add 24 and 13 in binary.

Solution:

(i) First, we convert the given numbers into binary numbers.

2	24		
2	12	0	LSB ↑
2	6	0	
2	3	0	
2	1	1	
	0	1	MSB

2	13		LSB ↑
2	6	1	
2	3	0	
2	1	1	
	0	1	MSB

(ii) Therefore, $(24)_{10} = (11000)_2$ Therefore, $(13)_{10} = (1101)_2$

Then, we perform the binary addition.

Carry			1				
$(24)_{10}$			1	1	0	0	0
$+ (13)_{10}$	=	+		1	1	0	1
$(37)_{10}$		1	0	0	1	0	1 ← Answer

1.21.1. Subtraction and Borrow

These two words will be used very frequently for the binary subtraction. For binary subtraction, we have to remember the following four cases given in Table 1.13.

Table 1.13. Four basic cases for binary subtraction

	A		B	Subtraction	Borrow	
Case 1 :	0	–	0	0	0	Same as decimal numbers
Case 2 :	1	–	0	1	0	
Case 3 :	1	–	1	0	0	
Case 4 :	0	–	1	1	(1)	← Borrow should be taken

Let us consider in Table 1.13. It is [0 – 1]. Hence, a logic 1 is borrowed. This will change the subtraction from [0 – 1] to [10 – 1] that means [• • – •] = • = 1.

Subtraction of Larger Binary Numbers

Let A = $(1\,1\,0\,1\,1)_2$ and B = $(1\,0\,1\,1\,0)_2$ are the two binary numbers. Let us obtain the result of the subtraction (A – B).

Let us solve an example.

A = $(1\,1\,0\,1\,1)_2$ and B = $(1\,0\,1\,1\,0)_2$.

(i) First, we subtract the LSBs.

Column of LSBs ↓

	A	1	1	0	1	1	LSB
–	**B**	1	0	1	1	0	LSB
						1	

Success often comes to those who dare and act. It seldom goes to the timid who are ever afraid of consequences.
—J. L. Nehru

(ii) We, now, subtract the bits of second column.

				Column 2 ↓	
	A	1	1	0	1	1
–	B	1	0	1	1	0
					0	1

(iii) Next, to subtract bits from column 3, we borrow 1 from column 4 :

				Column 3 ↓		
	Borrow			1		
	A	1	1	0	1	1
–	B	1	0	1	1	0
				1	0	1

Note that due to borrowed 1, the subtraction in column 3 takes place as under :

$(10)_2 - (1)_2 = 1$

(iv) Now, let us perform subtraction in column 4.

This has been borrowed from column 4 and hence it should be subtracted from column 4 as under :

			Column 4 ↓			
	Borrow			1		
	A	1	1	0	1	1
–	B	1	0	1	1	0
			–1			
		0	0	1	0	1 ← Answer

Thus, the subtraction in column 4 is 1 – 1 = 0.

(v) Lastly, we perform subtraction in column 5.

Subtraction in column 5 is straight forward 1 – 1 = 0.

Hence, the result of the subtraction is $(0\,0\,1\,0\,1)_2$.

EXAMPLE 1.48. Subtract the decimal numbers $(38)_{10}$ and $(29)_{10}$ by converting them into binary.

Solution:

(i) First, we convert $(38)_{10}$ and $(29)_{10}$ into their binary equivalents.

2	38		
2	19	0	LSB ↑
2	9	1	
2	4	1	
2	2	0	
2	1	0	
	0	1	MSB

Therefore, $(38)_{10} = (100110)_2$

2	29		LSB ↑
2	14	1	
2	7	0	
2	3	1	
2	1	1	
	0	1	MSB

Therefore, $(29)_{10} = (11101)_2$

(ii) Next, we perform the subtraction.

Borrow		1	1		1		
$(38)_{10}$		1	0	0	1	1	0
$-(29)_{10}$	−		1	1	1	0	1
$(09)_{10}$ →		0	0	1	0	0	1 ← Answer

1.21.2. Addition of Signed Numbers

(PTU, Jalandhar, Sem. Exam., 2008-09)

In the previous section, we have learnt how to represent signed numbers in three different forms. In this section, we will discuss the addition and subtraction of signed numbers. The 2's complement form of representation is the most widely used in computers and microprocessors-based systems. The addition of two numbers in signed-magnitude system follows the rules of ordinary arithmetic. This means that if the signs of numbers A and B are same, then we add their magnitudes and give the sum their common sign. If the signs are different, then, we subtract the smaller number from the bigger one and give the sign of the larger number to the result. The same procedure applies to the binary numbers in the signed magnitude representation. For example, (+ 16) + (– 25) = (– 25 + 16) = – 9 and this subtraction is performed by subtracting the smaller magnitude of 16 from the larger magnitude 25 and giving the sign of larger number to the subtraction.

This subtraction requires the comparison of signs and magnitudes of the two numbers. The same procedure is applicable to the binary numbers in signed magnitude representation. The rule for adding numbers in signed complement form does not need comparison or subtraction but it requires only the addition. This procedure is very simple and can be stated as under :

Procedure

The addition of two signed binary numbers with negative numbers represented in signed 2's complement form is obtained from the addition of two numbers, including the sign bits. A carry out of the sign bit position is discarded. This procedure can be explained with the help of following examples :

Case 1 : Addition of both positive numbers.

A = + 7		0	0 0 0 0 1 1 1	
B = + 12	+	0	0 0 0 1 1 0 0	
19		0	0 0 1 0 0 1 1	
		↑ Sign bit	Answer	

The answer is positive (sign bit is 0) and equal to 00010011, *i.e.*, 19.

Case 2 : Smaller number is negative.

Let A = – 7 and B = + 12 and their addition is to be performed.

We express (– 7) in the signed 2's complement form as under :

+ 7	=	0 0 0 0 0 1 1 1	← signed magnitude form
– 7	=	1 1 1 1 1 0 0 1	← 2's complement of + 7

Next, we perform the addition.

Carry		1 1 1 1	
A = – 7		1 1 1 1 1 0 0 1	
B = + 12	+	0 0 0 0 1 1 0 0	
+ 5	~~1~~	0 0 0 0 0 1 0 1	← Answer

Discard final carry ↑

The force that rules the world is conduct, whether it be a moral or immoral.

— **Nichola Murray**

Now, the final carry is discarded to obtain the correct answer.

sign (+)	0	0	0	0	0	1	0	1	= + 5
		Magnitude 5							

Case 3 : Bigger number negative.

Let A = + 7 and B = – 12

Carry				
+ 7		0 0	0 0 0 1 1 1	
– 12	+	1 1	1 1 0 1 0 0	← 2's complement of + 12
– 5		1 1	1 1 1 0 1 1	← Answer negative and in 2's complement form

Therefore, we bring the answer in its true form by subtracting 1 from the answer and then inverting all the bits to get the final answer as under :

The answer is

sign (–)	1	0	0	0	0	1	0	1	= – 5

Case 4 : Both numbers negative.

Carry		1 1 1	
– 7		1 1 1 1 1 0 0 1	
– 12	+	1 1 1 1 0 1 0 0	
– 19	~~1~~	1 1 1 0 1 1 0 1	← Answer

Discard carry ↑   2's complement of 19, i.e., – 19

1.22. BINARY SUBTRACTION USING 1's AND 2's COMPLEMENTS

The direct binary subtraction becomes complicated as the number size increases. Hence, we can represent the subtraction of A – B in the form of addition as A + (– B). We can represent number B (which is to be subtracted) in its 1's complement or 2's complement form and use addition instead of subtraction to get the result of A – B.

1.22.1. Subtraction using 1's Complement

The steps to be followed for subtraction $(A)_2 - (B)_2$ using 1's complement are as follows :

(i) First, we convert number to be subtracted $(B)_2$ to its 1's complement.

(ii) Then, we add $(A)_2$ and 1's complement of $(B)_2$ using the rules of binary addition.

(iii) If final carry is 1, then, we add it to the result of addition obtained in step (ii) to get the final result of $(A)_2 - (B)_2$. Note that if the final carry is 1, then, the subtraction is positive and is in its true form.

(iv) If the final carry produced in step 2 is 0, then the result obtained in step (ii) is negative and is in the 1's complement form and needs to be inverted.

The following examples will make the concept of subtraction using 1's complement quite clear. There may be two possible cases depending on the magnitude and sign of the numbers involved.

Case 1 : Number A and B, both positive and A > B.

Case 2 : A and B both positive and A < B.

Let us discuss these cases one-by-one.

Case 1 : A and B both positive and A > B.

EXAMPLE 1.49. Perform $(1011)_2 - (0100)_2$ using 1's complement method.

Solution: We have $A = (1011)_2$ and $B = (0100)_2$

(i) First, we obtain 1's complement of B as 1011.

(ii) Then, we add A and 1's complement of B.

A		1	0	1	1	
	+					
1's complement of B		1	0	1	1	
Final carry →	[1]	0	1	1	0	
+					1	Add the final carry
Final answer →		0	1	1	1	

Note that since final carry was produced, the answer is positive and is in the true form.

Case 2 : A and B both positive with A < B.

EXAMPLE 1.50. Subtract $(9)_{10}$ from $(4)_{10}$ using 1's complement method.

Solution: Given that

$$A = (4)_{10} = (0100)_2$$

$$B = (9)_{10} = (1001)_2$$

(i) First, we obtain 1's complement of $(9)_{10}$.

1's complement of $(9)_{10}$ or $(1001)_2$ is $(0110)_2$.

(ii) Then, we add $(4)_{10}$ and 1's complement of $(9)_{10}$.

A		0	1	0	0	
1's complement of B	+	0	1	1	0	
Zero final carry →	[0]	1	0	1	0	← Final answer

Note that as the final carry is zero, the answer is negative and is in its 1's complement form.

(iii) Then we invert the final answer.

Since the final answer is in 1's complement form, we have to invert it to obtain the true answer.

Thus, $$1010 \xrightarrow{\text{Invert}} (0101)_2 = (5)_{10}$$

Hence, the answer is $-(5)_{10}$.

1.22.2. Binary Subtraction using 2's Complement

If the subtraction of two binary numbers A and B is to be performed using the 2's complement, then the following steps are to be followed.

Steps to be followed:

(i) First, we add $(A)_2$ to the 2's complement of $(B)_2$.

(ii) If the carry is generated, then, the result is positive and is in its true form.

(iii) If the carry is not produced, then, the result is negative and is in its 2's complement form.

Important Point : Carry is always to be discarded.

Depending on the magnitude of numbers A and B, there may be two different possibilities as follows :

Case 1 : Both A and B are positive with A > B.

Case 2 : Both A and B are positive with A < B.

Case 1 : Both numbers positive with A > B.

The subtraction (A – B) for A > B is illustrated in Example 1.51.

EXAMPLE 1.51. Perform $(9)_{10} - (5)_{10}$ using 2's complement method.

Solution: (i) First, we obtain 2's complement of $(5)_{10}$.

Decimal	**Binary**	**2's complement**
$(5)_{10}$	$(0101)_2$	(1011)

(ii) Then, we add $(9)_{10}$ to 2's complement of $(5)_{10}$.

$(9)_{10}$		1	0	0	1
2's complement of $(5)_{10}$		1	0	1	1
Carry			1	1	
Discard Carry →	1	0	1	0	0 → $(4)_{10}$

Final carry indicates that the answer is positive and is in its true form.

Answer

Therefore, $(9)_{10} - (5)_{10} = (4)_{10}$ **Ans.**

Important Point: The final carry bit acts as a sign bit for the answer. If it is 1 then the answer is positive, and if it is 0, then, answer is negative.

Case 2 : A and B both positive with A < B.

EXAMPLE 1.52. Perform $(4)_{10} - (9)_{10}$ using the 2's complement method.

Solution: Let us convert both the numbers to binary.

$$(4)_{10} = 0100_2 \text{ and } (9)_{10} = 1001_2$$

(i) First, we obtain 2's complement of $(9)_{10}$.

Decimal	**Binary**	**2's complement**
$(9)_{10}$	$(1001)_2$	(1011)

(ii) Then, we add $(4)_{10}$ to 2's complement of $(9)_{10}$.

$(4)_{10}$:		0	1	0	0
+					
2's complement of $(9)_{10}$:		0	1	1	1
Carry :		1			
Final carry →	0	1	0	1	1

Answer is negative and is in the 2's complement form

0 carry indicates that the result is negative and is in its 2's complement form

(iii) Then, we convert the answer into its true form.

Answer :	1	0	1	1	In 2's complement form
Subtract 1 :	–			1	
	1	0	1	0	
Invert all bits :	0	1	0	1	Answer is in true form.

Thus, the answer is $-(0101)_2$, i.,e., $(-5)_{10}$.

1.22.3. Arithmetic Subtraction (Subtraction of Signed Numbers)

1. Subtraction of two signed binary numbers when the negative numbers are in the 2's complement form can be performed as under :

   First, we take the 2's complement of the subtrahend (including the sign bit) and add it to the minuend (including the sign bit). A carry out of the sign bit position should be discarded.

2. We can use this procedure because a subtraction is equivalent to an addition if the sign of the subtrahend is changed. This is demonstrated as under :

Minuend	$\rightarrow \pm A$	
Subtrahend	$\rightarrow -(+B)$	i.e., $(\pm A) - (+B) = (\pm A) + (-B)$
	$(\pm A) + (-B)$	
	And	$(\pm A) - (-B) = (\pm A) + (+B)$

3. We consider the subtraction (– 7) – (– 12) = + 5 in binary with eight bits. The numbers (– 7) and (– 12) are expressed in the 2's complement form as (1 1 1 1 1 0 0 1) and (1 1 1 1 0 1 0 0) respectively.
4. The subtraction of these numbers is changed to addition by taking the 2's complement of subtrahend (– 12) to (+ 12).
5. The 2's complement of – 12 = (0000 1100). Then, the addition is performed as under :

– 7		1	1	1	1	1	0	0	1
+ 12		0	0	0	0	1	1	0	0
+ 5	1	0	0	0	0	0	1	0	1
	Discard				Answer is + 5				

Discarding the final carry, we get the answer of + 5.

Important Conclusion : The binary numbers in the signed complement system are added and subtracted by the same basic addition and subtraction rules as unsigned numbers. Hence, the computers need only one common hardware circuit to handle the addition and subtraction. The user must interpret the results of such addition or subtraction differently depending on whether it is assumed that the numbers are signed or unsigned.

1.23. INTRODUCTION TO BINARY CODE

Usually, the digital data is represented, stored and transmitted as groups of binary digits (bits). The group of bits, also known as **binary code,** represent both numbers and letters of the alphabets as well as many special characters and control functions. They are classified as numeric or alphanumeric. Numeric codes are used to represent numbers. On the other hand, alphanumeric codes are used to represent characters: alphabetic letters and numerals. In these codes, a numeral is treated simply as another symbol rather than as a number or numeric value.

123.1. Classification of Binary Codes

The different binary codes can be classified as under :

1. Weighted codes
2. Non-weighted codes
3. Reflective codes
4. Sequential codes
5. Alphanumeric codes
6. Error Detecting and Correcting codes

1. Weighted Codes

In weighted codes, each digit position of the number represents a specific weight. For example, in decimal code, if number is 567, then, weight of 5 is 100, weight of 6 is 10 and weight of 7 is 1. In weighted binary codes, each digit has a weight 8, 4, 2 or 1.

The codes 8421, 2421 and 5211 are all weighted codes.

2. Non-weighted Codes

Non-weighted codes are not assigned with any weight to each digit position, i.e., each digit position within the number is not assigned fixed value. Excess-3 and gray codes are the non-weighted codes.

3. Reflective Codes

A code is said to be reflective when the code for 9 is the complement for the code for 0, 8 for 1, 7 for 2, 6 for 3, and 5 for 4. Note that the 2421, 5211 and excess-3 codes are reflective, whereas the 8421 code is not. Reflectivity is desirable in a code when the nine's complement must be found, such as in nine's complement subtraction.

4. Sequential Codes

In sequential codes, each succeeding code is one binary number greater than its preceding code. This greatly aids mathematical manipulation of data. The 8421 and excess-3 are sequential, whereas the 2421 and 5211 codes are not.

5. Alphanumeric Codes

The codes which consists of both numbers and alphabetic characters are called *alphanumeric codes.* Most of these codes, however, also represent symbols and various instructions necessary for conveying intelligible information. The most commonly used alphanumeric codes are : ASCII (American Standard Code for Information Interchange), EBCDIC (Extended Binary Coded Decimal Interchange Code) and Hollerith code.

6. Error Detecting and Correcting Codes

When the digital information in the binary form is transmitted from one circuit or system to another circuit or system, an error may occur. This means that a signal corresponding to 0 may change to 1 or vice-versa due to presence of noise. To maintain the data integrity between transmitter and receiver, extra bit or more than one bit are added in the data. These extra bits allow the detection and some times correction or error in the data. The data along with the extra bit/bits forms the code. Codes which allow only error detection are called *error detecting codes* and codes which allow error detection and correction are called *error detecting and correcting codes.*

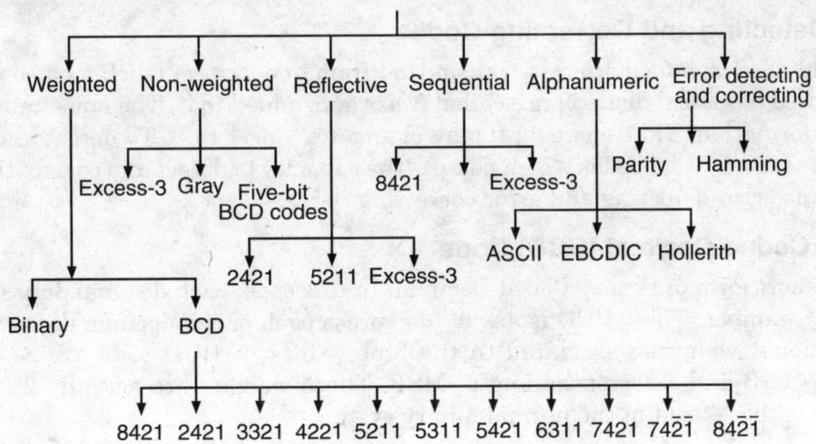

Fig. 1.25. Classification of various binary codes.

1.23.2. Weighted Binary Codes

As a matter of fact, weighted binary codes are those which obey the positional weight principle. Each position of the number represents a specific weight. Several systems of codes are used to express the decimal digits 0 through 9. These codes have been listed in figure 1.25. In fact, 8421, 2421, 3321 ... all are the weighted codes. In these codes, each decimal digit is represented by a group of four bits as shown below.

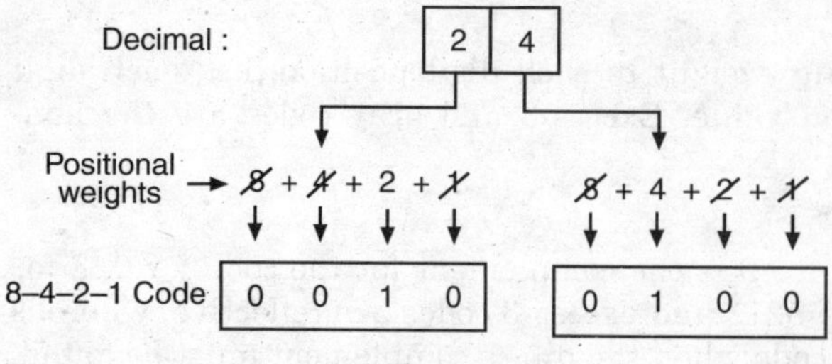

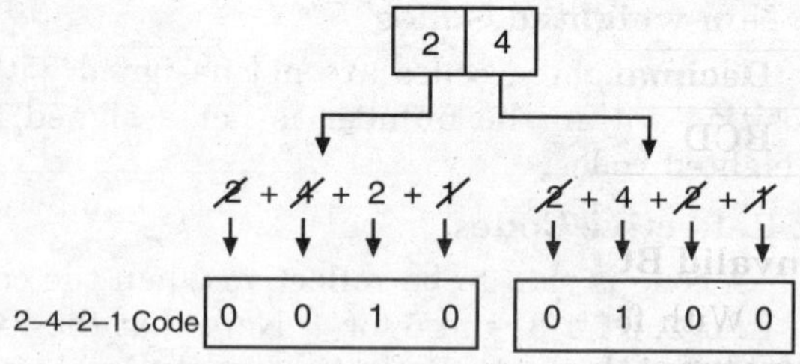

1.23.3. Non-Weighted Codes

In these codes, the positional weights are not assigned. The examples of non-weighted codes are excess-3 and gray codes.

1.23.4. Reflective Codes

A code is said to be reflective, when the code for 9 is the complement for 0, code for 8 is complement for 1, 7 for 2, 6 for 3 and 5 for 4. Reflectivity is desirable in a code when the 9's complement is required to be found, e.g., in 9's complement subtraction. Also, 2421, 5211 and excess-3 are the examples of reflective codes.

1.23.5. Sequential Codes

A code is said to be sequential when each succeeding code is one binary number greater than the preceding code. This greatly simplifies the mathematical manipulations. The 8421 and XS-3 are sequential codes.

1.23.6. Alphanumeric Codes

The alphanumeric codes are designed to represent numbers as well as alphabetic characters. Some of these codes are capable to representing some symbols and instructions as well. Examples of alphanumeric codes are : ASCII (American Standard Code for Information Interchange), EBCDIC (Extended Binary Coded Decimal Interchange Code) and Hollerith code.

1.23.7. Error Detecting and Correcting Codes

When digital data or information is transmitted from one system (such as a computer) to the other, an unwanted electrical disturbance called *Noise* gets added to it. The noise can force an *error* in the digital information. That means a 0 may change to 1 or 1 to 0. To detect and correct such errors we can use some special codes which possess the capacity to detect and correct the error. Such codes are called as error detecting and error correcting codes.

1.23.8. Binary Coded Decimal (BCD) Code

BCD is the short form of Binary Coded Decimal. In this code, each decimal digit is represented by a 4-bit binary number. Thus, BCD is a way to express each of the decimal digits with a binary code. The positional weights associated to the binary bits in BCD code are 8–4–2–1 with 1 corresponding to LSB and 8 corresponding to MSB. These weights are actually 2^3, 2^2, 2^1 and 2^0 which are same as those used in the normal binary system.

Conversion from decimal to BCD

The decimal digits 0 to 9 are converted into a BCD, exactly in the same way as binary. For example, decimal 4 corresponds to 0100 BCD which is same as binary. Similarly, the BCD numbers corresponding to the decimal numbers 0 to 9 are exactly same as the corresponding binary numbers. This has been illustrated in Table 1.14.

Table 1.14.

Decimal	0	1	1	3	4	5	6	7	8	10
BCD	0000	0001	0010	0011	0100	0101	0110	0111	1000	1001

BCD numbers are same as 4 bit binary numbers

Invalid BCD Codes

With four bits, we can represent sixteen numbers (0000 to 1111). But, in BCD code, only the first ten of these are used (0000 to 1001). The remaining six code combinations, i.e., 1010 to 1111 are invalid in BCD.

Conversion of big decimal numbers to BCD

To express any decimal number to BCD, we simply express each decimal digit with its equivalent 4 bit BCD code as illustrated below. Figure 1.26 shows the BCD codes for various decimal numbers.

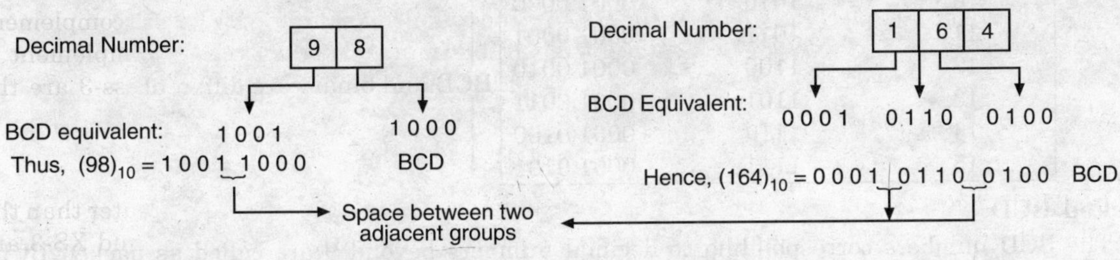

Fig. 1.26. Decimal to 8-4-2-1 BCD conversion

Smallest and largest number in BCD

The smallest number in BCD is (0000) and the largest one is (1001), i.e., 9. The next number to 9 will be $(10)_{10}$ which is expressed as (0001 0000) in BCD.

1.23.9. Comparison with Binary

1. BCD is less efficient than binary.

   Conversion of a decimal number 78 into BCD and binary is illustrated in figure 1.27.

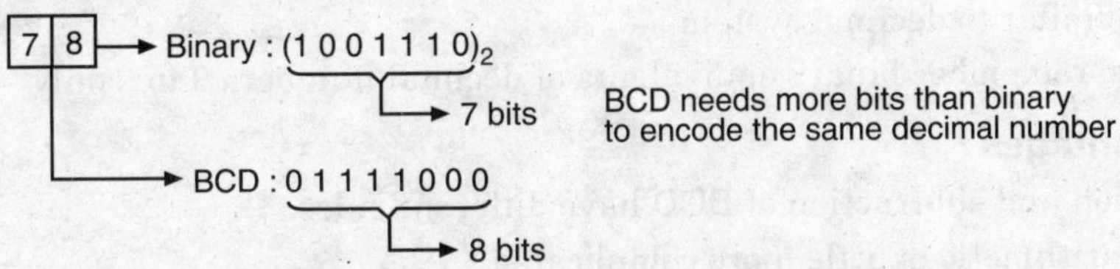

Fig. 1.27. BCD is less efficient than binary

   Figure1.27 illustrates that to encode the same decimal number, BCD needs more number of bits than binary. Hence, BCD is less efficient as compared to binary.

2. BCD arithmetic is more complicated than Binary arithmetic.
3. The advantage of a BCD code is that the conversion from Decimal to BCD or vice-versa is quite simple.
4. Table 1.15 shows the comparison of binary and BCD numbers from 0 to 15 decimal.

Table 1.15

Decimal	Binary	BCD	
0	0000	0000	
1	0001	0001	
2	0010	0010	
3	0011	0011	
4	0100	0100	BCD and Binary are same
5	0101	0101	
6	0110	0110	
7	0111	0111	
8	1000	1000	
9	1001	1001	
10	1010	0001 0000	
11	1011	0001 0001	
12	1100	0001 0010	BCD and binary are different
13	1101	0001 0011	
14	1110	0001 0100	
15	1111	0001 0101	

Packed BCD

The BCD numbers corresponding to decimal numbers beyond 9 are called as packed BCD. Some examples of packed BCD are presented in Table 1.16.

Table 1.16. Examples of packed BCD

Decimal	Packed BCD		
25		0010	0101
169	0001	0110	1001
523	0101	0010	0011

1.23.10. Advantages of BCD Codes

(i) It is very similar to decimal system.

(ii) We need to remember binary equivalents of decimal numbers 0 to 9 only.

1.23.11. Disadvantages

(i) The addition and subtraction of BCD have different rules.

(ii) The BCD arithmetic is little more complicated.

(iii) BCD needs more number of bits than binary to represent the same decimal number. Hence, BCD is less efficient than binary.

EXAMPLE 1.53. Convert the following decimal numbers to BCD.

(a) 35 (b) 174 (c) 2479

Solution:

(a) Decimal 3 5 → BCD 0011 0101

(b) 1 7 4 → 0001 0111 0100

(c) 2 4 7 9 → 0010 0100 0111 1001

1.24. BCD ARITHMETIC

In this article, let us understand the rules for BCD addition and subtraction.

1.24.1. BCD Addition

(PTU, Jalandhar, Sem. Exam., 2008-09)

In BCD addition, we have to deal with three different situations. Let us assume that two 4 bit BCD numbers A and B are being added. Then the three cases to be considered may be listed as under:

Case 1 : Sum is equal to or less than 9 and carry is 0.

Case 2 : Sum is greater than 9 and carry is 0.

Case 3 : Sum is less than or equal to 9 but carry is 1.

The addition is to be carried out as normal binary addition.

Case 1 : Sum equal to or less than 9 with carry 0.

The addition of $(2)_{10}$ and $(6)_{10}$ in BCD is shown in figure 1.28.

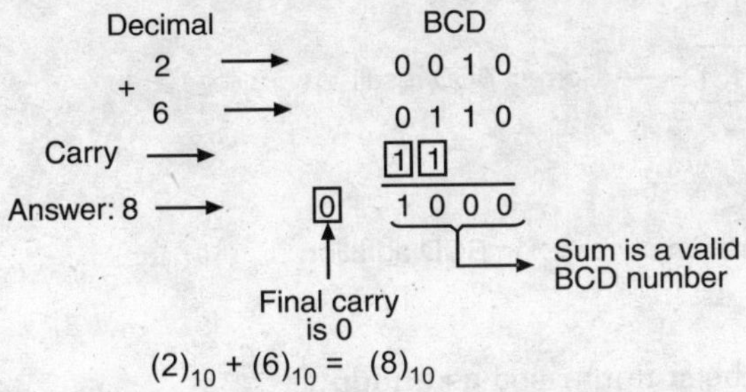

Important Conclusion

If sum is less than or equal to 9 with final carry equal to 0, then, the sum is in proper BCD form and requires no correction.

Fig. 1.28. (a) Illustration of case 1 in BCD addition

Case 2 : Sum greater than 9 but carry = 0.

In this case, the sum of the two BCD numbers is greater than 9 that means it is an invalid BCD number. But, the final carry is 0. Hence, we have to correct the sum by adding decimal 6 or BCD 0110 to it. After doing this, we get the correct BCD sum. Case 2 of BCD addition is illustrated in figure 1.29.

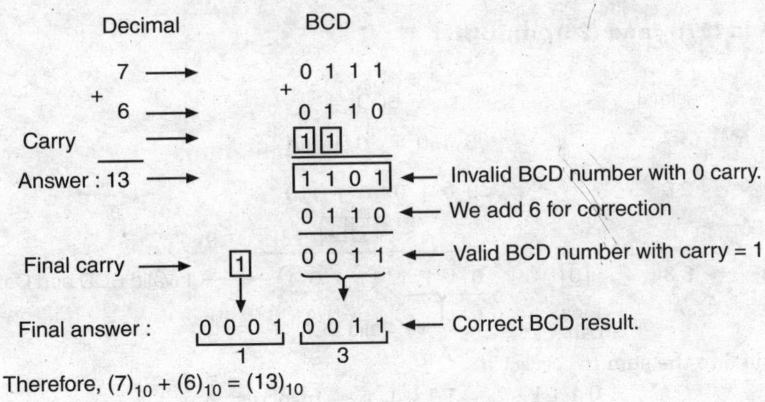

Fig. 1.28. (b) Illustration of case 2 in BCD addition

Case 3 : Sum less than or equal to 9 but carry = 1.

Here, the sum of two BCD numbers is less than or equal to 9, i.e., it is a valid BCD number, and final carry = 1. This inducates that the answer is wrong and needs correction. Thus, we add $(6)_{10}$ or (0110) BCD to the sum to correct the answer. Case 3 of BCD addition is illustrated in figure 1.30.

Decimal		BCD	
9	→	1 0 0 1	
+ 8	→	+ 1 0 0 0	
Carry		[1]	
Sum: 17	→	[1] 0 0 0 1	← Sum is valid BCD and carry = 1
		[0 0 0 1 0 0 0 1]	← Incorrect BCD result

Now, we add 6 to incorrect BCD result.

	0 0 0 1	0 0 0 1	← Incorrect BCD result
+	0 0 0 0	0 1 1 0	← We add 6
Carry →			
	0 0 0 1	0 1 1 1	← Correct BCD result
	1	7	

Thus, $(9)_{10} + (8)_{10} = (17)_{10}$

Fig. 1.28. (c) Illustration of case 3 in BCD addition

1. Summary of BCD addition

The BCD addition discussed so far may be summarised as under :

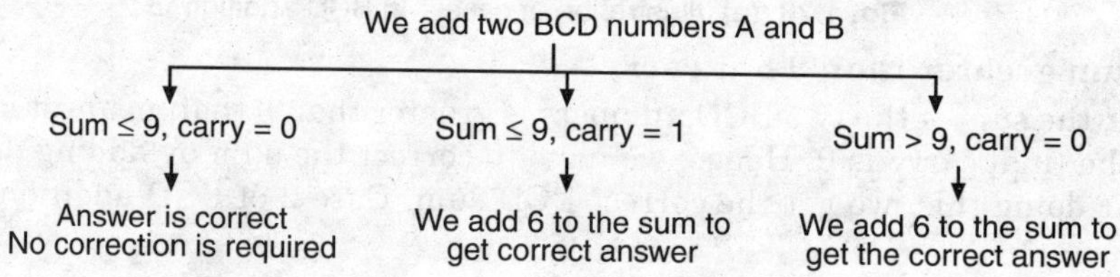

EXAMPLE 1.54. Add $(57)_{10}$ and $(26)_{10}$ in BCD.

Solution:

Decimal		BCD		
57		0 1 0 1	0 1 1 1	
+ 26		0 0 1 0	0 1 1 0	
Carry →			[1][1]	
Sum → 83	[0] Final carry	0 1 1 1 (Valid BCD)	1 1 0 1	← Invalid BCD and Carry = 0

We have to add 6 to the sum to correct it.

Thus,	0111	1101	Incorrect sum
+	0000	0110	We add 6 for correction
Carry	1111	1	
	1000	0011	Correct result
	8	3	

Therefore, $(57)_{10} + (26)_{10} = (83)_{10}$ **Ans.**

EXAMPLE 1.55. Add $(83)_{10}$ and $(34)_{10}$ in BCD.

Solution:

Decimal		BCD		
$(83)_{10}$		1000	0011	
+ $(34)_{10}$		+ 0011	0100	
	Carry			
$(117)_{10}$	Sum:	1011	0111	Sum > 9, Carry = 0
		Invalid BCD	Valid BCD	Hence, correction is required

Thus, we add 6 to the invalid BCD, for correction

	1011	0111	
+	0110	0000	6 is added only to the invalid BCD.
Carry	11		
	0001	0111	
[1] ↓	↓	↓	
0001	0001	0111	← Correct BCD result.
1	1	7	

Therefore, $(83)_{10} + (34)_{10} = (117)_{10}$ **Ans.**

EXAMPLE 1.56. Add $(569)_{10}$ and $(687)_{10}$ in BCD.

Solution:

569	0101	0110	1001	
+ 687	0110	1000	0111	
	0	1 ↖	111	
1256	1011	1111	10000	Incorrect answer
	Invalid BCD	Invalid BCD	Valid BCD with carry 1	

We add $(0110)_2$ to only the invalid BCD numbers to get correct answer.

	1011	1111	0000	Incorrect sum
+	0110	0110	0110	
	1111 ↖	11		
[1]	0010	10100	0110	
↓				
0001	0010	0101	0110	Correct BCD answer
1	2	5	6	

Therefore, $(569)_{10} + (687)_{10} = (1256)_{10}$ **Ans.**

1.24.2 BCD Subtraction

The subtraction of two BCD numbers A and B can be performed using one of the following methods.

1. Nine's complement method.
2. Ten's complement method

1. BCD subtraction using 9's complement

Nine's complement : The 9's complement of a BCD number can be obtained by subtracting it from 9. For example, 9's complement of 1 is 8. The 9's complement of various digits are given in Table 1.17.

Table 1.17 9's complement of various decimal digits

Decimal digit	0	1	2	3	4	5	6	7	8	9
9's complement	9	8	7	6	5	4	3	2	1	0

This is similar to the subtraction using 1's complement. It is performed as under :

(i) First, we obtain the 9's complement of number B.

(ii) Then, we add A and 9's complement of B.

(iii) If a carry is generated in step (ii), then we add it to the sum to obtain the final result. The carry is called as end around carry.

(iv) If carry is not produced then the result is negative and hence we take the 9's complement of the result.

EXAMPLE 1.57. Subtract $(3)_{10}$ from $(7)_{10}$ in BCD.

Solution:

(i) First, we obtain 9's complement of $(3)_{10}$.
9's complement of $(3)_{10}$ is $9 - 3 = 6$.

(ii) Then, we add 7 and nine's complement of 3.

Thus,

		7	
	+	6	Nine's complement of 3
End around carry	[1]	3	Sum
	+ →	1	We add carry to sum
		4	Final result is positive and is in its true form

Therefore, $(7)_{10} - (3)_{10} = (4)_{10}$ **Ans.**

EXAMPLE 1.58. Perform the subtraction $(4)_{10} - (7)_{10}$ using the 9's complement

Solution:

(i) First, we obtain 9's complement of $(7)_{10}$.
So, 9's complement of 7 is $(9 - 7) = (2)_{10}$.

(ii) Then, we add $(4)_{10}$ and 9's complement of $(7)_{10}$

$(4)_{10}$		0	1	0	0	BCD of 4
9's complement of $(7)_{10}$	+	0	0	1	0	
	[0]	0	1	1	0	Result

Final carry is 0. Hence, result is negative, hence, we take 9's complement of the result.

(iii) Next, we take 9's complement of the result.

9		1 0 0 1	
− 6	−	0 1 1 0	Result obtained in step (ii).
3		0 0 1 1	Final answer

Therefore, $(4)_{10} - (7)_{10} = (-3)_{10}$ **Ans.**

EXAMPLE 1.59. Perform $(83)_{10} - (21)_{10}$ using the 9's complement method.

Solution:

(i) First, we obtain 9's complement of $(21)_{10}$.

For this, we subtract each digit of 21 from 9.

Then,

	99	
−	21	Given number
	78	9's complement of 21

(ii) Then, we add $(83)_{10}$ and 9's complement of $(21)_{10}$.

```
(83)10          1 0 0 0   0 0 1 1
+ (78)10        0 1 1 1   1 0 0 0    9's complement of 21
   1    Carry
 161    Sum     1 1 1 1   1 0 1 1    Result is in invalid BCD form.
                Invalid   Invalid
                BCD       BCD
```

(iii) Next, we add $(6)_{10}$, i.e., $(0110)_2$ to each invalid BCD number.

```
      1 1 1 1            1 0 1 1   Result obtained in step (ii)
    + 0 1 1 0            0 1 1 0   We add (6)10 for correction
      1 1 1 1            1 1
  [1] 0 1 1 0         1  0 0 0 1
                      └── Carry is shifted to next higher digit
   └── Final carry is 1 which indicates that the result is positive and is
       in its true form.
```

(iv) Lastly, we add carry to the sum obtained in step (iii).

```
Final carry   [1]   0 1 1 0   0 0 0 1   Result obtained in step (iii)
               + └──────────────────→ 1   End around carry.
Carry :                             1
Sum :               0 1 1 0   0 0 1 0   Final answer is positive and is
                       6         2      in its true form.
```

Therefore, $(83)_{10} - (21)_{10} = (62)_{10}$ **Ans.**

EXAMPLE 1.60. Perform $(52)_{10} - (89)_{10}$ using the 9's complement.

Solution:

(i) First, we obtain 9's complement of $(89)_{10}$.

9's complement of $(89)_{10}$ = 99 – 89 = 10

(ii) Then, we add $(52)_{10}$ and 9's complement of $(89)_{10}$.

```
thus,       0 1 0 1   0 0 1 0    BCD of (52)10
        +   0 0 0 1   0 0 0 0    9's complement of (89)10
Carry :       1
Sum :       0 1 1 0   0 0 1 0    Final carry is 0, so, the sum is
                                 negative.
```

(iii) Next, we take 9's complement of sum.

```
    1 0 0 1   1 0 0 1   BCD of 99
  – 0 1 1 0   0 0 1 0   Sum
    1 1       1 1
    0 0 1 1   0 1 1 1   Final answer
       3         7
```

Therefore, $(52)_{10} - (89)_{10} = -(37)_{10}$ **Ans.**

2. Subtraction using 10's complement

The 10's complement is obtained by adding 1 to the 9's complement. The 10's complement can be used to perform the BCD subtraction as under :

(i) First, we obtain the 10's complement of subtrahend (number to be subtracted).

(ii) Then, we add the minuend to the 10's complement of subtrahend.

(iii) Then, we discard carry. If carry is 1 then the answer is positive and is in its true form.

(iv) If carry is not produced then the answer is negative. So, we take 10's complement to get the answer.

EXAMPLE 1.61. Perform the subtraction $(9)_{10} - (4)_{10}$ in BCD using 10's complement method.

Solution: (i) First, we obtain the 10's complement of $(4)_{10}$.

9's complement of 4 = 9 – 4 =	$(5)_{10}$
We add 1	+ 1
10's complement of 4	$(6)_{10}$

(ii) Then, we add $(9)_{10}$ and 10's complement of $(4)_{10}$.

$(9)_{10}$	1 0 0 1	
+ $(6)_{10}$	0 1 1 0	
Carry		
	1 1 1 1	Invalid BCD and carry = 0
(iii) We add $(6)_{10}$	+ 0 1 1 0	We add $(0110)_2$ for correction
	1 1	
Discard final carry [1]	0 1 0 0 (5)	Answer is positive and is in true BCD form

Therefore, $(9)_{10} - (4)_{10} = (5)_{10}$ **Ans.**

EXAMPLE 1.62. Perform $(3)_{10} - (8)_{10}$ in BCD using 10's complement method.

Solution:

(i) First, let us obtain 10's complement of $(8)_{10}$.

10's complement of $(8)_{10}$ is (9 – 8) + 1 = 2.

(ii) Then, we add 3 and 10's complement of 8.

$(3)_{10}$	0	0	1	1	
10's complement of 8 +	0	0	1	0	
		1			
	0	1	0	1	Sum is negative

Here, final carry is 0. Hence, sum is negative and is not in its true form.

(iii) Lastly, we obtain 10's complement of the sum.

For this, we subtract the sum from $(9)_{10}$

Thus,	1	0	0	1	
–	0	1	0	1	Sum
	1				
	0	1	0	0	
Add 1 +				1	
Final Result :	0	1	0	1	BCD 5

The result is negative.

Therefore, $(3)_{10} - (8)_{10} = (-5)_{10}$ **Ans.**

EXAMPLE 1.63. Perform $(54)_{10} - (22)_{10}$ in BCD using 10's complement.

Solution:

(i) First, let us obtain 10's complement of 22.

9's complement of 22	99 – 22 = 77
Add	+ 1
10's complement of 22	78

(ii) Next, we add $(54)_{10}$ and 10's complement of $(22)_{10}$.

$(54)_{10}$	0	1	0	1	0	1	0	0	
10's complement of $(22)_{10}$ +	0	1	1	1	1	0	0	0	
Carry	1	1	1						
Invalid BCD numbers	1	1	0	0	1	1	0	0	
(iii) Then, we add $(0110)_2$	0	1	1	0	0	1	1	0	
Carry	1		1		1				
Discard final carry [1]	0	0	1	1	0	0	1	0	Answer is positive and is in true BCD form
		3				2			

Therefore, $(54)_{10} - (22)_{10} = (32)_{10}$ **Ans.**

EXAMPLE 1.64. Perform $(22)_{10} - (54)_{10}$ in BCD using 10's complement.

Solution:

(i) First, let us obtain 10's complement of $(54)_{10}$.

We have $99 - 54 = 45 + 1 = 46$

Hence, 10's complement of 54 is 46

(ii) Then, we add $(22)_{10}$ and 10's complement of $(54)_{10}$.

$(22)_{10}$		0	0	1	0	0	0	1	0	
10's complement of $(54)_{10}$	+	0	1	0	0	0	1	1	0	$(46)_{10}$
						1	1			
		0	1	1	0	1	0	0	0	BCD for $(68)_{10}$

Carry = 0 and answer is invalid BCD form. Hence, answer is negative.

(iii) Next, we take 10's complement of the answer.

	1	0	0	1	1	0	0	1	BCD for $(99)_{10}$
–	0	1	1	0	1	0	0	0	Answer
	1	1							
	0	0	1	1	0	0	0	1	9's complement of answer
+								1	Add 1
							1		
	0	0	1	1	0	0	1	0	Final answer
		3				2			

Therefore, $(22)_{10} - (54)_{10} = -(32)_{10}$ **Ans.**

1.24.3. Signed-Magnitude BCD Numbers

The representation of signed decimal numbers in BCD is similar to the representation of signed numbers in binary. It is possible to use either the sign-magnitude system or signed-complement system. The sign of a decimal number is generally represented with four bits. The plus sign is denoted by four zeros and minus with the BCD equivalent of 9, i.e., 1001.

EXAMPLE 1.65. Represent the numbers + 52 and – 52 in the signed magnitude form.

Solution: The representation is shown in figure 1.29.

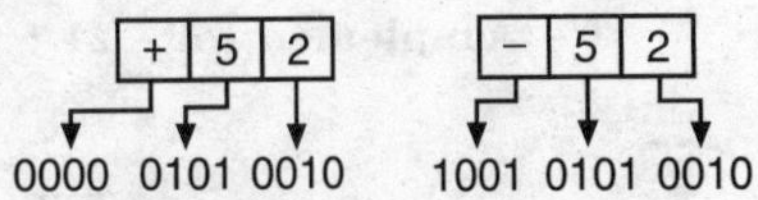

Fig. 1.29.

1.24.4. Signed-Complement BCD Numbers

The signed-magnitude system is generally not used in computers. The signed-complement system is generally preferred. The signed-complement system can be either the 9's or the 10's complement, but the 10's complement is the most preferred one. The procedure developed for the signed 2's complement system is also applicable to the signed 10's complement system for decimal numbers.

EXAMPLE 1.66. Represent + 240 and – 240 in the signed complement form.

Solution: 1. Representation of + 240.

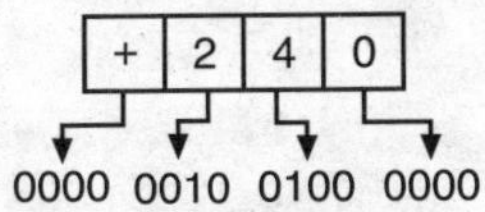

Fig. 1.30.

2. Representation of – 240.

(i) First, we represent – 240 in the signed magnitude form.

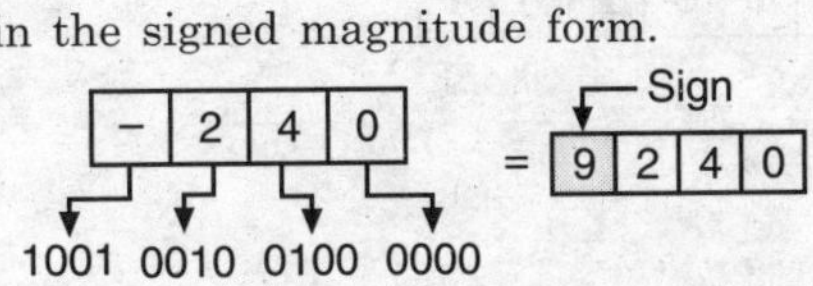

(ii) Next, we take 9's complement.

```
   9  9  9
9  2  4  0   ← Signed 9's complement
--------------
|9| 7| 5| 9|
```

No change → in the sign bit

(iii) Then, we obtain 10's complement.

```
  9  7  5  9
+          1
-------------
|9| 7| 6| 0|   ← Signed 10's complement of – 240
 ↑
Sign
```

Important Point: Here, addition is to be done by adding all digits, including the sign digits and discarding the end carry.

EXAMPLE 1.67. Add + 370 and – 240 using the signed 10's complement system.

Solution:

(i) First, we express both numbers in signed 10's complement form.

+ 370 = 0 370
– 240 = 9 760 } Signed 10's complement form

(ii) Then, we carry out subtraction.

```
+ 370 ⟶    0370
– 240 ⟶  + 9760
          ------
          ↗0130   ← Answer is in true form
Carry discarded
```

EXAMPLE 1.68. Convert the decimal integers + 21 and – 21 into 10's complement and 9's complement form. **(PTU, Jalandhar, Sem. Exam; 2002-03 (3.5 Marks)**

Solution:

9's complement of + 21 : 9's complement of 21 is 021.

10's complement of + 21 : 10's complement of 21 is 021.

9's complement of – 21 : 99 – 21 = 78

10's complement of – 21 :

```
  78  ←  9's complement of – 21
+  1  ←  We add 1
-----
  79  ←  10's complement of – 21
```

EXAMPLE 1.69. Convert the above two numbers in binary and express them in six bit (total) signed magnitude and 2's complement form. **(PTU, Jalandhar, Sem. Exam; 2002-03 (3.5 Marks)**

Solution: (i) Binary of + 21.

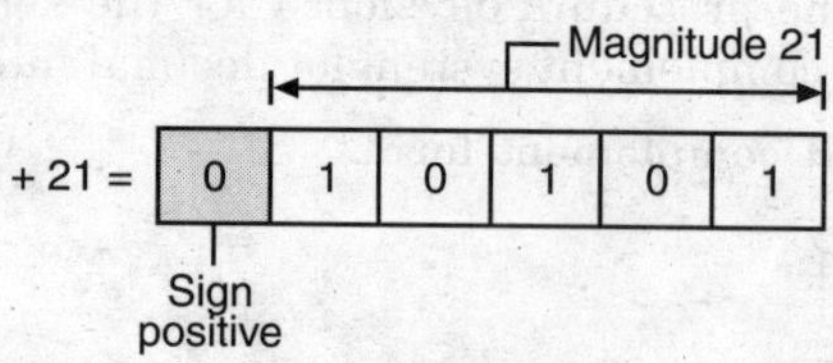

Binary of – 21

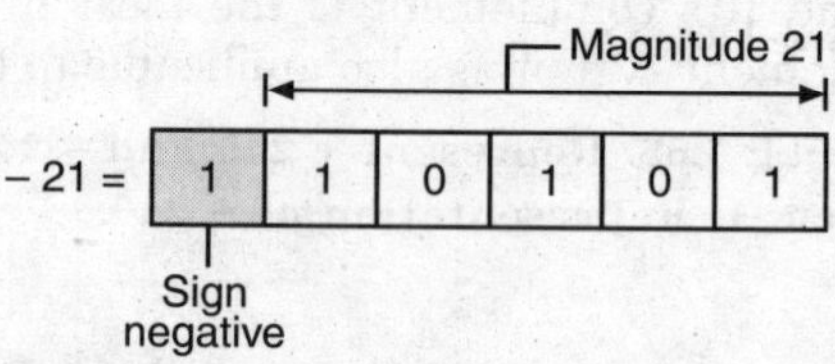

(ii) 2's complement of 21.

In the signed complemented form, the 2's complement of 21 is given by

0	1	0	1	0	1

Sign

(iii) 2's complement of – 21.

1	1	0	1	0	1	Binary – 21
1	0	1	0	1	0	Signed 1's complement representation
1	0	1	0	1	1	Signed 2's complement representation

Sign (leftmost bit)

EXAMPLE 1.70. Write 9's and 10's complement of the following numbers :

+ 9090, – 3578, + 136.8, – 136.8 **(PTU, Jalandhar, Sem. Exam; 2003-04)(6.5 Marks)**

Solution:

1. 9's complement of + 9090.

0000 1001 0000 1001 0000

2. 9's complement of – 3578.

Given number in sign magnitude form = 93578

```
      9  9  9  9
  – 9  3  5  7  8
  ---------------
    9  6  4  2  1   ← Ans.
Sign ┘
```

3. 9's complement of + 136.8.

0000 0001 0011 0110 1000

4. 9's complement of – 136.8.

```
      9  9  9 . 9
  – 9  1  3  6 . 8
  ----------------
    9  8  6  3 . 1   ← Ans.
Sign ┘
```

5. 10's Complement.

Given Number	**10's Complement**
+ 9090	09090
– 3578	96422
+ 136.8	0136.8
– 136.8	9864.1

1.24.5. Other 4-Bit BCD Codes

1.24.5.1. 2-4-2-1 Code

The 2-4-2-1 BCD is another self-complementing code. Unlike excess-3, it has the additional feature that its 4-bit code groups are weighted. In this case, the weights are 2-4-2-1, meaning that bit 1 and bit 3 have the same weight of 2. It is sometimes referred to as 2*-4-2-1 code, where the asterisk simply distinguishes one position with weight 2 from the other. Since two positions have the same weight, there are two possible bit patterns that could be used to represent some decimal digit, but only one of those patterns is actually assigned. Table 1.18 shows the code assignments for 2-4-2-1 code.

Table 1.18. 2-4-2-1 Code

Decimal digit	2-4-2-1 Code			
0	0	0	0	0
1	0	0	0	1
2	0	0	1	0
3	0	0	1	1
4	0	1	0	0
5	1	0	1	1
6	1	1	0	0
7	1	1	0	1
8	1	1	1	0
9	1	1	1	1

EXAMPLE 1.71. Convert the decimal digit 4 into 2-4-2-1 code.

Solution:

(1)	$(4)_{10}$ =	~~2~~	+ 4	+ ~~2~~	+ ~~1~~	← 2 - 4 - 2 - 1 weights
		↓	↓	↓	↓	
		0	1	0	1	← 2 - 4 - 2 - 1 Code word for $(4)_{10}$
(2)	OR $(4)_{10}$ =	2	+ ~~4~~	+ 2	+ ~~1~~	← 2 - 4 - 2 - 1 Weights
		↓	↓	↓	↓	
		1	0	1	0	← 2 - 4 - 2 - 1 Code word for $(4)_{10}$

Thus, in 2-4-2-1 code, we can have two different patterns to represent the same decimal digit. Practically, only one of them is assigned. The 2-4-2-1 codes for various decimal digits are shown in table 1.19. Table 1.19 shows two sets of 2-4-2-1 code words. Out of them, set-I does not exhibit the reflective property but set-II exhibits the reflective property. Hence, code words from set-II are generally preferred.

Table 1.19. 2-4-2-1 codes for various decimal digits

Decimal digit	Set I				Set II			
	2	4	2	1	2	4	2	1
0	0	0	0	0	0	0	0	0
1	0	0	0	1	0	0	0	1
2	0	0	1	0	0	0	1	0
3	0	0	1	1	0	0	1	1
4	0	1	0	0	0	1	0	0
5	0	1	0	1	1	0	1	1
6	0	1	1	0	1	1	0	0
7	0	1	1	1	1	1	0	1
8	1	1	1	0	1	1	1	0
9	1	1	1	1	1	1	1	1

Code words of set II exhibit the reflective property

Code words of set I do not exhibit the reflective property

1.24.5.2. Other Codes

There are various other weighted 4-bit BCD codes, each developed to have certain properties useful for special applications. Table 1.20 shows these codes, identified by the weights assigned to their bit positions. The 7421 code is somewhat different than the other codes. When 1 occurs in either of the two right most positions means that the weight of that position is subtracted, rather than added, to determine the decimal value.

Table 1.20.

Decimal	3 3 2 1	4 2 2 1	5 2 1 1	5 3 1 1	5 4 2 1	6 3 1 1	7 4 2 1	7 4 2 1
0	0 0 0 0	0 0 0 0	0 0 0 0	0 0 0 0	0 0 0 0	0 0 0 0	0 0 0 0	0 0 0 0
1	0 0 0 1	0 0 0 1	0 0 0 1	0 0 0 1	0 0 0 1	0 0 0 1	0 0 0 1	0 1 1 1
2	0 0 1 0	0 0 1 0	0 0 1 1	0 0 1 1	0 0 1 0	0 0 1 1	0 0 1 0	0 1 1 0
3	0 0 1 1	0 0 1 1	0 1 0 1	0 1 0 0	0 0 1 1	0 1 0 0	0 0 1 1	0 1 0 1
4	0 1 0 1	1 0 0 0	0 1 1 1	0 1 0 1	0 1 0 0	0 1 0 1	0 1 0 0	0 1 0 0
5	1 0 1 0	0 1 1 1	1 0 0 0	1 0 0 0	1 0 0 0	0 1 1 1	0 1 0 1	1 0 1 0
6	1 1 0 0	1 1 0 0	1 0 0 1	1 0 0 1	1 0 0 1	1 0 0 0	0 1 1 0	1 0 0 1
7	1 1 0 1	1 1 0 1	1 0 1 1	1 0 1 1	1 0 1 0	1 0 0 1	1 0 0 0	1 0 0 0
8	1 1 1 0	1 1 1 0	1 1 0 1	1 1 0 0	1 0 1 1	1 0 1 1	1 0 0 1	1 1 1 1
9	1 1 1 1	1 1 1 1	1 1 1 1	1 1 0 1	1 1 0 0	1 1 0 0	1 0 1 0	1 1 1 0

1.24.6. Excess-3 Code

Excess-3 code is a modified form of a BCD number. The Excess-3 code can be derived from the natural BCD code by adding 3 to each coded number. For example, decimal 12 can be represented in BCD as 0001 0010. Now, adding 3 to each digit we get Excess-3 code as 0100 0101 (12 in decimal). Table 1.21 shows excess-3 codes to represent single decimal digit.

Table 1.21. Excess-3 Code

Decimal digit	Excess-3 Code			
0	0	0	1	1
1	0	1	0	0
2	0	1	0	1
3	0	1	1	0
4	0	1	1	1
5	1	0	0	0
6	1	0	0	1
7	1	0	1	0
8	1	0	1	1
9	1	1	0	0

We have observed that in BCD subtraction, we have to compute 9's (or 10's) complement of the number before subtraction. In excess-3 code, we get 9's complement of a number by just complementing each bit. Because of this, excess-3 code is called *self-complementing code.*

EXAMPLE 1.72. Find the excess-3 code and its 9's complement for following decimal numbers.

(a) 592 **(b) 403**

Solution: (a) With reference to Table 1.20, we have

$$(592)_{10} = 1000\ \ 1100\ \ 0101$$

$$\text{9's complement of } (592)_{10} = 0111\ \ 0011\ \ 1010$$

(b) With reference to Table 1.20, we have

$$(403)_{10} = 0111\ \ 0011\ \ 0110$$

$$\text{9's complement of } (403)_{10} = 1000\ \ 1100\ \ 1001$$

1.24.6.1. Excess-3 Addition

To perform Excess-3 addition, we have to

(*i*) Add two Excess-3 numbers.

(*ii*) If carry = 1 → add 3 to the sum of two digits

= 0 → subtract 3

EXAMPLE 1.73.

(a)

					1	0	1	1	→ Excess-3 for 8
				+	1	0	0	1	→ Excess-3 for 6
				1	0	1	0	0	
Add 3	0	0	1	1	0	0	1	1	
	0	1	0	0	0	1	1	1	→ Excess-3 for 14
		(1)				(4)			

(b)

	0	1	0	0	→ Excess-3 for 1
+	0	1	0	1	→ Excess-3 for 2
	1	0	0	1	
Sub 3 −	0	0	1	1	
	0	1	1	0	→ Excess-3 for 3

1.24.6.3. Excess-3 Subtraction

To perform Excess-3 subtraction, we have to

(i) Complement the subtrahend

(ii) Add complemented subtrahend to minuend

(iii) If carry = 1, Result is positive. Add 3 and end around carry

(iv) If carry = 0, Result is negative. Subtract 3.

EXAMPLE 1.74.

(a) 8 – 5 **(b) 5 – 8**

Solution: (a) 8 – 5

	1 0 1 1	→ Excess-3 for 8
	+ 0 1 1 1	→ Complement of 5 in Excess-3
	1 0 0 1 0	
Add 3	0 0 1 1	
	0 1 0 1	
Add end around carry	→ 1	
	0 1 1 0	→ Excess-3 for 3

(b) 5 – 8	1 0 0 0	→ Excess-3 for 5
	+ 0 1 0 0	→ Complement of 8 in Excess-3
	1 1 0 0	
	– 0 0 1 1	
	1 0 0 1	→ Excess-3 for –3

EXAMPLE 1.75. Perform each of the following decimal additions in excess-3 code.

(a) 5 + 4 **(b)** 16 + 29 **(c)** 103 + 287

Solution:

(a)	5	1 0 0 0	Excess-3 for 5
	+ 4	+ 0 1 1 1	Excess-3 for 4
	9	1 1 1 1	No carry
		– 0 0 1 1	subtract 3
		1 1 0 0	Excess-3 for 9

(b)	16	0 1 0 0	1 0 0 1	Excess-3 for 16
	+ 29	+ 0 1 0 1	1 1 0 0	Excess-3 for 29
	45	1 0 0 1	1 0 1 0 1	
		1 ↵		Propagate carry
		1 0 1 0	0 1 0 1	
			+ 0 0 1 1	Add 3 to correct 0 1 0 1
		1 0 1 0	1 0 0 0	
		– 0 0 1 1		Subtract 3 to correct 1 0 1 0
		0 1 1 1	1 0 0 0	Excess-3 for 45

(c)					
103		0 1 0 0	0 0 1 1	0 1 1 0	Excess-3 for 103
+ 287	+	0 1 0 1	1 0 1 1	1 0 1 0	Excess-3 for 287
390		1 0 0 1	1 1 1 0 1	0 0 0 0	
			+ 1 ←		Propagate carry
		1 0 0 1	1 1 1 1	0 0 0 0	
			+	0 0 1 1	Add 3 to correct 0 0 0 0
		1 0 0 1	1 1 1 1	0 0 1 1	
		0 0 1 1	0 0 1 1		Subtract 3 to correct 1001 and 1111
		0 1 1 0	1 1 0 0	0 0 1 1	Excess-3 for 390

EXAMPLE 1.76. Perform the subtraction $(645)_{10} - (319)_{10}$ in Excess-3 using the 9's complement method.

Solution:

Excess-3 for 645 :	1 0 0 1	0 1 1 1	1 0 0 0	
Excess-3 for 319 :	0 1 1 0	0 1 0 0	1 1 0 0	
9's complement for 319 :	1 0 0 1	1 0 1 1	0 0 1 1	

645		1 0 0 1	0 1 1 1	1 0 0 0	
− 319	+	1 0 0 1	1 0 1 1	0 0 1 1	
326	1	0 0 1 0 1	0 0 1 0	1 0 1 1	Propagate carry and add end around carry
		1 ←	1	− + 1	
		0 0 1 1	0 0 1 0	1 1 0 0	
	+	0 0 1 1	0 0 1 1		Add 3 to correct 0011 and 0010
		0 1 1 0	0 1 0 1	1 1 0 0	
	−			0 0 1 1	Subtract 3 to correct 1001
		0 1 1 0	0 1 0 1	1 0 0 1	Excess-3 for 326

1.24.7. Gray Code

Gray code is a special case of unit-distance code. In unit-distance code, bit patterns for two consecutive numbers differ in only one bit position. These codes are also called cyclic codes. The Table 1.22 shows the bit patterns assigned for gray code from decimal 0 to decimal 15.

As shown in Table 1.22 for gray code any two adjacent code groups differ only in one bit position. The gray code is also called as reflected code. Notice that the two least significant bits for 4_{10} through 7_{10} are the mirror images of those for 0_{10} through 3_{10}. Similarly, the three least significant bits for 8_{10} through 15_{10} are the mirror images of those for 0_{10} through 7_{10}. In general, the n least significant bits for 2^n through $2^{n+1} - 1$ are the mirror images of those for 0 through $2^n - 1$. This observation gives us a method for counting in gray code.

Table 1.22. Gray Code

Decimal code	Gray code
0	0 0 0 0
1	0 0 0 1
2	0 0 1 1
3	0 0 1 0
4	0 1 1 0
5	0 1 1 1
6	0 1 0 1
7	0 1 0 0
8	1 1 0 1
9	1 1 0 1
10	1 1 1 1
11	1 1 1 0
12	1 0 1 0
13	1 0 1 1
14	1 0 0 1
15	1 0 0 0

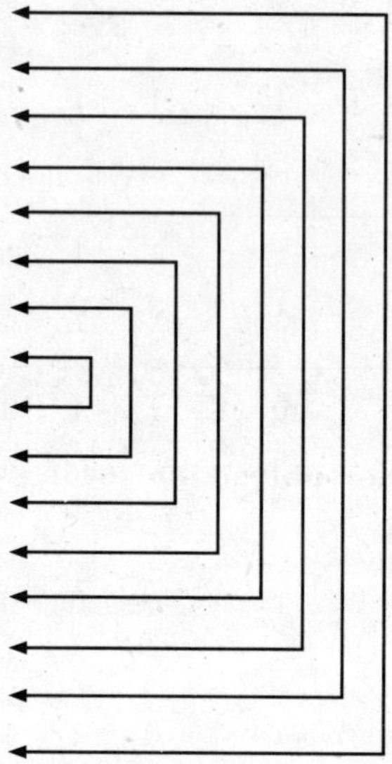

Another property of gray code is that the gray-coded number corresponding to the decimal number $2^n - 1$, for n, differ from gray coded 0 (0000) in one bit position only. For example, for n = 2, 3 and 4, we observe that

$$2^2 - 1 = 3_{10} = 0\ 0\ 1\ 0 \text{ in gray}$$

$$2^3 - 1 = 7_{10} = 0\ 1\ 0\ 0 \text{ in gray and}$$

$$2^4 - 1 = 15_{10} = 1\ 0\ 0\ 0 \text{ in gray}$$

Each differ from 0000 in one bit position only. This property places the gray code for the largest n-bit binary number at unit distance from 0.

1.24.7.1. Advantage of Gray Code

Let us consider an application where 3-bit binary code is provided to indicate position of the rotating disk. As shown in figure 1.31, brushes are used to indicate black and white portions on the disk. When the brushes are on the black portion, they output a 1. When on the white portion, they output a 0.

Now, let us consider what happens when the brushes are on the 111 sector and almost ready to enter the 000 sector. If one brush is slightly ahead of the other, say the 3rd brush, then the position is indicated by a 011 instead of a 111 or 000. This results a 180° error in the disk position. Since it is physically impossible to have all the brushes precisely aligned, some error will always be present at the edges of the sectors. If we use gray code to represent disk position then error due to improper brush alignment can be reduced. This is because the gray code assures that only one bit will change each time, the decimal number is incremented. Hence, in 3-bit code, error may occur due to one bit position. Other two bits positions of two adjacent sectors are always same and hence there is no possibility of error. Therefore, in 3-bit code, probability of error is reduced upto 66%. In case of 4-bit code, it is reduced upto 75%.

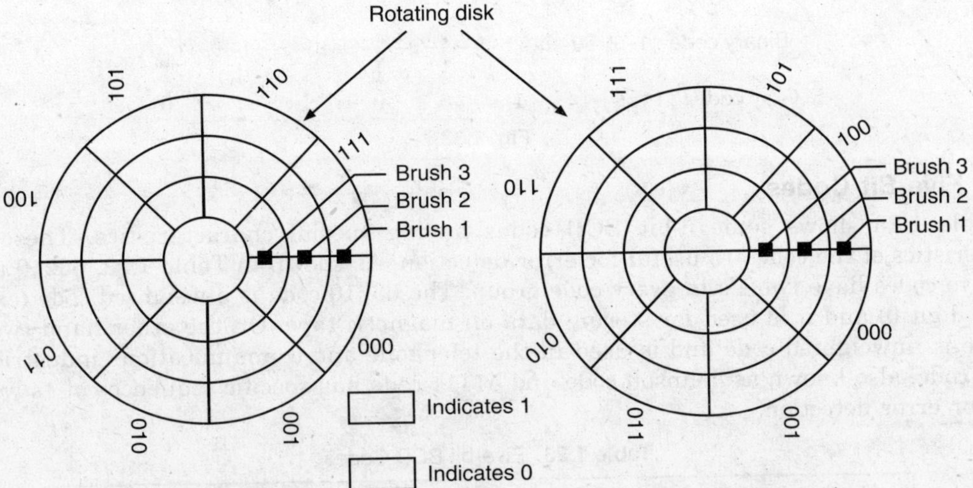

(a) Binary code (b) Gray code

Fig. 1.31. Position indication using binary and gray codes.

1.24.7.2. Gray-to-Binary Conversion

The gray to binary code conversion can be achieved using following steps :

(i) The most significant bit of the binary number is the same as the most significant bit of the gray code number. Hence, we write it down.

(ii) To obtain the next binary digit, perform an exclusive-OR-operation between the bit just written down and the next gray code bit. We write down the result.

(iii) Repeat step 2 until all gray code bits have been exclusive-ORed with binary digits. The sequence of bits that have been written down is the binary equivalent of the gray-code number.

EXAMPLE 1.77. Convert gray code 101011 into its binary equivalent.

Solution:

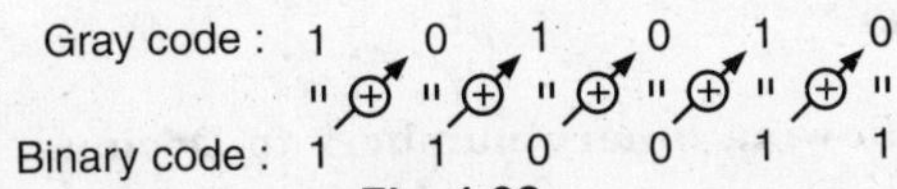

Fig. 1.32.

1.24.7.3. Binary to Gray Conversion

Let us represent binary number as

$B_1\ B_2\ B_3\ B_4 \dots B_n$ and its equivalent gray code as

$G_1\ G_2\ G_3\ G_4 \dots G_n$.

With this representation, gray code bits are obtained from the binary bits as under :

$$G_1 = B_1$$
$$G_2 = B_1 \oplus B_2$$
$$G_3 = B_2 \oplus B_3$$
$$G_4 = B_3 \oplus B_4$$
$$\vdots$$
$$G_n = B_{n-1} \oplus B_n$$

EXAMPLE 1.78. Convert 10111011 in binary into its equivalent gray code.

Solution:

Binary code : 1–⊕→0–⊕→1–⊕→1–⊕→1–⊕→0–⊕→1–⊕→1

↓ ↓ ↓ ↓ ↓ ↓ ↓ ↓

Gray code : 1 1 1 0 0 1 1 0

Fig. 1.33.

1.24.8. Five-Bit Codes

Table 1.23 shows some 5-bit BCD codes having special characteristics. These special characteristics of the code are useful for error detection. As shown in Table 1.23, 63210 and two-out-of-five codes have two 1's in every code group. The 63210 code is a weighted code (except for decimal digit 0) and it is used for storing data on magnetic tape. On the other hand two-out-of-five code is unweighted code and is used in the telephone and communications industries. Shift counter code, also known as Johnson code and 51111 code has specific sequences of 1s which are useful for error detection.

Table 1.23. Five-bit BCD Codes

Decimal	6 3 2 1 0	Two-out-of-five	Shift counter	5 1 1 1 1
0	0 0 1 1 0	0 0 0 1 1	0 0 0 0 0	0 0 0 0 0
1	0 0 0 1 1	0 0 1 0 1	0 0 0 0 1	0 0 0 0 1
2	0 0 1 0 1	0 0 1 1 0	0 0 0 1 1	0 0 0 1 1
3	0 1 0 0 1	0 1 0 0 1	0 0 1 1 1	0 0 1 1 1
4	0 1 0 1 0	0 1 0 1 0	0 1 1 1 1	0 1 1 1 1
5	0 1 1 0 0	0 1 1 0 0	1 1 1 1 1	1 0 0 0 0
6	1 0 0 0 1	1 0 0 0 1	1 1 1 1 0	1 1 0 0 0
7	1 0 0 1 0	1 0 0 1 0	1 1 1 0 0	1 1 1 0 0
8	1 0 1 0 0	1 0 1 0 0	1 1 0 0 0	1 1 1 1 0
9	1 1 0 0 0	1 1 0 0 0	1 0 0 0 0	1 1 1 1 1

FEW SOLVED PROBLEMS

EXAMPLE 1.79. Convert the following binary numbers to Decimal, Hexadecimal and Octal form:

(i) $(1\,0\,1\,1\,0\,1\,.\,1\,1\,0\,1)_2$ **(ii) $(1\,1\,0\,1\,1\,0\,1\,1\,.\,1\,0\,0\,1\,0\,1)_2$**

Solution: (i) $(1\,0\,1\,1\,0\,1\,.\,1\,1\,0\,1)_2$

Decimal Conversion

$$= 2^5 \times 1 + 0 \times 2^4 + 1 \times 2^3 + 1 \times 2^2 + 0 \times 2^1 + 1 \times 2^0 + 1 \times 2^{-1} + 1 \times 2^{-2} + 0 \times 2^{-3} + 1 \times 2^{-4}$$

$$= 32 + 8 + 4 + 1 + \frac{1}{2} + \frac{1}{4} + \frac{1}{16}$$

$$= 45 + 0.5 + 0.25 + 0.0625$$

$$= 45 + 0.8125 = 45.8125$$

Therefore, $(101101.1101)_2 = (45.8125)_{10}$ **Ans.**

Hex Conversion : 10 1101 . 1101

2 D . D $= (2D.D)_H$

Octal Conversion : 101 101 . 110 1

5 5 . 6 $4 = (55.64)_8$

(ii) $(1\,1\,0\,1\,1\,0\,1\,1\,.\,1\,0\,0\,1\,0\,1)_2$

Decimal Conversion

$$= 1 \times 2^7 + 1 \times 2^6 + 0 \times 2^5 + 1 \times 2^4 + 1 \times 2^3 + 0 \times 2^2 + 1 \times 2^1 + 1 \times 2^0 + 1 \times 2^{-1} + 0 \times 2^{-2} + 0 \times 2^{-3} + 1 \times 2^{-4} + 0 \times 2^{-5} + 1 \times 2^{-6}$$

$$= 128 + 64 + 0 + 16 + 8 + 0 + 2 + 1 + \frac{1}{2} + \frac{1}{16} + \frac{1}{64}$$

$$= 219 + 0.5 + 0.0625 + 0.015625 = 219 + 0.578125$$

Therefore, $(11011011.100101)_2 = (219.578125)_{10}$ **Ans.**

Hex Conversion

$(11011011.100101)_2 = (DB.94)_H$ (Hex)

D B 9 4

Octal Conversion

$(11011011.100101)_2 = (333.45)_8$ (Octal)

3 3 3 4 5

EXAMPLE 1.80. Convert $(2AC5.D)_H$ to decimal, octal and binary.

Solution: Hex to Decimal

$$(2AC5.D)_{16} = 2 \times 16^3 + 10 \times 16^2 + 12 \times 16^1 + 5 \times 16^0 + 13 \times \frac{1}{16}$$

$$= (10949.8125)_{10}$$

Hex to Binary

$(2AC5.D)_{16}$

$(0010 \quad 1010 \quad 1100 \quad 0101 \;.\; 1101)_2$

Binary to Octal

000	010	101	011	000	101	.	110	100
0	2	5	3	0	5	.	6	4

Therefore, we have $(2AC5.D)_{16} = (25305.64)_8$. **Ans.**

EXAMPLE 1.81. Express the following numbers in sign magnitude 1's and 2's complement

(i) –56 **(ii) 107**

Solution: (i) –56

56 =	0 1 1 1 0 0 0	
–56 =	1 0 0 0 1 1 1	1's Complement
	+ 1	
=	1 0 0 1 0 0 0	2's Complement

(ii) 107

107 =	0 1 1 0 1 0 1 1	
–107 =	1 0 0 1 0 1 0 0	1's Complement
	+ 1	
=	1 0 0 1 0 1 0 1	2's Complement

EXAMPLE 1.82. Using 2's complement perform the following : $(42)_{10} - (68)_{10}$.

Solution:

(i) First, let us convert $(42)_{10}$ to its binary equivalent.

(ii) Then, we convert $(68)_{10}$ to its binary equivalent.

(iii) Now, let us find 2's complement of 68.

Thus,

$(68)_{10}$ =	1 0 0 0 1 0 0	
	1 1	← Carry
	0 1 1 1 0 1 1	1's Complement
+	1	
	0 1 1 1 1 0 0	2's Complement

Next, we add $(1\ 0\ 1\ 0\ 1\ 0)_2$ and $(0\ 1\ 1\ 1\ 1\ 0\ 0)_2$

```
    1 1 1              ← Carry
    0 1 0 1 0 1 0
  + 0 1 1 1 1 0 0
    -------------
    1 1 0 0 1 1 0
```

We have, 1 1 0 0 1 1 0

```
  21
2)42         R
 -42
   0 ────→   0   LSB

  10
2)21
 -20
   1 ────→   1

   5
2)10
 -10
   0 ────→   0

   2
2)5
 -4
  1 ────→    1

   1
2)2
 -2
  0 ────→    0

   0
2)1
 -0
  1 ────→    1   MSB
```

≡

	Q	R	
2	42		↑ LSB
2	21	0	
2	10	1	
2	5	0	
2	2	1	
2	1	0	
	0	1	MSB

Here : Q : Quotient
R : Remainder

Since we have subtracted larger number from smaller number result is negative and hence it is necessary to take 2's complement of this result.

Lastly, let us find 2's complement of $(1\ 1\ 0\ 0\ 1\ 1\ 0)_2$

```
    1 1 0 0 1 1 0    Number
    0 0 1 1 0 0 1    1's complement
  + 0 0 0 0 0 0 1    2's complement
    -------------
    0 0 1 1 0 1 0    = (26)10
```

Since, we have taken 2's complement, result is $(-26)_{10}$.

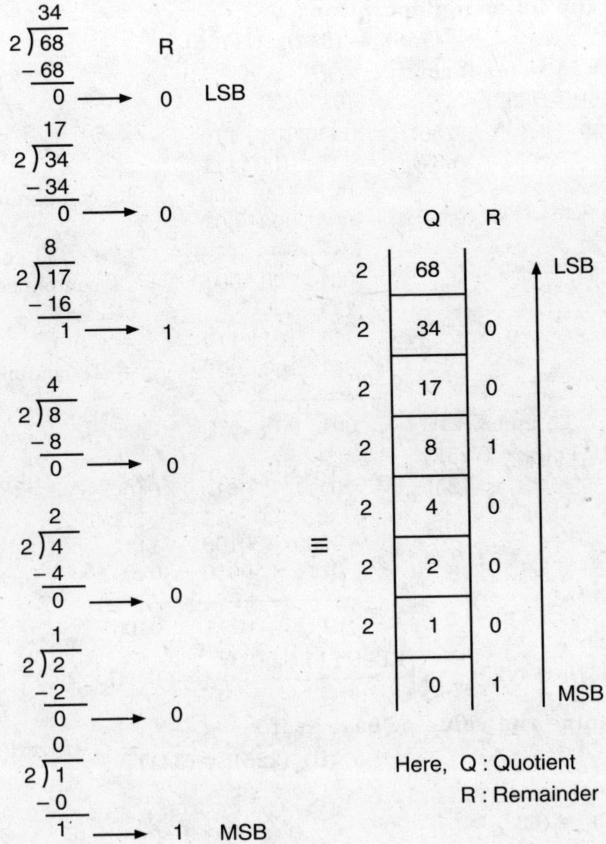

Here, Q : Quotient
R : Remainder

EXAMPLE 1.83. **Solve $(DDCC)_{16} + (BBAA)_{16} = (...)_{16}$.**

Solution: We have,

DDCC →		1101	1101	1100	1100
BBAA →	+	1011	1011	1010	1010
19976	11001	1001	0111	0110	

Result → $(19976)_{16}$ **Ans.**

EXAMPLE 1.84. **Solve $(723)_8 + (237)_8 = (...)_8$**

Solution: We have,

723	→		111	010	011
+ 237	→	+	010	011	111
1 162		1	001	110	010

Result → $(1162)_8$ **Ans.**

EXAMPLE 1.85. **Solve $(10101100)_2 - (00111000)_2$**

Solution: We have,

```
  10101100
- 00111000
----------
  01110100
```

Result → $(01110100)_2$. **Ans.**

* A life, which does not go into action, is a failure. — **Arnold J. Toynbee**

EXAMPLE 1.86. Perform the following operation :

$$(756)_8 - (637)_8 + (725)_{16}$$

Express the answer in octal form.

Solution: (i) We have, $(756)_8 - (637)_8$

$$\begin{array}{rrrrl}
(756)_8 = & 111 & 101 & 110 & \\
(637)_8 = & 110 & 011 & 111 & \\
 & 001 & 100 & 000 & \text{1's complement of } (637)_8 \\
+ & 000 & 000 & 001 & \\
\hline
 & 001 & 100 & 001 & \text{2's complement of } (637)_8 \\
 & \mathbf{111} & & & \leftarrow \text{carry} \\
 & 111 & 101 & 110 & \rightarrow (756)_8 \\
+ & 001 & 100 & 001 & \rightarrow \text{2's complement of } (637)_8 \\
\hline
\text{Ignore carry } 1 & 001 & 001 & 111 &
\end{array}$$

(ii) We have, $(001\ 001\ 111)_2 + (725)_{16}$

$$\begin{array}{rrrrl}
(725)_{16} = & 0111 & 0010 & 0101 & \\
 & & 1 & 111 & \leftarrow \text{carry} \\
 & 0000 & 0100 & 1111 & \\
+ & 0111 & 0010 & 0101 & \rightarrow (725)_{16} \\
\hline
 & 0111 & 0111 & 0100 &
\end{array}$$

Therefore, $(756)_8 - (637)_8 + (725)_{16} = (\underbrace{111}_{3}\underbrace{0\,11}_{5}\underbrace{110}_{6}\underbrace{100}_{4})_2 = (3564)_8$ **Ans.**

EXAMPLE 1.87. (a) Determine the value of base x, if :

(i) $(193)_x = (623)_8$ **(ii) $(225)_x = (341)_8$**

(iii) $(211)_x = (152)_8$

Solution: (i) We have, $(193)_x = (623)_8$

Converting octal into decimal, we have

$$6 \times 8^2 + 2 \times 8 + 3 = (403)$$

$$(623)_8 = (403)_{10}$$

Hence, $$(193)_x = (403)_{10}$$

$$1 \times x^2 + 9 \times x + 3 + x^0 = 403$$

or $$x^2 + 9x + 3 = 403$$

Therefore, $$x = 16 \text{ or } x = -25$$

Since, negative is not applicable, therefore

$$x = 16$$

Hence, $$(193)_{16} = (623)_8$$

(ii) We have, $(225)_x = (341)_8$

Converting octal into decimal, we get

$$3 \times 8^2 + 4 \times 8^1 + 1 \times 8^0 = 255_{10}$$

Hence, $$x = 10$$

(iii) We have, $(211)_x = (152)_8$

Converting octal into decimal, we obtain

$$2x^2 + x + 1 = 8^2 \times 1 + 5 \times 8 + 2$$

$$2x^2 + x + 1 = 106$$

Hence, $$2x^2 + x - 105 = 0$$

or $$(x - 7)(2x + 15) = 0$$

$$x = 7$$

Therefore, we have $(211)_7 = (152)_8$. **Ans.**

Knowledge is, indeed, that which, next to virtue, truly and essentially raises one man above another.

EXAMPLE 1.88. (a) Perform the subtraction using :

(1) 1's complement method **(2) 2's complement method**

(i) $(11010)_2 - (10000)_2$ **(ii) $(1000100)_2 - (1010100)_2$**

Solution: (1) 1's complement method

(i) $(1\,1\,0\,1\,0)_2 - (1\,0\,0\,0\,0)_2$

```
      1 1 0 1 0
    + 0 1 1 1 1      (one's complement of 1 0 0 0 0)
    -----------
    1 0 1 0 0 1
    +└────────→1
    -----------
        1 0 1 0
```

Therefore, $(1\,1\,0\,1\,0)_2 - (1\,0\,0\,0\,0)_2 = (0\,1\,0\,1\,0)_2$. **Ans.**

(ii) $(1\,0\,0\,0\,1\,0\,0)_2 - (1\,0\,1\,0\,1\,0\,0)_2$

```
      1 0 0 0 1 0 0
    + 0 1 0 1 0 1 1      (one's complement of 1 0 1 0 1 0 0)
    ---------------
      1 1 0 1 1 1 1
```

No carry, so answer is negative.

Therefore, 0 0 1 0 0 0 0 (one's complement of 1 1 0 1 1 1 1)

(2) 2's complement method

(i) $(1\,1\,0\,1\,0)_2 - (1\,0\,0\,0\,0)_2$

```
                  1 1 0 1 0
                + 1 0 0 0 0      (2's complement of 1 0 0 0 0)
                -----------
Ignore carry →    0 1 0 1 0
```

Therefore, $(1\,1\,0\,1\,0)_2 - (1\,0\,0\,0\,0) = (0\,1\,0\,1\,0)_2$.

(ii) $(1\,0\,0\,0\,1\,0\,0)_2 + (1\,0\,1\,0\,1\,0\,0)_2$

```
      1 0 0 0 1 0 0
    + 0 1 0 1 1 0 0      (2's complement of 1 0 1 0 1 0 0)
    ---------------
      1 1 1 0 0 0 0
```

Since final carry is zero, so the answer is negative and in 2's complement form.

Therefore, $(1\,0\,0\,0\,1\,0\,0)_2 - (1\,0\,1\,0\,1\,0\,0)_2 = (0\,0\,1\,0\,0\,0\,0)_2$. **Ans.**

EXAMPLE 1.89. Give the binary, BCD, excess-3, gray code representations of decimal numbers : 5, 8, 14.

Solution:

Decimal Numbers	Binary Code	BCD Code	Excess-3 Code	Gray Code
5	0 1 0 1	0 1 0 1	1 0 0 0	0 1 1 1
8	1 0 0 0	1 0 0 0	1 0 1 1	1 1 0 0
14	1 1 1 0	0 0 0 1 0 1 0 0	0 1 0 0 0 1 1 1	1 0 0 1

EXAMPLE 1.90. Convert each pair of decimal number to BCD and add.

(i) $(65)_{10} + (58)_{10}$

(ii) $(113)_{10} + (101)_{10}$.

Solution:

(i)

	Decimal		BCD (tens)	BCD (units)	
	65		0 1 1 0	0 1 0 1	BCD for 65
+	58	+	0 1 0 1	1 0 0 0	BCD for 58
	123		1 0 1 1	1 1 0 1	1101 > 9, so,
			+	0 1 1 0	Add 6
			1 0 1 1	1 0 0 1 1	
			+ 1 ←		Propagate carry to next digit
			1 1 0 0	0 0 1 1	1100 > 9, so,
		+	0 1 1 0		Add 6
		000 1	0 0 1 0	0 0 1 1	BCD for 123

(ii)

	Decimal	BCD	
	113	0 0 0 1 0 0 0 1 0 0 1 1	BCD for 113
+	101	0 0 0 1 0 0 0 0 0 0 0 1	BCD for 101
	214	0 0 1 0 0 0 0 1 0 1 0 0	BCD for 214

EXAMPLE 1.91. Convert each of the following decimal numbers to excess-3 code

(i) $(18)_{10}$ (ii) $(56)_{10}$

Solution: (i) 18_{10} = 0100 1011 in excess-3

(ii) 56_{10} = 1000 1001 in excess-3

EXAMPLE 1.92. Convert $(85.63)_{10}$ into its equivalent binary number.

Solution:

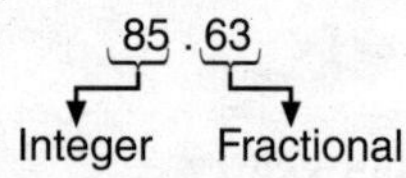

(i) First, let us separate integer and fractional parts as under :

(ii) Now, we convert the integer

Divisor	Number	Remainder	
2	85		
2	42	1	LSB
2	21	0	↑
2	10	1	↑
2	5	0	↑
2	2	1	↑
2	1	0	↑
	0	1	MSB

Therefore, $(85)_{10} = (1\,0\,1\,0\,1\,0\,1)_2$

(iii) Now, we convert the fractional part

$0.63 \times 2 = 1.26$ → 1 MSB

$0.26 \times 2 = 0.52$ → 0

$0.52 \times 2 = 1.04$ → 1

$0.04 \times 2 = 0.08$ → 0

$0.08 \times 2 = 0.16$ → 0 LSB

Therefore, $(0.63)_{10} = (1\,0\,1\,0\,0)_2$

(iv) Finally, we combine the results of steps 2 and 3.

Hence, $(85.63)_{10} = (1010101.10100)_2$ **Ans.**

EXAMPLE 1.93. Convert $(3000.45)_{10}$ into its equivalent octal number.

Solution:

(i) First, let us separate the integer and fractional parts.

Integer part = 3000 and fractional part = 0.45.

(ii) Then, we convert the integer part

8	3000	
8	375	0 LSD
8	46	7
8	5	6
	0	5 MSD

Column of quotients — Column of remainders

Therefore, $(3000)_{10} = (5670)_8$

(iii) Now, we convert the fractional part

Base	Product	Carry
$0.45 \times 8 = 3.60$		3 MSD
$0.60 \times 8 = 4.80$		4
$0.80 \times 8 = 6.40$		6
$0.40 \times 8 = 3.20$		3 LSD

Therefore, $(0.45)_{10} = (0.3463)_8$

Fig. 1.34.

(iv) Finally, we combine the results of steps 2 and 3.

Combining the results of steps 2 and 3 we get the answer.

Therefore, $(3000.45)_{10} = (5670.3463)_8$. **Ans.**

EXAMPLE 1.94. Convert $(2003.31)_{10}$ into its equivalent hex number.

Solution:

(i) First, we separate the integer and fractional parts as under :

2003 . 31 (Integer ← 2003; 31 → Fractional)

(ii) Then, we convert integer part.

			Hex
16	2003		
16	125	3	→ 3 LSD
16	7	13	→ D
	0	7	→ 7 MSD

Therefore, $(2003)_{10} = (7D3)_{16}$

(iii) Next, we convert the fractional part into hex.

Decimal fraction	× Base	Product	Recorded digit Decimal	Recorded digit Hex
0.31	× 16 =	4.96	4	4 MSD
0.96	× 16 =	15.36	15	F
0.36	× 16 =	5.76	5	5
0.76	× 16 =	12.16	12	C
0.16	× 16 =	2.56	2	2 LSD

Therefore, $(0.31)_{10} = (0.4F5C2)_{16}$

(iv) Finally, we combine the results of steps 2 and 3.

Thus, combining the results of steps 2 and 3, we get

$(2003.31)_{10} = (7D3.4F5C2)_{16}$ **Ans.**

FEW IMPORTANT EXAMPLES

EXAMPLE 1.95. Convert the decimal integers +21 and –21 into 10's complement and 9's complement.

(PTU, Jalandhar, Sem. Exam 2009-2010)

Solution: 9's complement of + 21 : 9's complement of 21 is 021.

10's complement of + 21 : 10's complement of 21 is 021.

9's complement of – 21 : 99 – 21 = 78

10's complement of – 21 :

$$\begin{array}{rll} & 78 & \leftarrow \quad \text{9's complement of} - 21 \\ + & 1 & \leftarrow \quad \text{Add 1} \\ \hline & 79 & \leftarrow \quad \text{10's component of} -21 \end{array}$$

EXAMPLE 1.96. Convert the above two numbers in binary and express them in six bit (total) signed magnitude and 2's complement.

(PTU, Jalandhar, Exam 2008-2009)

Solution: Binary of + 21 Binary of – 21

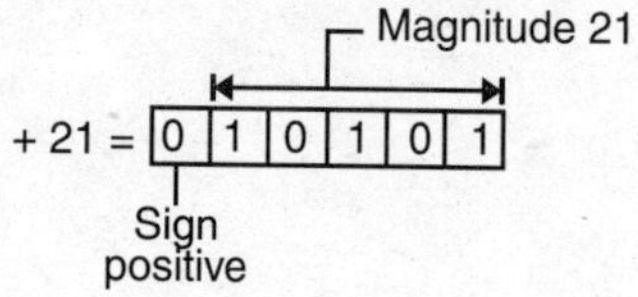

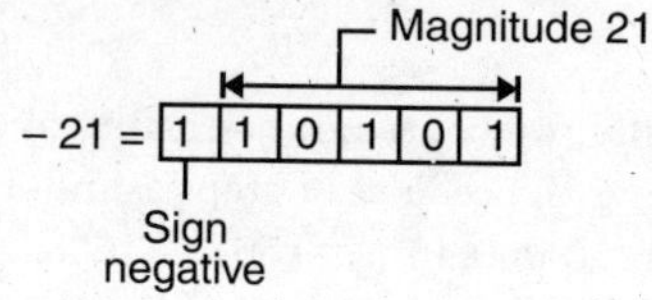

2's complement of 21

in the signed complement form, the 2's complement of 21 is given by

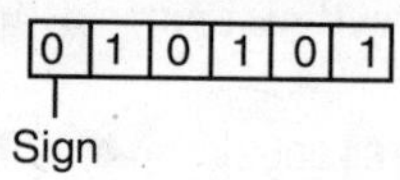

2's complement of – 21

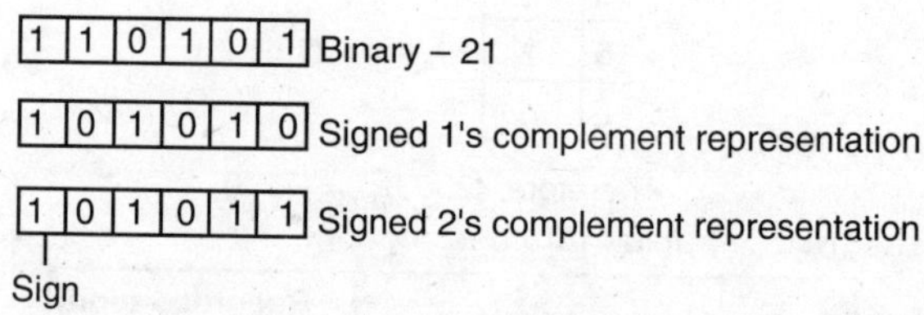

EXAMPLE 1.97. How is 2's complement advantageous for adder circuit design? Explain.

(PTU, Jalandhar, Sem. Exam 2007-08)

Solution: The 2's complement of the number to be subtracted helps us to obtain the subtraction by simply adding the two number. Hence, it is possible to use adder for carrying out subtraction.

2's complement, please go through the text.

EXAMPLE 1.98. Explain the floating-point data representation for decimal number and binary number. What are the advantages and disadvantages of having different values for base and radix in floating-point number format? Consider both the cases when the base is an integral power of the radix and when it is not.

Solution: Please go through the text.

1.25 ALPHANUMERIC CODES

(i) As a matter of fact, a binary digit or bit can represent only two symbols as it has only two states 0 or 1. But this is not enough for communication between two computers because there we need many more symbols for communication. These symbols are required to represent:

(a) 26 alphabets with capital and small letters

(b) Numbers from 0 to 9.

(c) Punctuation marks and other symbols

(ii) Therefore, instead of using only single binary bits, a group of bits is used as a code to represent a symbol. The number of bits used per code word will be dependent on the total number of symbols to be represented e.g. if 5 bits are used per code word then $2^5 = 32$ combination would be possible. If the word length is increased to 8, then, the numbr of combinations would be $2^8 = 256$, hence, an 8-bit code can represent 256 symbols.

(iii) A set of such code words is called as **code set**. The following three code stes are very commonly used for the data representation:
(a) American standard code for information interchange (ASCII)
(b) Extended Binary coded-decimal interchange code (EBCDIC)
(c) Five bit Baudot code.

ASCII code is a 7-bit code whereas EBCDIC is an 8-bit code. ASCII code is more commonly used world wide while EBCDIC is used primarily in large IBM computers.

1.25.1. ASCII (American Standard Code for Information Interchange)

It is defined by American national Standards Institute (ANSI). It is a 7 bit code with i.e. 128 possible combinations and all of them have defined meanings. The ASCII code set consists of 94 printable characters, SPACE and DELETE characters, and 32 control symbols. This has been shown in tables 1.24 and 1.25.

Table 1.24.

Bit numbers	7	0	0	0	0	1	1	1	1
	6	0	0	1	1	0	0	1	1
	5	0	1	0	1	0	1	0	1
4321	0	0	1	2	3	4	5	6	7
0000	0	NUL	DLE	SPACE	0	@	P		p
0001	1	SOH	DC1	!	1	A	Q	a	q
0010	2	STX	DC2	"	2	B	R	b	r
0011	3	ETX	DC3	#	3	C	S	c	s
0100	4	EOT	DC4	$	4	D	T	d	t
0101	5	ENQ	NAK	%	5	E	U	e	u
0110	6	ACK	SYN	&	6	F	V	f	v
0111	7	BEL	ETB	'	7	G	W	g	w
1000	8	BS	CAN	(	8	H	X	h	x
1001	9	HT	EM	)	9	I	Y	i	y
1010	A	LF	SUB	*	:	J	Z	j	z
1011	B	VT	ESC	+	;	K	[	k	{
1100	C	FF	FS	'	<	L	\	l	|
1101	D	CR	GS	–	=	M	]	m	}
1110	E	SO	RS		>	N	^	n	~
1111	F	SI	US	/	?	O	—	o	DEL

EXAMPLE 1.99. Using the Table 1.24, obtain the ASCII code word for the character K.

Solution: Considering Table 1.24, the co-ordinates of the character K are (4, B). Hence, the seven bit ASCII code for K is as under:

	7	6	5	4	3	2	1
ASCII Code :	1	0	0	1	0	1	1
	4			B			

Other symbols and their meanings

The control symbols are codes reserved for special functions. The control symbols are as listed in table 1.25. CR (Carriage Return) and LF (Line Feed) are the symbols used for basic operations of printers or displays. The symbols ACK (Acknowledgment) or NAK (Negative Aknowledgement) are used for the error control. And the symbols STX (Start to Text) and ETX (End of Text) are used for grouping of data character. The other symbols and their meanings are as listed in Table 1.25.

Table 1.25. Control symbols used in ASCII code set

ACK	Acknowledgment	FF	Form Feed
BEL	Bell	FS	File Separator
BS	Backspace	GS	Group Separator
CAN	Cancel	HT	Horizontal Tabulation
CR	Carriage Return	LF	Line feed
DC1	Device Control 1	NAK	Negative Acknowledgment
DC2	Device Control 2	NUL	Null
DC3	Device Control 3	RS	Record Separator
DC4	Device Control 4	SI	Shift-In
DEL	Delete	SO	Shift-Out
DLE	Data Line Escape	SOH	Start of Heading
EM	End of Medium	STX	Start of Text
ENQ	Enquiry	SUB	Substitute Character
EOT	End of Transmission	SYN	Synchronous Idle
ESC	Escape	US	Unit separator
ETB	End of Transmission Block	VT	Vertical Tabulation
ETX	End of Test		

1.25.2. EBCDIC (Extended Binary Coded Decimal Interchange Code)

This is an 8-bit code. However, all the possible 256 combinations are not used. There is no parity bit used to check error in the basic code set. The EBCDIC code set has been shown in Table 1.26.

Table 1.26. EBCDIC codes

Bit numbers	0	1	2	3	4	5	6	7	8	9	A	B	C	D	E	F
b_0	0	0	0	0	0	0	0	0	1	1	1	1	1	1	1	1
b_1	0	0	0	0	1	1	1	1	0	0	0	0	1	1	1	0
b_2	0	0	1	1	0	0	1	1	0	0	1	1	0	0	1	1
b_3	0	1	0	1	0	1	0	1	0	1	0	1	0	1	0	1
$b_4 - b_7$ 4567	0	1	2	3	4	5	6	7	8	9	A	B	C	D	E	F
0000 0	NUL	DLE				SP	&	–								0
0001 1	SOH	SBA					\		a	j			A	J		1
0010 2	STX	EUA		SYN					b	k	s		B	K	S	2
0011 3	ETX	IC							c	l	t		C	L	T	3
0100 4									d	m	u		D	M	U	4
0101 5	HT	NL							e	n	v		E	N	V	5
0110 6			ETB						f	o	w		F	O	W	6
0111 7			ESC	EOT					g	p	x		G	P	X	7
1000 8									h	q	y		H	Q	Y	8
1001 9		EM							i	r	z		I	R	Z	9
1010 A					¢	!	!									
1011 B						$		#								
1100 C		DUP		RA	<	*	%	@								
1101 D		SF	ENQ	NAK	(	)	—									
1110 E		FM			+	;	>	=								
1111 F		ITB		SUB	|	¬	?	"								

Code word Format

The 8-bit code word format of an EBCDIC code has been shown in figure 1.35.

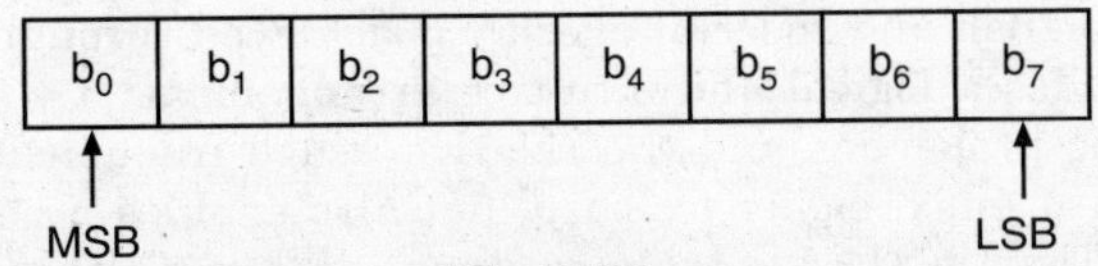

Fig. 1.35.

For example, the EBCDIC code for representing SP is under:

	b_0	b_1	b_2	b_3	b_4	b_5	b_6	b_7
SP =	1	0	1	0	0	0	0	0

1.25.3. Hollerith Code

This code is used in punched cards. Before discussing the code, let us see how the punched cards are organized.

Organization of a Card

(i) Each card has vertically oriented 80 columns and horizontally oriented 12 rows as shown in figure 1.36.

(ii) Columns are numbered through 0 to 80 and rows are numbered through 0 to 12.

(iii) EAch character is uniquely identified by rows punched in that column. For example, the letter A is a 12 – 1 punch (a punch in the 12 row and a punch in the 1 row).

(iv) The Hollerith code is a BCD and not natural binary oriented. Translation to and from EBCDIC is simple. Most large computers use punched cards. Hence, they use Hollerith code for their card readers and punches and FECDIC within the computer.

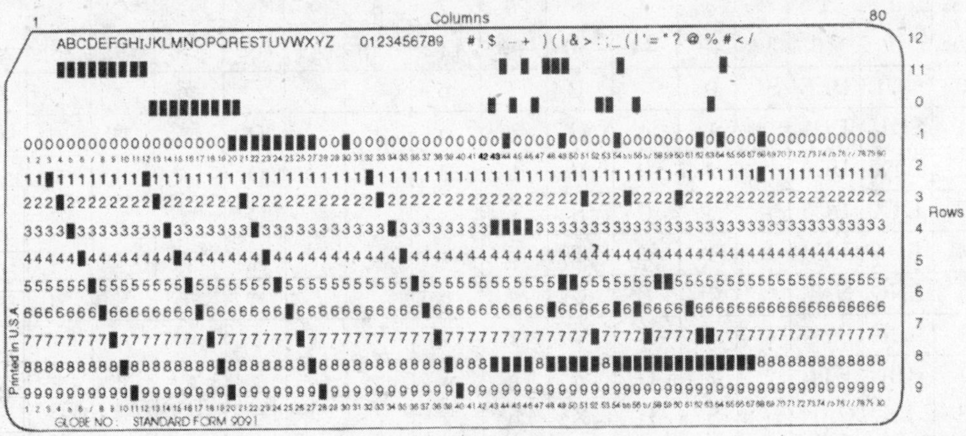

Fig 1.36. Hollerith code

1.25.4. Indian Script Code for Information Interchange (ISCII)

Since the 70s, different committees of the Department of Electronics and the Department of Official Language have been evolving different codes and keyboards which could cater to all the Indian scripts due to their common phonetic structure. Earlier efforts could not keep the ASCII code intact. The BIS standards IS 13194 :1991 confirms to the earlier standard IS 10402:1982, *8-bit* coded character set for information interchange. It is intended for use in all computer and communication media which allow usage of 7 or 8 bit characters. In an 8-bit environment, the lower 128 characters are the same as defined IS 10315:1982, 7-bit coded charactrer set for information interchange also known as **ASCII character set**.

The top 128 characters cater to all the ten Indian scripts based on the ancient Brahmi script. In a 7-bit environment, the control code SI can be used for invocation of ISCII code set, the control code SO can be used for reselection of the ASCII code set. The common INSCRIPT keyboard overlay allows typing of all the ten Indian scripts. This overlays fits on any existing English keyboard. Alternating between the English and Inscript overlay is achieved through the CAPSLOCK key. The INSCRIPT keyboard, provides a logical and intuitive arrangement of vowels and consonants. It is based both on the phonetic properties and the relative usage frequencies of the letters. Not only does this made the keyboard much easier to learn, but also enables a person to type subsequently in all the Indian Scripts. The ISCII code table is a super-set of all the characters required in the ten Brahmi based Indian scripts. An optimal keyboard overlay for these scripts is made posible by the phonetic nature of the alphabet. The differences between scripts primarily are in their written forms, where different combination rules get used.

There are manifold advantages in having a common code and keyboard for all the Indian scripts. Any software which allows ISCII codes to be used, can be used in any Indian script, enhancing its commercial viability. Furthermore, immediate transliteration between different Indian scripts becomes possible, just by changing the display modes. The 8-bit ISCII code retains the standard ASCII code, while the Indian script keyboard overlay is designed for the standard English can co-exist with Indian scripts. This approach also makes it feasible to used Indian scripts along with existing English computers and software, so long as 8-bit character codes are allowed. The ISCII code (Indian Script Code for Information Interchange), co-exists with the standard English code : It contains only the basic alphabet arranged in an order which confirms to that in most of the Indian dictionaries. As the display rendering depends entirely on the composition methodology, it is possible to view the same text in different styles prevalent in the script. It is possible to transliterate the text to Roman script, using additional signs on the letters (diacritic marks), for removing ambiguities. It is even possible to have the initial letters capitalised in the Romann text.

Table 1.27.

	Hex	0	1	2	3	4	5	6	7	8	9	A	B	C	D	E	F
Hex	Dec	0	16	32	48	64	80	96	112	128	144	160	176	192	208	224	240
0	0	NUL	DLE	SP	0	@	P	'	p				ओ	ढ	र	O	EXT
1	1	SOH	DC1	!	1	A	Q	a	q			O	ओं	ण	ल	O	0
2	2	STX	DC2	"	2	B	R	b	r			O	ऑ	त	ळ	O	१
3	3	ETX	DC3	#	3	C	S	c	s			O	क	थ	ऴ	O	२
4	4	EOT	DC4	$	4	D	T	d	t			अ	ख	द	व	ौ	३
5	5	ENQ	NAK	%	5	E	U	e	u			आ	ग	ध	श	Oा	४
6	6	ACK	SYN	&	6	F	V	f	v			इ	घ	न	ष	Oा	५
7	7	BEL	ETB	'	7	G	W	g	w			ई	ङ	ऩ	स	Oृ	
8	8	BS	CAN	(	8	H	X	h	x			उ	च	प	ह	Q	
9	9	HT	EM	)	9	I	Y	i	y			ऊ	छ	फ	INV	।	
A	10	LF	SUB	*	:	J	Z	j	z			ऋ	ज	ब	Oा		
B	11	VT	ESC	+	;	K	[	k	{			ऎ	झ	भ	f O		
C	12	FF	FS	,	<	L	\	|	|			ए	ञ	म	Oा		
D	13	CR	GS	–	=	M	]	m	}			ऐ	ठ	य	Oृ		
E	14	SO	RS	.	>	N	^	n	~			ऍ	ड	य़	Oृ		
F	15	SI	US	/	?	O	—	0	DEL			ऑ	ड	र	Ȯ	ATR	

1.26 CODE COMPARISON

1.26.1 Comparison of Binary and Gray Codes

S.No.	Parameter	Binary code	Gray code
1.	Type of code	Weighted code	Unweighted code
2.	Number of digits changing their value at a time.	Any number of digits can at a time change their value.	Only one digit can change its value at a time.
3.	Use in arithmetic	Possible	Not possible
4.	Use for representing a decimal number	We can use binary code to represent a decimal number	We can not use the gray code to represent a decimal number directly
5.	Application	In digit communication and in digital computers.	In A to D converters, speed measurements.

1.26.2 Comparison of Excess-3 code and Gray code

S.No.	Parameter	Excess – 3 code	Gray code
1.	Type of code	Unweighted code	Unweighted code
2.	Obtained from	Natural binary code by adding 3	Natural binary code
3.	Number of bits changing simultaneously	Any number of bits.	One bit only at a time.
4.	Application	For coding the binary information	In A to D converter, or speed measurement.

1.26.3. Comparison of BDC and Gray Codes

S.No.	Parameter	BCD code	Gray code
1.	Type of code	Un weighted code	Unweighted code
2.	Highest 4-bit	Equivalent to 9 decimal	Equivalent to 15 decimal
3.	Arithmetic operation	Possible	Not possible
4.	Number of bits changing at a time.	Can be more than 1.	Only 1.
5.	Applications	BCD counter, simplification arithmetic circuits.	A/D converter, speed measurement

GLOSSARY

1. **Active HIGH input :** Digital input which executes its function when a HIGH is present.
2. **Active LOW input :** Digital input which executes its function when a LOW is present.
3. **A/D converter :** Device for converting an analog voltage into a proportional digital quantity. Types include a microprocessor-compatible device with binary outputs or meter-type with decimal outputs.
4. **Alphanumeric :** Consisting of numbers, letters and other characters. ASCII is a common alphanumeric code.
5. **Analog :** A branch of electronics dealing with infinitely varying quantities. Also referred to as *linear electronics*.
6. **Analog to digital :** Conversion of an analog signal to a digital quantity such as binary.
7. **American Standard Code for information interchange :** One of the most widely used alphanumeric codes.

8. **Binary :** Base 2 number system using numbers 0 or 1.
9. **Binary-coded decimal :** A common code in which each decimal digit (0 – 9) is represented by a 4-bit group.
10. **Bit :** A single binary digit (0 or 1.) Useful in representing on-off switching in digital circuits. An acronym for *binary digit*.
11. **Digital :** Branch of electronics dealing with discrete signal levels. Signals are commonly HIGH or LOW and may be represented by binary numbers.
12. **Digital to analog :** Conversion of a digital signal to its analog equivalent, such as a voltage.
13. **Digitize :** To convert an analog signal into digital units or pulses.
14. **Dual-in-line package :** Popular packaging method for ICs.
15. **Even parity :** In data transmission, sending a parity bit that will make the number of 1s in a group even.
16. **Extended Binary-Coded :** An 8-bit alphanumeric code used mainly on mainframe Decimal Interchange Code computers.
17. **Floating input :** An input not held HIGH or LOW which may "float" either HIGH, LOW or in between. Can cause problems.
18. **Hexadecimal :** Base 16 number system using characters 0 through 9, A, B, C, D, E, and F. Used to represent binary numbers 0000 through 1111.
19. **IEEE :** Institute for Electrical and Electronic Engineers.
20. **Least significant bit :** The bit position in a binary number with the least weight.
21. **Nibble :** One half a byte. A 4-bit binary word.
22. **Octal :** Base 8 number system using characters 0 through 7.
23. **Odd parity :** In data transmission sending a parity bit that will make the number of 1s in a group odd.
24. **Parity :** A system used to detect errors in binary data transmission.
25. **Parity bit :** An extra bit sent with data bits to check for errors in transmission.
26. **Radix :** The base of a number.
27. **2s complement :** Notation commonly used to indicate sign and magnitude of a number using only 0s and 1s. To form 2s complement, we take 1s complement of binary and add 1. Helpful when using binary adders for binary subtraction.

SUMMARY

- Every day digital concepts are being applied to problems that could only be solved by analog methods several years ago.
- Nowadays, digital systems are used in wide variety of industrial and consumer products such as automated industrial machinery, pocket calculators, microprocessors, digital computers, digital watches, TV games, signal processing and so on. The areas of applications of digital systems have been increasing every day.
- The digital computer can be programmed to perform different tasks. The users are allowed to specify and change the program or the data according to the specific need. Because of this flexibility, general purpose digital computers can perform a variety of information processing tasks that range over a wide spectrum of applications.
- In the day-to-day life, we come across various types of signals and systems. The signals are broadly classified into following two categories :
  (i) Analog signals
  (ii) Digital signals
- These are the signals which can have infinite number of different magnitudes or values. They vary continuously with time. Sine wave, triangular wave etc. are the examples of analog signals.
- A signal is called as a digital signal if it has only a finite number of predetermined distinct magnitudes. Depending on the number of distinct magnitudes, the digital signals may be classified as under.
- Few examples of analog systems may be listed as under :
  (i) Filters
  (ii) Amplifiers
  (iii) Signal generators

(iv) Power supplied
(v) Motor speed controllers

- Comparison of Analog and Digital Systems

(i) Analog systems are less accurate.
(ii) Analysis of analog systems requires time domain analysis which requires sound mathematical background.
(iii) Analog systems are affected more by the unwanted electrical disturbance called noise.
(iv) Once the noise contaminates an analog signal, it is impossible to separate the signal from noise. The signal contaminated by noise will lead to degradation of quality.
(v) The performance of these systems degrades due to component ageing.
(vi) Their performance changes with change in temperature.
(vii) Less reliability.
(viii) Less versatility since no software is included.
(ix) Due to use of analog meters, it is difficult to take accurate reading.
(x) Storage and processing of data is not possible as they do not have memory.
(xi) Computerized control cannot be exercised.

- (i) Digital systems manipulate discrete elements of information.
(ii) Discrete elements are nothing but the digits such as 10 decimal digits, 26 letters of alphabets and so on.
(iii) Digital systems use physical quantities called signals to represent discrete elements.
(iv) In digital systems, the signals have two discrete values and are therefore said to be binary.
(v) A signal in digital system represents one binary digit, called a bit. The bit has a value either 0 or 1.
(vi) Discrete elements of information are represented with group of bits called binary codes. For example, the decimal digits 0 through 9 are represented in a digital system with 4-bit code called BCD (Binary Coded Decimal) code.
(vii) The digital systems like a digital computer is a programmable device which can be programmed to perform variety of tasks.
(viii) The digital systems like a digital computer is an interconnection of digital modules. The major units/modules of digital computer are a central processing unit, memory unit and input-output unit. The memory unit stores programs and data. The central processing unit performs arithmetic and logical operation on data stored in the memory specified by the program. Input device such as keyboards is used to store program and data into the memory and display device such as printer, monitor is used to print/display the result of operation.

- Digital systems are easy to design than analog systems, because digital design involves logic design and logic design does not require special maths skill and its behaviour can be visualized part by part.

  To visualize the behaviour of analog systems is somewhat difficult. It needs the special insights about the analog components such as capacitors, inductors and transistors used in the analog systems.

- When logic gates are connected together to produce a specified output for certain specified combinations of input variables, with no storage involved, the resulting circuit is called combinational logic. In combinational logic, the output variables are at all times dependent on the combination of input variables. A combinational circuit consists of input variables, logic gates, and output variables. The logic gates accept signals from the input variables and generate output signals. This process transforms binary information from the given input data to the required output data.

- There are many applications in which digital outputs are required to be generated in accordance with the sequence in which the input signals are received. This requirement cannot be satisfied using a combinational logic system. These applications require outputs to be generated that are not only dependent on the present input conditions but they also depend upon the past history of these inputs. The past history is provided by feedback from the output back to the input.

- The integrated circuits may be classified using following four levels of integration :

(i) Small Scale Integration (SSI)
(ii) Medium Scale Integration (MSI)

(iii) Large Scale Integration (LSI), and
(iv) Very Large Scale Integration (VLSI)

- Logic is defined as a statement which is true if some condition is satisfied and false if that condition is not satisfied.
  (i) A bulb turns ON, if we close the switch, otherwise it is OFF.
  (ii) You will pass with good marks, if you study well.
  (iii) I will miss the bus, if I am late.
  Thus, each logical statement has two possible outcomes. Such a logic which has only two possible outcomes is called as the Binary Logic. The two possible outcomes of a binary logic are represented by two voltage levels. The true outcome is denoted by a HIGH level and called as Logic 1 level and the false outcome is denoted by a LOW level and it is called as Logic 0 level.
- Number system is a basis for counting various items. On hearing the word 'number', all of us immediately think of the familiar decimal number system with its 10 digits : 0, 1, 2, 3, 4, 5, 6, 7, 8 and 9.
- As a matter of fact, modern computers communicate and operate with binary numbers which use only the digits 0 and 1. Let us consider decimal number 18. This number is represented in binary as 10010. In the example, if decimal number is considered, we require only two digit to represent the number, whereas if binary number is considered we require five digits. Therefore, we can say that, when decimal quantities are represented in the binary form, they take more digits. For large decimal numbers people have to deal with very large binary strings and therefore, they do not like working with binary numbers. This fact gave rise to three new number systems : Octal, Hexadecimal and Binary Coded Decimal (BCD). These number systems represent binary number in a compressed form. Therefore, these number systems are now widely used to compress long strings of binary numbers.
- To define any number system, we have to specify following aspects :
  (i) Base of the number system such as 2, 8, 10 or 16.
  (ii) The base decides the total number of digits available in that number system. For example, if base is 2, we have 2-digits (0 and 1), if base is 10 we have 10-digits (0 through 9) and so on. Therefore, total number of digits available in the number system is equal to the base of the number system.
  (iii) First digit in the number system is always zero (0) and last digit in the number system is always, base –1. For example, in decimal system, last digit is 9 (10 – 1).

SHORT QUESTIONS WITH ANSWER

Q.1. What is meant by radix (or base) of a number system?
Ans. Radix or base of a number system is the number of digits or distinct symbols it uses to represent various numbers.

Q.2. What is a weighted code ? Give one example.
Ans. In a weighted code the value of a digit in a 4-bit depends on its position such as in BCD code.

Q.3. What is non-weighted code?
Ans. In a non-weighted code, no definite weights are assigned to the four digit positions such as in excess-3 code.

Q.4. What is meant by LSB and MSB?
Ans. The right most bit of a binary number is called least significant bit (LSB) while the left most bit of a binary number is known as most-significant bit (MSB).

Q.5. What is (*a*) octal system (*b*) hexadecimal system?
Ans. The number systems with base (or radix) 8 and 16 are known as octal number system and hexadecimal number system respectively.

Q.6. What is meant by BCD code?
Ans. BCD code is a decimal system with each digit encoded in its binary equivalent of 4 bits.

Q.7. What is an excess-3 code?
Ans. Excess-3 code is derived from the natural BCD code by adding 3(0011) to each coded number.

Q.8. Why hexadecimal code is widely used in digital systems?
Ans. It is very convenient to enter binary data in a digital system using hex code.

Q.9. What is the difference between binary code and BCD?
Ans. BCD is not a number system like binary. It is a decimal system with each decimal digit encoded in its

binary equivalent. A straight binary code takes the complete decimal number and represents it in binary while the BCD code converts each decimal digit to binary individually.

Q.10. State the disadvantage of a 8-4-2-1 code.

Ans. BCD requires more bits than straight binary to represent decimal numbers of more than one digit. This is because BCD does not use all possible 4-bit groups and is, therefore, inefficient.

Q.11. Why is an excess-3 code called an unweighted code? What will be excess-3 code for decimal 5849.

Ans. Since no definite weights can be assigned to the four digit positions in an excess-3 code, it is an unweighted code.

$(5849)_{10} = (1000\ 1011\ 0111\ 1100)_{XS-3}$

Q.12. Explain the use of excess-3 code.

Ans. The key feature of the excess-3 code is that it is self-complementing code. It means that 1's complement of the coded number yields 9's complement of the number itself. The self-complementing property makes the excess-3 code useful in some arithmetic operations, because subtraction can be performed using 9's complement method and 1's complement can be easily produced with digital logic circuits by simply inverting each bit.

Q.13. What are the advantages of Gray code ?

Ans. In Gray code, if we go from any one decimal number to the next, only one bit of the Gray code changes. Because of this feature, an amount of switching is minimized and the reliability of the switching systems is improved.

Q.14. What is an alphanumeric code ?

Ans. The codes used for representing letters of alphabet, punctuation marks and other special characters along with numbers are called alphanumeric codes.

Q.15. Which is the most commonly used code for representing alphanumeric information ?

Ans. ASCII (American Standard Code for Information Interchange) is the most widely used alphanumeric code. It is a 7-bit code.

REVIEW QUESTIONS

1. Write the characteristics of digital systems.
2. What is meant by analog system ?
3. What is meant by digital system ?
4. Write are the salient features of digital systems over analog-systems ?
5. Write at least four examples where digital systems are preferred over analog systems.
6. What is meant by logic gates ?
7. Describe the combinational logic and sequential logic.
8. What is flip-flop ?
9. What are the salient features of using integrated circuits in digital design ?
10. Write a short note on the following :
    (a) Small scale integration (SSI) (b) Medium scale integration (MSI)
    (c) Large scale integration (LSI) (d) Very large scale integration (VLSI)
11. Convert $(725.25)_8$ to its decimal, binary and hexadecimal equivalent.
    [**Ans.** $(469.328125)_{10}$ $(111010101\ .\ 010101)_2$, $(E5.54)_{16}$]
12. Convert $(11001011\ .\ 01110)_2$ into decimal. [**Ans.** (203) . $(4375)_{10}$]
13. Convert decimal number 61.3 to binary. [**Ans.** $(111101\ .\ 010011001)_2$].
14. Convert octal number 574 to binary and decimal. [**Ans.** (i) $(101111100)_2$, (ii) $(380)_{10}$]
15. Convert A92H to octal. [**Ans.** 5222_8]
16. Convert the following numbers to hexadecimal :
    (i) $(360)_8$ (ii) $(2262)_{10}$
    (iii) $(10011\ .\ 1101)_2$ (iv) $(10.1)_2$.
    [**Ans.** (i) $(F0)_H$, (ii) $(16.9)_H$, (iii) $(13.D)_H$, (iv) $(2.8)_H$]
17. Convert the following numbers to their octal equivalents :
    (i) $(1100101011\ .\ 1110)_2$ (ii) 37.29_{10}
    (iii) $(672)_H$. [**Ans.** (i) $(1453.7)_8$, (ii) $(45.224)_8$, (iii) $(3162)_8$]
18. Using 2's complements method perform :
    (i) $(57)_{10} - (28)_{10}$ (ii) $(432)_{10} - (579)_{10}$.

2

BOOLEAN ALBEGRA

Syllabus

Boolean Algebra: Boolean postulates and laws – De-Morgan's Theorem, Principle of Duality, Boolean expression – Boolean function, Minimization of Boolean expressions – Sum of Products (SOP), Product of Sums (POS), Minterm, Maxterm, Canonical forms, Conversion between canonical forms, Karnaugh map Minimization, Quine-McCluskey method – Don't care conditions.

Chapter Outline

- Introduction
- Postulates of a Mathematical System
- Boolean (i.e., Binary Algebra)
- Boolean Postulates and Laws
- Boolean Laws
- De-Morgan's Theorems
- Boolean Expression and Boolean Function
- Classification of Digital Circuits
- SOP and POS Representations for Logic Expressions
- Minterm and Maxterm
- Various Boolean Expressions
- Simplification of Karnaugh-map (K-Map)
- Boolean Expressions with the Help of K-map
- How to Minimize SOP Expressions
- K-map Representation of POS Form
- How to Simplify Standard POS Form with the help of K-map
- EX-OR and EX-NOR Simplification of K-Maps
- Drawback of K-Map Technique
- Five and Six Variable K-maps
- Five variable K-map
- Quine-McCluskey Minimization Technique (i.e., Tablur Method)
- Implementation Using Only NAND or Only NOR Gates

2.1 INTRODUCTION

In order to deal mathematically with the digital circuits, we need to use the Boolean algebra. Boolean algebra may be defined with the help of the following aspects :

(i) A set of elements
(ii) A set of operators
(iii) A number of unproved axioms or postulates.

2.2 POSTULATES OF A MATHEMATICAL SYSTEM

The postulates of a mathematical system form the basic assumptions from which we can deduce the rules, theorems and also various properties of a system. The most important postulates may be listed as under :

(i) Closure
(ii) Associative law
(iii) Commutative law
(iv) Identity element
(v) Inverse
(vi) Distributive law.

2.3 BOOLEAN (i.e., BINARY) ALGEBRA

As a matter of fact, Boolean Algebra is used to analyze and simplify the digital, (i.e., logic) circuits. Because, it uses only the binary numbers, i.e., 0 and 1, it is also called as *Binary Algebra* or *Logical Algebra*. It was invented by George Boole in year 1854. The rules of Boolean Algebra are different from those of the conventional Algebra in the following manner :

(i) Symbols used in Boolean algebra (usually letters) do not represent numerical values.

(ii) Arithmetic operations (addition, subtraction, division etc.) are not performed in Boolean algebra. Also, there are no fractions, negative numbers, square, square root, logarithms, imaginary numbers etc.

(iii) The third and most important point is that Boolean Algebra allows only two possible values (0 to 1) for any variable.

(iv) Various logical operations along with their Boolean expressions have been listed in Table 2.1.

Table 2.1. Various logical operations

S. No.	Name of the gate	Logical operation	Boolean expression
1.	NOT gate	Inversion	$Y = \overline{A}$
2.	AND gate	Logical multiplication	$Y = AB$
3.	OR gate	Logical addition	$Y = A + B$
4.	NAND gate	NOT AND	$Y = \overline{AB}$
5.	NOR gate	NOT OR	$Y = \overline{A+B}$
6.	Exclusive OR	Addition/subtraction	$Y = A \oplus B$
7.	Exclusive NOR	NOT EX-OR	$Y = \overline{A \oplus B}$

Rules in Boolean algebra*

There are some rules which are to be followed while using a Boolean algebra, these rules may be described as under :

(i) Variables used can have only two values. Binary 1 for HIGH and Binary 0 for LOW.

(ii) Complement of a variable is represent by a overbar (–). Thus, complement of variable B is represented as $\overline{B}$. Therefore, if B = 0 then $\overline{B}$ = 1 if B = 1 then $\overline{B}$ = 0.

(iii) ORing of the variables is represented by a plus (+) sign between them. For example, ORing of A, B, C is represented as A + B + C.

(iv) Logical ANDing of the two or more variables is represented by writing a dot between them such as $A \cdot B \cdot C \cdot D \cdot E$. Sometimes, the dot may be omitted like ABCDE.

GEORGE BOOLE

George Boole was born in Lincon, England on Novermber 2, 1815. He was a self-taught mathematician who invented modern symbolic logic and pioneered the calculus of operators. Around 1850, George Boole created Boolean algebra, which underlies the theory of logic.

2.4 BOOLEAN POSTULATES AND LAWS

As discussed earlier, Boolean algebra is an algebraic structure which is defined by a set of elements and two binary operations namely (+) and (·) provided that the following postulates are satisfied.

1(a) **Closure with respect to operator (+)**

When operator (+), i.e., OR is used over two binary elements, it results in a unique binary element.

* Binary logic gets are the basic building blocks for all digital circuits.

1(b) **Closure with respect to operator (·)**

When operator (·), i.e., AND is used over two binary elements, it results in a unique binary element.

2(a) An identity element with respect to (+), designated by 0 :

$$A + 0 = 0 + A = A$$

2(b) An identity element with respect to (·), designated by 1 :

$$A \cdot 1 = 1 \cdot A = A$$

3(a) Commutative with respect to (+) :

$$A + B = B + A$$

3(b) Commutative with respect to (·) :

$$A \cdot B = B \cdot A$$

4(a) (·) is distributed over (+) :

$$A \cdot (B + C) = (A \cdot B) + (A \cdot C)$$

4(b) (+) is distributed over (·) :

$$A + (B \cdot C) = (A + B) \cdot (A + C)$$

5. For every binary element A, there exists an element called complement of A such that,

$$A + \overline{A} = 1 \text{ and } A \cdot \overline{A} = 0$$

Here, it may be noted that if A = 0 then $\overline{A} = 1$ if A = 1 then $\overline{A} = 0$.

6. In a set of binary elements, there always exist atleast two element A and B such that $A \neq B$, A = 0 then B = 1 and vice versa.

2.5 BOOLEAN LAWS

(PTU, Jalandhar, Sem. Exam., 2005-06)

In this article, let us discuss the following Boolean laws :

1. Commutative law
2. Associative law
3. Distributive law
4. AND law
5. OR law
6. INVERSION law.

DO YOU KNOW?

Because binary logic is used in all of today's digital computers and devices, the cost of the circuits that implement it is an important factor addressed by designers.

1. Commutative Law

Any binary operation which satisfies the following expression is called as commutative operation:

(i) $A \cdot B = B \cdot A$

(ii) $A + B = B + A$

Actually, the commutative law states that changing the sequence of the variables (inputs) does not have any effect on the output of a logic circuit. A commutative law may be expressed diagrammatically as shown in figure 2.1.

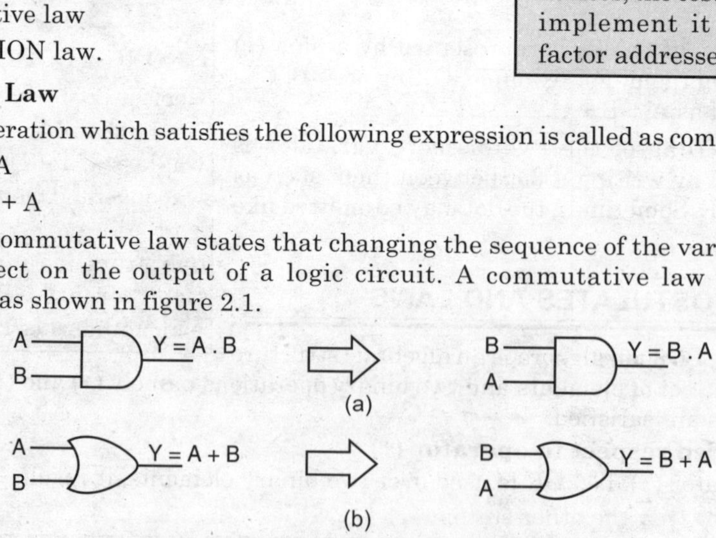

Fig. 2.1. Illustration of commutative law

The commutative law may be verified by following truth table :

Table 2.2. Verification of the commutative law

Inputs		A · B	B · A	A + B	B + A
A	B				
0	0	0	0	0	0
0	1	0	0	1	1
1	0	0	0	1	1
1	1	1	1	1	1

A · B = B · A   A + B = B + A

2. Associate Law

This law states that the order in which the logic operations are performed is irrelevant as their effect is the same. This means that we have

$$(A \cdot B) \cdot C = A \cdot (B \cdot C)$$

and
$$(A + B) + C = A + (B + C)$$

The associative law may be expressed diagrammatically as depicted in figure 2.2.

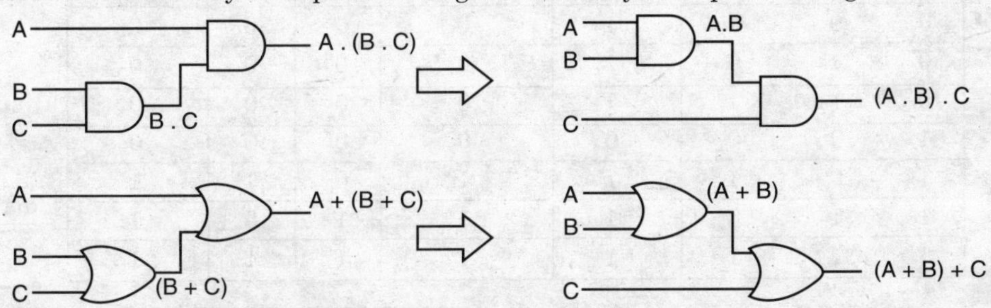

Fig. 2.2. Illustration of Associative law

An associative law can be verifed by considering the following truth table :

Table 2.3. Verification of the associative law

Inputs			B · C	A · (B · C)	A · B	(A · B) · C
A	B	C				
0	0	0	0	0	0	0
0	0	1	0	0	0	0
0	1	0	0	0	0	0
0	1	1	1	0	0	0
1	0	0	0	0	0	0
1	0	1	0	0	0	0
1	1	0	0	0	1	0
1	1	1	1	1	1	1

A · (B · C) = (A · B) · C

Similarly, we can verify the other statement, i.e.,

$$A + (B + C) = (A + B) + C$$

3. Distributive Law

The distributive law states that

$$A \cdot (B + C) = AB + AC$$

Figure 2.3 illustrates the distributive law.

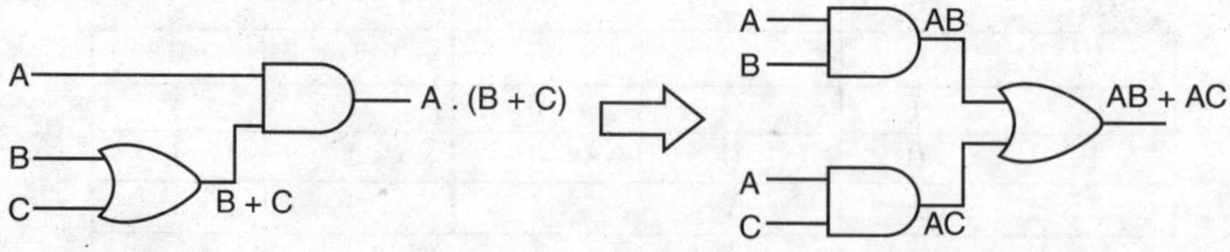

Fig. 2.3. Illustration of distributive law

The distributive law can be verified by using the following truth table :

Table 2.4. Verification of the distributive law

Inputs			B + C	A · (B + C)	AB	AC	AB + AC
A	B	C					
0	0	0	0	0	0	0	0
0	0	1	1	0	0	0	0
0	1	0	1	0	0	0	0
0	1	1	1	0	0	0	0
1	0	0	0	0	0	0	0
1	0	1	1	1	0	1	1
1	1	0	1	1	1	0	1
1	1	1	1	1	1	1	1

$A \cdot (B + C) = AB + AC$

4. AND Laws

AND laws use the AND operation, therefore, they are called as AND laws. The AND laws may be described as under :

(i) $A \cdot 0 = 0$, i.e., ANDing with a 0, always results in 0 output

(ii) $A \cdot 1 = A$, i.e., ANDing of A with a 1, results in A

(iii) $A \cdot A = A$, i.e., ANDing of an input with itself produces the same output

(iv) $A \cdot \overline{A} = 0$, i.e., ANDing of an input with its complement results in a 0 output.

Figure 2.4 illustrates AND laws diagrammatically.

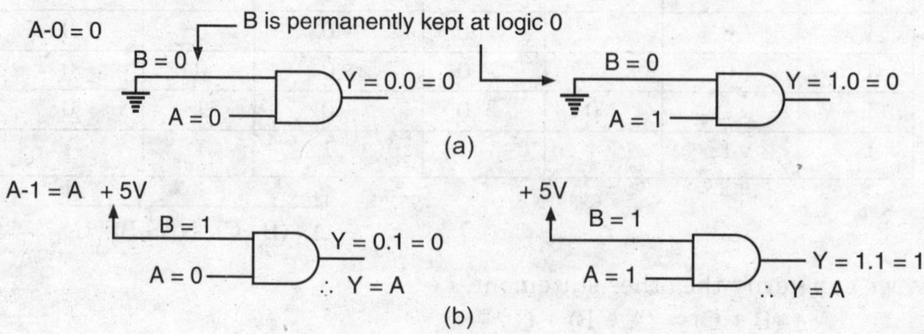

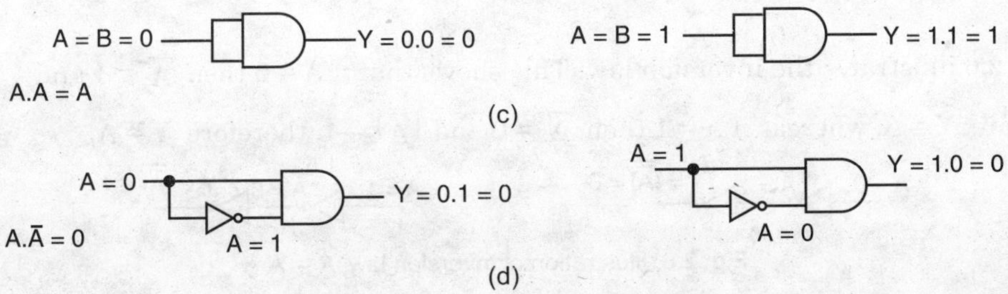

Fig. 2.4. Illustration of the AND laws

5. OR Laws

OR laws use the OR operation. Therefore, they are known as OR laws. The OR laws may be described as under :

> **DO YOU KNOW?**
>
> Finding simpler and cheaper, but equivalent, realizations of a circuit can reap huge pay offs in reducing the overall cost of the design.

(i) A + 0 = A, i.e., ORing an input with 0, results in the output equal to input.

(ii) A + 1 = 1, i.e., ORing an input with 1, always results in a high output.

(iii) A + A = A, i.e., ORing an input with itself, results in the same output.

(iv) A + $\overline{A}$ = 1, i.e., ORing an input and its complement always results in a high output.

The OR laws have been illustrated in figure 2.5.

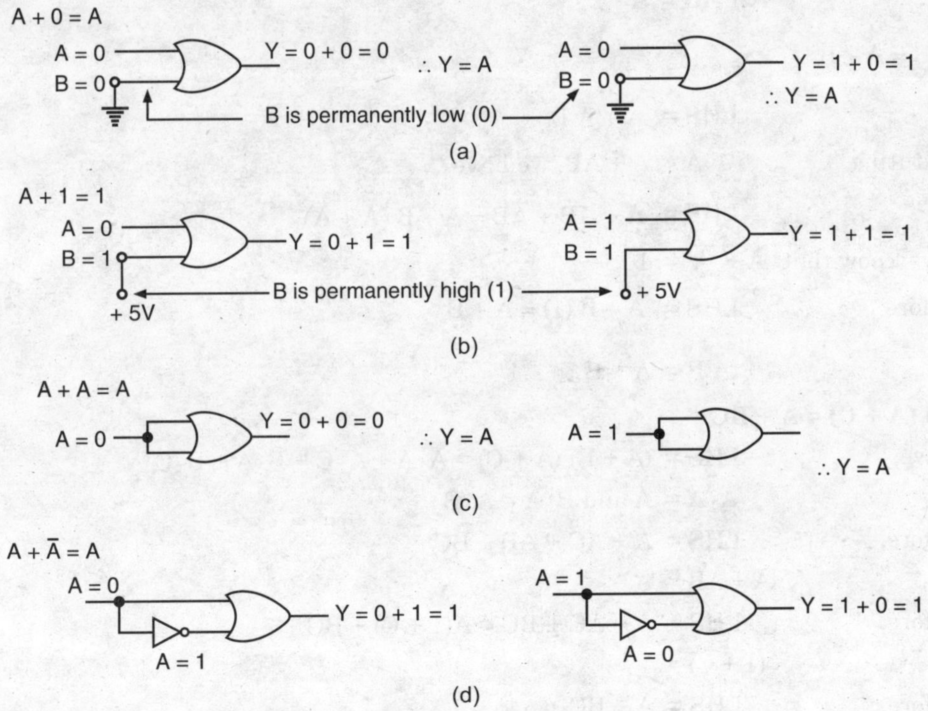

Fig. 2.5. Illustration of OR laws

6. INVERSION Law

This law uses the NOT operation. The inversion law states that double inversion of a variable results in the original variable itself. This means that

$$\bar{\bar{A}} = A$$

Figure 2.6 illustrates the inversion law. This shows that if A = 0 then $\bar{A}$ = 1 and $(\bar{\bar{A}}) = 0$

Therefore, Y = A, whereas if A = 1 then $\bar{A}$ = 0 and $(\bar{\bar{A}})$ = 1, therefore Y = A.

Fig. 2.6. Illustration of inversion law $\bar{A}$ = A

7. Other Important Rules

For the simplification of Boolean expressions, we take help of following additional rules :

(i) $A + AB = A$

(ii) $A + \bar{A}B = A + B$

(iii) $(A + B)(A + C) = A + BC$

Now, let us prove above rules by using the basic laws as under :

1. A + AB = A

LHS = $A + AB = A(1 + B)$

But $1 + B = 1$

Therefore, LHS = $A \cdot 1 = A$ $(\because A \cdot 1 = A)$

Hence, $A + AB = A$

2. $A + \bar{A}B = A + B$

LHS = $A + \bar{A}B$

Substituting $A = A + AB$, we have

LHS = $A + AB + \bar{A}B = A + B(A + \bar{A})$

But, we know that $A + \bar{A} = 1$

Therefore, LHS = $A + B(1) = A + B$

Hence, $A + \bar{A}B = A + B$

3. (A + B) (A + C) = A + BC

We have, LHS = $(A + B)(A + C) = A \cdot A + A \cdot C + B \cdot A + B \cdot C$

But $A \cdot A = A$ and $B \cdot A = A \cdot B$

Therefore, LHS = $A + AC + AB + BC$

But $A + AB = A$

Therefore, LHS = $A + AC + BC = A(1 + C) + BC$

But $(1 + C) = 1$

Therefore, LHS = $A + BC$

Hence, $(A + B)(A + C) = A + BC$ **Hence Proved.**

Table 2.5. Summary of All Boolean Laws

S. No.	Name of Law	Statement
1.	Commutative law	$A \cdot B = B \cdot A$ $A + B = B + A$
2.	Associative Law	$(A \cdot B) \cdot C = A \cdot (B \cdot C)$ $(A + B) + C = A + (B + C)$
3.	Distributive Law	$A \cdot (B + C) = AB + AC$
4.	AND Laws	$A \cdot 0 = 0$ $A \cdot 1 = A$ $A \cdot A = A$ $A \cdot \overline{A} = 0$
5.	OR Laws	$A + 0 = A$ $A + 1 = 1$ $A + A = A$ $A + \overline{A} = 1$
6.	Inversion Law	$\overline{\overline{A}} = A$
7.	Other Important Laws	$A + BC = (A + B)(A + C)$ $\overline{A} + AB = \overline{A} + B$ $\overline{A} + A\overline{B} = \overline{A} + \overline{B}$ $A + AB = A$ $A + \overline{A}B = A + B$

2.6 DUALITY PRINCIPLE

(PTU, Jalandhar, Sem. Exam., 2008-09)

An important property of Boolean algebra is known as the duality principle. It states that in a two valued Boolean algebra, the *dual* of an algebraic expression can be obtained simply by interchanging OR and AND operators and by replacing 1s by 0s and 0s by 1s. The duality principle has several applications in digital systems.

2.6.1 Duality Theorem

According to the duality theorem, the following conversions are possible in a given Boolean expression :

(i) We can change each AND operation to an OR operation.

(ii) We can change each OR operation to an AND operation.

(iii) We can complement any 1 or 0 appearing in the expression.

As a matter of fact, duality theorem is sometimes useful in creating new expressions from the given Boolean expressions. As an example, if the given expression is $A + 1 = 1$, then, we replace the OR (+) operation by AND ($\cdot$) operation and take complement of 1 to write the dual of the given relation as under :

$$A \cdot 0 = 0$$

The dual of $A(B + C) = AB + AC$ is given by

$$A + (B \cdot C) = (A + B) \cdot (A + C)$$

Important Point : It is possible to verify this theorem by constructing truth tables for both the sides of equations.

EXAMPLE 2.1. Find the dual of following Boolean expressions :

1. $A + AB = A$
2. $A + \overline{A}B = A + B$
3. $A + \overline{A} = 1$
4. $(A + B)(A + C) = A + BC$.

DO YOU KNOW?

Mathematical methods that simplify circuits rely primarily on Boolean algebra.

Solution:

1. $A + AB = A$

Let us replace (+) by (·) and (·) by (+) to obtain the dual of the given equation as under :

$A \cdot (A + B) = A$ **Ans.**

Similarly, the duals of the other expressions may be obtained as shown in Table 2.6.

Table 2.6

S.No.	Given expression	Dual of given expression
1.	$A + \overline{A}B = A + B$	$A \cdot (\overline{A} + B) = A \cdot B$
2.	$A + \overline{A} = 1$	$A \cdot \overline{A} = 0$
3.	$(A + B)(A + C) = A + BC$	$AB + AC = A \cdot (B + C)$

2.7 DeMORGAN'S THEOREMS

(PTU, Jalandhar, Sem. Exam., 2008-09)

The two important theorems have been suggested by DeMorgan. These two theorems are extremely useful in Boolean algebra. These may be described as under :

Theorem 1 : $\overline{AB} = \overline{A} + \overline{B}$; NAND = Bubbled OR

This theorem states that the complement of a product is equal to addition of the complements. Figure 2.7 (a) illustrates this rule. The left hand side (LHS) of this theorem represents a NAND gate with inputs A and B whereas the right hand side (RHS) of the theorem represents an OR gate with inverted inputs. This OR gate is known as *Bubbled OR*. Thus, we can state DeMorgan's first theorem as under :

NAND = Bubbled OR

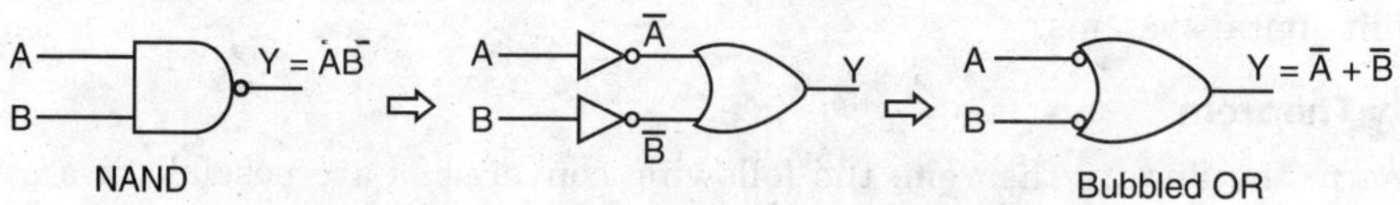

Fig 2.7. (a) Illustration of DeMorgan's first theorem

2.7.1 Verification of the First Theorem

The first theorem may be verified by writing a truth table as shown in figure 2.7 (*b*).

S.No.	A	B	$\overline{AB}$	$\overline{A}$	$\overline{B}$	$\overline{A} + \overline{B}$
1.	0	0	1	1	1	1
2.	0	1	1	1	0	1
3.	1	0	1	0	1	1
4.	1	1	0	0	0	0

Fig. 2.7. (b) Verification of first theorem $\overline{AB} = \overline{A} + \overline{B}$

Theorem 2 : $\overline{A + B} = \overline{A} \cdot \overline{B}$: NOR = Bubbled AND

This theorem states that the complement of a sum is equal to the product of complements.

Figure 2.8 illustrates this theorem. The LHS of this theorem represents a NOR gate with inputs A and B whereas the RHS represents an AND gate with inverted inputs. This AND gate is known as *Bubbled AND*. Therefore, we can state DeMorgan's second theorem as under :

NOR = Bubbled AND

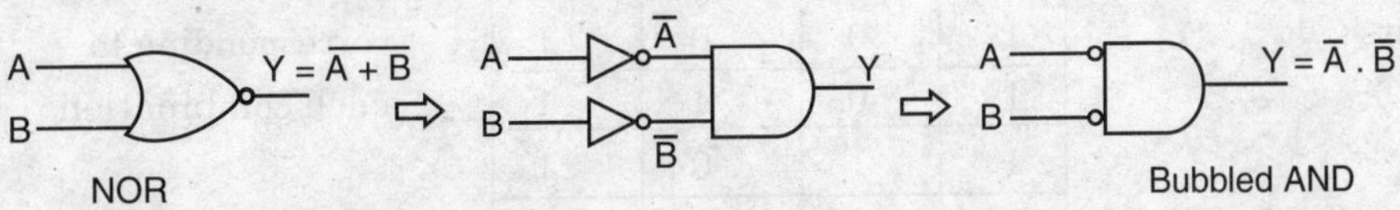

Fig. 2.8. Illustration of DeMorgan's second theorem

2.7.2 Verification of the Second Theorem

S.No.	A	B	$\overline{A+B}$	$\overline{A}$	$\overline{B}$	$\overline{A} \cdot \overline{B}$
1.	0	0	1	1	1	1
2.	0	1	0	1	0	0
3.	1	0	0	0	1	0
4.	1	1	0	0	0	0

Fig. 2.9. Truth table to verify DeMorgan's second theorem

2.8 BOOLEAN EXPRESSION AND BOOLEAN FUNCTION

As a matter of fact, Boolean algebra deals with binary variables and logic operations. A Boolean function is described by an algebraic expression called Boolean expression which consists of binary variables, the constants 0 and 1 and the logic operation symbols. Let us consider the following illustrative example :

$$\underbrace{F(A, B, C, D)}_{\text{Boolean function}} = \underbrace{A + B\overline{C} + ADC}_{\text{Boolean expression}} \quad ...(2.1)$$

Another example of Boolean function may be written as under:

$$F(A, B, C) = A + BC \quad ...(2.2)$$

The Boolean function can be represented by a **truth table**. A truth table can be defined as the list of combination of 0s and 1s assigned to the binary variables (A, B, C etc.) and a column that shows the value of the function F (A, B, C) for each input combination. The truth table for equation (2.2) has been shown in Table 2.7. The number of rows in a truth table is 2^n where n is the number of input variables (n = 3 for equation (2.2)). Therefore, there are $2^3 = 8$ possible combinations or rows in truth table.

AUGUSTUS DeMORGAN (1806-1871)

Augustus DeMorgan (1806-1871) was born in Madras Province, India. He taught mathematics at the University of London for 30 years, and published many texts on arithmetic, algebra, trigonometry, and calculus, and important treatises on the theory of probability and formal logic. DeMorgan contributed the method of changing from sum-of-products to product-of-sums.

Table 2.7. Truth table for F (A, B, C) = A + BC

8 possible combination of inputs

A	B	C	F
0	0	0	0
0	0	1	0
0	1	0	0
0	1	1	1
1	0	0	1
1	0	1	1
1	1	0	1
1	1	1	1

Values of the function F corresponding to each combination

Basically, Boolean function can be converted from an algebraic expression into a circuit diagram containing logic gates.

F = A + BC
→ AND gate with B and C inputs
→ OR gate with inputs A and BC

Figure 2.10 shows the logic diagram for F. This logic diagram shows that an AND gate is used with inputs B and C whereas an OR gate is used with inputs A and B · C (i.e., output of the AND gate).

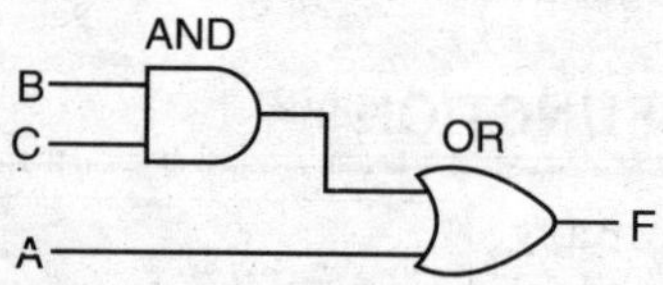

Fig. 2.10. Logic diagram or gate implementation of F = A + BC

> **DO YOU KNOW?**
>
> The postulates of a mathematical system form the basic assumptions from which it is possible to deduce the rules, theorems and properties of the system.

EXAMPLE 2.2. Prove the following Boolean expressions :

(i) $A + AB = A$ **(ii)** $A + \bar{A}B = A + B$

(iii) $(A + B)(A + C) = A + BC$

Solution: (i) LHS = $A + AB = A(1 + B)$

But $(1 + B) = 1$, therefore, LHS = $A \cdot 1$

But $A \cdot 1 = A$

Therefore, LHS = A = RHS

Hence, $A + AB = A$

(ii) LHS = $A + \bar{A}B = A + AB + \bar{A}B$ $(\because A = A + AB)$

Therefore, LHS = $A + B(A + \bar{A})$

But $(A + \bar{A}) = 1$

Therefore, LHS = $A + (B \cdot 1) = A + B$

Hence, LHS = RHS

Thus, $A + \bar{A}B = A + B$

(iii) LHS = $(A + B)(A + C)$

LHS = $AA + AC + BA + BC$

But $AA = A$, therefore, LHS = $A + AC + BA + BC$

Therefore, LHS = $A(1 + C) + AB + BC$

But $(1 + C) = 1$

Hence, $\text{LHS} = A + AB + BC = A(1 + B) + BC$

But $(1 + B) = 1$

Therefore, $\text{LHS} = A + BC$

Thus, $(A + B)(A + C) = A + BC$ **Hence Proved**

EXAMPLE 2.3. Prove that $(A+\bar{B})+AB)(A+\bar{B})(\bar{A}B)=0$

Solution: $\text{LHS} = (A+\bar{B}+AB)(A+\bar{B})(\bar{A}B)$

But $A + AB = A$

Therefore, $\text{LHS} = (A+\bar{B})(A+\bar{B})(\bar{A}B) = (AA + A\bar{B} + A\bar{B} + B\bar{B})(\bar{A}B)$

But $A \cdot A = A$ and $\bar{B} \cdot B = \bar{B}$ and $A\bar{B} + A\bar{B} = \bar{B}(A + A) = A\bar{B}$

Hence, $\text{LHS} = (A + A\bar{B} + \bar{B})(\bar{A}B) = [A(1+\bar{B})+\bar{B}](\bar{A}B)$

But $(1 + \bar{B}) = 1,$

Hence, $\text{LHS} = [(A \cdot 1)+\bar{B}](\bar{A}B)$

Therefore, $\text{LHS} = (A+\bar{B})(\bar{A}B)$ $(\because A \cdot 1 = A)$

$= A\bar{A}B + \bar{A}B\bar{B}$

But $A\bar{A} = 0$ and $B\bar{B} = 0$

Therefore, $\text{LHS} = 0 + 0 = 0$ **Hence Proved**

EXAMPLE 2.4. Prove that $A + \bar{A}B + AB = A + B$.

Solution: $\text{LHS} = A + \bar{A}B + AB = A + B(\bar{A} + A)$

But $\bar{A} + A = 1$

Therefore, $\text{LHS} = A + B = \text{RHS}$ **Hence Proved**

EXAMPLE 2.5. Simplify $ABCD + A\bar{B}CD$

Solution: We have, $Y = ABCD + A\bar{B}CD = ACD(B + \bar{B})$

But $(B + \bar{B}) = 1$

Therefore, $Y = ACD$ **Ans.**

EXAMPLE 2.6. Simplify the following expression :

$$\mathbf{Y} = \overline{\left(\overline{AB}+\bar{A}+AB\right)}$$

Solution: We have, $Y = \overline{\left(\overline{AB}+\bar{A}+AB\right)}$

But $\overline{AB} = \bar{A} + \bar{B}$ (DeMorgan's first theorem)

or $Y = \overline{\left(\bar{A}+\bar{B}+\bar{A}+AB\right)}$

Therefore, $Y = \overline{(\bar{A} + \bar{B} + \bar{A} + AB)}$

But $\bar{A} + \bar{A} = \bar{A}$

Therefore, $Y = \overline{\left(\bar{A}+\bar{B}+AB\right)}$

> **DO YOU KNOW?**
>
> A Boolean function can be transformed from an algebraic expression into a circuit diagram composed of logic gates connected in a particular structure.

Now, let us use DeMorgan's second theorem which states that,

$$\overline{A+B+C} = \bar{A}\cdot\bar{B}\cdot\bar{C}$$

Therefore, $Y = \bar{\bar{A}}\cdot\bar{\bar{B}}\cdot\bar{\bar{AB}}$

But $\bar{\bar{A}} = A$ and $\bar{\bar{B}} = B$

Therefore, $Y = A\cdot B\cdot\overline{AB}$

But $\overline{AB} = (\bar{A}+\bar{B})$ (DeMorgan's second theorem)

Therefore, $Y = A\cdot B\,(\bar{A}+\bar{B}) = A\bar{A}B + AB\bar{B}$

But $A\bar{A} = 0$ and $B\bar{B} = 0$

Therefore, $Y = 0\cdot B + A\cdot 0 = 0 + 0$ $(\because 0\cdot B = 0 \text{ and } A\cdot 0 = 0)$

Hence, $Y = 0$ **Ans.**

EXAMPLE 2.7. Simplify the following Boolean expressions :

(i) $A\bar{B}+\bar{A}B+\bar{A}\bar{B}+AB$ **(ii)** $A\bar{B}C+\bar{A}BC+ABC$

Solution: (i) We have, $Y = A\bar{B}+\bar{A}B+AB+\bar{A}\bar{B} = \bar{B}(A+\bar{A})+B(\bar{A}+A) = \bar{B}+B$ $(\because (\bar{A}+A)=1)$

Hence, $Y = 1$ $(\because \bar{B}+B=1)$

(ii) We have, $Y = A\bar{B}C+\bar{A}BC+ABC = AC(\bar{B}+B)+\bar{A}BC = AC+\bar{A}BC$ $(\because \bar{B}+B=1)$

$= C(A+\bar{A}B)$

Therefore, $Y = C(A+B)$ $(\because A+\bar{A}B = A+B)$

EXAMPLE 2.8. Prove the following Boolean identities :

(i) $A+\bar{A}B+AB = A+B$ **(ii)** $A\bar{B}+\bar{A}B+AB+\bar{A}\bar{B}=1$

Solution: (i) We have, $\text{LHS} = A+\bar{A}B+AB = A+B(\bar{A}+A) = A+B(1)$ $(\because A+\bar{A}=1)$

Therefore, $\text{LHS} = A+B = \text{RHS}.$

(ii) $\text{LHS} = A\bar{B}+\bar{A}B+AB+\bar{A}\bar{B} = A\bar{B}+\bar{A}\bar{B}+\bar{A}B+AB$

$= (A+\bar{A})\,\bar{B}+(\bar{A}+A)\,B = \bar{B}+B.$ $(\because \bar{A}+A=1)$

$= 1$ $(\because \bar{B}+B=1)$

Therefore, $\text{LHS} = 1$ **Hence Proved**

EXAMPLE 2.9. Simplify the following Boolean expression :

$$\mathbf{Y = (AB + C)\,(AB + D).}$$

Solution: $Y = (AB + C)(AB + D) = AB\,AB + ABD + ABC + CD$

But $A\cdot A = A$ and $B\cdot B = B$

Therefore, $Y = AB + ABD + ABC + CD = AB\,(1 + D + C) + CD$

But, $(1 + D + C) = 1$

Therefore, $Y = AB\,(1) + CD = AB + CD$ **Ans.**

EXAMPLE 2.10. Prove that $\mathbf{AB+ABC+A\bar{B}=A}$.

Solution: $\text{LHS} = AB+ABC+A\bar{B} = AB\,(1+C)+A\bar{B}$

But $(1 + C) = 1$

Therefore, $\text{LHS} = AB+A\bar{B} = A\,(B+\bar{B})$

But $B+\bar{B} = 1$

Therefore, $\text{LHS} = A = \text{RHS}$

Hence, $AB+ABC+A\bar{B} = A$ **Hence Proved**

2.9 CLASSIFICATION OF DIGITAL CIRCUITS (PTU, Jalandhar, Sem. Exam., 2007-08)

We have already discussed the logic operations and Boolean algebra. The Boolean theorems and DeMorgan's theorems are useful in manipulating the logic expressions. we can then realize the logical expressions using gates. The number of logic gates required for the realization of a logical expression must be reduced to the minimum possible value. This is possible if we can simplify the logical expressions. In this chapter, we shall discuss one of the simplification techniques called *Karnaugh map or K-map*.

> **INTERESTING TO KNOW?**
>
> Electronic Highways Intelligent Vehicle Highway Systems (VHS) are currently being developed. These electronic in Japan, drivers can access roadside beacons to get traffic information, individual communications, and directions. Monitoring systems that allow controllers to quickly reroute traffic from backed-up areas are also being developed.

2.9.1 Combinational Circuit Design

In this chapter, we shall discuss the design of combinational circuit. To design a combinational circuit, we have to specify the requirements of combinational circuits. Such requirements may be specified in the following ways :

(i) A set of statements

(ii) Boolean expressions or

(iii) Truth table.

From the specified requirements, we have to design to a circuit using a combination of logic gates or using some other circuits derived from the basic gates.

2.9.2 Various Approaches to the Combinational Circuit Design

There are two different approaches to the design of the combinational circuits as under :

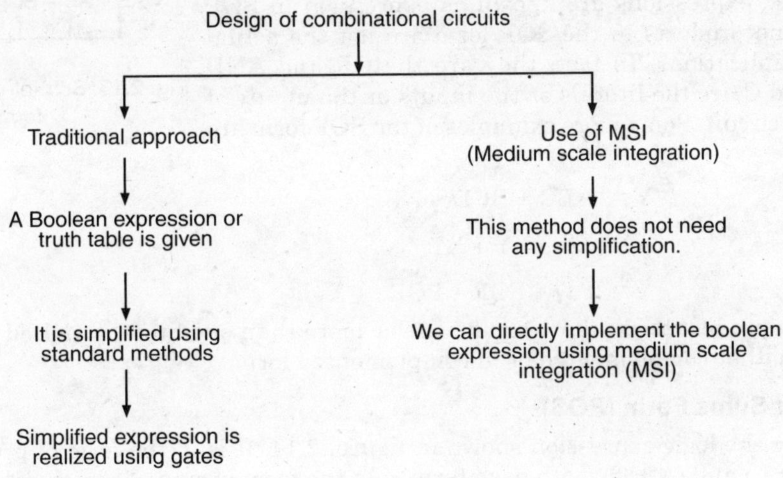

Fig. 2.11. Various approaches to combinational circuit design

2.9.3 How to Simplify the Boolean Functions

The various techniques used for simplifying the Boolean functions may be listed as under:

(i) Algebraic method

(ii) Karnaugh-map simplification.

(iii) Quine-McClusky method, and

(iv) Variable entered mapping (VEM) technique.

2.10 SOP AND POS REPRESENTATIONS FOR LOGIC EXPRESSIONS

Let us assume that the logic expression is

$$Y = AC + BC$$

Then it can be realized using basic gates as shown in figure 2.12. In this Boolean expression, Y is the result or output and A, B, C are called as literals.

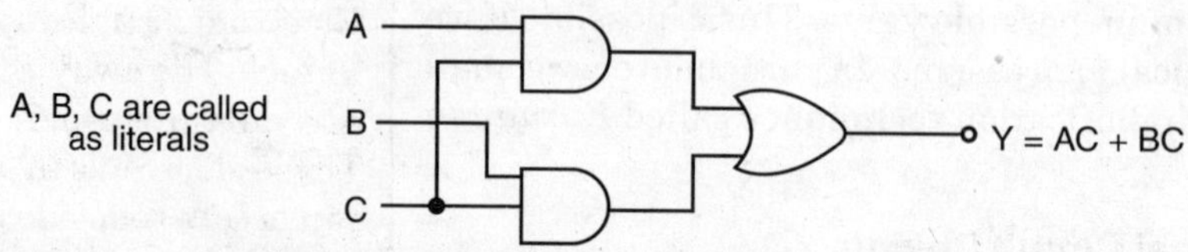

Fig. 2.12. Realization of given logic expression

When we realize the Boolean expression by using gates, each literal acts as an input as shown in figure 2.12. Any logic expression may be expressed in the following two standard forms:

(i) Sum-of-products form (SOP), and

(ii) Product-of-sums form (POS)

These two forms are suitable for reducing the given logic expression to its simplest form.

> **DO YOU KNOW?**
>
> When a Boolean expression is implemented with gates, each term requires a gate and each variable within a term, in complemented or uncomplemented form.

2.10.1 Sum-of-Products (SOP) Form

Let us consider logic expression shown in figure 2.13. It is in the form of sum of three terms AB, AC and BC with each individual term is product of two variables. Say $A \cdot B$ or $A \cdot C$ etc.

Therefore, such expressions are known as expression in SOP form. The sums and products in the SOP form are not the actual additions or multiplications. In fact, they are the OR and AND functions. A, B and C are the literals or the inputs or the inputs of the combinational circuit. Some more examples of the SOP form are as under :

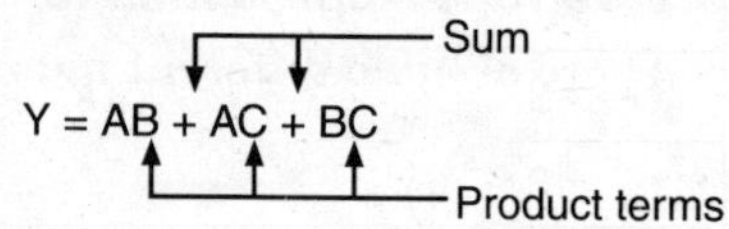

Fig. 2.13. Sum-of-products (SOP) form

$$Y = ABC + BCD + AB\overline{D}$$

$$A = XY + \overline{X}\overline{Y} + X\overline{Y}$$

$$Y = \overline{P}\,\overline{Q} + PQR + PQ\overline{R}$$

Hence, in each product term, there can be one or more than one literals ANDed together. The literals may be in either complemented or uncomplemented form.

2.10.2 Product-of-Sums Form (POS)

Let us consider the logic expression shown in figure. 2.14. It is in the form of product of three terms (A + B), (B + C) and (A + C), with each term is in the form of sum of two variables.

Such expressions are said to be in the product-of-sums (POS) form. Some other examples of POS form are as under :

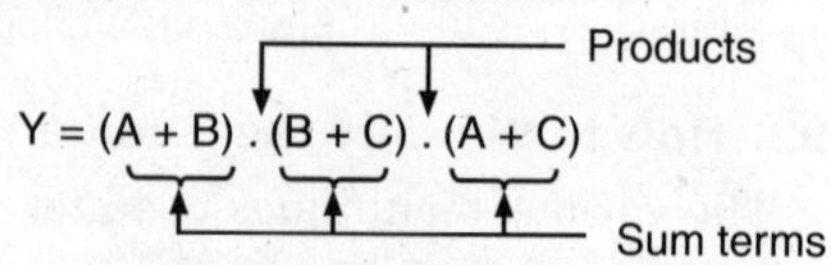

Fig. 2.14. Product-of-sums (POS) form

$$Y = (A + \overline{B} + \overline{C}) \cdot (A + B) \cdot (\overline{A} + C)$$

$$A = (\overline{X} + \overline{Y}) \cdot (X + Y + Z)$$

$$Y = (P + R) \cdot (P + \overline{Q}) \cdot (\overline{P} + R)$$

The literals are ORed together to form the sum terms and the sum terms are ANDed to get the expression in the POS form.

2.10.3 Standard or Canonical SOP and POS Forms

Let us consider the SOP and POS expressions shown in figure. 2.15.

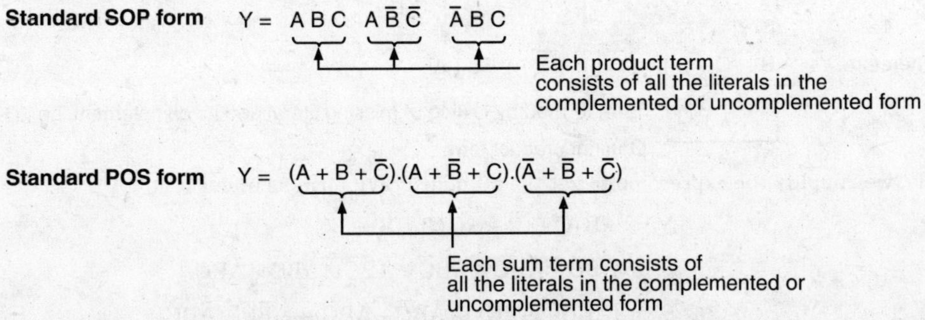

Fig. 2.15. Standard SOP and POS forms

According to figure 2.15 we can say that a logic expression is said to be in the standard (or canonical) SOP or POS form if each product term (for SOP); and sum term (for POS) consists of all the literals in their complemented or uncomplemented form. The examples of standard and non-standard SOP and POS expressions are given in Table 2.8.

Table 2.8. Standard and non-standard SOP amd POS forms

S.No.	Logical Expression	Type of expression
1.	$Y = AB + AB\bar{C} + \bar{A}BC$	Non-standard SOP
2.	$Y = AB + A\bar{B} + \bar{A}\bar{B}$	Standard SOP
3.	$Y = (\bar{A} + B) \cdot (A + B) \cdot (A + \bar{B})$	Standard POS
4.	$Y = (\bar{A} + B) \cdot (A + B + C)$	Non-standard POS

2.10.4. Conversion of a Logical Expression to Standard SOP or POS Form

Let us observe how we can convert the given SOP and POS expressions into the corresponding standard SOP and POS forms.

1. Conversion procedure from SOP to standard SOP form

The procedure to be followed for converting a general SOP expression into a standard SOP expression may be described as under :

(i) For each term, we find the missing literal.

(ii) Then, we AND term with the term formed by ORing the missing literal and its complement.

EXAMPLE 2.11. Convert the expression $Y = AB + A\bar{C} + BC$ into the standard SOP form.

Solution: Given expression $Y = AB + A\bar{C} + BC$.

First, let us find the missing literal for each term.

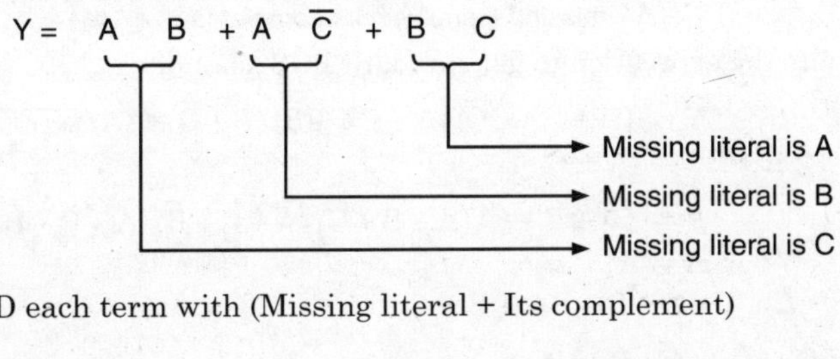

(ii) Secondly, we AND each term with (Missing literal + Its complement)

Complement — Missing literal — Missing literal — Complement

Therefore, $Y = AB\,.\,(C+\bar{C}) + A\bar{C}\,(B+\bar{B}) + BC\,(A+\bar{A})$

→ Term formed by ORing of missing literal and its complement, i.e., C and $\bar{C}$

→ Original product term

Lastly, we simplify the expression to get the standard SOP form as under

$$Y = AB(C+\bar{C}) + AC(B+\bar{B}) + BC(A+\bar{A})$$

$$= ABC + AB\bar{C} + AB\bar{C} + A\bar{B}\bar{C} + ABC + \bar{A}BC$$

$$= (ABC + ABC) + (AB\bar{C} + AB\bar{C}) + A\bar{B}\bar{C} + \bar{A}BC$$

But $\qquad A + A = A$

Therefore, $\qquad (ABC + ABC) = ABC \text{ and } (AB\bar{C} + AB\bar{C}) = AB\bar{C}$

Therefore, we get the following standard SOP form :

$$Y = ABC + AB\bar{C} + A\bar{B}\bar{C} + \bar{A}BC$$

→ Each term consists of all the literals

2. Conversion from POS to standard POS form

The conversion of given POS expression into standard POS form can be obtained by following the steps given below :

(i) For each term, we find the missing literal.

(ii) Then, we OR each term with the formed by ANDing the missing literal in that term with its complement.

(iii) Lastly, we simplify the expression to get the standard POS form.

EXAMPLE 2.12. Convert the expression $Y = (A + B)(A + C)(B + \bar{C})$ into standard POS form.

Solution: (i) First, let us find the missing literal for each term as under :

$$Y = (A+B)\;(A+C)\;(B+\bar{C})$$

→ Missing literal is A

→ Missing literal is B

→ Missing literal is C

Secondly, we OR each term with (Missing literal or its complement).

Therefore, we have $\quad Y = (A + B + C\bar{C})\,.\,(A + C + B\bar{B})\,.\,(B + \bar{C} + A\bar{A})$

→ Missing literal ANDed with its complement

→ This term is ORed with the term formed by ANDing the missing literal with its complement.

(iii) Lastly, we simplify the expression to get standard POS as under :

$$Y = (A+B+C\bar{C})(A+C+B\bar{B})(B+\bar{C}+A\bar{A})$$

But $\qquad A + BC = (A + B)(A + C)$

Therefore, $\qquad Y = (A+B+C)(A+B+\bar{C})(A+C+B)(A+C+\bar{B})(B+\bar{C}+A)\;(B+\bar{C}+A)$

But $\qquad A \cdot A = A$

Therefore, $\quad (A + B + C)(A + C + B) = (A + B + C)$

and $(A+B+\bar{C})(B+\bar{C}+A) = (A+B+\bar{C})$

Therefore, $Y = \underbrace{(A+B+C)(A+B+\bar{C})(A+\bar{B}+C)(\bar{A}+B+\bar{C})}_{\text{Each term consists of all the literals. Hence, this is the standard POS form.}}$

EXAMPLE 2.13. Convert the following expressions into their standard SOP or POS forms :

(i) $Y = AB + AC + BC$

(ii) $Y = (A+B)(\bar{B}+C)$

(iii) $Y = A + BC + ABC$

Solution: (i) $Y = AB + AC + BC$

Therefore, $Y = AB(C+\bar{C}) + AC(B+\bar{B}) + BC(A+\bar{A})$

$= AB(C+\bar{C}) + AC(B+\bar{B}) + BC(A+\bar{A})$

$= \underbrace{ABC+ACB\ BCA+AB\bar{C}+AC\bar{B}+BC\bar{A}}$

Therefore, $Y = ABC + AB\bar{C}+A\bar{B}C+\bar{A}BC$ $(\because A+A=A)$

This is the required expression in standard POS form.

(ii) $Y = (A+B)(\bar{B}+C) = (A+B+C\bar{C})(\bar{B}+C+A\bar{A})$

But, $A+BC = (A+B)(A+C)$

Therefore, $Y = (A+B+C)(A+B+\bar{C})(\bar{B}+C+A)(\bar{B}+C+\bar{A})$

This is in the standard POS form.

(iii) $Y = A + BC + ABC$

$= A(B+\bar{B})(C+\bar{C}) + BC(A+\bar{A}) + ABC$

$= ABC + AB\bar{C} + A\bar{B}C + A\bar{B}\bar{C} + BCA + BC\bar{A} + ABC$

$= (ABC + BCA + ABC) + AB\bar{C} + A\bar{B}C + A\bar{B}\bar{C} + BC\bar{A}$

But, $A + A = A$ so $(ABC + BCA + ABC) = ABC$

Therefore, $Y = ABC + AB\bar{C} + A\bar{B}C + A\bar{B}\bar{C} + BC\bar{A}$

This expression is in the standard SOP form.

2.11 MINTERM AND MAXTERM DESIGNATION **(PTU, Jalandhar, Sem. Exam., 2004-05)**

Minterm

Each individual term in the standard SOP form is called as *minterm*. This has been shown in figure 2.16.

Maxterm

Each individual term in the standard POS form is called as *maxterm*. This has been shown in figure 2.16.

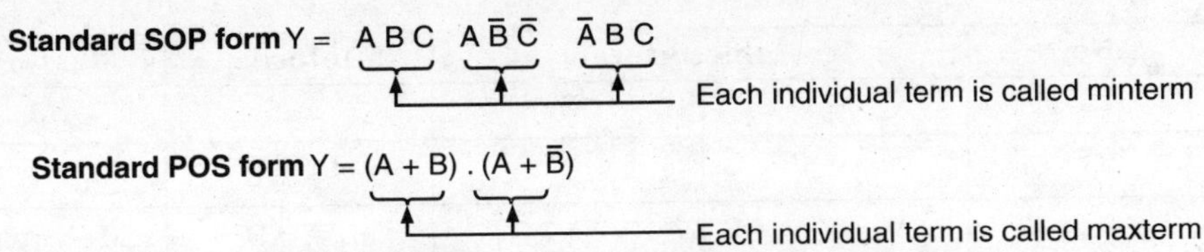

Fig. 2.16. Concept of maxterm and minterm

The concept of minterm and maxterm allows us to introduce a very convenient shorthand notation to express logic functions. Table 2.9 gives the minterms and maxterms for a three variable/literal logic function. Let Y be the output and A, B, C be the variables. The number of minterms and maxterms is $2^3 = 8$. In general, for n number of variables, the number of minterms or maxterms will be 2^n.

Table 2.9. Minterms and Maxterms for three variables

Variables			Minterms	Maxterms
A	B	C	m_i	M_i
0	0	0	$\overline{A}\overline{B}\overline{C} = m_0$	$A + B + C = M_0$
0	0	1	$\overline{A}\overline{B}C = m_1$	$A + B + \overline{C} = M_i$
0	1	0	$\overline{A}B\overline{C} = m_2$	$A + \overline{B} + C = M_2$
0	1	1	$\overline{A}BC = m_3$	$A + \overline{B} + \overline{C} = M_3$
1	0	0	$A\overline{B}\overline{C} = m_4$	$\overline{A} + B + C = M_4$
1	0	1	$A\overline{B}C = m_5$	$\overline{A} + B + \overline{C} = M_5$
1	1	0	$AB\overline{C} = m_6$	$\overline{A} + \overline{B} + C = M_6$
1	1	1	$A\,B\,C = m_7$	$\overline{A} + B + C = M_7$

Each minterm is represented by m_i where $i = 0, 1, 2^n - 1$ and maxterm is represented by M_i.

2.11.1 How to write Minterms and Maxterms ?

1. Writing minterm for a particular combination of ABC

Let ABC = 011. Let us assume that A, B, C are input to an AND gate. We want the output of the AND gate to be 1. For that, all its inputs must be 1. Hence, we complement the input which is 0, i.e., A in this case.

Therefore, Minterm = $\overline{A}BC$ corresponding to ABC = 011

Similarly, other minterms may be obtained.

> **DO YOU KNOW?**
>
> Although one can use some knowledge from ordinary algebra to deal with Boolean algebra, the beginner must be careful not to substitute the rules of ordinary algebra where they are not applicable.

2. Writing maxterm for a particular combination of ABC

Let ABC = 011. Let us assume that ABC are inputs to an OR gate. We want the output of the OR gate to be 0. Hence, all the inputs to the OR gate must be 0s. Therefore, we invert the inputs which are 1s. i.e., B and C in this case.

Therefore, Maxterm = $(A + \overline{B} + \overline{C})$ corresponding to ABC = 0 1 1

Similarly, other maxterms may be obtained.

EXAMPLE 2.14. For the truth table of two variables, write the minterms and maxterms.

Solution: Table 2.10 shows the required minterms and maxterms.

Table 2.10. Solution to Example 2.14

S.No.	Variables/literals		Minterms	Maxterms
1.	A	B	m_i	M_i
2.	0	0	$m_0 = \overline{A}\overline{B}$	$M_0 = A + B$
3.	0	1	$m_1 = \overline{A}B$	$M_1 = A + \overline{B}$
4.	1	0	$m_2 = A\overline{B}$	$M_2 = \overline{A} + B$
5.	1	1	$m_3 = A\,B$	$M_3 = \overline{A} + \overline{B}$

* The force that rules the world is conduct, whether it be a moral or immoral. —**Nichola Murray**

2.11.2 Representation of Logical Expressions with the help of Minterms and Maxterms

We can represent the logical expressions using the minterms and maxterms as under :

(i) We have, $$Y = \underbrace{ABC}_{m_7} + \underbrace{\bar{A}BC}_{m_3} + \underbrace{A\bar{B}\bar{C}}_{m_4}$$ ← Given logic expression ← Corresponding minterms

Therefore, $Y = m_7 + m_3 + m_4$

or $Y = \Sigma m\,(3, 4, 7)$ ← other way of representation

where Σ denotes sum of products.

(ii) We have, $$Y = \underbrace{\left(A+\bar{B}+C\right)}_{M_2}\ \underbrace{(A+B+C)}_{M_0}\ \underbrace{\left(\bar{A}+\bar{B}+C\right)}_{M_6}$$ ← Given logic expression ← Corresponding minterms

Therefore, $Y = M_2 \quad M_0 \quad M_6$

or $Y = \Pi M\,(0, 2, 6)$ ← Other way of representation.

where Π denotes product of sums.

2.11.3 How to write SOP and POS forms for a given Truth Table

We know that truth table representation of a logic expression is possible. As an example, the expression $Y = AB + \bar{A}\bar{B}$ which is the Boolean expression for an EX-NOR gate can also be represented using a truth table. Hence, it is possible to obtain the logic expression in the standard SOP or POS form if a truth table is given to us.

2.11.4 How to write standard SOP Expression for a given Truth Table

The procedure to be followed for writing the standard SOP expression is as under :

(i) First, we consider only those combinations of inputs which correspond to Y = 1.

(ii) Then, we write a product term for each such combination.

(iii) Lastly, we OR all these product terms to get the standard SOP form.

EXAMPLE 2.15. From the truth Table 2.11, obtain the logical expression in the standard SOP form.

Table 2.11. Given truth table

A	B	Y
0	0	0
0	1	1
1	0	1
1	1	0

Solution: (i) First, Let us consider only those combinations of inputs which corresponds to Y = 1.

Table 2.12. Truth table of an EX-OR gate

	A	B	Y	
①	0	0	0	← Y = 0, So do not consider this entry
②	0	1	1	← Y = 1, So these two entries must be considered
③	1	0	1	←
④	1	1	0	← Y = 0, So don't consider

(ii) For the second and the third entries in Table 2.13, we write the product terms.

Life is a continuous process of adjustment. — Indira Gandhi

Table 2.13

	A	B	Y	
①	0	0	0	X
②	0	1	1	→ $Y_1 = \bar{A}B$
③	1	0	1	→ $Y_2 = A\bar{B}$
④	1	1	0	X

Boolean expressions in the product forms

(iii) Next, we OR (add) all the product terms to write the final expression in standard SOP form as under :

$$Y = Y_1 + Y_2$$

or $$Y = \bar{A}B + A\bar{B}$$

This is the required expression in the standard SOP form. It can also be represented as under :

$$Y = \bar{A}B + A\bar{B} = m_1 + m_2$$

where m_1 is the minterm corresponding to AB = 01 and m_2 is the minterm corresponding to AB = 10. The expression can also expressed as under :

$$Y = \Sigma\, m\,(1, 2)$$

EXAMPLE 2.16. For the truth table given in Table 2.14, write the logic expression in the standard SOP form.

Solution:

Table 2.14. Given truth table

A	B	C	Y	
0	0	0	0	
0	0	1	1	← $\bar{A}\,\bar{B}\,C\ (m_1)$
0	1	0	0	
0	1	1	0	
1	0	0	1	← $A\,\bar{B}\,\bar{C}\ (m_4)$
1	0	1	0	
1	1	0	0	
1	1	1	1	← $A\,B\,C\ (m_7)$

First, we obtain product terms corresponding to combinations of inputs for which Y = 1

(ii) Now, we OR (add) all the product terms.

Thus, ORing all the product terms, we obtain,

$$Y = \bar{A}\bar{B}C + A\bar{B}\bar{C} + A\,B\,C \quad \textbf{Ans.}$$

This is the required logic expression is standard SOP form.

This expression may also be written as under :

$$Y = m_1 + m_4 + m_7 = \Sigma\, m\,(0, 4, 7).$$

2.11.5. How to write a standard POS Expression for a given Truth Table

We follow the procedure given below to get the expression in standard POS form as under :

(i) First, we consider only those combinations of inputs which produce a low output (Y = 0)

(ii) Then, we write the maxterms only for such combinations.

(iii) Finally, we AND these maxterms to obtain the expression in standard POS form.

EXAMPLE 2.17. Write the logic expression in standard POS form for the truth table given in Table 2.15.

Solution:

Table 2.15. Given truth table

A	B	C	Y	
0	0	0	0	← $A + B + C\ (M_0)$
0	0	1	1	
0	1	0	1	
0	1	1	0	← $A + \bar{B} + \bar{C}\ (M_3)$
1	0	0	1	
1	0	1	0	← $\bar{A} + B + \bar{C}\ (M_5)$
1	1	0	0	← $\bar{A} + \bar{B} + C\ (M_6)$
1	1	1	1	

First, we write maxterms for the combinations of input which produce Y = 0

(ii) Then, we AND (take product of) all the maxterms to get standard POS form.

Now, ANDing (taking product of) all the maxterms written in (i), we obtain

$$Y = (A + B + C) \cdot (A + \bar{B} + \bar{C})(\bar{A} + B + \bar{C})(\bar{A} + \bar{B} + C)$$

This is the required logic expression in the standard POS form.

This expression may also be written as under :

$$Y = M_0 \cdot M_3 \cdot M_5 \cdot M_6$$

or $$Y = \Pi\ (0, 3, 5, 6)$$

2.11.6 Conversion from SOP to POS and POS to SOP form

It is necessary to understand that the SOP and POS forms written for the same truth table are always logically equivalent. This point can be proved by solving the next example.

EXAMPLE 2.18. For the given truth Table 2.16, write the logical expressions in the standard SOP and POS forms and prove their equivalence.

Solution:

Table 2.16. Given truth table

A	B	C	Y	
0	0	0	0	← $A + B + C$
0	0	1	1	← $\bar{A}\ \bar{B}\ C$
0	1	0	0	← $A + \bar{B} + C$
0	1	1	0	← $A + \bar{B} + \bar{C}$
1	0	0	1	← $A\ \bar{B}\ \bar{C}$
1	0	1	0	← $\bar{A} + B + \bar{C}$
1	1	0	0	← $\bar{A} + \bar{B} + C$
1	1	1	1	← ABC

First, we obtain minterms and maxterms for the combinations of inputs producing Y = 1 and 0 respectively.

(ii) Then, we write the standard SOP and POS expressions.

Standard SOP form : $Y = \bar{A}\bar{B}C + A\bar{B}\bar{C} + ABC$

Standard POS form : $Y = (A + B + C)(A + \bar{B} + C)(A + \bar{B} + \bar{C})(\bar{A} + B + \bar{C})(\bar{A} + \bar{B} + \bar{C})$

(iii) Next, we prove the equivalence between SOP and POS forms.

Let us consider the standard POS form as under :

$$Y = (A + B + C)(A + \bar{B} + C)(A + \bar{B} + \bar{C})(\bar{A} + B + \bar{C})(\bar{A} + \bar{B} + C)$$

But, $(A+B+C)(A+\bar{B}+\bar{C}) = (A+C+B\cdot\bar{B}) = (A+C)$

Therefore, $Y = (A+C)(A+\bar{B}+\bar{C})(\bar{A}+B+\bar{C})(\bar{A}+\bar{B}+C)$

Multiplying the terms (1 × 2) and (3 × 4), we get

$$Y = [AA + A\bar{B} + A\bar{C} + AC + \bar{B}C + C\bar{C}]$$
$$[A\bar{A} + \bar{A}\bar{B} + \bar{A}C + \bar{A}B + B\bar{B} + BC + \bar{A}\bar{C} + \bar{B}\bar{C} + C\bar{C}]$$

But $\quad AA = A, C\bar{C} = 0, \bar{A}A = A, B\bar{B} = 0, C\bar{C} = 0$

Therefore, $Y = [A + A\bar{B} + A\bar{C} + AC + \bar{B}C]\ [(\bar{A} + \bar{A}\bar{B} + \bar{A}C + \bar{A}B + \bar{A}\bar{C} + \bar{B}\bar{C}]$

$$= [A(1+\bar{B}) + A(\bar{C}+C) + \bar{B}C][(1+\bar{B})\bar{A} + \bar{A}(C+\bar{C}) + \bar{A}B + BC + \bar{B}\bar{C}]$$

But $\quad (1+\bar{B}) = 1, (C+C) = 1, (1+\bar{B}) = 1$

Therefore, $Y = [A + A + \bar{B}C]\ [\bar{A} + \bar{A}B + BC + \bar{B}\bar{C}]$

But $\quad A + A = A$ and $\bar{A} + \bar{A} = \bar{A}$

Therefore, $Y = (A+\bar{B}C)[\bar{A} + \bar{A}B + BC + \bar{B}\bar{C}] = (A+\bar{B}C)[\bar{A}(1=B) + BC]$

But $\quad (1 + B) = 1$

Therefore, $Y = (A+\bar{B}C)(\bar{A} + BC + \bar{B}\bar{C}) = A\bar{A} + ABC + A\bar{B}\bar{C} + (\bar{B}C\cdot BC) + (\bar{B}C\cdot\bar{B}\bar{C})$

But $\quad A\bar{A} = 0, (\bar{B}C\cdot BC) = 0$ and $(\bar{B}C\cdot BC) = 0$

Therefore, $Y = ABC + A\bar{B}\bar{C} + \bar{A}\bar{B}C$ ← This is same as standard SOP form. Thus, the equivalence between SOPand POS forms is proved.

(iv) Now, let us express output in terms of minterms and maxterms.

The output can be expressed in terms of minterms and maxterms as under :

$$Y = m_1 + m_4 + m_7 \quad \text{(in terms of minterms)}$$
$$= \Sigma\, m\,(1, 4, 7)$$

and

$$Y = M_0 \cdot M_2 \cdot M_3 \cdot M_5 \cdot M_6 \quad \text{(in terms of maxterms)}$$
$$= \Pi\, M\,(0, 2, 3, 5, 6)$$

Because of equivalence between SOP and POS forms, we can write

$$\Sigma\, m\,(1, 4, 7) = \Pi\, M\,(0, 2, 3, 5, 6) \quad ...(i)$$

2.11.7 Complementary Relationship between Minterms and Maxterms

From equation (i), we can conclude that the relationship between the expressions expressed using minterms and maxterms is complementary relationship. We can exploit this complementary relationship to write the expression in terms of maxterms if the expression in terms of minterms is known and vice versa. As an example, if a SOP expression for 4-variables is given by

$$Y = \Sigma\, m\,(0, 1, 3, 5, 6, 7, 11, 12, 15)$$

Then we can get the equivalent POS expression using the complementary relationship as follows :

$$Y = \Pi\, M\,(2, 4, 8, 9, 10, 13, 14)$$

2.12. VARIOUS METHODS TO SIMPLIFY BOOLEAN EXPRESSIONS

Let us discuss the following methods to simplify Boolean expressions :

(i) Algebraic simplification

(ii) Simplification using Karnaugh maps.

(iii) Simplification using Quine McClusky method

2.12.1 Standard Procedure for Algebraic Simplification

Earlier, we have discussed the Boolean laws and DeMorgan's theorems. Simply, we can use them do simplify the given Boolean expression in the following way. The most important thing is to convert the given expression into SOP form.

(*i*) First, we bring the given expression into the SOP form by using the Boolean laws and DeMorgan's theorems.

(*ii*) Then, we simplify this SOP expression by checking the product terms for common factors.

DO YOU KNOW?

An alternative procedure for deriving the minterms of a Boolean function is to obtain the truth table of the function directly from the algebraic expression and then read the minterms from the truth table.

EXAMPLE 2.19. Simplify the expression given below.

$$\mathbf{Y = AB + (AB)(\bar{A} + B).}$$

Solution: First, let us bring the given expression in SOP form.

Given expression, $\quad Y = AB + (A + B)(\bar{A} + B) = AB + (A\bar{A} + AB + \bar{A}B + BB)$

Then, let us search for common factors and simplify :

$$Y = AB + A\bar{A} + AB + \bar{A}B + BB = AB + AB + BB + A\bar{A} + \bar{A}B$$

But $\quad AB + AB = AB,\ BB = B \text{ and } A\bar{A} = 0$

Therefore, $\quad Y = AB + B + \bar{A}B = B(A + 1) + \bar{A}B$

But $\quad (A + 1) = 1$

Therefore, $\quad Y = B + \bar{A}B = B(1 + \bar{A}) = B \qquad [\because\ (1 + A) = 1]$

or $\quad Y = B$ **...Ans.**

This is the required simplified expression

EXAMPLE 2.20. For the logic circuit shown in figure 2.17, write the Boolean expression and simplify it.

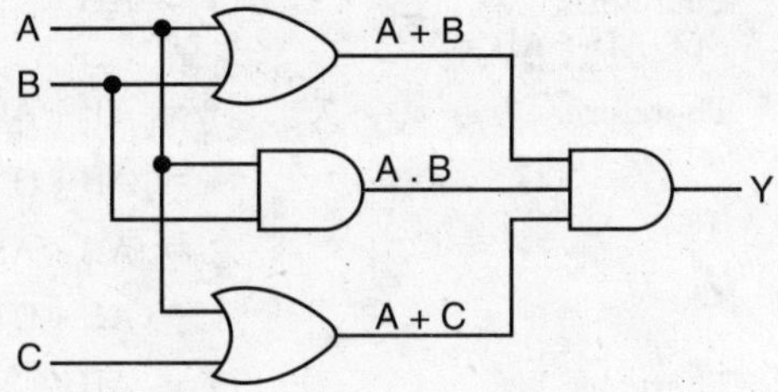

Fig. 2.17. Given logic circuit

Soluton: First, we write the Boolean expression.

The expression for output of the given logic circuit is given by,

$$Y = (A + B)(AB)(A + C)$$

Then, we bring the expression in SOP form.

Multiplying the terms, we get the expression into SOP form as under :

$$Y = (A + B)(AAB + ABC)$$

But AA = A, therefore, $\quad Y = (A + B (AB + ABC) = AAB + AABC + BAB + BABC$

$$= AB + ABC + AB + ABC = AB + AB + ABC + ABC$$

But, $AB + AB = AB$ and $ABC + ABC = ABC$

Therefore, $\quad Y = AB + ABC = AB(1 + C)$

or $\quad Y = AB \qquad (\because\ 1 + C = 1)$

This is the required simplified expression.

EXAMPLE 2.21. Simply the following three variable Boolean expression.

$$\mathbf{Y = \Sigma\, m\, (2, 4, 6)}$$

Solution: The given expression can be written in SOP form as under :

$$Y = m_2 + m_4 + m_6 = \bar{A}B\bar{C} + A\bar{B}\bar{C} + AB\bar{C} = \bar{A}B\bar{C} + A\bar{C}(\bar{B} + B)$$

But $\quad \bar{B} + B = 1,$

Therefore, $Y = \bar{A}B\bar{C} + A\bar{C}$

or $\quad Y = \bar{C}(\bar{A}B + A)$

But, $A + BC = (A + B)(A + C)$, hence, $(A + \bar{A}B) = (A + \bar{A})(A + B)$

Therefore, $\quad Y = \bar{C}(A + \bar{A})(A + B) = \bar{C}(1)(A + B) = \bar{C}(A + B)$

This is the required simplified expression.

EXAMPLE 2.22. Simplify the following three variable logic expression :

$$Y = \Pi M(1, 3, 5).$$

Solution: The given expression can be expressed in terms of maxterms as under :

$$Y = M_1 \cdot M_3 \cdot M_5 = (A + B + \bar{C})(A + \bar{B} + \bar{C})(\bar{A} + B + \bar{C})$$

First, we bring the expression into SOP form.

This means that $Y = (A + B + \bar{C})(A + \bar{B} + \bar{C})(\bar{A} + B + \bar{C})$

$$= (A + B + \bar{C})(A\bar{A} + AB + A\bar{C} + \bar{A}\bar{B} + \bar{B}B + \bar{B}\bar{C} + \bar{A}\bar{C} + B\bar{C} + \bar{C}\bar{C})$$

But $A\bar{A} = 0$, $B\bar{B} = 0$ and $\bar{C}\bar{C} = \bar{C}$

Therefore, $Y = (A + B + \bar{C})(AB + A\bar{C} + \bar{A}\bar{B} + \bar{B}\bar{C} + \bar{A}\bar{C} + B\bar{C} + \bar{C})$

$$= (A + B + \bar{C})[AB + A\bar{C} + \bar{A}\bar{C} + \bar{A}\bar{B} + \bar{B}\bar{C} + B\bar{C} + \bar{C})$$

$$= (A + B + \bar{C})[AB + \bar{C}(A + \bar{A}) + \bar{A}\bar{B} + \bar{C}(\bar{B} + B) + \bar{C}]$$

But, $(A + \bar{A}) = 1$, $(\bar{B} + B) = 1$

$$Y = (A + B + \bar{C})[AB + \bar{C} + \bar{A}\bar{B} + \bar{C} + \bar{C}]$$

But, $\bar{C} + \bar{C} = \bar{C}$

$$Y = (A + B + \bar{C})(AB + \bar{A}\bar{B} + \bar{C} + \bar{C}) = (A + B + \bar{C})(AB + \bar{A}\bar{B} + \bar{C})$$

$$= AAB + A\bar{A}\bar{B} + A\bar{C} + ABB + \bar{A}\bar{B}B + B\bar{C} + AB\bar{C} + \bar{A}\bar{B}\bar{C} + \bar{C}\bar{C}$$

But, $AAB = AB$, $A\bar{A}\bar{B} = 0$, $ABB = AB$, $\bar{A}\bar{B}B = 0$ and $\bar{C}\bar{C} = \bar{C}$

Therefore, $Y = AB + A\bar{C} + AB + B\bar{C} + AB\bar{C} + \bar{A}\bar{B}\bar{C} + \bar{C}$

But, $AB + AB = AB$

Therefore, $Y = AB + A\bar{C} + B\bar{C} + AB\bar{C} + \bar{A}\bar{B}\bar{C} + \bar{C} = AB + AB\bar{C} + A\bar{C} + B\bar{C} + \bar{A}\bar{B}\bar{C} + \bar{C}$

$$= AB + (1 + \bar{C}) + A\bar{C} + B\bar{C} + \bar{A}\bar{B}\bar{C} + \bar{C}$$

$$= AB + A\bar{C} + B\bar{C} + \bar{A}\bar{B}\bar{C} + \bar{C} \qquad [\because 1 + \bar{C} = 1]$$

$$= AB + \bar{C}(A + B + \bar{A}\bar{B} + 1)$$

Therefore, $Y = AB + \bar{C}$ $\qquad [\because (A + B + \bar{A}\bar{B} + 1) = 1]$

or $Y = AB + \bar{C}$ **Ans.**

This is the required simplified expression

2.12.2 Drawbacks of Algebraic Method of Simplification

(i) In the algebraic method of simplification, we require to write lengthy expressions, find the common terms, manipulate the expressions etc. Hence, it is a time consuming work.

(ii) Further, sometimes, we are not sure whether the further simplification is really possible or not.

2.12.3 Gate-Level Minimization

As a matter of fact, gate-level minimization refers to the design task of finding an optimal gate-level implementation of the Boolean functions describing a digital circuit. This task is well understood, but, it is difficult to execute by manual techniques when the logic has more than a few inputs. Fortunately, computer-based logic synthesis tools can minimize a large set of Boolean equations efficiently and quickly. Further, it is important that a designer understand the underlying mathematical description and solution of the problem.

2.13 SIMPLIFICATION OF KARNAUGH-MAP (K-MAP) (PTU, Jalandhar, Sem. Exam., 2007-08)

This is another simplification technique to reduce the Boolean expression. It overcomes all the drawbacks of the algebraic simplification technique. K-map (short form of Karnaugh-map) is a graphical method of simplifying a Boolean expression. K-map is a graphical chart which contains boxes. The information contained in a truth table or available in the SOP or POS form is represented on a K-map. The K-map method is a systematic approach to simplification of Boolean expression. K-maps can be written for 2, 3, 4 ..... upto 6 variables. Beyond that the K-map technique becomes very cumbersome.

Important Point: K-map is ideally suitable for designing the combinational logic circuits using either a SOP method or a POS method.

2.13.1 The Structure of Karnaugh-Map (K-Map)

The structure of a 2 input (variable) Karnaugh map is shown in figure 2.18. The K-map comprises a box for every line in the truth table. For a 2-input combinational circuit, there are 4-lines in the truth table, so there will be 4-boxes in the 2 variables K-map. For a 3-variable K-map, there will be 8 boxes, for 4-variable map, there will be 16 boxes and so on. The 0's and 1's written on top or sides of the boxes represent the values of the corresponding variables. In truth tables, the values of inputs follows a standard binary sequence (00, 01, 10, 11). But in K-maps, the input values are ordered in a different sequence (00, 10, 11, 10). This is as per the gray code and not binary code. So that as we move along a row or column only one variable will change its value.

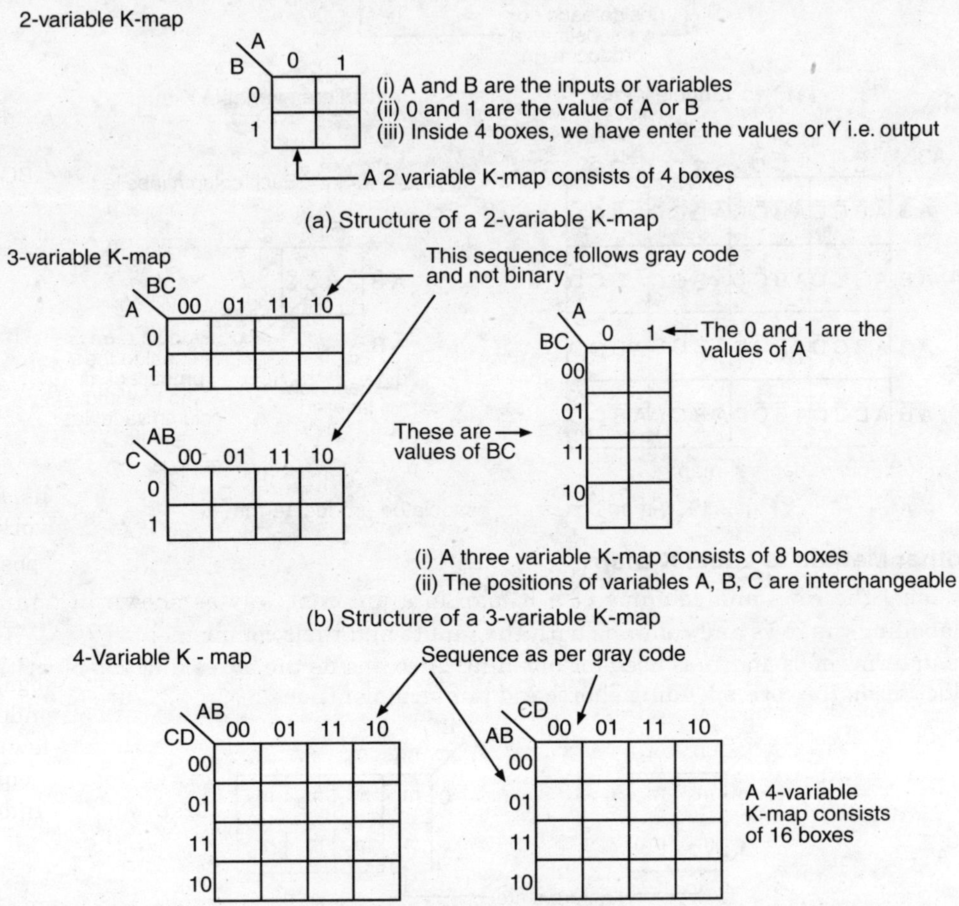

(a) Structure of a 2-variable K-map

(b) Structure of a 3-variable K-map

(c) Structure of a 4-variable K-map

Fig. 2.18. Structure of 2, 3, 4 variable K-map

* Change is not made without inconvenience even from worse to better. —**Richard Hooker**

2.13.2 How to Label K-Map

(PTU, Jalandhar, Sem. Exam., 2004-05)

The boxes in a K-map are to be filled with the values for output Y corresponding to different combinations of inputs as shown in figure 2.19. Each row and column of a K-map is labelled with a variable, or a group of variables or their complements. As depicted in figure 2.19 (a). This box corresponds to the first row which is labelled by $\overline{A}$ and first column labelled by $\overline{B}$. Hence, the product term written inside this box is $\overline{A}\overline{B}$ as shown in Fig. 2.19 (a).

Another example is the shaded box shown in figure 2.19 (c). The first row is labelled with $\overline{A}\overline{B}$ and the first column with $\overline{C}\overline{D}$. Hence, the product term written in this box is $\overline{A}\overline{B}\overline{C}\overline{D}$. Similarly, the other product terms have been inserted in the remaining boxes.

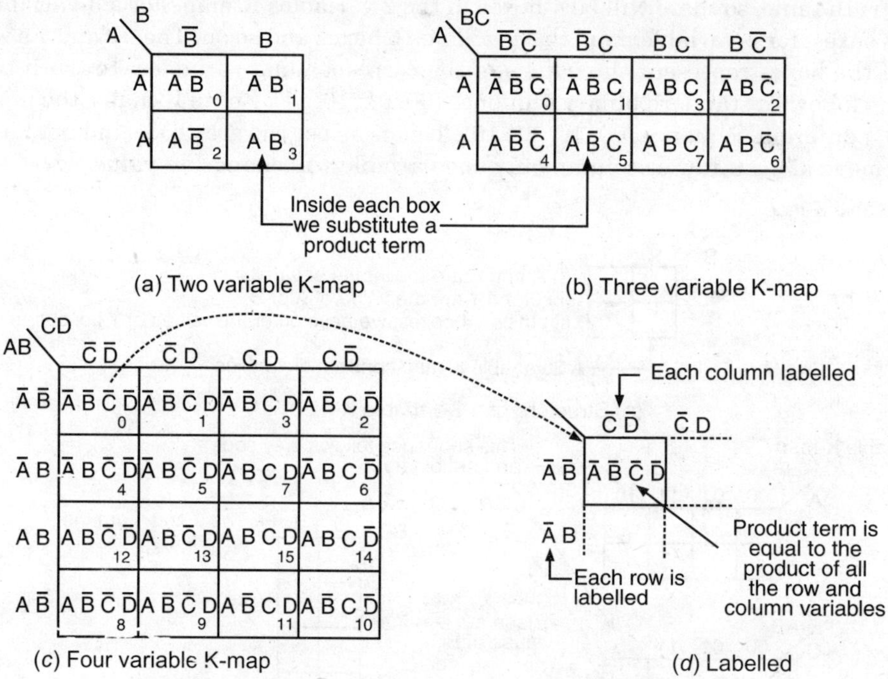

(a) Two variable K-map (b) Three variable K-map

(c) Four variable K-map (d) Labelled

Fig. 2.19. K-map boxes and associated product terms

2.13.3 Another Method to Label K-map

We can label the rows and columns of a K-map in a different way as shown in figure 2.20. Instead of labelling the rows and columns with the inputs and their complements (A, $\overline{A}$, A$\overline{B}$ etc.), their values in terms of 0s and 1s is used for labelling. Also, inside the boxes, instead of writing the actual product term, the corresponding shorthand minterm notations m_0, m_1 ..... are entered.

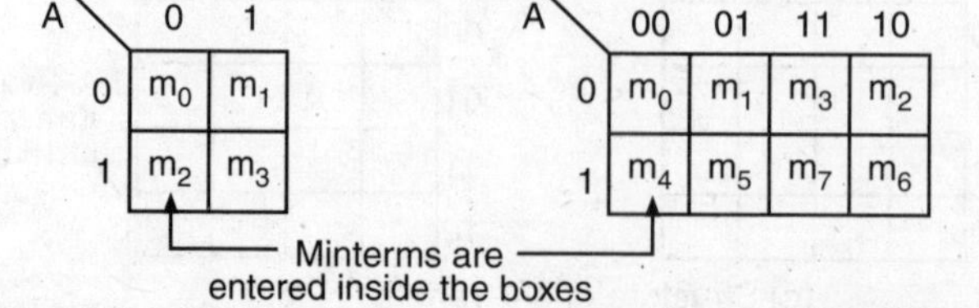

(a) Two variable K-map (b) Three variable K-map

* The happiness of every country depends upon the character of its people rather than the form of its government. — **Thomas C. Haliburton**

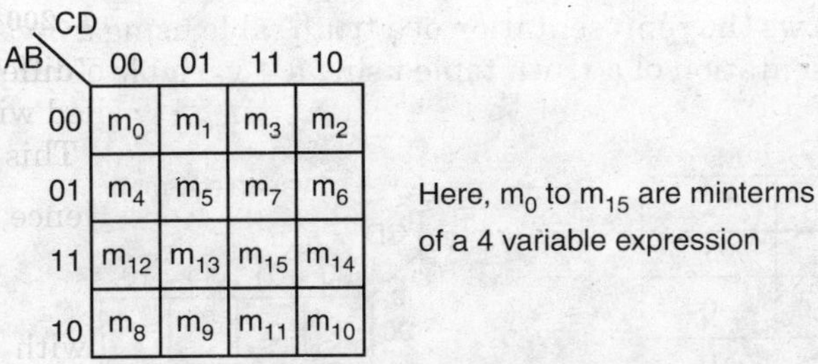

(c) Four variable K-map

Fig. 2.20. Alternative way to label K-map

> **DO YOU KNOW?**
>
> The two canonical forms of Boolean algebra are basic forms that one obtains from reading a given function from the truth table.

2.13.4 Relation Between a Truth Table and K-map Entries

Whenever the K-map is to be practically used for simplification, the entries inside the boxes are to be done from the given truth table. In this subarticle, let us discuss how to represent the given truth table on a Karnaugh-map.

Figure 2.21 (*a*) illustrates the mapping of a truth table on the K-map.

According to figure 2.21 (*a*), we conclude that inside the boxes of the K-map, we have to enter the values of output *Y* corresponding to various

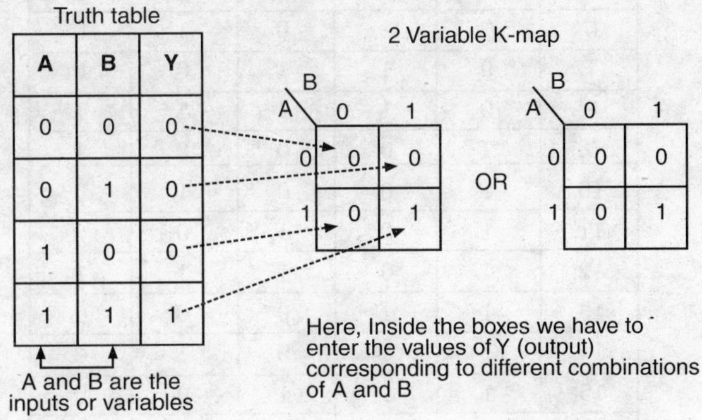

Fig. 2.21. Relation between a truth table and K-maps for 2 variables.

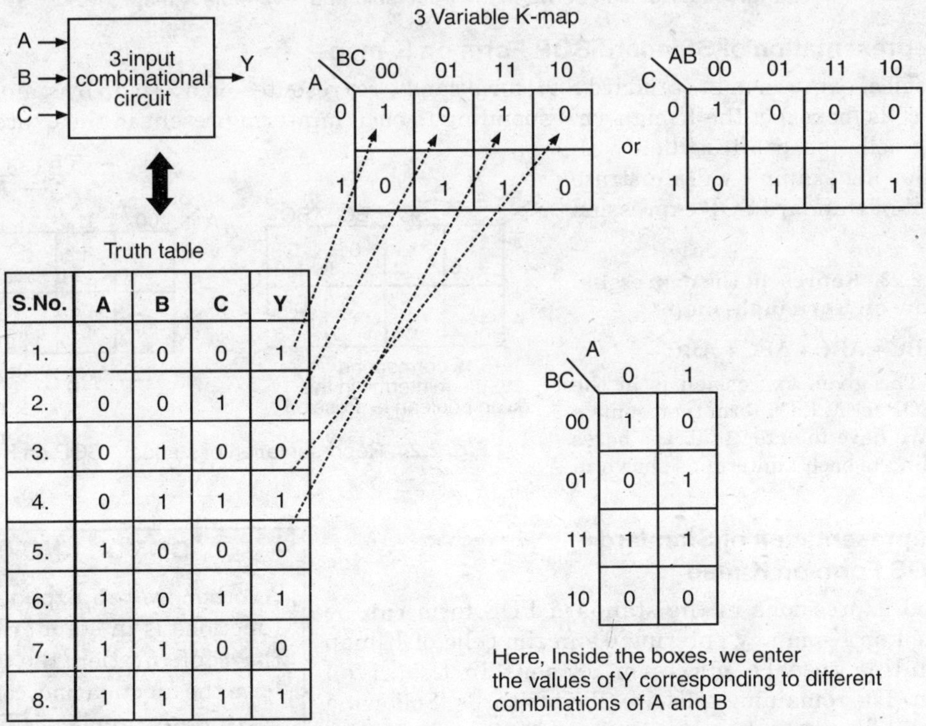

Fig. 2.22. Relation between a truth table and K-map for 3-variables

combinations of A and B. Figure 2.21 (b) shows the representation of a truth table using a 3-variable K-map and figure 2.21 (c) shows the representation of a truth table using a 4-variable K-map.

Truth table

S.No.	A	B	C	D	Y
1.	0	0	0	0	1
2.	0	0	0	1	0
3.	0	0	1	0	0
4.	0	0	1	1	1
5.	0	1	0	0	0
6.	0	1	0	1	0
7.	0	1	1	0	0
8.	0	1	1	1	1
9.	1	0	0	0	1
10.	1	0	0	1	0
11.	1	0	1	0	0
12.	1	0	1	1	1
13.	1	1	0	0	0
14.	1	1	0	1	0
15.	1	1	1	0	0
16.	1	1	1	1	1

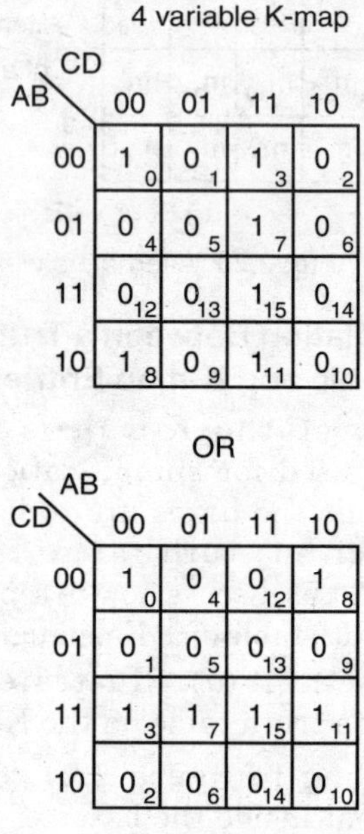

Fig. 2.23. Relation between the truth table and 4-variable K-map

2.13.5 Representation of Standard SOP Form on K-map

The logical expression in standard SOP form can be represented on a K-map by simply entering 1's in the cells (boxes) of the K-map corresponding to each minterm present in the expression. The remaining cells (boxes) are filled with zeros. Following example will illustrate the concept of standard SOP expression of K-map.

EXAMPLE 2.23. Represent the expression given below on Karnaugh- map.

$$Y = \bar{A}\bar{B}\bar{C} + \bar{A}\bar{B}C + A\bar{B}\bar{C} + ABC.$$

Solution: The given expression is in the standard SOP form. Each term represents a minterm. We have to enter 1s in the boxes corresponding to each minterm as shown in figure 2.24.

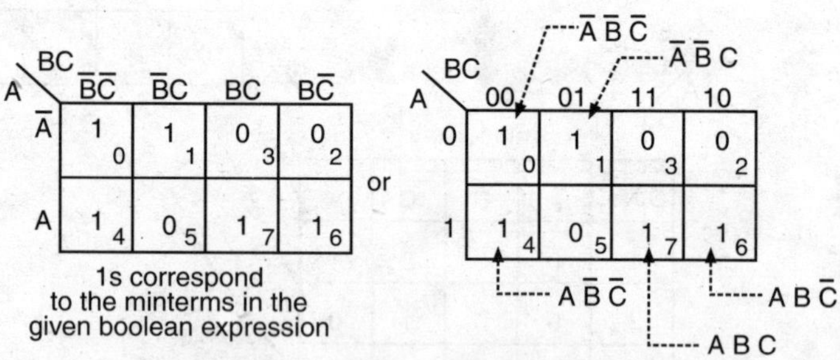

Fig. 2.24. Representation of standard SOP on K-map.

2.13.6 Representation of Standard POS Form on K-map

Logical expressions in the standard POS form can be represented on K-map by entering 0's in the cells of K-map corresponding to each maxterm present in the given expression. The remaining cells are filled with 1's. Following example will illustrate the procedure.

DO YOU KNOW?

Another way to explain Boolean functions is in standard form. In this configuration, the terms that form the function may contain one, two, or any number of literals.

EXAMPLE 2.24. Represent the following standard POS Expression on Karnaugh map.

$$Y = (A + B + C)(A + \bar{B} + C)(\bar{A} + \bar{B} + C).$$

Solution: Each term in the given logical expression is a maxterm. Enter a 0 corresponding to each maxterm as shown in figure 2.25. The given expression has three maxterms as under :

$$(A + B + C) = M_0,\ (A + \bar{B} + \bar{C}) = M_2,\ (\bar{A} + \bar{B} + C) = M_6$$

Hence, we have to write the structure of a 3 variable K-map as usual and enter 0's at M_0, M_2 and M_6 as shown in figure 2.25.

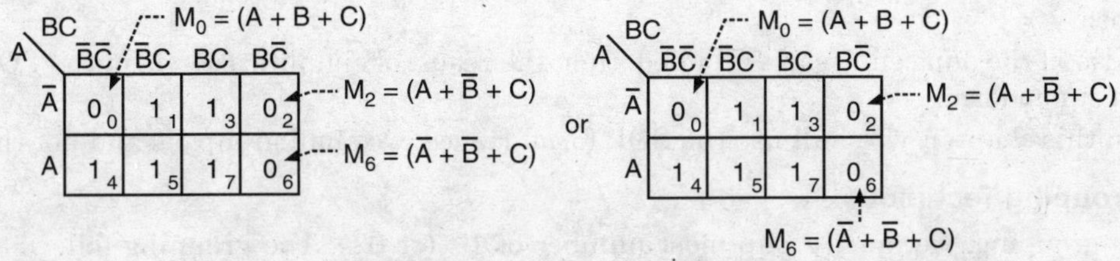

Fig. 2.25. Representation of standard POS on K-map

2.14. HOW TO SIMPLIFY BOOLEAN EXPRESSIONS WITH THE HELP OF K-MAP

As a matter of fact, the simplification of Boolean expressions using K-map is based on combining or grouping the terms in the adjacent cells (or boxes) of a K-map. Two cells of a K-map are said to be adjacent if they differ in only one variable as shown in figure 2.26.

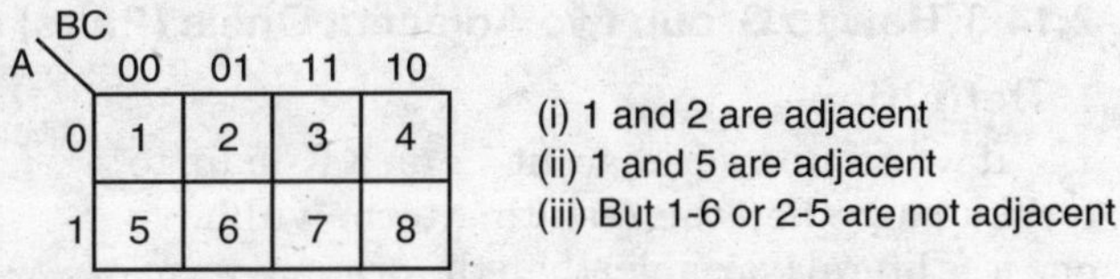

Fig. 2.26. Adjacent or non-adjacent cells

Here, it may be noted that the cells on left or right side or at the bottom top of a cells are adjacent cells. But the cells connected diagonally are not the adjacent ones. The left most cells are adjacent to their corresponding right cells as shown in figure 2.27 (*a*). Also, the top cells are adjacent to their corresponding bottom cells.

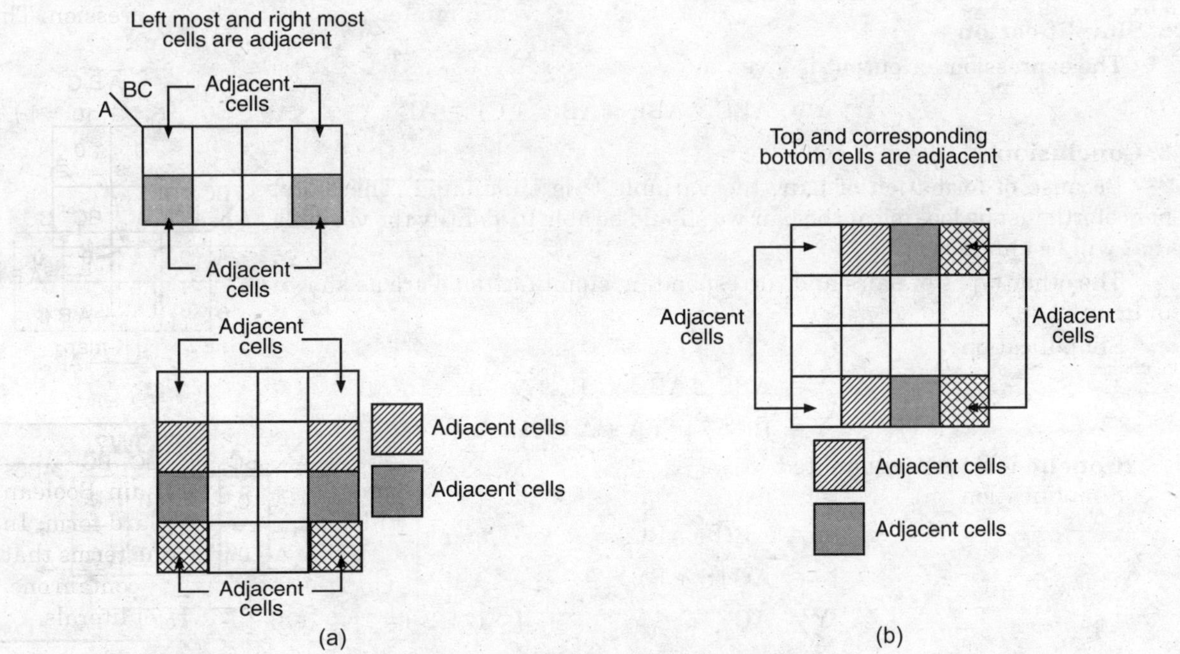

Fig. 2.27. Illustration of adjacent cells

2.14.1. Simplification Procedure

(i) Once, we plot the logic function or truth table on a Karnaugh-map, we have to use the grouping technique for simplifying the logical function.

(ii) Grouping means combining the terms in the adjacent cells.

(iii) The grouping of either adjacent 1's, or adjacent 0's results in the simplification of Boolean expression.

(iv) If we group the adjacent 1's, then the result of simplification is in SOP (sum of products) form.

(v) Also, if the adjacent 0's are grouped, then the result of simplification is in the POS (product of sums) form.

(vi) In this chapter, we shall use the SOP form. Hence, we shall group 1's and not the 0's.

2.14.2 Grouping Technique

While grouping, we must group most number of 1's (or 0's). The grouping follows the binary rule, i.e., we can group 1, 2, 4, 8, 16, 32, ... number of 1's or 0's. We cannot group 3, 5, 7, ... number of 1's or 0's.

Pairs : A group of two adjacent 1's or 0's is called as a pair.

Quad : A group of four adjacent 1's or 0's is called as a quad.

Octet : A group of eight adjacent 1's or 0's is called as octet.

2.14.3 How to Group two Adjacent One's (Pairs) on K-map

1. Definition

If we group two adjacent 1's on a K-map, to form a pair, then the resulting term will have one less literal (variable) than the original term. This means that by pairing two adjacent 1's, we can eliminate one variable. The grouping of two adjacent 1's and elimination of one variable due to pairing is illustrated in figure 2.28.

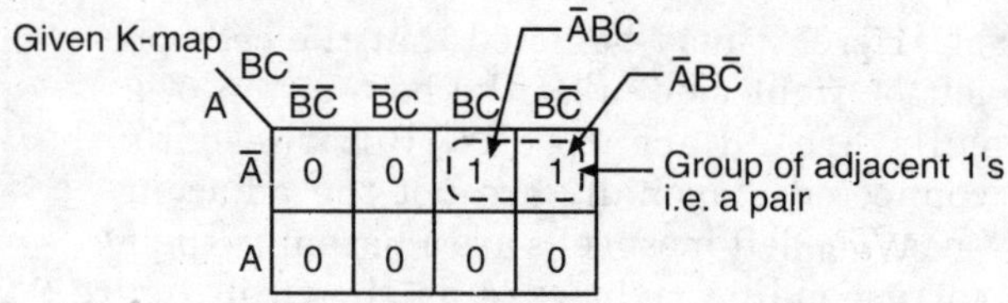

Fig. 2.28.

2. Simplification

The expression for output is given by,

$$Y = \bar{A}BC + \bar{A}B\bar{C} = \bar{A}B(C+\bar{C}) = \bar{A}B \qquad (\because C+\bar{C}=1)$$

3. Conclusion

Because of formation of pair, the variable C is eliminated. Therefore, henceforth just by looking at the pair we should be able to identify the variable that will be eliminated.

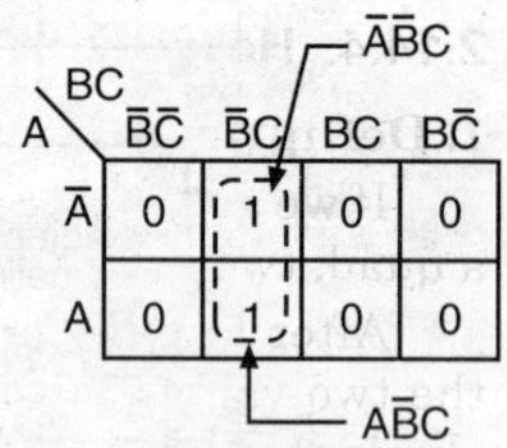

The other types of pairs and corresponding simplifications are as shown in figure 2.29.

Simplification

$$Y = \bar{A}\bar{B}C + A\bar{B}C = \bar{B}C(\bar{A}+A)$$

$$Y = \bar{B}C \qquad [\because (\bar{A}+A)=1]$$

Conclusion : A is eliminated

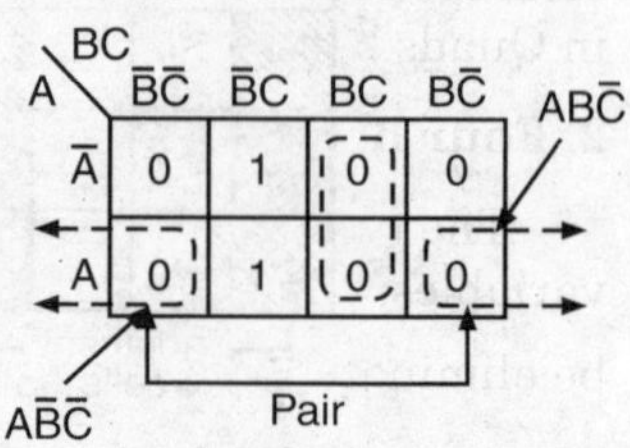

Simplification

$$Y = AB\bar{C} + A\bar{B}\bar{C}$$

$$= A\bar{C}(B+\bar{B})$$

$$Y = A\bar{C} \qquad [\because B+\bar{B})=1]$$

Conclusion : B is eliminated

Simplification

$$Y = \bar{A}\bar{B}\bar{C}D + A\bar{B}\bar{C}D$$

$$= \bar{B}\bar{C}D(\bar{A} + A)$$

$$Y = \bar{B}\bar{C}D$$

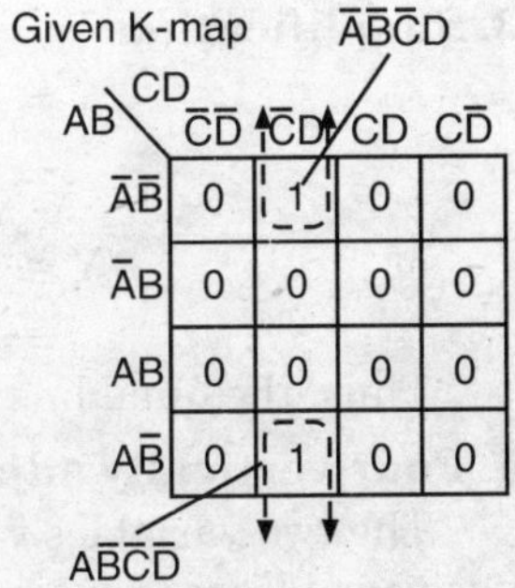

Conclusion : A is eliminated

Given K-map Overlapping is possible $\bar{A}$ can be shared by two pairs.

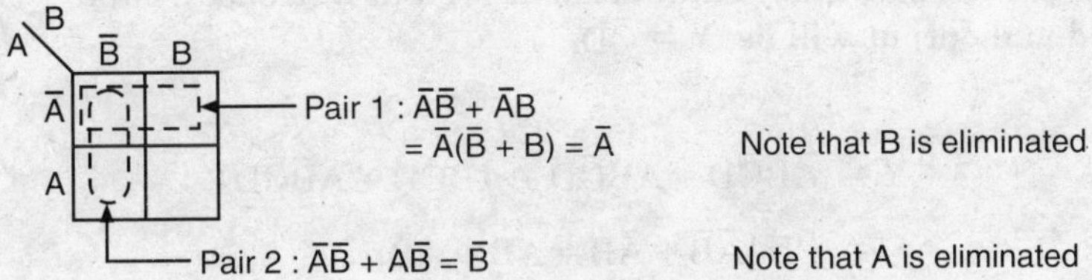

Conclusion : Due to overlapping of two pairs, two variables are eliminated.

Given K-map

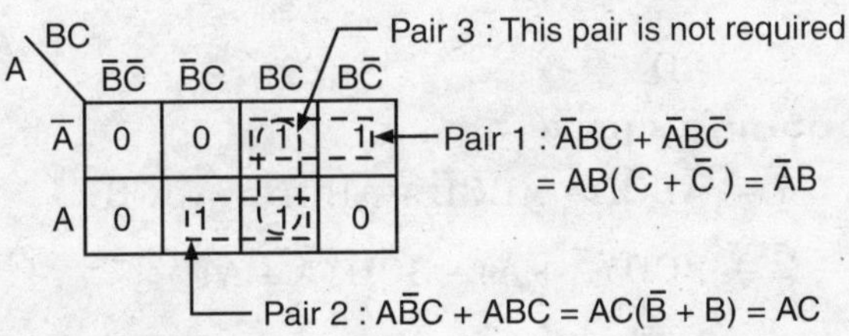

We add the two product terms to get,

$$Y = \bar{A}B = AC$$

Fig. 2.29. Different types of pairs and the corresponding simplifications.

Important Point. In this K-map, three pairs were possible to be formed. However, only two pairs are sufficient to include all the 1's present in the K-map. Then, the third pair is not required.

2.14.4. How to Group Four Adjacent Ones (Quad) on K-map

1. Definition

If we group four 1's from the adjacent cells of K-map, then the group is called as a Quad. In such a quad, two variables asssociated with the minterms are same and the other two are not the same.

After forming a Quad, the simplification takes place in such a way that the two variabels which are not same will be eliminated. Therefore, a quad eliminates two variables. Figure 2.30 shows various types of Quads and the corresponding mathematical simplification. Note that overlapping is possible in Quads.

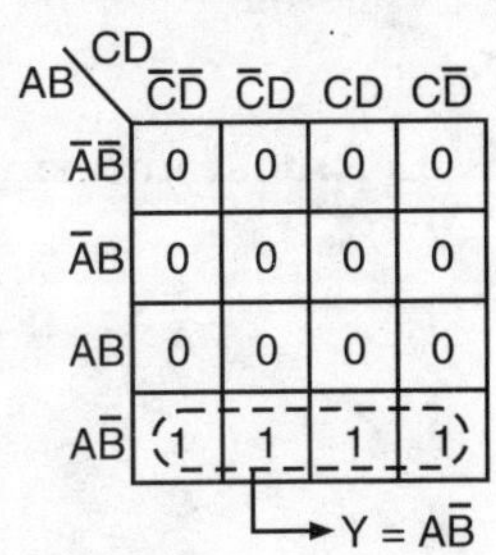

Fig. 2.30.

2. Four horizontally adjacent ones

The two variables which are small in all minterms are A and $\bar{B}$. And the variables which are not same in all minterms are C and D. Also, C and D will be eliminated and the output will be $Y = A\bar{B}$.

3. Simplification

$$Y = A\bar{B}\bar{C}\bar{D} + A\bar{B}\bar{C}D + A\bar{B}CD + A\bar{B}C\bar{D} = A\bar{B}\,[\bar{C}\bar{D} + \bar{C}D + CD + C\bar{D}]$$

$$= A\bar{B}\,[\bar{C}\,(\bar{D} + D) + C\,(D + \bar{D})]$$

or $$Y = A\bar{B}\,[\bar{C} + C]$$

$$= A\bar{B}$$

Hence Proved.

Thus, the quad has eliminated two variables.

4. Four vertically adjacent ones

The two variables which are same in all the minterms are $\bar{C}$ and D. The two variables which are not same in all minterms are A and B. Hence, A and B will be eliminated and output will be $Y = \bar{C}D$.

AB \ CD	$\bar{C}\bar{D}$	$\bar{C}D$	CD	$C\bar{D}$
$\bar{A}\bar{B}$	0	1	0	0
$\bar{A}B$	0	1	0	0
AB	0	1	0	0
$A\bar{B}$	0	1	0	0

→ $Y = \bar{C}D$

Fig. 2.31.

5. Simplification

$$Y = \bar{A}\bar{B}\bar{C}D + \bar{A}B\bar{C}D + AB\bar{C}D + A\bar{B}\bar{C}D$$

$$= \bar{C}D\,[\bar{A}\bar{B} + \bar{A}B + AB + A\bar{B}]$$

$$= \bar{C}D\,[\bar{A}\,(\bar{B} + B) + A\,(B + \bar{B})]$$

$$= \bar{C}D\,[\bar{A} + A]$$

$$= \bar{C}D$$

6. Four adjacent ones forming a square

AB \ CD	$\bar{C}\bar{D}$	$\bar{C}D$	CD	$C\bar{D}$
$\bar{A}\bar{B}$	0	0	0	0
$\bar{A}B$	1	1	0	0
AB	1	1	0	0
$A\bar{B}$	0	0	0	0

→ $Y = B\bar{C}$

$$Y = \bar{A}B\bar{C}\bar{D} + AB\bar{C}\bar{D} + \bar{A}B\bar{C}D + \bar{A}B\bar{C}D$$

$$= BCD\,(\bar{A} + A) + B\bar{C}D\,(\bar{A} + A)$$

$$= B\bar{C}\bar{D} + B\bar{C}D$$

$$= B\bar{C}\,(\bar{D} + D)$$

Therefore, $$Y = B\bar{C} \text{ (A and D are eliminated)}$$

7. Top and bottom 1's forming a Quad

AB \ CD	$\bar{C}\bar{D}$	$\bar{C}D$	CD	$C\bar{D}$
$\bar{A}\bar{B}$	0	1	1	0
$\bar{A}B$	0	0	0	0
AB	0	0	0	0
$A\bar{B}$	0	1	1	0

→ $Y = \bar{B}\,D$

$$Y = \bar{A}B\bar{C}\bar{D} + AB\bar{C}\bar{D}$$

$$= \bar{A}BC\bar{D} + ABC\bar{D}$$

$$= \bar{A}\bar{B}D\,(\bar{C} + C) + A\bar{B}D\,(\bar{C} + C)$$

$$= \bar{A}\bar{B}D + A\bar{B}D$$

Therefore, $$Y = \bar{B}D\,(\bar{A} + A)$$

$$= \bar{B}D \text{ (A and C are eliminated)}$$

8. Leftmost and rightmost 1's forming a Quad

AB \ CD	$\bar{C}\bar{D}$	$\bar{C}D$	CD	$C\bar{D}$
$\bar{A}\bar{B}$	0	0	0	0
$\bar{A}B$	1	0	0	1
AB	1	0	0	1
$A\bar{B}$	0	0	0	0

$Y = B\bar{D}$ (A and C are eliminated)

$$Y = \bar{A}B\bar{C}D + AB\bar{C}\bar{D} + \bar{A}BC\bar{D} + ABC\bar{D}$$

$$= B\bar{C}D\,(\bar{A} + A) + BC\bar{D}\,(\bar{A} + A)$$

$$= B\bar{C}\bar{D} + BC\bar{D}$$

$$Y = B\bar{D}\,(\bar{C} + C)$$

$$= B\bar{D} \text{ (A and C are eliminated)}$$

9. 1's corresponding to corners forming a Quad

$$Y = \bar{A}\bar{B}\bar{C}\bar{D} + \bar{A}\bar{B}C\bar{D} + A\bar{B}\bar{C}\bar{D} + A\bar{B}C\bar{D}$$

$$= \bar{B}\bar{C}\bar{D}\,(\bar{A} + A) + \bar{B}C\bar{D}\,(\bar{A} + A)$$

$$= \bar{B}\bar{C}\bar{D} + \bar{B}C\bar{D}$$

$$= \bar{B}\bar{D} + (\bar{C} + C)$$

Therefore, $Y = \bar{B}\,\bar{D}$ (A and C are eliminated)

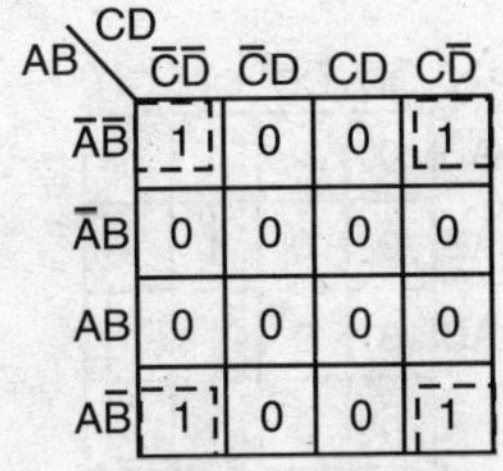

Y = B̄D̄ (A and C are eliminated)

10. Overlapping of Quads and pairs

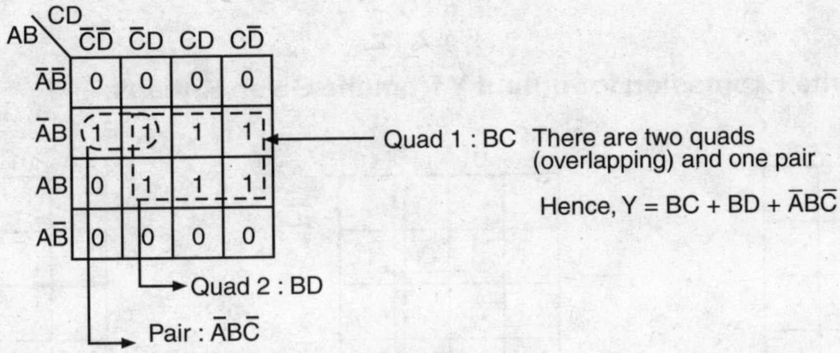

Fig. 2.32.

2.14.5. How to Group Eight Adjacent Ones (octet) on K-map

It is possible to form a group of eight adjacent ones. Such a group is called as octet. When an octet is formed, three variables will change and only one variable will remain same in all the minterms. The three variables that change will be eliminated and the variable which does not change will appear as output. Thus, octet eliminates three variables. Figure 2.33 shows various types of octets and the corresponding output.

Thus, we have $Y = \bar{A}\bar{B}\bar{C}\bar{D} + \bar{A}\bar{B}\bar{C}D + \bar{A}\bar{B}CD + \bar{A}\bar{B}C\bar{D} + \bar{A}B\bar{C}\bar{D} + \bar{A}B\bar{C}D + \bar{A}BC\bar{D}$

Therefore, $Y = \bar{A}\bar{B}\bar{C}\,(\bar{D} + D) + \bar{A}\bar{B}C\,(D + \bar{D}) + \bar{A}B\bar{C}\,(\bar{D} + D) + \bar{A}BC\,(D + \bar{D})$

or $Y = \bar{A}\bar{B}\,(\bar{C} + C) + \bar{A}B\,(\bar{C} + C)$

or $Y = \bar{A}\,(\bar{B} + B) = \bar{A}$

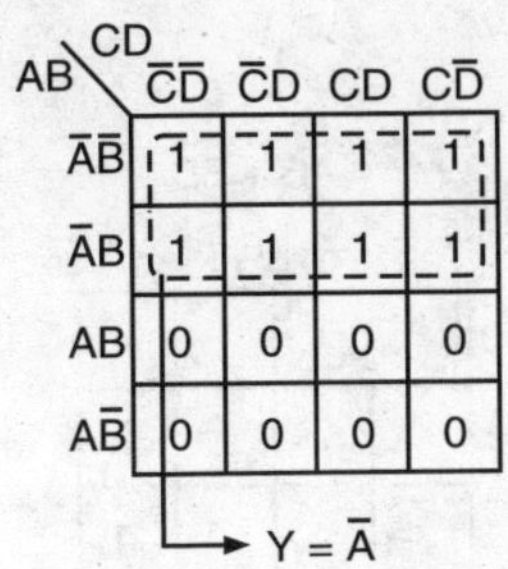

Fig. 2.33. (a) $Y = \bar{A}$

The only variable that remains same is $\bar{A}$. Hence, it appears as output.

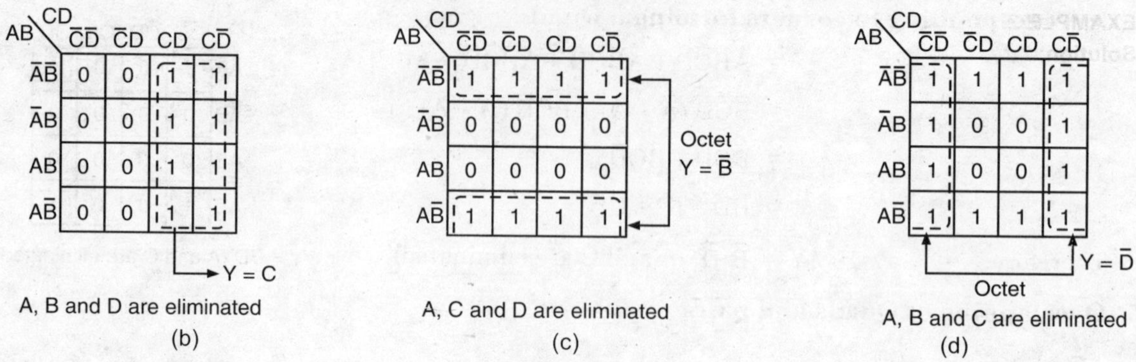

Fig. 2.33.

2.14.6. How to write Expression for output Y from the Given K-maps

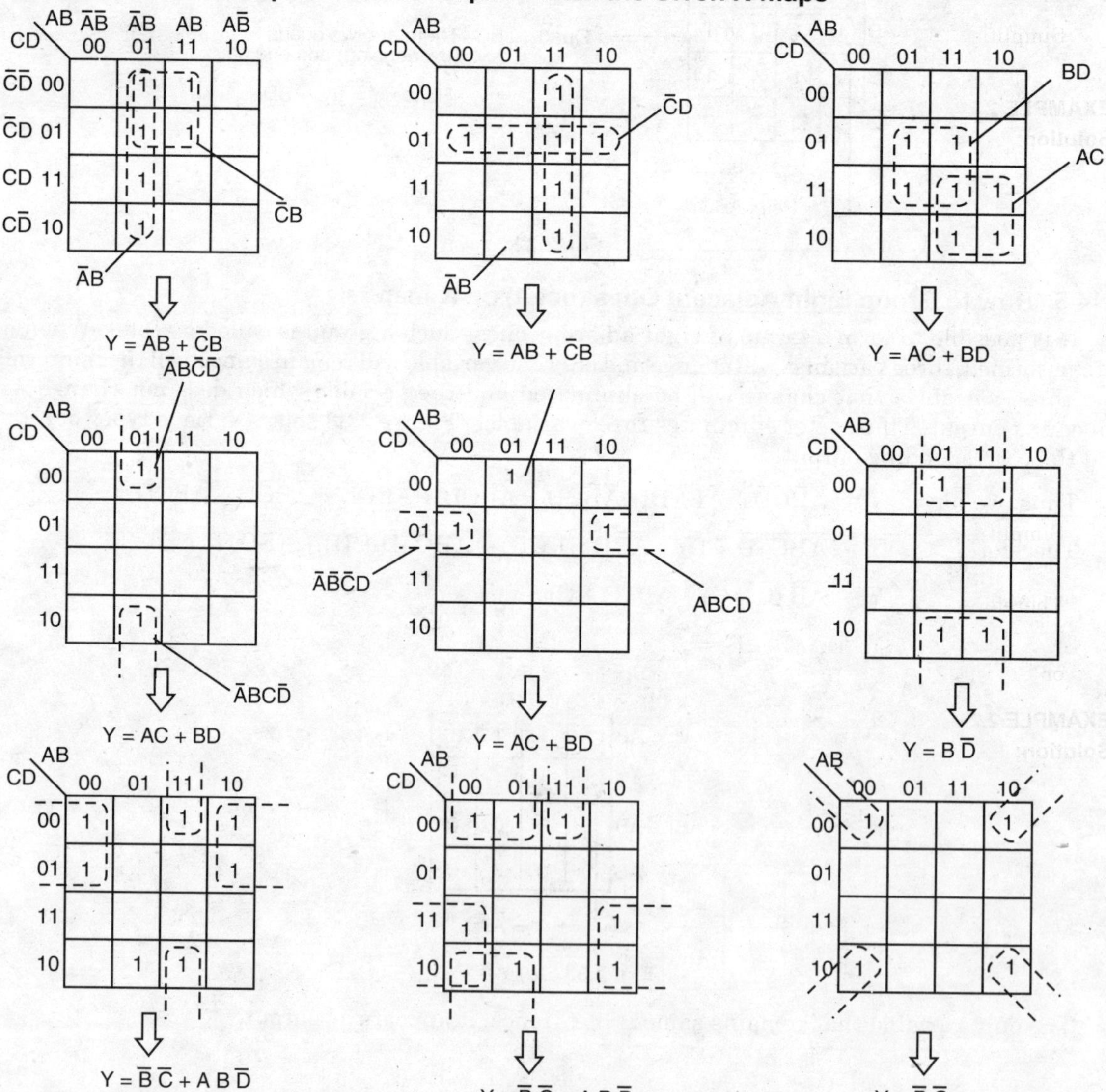

Fig. 2.34. Some additional K-map grouping possibilities

EXAMPLE 2.25. For the K-map shown in figure 2.35 write the simplified Boolean expression.

Solution:

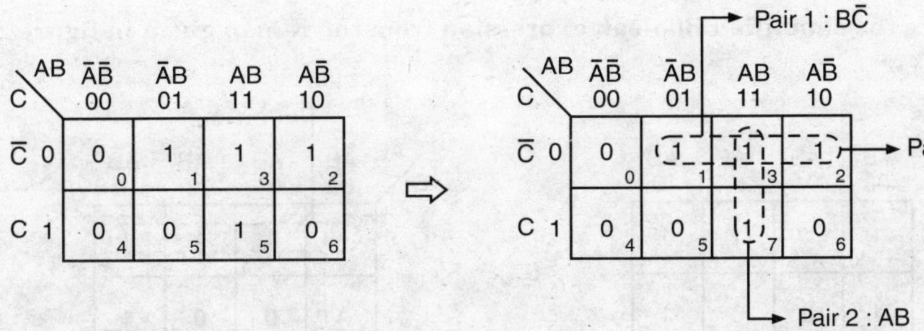

(a) Given K-map (b) Grouping

Fig. 2.35.

Simplified Boolean expression : $Y = \underset{(1)}{B\bar{C}} + \underset{(2)}{AB} + \underset{(3)}{A\bar{C}}$ **Ans.**

EXAMPLE 2.26. Write down the simplified Boolean expression for the K-map shown in figure 2.36.

Solution:

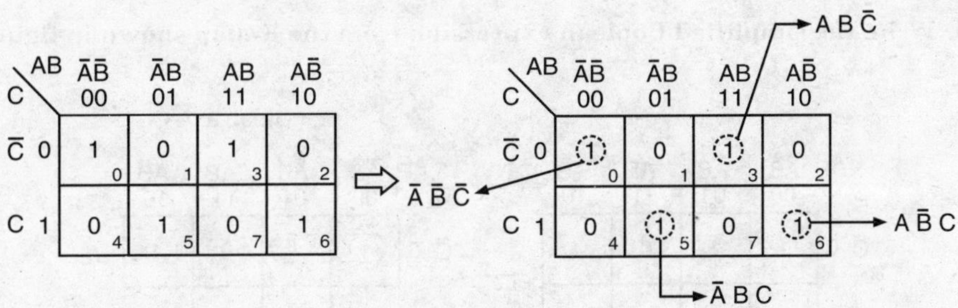

(a) Given K-map (b) Simplification

Fig. 2.36.

Simplified Boolean expression : $Y = \bar{A}\bar{B}\bar{C} + AB\bar{C} + \bar{A}BC + A\bar{B}C = \bar{A}\bar{B}\bar{C} + \bar{A}BC + AB\bar{C} + A\bar{B}C$

Therefore, $$Y = \bar{A}\underbrace{\left(\bar{B}\bar{C} + BC\right)}_{\text{EX-NOR}} + A\underbrace{\left(B\bar{C} + \bar{B}C\right)}_{\text{EX-OR}}$$

or $$Y = \bar{A}\,\overline{(B \oplus C)} + A\,(B \oplus C)$$ **Ans.**

EXAMPLE 2.27. For the Karnaugh-map shown in figure 2.37, obtain the simplified Boolean expression.

Solution:

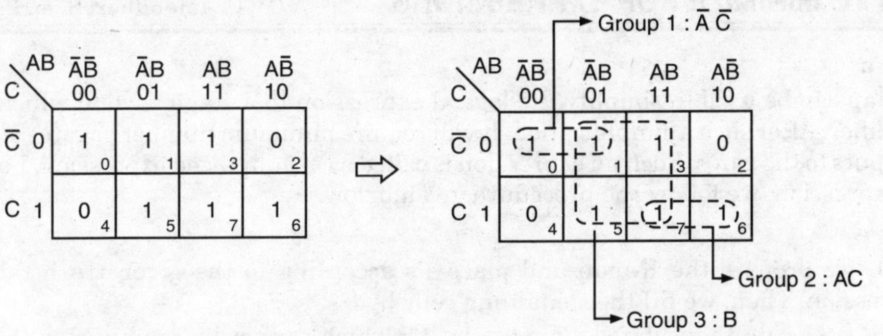

(a) Given K-map (b) Simplification

Fig. 2.37.

Simplified Boolean expression : $Y = \overline{A}\overline{C} + AC + B$ **Ans.**

(1) (2) (3)

EXAMPLE 2.28. Write the simplified Boolean expression from the K-map given in figure 2.38.

Solution :

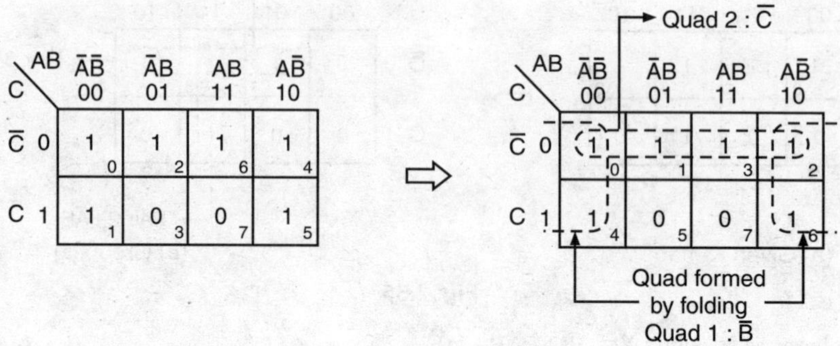

(a) Given K-map (b) Simplification

Fig. 2.38.

Simplified Boolean Equation : $Y = \overline{B} + \overline{C}$

EXAMPLE 2.29. Write the simplified Boolean expression from the K-map shown in figure 2.39.

Solution :

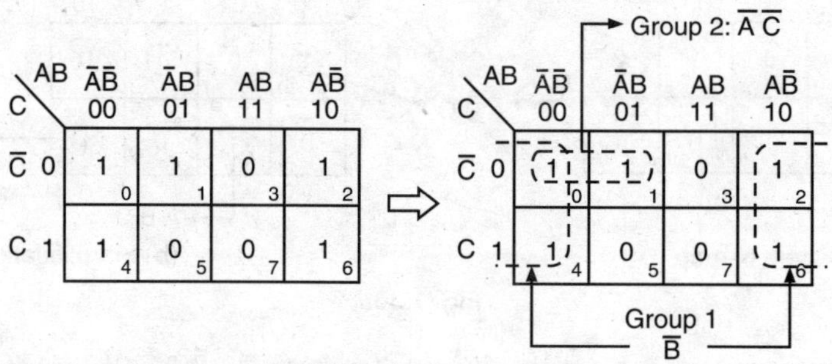

(a) Given K-map (b) Simplification

Fig. 2.39.

Simplified Boolean expression : $Y = \overline{B} + A\overline{C}$ **...Ans.**

$\downarrow$ $\downarrow$

(1) (2)

2.15. HOW TO MINIMIZE SOP EXPRESSIONS

(PTU, Jalandhar, Sem. Exam., 2004-05)

1. Definition

The K-map can be used to simplify the logical expression to a stage beyond which it cannot be further simplifier. After such a simplification, it will require minimum number of gates with minimum number of inputs to the gates. Such an expression is called as a minimized expression. For minimizing the logical expression, we follow the procedure given below.

2. Procedure

(i) First, we prepare the K-map and place 1's according to the given truth table or logical expression. Then, we fill the remaining cells by 0's.

(ii) Then, we locate the isolated 1's, i.e., the 1's which cannot be combined with any other 1. We encircle such 1's.

(iii) We identify the 1's which can be combined to form a pair in only one way and encircle them.
(iv) After that we identify the 1's which can form a quad in only one way and encircle them.
(v) Next, we identify the 1's which can form an octet in only one way and encircle them.
(vi) Lastly, after identifying the pairs, quads and octets, we check if any 1 is yet to be encircled. If yes then encircle them with each other or with the already encircled 1's.

Important Note: Note that the number of groups should be minimum. Also note that any 1 can be included any number of times without affecting the expression. Let us solve some examples on this to make the concept obvious.

EXAMPLE 2.30. A logical expression in the standard SOP form is as under :

$$Y = \bar{A}\bar{B}\bar{C} + \bar{A}B\bar{C} + \bar{A}BC + A\bar{B}C$$

Minimize it using the K-map technique.

Solution: First, we prepare the K-map and place 1's and 0's as shown in figure 2.40.

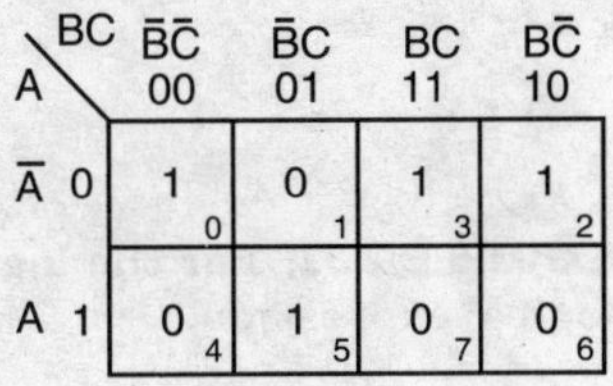

Fig. 2.40.

$$\bar{A}B\bar{C} = m_2,\ \bar{A}\bar{B}\bar{C} = m_0,\ ABC = m_3,\ A\bar{B}C = m_3$$

Then, we locate the isolated 1's and encircle them.

(a) (b)

Fig. 2.41.

Lastly, we identify the pairs.

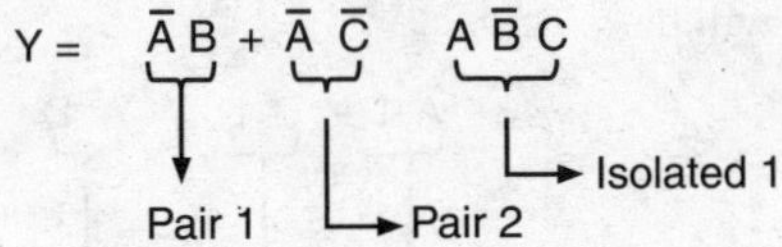

This is the required minimized expression.

EXAMPLE 2.31. The logical expression representing a logic circuit is Y = Σm (0, 1, 2, 5, 13, 15). Draw the K-map and find the minimized logical expression.

Solution: From the given expression, it is obvious that the number of variables is 4.

$$Y = m_0 + m_1 + m_2 + m_5 + m_{13} + m_{15}$$

First, let us prepare the K-map and place 1's and zeros as shown in figure 2.42.

Then, we write the minimized logical expression.

As shown in figure 2.42, three pairs are formed. There are no isolated 1's, no quads, no octets and no overlaps. The simplified expression will be

$$Y = \underbrace{\bar{A}\bar{B}\bar{D}}_{\text{Pair 1}} + \underbrace{\bar{A}\bar{C}D}_{\text{Pair 2}} + \underbrace{ABD}_{\text{Pair 3}}$$

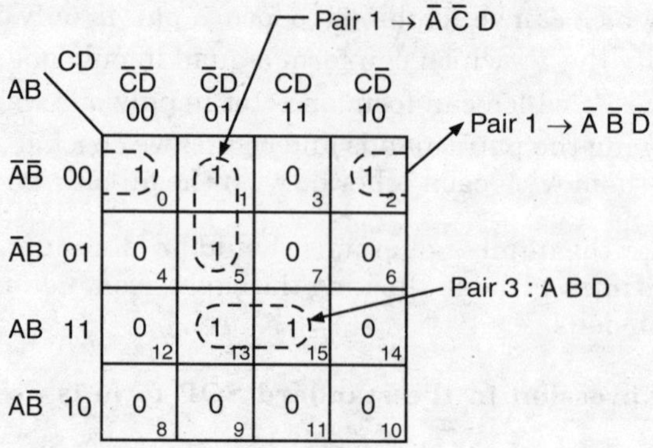

Fig. 2.42.

EXAMPLE 2.32. For the logical expression given below, draw the K-map and obtain the simplified logical expression.

$$Y = \Sigma m\,(1, 5, 7, 9, 11, 13, 15).$$

Realize the minimized expression using the basic gates.

Solution: The given expression is,

$$Y = m_1 + m_5 + m_7 + m_9 + m_{11} + m_{13} + m_{15}$$

It can be expressed on K-map as shown in figure 2.43.

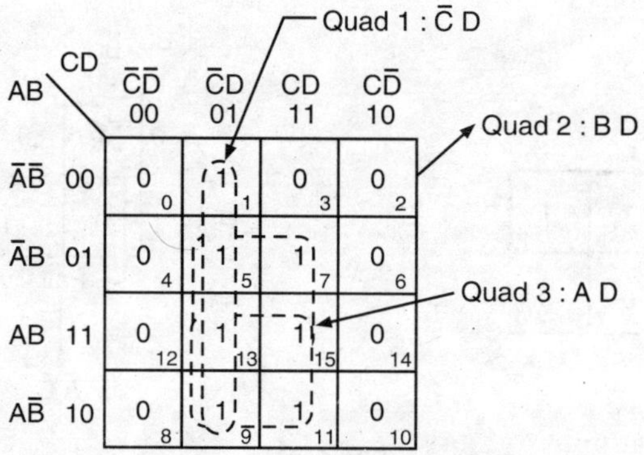

Fig. 2.43.

The simplified logical expression will be

$$Y = \underbrace{\bar{C}D}_{\text{Quad 1}} + \underbrace{BD}_{\text{Quad 2}} + \underbrace{AD}_{\text{Quad 3}} \qquad \text{...(i)}$$

$$\text{Minimized expression } Y = D\,(\bar{C} + B + A) \qquad \text{...}(ii)$$

Minimized expression : $Y = D\,(\bar{C} + B + A)$

Important Note : Here, it may be noted that to realize equation (i) we will require three AND gates and two OR gates and an inverter. Whereas to realize equation (ii) we require one three input OR gate and one inverter. Therefore, equation (ii) is the minimized logical expression.

Realization

Equation (ii) can be realized as shown in figure 2.44.

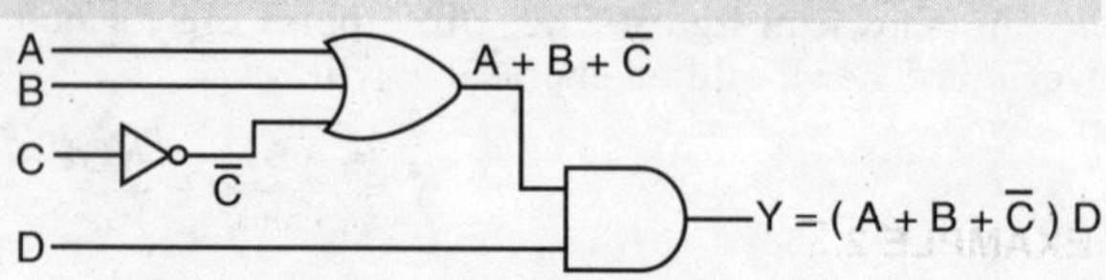

Fig. 2.44. Realization with minimum number of gates

EXAMPLE 2.33. Minimize the following Boolean expression using K-map and realize it using the basic gates.

$$Y = \Sigma m\,(1, 3, 5, 9, 11, 13)$$

Solution: The given expression can be expressed in terms of the minterms as under :

$$Y = m_1 + m_3 + m_5 + m_9 + m_{11} + m_{13}$$

The corresponding K-map is shown in figure 2.45.

The logical expression is given by,

$$Y = \underbrace{\bar{B}\,D}_{\text{Quad 1}} + \underbrace{\bar{C}\,D}_{\text{Quad 2}} \qquad \text{...(i)}$$

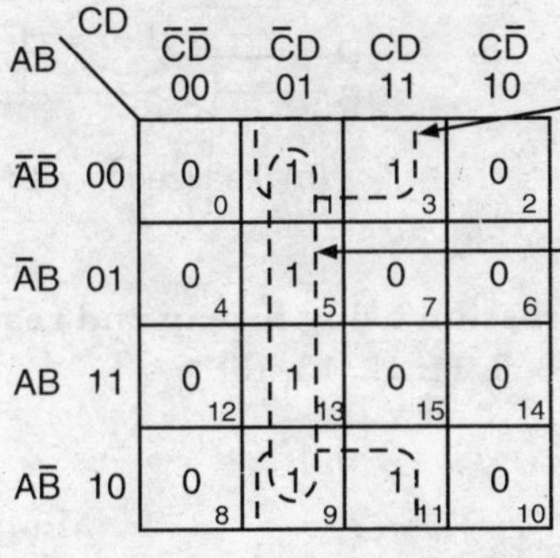

Fig. 2.45.

Therefore, $Y = D(\bar{B} + \bar{C})$

This is the minimized logical expression.

Realization

The minimized expression is realized as shown in figure 2.46.

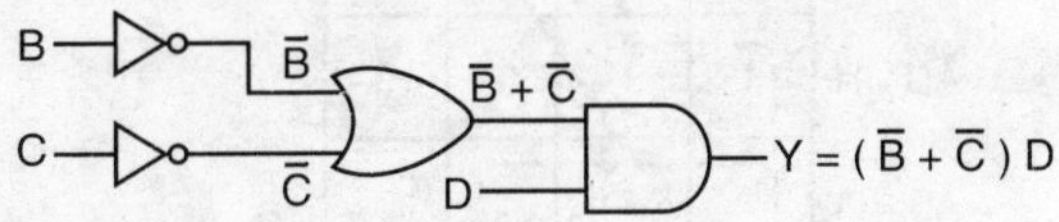

Fig. 2.46. Realization with minimum number of gates

EXAMPLE 2.34. Using K-map, realize the following expression using minimum number of gates.

$$Y = \Sigma m\,(1, 2, 3, 4, 5, 7, 9, 11, 13, 15)$$

Solution: The required K-map is shown in figure 2.47 (a) and the realization using minimum gates is shown in figure 2.47 (b).

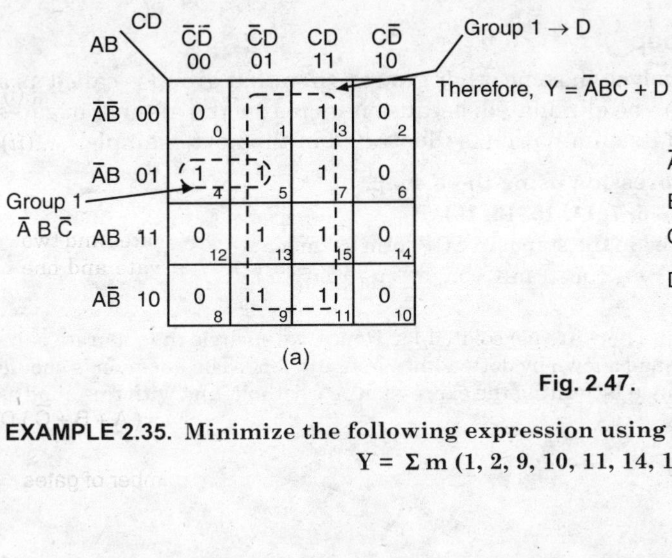

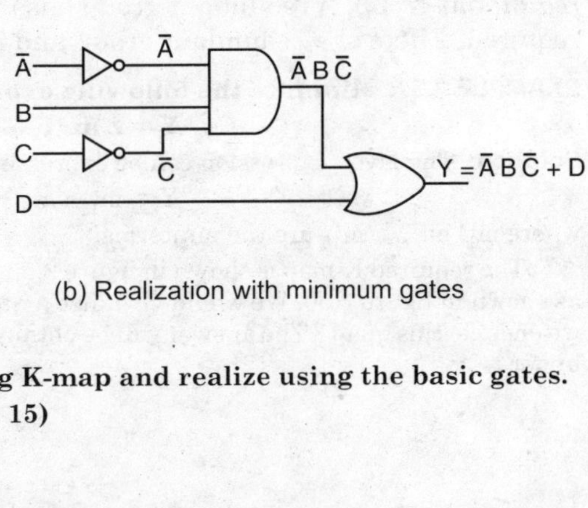

Fig. 2.47.

EXAMPLE 2.35. Minimize the following expression using K-map and realize using the basic gates.

$$Y = \Sigma m\,(1, 2, 9, 10, 11, 14, 15)$$

Solution:

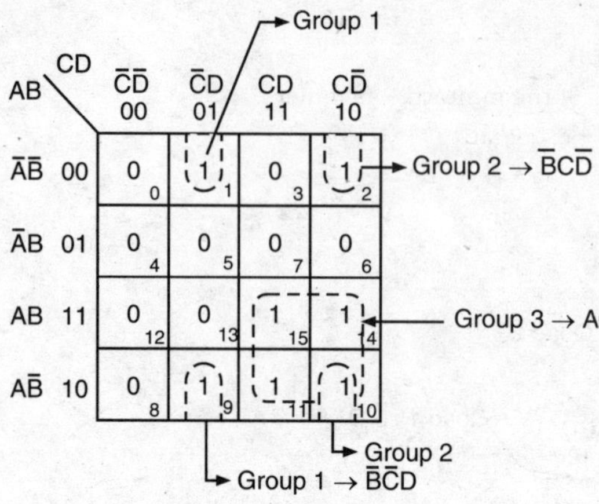

(a) K-map simplification

Minimized expression : $Y = \overline{B}\,\overline{C}D + \overline{B}C\overline{D} + AC$

$= \overline{B}\,(\overline{C}D + C\overline{D})\ AC$

EX OR gate

Therefore, $Y = \overline{B}\,(C \oplus D) + AC$

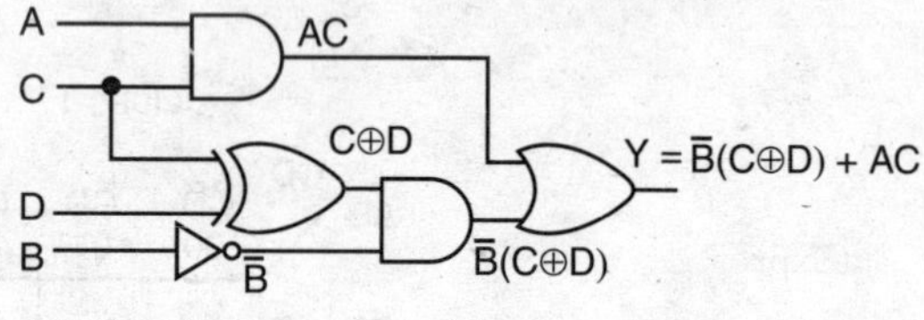

(b) Realization with minimum number of gates

Fig. 2.48.

EXAMPLE 2.36. Minimize the following expression using K-map and realize it using the basic gates.

$$Y = \Sigma m\,(4, 5, 8, 9, 11, 12, 13, 15)$$

Solution:

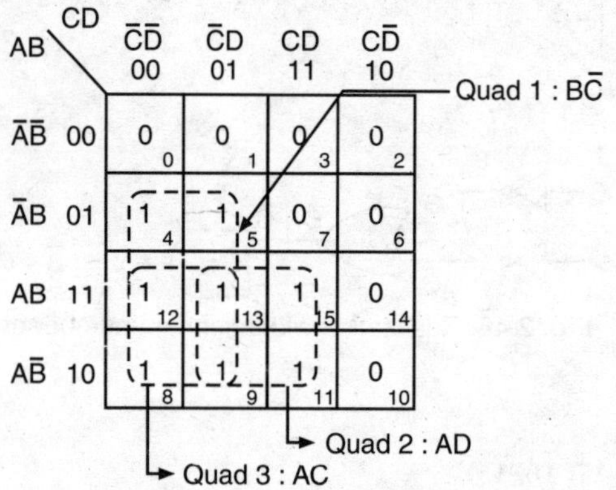

(a) K-map simplification

Minimized expression :

$Y = B\overline{C} + AD + A\overline{C}$

$= \overline{C}(A + B) + AD$

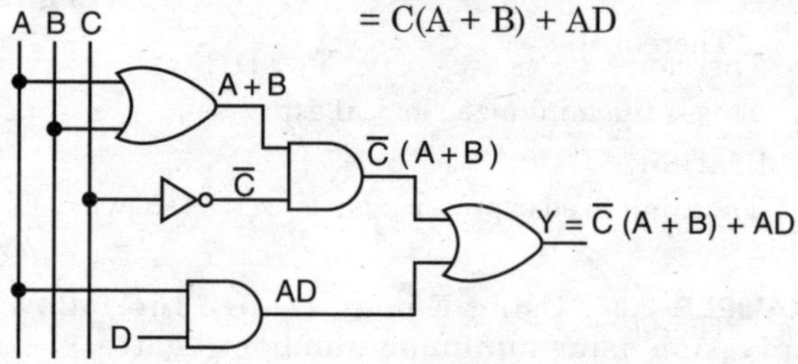

(b) Realization using minimum number of gates

Fig. 2.49.

2.15.1. Elimination of a Redundant Group

If all the 1's in a group are already involved in some other groups, then that group is called as a *redundant group*. A redundant group has to be eliminated, because it increases the number of gates required. Effect of a redundant group and its elimination is illustrated in the next example.

EXAMPLE 2.37. Minimize the following expression using the K-map.

$$Y = \Sigma m\,(1, 5, 6, 7, 11, 12, 13, 15)$$

Solution: The given expression can be expressed in the standard SOP form as under :

$$Y = m_1 + m_5 + m_6 + m_7 + m_{11} + m_{12} + m_{13} + m_{15}$$

where m_1, m_5 ..... m_{15} are the minterms.

The required K-map is shown in figure 2.50. There are no isolated 1's. Hence, we encircle the separate pairs as shown in figure 2.50. We would visualize a quad shown by dotted lines in figure 2.50. The question is should we encircle this quad ? The answer will be obtained by writing the expression of Y without and with this quad as under :

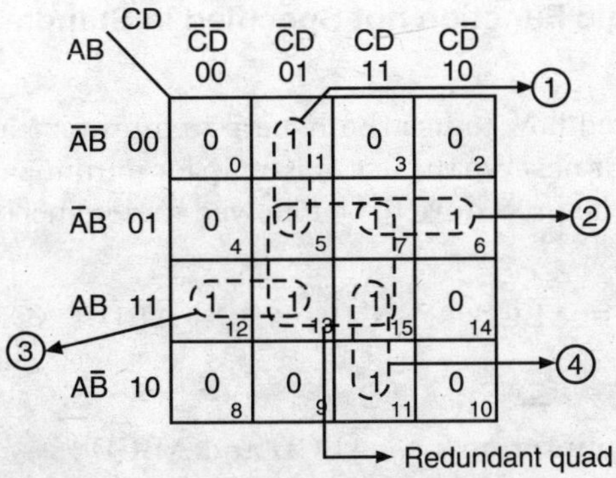

Fig. 2.50.

Expression without quad : $Y = \bar{A}\bar{C}D + \bar{A}BC + AB\bar{C} + ACD$

(1) (2) (3) (4)

Expression with quad : $Y = \bar{A}\bar{C}D + \bar{A}BC + AB\bar{C} + ACD + BD$ → Redundant quad

(1) (2) (3) (4)

Conclusion

If we encircle the quad, then the expression for output consists of an additional term. So quad should not be encircled, it is called as the redundant group. Hence, the correct K-map is shown in figure 2.51.

Therefore,

$$Y = \bar{A}\bar{C}D + \bar{A}BC + AB\bar{C} + ACD$$

$$= \bar{A}\bar{C}D + ACD + \bar{A}BC + AB\bar{C}$$

$$= \underbrace{D\left(\bar{A}\bar{C} + AC\right)}_{\text{EX-NOR}} + \underbrace{B\left(\bar{A}C + A\bar{C}\right)}_{\text{EX-OR}}$$

Therefore,

$$Y = D\,(\overline{A \oplus C}) + B\,(A \oplus C)$$

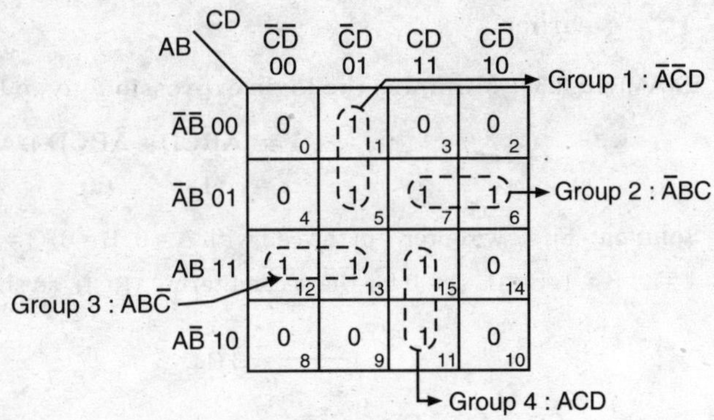

Fig. 2.51.

EXAMPLE 2.38. Minimize the following expression using K-map.

$$Y = \Sigma m\ (0, 1, 2, 5, 13, 15)$$

Solution: The required K-map is shown in figure 2.52.

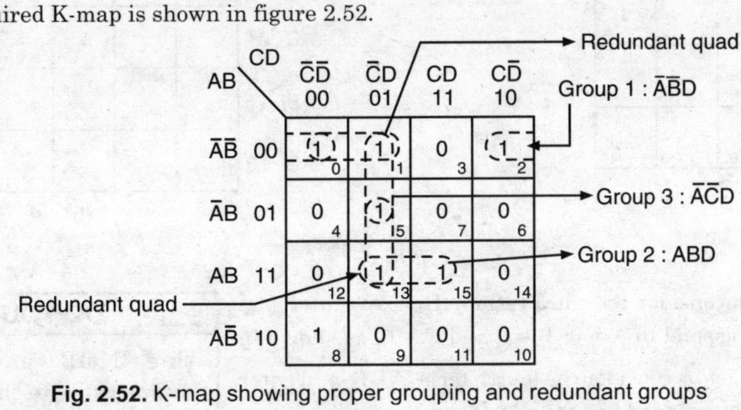

Fig. 2.52. K-map showing proper grouping and redundant groups

The groups shown by dotted lines in figure 2.52, are redundant. They should not therefore be considered.

Hence, minimised expression $Y = \bar{A}\bar{B}\bar{D} + ABD + \bar{A}\bar{C}D$

2.15.2. Minimization of Logic Function not Specified in Standard SOP Form

1. Basic Concept

Till now, we have observed how to use the K-map to minimize an expression which is given in standard SOP form. Now, let us observe the use of K-map for minimization when the given expression is not in the standard form. The procedure to be followed under such conditions may be described as under :

The given expression : $Y = \bar{A}\bar{B}CD + \bar{A}BC\bar{D} + \bar{A}\bar{B}\bar{C} + \bar{A}\bar{B}\bar{D} + A\bar{C} + \bar{B}$

2. Procedure

(i) First, we enter 1 for minterms (i.e., $\bar{A}\bar{B}CD$ and $\bar{A}BC\bar{D}$)

(ii) Secondly, we enter a pair of 1's for each terms with one less variable then total (i.e., for $\bar{A}\bar{B}\bar{C}$, $\bar{A}\bar{B}\bar{D}$)

(iii) Next, we enter four adjacent 1's for each terms with two less variables than total (i.e., $A\bar{C}$)

(iv) Then, we repeat for the other terms in a similar way.

(v) Lastly, once the K-map is obtained, the minimization procedure is same as the one discussed earlier.

EXAMPLE 2.39. Minimize the logic expression given below :

$$Y = \underset{(1)}{\bar{A}\bar{B}CD} + \underset{(2)}{\bar{A}BC\bar{D}} + \underset{(3)}{\bar{A}\bar{B}\bar{C}} + \underset{(4)}{\bar{A}\bar{B}\bar{D}} + \underset{(5)}{A\bar{C}} + \underset{(6)}{\bar{B}}$$

Solution: First, we enter 1 in the cell with A = 0, B = 0, C = 1, D = 1 for the first term $\bar{A}\bar{B}CD$ and in the cell with A = 0, B = 1. C = 1, D = 0 for the second term $\bar{A}BC\bar{D}$ as shown in figure 2.53 (a).

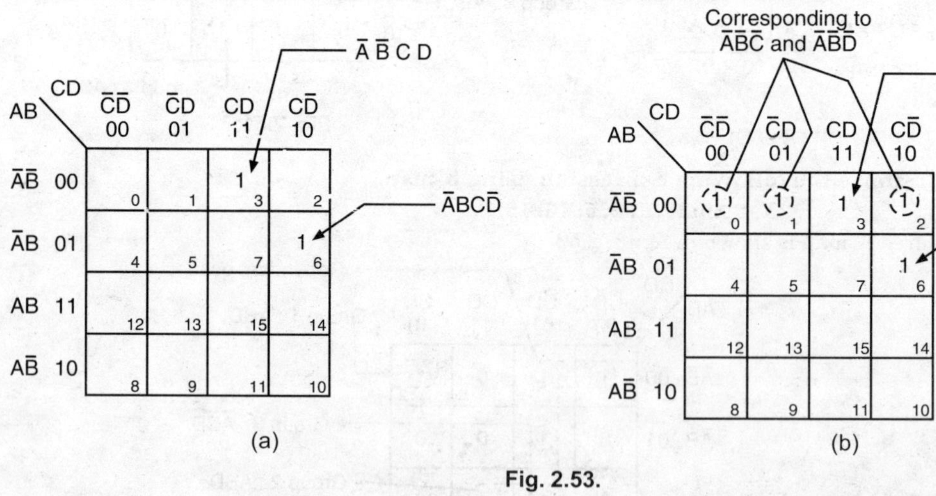

Fig. 2.53.

Secondly, let us consider the third term $\bar{A}\bar{B}\bar{C}$. We enter 1 in two cells which correspond to A = 0, B = 0 and C = 0, as shown in figure 2.53 (b). Now, let us consider the fourth term $\bar{A}\bar{B}\bar{D}$ and enter 1 in two cells which correspond to A = 0, B = 0 and D = 0 as shown in figure 2.53 (b).

DO YOU KNOW?

The NOR gate is another universal gate that can be used to implement any Boolean function.

Now, we consider the fifth term $A\bar{C}$. We enter 1 in four cells corresponding to A = 1 and C = 0, as shown in figure 2.54 (a).

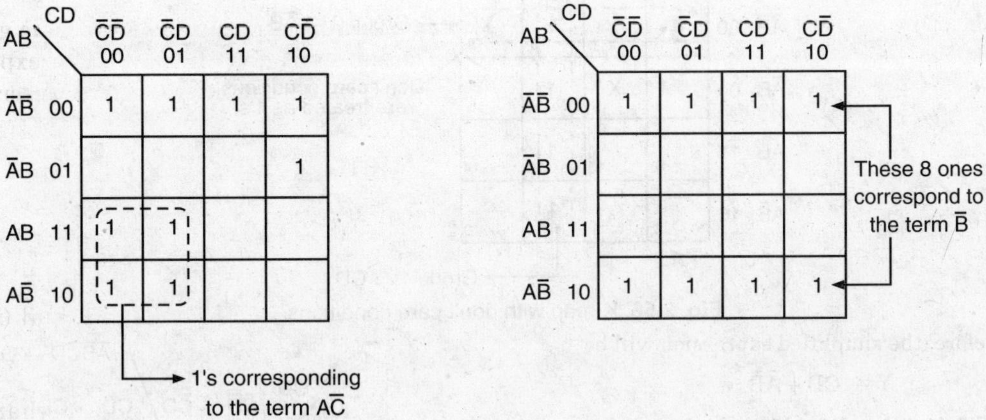

(a) (b)

Fig. 2.54

Finally corresponding to the sixth term, we enter eight 1's corresponding to B = 0 as shown in figure 2.54 (b).

Figure 2.55 shows the final K-map. We use the normal simplification techniques to simplify this K-map and get the minimized expressions. Thus, the simplified expression will be

$$Y = \bar{B} + A\bar{C} + \bar{A}C\bar{D}$$

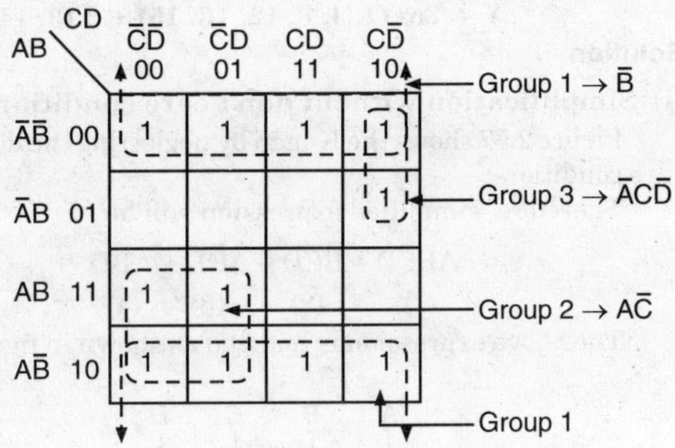

Fig. 2.55. The Final K-map

2.15.3. Don't Care Conditions

For SOP form, we enter 1's corresponding to the combinations of input variables which produce a high output. Also, we enter 0's in the remaining cells of the K-map. For the POS form, we enter 0's corresponding to the combinations of inputs which product a low output and enter 1's in the remaining cells of the K-map. However, it is not always true that the cells not containing 1's (in SOP) will contain 0's, because some combinations of input variable do not occur. Also, for some functions, the outputs corresponding to certain combinations of input variables *do not matter*. In such situations, we have a freedom to assume a 0 of 1 as output for each of these combinations. These conditions are known as the *Don't care conditions* and in the K-map it is represented as × (cross) mark in the corresponding cell.

Important Point. The don't care conditon (×) may be assumed to be 0 or 1 as per the requirement for simplification.

EXAMPLE 2.40. Simplify the expression given below using K-map. The don't care conditions are indicated by d( ).

$$\mathbf{Y = \Sigma m\,(1, 3, 7, 11, 15) + d\,(0, 2, 5)}$$

Solution: The given expression is

$$Y = \underbrace{m_1 + m_3 + m_7 + m_{11} + m_{15}}_{\text{Regular minterms, so we enter 1's}} + \underbrace{d(0,2,5)}_{\text{Don't care conditions so, we enter} \times \text{marks}}$$

Figure 2.56 shows the required K-map.

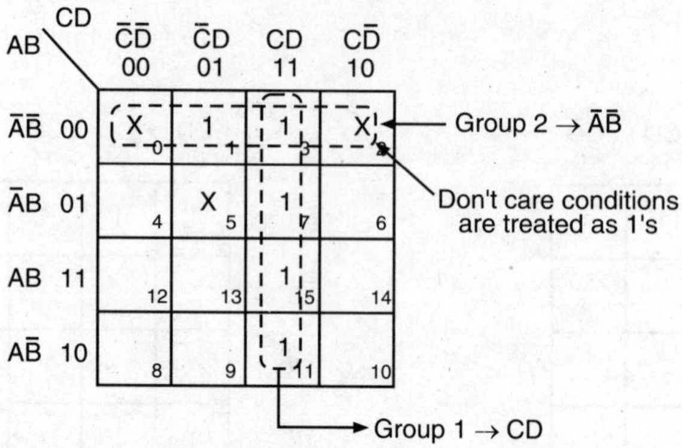

Fig. 2.56. K-map with don't care conditions

Therefore, the simplified expression will be

$$Y = CD + \bar{A}\bar{B}$$

EXAMPLE 2.41. Simplify the following expression once by considering the don't care conditions and once by ignoring the don't care condition.

$$Y = \Sigma m\,(1, 4, 8, 12, 13, 15) + d\,(3, 14)$$

Solution:

(i) Simplification without don't care condition

Figure 2.57 shows the K-map by neglecting the don't care condition

Therefore, simplified expression will be

$$Y = \underset{(1)}{\bar{A}\bar{B}\bar{C}D} + \underset{(2)}{B\bar{C}\bar{D}} + \underset{(3)}{A\bar{C}\bar{D}} + \underset{(4)}{ABD}$$

The above expression is realized as shown in figure 2.58.

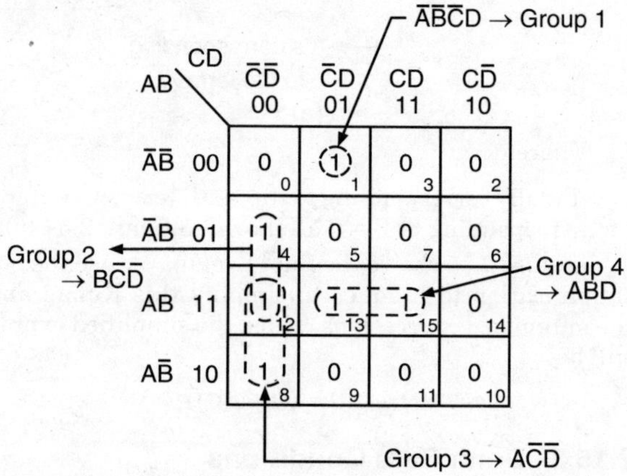

Fig. 2.57. Simplification neglecting don't care condition

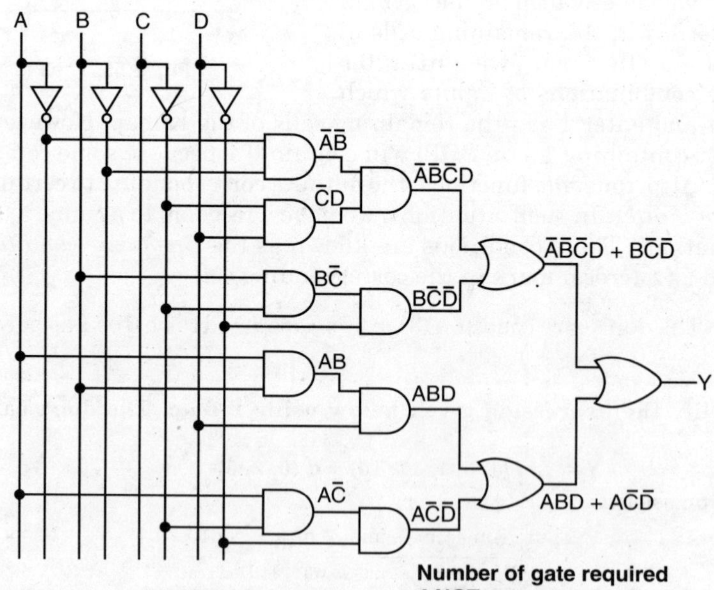

Number of gate required
4 NOT gates
9 AND gates
3 OR gates

Fig. 2.58. Realization without don't care condition

(ii) Simplification with don't care conditions

Figure 2.59 shows the K-map including the don't care conditions.

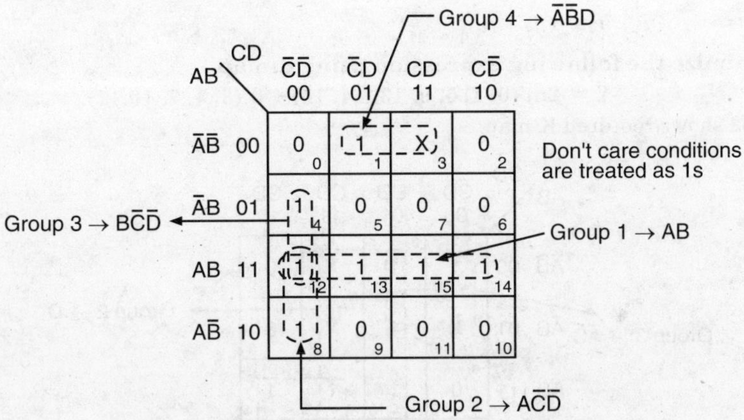

Fig. 2.59. Simplification using don't care conditions

Therefore, simplified expression will be $Y = AB + A\bar{C}\bar{D} + B\bar{C}\bar{D} + \bar{A}\bar{B}D$

(1) (2) (3) (4)

or $$Y = AB + \bar{C}\bar{D}(A + B) + \bar{A}\bar{B}D$$

The above expression is realized as shown in figure 2.60.

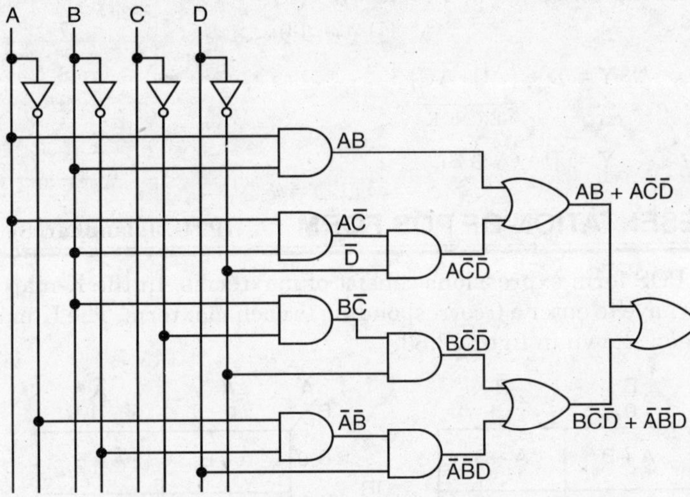

Fig. 2.60. Realization using the don't care conditions

The realization of this Boolean expression shall require 4 NOT gates, 7 AND gates and 3 OR gates.

Important Conclusion

Figure 2.60 illustrates that we have saved 2 AND gates by using the don't care conditions.

EXAMPLE 2.42. Minimize the SOP expression given below using K-map.

$$Y = \Sigma m(1, 3, 7, 11, 15) + \Sigma d(0, 2, 5, 8, 14)$$

Solution : Figure 2.61 shows the required K-map with don't care conditions.

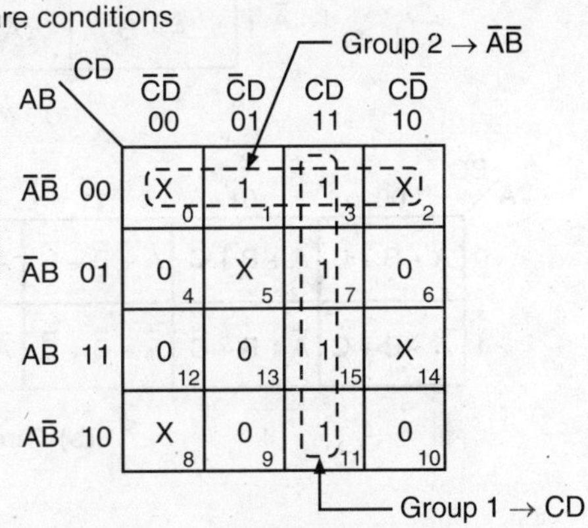

Fig. 2.61.

(a) 4 (b) 3 (c) 2 (d) 10

Therefore, the simplified expression is

$$Y = CD + \overline{A}\,\overline{B}$$

EXAMPLE 2.43. Minimize the following expression using K-map.

$$\mathbf{Y = \Sigma m\ (0, 1, 5, 9, 13, 14, 15) + d\ (3, 4, 7, 10, 11).}$$

Solution: Figure 2.62 shows required K-map.

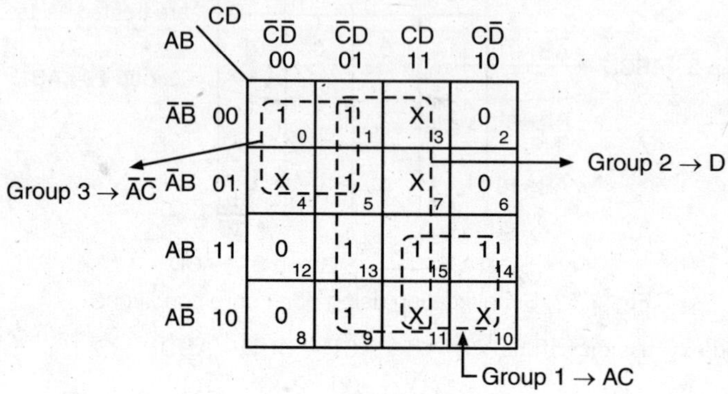

Fig. 2.62.

Therefore, the simplified expression will be $Y = AC + \overline{D} + \overline{A}\,\overline{C}$
(1) (2) (3)

or $$Y = D + \underbrace{(AC + \overline{A}\,\overline{C})}_{\text{EX-NOR}}$$

Hence, we have $$Y = D + (A \odot C)$$ **Ans.**

2.16. K-MAP REPRESENTATION OF POS FORM

(PTU, Jalandhar, Sem. Exam., 2004-05)

As a matter of fact, the POS form expressions consist of maxterms. In the K-map corresponding to POS form expression, we have to enter a 0 corresponding to each maxterm. The K-maps corresponding to the POS form have been shown in figure 2.63.

A \ B	B 0	$\overline{B}$ 1
A 0	$A + B$	$A + \overline{B}$
$\overline{A}$ 1	$\overline{A} + B$	$\overline{A} + \overline{B}$

OR

B \ A	A 0	$\overline{A}$ 1
B 0	$A + B$	$\overline{A} + B$
$\overline{B}$ 1	$A + \overline{B}$	$\overline{A} + \overline{B}$

(a) Two variable K-map for POS form

A \ BC	00	01	11	10
0	$A + B + C$	$A + B + \overline{C}$	$A + \overline{B} + \overline{C}$	$A + \overline{B} + C$
1	$\overline{A} + B + C$	$\overline{A} + B + \overline{C}$	$\overline{A} + \overline{B} + \overline{C}$	$\overline{A} + \overline{B} + C$

OR

C \ AB	00	01	11	10
0	$A + B + C$	$A + \overline{B} + C$	$\overline{A} + \overline{B} + C$	$\overline{A} + B + C$
1	$A + B + \overline{C}$	$A + \overline{B} + \overline{C}$	$\overline{A} + \overline{B} + \overline{C}$	$\overline{A} + B + \overline{C}$

(b) Three variable K-map for POS form

AB \ CD	00	01	11	10
00	$A+B+C+D$	$A+B+C+\bar{D}$	$A+B+\bar{C}+\bar{D}$	$A+B+\bar{C}+D$
01	$A+\bar{B}+C+D$	$A+\bar{B}+C+\bar{D}$	$A+\bar{B}+\bar{C}+\bar{D}$	$A+\bar{B}+\bar{C}+D$
11	$\bar{A}+\bar{B}+C+D$	$\bar{A}+\bar{B}+C+\bar{D}$	$\bar{A}+\bar{B}+\bar{C}+\bar{D}$	$\bar{A}+\bar{B}+\bar{C}+D$
10	$\bar{A}+B+C+D$	$\bar{A}+B+C+\bar{D}$	$\bar{A}+B+\bar{C}+\bar{D}$	$\bar{A}+B+\bar{C}+D$

(c) Four variable K-map for POS form

Fig. 2.63.

The entries in the K-map can be shown in terms of the maxterms M_0, M_1, ... etc., as shown in figures 2.64 and 2.65.

A \ B	0	1
0	M_0	M_1
1	M_2	M_3

(a)

A \ BC	00	01	11	10
0	M_0	M_1	M_3	M_2
1	M_4	M_5	M_7	M_6

(b)

Fig. 2.64.

AB \ CD	00	01	11	10
00	M_0	M_1	M_3	M_2
01	M_4	M_5	M_7	M_6
11	M_{12}	M_{13}	M_{15}	M_{14}
10	M_8	M_9	M_{11}	M_{10}

Fig. 2.65. K-map in terms of maxterms

2.17. HOW TO SIMPLIFY STANDARD POS FORM WITH THE HELP OF K-MAP

1. Definition

Till now we have focussed on minimizing the SOP form. But, now, let us consider the other form, i.e., the POS form. The procedure for minimizing the POS form may be described as under :

2. Minimization procedure

(i) The given POS expression consists of maxterms.

(ii) Corresponding to every maxterm, we enter a 0 in K-map.

(iii) Then, we enter 1's in the remaining cells of K-map.

(iv) We encircle/group 0's instead of 1's for carrying out the simplification.

(v) Rules of simplification are exactly the same as those for the SOP form.

EXAMPLE 2.44. Obtain the expression in the POS form for the K-map given in figure 2.66.

A \ BC	00	01	11	10
0	1	1	0	0
1	1	1	0	1

Group1 $\rightarrow (A+\bar{B}+\bar{C})\,(A+\bar{B}+C)$

Group 2 $\rightarrow (A+\bar{B}+\bar{C})\,(\bar{A}+\bar{B}+\bar{C})$

Fig. 2.66. Given K-map

Solution: We have Group 1 $\rightarrow (A+\bar{B}+\bar{C})\,(\bar{A}+\bar{B}+C)$

$$= AA + A\overline{B} + AC + A\overline{B} + B\overline{B} + BC + A\overline{C} + B\overline{C} + C\overline{C}$$

But $\quad AA = A, B\overline{B} = \overline{B}, C\overline{C} = 0, A\overline{B} + A\overline{B} = A\overline{B}$

Therefore, $\quad \text{Group 1} = A + A\overline{B} + AC + \overline{B} + \overline{B}C + A\overline{C} + \overline{B}\overline{C} = A(1+\overline{B}) + A(C+\overline{C}) + \overline{B}(1+C) + \overline{B}\overline{C}$

$$= A + A + \overline{B} + \overline{B}\overline{C} = A + \overline{B}(1+\overline{C})$$

Then, $\quad \text{Group 1} = A + \overline{B}$

Important Conclusion. When a pair of 0s is formed, the variable which changes will get eliminated, e.g., C changes for the pair of 0's in group 1. Hence, it is eliminated.

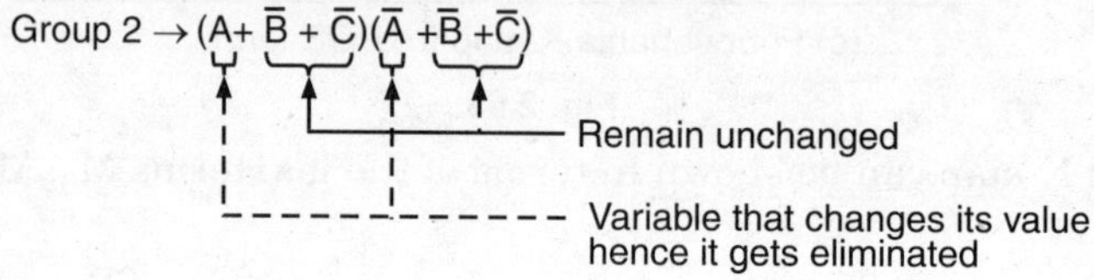

Also, $\quad \text{Group 2} \rightarrow \overline{B} + \overline{C}$

Therefore, final minimized expression is as under :

$$Y = \underset{(1)}{(A + \overline{B})} \cdot \underset{(2)}{(\overline{B} + \overline{C})}$$

EXAMPLE 2.45. Minimize the following standard POS expression using K-map.

$$\mathbf{Y = \Pi\, M\,(0, 2, 3, 5, 7)}$$

Solution: The given expression is in the POS form. It can be written in terms of the maxterms as under :

$$Y = M_0\, M_2\, M_3\, M_5\, M_7$$

Figure 2.67 (a) shows the terms variable K-map for POS form. Now, we enter a 0 corresponding to each maxterm in the given expression as shown in figure 2.67 (b).

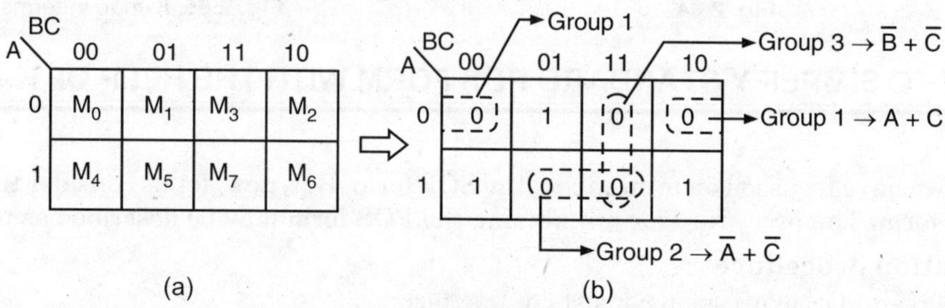

Fig. 2.67.

Hence, three groups are formed as shown in figure 2.67 (b).

Therefore, minimized expression is as under :

$$Y = (A + C)\,(\overline{A} + \overline{C}) + (\overline{B} + \overline{C})$$

EXAMPLE 2.46. Obtain the expression for the output in the POS form for the K-map shown in figure 2.68.

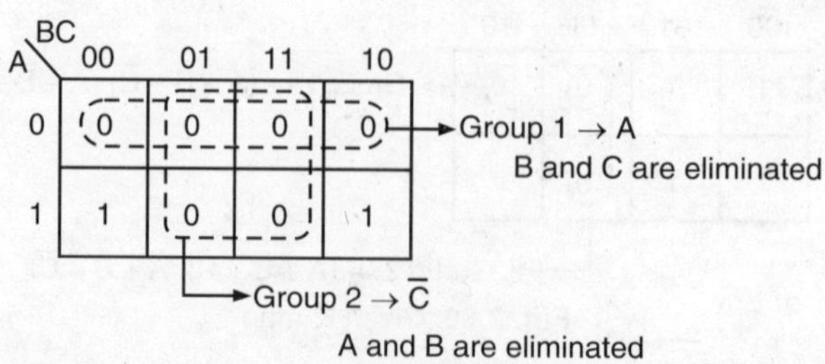

Fig. 2.68.

Solution: Here, it may be noted that two quads have been encircled in figure 2.68. When a quad is encircled, two variables which change will get eliminated. Expression for output in the POS form is as under :

$$Y = A \cdot \overline{C}$$

(A → Group 1, $\overline{C}$ → Group 2) **Ans.**

EXAMPLE 2.47. Write the simplified expression for output in the POS form the following expression:

$$\mathbf{Y = \Pi\, M\, (0, 2, 3, 7)}$$

Solution: The given expression is,

$$Y = M_0\, M_2\, M_3\, M_7$$

In the K-map, we enter 0's corresponding to these maxterms and enter 1's in the remaining cells as shown in figure 2.69 (a). Then, we group the zeros as shown in figure 2.69 (b) for the further simplification.

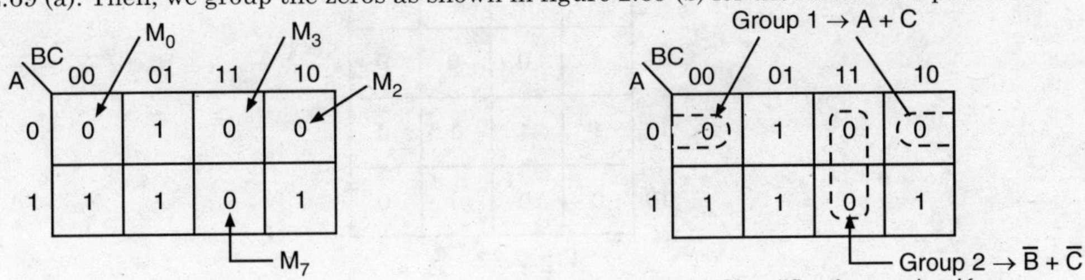

(a) (b) Simplifications using K-map

Fig. 2.69.

Therefore, the simplified expression using K-map will be as under :

$$Y = (A + C)\,(\overline{B} + \overline{C})$$ **Ans.**

EXAMPLE 2.48. Write the expression for output in the POS form for the K-map shown in figure 2.70.

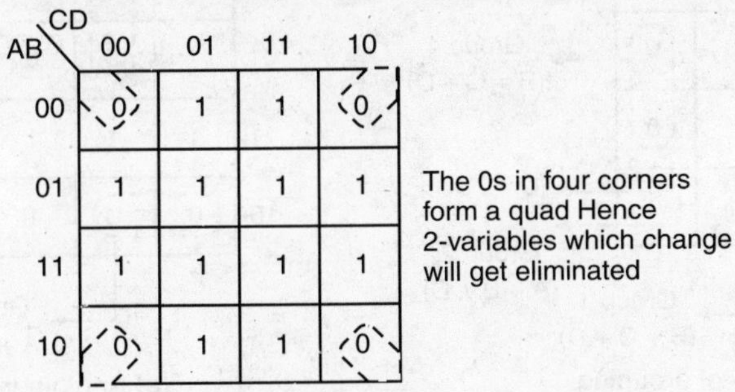

Fig. 2.70.

Solution: Therefore, expression for output will be Y = B + D.

EXAMPLE 2.49. For the K-map shown in figure 2.71, write the expression for output in the POS form.

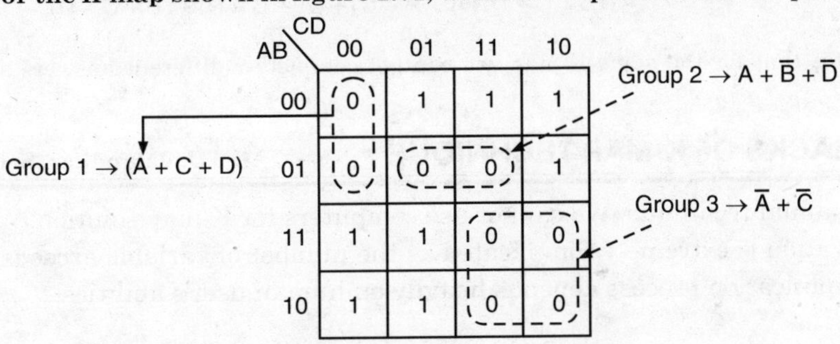

Fig. 2.71.

Solution: Therefore, the expression for output will be

$$Y = \underset{(1)}{(A + C + D)}\,\underset{(2)}{(A + \overline{B} + \overline{D})} + \underset{(3)}{(\overline{A} + \overline{C})}$$

Important Point. Sometimes there are more than one correct ways of encircling the 0's and the answers obtained as a result of all of them are correct.

EXAMPLE 2.50. Group the 0s given in the K-map of figure 2.72 in different ways to obtain the expression for output in the POS form.

AB \ CD	00	01	11	10
00	1	0	1	1
01	1	0	0	0
11	1	1	1	0
10	0	0	1	0

Fig. 2.72.

Solution: Figure 2.73 (a) and (b) show two different but valid ways of grouping 0's. The corresponding expressions for the output in the POS form are obtained after that.

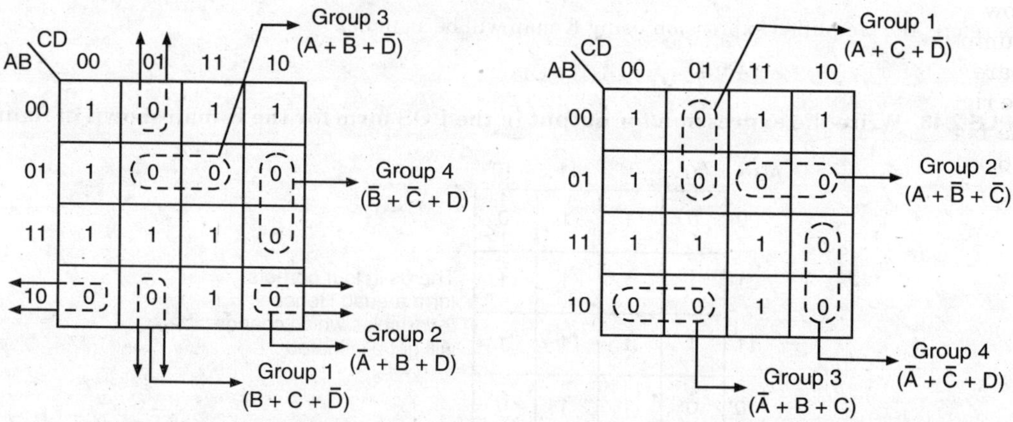

(a) One way of grouping (b) Another way of grouping

Fig. 2.73.

According to the grouping shown in figure 2.73, we obtain

$$Y = \underset{(1)}{(B + C + \overline{D})}\,\underset{(2)}{(\overline{A} + B + D)} + \underset{(3)}{(A + \overline{B} + \overline{D})} + \underset{(4)}{(\overline{B} + \overline{C} + D)}$$

This illustrates that for the same K-map, we can get completely different answers and all of them are correct.

2.18. DRAWBACKS OF K-MAP TECHNIQUE **(PTU, Jalandhar, Sem. Exam., 2008-09)**

(i) It is a manual technique. We cannot use computers for K-map solution.

(ii) Minimization is extremely complicated as the number of variable exceeds 6.

(iii) The simplification process depends heavily on human user's abilities.

2.19. FIVE AND SIX VARIABLE K-MAPS

Till now, we have discussed K-maps with 2, 3 and 4 variables. However, it is possible to obtain K-maps with five or even 6 variables.

2.19.1 A Five Variable K-map

1. Definition

The number of cells in a five variable K-map is 32 (i.e., 5). But all these 32 cells do not appear in the same K-map, because it is difficult to identify the adjacent cells for grouping. Therefore, two 16-cell K-maps are used as shown in figure 2.74. The five variables are A, B, C, D, E. The first K-map contains entries corresponding to A and B C D E whereas the other one contains entries corresponding to $\overline{A}$ and B C D E.

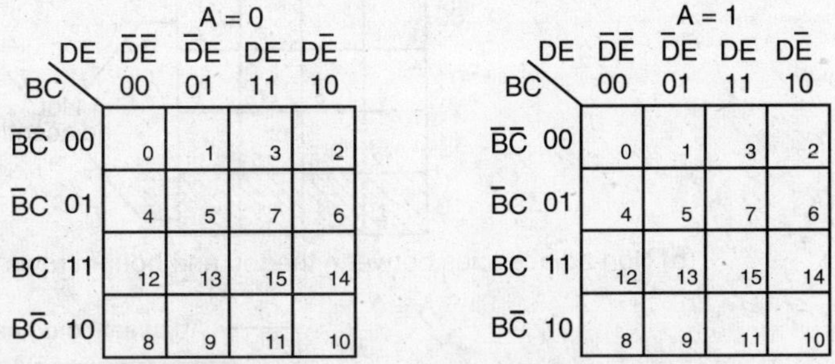

Fig. 2.74. Structure of a five variable K-map

2. How to decide the adjacent 1's or 0's ?

In a five variable K-map, the two four variable maps shown in figure 2.74 are to be considered as being superimposed on one another. They are not to be imagined as being *hinged* or mirror imaged. Every cell in the first 16 cell K-map is adjacent to the corresponding cell in the second 16 cell K-map because only 1 variable (A) changes between them. Applying the same principle, we can say that every row and every column in the first 16 cell K-map becomes adjacent to the corresponding rows and columns in the second 16 cell K-map. The adjacencies between the entries in each 4 variable K-map are visualized in a normal way. But the leftmost column of the first K-map is not adjacent with the rightmost column of the second K-map. Also, the top row of the first K-map is not adjacent with the bottom row of the second K-map. The adjacencies have been illustrated in figures 2.75 and 2.76.

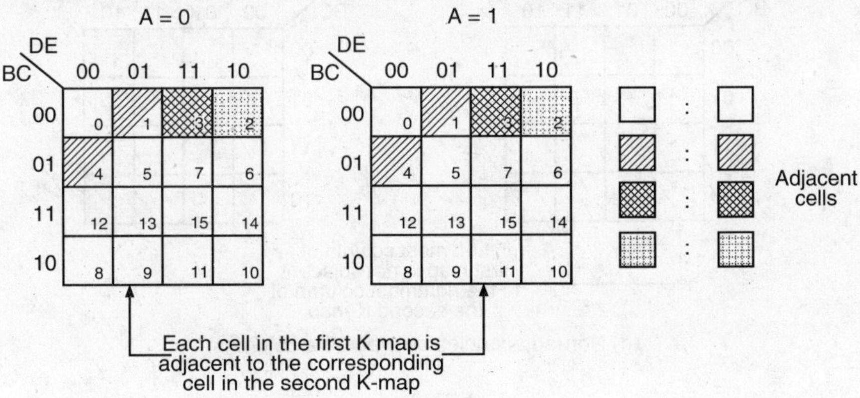

Fig. 2.75.

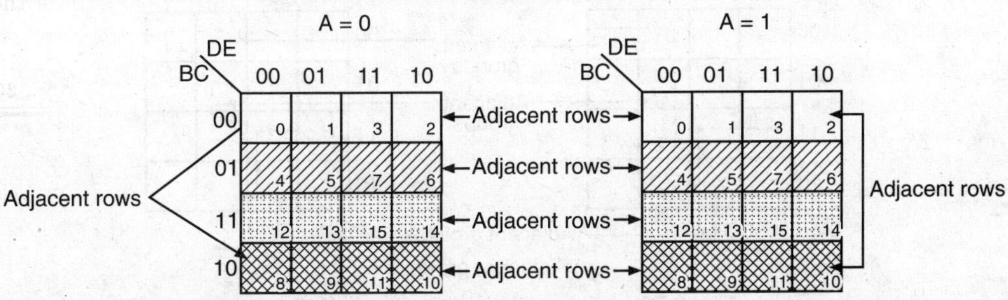

(a) Adjacencies between the rows of the two K-maps in a 5 variable K-map

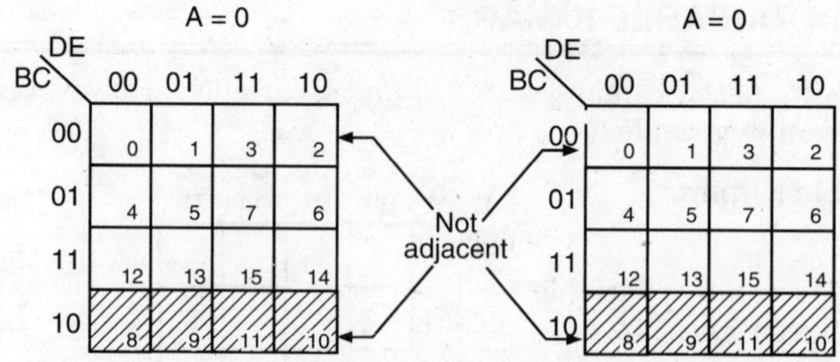

(b) Non-adjacencies between the top and bottom rows of the two K-maps in a 5 variable K-map

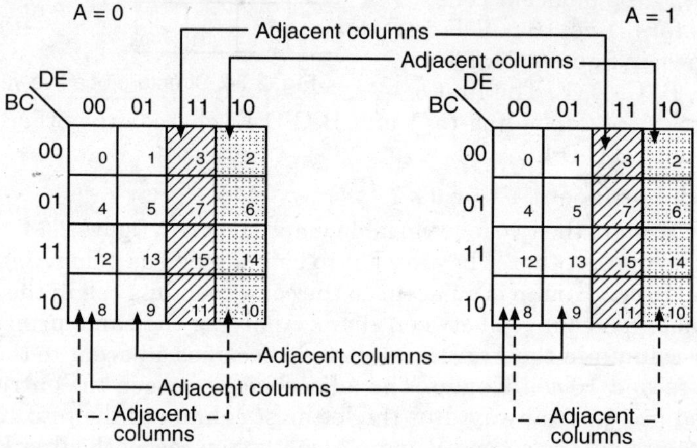

(c) Adjacencies between the columns of first K-map with the second K-map in a five variable K-map

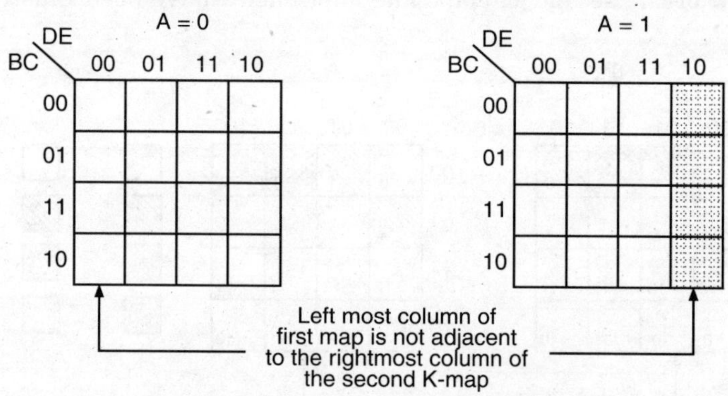

(d) Non-adjacencies between the columns

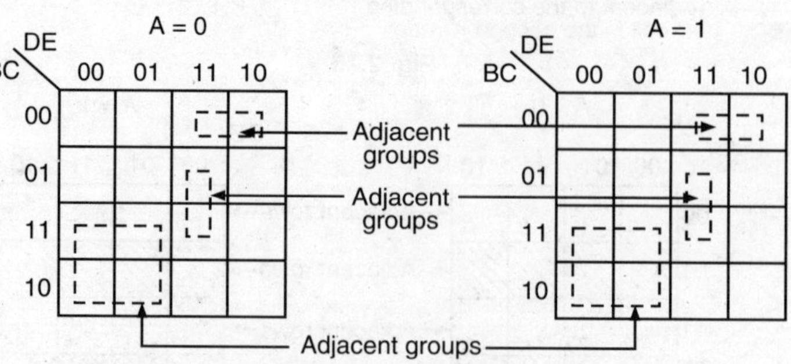

(e) Adjacent groups

Fig. 2.76. Adjacencies in a 5 variable K-map

EXAMPLE 2.51. Simplify the following Boolean function using K-map.

$$f(A, B, C, D, E) = \Sigma m (0, 2, 4, 6, 9, 11, 13, 15, 17, 21, 25, 27, 29, 31).$$

Solution: Corresponding to each minterm in the given expression, we substitute a 1 and in the remaining cells and enter 0's as shown in figure 2.77.

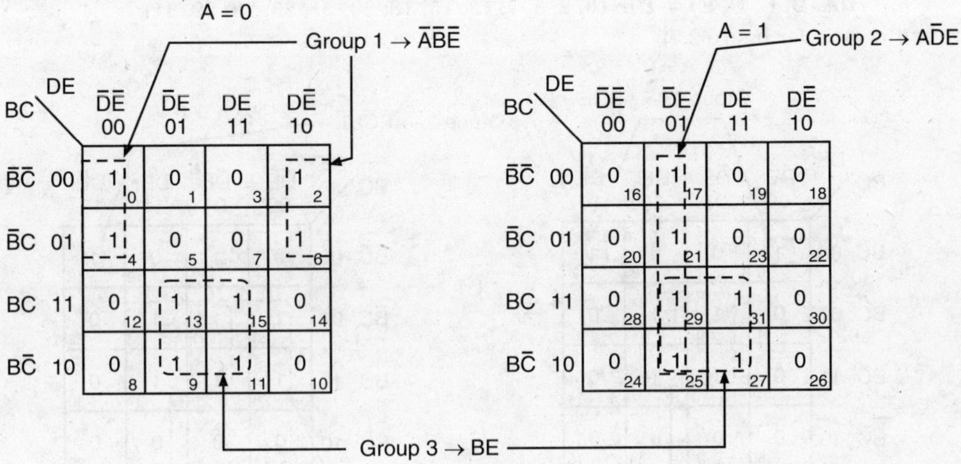

Fig. 2.77.

Let us consider group 3 of figure 2.77. The two quads are adjacent to each other to form an octet. Also, three variables A, C and D are eliminated.

Therefore, $f(A, B, C, D, E) = \underset{(1)}{\bar{A}\bar{B}\bar{E}} + \underset{(2)}{A\bar{D}E} + \underset{(3)}{BE}$ **Ans.**

INTERESTING TO KNOW?

Transistors as Far as the Eye Can See Sun Microsystems is developing a processor which may contain more than 30 million transistors in use each day worldwide. This amounts to about 40 million transistors in use for each person on earth.

EXAMPLE 2.52. Simplify the following expression using K-map :

f (A, B, C, D, E) = Σ m (0, 1, 7, 9, 11, 13, 15, 16, 17, 23, 25, 27)

Solution: Let us consider figure 2.78.

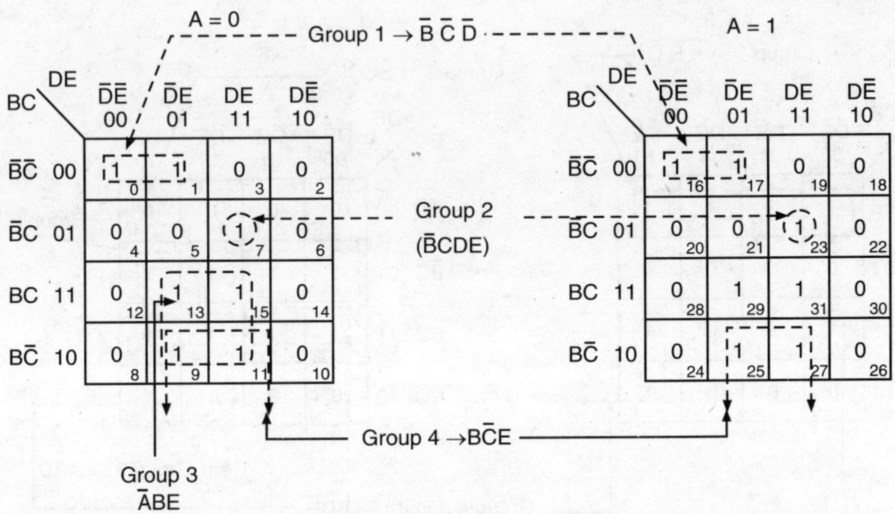

Fig. 2.78.

Therefore, $f(A, B, C, D, E) = \underset{(1)}{\bar{B}\bar{C}\bar{D}} + \underset{(2)}{\bar{B}CDE} + \underset{(3)}{\bar{A}BE} + \underset{(4)}{B\bar{C}E}$

We can further simplify it as under :

$$f(A, B, C, D, E) = \bar{B}(\bar{C}\bar{D} + CDE) + BE(\bar{A} + \bar{C})$$

EXAMPLE 2.53. Simplify the following expression using K-map :

$$\mathbf{f(A, B, C, D, E) = \Sigma\, m\, (0, 2, 5, 7, 13, 15, 18, 20, 21, 23, 28, 29, 31)}$$

Solution: Let us consider figure 2.79.

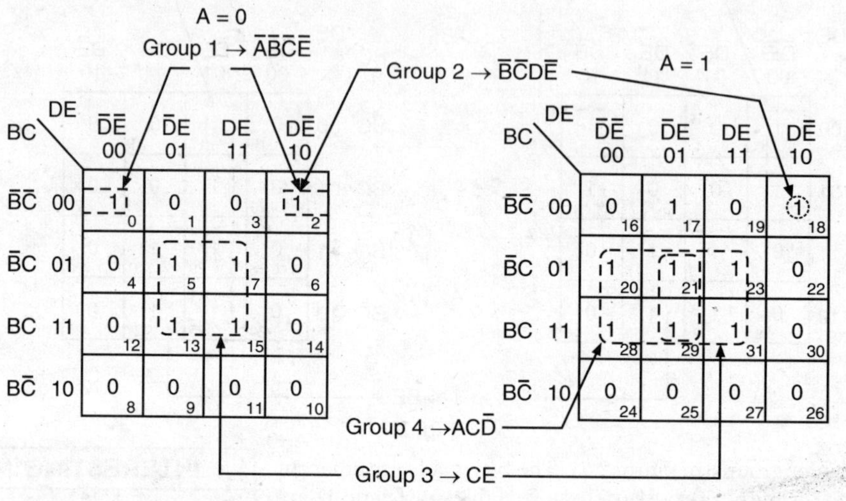

Fig. 2.79.

Therefore, $f(A, B, C, D, E) = \bar{A}\bar{B}\bar{C}\bar{E} + \bar{B}\bar{C}D\bar{E} + CE + AC\bar{D} = \bar{B}\bar{C}\bar{E} + (\bar{A} + D) + CE + AC\bar{D}.$

EXAMPLE 2.54. Reduce the following expression using 5 variable K-map :

$$\mathbf{f(A, B, C, D, E) = \Sigma\, m\, (0, 1, 2, 3, 10, 12, 13, 14, 20, 21, 22, 23, 26, 27, 28, 29, 30, 31)}$$

Solution: We consider figure 2.80 for solution

Therefore, $f(A, B, C, D, E) = \underset{(1)}{\bar{A}\bar{B}\bar{C}} + \underset{(2)}{AC} + \underset{(3)}{ABD} + \underset{(4)}{BD\bar{E}} + \underset{(5)}{BC\bar{D}}$

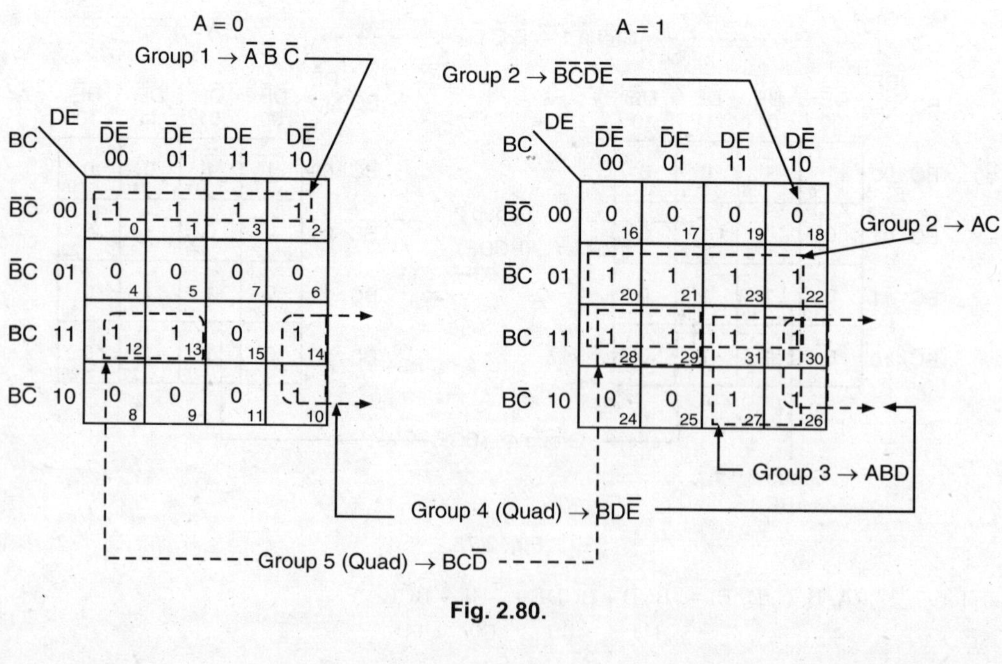

Fig. 2.80.

2.19.2. A Six Variable K-Map

Figure 2.81 shows the structure of a six variable K-map. It consists of four 4-variable maps and 64 cells (2^6). The adjacencies between the entries in each four variable map are visualised in the normal way. The four 16-cells K-maps correspond to four combinations of variables A and B, i.e., $\overline{A}\,\overline{B}$, $\overline{A}$ B, AB and A$\overline{B}$. The adjacencies between the four variable maps are visualized as groupings in a two variable map.

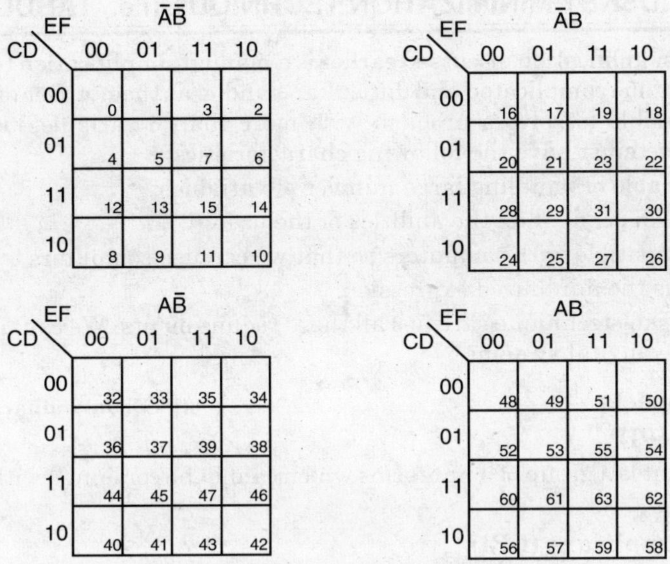

Fig. 2.81. Structure of a six variable K-map

EXAMPLE 2.55. Simplify the following six variable expression using K-map.

f(A, B, C, D, E, F) = Σ m (0, 5, 7, 8, 9, 13, 23, 24, 25, 28, 29, 37, 40, 41, 42, 43, 55, 56, 57, 60, 61)

Solution: Let us consider figure 2.82 for the solution of this problem.

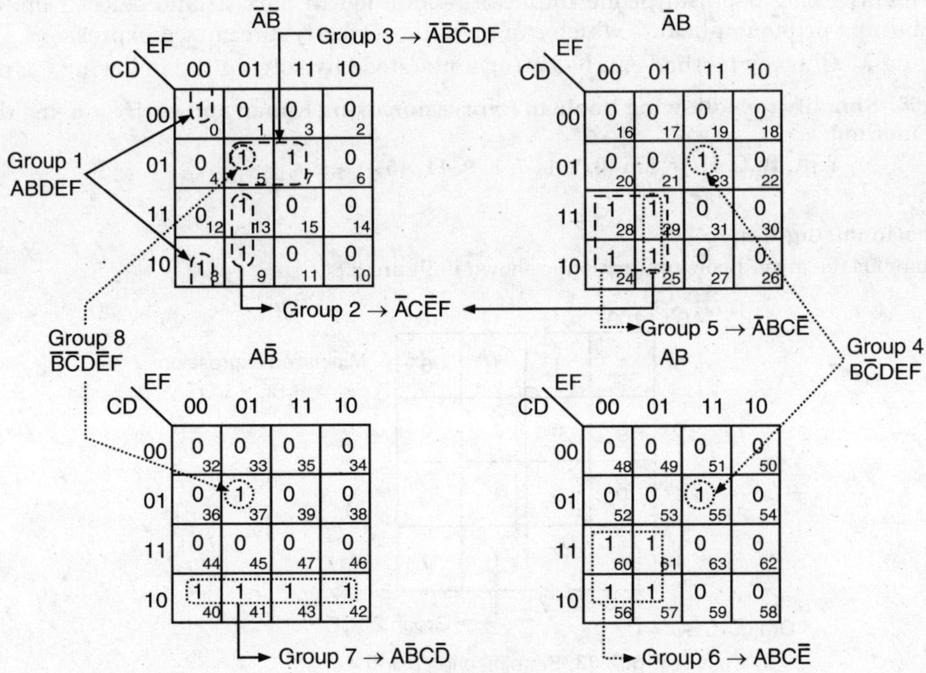

Fig. 2.82.

Therefore, the expression for output will be,

$$Y = \underset{(1)}{\bar{A}\bar{B}\bar{D}\bar{E}\bar{F}} + \underset{(2)}{\bar{A}C\bar{E}F} + \underset{(3)}{\bar{A}\bar{B}\bar{C}DF} + \underset{(4)}{B\bar{C}DEF}$$

$$+ \underset{(5)}{\bar{A}BC\bar{E}} + \underset{(6)}{ABC\bar{E}} + \underset{(7)}{A\bar{B}C\bar{D}} + \underset{(8)}{\bar{B}\bar{C}D\bar{E}F}$$

2.20. QUINE-McCLUSKEY MINIMIZATION TECHNIQUE (i.e., TABULAR METHOD)

The K-map method of simplification discussed earlier is a manual simplification technique. In addition to that, the K-maps become complicated and difficult to handle, as the number of variables increases, it is practically impossible to solve a problem with more than 6 variables using K-map. A logic minimization technique must have the following characteristics :

(i) It must be capable of handling large number of variables.

(ii) It must not be dependent on the abilities of the user.

(iii) It must be compatible with computers so that we can use computers to obtain the solution.

(iv) It must give us the minimized expression.

The Quine-McCluskey technique satisfies all these requirements. We can use this technique for designing very complex digital systems.

2.20.1. Important Aspects

(PTU, Jalandhar, Sem. Exam; 2009-10)

1. Prime Implicant (PI)

The prime implicant is a group of a minterms which cannot be combined with any other minterm or groups.

2. Essential Prime Implicant (EPI)

The essential prime implicant is a prime implicant in which one or more minterms are unique, i.e., it contains at least one minterm which is not contained in any other prime implicant. The Quine-McCluskey technique consists of two parts:

(i) First is to find all the prime implicants by an exhaustive search.

(ii) To identify the essential prime implicants obtained in part (i) and select from them the remaining prime implicants which can give the perfectly minimized expression.

The Quine-McCluskey method can be better understood by solving the following examples.

EXAMPLE 2.56. Simplify the following Boolean Expression using K-map and verify it using the Quine-McCluskey method.

Y (A, B, C, D) = Σm (0, 1, 3, 7, 8, 9, 11, 15).

Solution:

1. Simplification using K-map

The K-map for the given Boolean function is shown in figure 2.83.

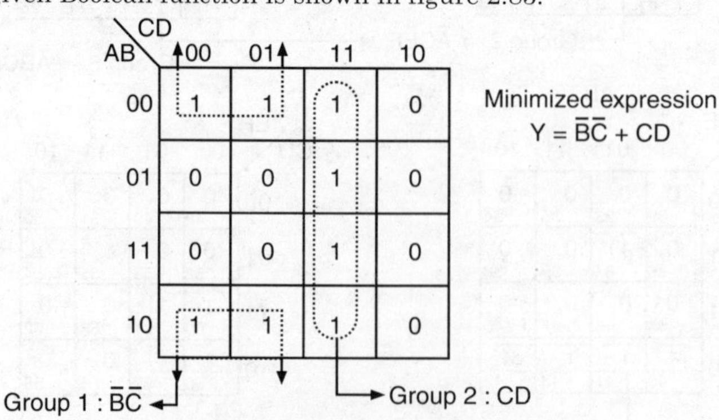

Fig. 2.83. Simplification using K-map

2. Simplification using Quine-McCluskey method

(i) First, we arrange all the minterms according to number of 1's contained and form the groups having no ones, one 1, two 1's, three 1's and so on, as shown in Table 2.17

Table 2.17.

Group	Minterm	A	B	C	D		
		Representation in binary form					
1	m_0	0	0	0	0	√	Zero 1
2	m_1	0	0	0	1	√	One 1
	m_8	1	0	0	0	√	
3	m_3	0	0	1	1	√	Two 1's
	m_9	1	0	0	1	√	
4	m_7	0	1	1	1	√	Three 1's
	m_{11}	1	0	1	1	√	
5	m_{15}	1	1	1	1	√	Four 1's

(ii) Now, we compare each minterm in group n in which each minterm in group (n + 1), and identify the matched pairs. Then, we put a check (√) mark on each matched pair as shown in Table 2.17. A matched pair is a pair of minterms which differ only in one variable, e.g., m_0 and m_4 is a matched pair because they differ only in the third position. After comparing each minterm in the nth group with every term in the $(n + 1)^{th}$ group, another table, i.e., Table 2.17 is prepared. This table lists the results of all the matching (all the matched terms) alongwith new terms generated as a result of matching of the corresponding terms. For example, let us consider the first entry in Table 2.17. The matched pair is $m_0 - m_1$.

		A	B	C	D
m_0	:	0	0	0	0
m_1	:	0	0	0	1

← Only these bits are different

Here, A, B and C are equal and m_0 and m_1 differ in only D. Hence, the new term is written in front of $m_0 - m_1$ as 0 0 0 –

Here, it may be noted that a (–) is put is place of D which indicates that D gets eliminated. Similarly, the other entries are made. Note that pairs can not be formed between minterms 8 and 3 or 9 and 7 in Table 2.18.

Table 2.18. Combination of minterms into groups of two

Group	Minterms (Matched pairs)	A	B	C	D	
		Binary Representation				
0	$m_0 - m_1$	0	0	0	–	✓
	$m_0 - m_8$	–	0	0	0	✓
1	$m_1 - m_3$	0	0	–	1	✓
	$m_1 - m_9$	–	0	0	1	✓
	$m_8 - m_7$	1	0	0	–	✓
2	$m_3 - m_7$	0	–	1	1	✓
	$m_3 - m_{11}$	–	0	1	1	✓
	$m_9 - m_{11}$	1	0	–	1	✓
3	$m_7 - m_{15}$	–	1	1	1	✓
	$m_{11} - m_{15}$	1	–	1	1	✓

First entry →

New terms generated from the matched pairs of minterms

Matched pairs of minterms obtained from Table 2.17

(iii) Now, we compare all the pairs of minterms in Table 2.18 with those in the adjacent groups, to find if we can form groups of four minterms. For example, we compare pair $(m_0 - m_1)$ of group 0 with each pair in group. The pair $(m_0 - m_8)$ of group 0 is compared with each pair in group 1. The matching between two pairs of minterms in the adjacent groups is obtained if the dashes are in the same position and only one variable differs. Such matched pairs are tick marked (√) in Table 2.17. The match pairs are also shown by dotted lines, in Table 2.18. For example, $(m_0 - m_1)$ of group 0 matches with $(m_8 - m_9)$ of group 1 or $(m_0 - m_8)$ of group 0 matches with $(m_1 - m_9)$ of group 1 and so on. Then, we prepare a new table as shown in Table 2.19, which contains the pairs of pairs. For example, $m_0 - m_1 - m_8 - m_9$, i.e., groups of four minterms. In front of each group of four minterms, the new terms are written with dashes (–). By comparing the minterm pairs of Table 2.18, bit-by-bit, we put the dashes (–) where the bits do not match.

DO YOU KNOW?

Quine-McCluskey (QM) method is more effective since a digital computer can be used to simplify the given Boolean switching function.

Table 2.19. Combination of minterms into groups of four

Group	Minterms	A	B	C	D	
		Binary representation				
0	$m_0 - m_1 - m_8 - m_9$	–	0	0	–	$\overline{B}\,\overline{C}$
	$m_0 - m_8 - m_1 - m_9$	–	0	0	–	
1	$m_1 - m_3 - m_9 - m_{11}$	–	0	–	1	$\overline{B}\,D$
	$m_1 - m_9 - m_3 - m_{11}$	–	0	–	1	
2	$m_3 - m_7 - m_{11} - m_{15}$	–	–	1	1	CD
	$m_3 - m_{11} - m_7 - m_{15}$	–	–	1	1	

(iv) We repeat the procedure of grouping. Hence, we see if we can group the Quads of minterms in the adjacent groups of Table 2.19 to obtain groups of eight minterms. There are no such matchings in Table 2.18. Hence, the process of grouping will end here.

(v) Then, we collect all the nonchecked (non √ marked) terms from Tables 2.17, 2.18 and 2.19. These are the prime implicants (PI) and they will appear in the simplified Boolean expression of Y. There are no such terms in Tables 2.17 and 2.18. But, there are three such terms in Table 2.19 as $\overline{B}\,\overline{C}$, $\overline{B}\,D$ and CD marked on that table.

Therefore, $$Y(A, B, C, D) = \overline{B}\,\overline{C} + \overline{B}D + CD \qquad \text{...(i)}$$

(vi) Now, we prepare the prime-implicant table as shown in Table 2.20.

Table 2.20. Prime implicants table

PI	Decimal numbers corresponding to P-I	1	3	7	8	9	11	15
		Given minterms						
$\overline{B}\,\overline{C}$	0, 1, 8, 9	X			⊗	X		
$\overline{B}D$	1, 3, 9, 11	X	X			X	X	
CD	3, 7, 11, 15		X	⊗			X	⊗

Obtained from Table 2.19

This table consists of the minterms contained in the given function, the PI terms included in equation (i) and the decimal numbers of the minterms corresponding to these prime implicants.

A (x) is put in each row under the minterms contained in a particular PI. Then, we find the minterms that contain only one (x) in its column and encircle these cross marks as shown in Table 2.19 and put a tick mark (√) in front of the corresponding PIs, i.e., $\overline{B}\,\overline{C}$ and CD. The encircled minterms 0 and 8 are contained in only one

DO YOU KNOW?

K-map and VEM methods are graphical methods, while QM, method is a tabular method of minimization.

PI, i.e., $\overline{B}\,\overline{C}$ whereas the minterms 7 and 15 are contained by only PI, CD. Hence, the prime implicants $\overline{B}\,\overline{C}$ and CD are called as the essential prime implicants (EPI). Now, we consider the other minterms (those which are not circle). The minterms 1 and 9 are contained in $\overline{B}\,\overline{C}$ and minterms 3 and 11 are contained in CD. Thus, all the minterms in the given function are included in the two EPIs. Hence, the minimized expression is given by,

$$Y = (A, B, C, D) = \overline{B}\,\overline{C} + CD \qquad \textbf{Ans.}$$

Importat Point. The answer obtained by the Quine-McCluskey method is exactly the same as that obtained by the K-map method.

EXAMPLE 2.57. Minimize the following logic function using K-map and verify the answer using the Quine-McCluskey method.

$$\mathbf{Y\ (A, B, C, D) = \Sigma\, m\ (0, 1, 2, 3, 5, 7, 8, 9, 11, 14)}$$

Solution:

1. Minimization using K-map

The K-map for the given logic function is shown in figure 2.84.

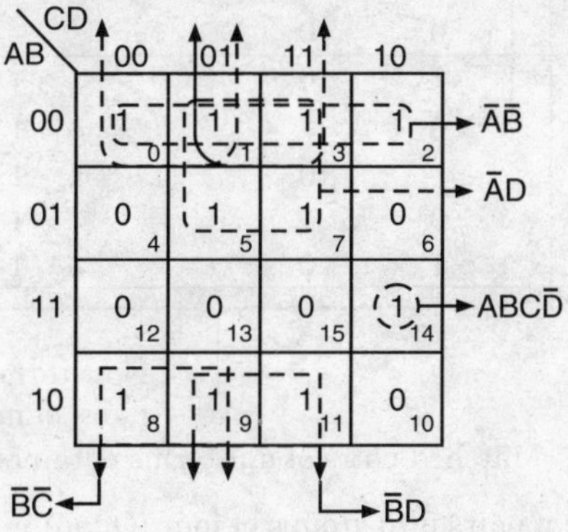

Therefore, simplified expression : $Y\ (A, B, C, D) = ABC\overline{D} + \overline{B}\,\overline{C} + \overline{B}D + \overline{A}D + \overline{A}\,\overline{B}$

Fig. 2.84.

2. Quine–McCluskey Method

(i) First, we arrange all minterms according to the number of 1's.

The minterms are arranged according to the number of 1's and groups are formed as no ones, one 1, two 1s etc. as shown in Table 2.21.

Table 2.21.

	Group	Minterm	Representation in binary form				
			A	B	C	D	
No 1′s	0	0	0	0	0	0	√
One 1	1	1	0	0	0	1	√
		2	0	0	1	0	√
		8	1	0	0	0	√
Two 1′s	2	3	0	0	1	1	√
		5	0	1	0	1	√
		9	1	0	0	1	√
Three 1′s	3	7	0	1	1	1	√
		11	1	1	0	1	√
		14	1	1	1	0	→

Minterm 0 forms a

Minterm 14 does not appear in any pair so it is not √ marked.

(ii) Then, we combine the minterms into a group of two.

Table 2.22 shows the matched pairs of minterms in the adjacent groups of Table 2.21 which differ at only one location (bit position) with respect to each other. We place (√) mark on the matched pairs in Table 2.21. The bit position where the minterms differ are represented by dashes (–) in the new terms written in front of matched pairs.

Table 2.22. Combination of minterms into grouped of two

Group	Minterms (matched pairs)	A	B	C	D	
0	0 – 1	0	0	0	–	√
	0 – 2	0	0	–	0	√
	0 – 8	–	0	0	0	√
1	1 – 3	0	0	–	1	√
	1 – 5	0	–	0	1	√
	1 – 9	–	0	0	1	√
	2 – 3	0	0	1	–	√
	8 – 9	1	0	0	–	√
2	3 – 7	0	–	1	1	√
	3 – 11	–	0	1	1	√
	5 – 7	0	1	–	1	√
	9 – 11	1	0	–	1	√

Matched quads of minterms

Now terms generated from the matched pairs of minterms

Matched pairs of minterms obtained from Table 2.21.

(iii) Next, we combine the minterm pairs into groups of four (Quad).

We group the minterm pairs of Table 2.22 (adjacent groups) to form minterm quads as shown in Table 2.23 and we place a (√) mark on the matched pairs in Table 2.22.

Table 2.23. Combination of minterms into groups of four

Group	Minterm (Group of four)	A	B	C	D
0	0 – 1 – 2 – 3	0	0	–	–
	0 – 1 – 8 – 9	–	0	0	–
	0 – 2 – 1 – 3	0	0	–	–
	0 – 8 – 1 – 9	–	0	0	–
1	1 – 5 – 3 – 7	0	–	–	1
	1 – 9 – 3 – 11	–	0	–	1
	1 – 3 – 5 – 7	0	–	–	1
	1 – 3 – 9 – 11	–	0	–	1

All these are prime implicants: $\bar{A}\,\bar{B}$, $\bar{B}\,\bar{C}$, $\bar{A}\,D$, $\bar{B}\,D$

No further grouping is possible after this. Hence, the process of grouping stops here.

(iv) Next, we collect all non-checked terms.

We collect all the non-checked (non-tickmarked) terms form Table 2.21, 2.22 and 2.23 because they are the prime-implicants (PI).

In Table 2.21, the minterm 14 is non-tickmarked so $AB\bar{C}\bar{D}$ is a PI. In Table 2.22, we don't have any such term and in Table 2.23, we have four such terms, i.e., $\bar{A}\bar{B}$, $\bar{B}\bar{C}$, $A\bar{D}$ and $\bar{B}D$.

Therefore, $Y(A, B, C, D) = ABC\bar{D} + \bar{A}\bar{B} + \bar{B}\bar{C} + \bar{A}D + \bar{B}D$ ...(i)

(v) Then, we pepare the PI table and obtain the EPIs.

The PI table is shown in Table 2.24.

Table 2.24 Prime implicants (PI) table

PI	Decimal number corresponding to PI	Given minterms									
		0	1	2	3	5	7	8	9	11	14
ABCD ✓	14										(X)
$\bar{A}\bar{B}$ ✓	0, 1, 2, 3	X	X	(X)	X						
$\bar{B}\bar{C}$ ✓	0, 1, 8, 9	X	X					(X)	X		
$\bar{A}D$ ✓	1, 5, 3, 7	X			X	(X)	(X)				
$\bar{B}D$ ✓	1, 3, 9, 11		X		X				X	(X)	

In the PI table, we find the columns containing only 1 cross (×) and encircle those (×) marks. We put (√) mark in front of the corresponding PIs. These (√) marked prime implicants in Table 2.24 are the essential prime implicants (EPIs). Hence, the simplified expression for Y will be

$$Y(A, B, C, D) = ABC\bar{D} + \bar{A}\bar{B} + \bar{B}\bar{C} + \bar{A}D + \bar{B}D$$ **Ans.**

Note that this expression is same as the one obtained with K-map simplification.

EXAMPLE 2.58. Obtain the minimal expression using Quine–McCluskey method.

$$\mathbf{f(A, B, C, D) = \Sigma\, m\, (1, 5, 6, 12, 13, 14) + d\, (2, 4)}$$

Solution:

(i) First, we group the minterms according to number of 1s.

Table 2.25.

Group	Minterm	Binary Representation				
		A	B	C	D	
0	1	0	0	0	1	√
	2 *	0	0	1	0	√
	4 *	0	1	0	0	√
1	5	0	1	0	1	√
	6	0	1	1	0	√
	12	1	1	0	0	√
2	13	1	1	0	1	√
	14	1	1	1	0	√

* Represents the don't care terms.

(ii) Then, we group the minterms to form pairs.

Table 2.26.

Group	Minterm pair	A	B	C	D	
		Binary Representation				
0	1, 5	0	–	0	1	→ $\bar{A}\bar{C}D$ Prime imp-licants (non marked terms)
	2*, 6	0	–	1	0	→ $\bar{A}C\bar{D}$
	4*, 5	0	1	0	–	√
	4*, 6	0	1	–	0	√
	4*, 12	–	1	0	0	√
1	5, 13	–	1	0	1	√
	6, 14	–	1	1	0	√
	12, 13	1	1	0	–	√
	12, 14	1	1	–	0	√

(iii) Then, we group the minterms to form groups of four.

Table 2.27.

Group	Minterm Quads	A	B	C	D	Prime implicants (non √ marked terms.)
		Binary Representation				
0	4, 5, 12, 13	–	1	0	–	$B\bar{C}$
	4, 6, 12, 14	–	1	–	0	$B\bar{D}$
	4, 12, 5, 13	–	1	0	–	$B\bar{C}$
	4, 12, 6, 14	–	1	–	0	$B\bar{D}$

(iv) Then, we group the prime implicants.

The non-tick (√) marked terms from Tables 2.25, 2.26 and 2.27 are the prime implicants (PI). They are likely to appear in the simplified expression for output as under :

$$f(A, B, C, D) = \bar{A}\bar{C}D + \bar{A}\bar{C}D + \bar{A}C\bar{D} + B\bar{C} + B\bar{D} \qquad ...(i)$$

(v) Then, we prepare the table of prime implicants.

Table 2.28. Table of PI

Prime implicants	Decimal number (minterms)	1	5	6	12	13	14
		Given minterms					
$\bar{A}\bar{C}D$ ✓	1, 5	(X)	X				
$\bar{A}C\bar{D}$	2, 6			X			
$B\bar{C}$ ✓	4, 5, 12, 13		X		X	(X)	
$B\bar{D}$ ✓	4, 6, 12, 14				X		(X)

← Don't care terms are not listed in this table

The encircled crosses (×) represent the essential prime implicants (EPIs). They are $\bar{A}\bar{C}D$, $B\bar{C}$, $B\bar{D}$ and they cover the minterms 1, 5, 6, 12, 13 and 14, i.e., all the given minterms (we do not worry about 2 and 4 which are don't cares).

Therefore, $f(A, B, C, D) = \bar{A}\bar{C}D + B\bar{C} + B\bar{D}$ **Ans.**

Cross checking using K-map

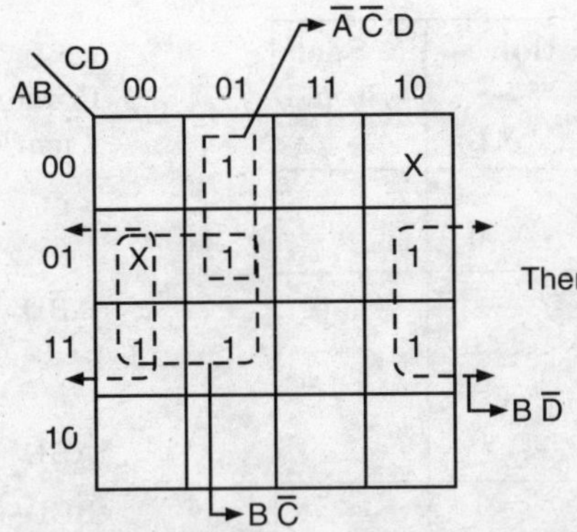

Therefore, f (A B C D) = $\overline{A}\,\overline{C}\,D + B\,\overline{C} + B\,\overline{D}$

Thus the answers using both the methods match

Fig. 2.85.

EXAMPLE 2.59. Realize the following logic function in SOP form using Quine–McCluskey Method.

$$f(A, B, C, D) = \Pi M (2, 7, 8, 9, 10, 12)$$

Solution: Here, it may be noted that the given function is in the POS form or maxterm form and we want the solution in the SOP form. Hence, we have to first convert the given expression into SOP form using the complementary relationship between the POS and SOP form.

Hence,

$$f(A, B, C, D) = \Pi M (2, 7, 8, 9, 10, 12)$$
$$= \Sigma m (0, 1, 3, 4, 5, 6, 11, 13, 14, 15)$$

(i) First, we group the minterms according to number of 1s.

Table 2.29.

Group	Minterms	Binary representation				See Step 2	
		A	B	C	D		
0	0	0	0	0	0	√	No ones
1	1	0	0	0	1	√	One 1
	4	0	1	0	0	√	
2	3	0	0	1	1	√	Two 1s
	5	0	1	0	1	√	
	6	0	1	1	0	√	
3	11	1	0	1	1	√	Three 1s
	13	1	1	0	1	√	
	14	1	1	1	0	√	
4	15	1	1	1	1	√	Four 1s

(ii) Now, we group the minterms to form pairs.

DO YOU KNOW?

The standard form of expressing Boolean functions results in a two-level implementation. However, there are occasions, when the design of digital systems results in getting structures with three or more levels.

Table 2.30.

Group	Minterm pairs	Binary Representation A	B	C	D	See Step 3	Prime implicants (non √ marked terms)
0	0 – 1	0	0	0	–	√	
	0 – 4	0	–	0	0	√	
1	1 – 3	0	0	–	1		← $\bar{A}\bar{B}D$
	1 – 5	0	–	0	1	√	
	4 – 5	0	1	0	–	√	
	4 – 6	0	1	–	0		← $\bar{A}B\bar{D}$
2	3 – 11	–	0	1	1		→ $\bar{B}CD$
	5 – 13	–	1	0	1		→ $B\bar{C}D$
	6 – 14	–	1	1	0		→ $BC\bar{D}$
3	11 – 15	1	–	1	1		→ ACD
	13 – 15	1	1	–	1		→ ABD
	14 – 15	1	1	1	–		→ ABC

(iii) Then, we group the minterms to form quads (i.e., groups of 4).

Table 2.31.

Group	Minterms quads	Binary Representation A	B	C	D	
0	0, 1, 4, 5	0	–	0	–	$\bar{A}\bar{C}$ – Prime implicant
	0, 4, 1, 5	0	–	0	–	

(iv) Then, we collect all the prime implicants (PI).

All the nontick marked terms in Table 2.29, 2.30 and 2.31 are the prime implicants (PI), so they will appear in the expression for output as under :

$$f(A, B, C, D) = \bar{A}\bar{B}D + \bar{A}B\bar{D} + \bar{B}CD + B\bar{C}D + BC\bar{D} + ACD + ABC + \bar{A}\bar{C} + ABD \quad \text{...(i)}$$

(v) Now, let us prepare the PI table.

Table 2.32.

PI term	Decimal equivalent	Given minterms 0	1	3	4	5	6	11	13	14	15
$\bar{A}\bar{B}D$	1, 3		X	X							
$\bar{A}B\bar{D}$	4, 6				X		X				
$\bar{B}CD$ ✓	3, 11			X				X			
$B\bar{C}D$	5, 13					X			X		
$BC\bar{D}$ ✓	6, 14						X			X	
ACD	11, 15							X			X
ABD ✓	13, 15								X		X
ABC	14, 15									X	X
$\bar{A}\bar{C}$ ✓	0, 1, 4, 5	ⓧ	X		X	X					

The encircle term in Table 2.32 is called as the essential prime implicant (EPI) which covers the minterms 0, 1, 4 and 5. This term will appear in the minimized expression. The remaining minterms are 3, 6,, 11, 13, 14,

15, so, we find the terms which cover these minterms. From Table 2.32, such terms are $\bar{B}CD$ which covers the minterms 3 and 11, $BC\bar{D}$ which covers 6, 14 and ABD which covers 13 and 15. Hence, these terms also will appear in the minimized expression. Hence, the minimized expression in SOP form is given by

$$f(A, B, C, D) = \bar{A}\bar{C} + \bar{B}CD + BC\bar{D} + ABD \quad \textbf{Ans.}$$

Cross checking using K-map

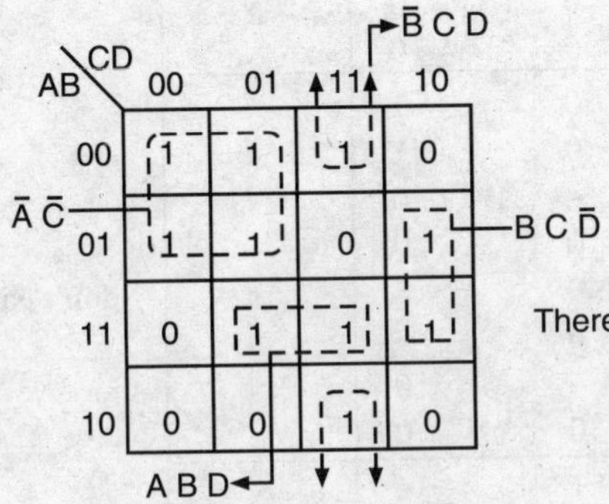

Therefore, f (A, B, C, D) = $\bar{A}\bar{C} + \bar{B}CD + BC\bar{D} + ABD$

Fig. 2.86.

Thus, the results obtained by the two methods are exactly identical.

EXAMPLE 2.60. Determine the minimal expression for the following function using Quine–McCluskey method.

$$f(A, B, C, D) = \Sigma m\ (0, 1, 4, 6, 8, 9, 10, 12) + d(5, 7, 14).$$

Solution:

(i) First, we arrange the minterms according to number of 1's.

Table 2.33.

Group	Minterm	Binary representation				See Step 2	
		A	B	C	D		
0	0	0	0	0	0	√	Zero 1's
1	1	0	0	0	1	√	One 1
	4	0	1	0	0	√	
	8	1	0	0	0	√	
2	6	0	1	1	0	√	Two 1's
	9	1	0	0	1	√	
	10	1	0	1	0	√	
	12	1	1	0	0	√	
	5*	0	1	0	1	√	
3	7*	0	1	1	1	√	Three 1's
	14*	1	1	1	0	√	

* don't care terms

(ii) Then, we group the minterms into groups of two.

Table 2.34.

Group	Pair of Minterm	Binary representation				See Step 2
		A	B	C	D	
0	0 – 1	0	0	0	–	√
	0 – 4	0	–	0	0	√
	0 – 8	–	0	0	0	√
1	1 – 9	–	0	0	1	√
	1 – 5	0	–	0	1	√
	4 – 6	0	1	–	0	√
	4 – 12	–	1	0	0	√
	4 – 5	0	1	0	–	√
	8 – 9	1	0	0	–	√
	8 – 10	1	0	–	0	√
	8 – 12	1	–	0	0	√
2	6 – 7 *	0	1	1	–	√
	6 – 14 *	–	1	1	0	√
	10 – 14 *	1	–	1	0	√
	12 – 14 *	1	1	–	0	√
	5 * – 7 *	0	1	–	1	√

* don't care terms

(iii) Now, we group the minterms into groups of four.

Table 2.35.

Group	Quad of minterms	Binary Representation				Prime implicants
		A	B	C	D	
0	0, 1, 4, 5*	0	–	0	–	$\overline{A}\,\overline{C}$
	0, 1, 8, 9	–	0	0	–	$\overline{B}\,\overline{C}$
	0, 4, 1, 5*	0	–	0	–	$\overline{A}\,\overline{C}$
	0, 4, 8, 12	–	–	0	0	$\overline{C}\,\overline{D}$
	0, 8, 1, 9	–	0	0	–	$\overline{B}\,\overline{C}$
	0, 8, 4, 12	–	–	0	0	$\overline{C}\,\overline{D}$
1	4, 6, 12, 14	–	1	–	0	$B\overline{D}$
	4, 6, 5, 7	0	1	–	–	$A\overline{B}$
	4, 12, 6, 14	–	1	–	0	$B\overline{D}$
	4, 5, 6, 7	0	1	–	–	$A\overline{B}$
	8, 10, 12, 14	1	–	–	0	$A\overline{D}$
	8, 12, 10, 14	1	–	–	0	$A\overline{D}$

* don't care terms

No further grouping is possible. Hence, from the Tables 2.33, 2.34 and 2.35, we list the terms which are not (√) marked. They are prime implicants (PI) and hence will appear in the expression for Y as under :

$$Y = \overline{A}\,\overline{C} + \overline{B}\,\overline{C} + \overline{C}\,\overline{D} + B\overline{B} + \overline{A}B + A\overline{D}$$

(iv) Then, we prepare the table the prime implicants.

Table 2.36 shows the table of prime implicants

Table 2.36. Table of PI

Prime Implicant		Decimal numbers	Given minterms							
			0	1	4	6	8	9	10	12
$\bar{A}\bar{C}$		0, 1, 4, 5*	X	X	X					
$\bar{B}\bar{C}$	✓	0, 1, 8, 9	X	X			X	⊗		
$\bar{C}\bar{D}$		0, 4, 8, 12	X		X		X			X
$B\bar{D}$	✓	4, 6, 12, 14*			X	X				X
$\bar{A}B$		4, 6, 5*, 7*			X	X				
$A\bar{D}$	✓	8, 10, 12, 14*					X		⊗	X

The encircled crosses represent the EPIs. Hence, the terms $\bar{B}\bar{C}$ and $A\bar{D}$ are the EPIs. They cover the terms 0, 1, 8, 9 and 10, 12. Now, we check the remaining minterms, i.e., 4 and 6. both these are covered by $B\bar{D}$, so, we have to include $B\bar{D}$.

Hence, $$Y = \bar{B}\bar{C} + A\bar{D} + B\bar{D}$$ **Ans.**

EXAMPLE 2.61. Use the Quine-McCluskey method of minimization and find the expression for the following function :

$$\mathbf{f(A, B, C, D) = \Sigma m(2, 4, 8, 11, 15) + d(1, 10, 12, 13)}$$

Solution: We will treat the don't care conditions to be the minterms producing a high output so the given function gets modified to

$$f(A, B, C, D) = \Sigma m(1, 2, 4, 8, 10, 11, 12, 13, 15)$$

(i) First, we arrange the minterms according to number of 1's.

Table 2.37.

Group	Minterms	A	B	C	D		Unchecked term
	1	0	0	0	1		One 1
0	2	0	0	1	0	√	
	4	0	1	0	0	√	
	8	1	0	0	0	√	Two 1's
1	10	1	0	1	0	√	
	12	1	1	0	0	√	Three 1's
2	11	1	1	1	1	√	
	13	1	1	0	1	√	Four 1's
3	15	1	1	1	1	√	

(ii) Then, we group the minterms into groups of two.

Table 2.38.

Group	Pair of minterms	Binary Representation A	B	C	D	Prime implicants.
0	2 – 10	–	0	1	0	$\bar{B}C\bar{D}$
	4 – 12	–	1	0	0	$B\bar{C}\bar{D}$
	8 – 10	1	0	–	0	$A\bar{B}\bar{D}$
	8 – 12	1	–	0	0	$A\bar{C}\bar{D}$
1	10 – 11	1	0	1	–	$A\bar{B}C$
	12 – 13	1	1	0	–	$AB\bar{C}$
2	11 – 15	1	–	1	1	ACD
	13 – 15	1	1	–	1	ABD

From Tables 2.37 and 2.38, we list the terms which are not √ marked, they are the prime implicants (PI). Therefore,

$$Y(A, B, C, D) = \bar{A}\bar{B}\bar{C}D + \bar{B}C\bar{D} + B\bar{C}\bar{D} + A\bar{B}\bar{D} + A\bar{C}\bar{D} + A\bar{B}C + A\bar{B}C + AB\bar{C} + ACD + ABD$$

(iii) Then, we prepare the table of prime implicants.

Table 2.39. Table of PI

PI Term	Decimal numbers	Given minterms 2	4	8	11	15
$\bar{A}\bar{B}\bar{C}D$	1					
$\bar{B}C\bar{D}$ ✓	2 - 10	ⓧ				
$B\bar{C}\bar{D}$	4 - 10		ⓧ			
$A\bar{B}\bar{D}$	8 - 10			X		
$A\bar{C}\bar{D}$ ✓	8 - 12			X		
$A\bar{B}C$	10 - 11				X	
$AB\bar{C}$	12 - 13					
ACD ✓	11 - 15				X	X
ABD	13 - 15					X

We encircle only those crosses which are alone in their columns and tickmark the corresponding PI's. These are the essential prime implicants.

Hence, $Y = \bar{B}C\bar{D} + B\bar{C}\bar{D}$

These terms cover the minterms. Now, we check the remaining minterms, i.e., 8, 11 and 15. The PIs $A\bar{C}\bar{D}$ and ACD cover these terms. Hence, they must also be included.

Hence, $Y = \bar{B}C\bar{D} + B\bar{C}\bar{D} + A\bar{C}\bar{D} + ACD$ **Ans.**

Cross checking

Let us now cross check using K-map.

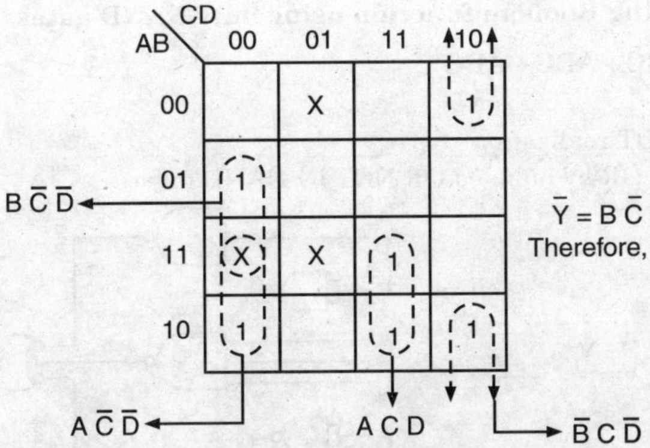

$\bar{Y} = B\bar{C}\bar{D} + A\bar{C}\bar{D} + ACD + \bar{B}C\bar{D}$

Therefore, we get the same expression

Fig. 2.87.

2.21. IMPLEMENTATION USING ONLY NAND OR ONLY NOR GATES

The SOP expressions are implemented using the AND and OR gates. Let us consider the SOP expression given below :

$$Y = AB\bar{C} + ABC + \bar{A}BC$$

Now, let us observe its implementation using AND-OR-NOT gates as shown in figure 2.88.

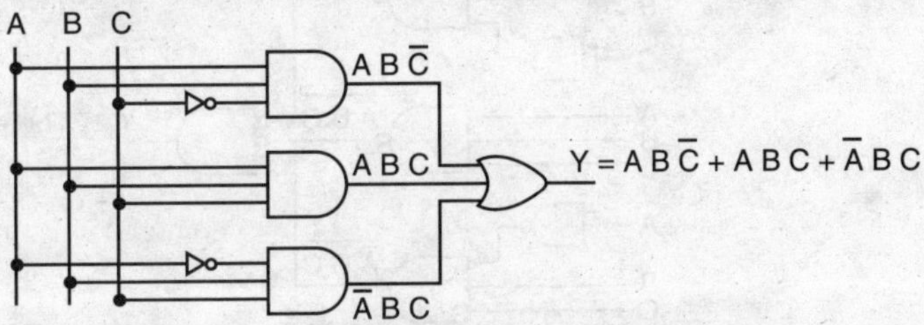

Fig. 2.88. Realization using AND-OR-NOT gates

From the implementation, it is obvious that the implementation procedure is as under :

(i) First, we implement all the product terms using AND gates.

(ii) Logically, we OR all the product terms to get the output.

However, many times, it is convenient to implement the given circuit with only NAND or only NOR gates. Hence, it is essential to convert the AND-OR-NOT circuit to a NAND-NAND or NOR-NOR circuit.

> **DO YOU KNOW?**
>
> NAND and NOR gates are easier to fabricate with electronic components and are the basic gates used in all IC digital logic families.

2.21.1 NAND-NAND Implementation

1. Alternative Representation of a NAND gate

According to DeMorgan's first theorem, we have

NAND = Bubbled OR

This has been shown in figure 2.89. Hence, a NAND gate can also be represented as a bubbled OR gate.

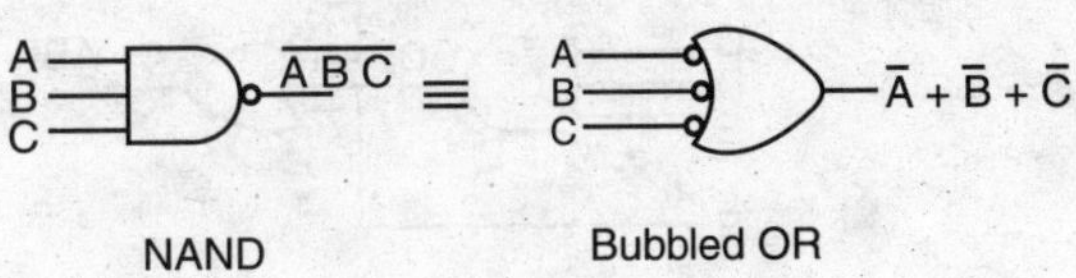

Fig. 2.89. Alternative representation of a NAND gate

This alternative representation is used for converting the AND-OR-NOT logic to NAND-NAND logic.

2. Procedure

(i) First, we simplify the given logical expression and convert it in the SOP form.

(ii) Then, we draw the AND-OR-NOT realization.

(iii) Then, we replace every AND gate by a NAND, every OR gate by a bubbled OR gate and NOT gate by a NAND inverter.

(iv) Finally, we draw the circuit using only NAND gates.

EXAMPLE 2.62. Implement the following Boolean function using only NAND gates.

$$Y = AB\bar{C} + ABC + \bar{A}BC$$

Solution:

(i) First, let us draw the AND-OR-NOT realization

(ii) Next, we replace AND by NAND, OR by bubbled OR NOT by NAND inverter

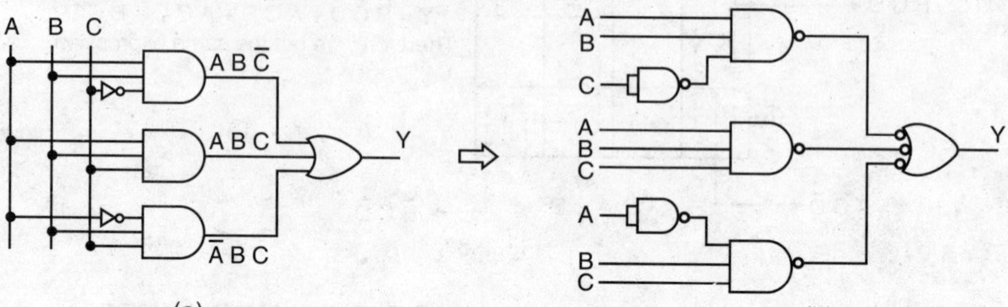

(a) (b)

Lastly, we draw the circuit using only NAND gates.

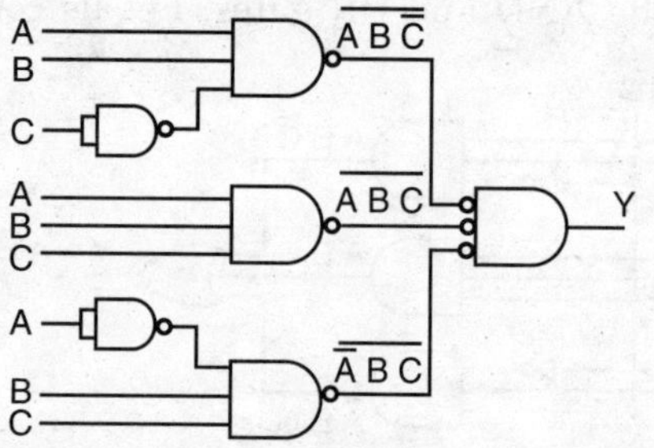

Therefore, $Y = \overline{[(\overline{A B \bar{C}})\ (\overline{A B C})\ (\overline{\bar{A} B C})]}$

$= A B \bar{C} + A B C + \bar{A} B C$

(c)

Fig. 2.90. Implementation using only NAND

EXAMPLE 2.63. Implement the following Boolean expression using only NAND gates.

$$Y = AB + CDE + F$$

Solution:

(i) First, we carry out the realization using basic gates.

(ii) Secondly, we replace AND → NAND, OR → Bubbled OR, NOT → NAND inverter

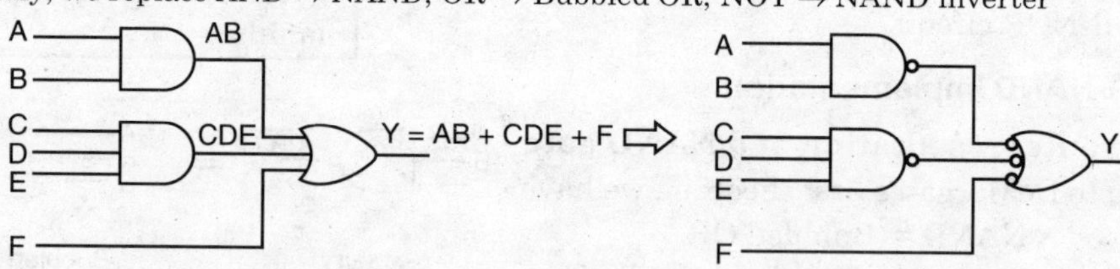

(a) (b)

(iii) Lastly, we draw the circuit using only NAND gates.

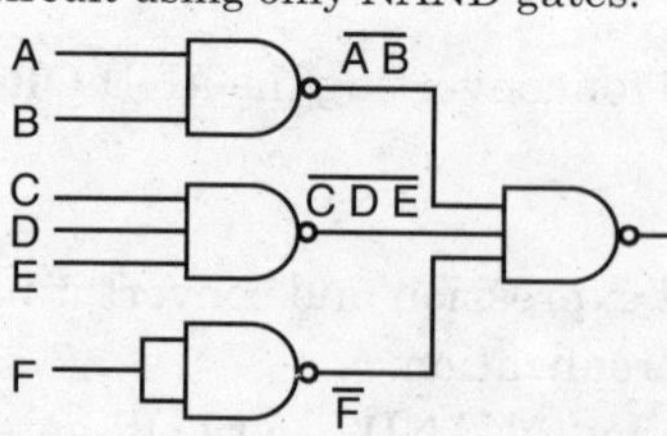

$Y = \overline{(\overline{AB})\ (\overline{CDE})\ (\overline{F})}$

$= AB + CDE + F$

(c)

Fig. 2.91. Realization using only NAND gates

EXAMPLE 2.64. Implement the following expression using NAND-NAND logic.

$$Y = \Sigma m\,(0, 1, 5).$$

Solution: (i) First, we write the K-map.

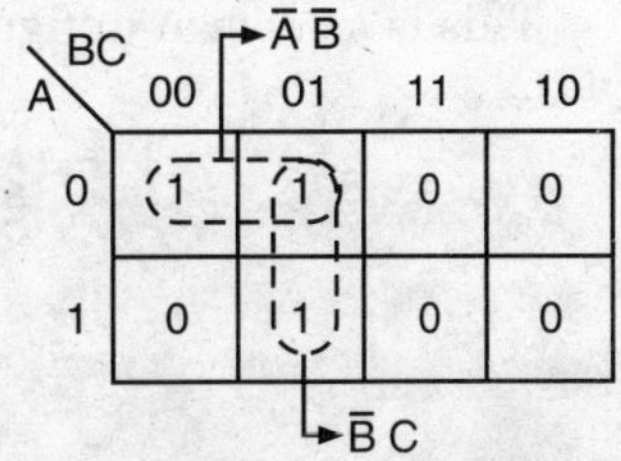

Fig. 2.92. K-map for given expression

Therefore, $Y = \bar{A}\bar{B} + \bar{B}C$

(ii) Next, we implement using AND-OR-NOT logic.

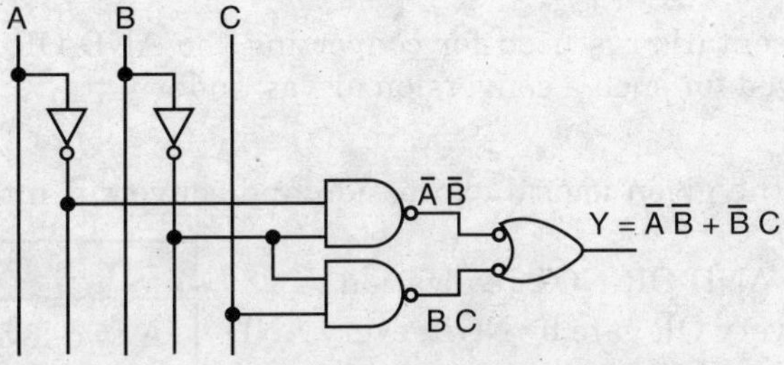

Fig. 2.93. AND-OR-NOT implementation

(iii) Now, we convert AND-OR-NOT into NAND-NAND logic :

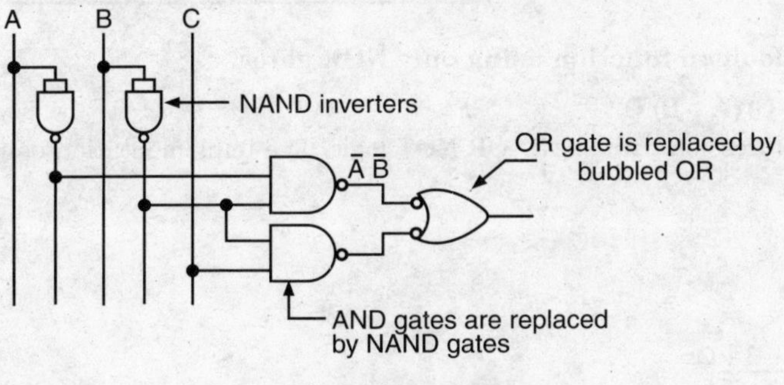

Fig. 2.94.

> **DO YOU KNOW?**
>
> The NOR operation is the dual of the NAND operation. Therefore, all procedures and rules for NOR logic are the duals of the corresponding procedures and rules developed for NAND logic.

Now, we replace every AND by NAND, every OR by a bubbled OR and every inverter by a NAND inverter to get the NAND-NAND logic shown in figure 2.94.

(iv) Finally, we draw the circuit using only NAND gates.

The circuit using only NAND gates is shown in figure 2.95.

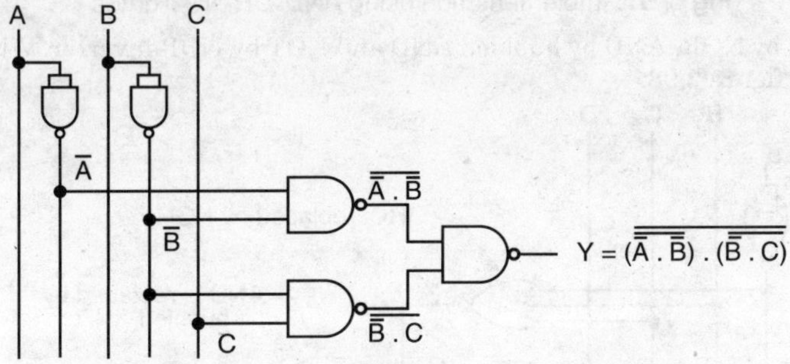

Fig. 2.95. Implementation using NAND-NAND logic

2.21.2 NOR-NOR Implementation

1. Alternative representation of a NOR gate

According to DeMorgan's second theorem, we have

NOR = Bubbled AND

This is as shown in figure 2.96. Hence, a NOR gate can also be represented as a bubbled AND gate.

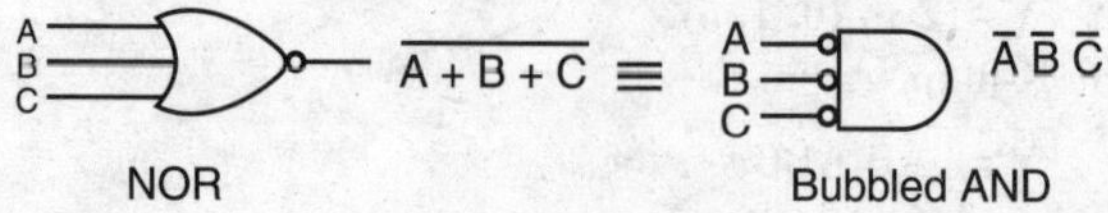

Fig. 2.96. Alternative representation of a NOR gate

This alternative representation is used for converting the AND-OR-NOT logic to NOR-NOR logic. The steps to be followed for such a conversion are as under :

2. Procedure

(i) First, we simplify the given logical expression and convert it into product of sums (POS) form.

(ii) Then, we draw the AND-OR-NOT realization

(iii) Then, we replace every OR gate by NOR, every AND gate by a bubbled AND gate and every inverter by a NOR inverter.

(iv) Finally, we draw the final circuit using only the NOR gates.

DO YOU KNOW?

A two-level implementation with NOR gates requires that the function be simplified into product-of-sums form.

EXAMPLE 2.65. Implement the following Boolean function using only NOR gates.

$$F = (\bar{A}+B+C)(A+B)D$$

Solution: (i) First, we implement the given expression using AND-OR-NOT logic. The implementation using AND-OR-NOT gates is shown in figure 2.97.

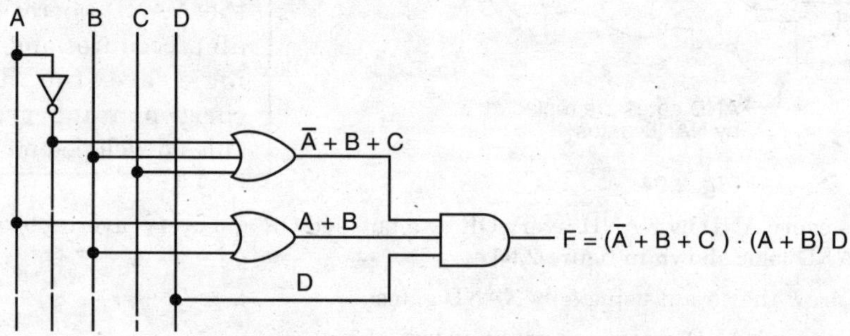

Fig. 2.97. Implementation using AND-OR-NOT gates

Next, we Replace OR by NOR, AND by bubbled AND and NOT by NOR inverter. With these replacements, the circuit is as shown in figure 2.98.

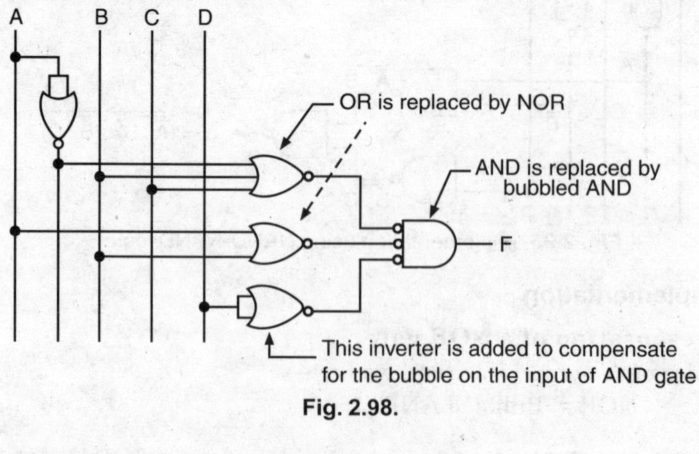

Fig. 2.98.

Then, we implement the circuit using only NOR gates : The NOR-NOR implementation is shown in figure 2.99.

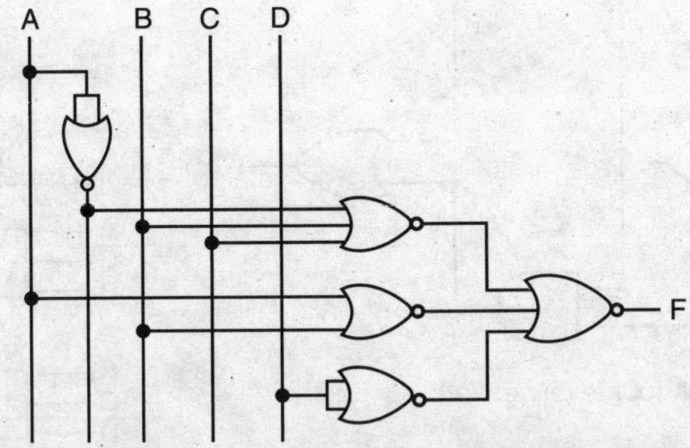

Fig. 2.99. Implementation using the NOR-NOR logic

EXAMPLE 2.66. Implementation the following POS expression using NOR-NOR logic.

$$F = \Pi M(0, 1, 2, 4, 6)$$

Solution: (i) First, we simplify the given function using K-map. The given Boolean function is,

$$F = M_0 M_1 M_2 M_4 M_6$$

Hence, the corresponding K-map is as shown in figure 2.100.

Therefore, simplified Boolean equation is as under :

$$F = (A + B) C$$

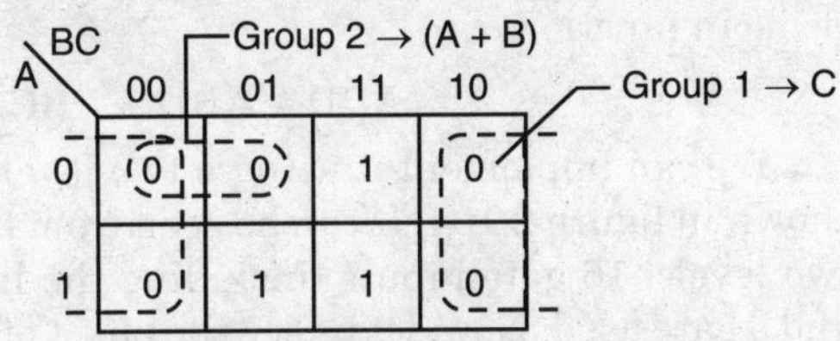

Fig. 2.100.

(ii) Next, we implement this Boolean expression using AND-OR gates.

The implementation using AND-OR gates has been shown in figure 2.101 (a).

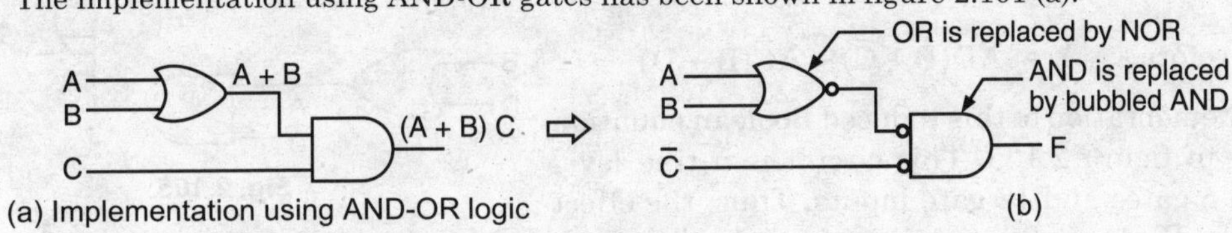

Fig. 2.101.

(iii) Then, we replace OR by NOR, AND by bubbled AND and NOT by NOR inverter.

These replacement have been shown in figure 2.101 (b).

(iv) Lastly, we implement the circuit using only NOR gates.

The NOR-NOR implementation is shown in figure 2.102.

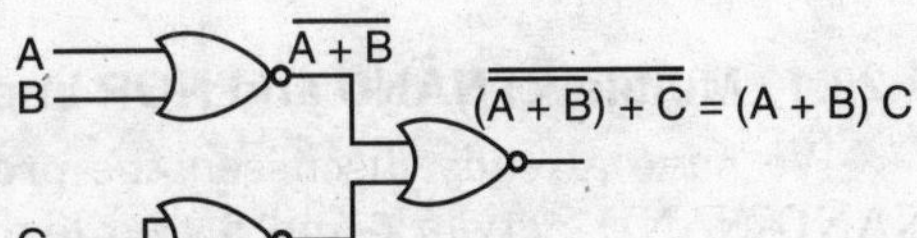

Fig. 2.102 Implementation using NOR-NOR logic

Cross checking

Referring equation (i), we obtain,

$$F = (A+B)\,C = \overline{\overline{(A+B)\cdot C}} = \overline{\overline{(A+B)}+\overline{C}} \quad \text{...(ii)}$$

We get equation (ii) at the output of NOR-NOR logic shown in figure 2.102.

2.22. MULTILEVEL GATE IMPLEMENTATIONS

We have already discussed the two level AND-OR or OR-AND implementation. We have seen that each variable in the given expression acts as an input in the implemented circuit. Note that the complements of inputs also are treated as inputs. Hence, inverters used on the input side are not treated as a separate level. A logic circuit can have more number of levels than two as shown in figure 2.105. Such networks are called as multilevel networks.

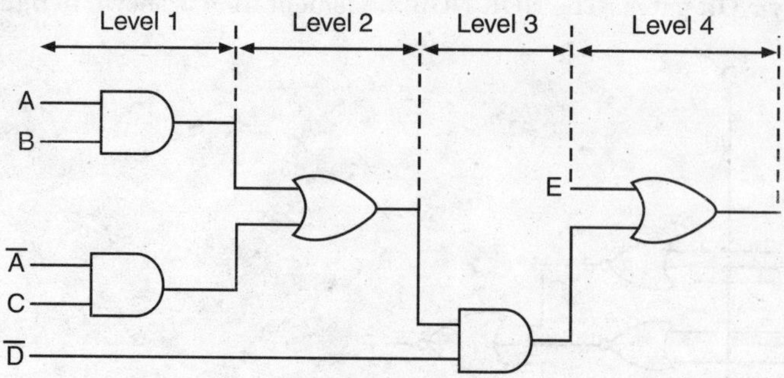

Fig. 2.103. A four level network

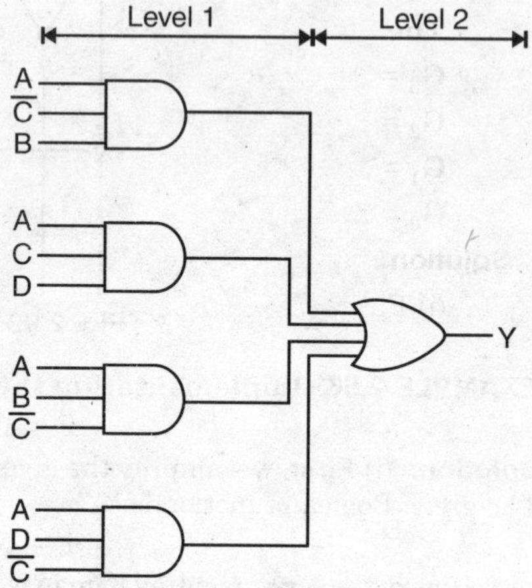

Fig. 2.104.

Effect of simplification

Now let us see the effect of simplification of a Boolean equation on the number of levels of a network. Let us consider a Boolean function

$$Y = A\overline{C}D + ABD + AB\overline{C} + AD\overline{C}$$

We can implement it using a two level AND-OR network as shown in figure 2.104. It can be seen from figure 2.104, there are two levels, 16 gate inputs (including the inputs of the OR gate) and 5 gate used. Now, let us see the effect of simplification of the given Boolean function.

$$Y = A\overline{C}D + ABD + AB\overline{C} + AD\overline{C}$$

Therefore, $\quad Y = AD(B + \overline{C}) + A\overline{C}(B + D)$

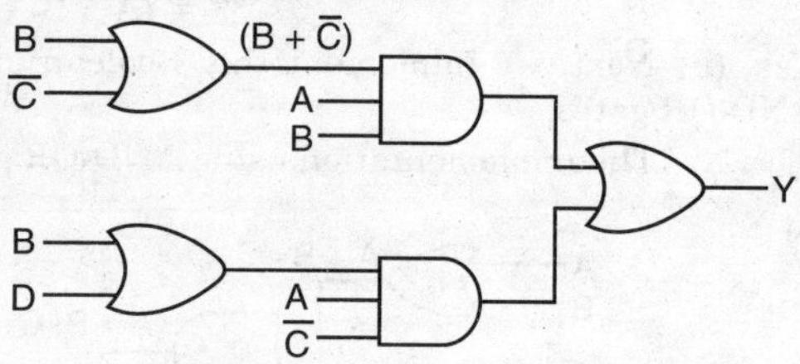

Fig. 2.105.

The implementation of this reduced Boolean equation is as shown in figure 2.105. This circuit is a tree level circuit, with 5 gates and 12 gate inputs. Thus, the effect of reduction in Boolean equation is increase in the number of levels and reduction in the number of gate inputs.

2.22.1. Multilevel NAND and NOR Implementations

We have already discussed the procedure for converting a two level AND-OR network to a NAND-NAND network and a two level OR-AND network to a NOR-NOR network. The same principle can be applied to the multilevel AND-OR or OR-AND network to convert them into multilevel NAND-NAND or NOR-NOR networks respectively.

2.23. MULTIPLE OUTPUT IMPLEMENTATIONS

(PTU, Jalandhar, Sem. Exam., 2007-08)

Some digital circuits produce more than one outputs from the same set of inputs as shown in figure 2.106. For such circuits each output is related to the inputs through a separate Boolean expression. We have to solve (simplify) each of these expressions individually. The simplification can be carried out using one of the following methods :

(i) K-map

(ii) Quine-McClusky method.

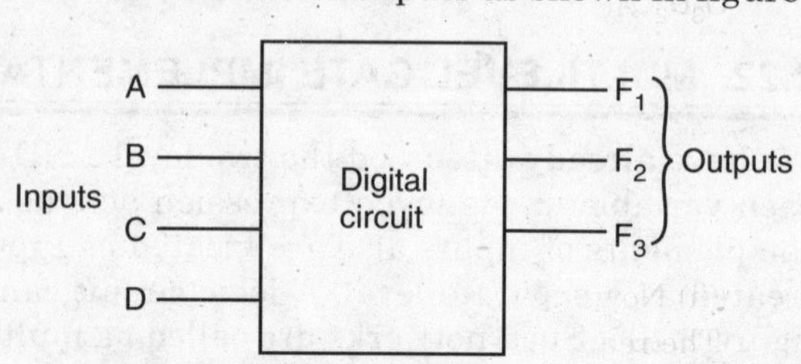

Fig. 2.106. Multiple output digital circuit

The simplified expressions for various outputs may have some common terms, which can be shared by other output functions.They reduces the required number of gates. Example 2.67 illustrates the concept of multiple output implementation.

EXAMPLE 2.67. Implement a combinational circuit that has four inputs D_3, D_2, D_1, D_0 and four outputs $G_3 \ldots G_0$.

The SOP equations relating them are as follows :

$$G_3 = \Sigma m(8, 9) + d\,(10, 11, 12, 13, 14, 15)$$
$$G_2 = \Sigma m(4, 5, 6, 7, 8, 9) + d\,(10, 11, 12, 13, 14, 15)$$
$$G_1 = \Sigma m(2, 3, 4, 5) + d\,(10, 11, 12, 13, 14, 15)$$
$$G_0 = \Sigma m(1, 2, 5, 6, 9) + d\,(10, 11, 12, 13, 14, 15)$$

Solution :

(i) First, we write K-map for each output and get simplified equation.

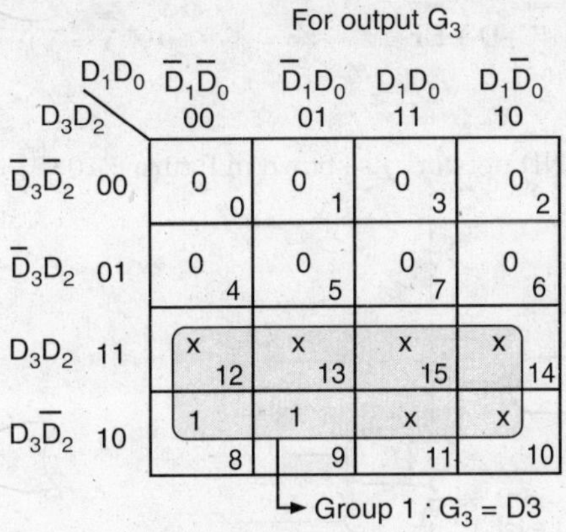

(a) K-map for G_3

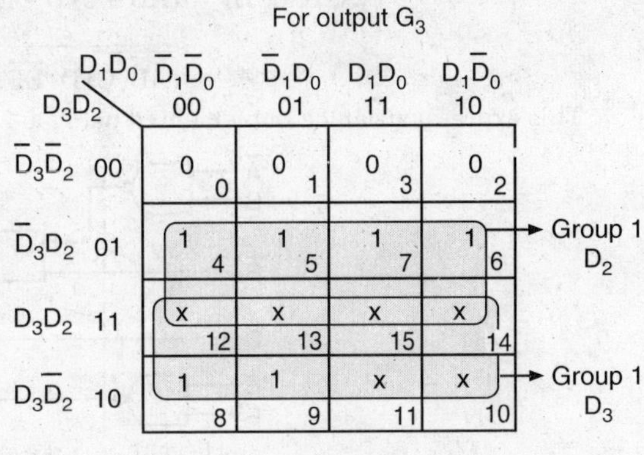

(b) K-map for G_2

Therefore, $G_1 = D_2\bar{D}_1 + \bar{D}_2 D_1 = (D_2 \oplus D_1)$

Therefore, $G_0 = (\bar{D}_1 D_0 + D_1 \bar{D}_0) = (D_1 \oplus D_0)$

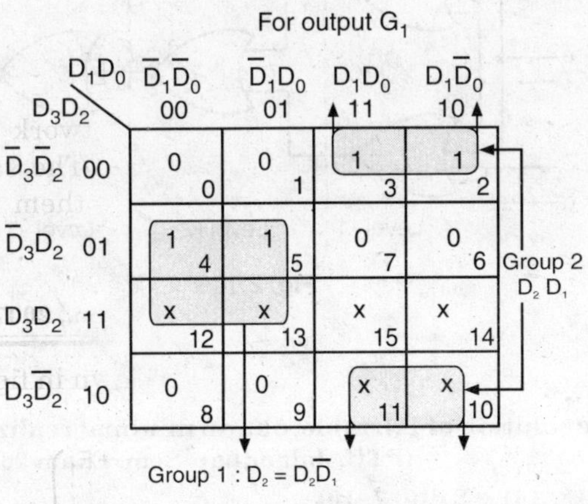

(c) K-map for G_1

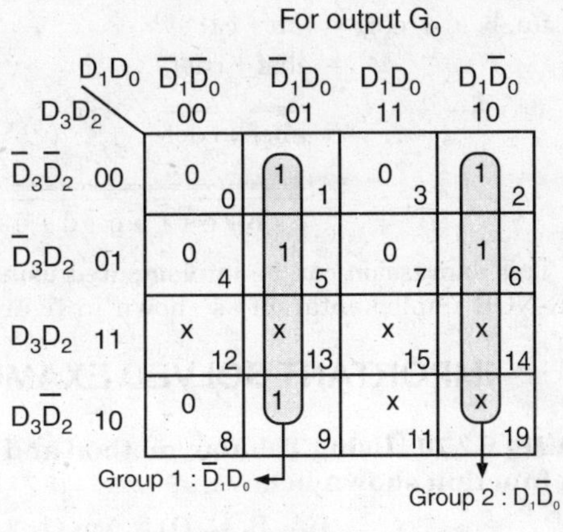

(d) K-map for G_0

Fig. 2.107.

(ii) Now, let us realize the required circuit.

The required circuit is shown in figure 2.107 (e).

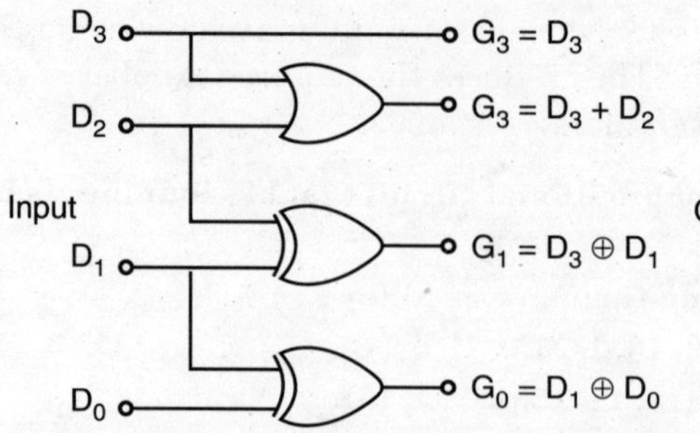

Fig. 2.107. (e) BCD to Gray code converter

EXAMPLE 2.68. Obtain a 4 level NAND network for f (A, B, C, D) = $(\bar{A}B + C)\,D + EF$.

Solution : Given expression is,

$$f\,(A, B, C, D) = (\bar{A}B + C)\,D + EF = \overline{\overline{(\bar{A}B + C)\,D + EF}} \qquad (\because\ \overline{\overline{Y}} = Y)$$

$$= \overline{\overline{(\overline{\overline{\bar{A}B}.\,C})\,D}.\,\overline{EF}}$$

This expression can be implemented using a 4 level NAND network as shown in figure 2.108.

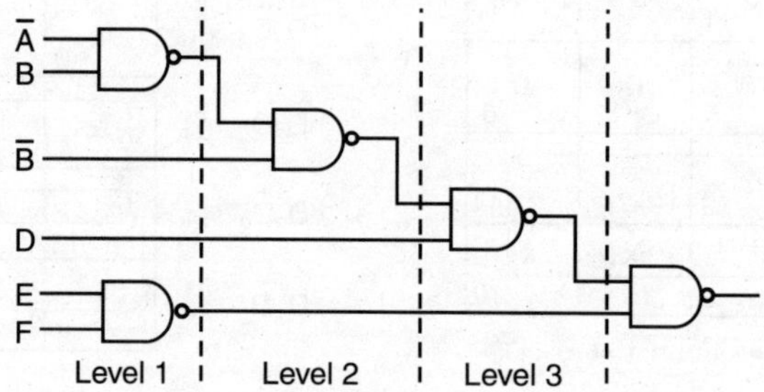

Fig. 2.108.

EXAMPLE 2.69. Obtain 3-level NOR-NOR implementation of f(a, b, c, d, e, f) = (ab + cd)ef.

Solution : Given expression is,

$$f(a, b, c, d, e, f) = (ab + cd)\,ef$$
$$= abef + cdef$$
$$= \overline{\overline{abef}} + \overline{\overline{cdef}} \qquad (\because\ \overline{\overline{Y}} = Y)$$
$$= \overline{\overline{\overline{\bar{a} + \bar{b} + \bar{e} + \bar{f}} + \overline{\bar{c} + \bar{d} + \bar{e} + \bar{f}}}}$$

This expression can be implemented using 3-level NOR-NOR implementation as shown in figure 2.109.

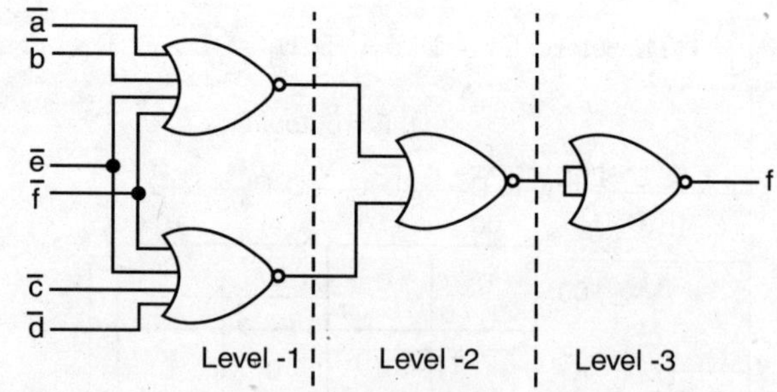

Fig. 2.109.

IMPORTANT SOLVED EXAMPLES

EXAMPLE 2.70. Using Tabular method and algebraic solution of P.I. table, obtain minimal realization of a function shown below : (PTU, Jalandhar, Sem. Exam 2008-09)

$$\mathbf{f(A, B, C, D) = \Sigma m\,(1, 2, 3, 9, 12, 13, 14) + \Sigma\, d(0, 7, 10).}$$

Solution: The given function is,

$$f(A, B, C, D) = \Sigma m\,(1, 2, 3, 9, 12, 13, 14) + \Sigma\, d(0, 7, 10)$$

(i) First, we group the minterms according to the number of 1's.

Group	Minterm	Representation in binary form				
		A	B	C	D	
0	0*	0	0	0	0	√
1	1	0	0	0	1	√
	2	0	0	1	0	√
2	3	0	0	1	1	√
	9	1	0	0	1	√
	10*	1	0	1	0	√
	12	1	1	0	0	√
3	7*	0	1	1	1	√
	13	1	1	0	1	√
	14	1	1	1	0	√

(ii) Secondly, we group the minterms to form pairs.

Group	Minterm	Representation in binary form				
		A	B	C	D	
0	0*, 1	0	0	0	–	√
	0*, 2	0	0	–	0	√
1	1, 3	0	0	–	1	√
	1, 9	–	0	0	1	$\overline{B}\,\overline{C}D$
	2, 3	0	0	1	–	√
	2, 10*	–	0	1	0	$\overline{B}C\overline{D}$
2	3, 7*	0	–	1	1	$\overline{A}CD$
	9, 13	1	–	0	1	$A\overline{C}D$
	10*, 14	1	–	1	0	$AC\overline{D}$
	12, 13	1	1	0	–	$AB\overline{C}$
	12, 14	1	1	–	0	$AB\overline{D}$

(iii) Next, we group the minterms to form groups of four.

Group	Minterm Quad	Binary representation				
		A	B	C	D	
0	0*, 1, 2, 3	0	0	–	–	$\overline{A}\,\overline{B}$
	0*, 2, 1, 3	0	0	–	–	

(iv) Then, we group the prime implicants.

$$f(A, B, C, D) = \overline{B}\,\overline{C}D + \overline{B}C\overline{D} + \overline{A}CD + A\overline{C}D + AC\overline{D} + AB\overline{C} + AB\overline{D} + \overline{A}\,\overline{B}$$

(v) Finally, we prepare the table of prime implicants.

Prime implicants	Decimal numbers (Minterms)		Given Minterms						
			1	2	3	9	12	13	14
$\overline{A}\overline{B}$	0, 1, 2, 3	√	×	×	×				
$\overline{B}\overline{C}D$	1,9		×			×			
$\overline{B}C\overline{D}$	2, 10			×					
$\overline{A}CD$	3, 7				×				
$A\overline{C}D$	9, 13	√				×		×	
$AC\overline{D}$	10, 14								×
$AB\overline{C}$	12, 13						×	×	
$AB\overline{D}$	12, 14	√					×		×
			√	√	√	√	√	√	√

Therefore, the simplified expression for the given function will be,

$$f(A, B, C, D) = \overline{A}\overline{B} + A\overline{C}D + AB\overline{D}.$$

EXAMPLE 2.71. Minimize the following using Tabular method :

$f(x, y, z, w, p) = \Sigma m\ (13, 15\ 17\ 18\ 19\ 20\ 21\ 23\ 25\ 27\ 29\ 31) + \Sigma\ d(1, 2, 12, 24)$

(PTU, Jalandhar, Sem. Exam. 2007-08)

Solution: The given function is,

$$f(x, y, z, w, p) = \Sigma m\ (13, 15, 17, 18, 19, 20, 21, 23, 25, 27, 29, 31) + \Sigma\ d(1, 2, 12, 24)$$

(i) First, we group the minterms according to the number of 1's.

Group	Minterm	Binary representation					
		x	y	z	w	p	
1	1*	0	0	0	0	1	√
	2*	0	0	0	1	0	√
2	12*	0	1	1	0	0	√
	17	1	0	0	0	1	√
	18	1	0	0	1	0	√
	20	1	0	1	0	0	√
	24*	1	1	0	0	0	√
3	13	0	1	1	0	1	√
	19	1	0	0	1	1	√
	21	1	0	1	0	1	√
	25	1	1	0	0	1	√
4	15	0	1	1	1	1	√
	23	1	0	1	1	1	√
	27	1	1	0	1	1	√
	29	1	1	1	0	1	√
5	31	1	1	1	1	1	√

(ii) Next, we group the minterms to form minterm pairs.

Group	Minterm Pairs	Binary representation					
		x	y	z	w	p	
1	1*, 17	–	0	0	0	1	$\bar{y}\bar{z}\bar{w}p$
	2*, 18	–	0	0	1	0	$\bar{y}\bar{z}w\bar{p}$
2	12*, 13	0	1	1	0	–	$\bar{x}yz\bar{w}$
	17, 19	1	0	0	–	1	√
	17, 21	1	0	–	0	1	√
	17, 25	1	–	0	0	1	√
	18, 19	1	0	0	1	–	$x\bar{y}\bar{z}w$
	20, 21	1	0	1	0	–	$x\bar{y}z\bar{w}$
	24*, 25	1	1	0	0	–	$xy\bar{z}\bar{w}$
3.	13, 15	0	1	1	–	1	√
	19, 23	1	0	–	1	1	√
	19, 27	1	–	0	1	1	√
	21, 23	1	0	1	–	1	√
	21, 29	1	–	1	0	1	√
	25, 27	1	1	0	–	1	√
	25, 29	1	1	–	0	1	√
4	15, 31	–	1	1	1	1	√
	23, 31	1	–	1	1	1	√
	27, 31	1	1	–	1	1	√
	29, 31	1	1	1	–	1	√

(iii) Then, we group the minterms to form minterm quads.

Group	Minterm Quads	Binary representation				
		x	y	z	w	p
2	17, 19, 21, 23	1	0	–	–	1
	17, 19, 25, 27	1	–	0	–	1
	17, 21, 19, 23	1	0	–	–	1
	17, 21, 25, 29	1	–	–	0	1
	17, 25, 19, 27	1	–	0	–	1
	17, 25, 21, 29	1	–	–	0	1
3	13, 15, 29, 31	–	1	1	–	1
	13, 29, 15, 31	–	1	1	–	1
	19, 23, 27, 31	1	–	–	1	1
	21, 23, 29, 31	1	–	1	–	1
	21, 29, 23, 31	1	–	1	–	1
	25, 27, 29, 31	1	1	–	–	1
	25, 29, 27, 31	1	1	–	–	1

This table can be rearranged written eliminating the similar terms are under :

Group	Minterm Quads	Binary Representation					
		x	y	z	w	p	
2	17, 19, 21, 23	1	0	–	–	1	√
	17, 19, 25, 27	1	–	0	–	1	√
	17, 21, 25, 29	1	–	–	0	1	√
3	13, 15, 29, 31	–	1	1	–	1	yzp
	19, 23, 27, 31	1	–	–	1	1	√
	21, 23, 29, 31	1	–	1	–	1	√
	25, 27, 29, 31	1	1	–	–	1	√

(iv) Now, we group the minterm quads to form octets

Group	Minterm Octets	Binary Representation					
		x	y	z	w	p	
	17, 19, 21, 23, 25, 27, 29, 31	1	–	–	–	1	xp
	17, 19, 25, 27, 21, 23, 29, 31	1	–	–	–	1	
	17, 21, 25, 29, 19, 23, 27, 31	1	–	–	–	1	

(v) Finally, we form the table of Prime implicants.

Prime implicants	Decimal numbers	Given minterms											
		13	15	17	18	19	20	21	23	25	27	29	31
$\overline{y}\,\overline{z}\,\overline{w}\,p$	1, 17			X									
$\overline{y}\,\overline{z}\,w\,\overline{p}$	2, 18				X								
$\overline{x}\,yz\,\overline{w}$	12, 13	X											
$x\,\overline{y}\,\overline{z}\,w$	18, 19 √				X	X							
$x\,\overline{y}\,z\,\overline{w}$	20, 21 √						Ⓧ	X					
$xy\,\overline{z}\,\overline{w}$	24, 25									X			
yzp	13, 15, 29, 31 √	X	Ⓧ									X	X
xp	17, 19, 21, 23, 25, 27, 29, 31 √			X		X		X	Ⓧ	X	Ⓧ	X	X
			√	√	√	√	√	√	√	√	√	√	√

The simplified expression for given function is

$$f(x, y, z, w, p) = x\,\overline{y}\,\overline{z}\,w + x\,\overline{y}\,z\,\overline{w} + y\,z\,p + z\,p\,.$$

Ans

EXAMPLE 2.72. Minimize the function in Example 2.68, using K-map in POS form.

(PTU, Jalandhar, Sem. 2005-06)

Solution: The given function is,

$$f(x, y, z, w, p) = \Sigma\, m\,(13, 15, 17, 18, 19, 20, 21, 23, 25, 27, 29, 31) + \Sigma\, d(1, 2, 12, 24)$$

The K-map for this function is as shown in figure 2.110.

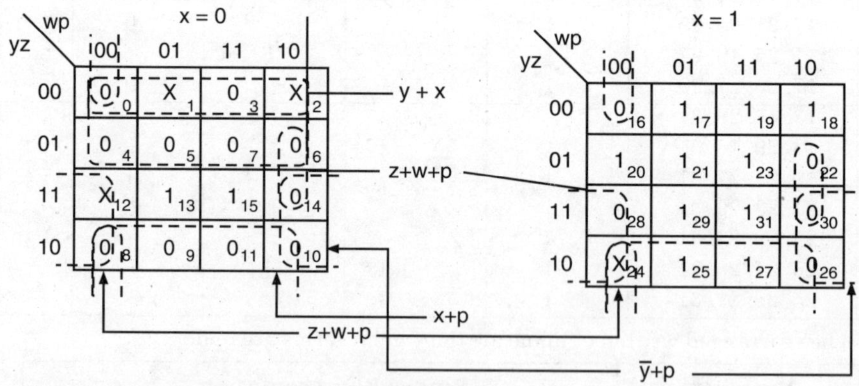

Fig. 2.110.

The simplified expression in POS form is,

$$f(x, y, z, w, p) = (y + x)\,(x + p)\,(x + z)\,(z + w + p)\,(\overline{y} + p)\,.$$

Ans.

EXAMPLE 2.73. Obtain a reduced expression for the following multiple output system, given as F_1 (A, B, C, D) = Σm (0, 1, 2, 5, 7, 8, 9, 10, 13, 15)
F2 (A, B, C, D) = Σm (0, 1, 2, 8, 10, 11, 14 ,15) and realise the minimized function using NOR gates only.

(PTU, Jalandhar, Sem. Exam., 2007-08)

Solution : Minimize F_1 and F_2 for POS expression using K-maps for both functions separately and combining 0s.

$$F_1\,(A, B, C, D) = \Sigma m\,(0, 1, 2, 5, 7, 8, 9, 10, 13, 15)$$

From K-map shown in figure 2.111, we have

$$\bar{F}_1 = B\bar{D} + \bar{B}CD$$

$$F_1 = \bar{\bar{F}}_2 = \overline{B\bar{D} + \bar{B}CD} = \overline{B\bar{D}}\,.\,\overline{\bar{B}CD} = (\bar{B} + D)(B + \bar{C} + \bar{D}).\quad \textbf{Ans.}$$

AB \ CD	$\bar{C}\bar{D}$	$\bar{C}D$	CD	$C\bar{D}$
$\bar{A}\bar{B}$	(0) 1	(1) 1	(3) 0	(2) 1
$\bar{A}B$	(4) 0	(5) 1	(7) 1	(6) 0
AB	(12) 0	(13) 1	(15) 1	(14) 0
$A\bar{B}$	(8) 1	(9) 1	(11) 0	(19) 1

Fig. 2.111.

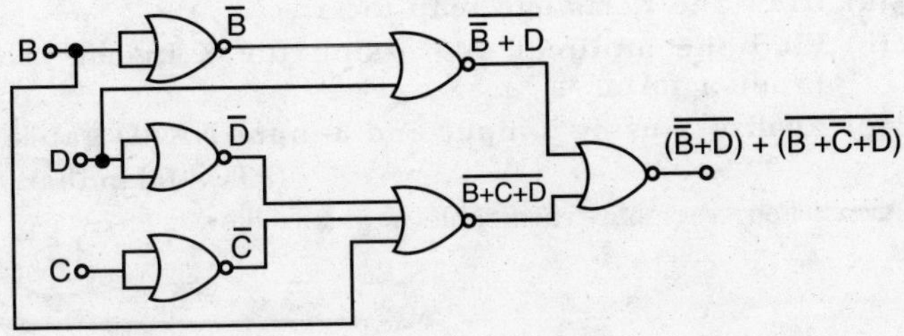

Fig. 2.112.

$$F_2\,(A, B, C, D) = \Sigma m\,(0, 1, 2, 8, 10, 11, 14, 15)$$

From K-map shown in figure 2.112, we have

$$\bar{F}_2' = B\bar{C} + \bar{A}B + \bar{A}CD + A\bar{C}D$$

$$F_2 = \bar{\bar{F}}_2 = \overline{B\bar{C} + \bar{A}B + \bar{A}CD + A\bar{C}D} = (\bar{B} + C)(A + \bar{B})(A + \bar{C} + \bar{D})(\bar{A} + C + \bar{D})\quad \textbf{Ans.}$$

AB \ CD	$\bar{C}\bar{D}$	$\bar{C}D$	CD	$C\bar{D}$
$\bar{A}\bar{B}$	(0) 1	(1) 1	(3) 0	(2) 1
$\bar{A}B$	(4) 0	(5) 0	(7) 0	(6) 0
AB	(12) 0	(13) 0	(15) 1	(14) 1
$A\bar{B}$	(8) 1	(9) 0	(11) 1	(10) 1

Fig. 2.113.

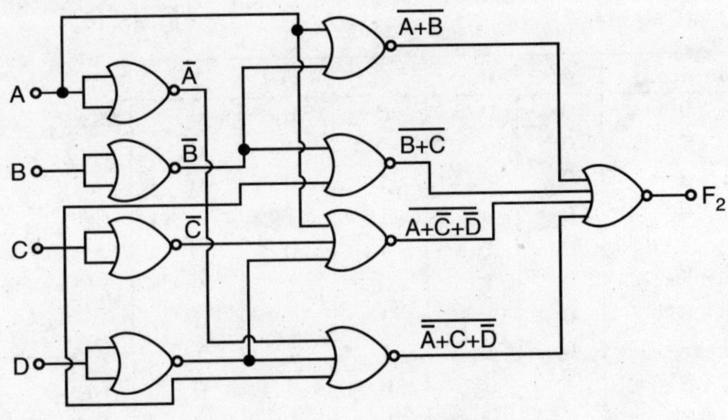

Fig. 2.114.

EXAMPLE 2.74. A truth table has output 1s for these inputs

$$\bar{A}\bar{B}CD = 0011$$

$$\bar{A}BC\bar{D} = 0110$$

$$A\bar{B}\bar{C}\bar{D} = 1000$$

$$A\bar{B}\bar{C}\bar{D} = 1000$$

$$AB\bar{C}\bar{D} = 1100$$

and 0s for other inputs.

Draw the Karnaugh-map and find the simplified Boolean equation for truth table.

(PTU, Jalandhar, Elec. Engineering-II 1992)

Solution : K-map for the given truth table is shown in figure 2.117. From K-map, the simplified Boolean expression comes out to be

$\bar{A}\bar{B}CD + \bar{A}BC\bar{D} + A\bar{C}\bar{D}$. **Ans.**

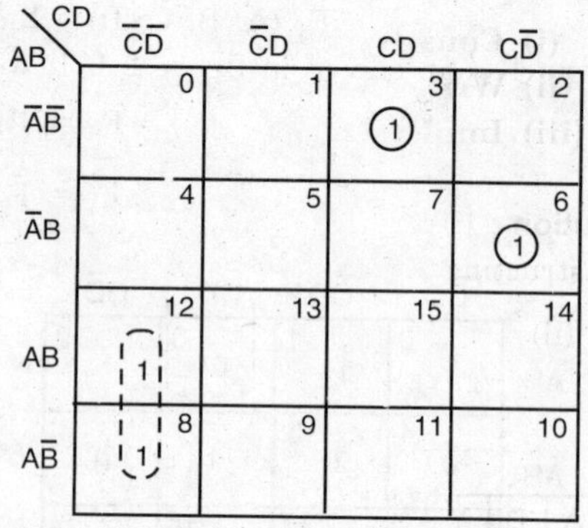

Fig. 2.115.

EXAMPLE 2.75. A digital system has four bits of a 4-bit word 'ABCD' as inputs. The output Y is equal to 1 when any two adjacent bits are 1, or any three or all four bits are 1.

(a) Draw the Karnaugh map for Y.

(b) Find the minimal expression for Y in the 'Sum-of Product form.'

(c) Realise Y using 2-input and 3-input NAND gates only.

(PTU, Jalandhar, 1992)

Solution : Four-variable truth table is given below :

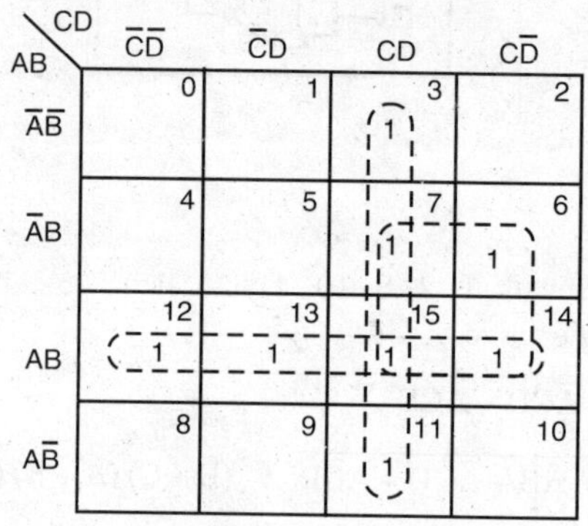

Fig. 2.116.

Decimal	A	B	C	D	Y
0	0	0	0	0	0
1	0	0	0	1	0
2	0	0	1	0	0
3	0	0	1	1	1
4	0	1	0	0	0
5	0	1	0	1	0
6	0	1	1	0	1
7	0	1	1	1	1
8	1	0	0	0	0
9	1	0	0	1	0
10	1	0	1	0	0
11	1	0	1	1	1
12	1	1	0	0	1
13	1	1	0	1	1
14	1	1	1	0	1
15	1	1	1	1	1

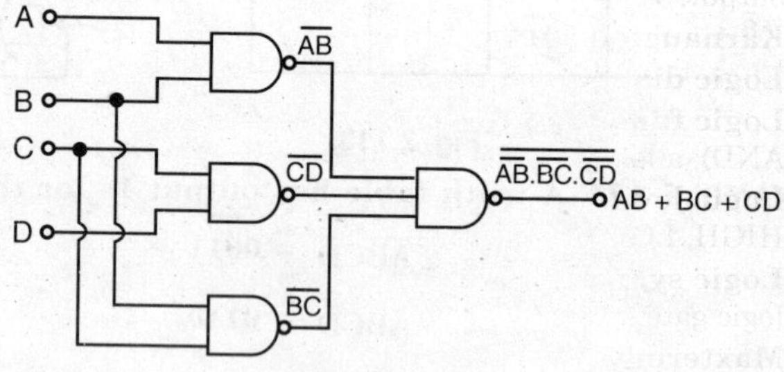

Fig. 2.117.

K-map is shown in figure 2.116.

From K-map shown in figure 2.116, we have

$$Y = AB + CD + BC$$

To implement with NAND gate double complement Y

$$\bar{\bar{Y}} = \bar{\bar{A}}\,\bar{\bar{B}} + \bar{\bar{C}}\,\bar{\bar{D}} + \bar{\bar{B}}\,\bar{\bar{C}} = \overline{AB}\,.\,\overline{BC}\,.\,\overline{CD} \text{ Ans.}$$

EXAMPLE 2.76. An equality detector gives the output Y = 1, if both the inputs A and B are either 1 or 0.

(i) Construct the truth table.
(ii) Write the Boolean expression of Y.
(iii) Implement the circuit using NAND gates only.

(U.P.S.C. I.E.S., Electrical Engineering II, 2004)

Solution : (i) Truth table is given in figure 2.118.
Constructing K-map, we have Boolean equation for Y as

(ii) $$Y = \bar{A}\bar{B} + AB$$

A	B	Y
0	0	1
1	1	1
0	1	0
1	0	0

Fig. 2.118.

A \ B	0	1
0	1	0
1	0	1

Fig. 2.119.

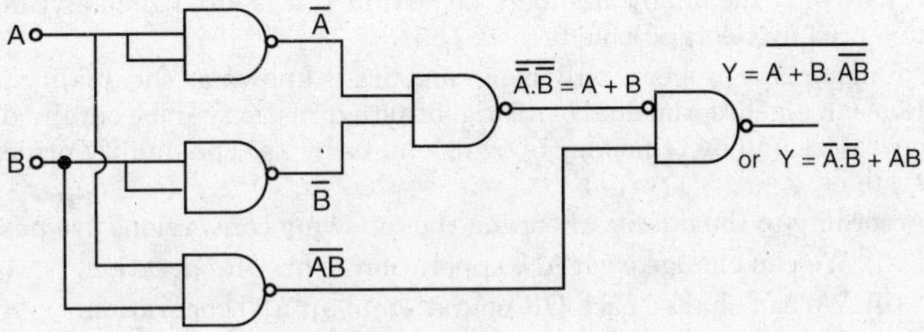

Fig. 2.120.

GLOSSARY

1. **2s complement :** Method of subtraction using a 2s complement subtrahend added subtraction to the minuend. Used so adders can be used to perform subtraction.
2. **Boolean algebra :** Mathematical system for representing logical statements. Very useful in digital electronics.
3. **Boolean expression :** Mathematical representation of a logic function. Function could also be described using a truth table of logic circuit diagram.
4. **Bubble :** On logic symbol, it means an active LOW input or output.
5. **Enable :** To activate a function or input to a digital circuit. The opposite of disable.
6. **Gate :** Basic combinational logic device which performs a specific logic function (AND, OR, NOT, NAND, NOR).
7. **Generic array logic :** A specific programmable logic device (PLD) with an array of ANDs that can be reprogrammed, a fixed array of ORs gates.
8. **GND :** Label for negative of power supply in TTL ICs and some CMOS ICs. Common ground.
9. **Increment :** To increase the count by 1.
10. **Input/Output :** A connection to a digital device that can be programmed to serve as either an input or output. Very common on many complex devices including microcontrollers.
11. **Karnaugh-map :** A graphic method of reducing Boolean expressions to simpler forms.
12. **Logic diagram :** A schematic showing interconnected between logic devices like gates, flip-flops, etc.
13. **Logic function :** The logical task needed to be performed. It might be represented by the name (such as AND), a logic symbol, a Boolean expression (such as AB = Y), and/or a truth table.
14. **Logic levels :** In digital electronics, voltage ranges at which inputs to digital devices interpret signal as HIGH, LOW or undefined. Voltage thresholds may be different for various logic families.
15. **Logic symbols :** Two systems are used in the U.S. Traditional representations using the unique shaped logic gate symbols. The newer IEEE symbols using rectangle boxes.
16. **Maxterm Boolean expression :** See product-of-sums.
17. **Product-of-sums :** The form of a Boolean expression that looks like this : (A + B) (C + D) = Y. Implemented using an OR/AND logic diagram. Also called a maxterm Boolean expression.
18. **Sum of products :** The form of Boolean expression that looks like this : AB + CD = Y. Implemented using an AND/OR logic diagram. Also called a minterm Boolean expression.

SUMMARY

- In order to deal mathematically with the digital circuits, we need to use the Boolean algebra. Boolean algebra may be defined with the help of the following aspects :

(i) A set of elements (ii) A set of operators
(iii) A number of unproved axioms or postulates.

- The postulates of a mathematical system form the basic assumptions from which we can deduce the rules, theorems and also various properties of a system. The most important postulates may be listed as under :
  (i) Closure (ii) Associative law
  (iii) Commutative law (iv) Identity element
  (v) Inverse (vi) Distributive law.
- As a matter of fact, Boolean Algebra is used to analyze and simplify the digital (i.e., logic) circuits. Because, it uses only the binary numbers i.e., 0 and 1, it is also called as Binary Algebra or Logical Algebra. It was invented by George Boole in year 1854.
- An important property of Boolean algebra is known as the duality principle. It states that in a two valued Boolean algebra, the dual of an algebraic expression can be obtained simply by interchanging OR and AND operators and by replacing 1s by 0s and 0s by 1s. The duality principle has several applications in digital systems.
- According to the duality theorem, the following conversions are possible in a given Boolean expression :
  (i) We can change each AND operation to an OR operation.
  (ii) We can change each OR operation to an AND operation.
  (iii) We can complement any 1 or 0 appearing in the expression.
- As a matter of fact, duality theorem is sometimes useful in creating new expressions from the given Boolean expressions. As an example, if the given expression is A + 1 = 1, then, we replace the OR (+) operation by AND (·) operation and take complement of 1 to write the dual of the given relation as under :

$$A \cdot 0 = 0$$

The dual of A(B + C) = AB + AC is given by

$$A + (B \cdot C) = (A + B) \cdot (A + C).$$

- The two important theorems have been suggested by DeMorgan. These two theorems are extremely useful in Boolean algebra. These may be described as under :

Theorem 1 : $\overline{AB} = \overline{A} + \overline{B}$; NAND = Bubbled OR

This theorem states that the complement of a product is equal to addition of the complements.

- **Theorem 2 :** $\overline{A+B} = \overline{A} \cdot \overline{B}$: NOR = Bubbled AND

This theorem states that the complement of a sum is equal to product of complements.

- The procedure to be followed for converting a general SOP expression into a standard SOP expression may be described as under :
  (i) For each term, we find the missing literal.
  (ii) Then, we AND term with the term formed by ORing the missing literal and its complement.
- The conversion of given POS expression into standard POS form can be obtained by following the steps given below :
  (i) For each term, we find the missing literal.
  (ii) Then, we OR each term with the formed by ANDing the missing literal in that term with its complement.
  (iii) Lastly, we simplify the expression to get the standard POS form.
- We know that truth table representation of a logic expression is possible. As an example, the expression Y $= AB + \overline{A}\overline{B}$ which is the Boolean expression for an EX-NOR gate can also be represented using a truth table. Hence, it is possible to obtain the logic expression in the standard SOP or POS form if a truth table is given to us.
- As a matter of fact, gate-level minimization refers to the design task of finding an optimal gate-level implementation of the Boolean functions describing a digital circuit. This task is well understood, but is difficult to execute by manual techniques when the logic has more than a few inputs. Fortunately, computer-based logic synthesis tools can minimize a large set of Boolean equations efficiently and quickly. Further, it is important that a designer understand the underlying mathematical description and solution of the problem.
- This is another simplification technique to reduce the Boolean expression. It overcomes all the drawbacks of the algebraic simplification technique. K-map (short form of Karnaugh map) is a graphical method of simplifying a Boolean expression. K-map is a graphical chart which contains boxes. The information contained

in a truth table or available in the SOP or POS form is represented on a K-map. The K-map method is a systematic approach to simplification of Boolean expression. K-maps can be written for 2, 3, 4 ..... upto 6 variables. Beyond that the K-map technique becomes very cumbersome.

- The logical expression is standard SOP form can be represented on a K-map by simply entering 1's in the cells (boxes) of the K-map corresponding to each minterm present in the expression. The remaining cells (boxes) are filled with zeros. Following example will illustrate the concept of standard SOP expression of K-map.
- If we group four 1's from the adjacent cells of K-map, then the group is called as a Quad. In such a quad, two variables asssociated with the minterms are same and the other two are not the same.
- If all the 1's in a group are already involved in some other groups, then that group if called as a redundant group. A redundant group has to be eliminated, because it increases the number of gates required.
- For SOP form, we enter 1's corresponding to the combinations of input variables which produce a high output. Also, we enter 0's in the remaining cells of the K-map. For the POS form, we enter 0's corresponding to the combinations of inputs which product a low output and enter 1's in the remaining cells of the K-map. However, it is not always true that the cells not containing 1's (in SOP) will contain 0's, because some combinations of input variable do not occur. Also, for some functions, the outputs corresponding to certain combinations of input variables do not matter. In such situations, we have a freedom to assume a 0 of 1 as output for each of these combinations. These conditions are known as the Don't care conditions and in the K-map it is represented as × (cross) mark in the corresponding cell.

SHORT QUESTIONS WITH ANSWER

Q. 1. What are the postulates of a mathematical system?

Ans. The postulates of a mathematical system form the basic assumptions from which we can deduce the rules, theorems and also various properties of a system. The most important postulates may be listed as under :

(i) Closure (ii) Associative law
(iii) Commutative law (iv) Identity element
(v) Inverse (vi) Distributive law.

Q.2. How Boolean Algebra differs with an ordinary algebra ?

Ans. As a matter of fact, Boolean Algebra is used to analyze and simplify the digital (i.e., logic) circuits. Because, it uses only the binary numbers i.e., 0 and 1, it is also called as Binary Algebra or Logical Algebra. It was invented by George Boole in year 1854. The rules of Boolean Algebra are different from those of the conventional Algebra in the following manner :

(i) Symbols used in Boolean algebra (usually letters) do not represent numerical values.
(ii) Arithmetic operations (addition, subtraction, division etc.) are not performed in Boolean algebra. Also, there are no fractions, negative numbers, square, square root, logarithms, imaginary numbers etc.
(iii) The third and most important point is that Boolean Algebra allows only two possible values (0 to 1) for any variable.

Q.3. What are Rules in Boolean Algebra ?

Ans. There are some rules which are to be followed while using a Boolean algebra, these rules may be described as under :

(i) Variables used can have only two values. Binary 1 for HIGH and Binary 0 for LOW.
(ii) Complement of a variable is represent by a overbar (–). Thus, complement of variable B is represented as $\overline{B}$. Therefore, if B = 0 then $\overline{B} = 1$ if B = 1 then $\overline{B} = 0$.
(iii) ORing of the variables is represented by a plus (+) sign between them. For example ORing of A, B, C is represented as A + B + C.
(iv) Logical ANDing of the two or more variables is represented by writing a dot between them such as $A \cdot B \cdot C \cdot D \cdot E$. Sometimes, the dot may be omitted like ABCDE.

Q. 4. Explain about Duality Principle.

Ans. An important property of Boolean algebra is known as the duality principle. It states that in a two valued Boolean algebra, the dual of an algebraic expression can be obtained simply by interchanging OR and AND operators and by replacing 1s by 0s and 0s by 1s. The duality principle has several applications in digital systems.

Q.5. What is meant by Duality Theorem ?

Ans. According to the duality theorem, the following conversions are possible in a given Boolean expression :

(i) We can change each AND operation to an OR operation.

(ii) We can change each OR operation to an AND operation.

(iii) We can complement any 1 or 0 appearing in the expression.

As a matter of fact, duality theorem is sometimes useful in creating new expressions from the given Boolean expressions. As an example, if the given expression is A + 1 = 1, then, we replace the OR (+) operation by AND (·) operation and take complement of 1 to write the dual of the given relation as under :

$$A \cdot 0 = 0$$

The dual of A(B + C) = AB + AC is given by

$$A + (B \cdot C) = (A + B) \cdot (A + C).$$

Q.6. How can you simplify the Boolean Functions ?

Ans. The various techniques used for simplifying the Boolean functions may be listed as under:

(i) Algebraic method

(ii) Karnaugh-map simplification.

(iii) Quine-McCluskey method, and

(iv) Variable entered mapping (VEM) technique.

Q.7. What are the various methods to Simplify Boolean Expression ?

Ans. There are following methods to simplify Boolean expressions :

(i) Algebraic simplification

(ii) Simplification using Karnaugh maps.

(iii) Simplification using Quine-McCluskey method.

Q.8. What is Boolean algebra?

Ans. Boolean algebra is a mathematical system of logic in which truth functions are expressed as symbols and then these symbols are manipulated to arrive at conclusion.

Q.9. What are the basic logic elements?

Ans. Basic logic elements are NOT gate, AND gate, OR gate and the flip-flop.

Q.10. What is a truth table?

Ans. Truth table is a table that gives outputs for all possible combinations of inputs to a logic circuit.

Q.11. Define positive logic and negative logic.

Ans. If the higher of the two voltages represents a 1 and the lower voltage represents a, 0 the logic is called a positive logic. On the other hand, if the lower voltage represents a 1 and the higher voltage a 0, we have a negative logic.

Q.12. What is pulse logic system ?

Ans. A logic system in which a bit is recognised by the presence or absence of a pulse is called a pulse or dynamic logic system.

Q.13. Why is a two-input NAND gate called universal gate?

Ans. NAND gate is called universal gate because any digital system can be implemented with the NAND gate. Sequential and combinational circuits can be constructed with these gates because element circuits like flip-flop can be constructed from two NAND gates connected back-to-back. NAND gates are common in hardware because they are easily available in the ICs form. A NAND gate is in fact a NOT-AND gate. It can be obtained by connecting a NOT gate in the output of an AND gate.

Q.14. What is associate law?

Ans. Associate law is a law of addition and multiplication and according to this law grouping of the variable in the ORing or ANDing or several variables is immaterial and the results obtained are the same.

Q.15. Distinguish between sum-of-products and product-of-sums.

Ans. Sum-of-products (SOP) is a logic expression in the form of ANDed terms ORed together where as product-of-sums (POS) is a logic expression in the form of ORed terms ANDed together.

Q.16. What is a minterm ?

Ans. A term containing literals corresponding to all the variables in ANDed form is known as a minterm.

Q.17. What are maxterms?

Ans. A term containing literals corresponding to all the variables in ORed form is known as a maxterm.

Q.18. How many output functions are possible when we have n input variables?

Ans. For n binary variables we can have 2^n distinct minterms. Since the function can be either 0 or 1 for each minterm, and since there are 2^n minterms, one can calculate the possible functions that can be formed with n variables to be 2^n. Thus, for n = 3

Number of minterms = $2^n = 2^3 = 8$

and number of possible functions = $2.2^n = 2 \times 2^3 = 16$.

Q.19. What is the difference between canonical form and standard form?

Ans. In canonical form, a Boolean function is expressed as sum of minterms or product of maxterms and is obtained by reading a function from the truth table. Canonical forms may or may not contain the least numbers of literals because by definition, each maxterm or minterm must contain all variables, complemented or uncomplemented. In standard form, the terms of function may contain one, two or any number of literals. Standard form is expressed as sum of products or product of sums.

Q.20. What is the use of DeMorgan's theorem?

Ans. DeMorgan's theorems are extremely useful in simplification of Boolean expressions in which a sum or product of variables is inverted.

Q.21. What is Karnaugh-map?

Ans. Karnaugh-map is essentially a geometrical pattern of squares that carry the labels of the logic variables in both normal and complemented forms.

Q.22. What is the use of a Karnaugh-map?

Ans. Karnaugh-map is used for simplification and manipulation of Boolean expressions.

Q.23. State the limitation of Karnaugh's-map?

Ans. The limitation of Karnaugh's-map is that beyond 6 variables it becomes very cumbersome and so it is generally used up to 6 variables.

REVIEW QUESTIONS

1. Explain the following terms :
   (i) Product term (ii) Sum term
2. Explain the different methods used to simplify the Boolean function.
3. Explain the term SOP and POS related to Boolean function.
4. Convert the following equation into standard POS form $Y = (A + B)(A + C)(B + \bar{C})$.
5. Explain the K-map reduction technique.
6. State and prove DeMorgan's first and second theorem.
7. Explain the rules of Boolean algebra.
8. Explain the term SOP and POS related to Boolean function.
9. Convert the following equation into standard POS form $Y = (A + B)(A + C)(B + \bar{C})$.
10. Explain the K-map reduction technique.

NUMERICAL PROBLEMS

1. Using Tabular method and algebraic solution of P.I. table, obtain minimal realization of a function shown below :
   $$f(A, B, C, D) = \Sigma(1, 2, 3, 9, 12, 13, 14) + \Sigma\phi(0, 7, 10)$$
2. What are prime implicants and explain their role in Boolean function representation in its minimal form.
3. Minmize the following using Tabular method :
   $$f(x, y, z, w, p) = \Sigma m(13, 15, 17, 18, 19, 20, 21, 23, 25, 27, 29, 31) + \Sigma\, d(1, 2, 12, 24)$$
4. Minimize the function in Problem 3, using K-map is SOP form.
5. Minmize the function in Problem 3, using K-map in POS form.

❑❑❑

3

LOGIC GATES

Syllabus

Logic Gates: AND, OR, NOT, NAND, NOR, Exclusive-OR and Exclusive-NOR Gates.

Chapter Outline

- Introduction
- Logic Operators
- Logic Gates
- Universal Gates

3.1 INTRODUCTION

As a matter of fact, logic gates are the basic elements that make up a digital system. The electronic gate is a circuit that is able to operate on a number of binary inputs in order to perform a particular logical function. In other words, we can say that logic gates are the devices used as basic building blocks of all the digital circuits. The basic logic gates are NOT, AND and OR along with NOR, NAND, EX–OR etc. We have to use different laws, rules and theorems to analyze the digital circuits. The Boolean algebra developed by Charles Boole way back in 1884 is used for representing, simplifying and analyzing the logic circuits. The gate is a digital circuit with one or more input voltages but only one output voltage. By connecting the different gates in different ways, we can build circuits that can perform arithmetic and other functions associated with the human brain. Because they simulate mental processes, gates are often called as logic circuits. The operation of a logic gate can be easily understood with the help of a truth table. A truth table is a table which shows all the input-output possibilities of the inputs.

3.2 LOGIC OPERATORS

In the decimal system which is used in our day–to–day life, the arithmetic operations such as addition, subtraction, multiplication, division, square, square root etc. are used to solve arithmetic equations.

Similarly, to solve or simplify the logical expressions, used in digital circuits, we need to use logical operators. The three main logic operators may be listed as under :

(i) AND operator.

(ii) OR operator.

(iii) NOT operator (i.e., invert).

Let us discuss all these operators one by one in details.

3.2.1 AND Operator

The AND operator represents logical multiplication. It is denoted by a dot between the two variables to be multiplied i.e.,

$$A \,.\, B \rightarrow \text{Logical multiplication}$$

However, sometimes, this dot is not used and we denote the logical multiplication of A and B as AB. Thus, if A is to be multiplied with B in a logic circuit then the multiplication is represented as A . B, AB and it is to be read as A AND B.

3.2.2 OR Operator

The OR operator represents logical addition. It is denoted by a (+) sign between the two variables to be added. If A and B are to be added in a logic circuit it is represented as under:

$$A + B \rightarrow \text{Logical addition}$$

And, it is to be read as A OR B.

3.2.3 NOT Operator (i.e., Inverter)

The NOT operator represents a logical inversion or complementing. The NOT operation is denoted by a bar (–) over the variable to be inverted. For example, if A is to be inverted then the process of inversion is denoted by $\overline{A}$ and it is to be read as NOT A.

$$\overline{A} = \text{NOT A}$$

AUGUSTUS DeMORGAN (1806-1871)

Augustus DeMorgan (1806-1871) was born in Madras Province, India. He taught mathematics at the University of London for 30 years, and published many texts on arithmetic, algebra, trigonometry, and calculus, and important treatises on the theory of probability and formal logic. DeMorgan contributed the method of changing from sum-of-products to product-of-sums.

DO YOU KNOW?

Logic gates have two or more inputs and only one output except for the NOT gate, which has only one input. The output signal appears only for certain combinations of the input signals.

3.3 LOGIC GATES

(PTU, Jalandhar, Sem. Exam., 2007-08)

1. Definition

Logic gates are the basic building blocks of any digital system. It is an electronic circuit having one or more than one inputs and only one output. The relationship between the input and the output is based on a certain logic. Based on this logic, the gates are named as NOT gate, AND gate, OR gate, NAND gate, NOR gate etc.

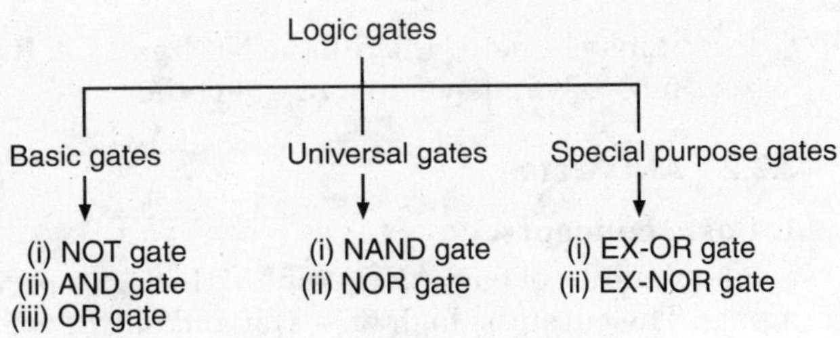

Fig. 3.1. Classification of gates

2. Truth table

The operation of a logic gate can be best understood with the help of a table called *Truth Table*. The truth table consists of all the possible combinations of the inputs and the corresponding state of output of a logic gate.

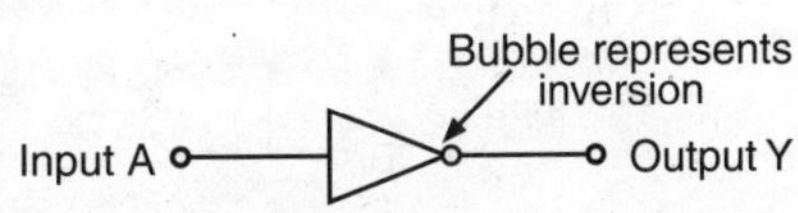

Boolean expression : $Y = \overline{A}$

(a) Symbol

3. Boolean expression

The relationship between the inputs and the outputs of a

gate can be expressed mathematically by means of the Boolean Expression. Let us now discuss the operation of various logic gates.

4. Classification of logic gates.

As shown in figure 3.1, the logic gates are classified into three categories namely, the basic gates, the universal gates and the special purpose gates.

3.3.1 NOT Gate or Inverter

1. Basic concepts

The NOT gate or Inverter is a logic gate having one input (A) and one output (Y). Its symbol and truth table are shown in figure 3.2.

Truth table

Input A	Output Y
0	1
1	0

(b)

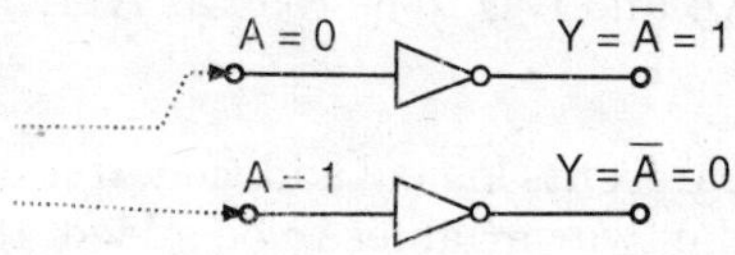

(c)

Fig. 3.2. Symbol, truth table and equivalent circuit of a NOT gate

2. Few points about the NOT gate

(i) The NOT gate is also known as an Inverter because its output is the inverted version or complement of its input. This has been shown in the truth table of NOT gate.

(ii) The bubble (o) in the symbol of a NOT gate indicates inversion.

(iii) The operation of a NOT gate for various input combinations has been shown in figure 3.2 (c).

3. Standard Package

As a matter of fact, the NOT gate is available in the monolithic integrated circuit form. Also, IC 7404 is a 14 pin TTL integrated circuit with six inverters inside. It is therefore called as a hex inverter. The pin configuration of IC 7404 has been shown in figure 3.3.

Here, it may be noted that all the NOT gates in IC 7404 are completely independent of each other.

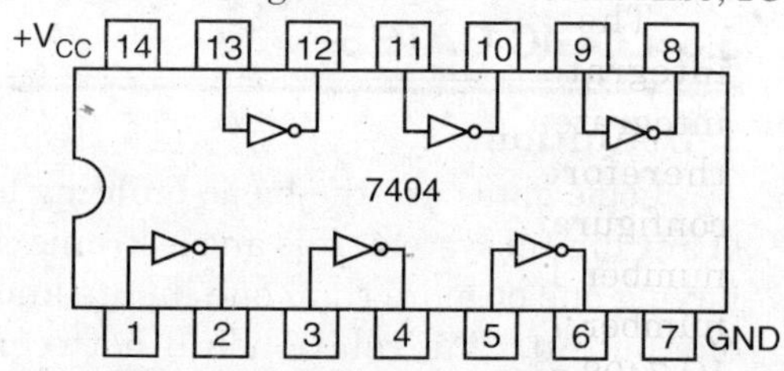

Fig. 3.3. Pin configuration of TTL IC 7404

3.3.2 AND Gate

1. Basic concepts

As a matter of fact, AND is one of the logic operators. It performs the logical multiplication on its inputs. The output is high (Y = 1) if and only if all the inputs to the gate are high (1). The output is low (0), if atleast one of the inputs is low (0). AND gate can have two or more inputs and only one output.

2. Symbols

Two input AND gate

(i) Boolean equation $Y = A \cdot B$

Three input AND gate

(ii) Boolean equation $Y = A \cdot B \cdot C$

(a)

Those who dare to fail miserably can achieve greatly. — **Robert F. Kennedy**

3. Truth table

S.No.	Inputs		Output
	A	B	Y
1.	0	0	0
2.	0	1	0
3.	1	0	0
4.	1	1	1

(b) Truth table of a two input AND gate

S.No.	Inputs			Output Y
	A	B	C	
1	0	0	0	0
2	0	0	1	0
3	0	1	0	0
4	0	1	1	0
5	1	0	0	0
6	1	0	1	0
7	1	0	0	0
8	1	1	1	1

(c) Truth table for a 3-input AND gate

Fig. 3.4. Symbols, Boolean expression and truth tables for an AND gate

4. Boolean expressions

(i) The expression relating the inputs and output of a gate is called as the Boolean expression. The Boolean expression for an AND gate is,

$$Y = A \, . \, B$$

where the dot between A and B represents multiplication.

(ii) By substituting different values of A and B into the Boolean expression, we can get the corresponding state of output. This is how we can verify the truth table of the gate.

DO YOU KNOW?

The logic gates are the building blocks of hardware which are available in the form of various IC families.

5. Standard package

The AND gate is available in the monolithic integrated circuit form. IC 7408 is a 14 pin TTL integrated circuit with four AND gates inside. It is therefore called as Quad AND gate IC. The pin configuration of IC 7408 is shown in figure 3.5. Pin number 14 is to be connected to + 5 V supply and pin number 7 is to be grounded. The four AND gates inside IC 7408 are completely independent of each other.

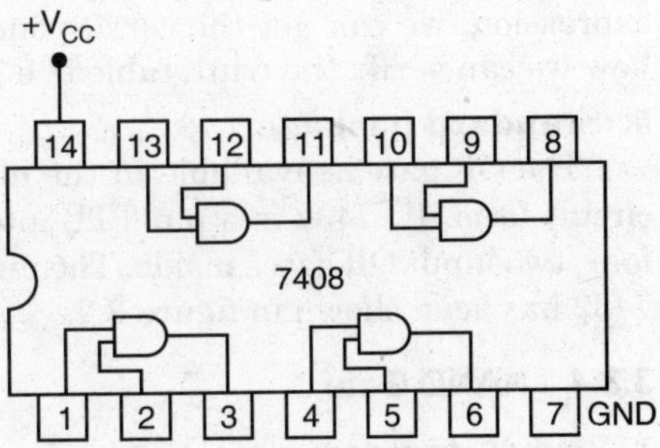

Fig. 3.5. Pin configuration of TTL IC 7408

3.3.3 OR Gate

1. Definition

As discussed earlier, OR is one of the logic operators.

An OR gate performs the logical addition on its inputs therefore its output will be high (1) if any one or both the inputs are high (1). Its output will be low (0) if and only if both the inputs are simultaneously low (0).

2. Symbols

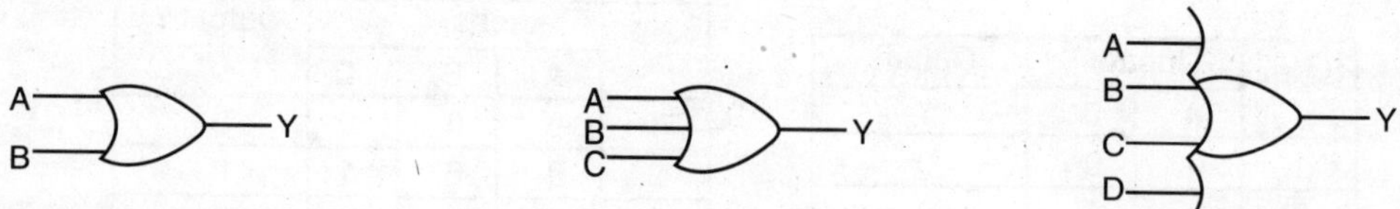

Two input OR gate
Boolean equation
$Y = A + B$
(a)

Three input OR gate
Boolean equation
$Y = A + B + C$
(b)

Four input OR gate
Boolean equation
$Y = A + B + C + D$
(c)

Fig. 3.6.

3. Truth table

Truth table for two input OR gate

S.No.	Inputs		Output
	A	B	Y
1.	0	0	0
2.	0	1	1
3.	1	0	1
4.	1	1	1

Truth table for three input OR gate

S.No.	Inputs			Output Y
	A	B	C	
1.	0	0	0	0
2.	0	0	1	1
3.	0	1	0	1
4.	0	1	1	1
5.	1	0	0	1
6.	1	0	1	1
7.	1	1	0	1
8.	1	1	1	1

4. Boolean Expression

The Boolean expression for a two input OR gate is,

$$Y = A + B$$

By substituting different values of A and B into this expression, we can get the corresponding output. This is how we can verify the truth table of a gate.

5. Standard package

The OR gate is available in the monolithic integrated circuit form. IC 7432 is a pin TTL integrated circuit with four, two input OR gates inside. The pin configuration of IC 7432 has been shown in figure 3.7.

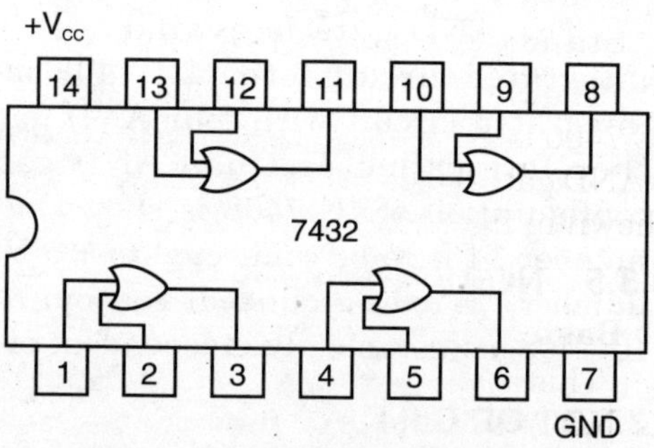

Fig. 3.7. Pin configuration of IC 7432

3.3.4 NAND Gate

1. Basic Concepts

The term NAND can be split as NOT–AND which means that the NAND operation can be implemented with the **combination of an AND gate and a NOT gate i.e., inverter**. Thus a NAND gate is equivalent to an AND gate followed by an inverter as shown in figure 3.8 (b). The symbol of a two input NAND gate is shown in figure 3.8 (a) where a bubble (o) on the output side represents inversion. The truth table of a two input NAND gate is shown in figure 3.8 (c), which shows that the output is low (0) if and only if both the inputs are high (1) simultaneously. For all other input combinations the output voltage will be high (1). A NAND gate is also known as an *Universal Gate* because we can construct AND, OR and NOT gates using only NAND gates.

2. Symbols

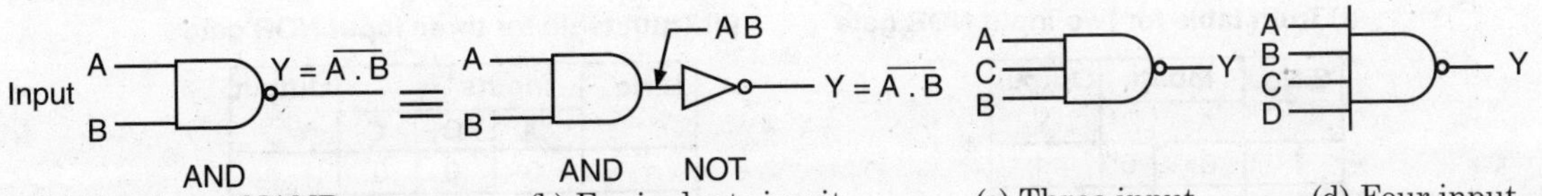

(a) Two input NAND gate (b) Equivalent circuit (c) Three input NAND gate (d) Four input NAND gate

Fig. 3.8.

3. Truth tables

(i) Truth tables for two input NAND gate

S.No.	Inputs		Output
	A	B	Y
1.	0	0	1
2.	0	1	1
3.	1	0	1
4.	1	1	0

(ii) Truth table for three input NAND gate

S.No.	A	B	C	Y
1.	0	0	0	1
2.	0	0	1	1
3.	0	1	0	1
4.	0	1	1	1
5.	1	0	0	1
6.	1	0	1	1
7.	1	1	0	1
8.	1	1	1	0

4. Boolean expression

The Boolean expression for the NAND gate is,

$$Y = \overline{A.B}$$

Here, A . B represents an AND gate while the bar represents an inversion. By substituting various combinations of the inputs, it is possible to obtain the state of output. This is how we can verify the truth table.

5. Standard package

The NAND gate is available in the monolithic IC form. IC 7400 is a 14 pin TTL integrated circuit four, two input NAND gates inside. The pin configuration of IC 7400 has been shown in figure 3.9.

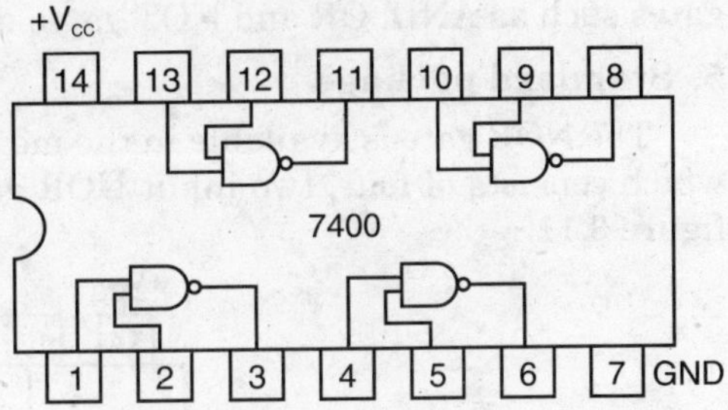

Fig. 3.9. Pin configuration of IC 7400

DO YOU KNOW?

Each gate has a distinct logic symbol and its operation can be described by means of an algebraic function.

3.3.5 NOR Gate

1. Basic concepts

This is another universal gate. The word NOR can be split as NOT–OR which means that a NOR operation can be implemented with the combination of an OR gate and a NOT gate *i.e.*, inverter. Thus, a NOR gate is equivalent to an OR gate followed by an inverter as shown in figure 3.10. The symbol of a two input NOR gate is shown in figure 3.10, where bubble (o) on the output side represents inversion. Also, the output of a NOR gate is high (1) if and only if all its inputs are low (0) simultaneously. The output of a NOR gate is low (0) if any one or all the inputs are at high (1) level.

2. Symbols

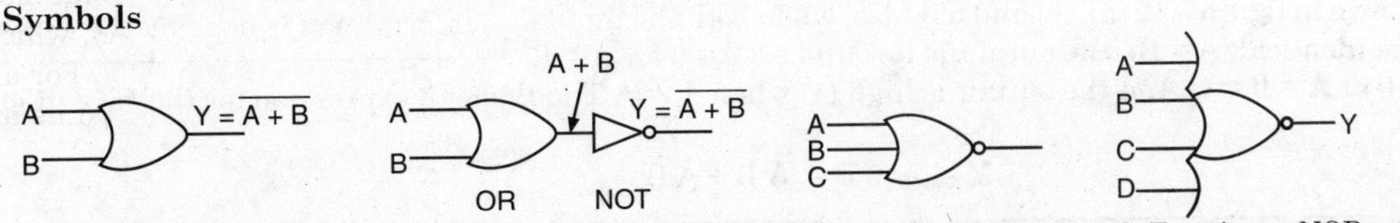

(a) Two input NOR gate (b) Equivalent circuit (c) Three input NOR gate (d) Four input NORgate

Fig. 3.10.

3. Truth table

(i) Truth table for two input NOR gate

S.No.	Inputs		Output
	A	B	Y
1.	0	0	1
2.	0	1	0
3.	1	0	0
4.	1	1	0

(ii) Truth table for three input NOR gate

S.No.	Inputs			Output
	A	B	C	Y
1.	0	0	0	1
2.	0	0	1	0
3.	0	1	0	0
4.	0	1	1	0
5.	1	0	0	0
6.	1	0	1	0
7.	1	1	0	0
8.	1	1	1	0

4. Boolean expression

The Boolean expression for the NOR gate is given by,

$$Y = \overline{A+B}$$

Here, A + B represents an OR gate whereas the bar (overhead line) represents the inversion, or complement. By substituting various combinations of the inputs A and B into the Boolean expression, we get the corresponding output. This is how we obtain the truth table of the NOR gate. The NOR gate like NAND gate is called as an universal gate because it is possible to construct all the basic gates such as AND, OR and NOT gates using the NOR gates only.

5. Standard package

The NOR gate is available in the monolithic integrated circuit form. IC 7402 is a 14 pin TTL IC which consists of four, two input NOR gates. The pin configuration of IC 7402 has been shown in figure 3.11.

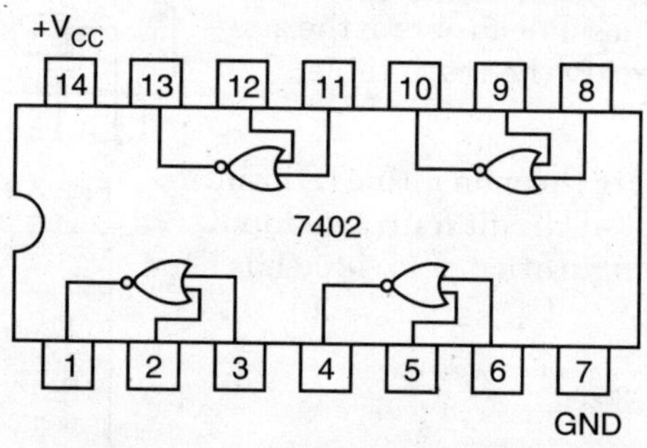

Fig. 3.11. Pin configuration of IC 7402

3.3.6 EX-OR Gate

(PTU, Jalandhar, Sem. Exam., 2004-05)

1. Definition

The exclusive-OR gate is abbreviated as EX–OR gate or sometimes as X-OR gate. An EX-OR gate can have two or more than two input terminals and one output terminal as shown in figure 3.12 (a), (b) and (c). Also, when both the inputs are identical (A = B), the output is low (0) i.e., Y = 0 for A = B = 0 or A = B = 1. And the output is high (1) when A ≠ B. The Boolean expression for the EX-OR gate is,

DO YOU KNOW?

The manipulation of binary information is done by the gates.

$$Y = A \oplus B = \overline{A}B + A\overline{B}$$

Action may not always bring happiness, but there is no happiness without action. — **Disraeli**

2. Symbols

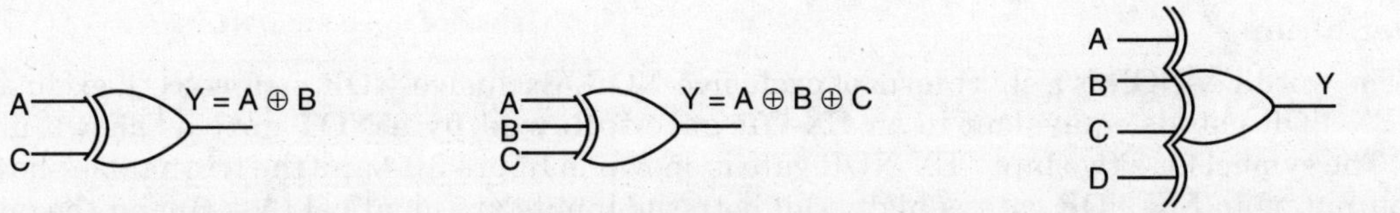

(a) Two input EX–OR gate (b) Three input EX–OR gate (c) Four input EX–OR gate

Fig. 3.12.

3. Truth table

(i) Truth table for two input EX-OR

S.No.	A	B	Y
1.	0	0	0
2.	0	1	1
3.	1	0	1
4.	1	1	0

(ii) Truth table for three input EX-OR

S.No.	A	B	C	Y
1.	0	0	0	0
2.	0	0	1	1
3.	0	1	0	1
4.	0	1	1	0
5.	1	0	0	1
6.	1	0	1	0
7.	1	1	0	0
8.	1	1	1	1

4. Standard Package

IC 7486 consists of four two input EX–OR gates. The pin configuration of IC 7486 has been shown in figure 3.13

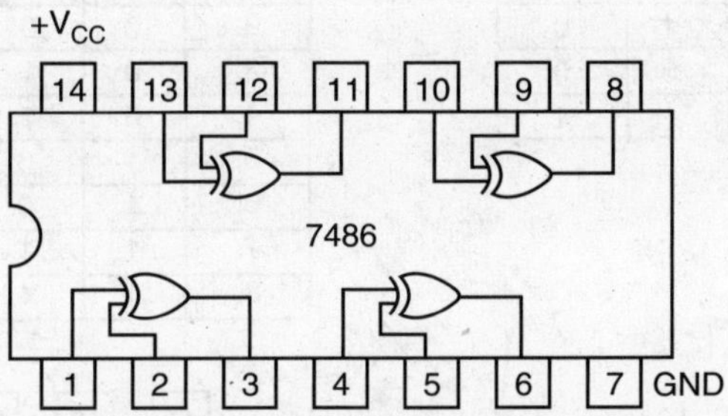

Fig. 3.13. Pin configuration of IC 7486

5. Multiple input EX–OR gate

EX-OR gates having more than two inputs and one output are available in market. Symbols of such EX-OR gates have been shown in figure 3.13. Also, the truth table of a three input EX-OR gate is shown in figure 3.13. This truth table shows that the output of an EX-OR gate is high (1) if odd number of inputs are high (1).

6. Applications of EX-OR gate

Some of the applications of an EX–OR gate may be listed as under :

(i) As a magnitude comparator.

(ii) In binary to gray code convertor.

(iii) In adder and subtractor circuits.

(iv) In parity generator.

(v) As a modulo–2 adder.

3.3.7 The EX-NOR Gate

1. Definition

The word EX-NOR is a short form of exclusive-NOR. Exclusive-NOR means NOT-exclusive OR, SO, EX-NOR gate is equivalent to an **EX-OR gate followed by a NOT gate** as shown in figure 3.14. The symbol for a two input EX-NOR gate is shown in figure 3.14 and the truth table shows that the output of an EX–NOR gate is high (1) if both the inputs are identical (A = B) and the output is low (0) if the inputs are not identical (A ≠ B). The Boolean expression for the EX–NOR gate is,

$$Y = \overline{A \oplus B} = \bar{A}\,\bar{B} + AB.$$

2. Symbols

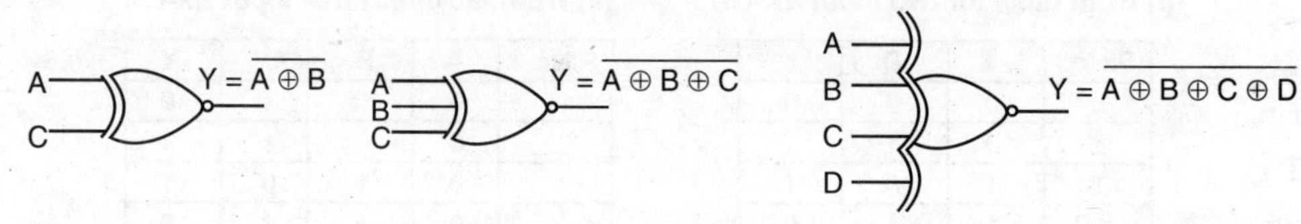

(a) Two input EX-NOR gate (b) Three input EX-NOR gate (c) Four input EX-NOR gate

Fig. 3.14.

3. Truth table

(i) Truth table for a two input EX-NOR gate

S.No.	Inputs		Output
	A	B	Y
1.	0	0	1
2.	0	1	0
3.	1	0	0
4.	1	1	1

(ii) Truth table of a three input EX-NOR gate

S.No.	Inputs			Outputs
	A	B	C	Y
1.	0	0	0	1
2.	0	0	1	0
3.	0	1	0	0
4.	0	1	1	1
5.	1	0	0	0
6.	1	0	1	1
7.	1	1	0	1
8.	1	1	1	0

4. Standard package

The EX-NOR gate is available in the IC form in the market. The TTL IC 74266 contains four, two input EX-NOR gates. It is a 14 pin quad-two input EX-NOR gate IC in dual in line package. The pin configuration for the same has been shown in figure 3.15.

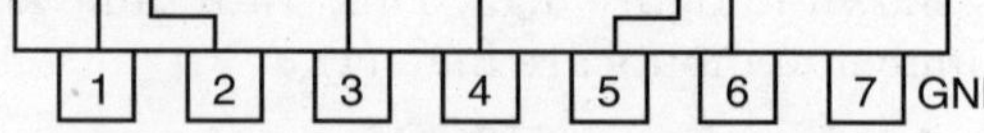

Fig. 3.15. Pin configuration of IC 74266

5. Multiple input EX-NOR gate

The EX-NOR gate with more than two inputs are also available. The symbol of a three input EX-NOR gate has been shown in figure 3.14. Looking at the truth table, we conclude that the output of an EX–NOR gate is high (1) when even number of inputs are high (1) or when all inputs are zeros.

6. Applications of EX-NOR gate

(i) As even parity generator.

(ii) As a comparator.

(iii) To check the even parity.

What we anticipate seldom occurs; what we least expected generally happens .

—**BENJAMIN DISRAELI.**

3.4. UNIVERSAL GATES

(PTU, Jalandhar, Sem. Exam., 2004-05)

The NAND and NOR gates are called as Universal Gates because it is possible to implement any Boolean expression with the help of only NAND or only NOR gates. Hence, a user can build any combinational circuit with the help of only NAND gates or only NOR gates. This is a great advantage because a user will have to make a stock of only NAND or NOR gate ICs.

3.4.1 NAND Gate as a Universal Gate

In this article, let us illustrate the conversion of each logic function into an expression using only NAND gates.

As discussed earlier, a NAND gate is expressed mathematically as under :

$$Y = \overline{A \cdot B}$$

Hence, we have to bring the given Boolean expression into this form.

1. NOT gate (Inverter) using NAND gate

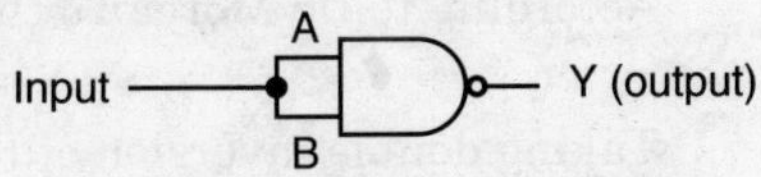

Fig. 3.16. NOT gate using NAND gate

Figure 3.16 shows the realization of a NOT gate (inverter) using a two input NAND gate. As both the inputs of the NAND are connected together, we can write that

$$\text{Input} = A = B$$

Hence, output is given by,

$$Y = \overline{A \cdot B}$$

$$Y = \overline{A \cdot A} \qquad \text{(since A = B = A)}$$

But, $A \,.\, A = A$

Therefore, $Y = \bar{A}$

This expression is the Boolean expression for an inverter. Hence, figure 3.16 is an inverter.

2. AND gate using NAND gate

The Boolean expression for an AND gate is,

$$Y = A \,.\, B$$

Taking double inversion of RHS, we get,

$$Y = \overline{\overline{A \cdot B}}$$

$$Y = A \,.\, B = \overline{\overline{A \cdot B}} \qquad \text{...(3.1)}$$

Equation (3.1) can be realized using only NAND gates as shown in figure 3.17.

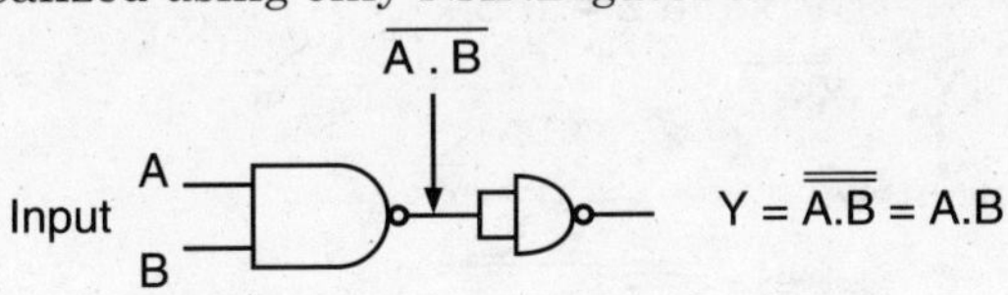

Fig. 3.17. AND gate using NAND gate

3. OR gate using NAND gate

The Boolean expression for OR gate is,

$$Y = A + B$$

Taking double inversion on RHS, we get,

$$Y = \overline{\overline{A + B}}$$

But as per De Morgan's theorem, we know that

$$\overline{A + B} = \bar{A} \,.\, \bar{B}$$

Therefore, $Y = \overline{\bar{A} \cdot \bar{B}}$

This is the required expression.

So, $Y = A + B = \overline{\bar{A} \cdot \bar{B}}$ and figure 3.18 shows the realization.

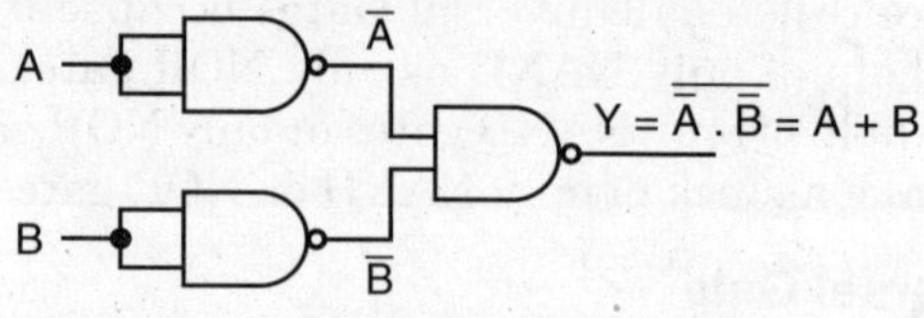

Fig. 3.18. OR gate using NAND gates

4. NOR gate using NAND gate

The Boolean expression for NOR gate is given by

$$Y = \overline{A+B}$$

According to De Morgan's theorem, we have

$$Y = \bar{A} \cdot \bar{B}$$

Taking double inversion of RHS, we write

$$Y = \overline{\overline{\bar{A} \cdot \bar{B}}}$$

This is the required expression,

Figure 3.19 shows the realization of NOR gate using NAND gates.

Therefore, $\overline{A+B} = \overline{\overline{\bar{A} \cdot \bar{B}}}$

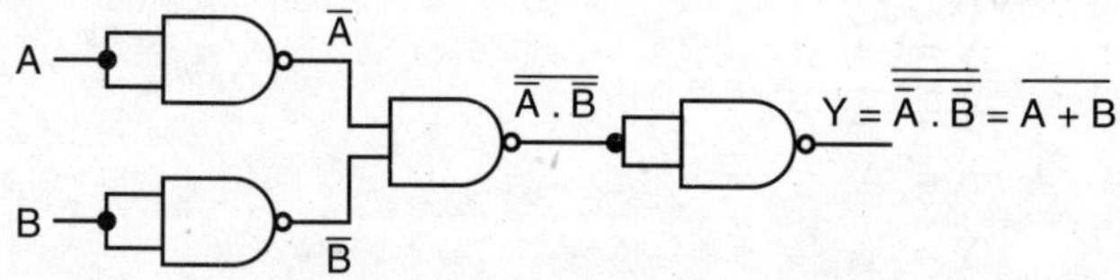

Fig. 3.19. NOR gate using only NAND gates

5. EX-OR gate using NAND gates only

The Boolean expression for EX-OR gate is,

$$Y = A \oplus B = \bar{A}B + A\bar{B}$$

Taking double inversion of RHS, we get

$$Y = \overline{\overline{\bar{A}B + A\bar{B}}}$$

Let $X = \bar{A}B$ and $Z = A\bar{B}$.

Hence, $Y = \overline{\overline{X+Z}}$

Using De Morgan's theorem, we have $\overline{X+Z} = \bar{X} . \bar{Z}$

Therefore, we have $Y = \overline{\bar{X} \cdot \bar{Z}} = \overline{(\overline{\bar{A}B}) \cdot (\overline{A\bar{B}})}$

> **DO YOU KNOW?**
>
> The relationship between input and output variables of each gate can be represented in a tabular form called a truth table.

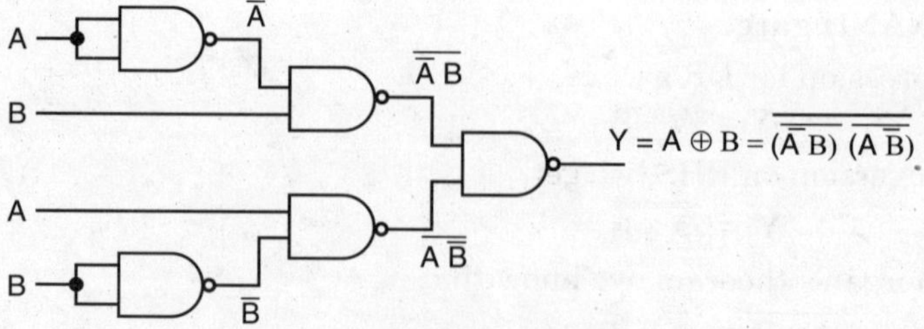

Fig. 3.20. EX-OR gate using NAND gates only

This is the required expressions for EX-OR gate using only NAND gates. Figure 3.20 shows the realization.

6. EX-NOR using NAND gates only

The Boolean expression for EX–NOR is given by

$$Y = A \odot B = \bar{A}\bar{B} + AB = X + Z$$

Taking double inversion of RHS, we get

$$Y = \overline{\overline{X+Z}} = \overline{\bar{X} \cdot \bar{Z}} \quad \text{(applying De Morgan's theorem)}$$

Therefore, we have $\quad Y = \overline{\left(\overline{\bar{A}\bar{B}}\right)\left(\overline{AB}\right)}$

This is the required expression for EX-NOR in terms of only NAND gates. Figure 3.21 shows the realization.

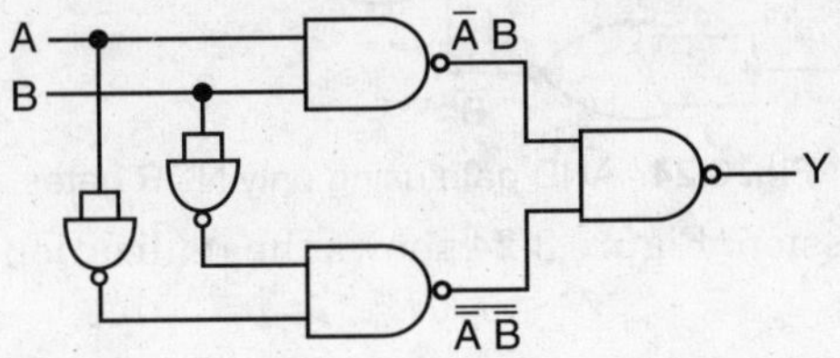

Fig. 3.21. EX–OR gate using only NAND gates

3.4.2 NOR gate as a Universal Gate

In this article, let us illustrate the implementation of every logic operation using only NOR gates. In order to do so, the Boolean expression of the given logic circuit must be first converted into the NOR format which is,

$$Y = \overline{A+B}$$

The implementation of various logic function using only NOR gates is illustrated in various figures as follows :

1. NOT gate (inverter) using NOR gate

Figure 3.22 shows the realization of a NOT gate using only NOR gate. Since both the inputs of the NOR gate are connected together, we can write,

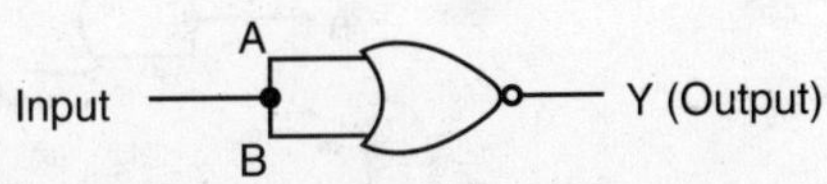

Fig. 3.22. NOT gate using NOR gate

$$A = B = A$$

Therefore, output of NOR will be,

$$Y = \overline{A+B} = \overline{A+A}$$

Since $\quad A + A = A$

Therefore, $\quad Y = \bar{A}$

This is the Boolean expression of an inverter. Hence, figure 3.22 represents an inverter.

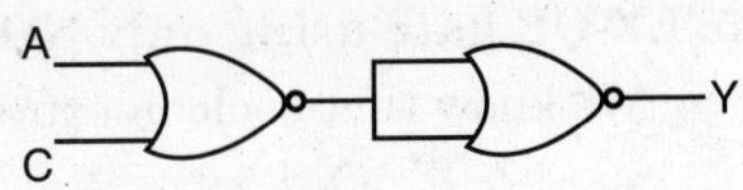

Fig. 3.23. OR gate using only NOR gates

2. OR gate using NOR gates only

We know that the boolean expression for an OR gate is given by

$$Y = A + B$$

Taking double inversion of RHS, we get

$$Y = \overline{\overline{A+B}}$$

This is the required expression

Figure 3.23 shows the realization of OR gate using only NOR gates.

Therefore, $Y = A + B = \overline{\overline{A+B}}$

3. AND gate using only NOR gates

We know that the boolean expression for an AND gate is given by

$$Y = A \,.\, B$$

Taking double inversion of RHS, we obtain

$$Y = \overline{\overline{A \cdot B}}$$

Therefore, $Y = \overline{\bar{A}+\bar{B}}$ (Applying De Morgan's theorem)

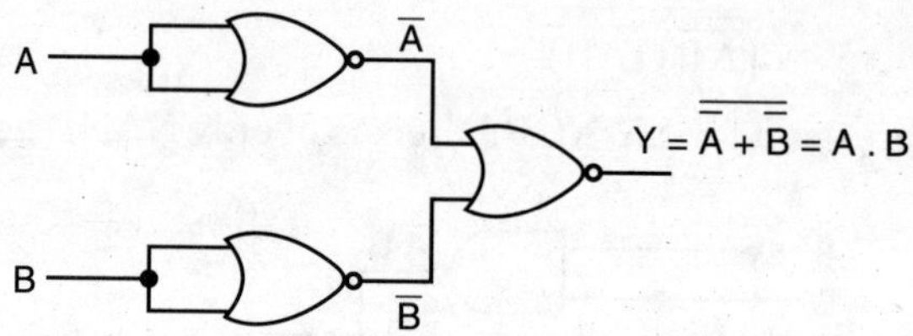

Fig. 3.24. AND gate using only NOR gates

This is the required expression. Figure 3.24 shows the realization of AND gate using NOR gates only.

4. NAND gate using only NOR gates

We know that boolean expression for a NAND gate is given by

$$Y = \overline{A \cdot B}$$

$$= \bar{A} + \bar{B}$$ (Applying De Morgan's theorem)

Taking double inversion of RHS, we obtain

$$Y = \overline{\overline{\bar{A}+\bar{B}}}$$

This is the required expression. Figure 3.25 shows the realization of NAND gate using only NOR gates.

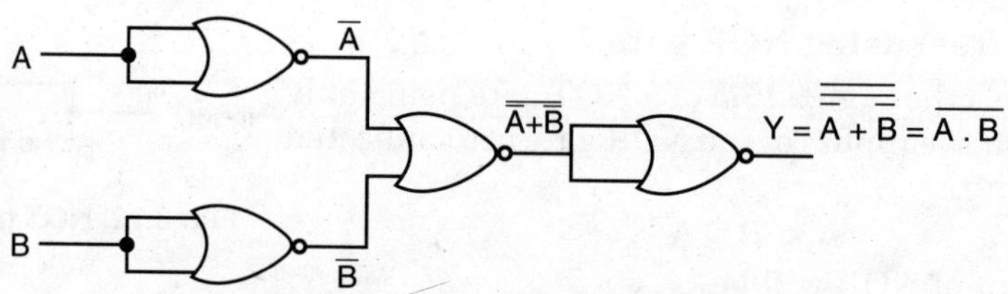

Fig. 3.25. NAND gate using only NOR gates

5. EX-OR gate using only NOR gates

We know that boolean expression for a EX-OR gate is

$$Y = A \oplus B = \bar{A}B + A\bar{B}$$

Let $X = \bar{A}B$ and $Z = A\bar{B}$

Therefore, $Y = X + Z$

Taking double inversion of RHS, we obtain,

$$Y = \overline{\overline{X+Z}}$$

Hence, $Y = \overline{\bar{X} \cdot \bar{Z}}$ (Applying De Morgan's theorem)

A useless life is an early death. – **GOETHE.**

or $$X = \overline{(\overline{A}B)}.\overline{(A\overline{B})}$$

But, $$\overline{\overline{A}B} = (A + \overline{B})$$

and $$\overline{A\overline{B}} = (\overline{A} + B)$$ (Applying De Morgan's theorem)

Therefore, $$Y = \overline{(A+\overline{B})\cdot(\overline{A}+B)}$$

or $$Y = \overline{(A+\overline{B})\cdot(\overline{A}+B)}$$ (Applying De Morgan's theorem)

Taking double inversion of RHS, we obtain,

$$Y = \overline{\overline{(A+\overline{B})}+\overline{(\overline{A}+B)}}$$

This is the required expression. Figure 3.26 shows the realization of EX-OR gate using only NOR gates.

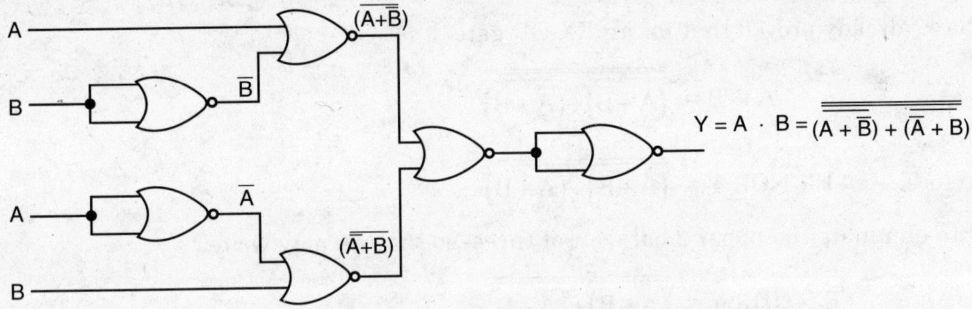

Fig. 3.26. EX-OR gate using only NOR gates

6. EX-NOR gate using only NOR gates

We know that the boolean expression for an EX-NOR gate is given by,

$$Y = \overline{A}\,\overline{B} + AB = X + Z$$

Taking double inversion of RHS, we obtain,

$$Y = \overline{\overline{\overline{A}\,\overline{B} + AB}}$$

or $$Y = \overline{\overline{\overline{A}\,\overline{B}+AB}}$$ (Applying De Morgan's theorem)

Therefore, $$Y = \overline{(A+B)\cdot(\overline{A}+\overline{B})}$$ (Applying De Morgan's theorem)

or $$Y = \overline{(A+B)}+\overline{(\overline{A}+\overline{B})}$$ (Applying De Morgan's theorem)

Taking double inversion of RHS, we obtain,

$$Y = \overline{\overline{(A+B)}+\overline{(\overline{A}+\overline{B})}}$$

This is the required expression. Figure 3.27 shows the realization of an EX-NOR gate using only NOR gates.

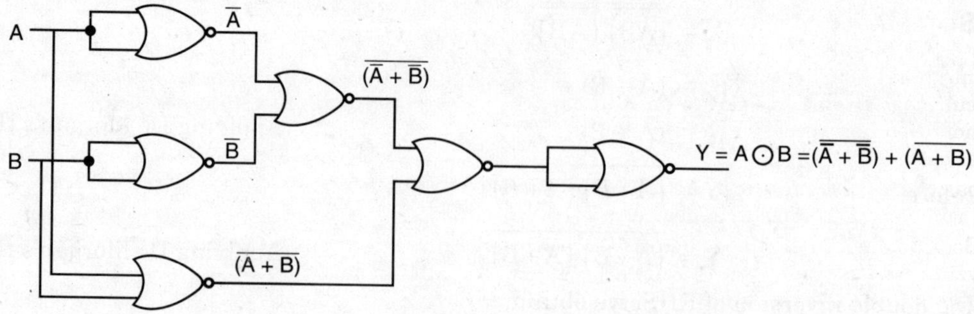

Fig. 3.27. EX-NOR gate using only NOR gates

Alternative Method

We know that EX-NOR function is given by $Y = A \odot B = \overline{A \oplus B} = \overline{\text{EX-OR}}$

We have already proved that for an EX-OR gate

$$A \oplus B = \overline{\overline{\overline{(A+\overline{B})}+\overline{(\overline{A}+B)}}}$$

Therefore, EX-NOR $$Y = \overline{\overline{\overline{\overline{(A+\overline{B})}+\overline{(\overline{A}+B)}}}}$$

We can eliminate the upper 2 bars out of three, so that we may write,

$$\text{EX–NOR } Y = \overline{\overline{(A+\overline{B})}+\overline{(\overline{A}+B)}}$$

This is the required expression and has been is shown in figure 3.28.

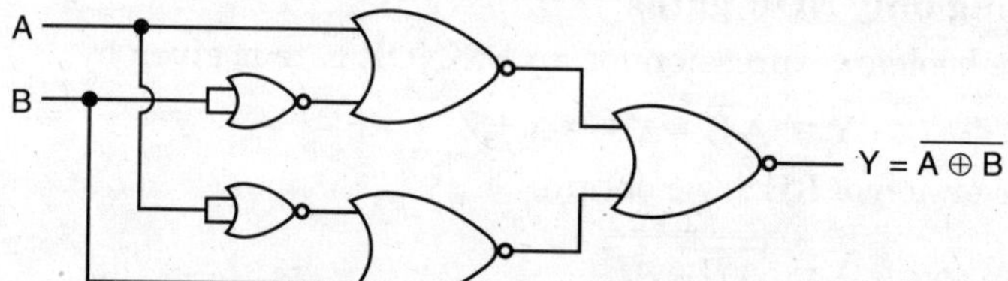

Fig. 3.28. EX–NOR gate using only NOR gates

GLOSSARY

1. **AND gate :** Basic combinational logic device where all inputs must be HIGH for the output to be HIGH.
2. **NAND gate :** Basic combinational logic device where all inputs must be HIGH for the output to be LOW. A not AND circuit.
3. **NOR gate :** Basic combinational logic device where all inputs must be LOW for the output to be HIGH. A not OR circuit.
4. **NOT :** Basic combinational logic device where the output is always the opposite from the input. Also called an inverter.
5. **OR gate :** Basic combinational logic device where the output goes HIGH when any or all inputs are HIGH.
6. **Truth table :** Tabular listing of all inputs and resultant output conditions for a logic function or circuit.
7. **XNOR gate :** Basic combinational logic device where an even number of HIGH inputs generates a HIGH output. A not XOR gate.
8. **XOR gate :** Basic combinational logic device where an odd number of HIGH inputs generates a HIGH output.

SUMMARY

- As a matter of fact, logic gates are the basic elements that make up a digital system. The electronic gate is a circuit that is able to operate on a number of binary inputs in order to perform a particular logical function. In other words, we can say that logic gates are the devices used as basic building blocks of all the digital circuits.
- The basic logic gates are NOT, AND and OR along with NOR, NAND, EX–OR etc. We have to use different laws, rules and theorems to analyze the digital circuits.
- The Boolean algebra developed by Charles Boole way back in 1884 is used for representing, simplifying and analyzing the logic circuits.
- The gate is a digital circuit with one or more input voltages but only one output voltage. By connecting the different gates in different ways, we can build circuits that can perform arithmetic and other functions associated with the human brain.
- In the decimal system which is used in our day–to–day life, the arithmetic operations such as addition, subtraction, multiplication, division, square, square root etc. are used to solve arithmetic equations.
- Similarly, to solve or simplify the logical expressions, used in digital circuits, we need to use logical operators. The three main logic operators may be listed as under :

  (i) AND operator. (ii) OR operator.
  (iii) NOT operator (i.e., invert).
- The AND operator represents logical multiplication. It is denoted by a dot between the two variables to be multiplied i.e.,

  $$A \, . \, B \rightarrow \text{Logical multiplication}$$

  However, sometimes, this dot is not used and we denote the logical multiplication of A and B as AB. Thus, if A is to be multiplied with B in a logic circuit then the multiplication is represented as A . B, AB and it is to be read as A AND B.
- The OR operator represents logical addition. It is denoted by a (+) sign between the two variables to be added. If A and B are to be added in a logic circuit it is represented as under

  $$A + B \rightarrow \text{Logical addition}$$

  And, it is to be read as A OR B.
- The NOT operator represents a logical inversion or complementing. The NOT operation is denoted by a bar (–) over the variable to be inverted. For example, if A is to be inverted then the process of inversion is denoted by $\overline{A}$ and it is to be read as Not A.

  $$\overline{A} = \text{NOT A}$$
- Logic gates are the basic building blocks of any digital system. It is an electronic circuit having one or more than one inputs and only one output. The relationship between the input and the output is based on a certain logic. Based on this logic, the gates are named as NOT gate, AND gate, OR gate, NAND gate, NOR gate etc.
- The operation of a logic gate can be best understood with the help of a table called Truth Table. The truth table consists of all the possible combinations of the inputs and the corresponding state of output of a logic gate.
- The relationship between the inputs and the outputs of a gate can be expressed mathematically by means of the Boolean Expression.
- AND is one of the logic operators. It performs the logical multiplication on its inputs. The output is high (Y = 1) if and only if all the inputs to the gate are high (1). The output is low (0), if atleast one of the inputs is low (0). AND gate can have two or more inputs and only one output.
- OR is one of the logic operators. An OR gate performs the logical addition on its inputs therefore its output will be high (1) if any one or both the inputs are high (1). Its output will be low (0) if and only if both the inputs are simultaneously low (0).
- The term NAND can be split as NOT–AND which means that the NAND operation can be implemented with the **combination of an AND gate and a NOT gate i.e., inverter**.
- This is another universal gate. The word NOR can be split as NOT–OR which means that a NOR operation can be implemented with the combination of an OR gate and a NOT gate i.e., inverter.
- Some of the applications of an EX–OR gate may be listed as under :

  (i) As a magnitude comparator. (ii) In binary to gray code convertor.

(iii) In adder and subtractor circuits. (iv) In parity generator.
(v) As a modulo–2 adder.

- The NAND and NOR gates are called as Universal Gates because it is possible to implement any Boolean expression with the help of only NAND or only NOR gates. Hence, a user can build any combinational circuit with the help of only NAND gates or only NOR gates. This is a great advantage because a user will have to make a stock of only NAND or NOR gate ICs.

SHORT QUESTIONS WITH ANSWER

Q.1. What is meant by Logic Operators.

Ans. In the decimal system which is used in our day–to–day life, the arithmetic operations such as addition, subtraction, multiplication, division, square, square root etc. are used to solve arithmetic equations.

Q.2. What is meant by AND Operator.

Ans. The AND operator represents logical multiplication. It is denoted by a dot between the two variables to be multiplied *i.e.,*

$$A \cdot B \rightarrow \text{Logical multiplication}$$

However, sometimes, this dot is not used and we denote the logical multiplication of A and B as AB. Thus, if A is to be multiplied with B in a logic circuit then the multiplication is represented as A . B, AB and it is to be read as A AND B.

Q.3. What is meant by OR Operator.

Ans. The OR operator represents logical addition. It is denoted by a (+) sign between the two variables to be added. If A and B are to be added in a logic circuit it is represented as under

$$A + B \rightarrow \text{Logical addition}$$

And, it is to be read as A OR B.

Q.4. What do you mean by Logic Gates.

Ans. Logic gates are the basic building blocks of any digital system. It is an electronic circuit having one or more than one inputs and only one output. The relationship between the input and the output is based on a certain logic. Based on this logic, the gates are named as NOT gate, AND gate, OR gate, NAND gate, NOR gate etc.

Q.5. What is an OR gate.

Ans. OR is one of the logic operators. An OR gate performs the logical addition on its inputs therefore its output will be high (1) if any one or both the inputs are high (1). Its output will be low (0) if and only if both the inputs are simultaneously low (0).

Q.6. Write the applications of EX-NOR gate.

Ans. (i) As even parity generator.
(ii) As a comparator.
(iii) To check the even parity.

Q.7. Explain about Universal gates.

Ans. The NAND and NOR gates are called as Universal Gates because it is possible to implement any Boolean expression with the help of only NAND or only NOR gates. Hence, a user can build any combinational circuit with the help of only NAND gates or only NOR gates. This is a great advantage because a user will have to make a stock of only NAND or NOR gate ICs.

Q.8. What is an inverter?

Ans. An inverter is a logic gate whose output is the inverse or complement of its input.

Q.9. What are the universal logic gates?

Ans. Universal gate is a gate that can perform all the basic logical operations such as NAND and NOR gates.

Q.10. What are basic logic gates and what are universal logic gates?

Ans. Basic logic gates are AND gate, OR gate, and NOT (INVERTER) gate. Universal logic gates are NAND and NOR gates.

Q.11. What is the speciality of NAND and NOR gates?

Ans. The speciality of NAND and NOR gates is that they are universal gates and can perform all the basic logical operations.

Q.12. Why NAND-NAND realization is preferred over AND-OR realization?

Ans. NAND-NAND realization needs only one type of gate (NAND), that minimizes IC package counter.

Q.13. What is the logic signal required to inhibit a NAND gate?

Ans. Logic 0 is the logic signal required to inhibit a NAND gate and 1 is its output, irrespective of all other inputs, when inhibited.

REVIEW QUESTIONS

1. Describe with circuit diagrams a two input EX-NOR gate using only NAND gates.
2. Describe with circuit diagrams a two input EX-NOR gate using only NOR gates.
3. What are the applications EX-NOR gate? Explain.
4. Define 'truth table'. What is mean by a logic gate ? Classify the logic gates.
5. Describe the AND gate. Specify its symbol and write its truth table.
6. Describe the OR gate. Specify its symbol and write its truth table.
7. Describe the EX-OR gate. Specify its symbol and write its truth table.
8. Describe the EX-NOR gate. Specify its symbol and write its truth table.
9. Describe the NAND gate. Specify its symbol and write its truth table.
10. Describe the NOR gate. Specify its symbol and write its truth table.
11. Describe and mention the universal gates.
12. Draw the equivalent circuit of all gates using NAND gate ?
13. Draw an equivalent circuit of all gates using NOR gate ?

❑❑❑

4

LOGIC FAMILIES

Syllabus

Logic Families: Implementations of TTL, MOS, CMOS, ECL and their characteristics.

4.1 INTRODUCTION

As a matter of fact, various digital ICs available in market belong to various types. These types are known as families. Based on the components and devices, internally used, the digital IC families are named as RTL (Resistor transistor logic). TTL (Transistor Transistor Logic), DTL (Diode Transistor Logic). CMOS etc.

4.1.1. Switching Elements

The digital circuits are in the integrated circuit (IC) form the elements used for building such circuits are diodes, resistors, transistors and MOSFETs. The diodes, transistors and MOSFETs are used as switches. Hence, they are known as switching elements. All these devices are expected to operate at very high frequencies because all these ICs are designed to operate a high frequencies.

4.1.2. Diode as Switching Element

1. Basic Aspects

When a diode is driven from reverse condition to forward condition or in the opposite direction, it does not change its state instantaneously. Instead it takes a finite time to switch from OFF to ON state or from ON to OFF state. This time consists of a transient time and the time interval elapsed before the diode recovers to its steady state. The diode characteristics corresponding to the interval in which the change of state is taking place is called as the **switching characteristics**. Diodes are used in applications such as logic gates. In such applications, diodes are used as switches and not for the purpose of rectification. The switching characteristics and the parameters derived from it are useful for such

Chapter Outline

applications. In the switching applications, we define two time durations namely the turn ON time and the turn OFF time of the diode. The time required to turn OFF is longer than the time required to turn ON a diode. Therefore, the turn OFF time and turn OFF characteristics is more important.

DENNIS C. HAYES

In 1978, 28-year-old Dennis C. Hayes formed what became Hayes Microcomputer Products, Inc. When he began, Hayes hand-assembled and soldered products on a borrowed dining room table in his home. Then, in 1981 the Hayes™ Smartmodem™, which was easily integrated into the computer environment, started a communications revolution.

Forward and Reverse Recovery Times

Now, let us define two important delays related to the switching of a diode. They are as under:

(i) Forward recovery time t_{rf}

(ii) Reverse recovery time t_{rr}

(i) Forward Recovery Time t_{rf}

The forward recovery time (t_{rf}) is defined as the time required for the diode current to change from 10 to 90 percent of its final value when the diode is switched from OFF to ON state. In practice, the forward recovery time is very short as compared to the reverse recovery time. Hence, we can neglect it. Hence, the turn on time of a diode is always shorter than its turn off time. Manufacturers normally specify the reverse recovery time t_{rr} of a diode.

Storage and Transition Times

If the conducting diode is suddenly reverse biased by applying the voltage waveform shown in figure 4.1 (a), then the sequence of events will take place.

The sequence of events are as under:

The voltage waveform in figure 4.1(a) is assumed to be applied at the input of the diode circuit of figure 4.1 (b) so as to reverse bias the conducting diode.

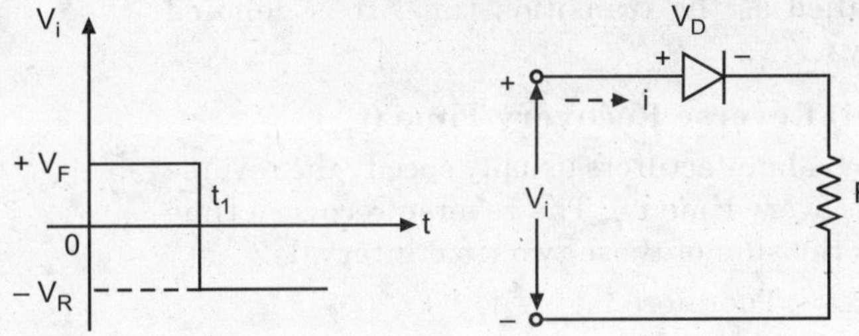

(a) Applied voltage at the input of diode circuit to reverse bias it

(b) Simple Diode circuit

Fig. 4.1.

Interval 0 – t_1

Now, we consider the waveforms shown in figure 4.2. From t = 0 to t = t_1, the input voltage $V_i = V_F$ and the diode remains forward biased. The resistance R_L is assumed to be large as compared to the resistance of diode

Therefore, current $$i = +\frac{V_F}{R_L} = +I_F$$

At t = t_1

At t = t_1, the input voltage reverses abruptly from + V_F to – V_R

Due to the presence of large number of minority carrier on either sides, the current will simply reverse. The new value of reverse current is,

$$i = -\frac{V_R}{R_L} = -I_R$$

Interval t_1 – t_2

The diode current will remain constant at – I_R from t_1 to t_2. At instant t = t_2 as shown in figure 4.2, the injected minority carrier density at x = 0 has reached its equilibrium state. At t = t_2, when the excess minority carriers in the immediate neighborhood of the junction have been swept back across the junction, the diode voltage begins to reverse and the magnitude of diode current starts decreasing. The interval t_1 to t_2 is the duration required for the stored minority charge to become zero and it is called as the **storage time t_s**.

Definition of Storage Time T_s

The storage time is defined as the time required for the stored minority charge to become zero. It is denoted by t_s.

After $t = t_2$

As shown in figure 4.2, the diode voltage starts becoming negative and increases towards the maximum value $-V_R$. The magnitude of reverse current decreases towards I_o. This is an indication that the diode has fully recovered.

Definition of Transition Time t_t

The time elapsed between $t = t_2$ and the time when diode is completely recovered is called as the transition time. It is denoted by t_t.

(ii) Reverse Recovery Time (t_{rr})

Manufacturers usually specify the reverse recovery time t_{rr}. The reverse recovery time is the sum of these two time intervals.

Therefore, $t_{rr} = t_s + t_t$

Where, t_{rr} = reverse recovery time,

t_s = storage time,

t_t = transition time

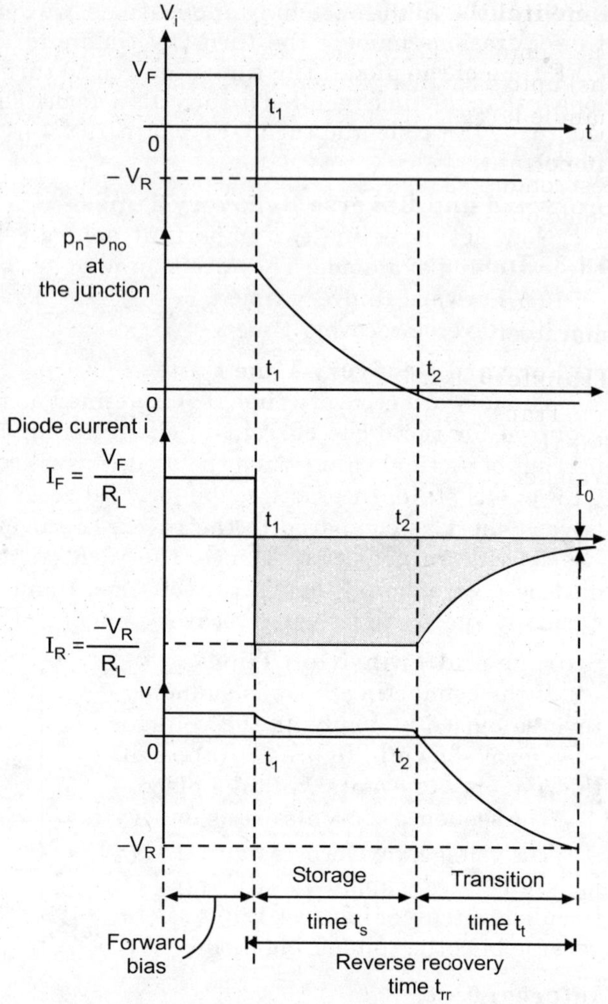

Fig. 4.2. Diode switching waveforms

In words, we can define the reverse recovery time as the time interval from the current reversal at $t = t_1$ until the diode has recovered completely.

Importance of Diode Switching Times

(i) The reverse recovery time decides the time required to switch the diode off. Hence, it decides, how fast or at what maximum rate we can switch the diode on and off.

(ii) Thus, reverse recovery time decides the highest switching frequency of a diode.

(iii) The reverse recovery time depends on the transition capacitance of the diode. So the transition capacitance of switching diodes is small.

(iv) The maximum switching frequency is given by,

$$f_{max} = \frac{1}{10\, t_{rr}}$$

Therefore, as t_{rr} decreases, f_{max} will increase.

4. Switching Diodes

Commercial switching type diodes are available with time t_{rr} ranging from 1 nanosecond (ns) upto as high as 1 microsecond (μ s). Higher values of t_{rr} correspond to the diodes intended to handle larger currents.

IN 4153 is a switching diode. It has a reverse recovery time of a few nanoseconds under the test conditions. IN 4153 is high speed switching diode for computer and general purpose applications. Some of the important features of this switching diode are as follows.

4.1.3. Transistor (BJT) as a Switch

Before discussing about how to use the transistor as a switch, let us study its transfer characteristics.

Transfer Characteristics of BJT

Transfer characteristics is the graph of controlled quantity versus the controlling quantity. For the transistor to operate as a switch, it is operated in CE configuration. So, let us see the transfer characteristics of a transistor in CE configuration.

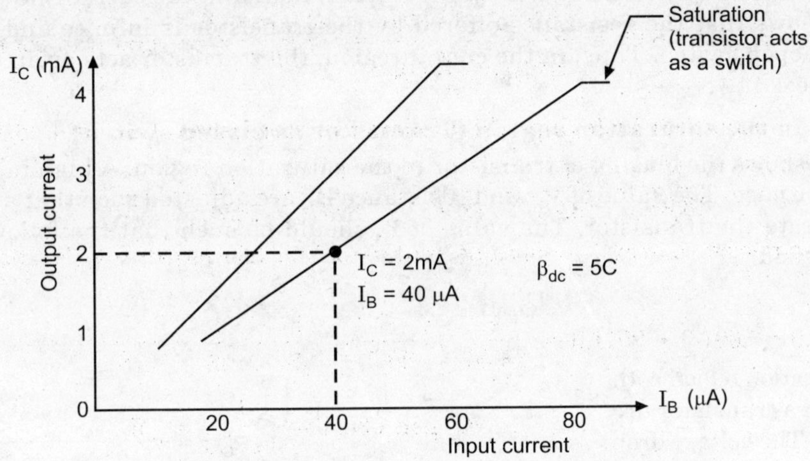

Fig. 4.3. Transfer characteristics for CE configuration

Transfer characteristics for a CE configuration is the graph of output current (I_C) versus input current (I_B) at a constant value of V_{CE}. It is as shown in figure 4.3. Transfer characteristics is also called as current gain characteristics. We can measure the current gain of the transistor directly from this characteristics. The characteristics in figure 4.3 shows that the relation between I_C and I_B is linear. This portion of characteristic is used for the amplification application. To operate the transistor as a switch, we have to operate it in its saturation region as shown in figure 4.3.

4.1.4. Transistor as a Switching Element

The transistor can be used for two types of applications i.e,. amplification and switching. For the amplification applications a transistor is biased in its active region. Whereas for switching applications it is biased to operate in the saturation (fully on) or cutoff (fully off) regions.

1. Transistor in Cutoff Region (Transistor as an Open Switch)

In the cutoff region both the junctions of a transistor are reverse biased and a very small reverse current flows through the transistor. The voltage drop across the transistor (V_{CE}) is high. Thus, in the cutoff region the transistor is equivalent to an open switch.

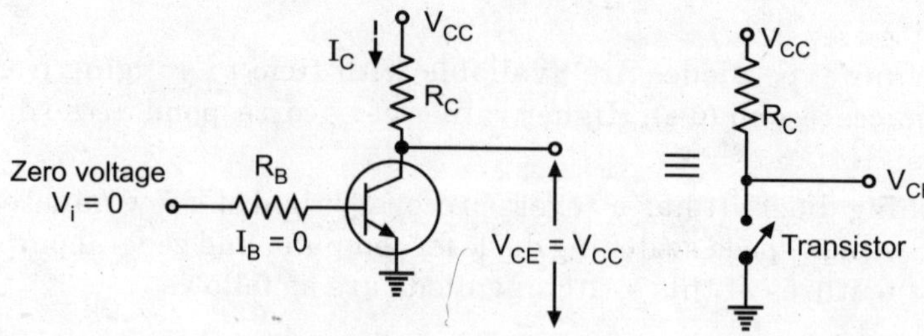

Fig. 4.4. Transistor in cutoff is equivalent to an open switch

The input voltage is zero, so, $I_B = 0$ and $I_C = 0$. Applying KVL to the collector circuit of figure 4.4, we get

$$V_{CC} = I_C R_C + V_{CE}$$

But $\quad I_C = 0$

Therefore, $\quad V_{CE} = V_{CC}$

The voltage across the transistor is $V_{CE} = V_{CC}$ i.e., high and the current flowing through it $I_C = 0$. This shows that the resistance offered by the transistor is infinite and the transistor operates as an open switch. Thus, in the cut off region, the transistor acts as an open switch as shown in figure 4.4.

2. Transistor in the saturation region (Transistor as closed switch)

Figure 4.5 shows the biasing of transistor in the saturation region. A high input voltage V_{in} is applied at the base. The value of V_{in} and resistance R_B are adjusted such that a large I_B flows. This will saturate the transistor. The value of I_B should be such that the following condition should be satisfied.

$$I_B \geq \frac{I_{C(sat)}}{\beta}$$

In the saturation region both the junctions of a transistor are forward biased. The voltage drop across the transistor (V_{CE}) is very small of the order of 0.2 V to 1V depending on the type of transistor and collector current is very large.

A low (almost 0) voltage drop across the transistor and a large current through it indicates that the transistor is equivalent to a closed switch.

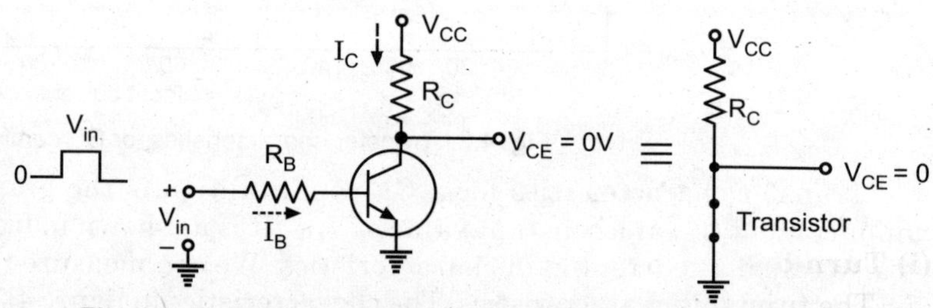

Fig. 4.5. Transistor in saturation is equivalent to a close switch

3. Switching Waveforms of an Ideal Transistor Switch

Let us consider the circuit diagram of figure 4.6 which shows the biasing arrangement for a transistor to operate as a switch. A rectangular waveform is applied at the base and the transistor is assumed to be an ideal one. Hence, the transistor will change its state from ON to OFF or vice versa instantaneously.

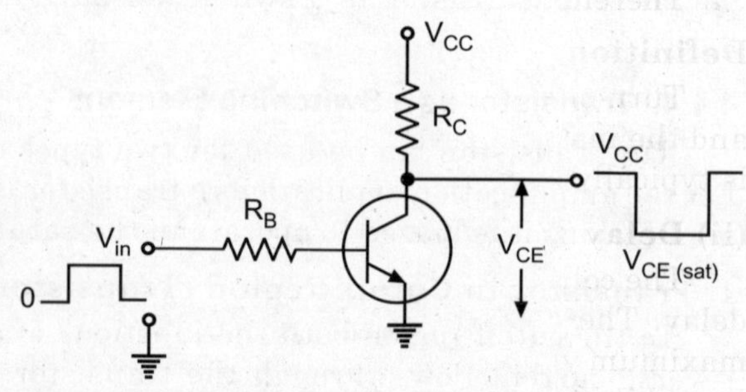

Fig. 4.6. Circuit diagram showing transistor biased to operate as a switch

When the input is low (0), the transistor is off, it acts as an open switch and the output voltage is $V_{CE} = V_{CC}$. When the input voltage goes high, the transistor is driven into saturation. It acts as a closed switch and the output voltage $V_{CE} = V_{CE(sat)} \approx 0$ V. The input and output voltage waveforms are shown in figure 4.7 which shows that the transistor state change takes place instantly.

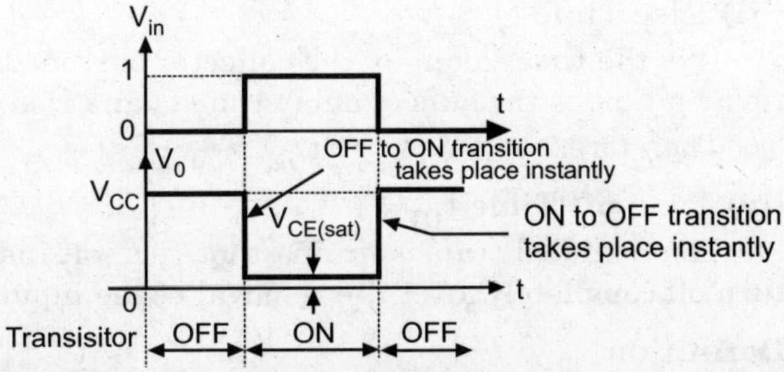

Fig. 4.7. Input output waveforms for transistor switching of an ideal transistor

4. Transistor Switching Times

The practical transistor does not switch its state instantaneously. It needs a finite time to turn on and to turn off completely. So, consider a practical transistor being driven by a pulse waveform (figure 4.6). The input voltage and collector current waveforms for a non ideal transistor is shown in figure 4.8. Various delays and time intervals are involved in the switching waveforms of the transistor in figure 4.8.

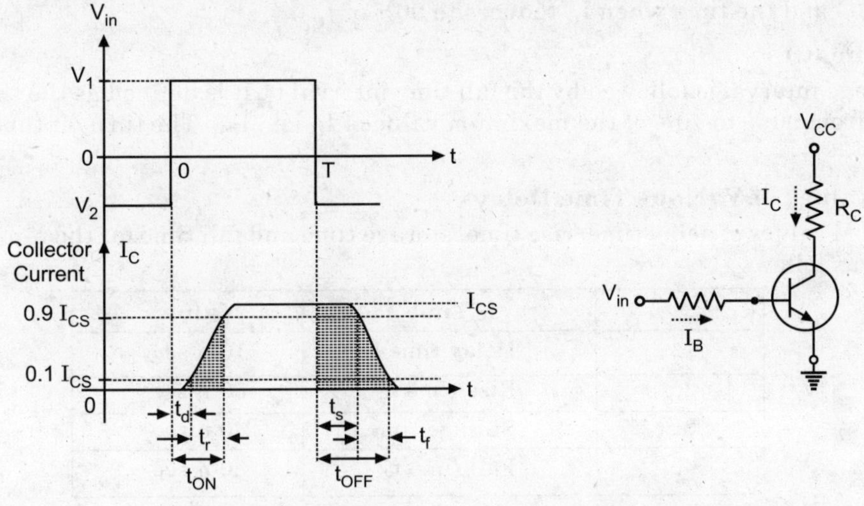

Fig. 4.8. Switching waveforms for the non ideal transistor

(i) Turn on Time

The turn on time t_{ON} is the sum of two time intervals namely the delay time t_d and the rise time t_r.

Therefore, $$t_{ON} = (t_d + t_r)$$

Definition of Turn on Time t_{ON}

Turn on time is the time delay between the instant at which a high base voltage is applied and the instant at which the collector current reaches 90% of its maximum value. Turn on time is typically a few microseconds.

(ii) Delay Time t_d

The collector current I_C does not immediately respond to the input signal instead there is a delay. The delay time t_d is the time required for the collector current to attain 10% of the maximum value of I_C i.e., I_{CS}, after the application of input signal. I_{CS} is the saturation value of I_C and it is given by,

$$I_{CS} = \frac{V_{CC}}{R_C}$$

(iii) Rise Time t_r

It is the time taken by the collector current I_C to increase from 0.1 I_{CS} to 0.9 I_{CS}. The total turn on time is the sum of delay time t_d and rise time t_r.

Therefore, $$t_{ON} = t_d + t_r$$

(iv) Turn off Time t_{OFF}

The nonideal transistor does not turn off instantaneously instead it takes a finite time to turn off completely after the removal of the input signal.

Definition

The turn off time (t_{OFF}) is defined as the time taken by the collector current to reduce from I_{CS} to 10% of I_{CS} after switching the input voltage from V_1 to V_2. The turn off time is the sum of two components namely the storage time (t_s) and the fall time (t_f).

Therefore, $$t_{OFF} = t_s + t_f$$

(v) Storage Time (t_s)

It is defined as the time interval which elapses between the transition of input waveform (from V_1 to V_2) and the time when I_C reduces to 90% of I_{CS}.

(vi) Fall Time (t_f)

The storage interval is followed by the fall time interval t_f. It is defined as the time required for I_C to fall from 90 % to 10% of the maximum value of I_C i.e., I_{CS}. The turn off time is the sum of t_s and t_f.

5. Typical Values of Various Time Delays

The typical values of delay time, rise time, storage time and fall time for the n-p-n transistor 2N 3830 are as follows :

S. No.	Time	Value
1.	Delay time t_d	10 n sec
2.	Rise time t_r	50 n sec
3.	Storage time t_s	40 n sec
4.	Fall time t_f	30 n sec

6. Importance of Turn ON and Turn OFF Time

The turn on and turn off times are important in deciding the maximum switching frequency of the transistor. In order to operate the transistor at a high frequency, the turn on and turn off times should be as short as possible.

4.1.5. MOSFET as a Switching Element

The MOS-logic family uses MOSFET as the basic device the way TTL uses BJT. The MOSFETs are of two types namely the depletion MOSFETs and the enhancement type MOSFETs. In logic circuits, the enhancement MOSFETs are used. We use the E-MOSFET as a switch.

Types of E-MOSFETs

The E-MOSFETs are of two types namely

(i) P channel enhancement MOSFETs (PMOS)

(ii) N channel enhancement MOSFETs (NMOS)

The symbols of PMOS and NMOS are shown in figure 4.9 which shows that MOSFET is a three terminal device. Their names are Drain (D), Gate (G) and source (S). These are similar to collector, base and emitter. Gate is a control terminal and voltage V_{GS} controls the magnitude of drain current I_D.

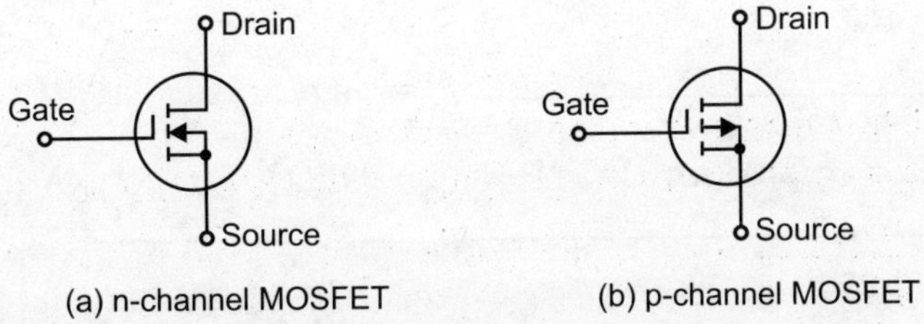

Fig. 4.9. Circuit symbols of E-MOSFETs

4.1.6. Characteristics of N-channel Enhancement Type MOSFET

The drain characteristics and transfer characteristics of a N-channel enhancement MOSFET are as shown in figure 4.10 (b) and (a) respectively. Here, it may be noted that the drain current is zero for $V_{GS} \le V_T$. The transfer characteristics shown in figure 4.10 (a) is very much different from those obtained earlier. It is now totally in the positive V_{GS} region and remains zero till $V_{GS} = V_T$. V_T is sometimes called as $V_{GS(OFF)}$. The relation between drain current and V_{GS} is given by the following equation,

$$I_D = K (V_{GS} - V_T)^2$$

This is a non-linear relation and is valid only for $V_{GS} > V_T$. The 'K' is a constant and its value depends on construction of the device.

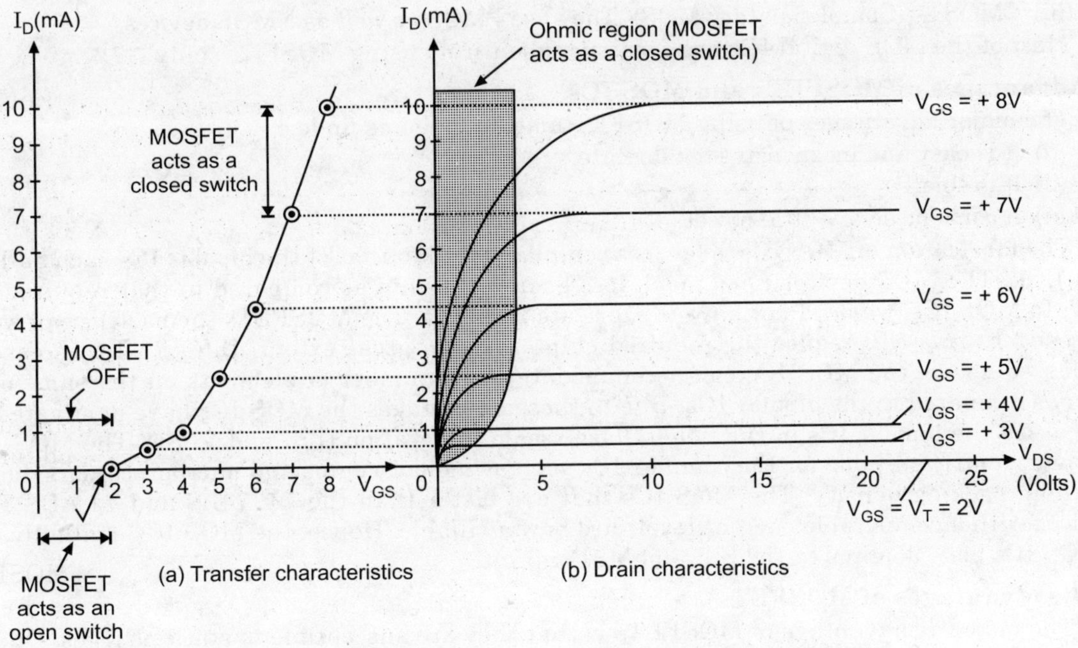

Fig. 4.10. Characteristics of a n-channel enhancement MOSFET

4.1.7. Characteristics of P-channel Enhancement type MOSFET

The drain characteristics of P-channel enhancement MOSFET are as shown in figure 4.11 (a) Note that the drain current increases with increase in the negative gate to source voltage (V_{GS}). The transfer characteristics is as shown in figure 4.11 (b). Here, it may be noted that it is exactly the mirror image of the transfer characteristic of an n-channel enhancement MOSFET. The drain current I_D remains zero upto the voltage $V_{GS} > V_T$ and increases as negative V_{GS} becomes greater than V_T as shown in figure 4.11 (b).

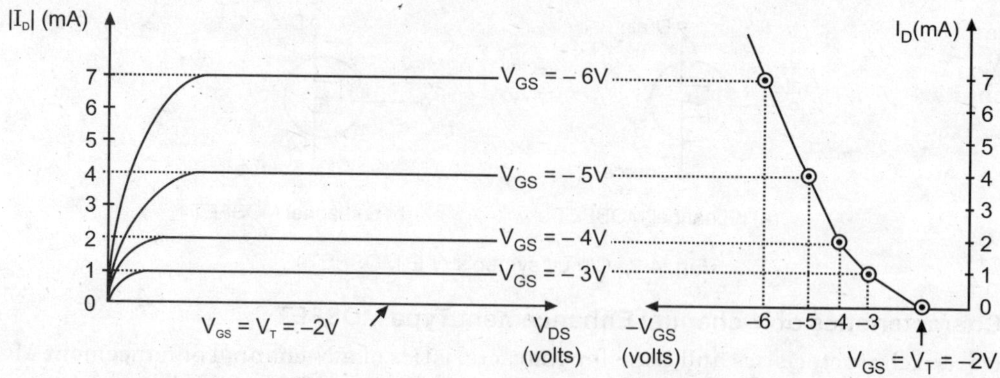

(a) Drain characteristics of p-channel enhancement MOSFET

(b) Transfer characteristics of p-channel enhancement MOSFET

Fig. 4.11.

4.1.8. MOS Technology

1. Classification

The digital circuits using MOSFET can be classified into three groups as under:

(i) PMOS — These circuits use only P channel E-MOSFETs

(ii) NMOS — These circuits use only N channel E-MOSFETs

(iii) CMOS — (Complementary MOS)–They use PMOS as well as NMOS devices.

Most of the MOS digital ICs are constructed entirely using MOSFETs only.

2. Advantages of MOSFETs and MOS ICs

The major advantages of using MOSFET may be listed as under:

(i) It is easy and inexpensive to fabricate

(ii) It is small

(iii) It consumes very little power.

The fabrication of MOS ICs is far less complicated than that of the bipolar ICs such as TTL and ECL. The MOS devices need much less space on a chip as compared to that required for BJT. In MOS ICs, we need not have to fabricate resistors. The MOSFETs themselves can work as resistor. This will reduce the required chip area to a great extent. Due to all the reasons discussed above, the MOS ICs can accommodate more number of elements on the same chip area as compared to the bipolar ICs. Due to these advantages the MOS ICs have dominated or surpassed the bipolar ICs in the field of large scale integration (LSI and VLSI). The MOS ICs are particularly suitable for the complex ICs such as memory chips and microprocessors due to the high packing density. The MOS ICs that are faster than the 74, 74LS and 74 ALS TTL devices with a comparable current levels are now available. Hence, the MOS ICs, specially the CMOS ICs have dominated the SSI and MSI market.

3. Disadvantages of MOSFETs

The major disadvantage of MOSFETs is that they are susceptible to static charges. Due to accumulated static charges, MOSFETs can get damaged. By using the proper handling procedures, it is possible to minimize the possibility of damage. But still TTL proves to be more reliable and durable for laboratory work.

4.1.9. The MOSFET as a Switch

1. Basic Aspects

Let us observe how can we operate the n-channel MOSFET as a switch as shown in figure 4.21(a).

For n-channel MOSFET the drain to source voltage is always made positive. The gate to source voltage V_{GS} is the input voltage and it is used for controlling the resistance between drain and source (channel resistance). Hence, V_{GS} decides whether the device is ON or OFF.

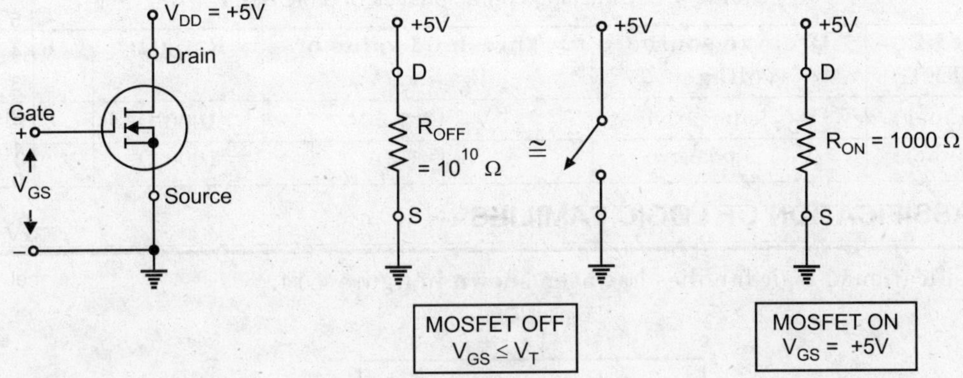

(a) Biasing of MOSFET (b) MOSFET OFF (c) MOSFET ON

Fig. 4.12.

2. MOSFET OFF

If $V_{GS} \le V_T$ then the MOSFET remains OFF. Drain current is almost equal to zero. Hence, the device offers a very large resistance of the order of 10^{10} Ω. So, MOSFET is equivalent to an open switch as shown in figure 4.13 (b). MOSFET remains of as long as V_{GS} is less then V_T.

3. MOSFET ON

As V_{GS} is made positive and increased, it reaches a voltage called "threshold voltage" (V_T). Typical value of V_T is 1.5 V. As soon as V_{GS} becomes higher than V_T, the MOSFET will be turned ON. When V_{GS} = + 5 V, the MOSFET becomes completely ON and its channel resistance reaches to about 1000 Ω. Hence, the MOSFET is equivalent to a resistance of value R_{ON} = 1000 Ω as shown in figure 4.13 (c).

4.1.10. P-channel MOSFET as a Switch

The P-channel MOSFET can also be used as a switch by following the same principle as that for an n-channel MOSFET. The only difference is that all the voltages will be negative as shown in figure 4.13 (a). The p-channel MOSFET will be OFF if V_{GS} = 0 Volts and R_{OFF} = 10 Ω and is equivalent to an open switch. The threshold voltage V_T for a p-channel MOSFET is negative. It

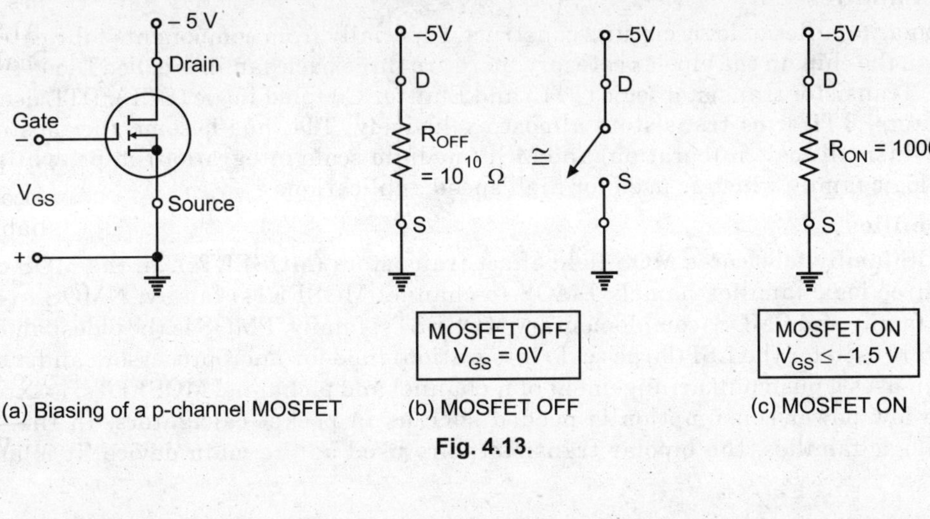

(a) Biasing of a p-channel MOSFET (b) MOSFET OFF (c) MOSFET ON

Fig. 4.13.

is typical value is – 1.5 V. The p-type MOSFET will conduct fully if V_{GS} = – 5 V. Then the on state resistance offered by the MOSFET is R_{ON} = 1000 Ω. The switching characteristics of the P and N type MOSFETs are listed in table 4.1.

Table 4.1. Switching characteristics of a MOSFET

Type of MOSFET	Drain to source voltage	Threshold value of V_{GS} *i.e.*, V_T	R_{ON} (Ω)	R_{OFF} (Ω)
P channel	V_{DS} is negative	V_T = – 1.5 V	1000 Ω	10^{10} Ω
N channel	V_{DS} is positive	V_T = 1.5 V	1000 Ω	10^{10} Ω

4.2 CLASSIFICATION OF LOGIC FAMILIES

The classification of logic families has been shown in figure 4.14.

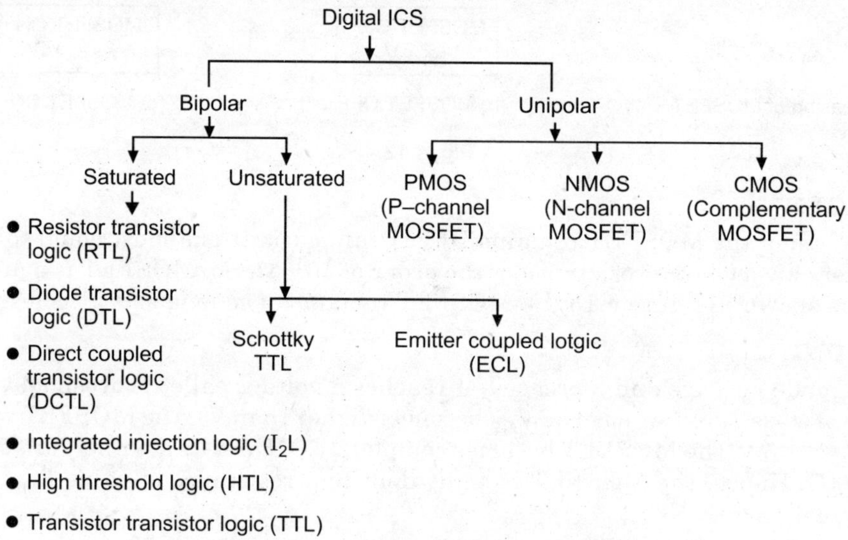

Fig. 4.14. Classification of logic families

The two basic techniques for manufacturing ICs are as under:

1. Bipolar technology
2. Metal oxide semiconductor (MOS) technology.

1. Bipolar Families

The bipolar families of logic circuits construct, especially from components fabricate bipolar transistors on the chip. In the bipolar category there are three basic families called Diode transistor logic (DTL). Transistor transistor logic (TTL) and Emitter Coupled Logic (ECL). DTL uses diodes and transistors, TTL uses transistors almost exclusively. TTL has become the most popular family in SSI (small scale integration) and MSI (medium scale integration) chips, while ECL is the fastest logic family which is used for high speed applications.

2. MOS Families

The MOS family fabricates MOS field effect transistors (MOSFETs). In the MOS category there are three logic families namely PMOS (p-channel MOSFETs) family, NMOS (n-channel MOSFETs) family and CMOS (complementary MOSFETs) family. PMOS is the oldest and slowest type. NMOS is used for the LSI (large scale integration) field for microprocessors and memories. CMOS which uses a push pull arrangement of n-channel and p-channel MOSFETs, is extensively used where low power consumption is needed such as in pocket calculators. In the "Bipolar saturated" logic families, the bipolar transistors are used as the main device. It is used as a switch and operated in the saturation or cutoff regions. TTL is an example of saturated bipolar logic. In the unsaturated bipolar logic, the bipolar transistors are not driven into hard saturation. This increases the speed of operation. So the unsaturated bipolar ICs such as schottky TTL and ECL (emitter coupled logic) are much faster as compared to TTL. All these ICs are fabricated on Silicon chips using different fabrication technologies.

4.2.1. Positive and Negative Logic

The digital circuits will have only two distinct output voltage levels. These two voltage levels will be used to represent the logic 0 and logic 1 outputs of the logic circuit as shown in figure 4.15.

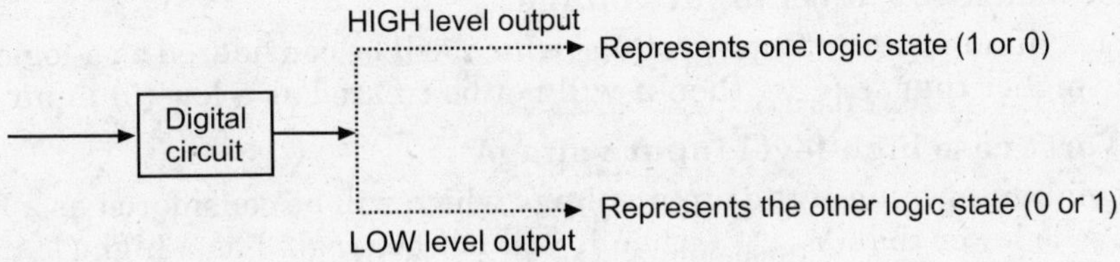

Fig. 4.15.

The high and low voltage levels of the output voltage can be assigned to represent logic 0 or logic 1 levels by using following two different types of logic :

1. Positive logic
2. Negative logic

1. Positive Logic

A **LOW** voltage level represents **logic 0** state and a comparatively **HIGH** output voltage level represents **logic 1** state, as shown in figure 4.16 (a). For example, 0 volts represent a logic 0 state and + 5 V represent logic 1. This is called as positive logic.

Positive logic : Logic 0 (LOW) = 0 V
Logic 1 (HIGH) = + 5 V.

2. Negative Logic

A LOW voltage level represents logic 1 state and a **HIGH** output voltage level represents logic 0 state, as shown in figure 4.16 (b). For example, 0 volts represent a logic 1 state and + 5 V represent logic 0 state. This is called as *negative logic*.

Negative Logic : Logic 0 (LOW) = + 5 V
Logic 1 (HIGH) = 0 V.

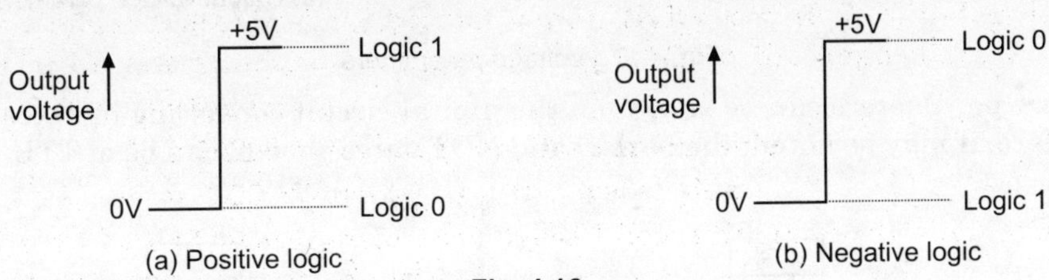

Fig. 4.16.

Important Point : In this chapter, we shall consider only the positive logic. Also we will assume the logic 0 level corresponds to 0 Volts and logic 1 level corresponds to + 5 V.

4.3 DIGITAL IC TERMINOLOGY

Even though there are various logic families, the general characteristics, their definitions, nomenclature and terminologies used for all of them have been standardized. So let us discuss some of the most important general characteristics first and then discuss them for some particular families.

4.3.1. Voltage and Current Parameters

Voltage parameters (Threshold Levels)

Ideally, the input voltage levels of 0 V and + 5 V (for TTL) are called as logic 0 and 1 levels respectively. However practically we won't always obtain voltage levels matching exactly to these values. Therefore, it is necessary to define the worst case input voltages.

(i) $V_{IL(max)}$-Worst case low level input voltage

This is the maximum value of input voltage which will be considered as a logic 0 level. If the input voltage is higher than $V_{IL(max)}$ then it will not be treated as a low (0) input level.

(ii) $V_{IH(min)}$- Worst case high level input voltage

This is the minimum value of the input voltage which will be considered as a logic 1 level. If the input voltage is lower than $V_{IH(min)}$, then it will not be treated as a high (1) input.

(iii) $V_{OH(min)}$-Worst case high level output voltage

This is the minimum value of the output voltage which will be considered as a logic HIGH (1) level. If the output voltage is lower than this level then it won't be treated as a HIGH (1) output.

(iv) $V_{OL(max)}$-Wrost case high level output voltage

This is the maximum value of the output voltage which will be considered as a logic LOW (0) level. If the output voltage is higher than this value when it won't be treated as a LOW (0) output. All the voltage parameters have been shown in figure 4.17.

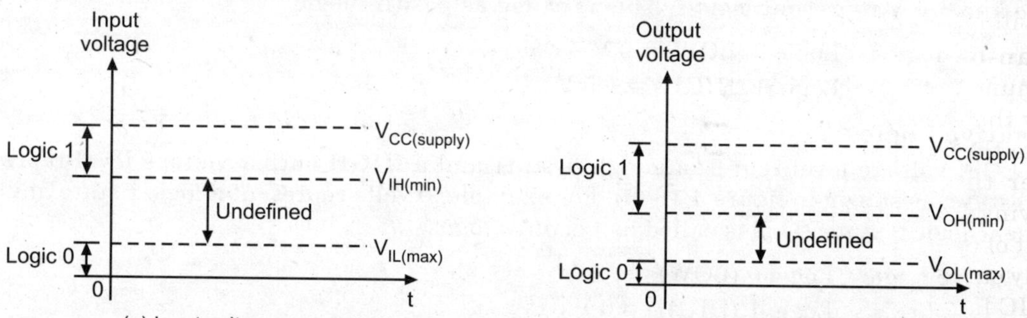

(a) Input voltage parameters (b) Output voltage parameters

Fig. 4.17. Voltage parameters

The voltage parameters can be shown on the digital circuit consisting of gates as shown in figure 4.18. Here, it may be noted, the NAND and NOT gates shown, can be of TTL, ECL, CMOS or any other type.

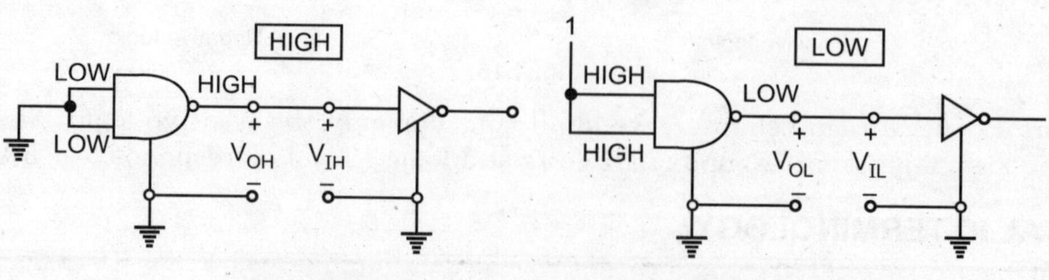

Fig. 4.18. Voltage parameters on a logic circuit

Current Parameters

(a) **I_{IL}-Low level input current:** It is the current that flows into the input when a low level input voltage in the specified range is applied.

(b) **I_{IH}-High level input current:** It is the current that flows into the input when a high level voltage input voltage in the specified range is applied.

(c) **I_{OL}-Low level output current :** This is the current that flows from the output when the output voltage is in the specified low (0) voltage range and a specified load is applied.

(d) **I_{OH}-High level output current :** This is the current flowing from the output when the output voltage is in the specified HIGH (1) voltage range and a specified load is applied.

If the output current flow into the output terminal then it is called as a sinking current and if the output current flows away from the output terminal then it is called as a sourcing current. The current parameters have been displayed on the logic circuit shown in figure 4.19.

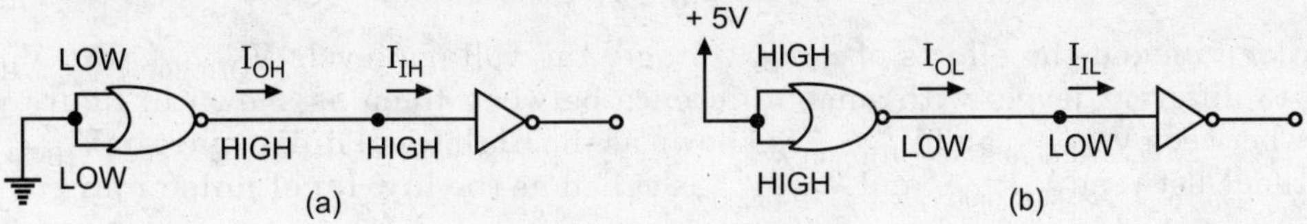

Fig. 4.19. Current parameters

Here, it may be noted that the actual current directions can be opposite to those shown in figure 4.19, depending on the logic family. In most data books, current flowing into a node or device is considered positive and current flowing out of node or device is considered negative.

4.3.2. Fan-in and Fan-out

Fan-in : Fan-in is defined as the number of inputs a gate has. For example a two input gate will have a fan-in equal to 2.

Fan-out : Fan-out is defined as the maximum number of inputs of the same IC family that a gate can drive without falling outside the specified output voltage limits. Higher the fan out higher the current supplying capacity of a gate. For example, a fan out of 5 indicates that the gate can drive (supply current to) at the most 5 input of the same IC family.

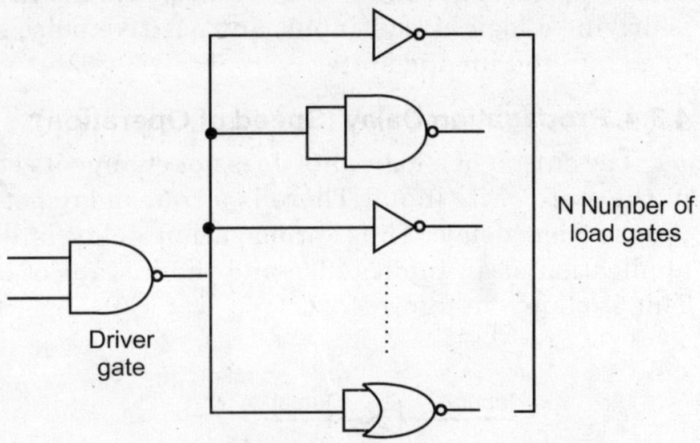

Fig. 4.20. Fan-out

The concept of Fan-out will be more clear if you refer to figure 4.20. As shown in figure 4.20, the fan-out of the driver gate which is driving N number of gates is N. Fan-out is also called as the **loading factor**. If the specified fan out of a gate is 10 then we should not load it with more than 10 gates. Because then the output logic level voltages can not be guaranteed. Fan-out depends on the nature of input devices that are connected to an output. Unless a different logic family is specified as the load device, fan out is assumed to be referred to load devices of same family as the driving device.

4.3.3. Noise Margin

To understand the meaning of the term Noise Margin or Noise Immunity, let us consider the input and output voltage profiles shown in figure 4.21. Noise is an unwanted electrical disturbance which may induce some voltage in the connecting wires used between two gates or from a gate output to load. **Noise immunity** is defined as the ability of a logic circuit to tolerate the noise without causing any unwanted changes in the output. A quantitative measure of noise immunity is called as **noise margin**.

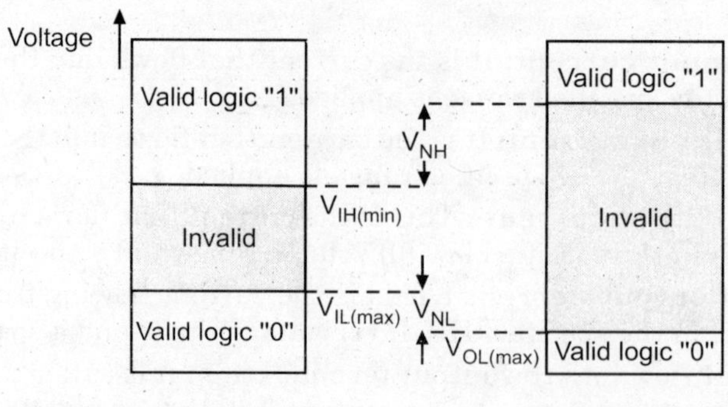

(a) Input profile (b) Output profile

Fig. 4.21.

In order to avoid the effects of noise voltage, the voltage levels $V_{OH(min)}$ and $V_{IH(min)}$ are adjusted to different levels with some difference between them as shown in figure 4.21. The difference between $V_{OH(min)}$ and $V_{IH(min)}$ is known as the **high level noise margin** V_{HN}. Similarly, the difference between $V_{IL(max)}$ and $V_{OL(max)}$ is called as the **low level noise margin** V_{NL}.

High level noise margin, $V_{NH} = V_{OH(min)} - V_{IH(min)}$

Low level noise margin, $V_{NL} = V_{IL(max)} - V_{OL(max)}$

When a high logic output is driving a logic circuit input, any negative noise spike greater than V_{NH} can cause the voltage to drop into the invalid range. Similarly, when a low logic output is driving a logic circuit input, any positive noise spikes greater than V_{NL} can cause the voltage to go into the invalid state.

4.3.4. Propagation Delay (Speed of Operation)

The output of a logic gate does not change its state instantaneously in response to the change in the state of its input. There is a time delay between these two events, which is called as the propagation delay. Thus, propagation delay is defined as time delay between the instant of application of an input pulse and the instant of occurrence of the corresponding output pulse. This is shown in figure 4.22.

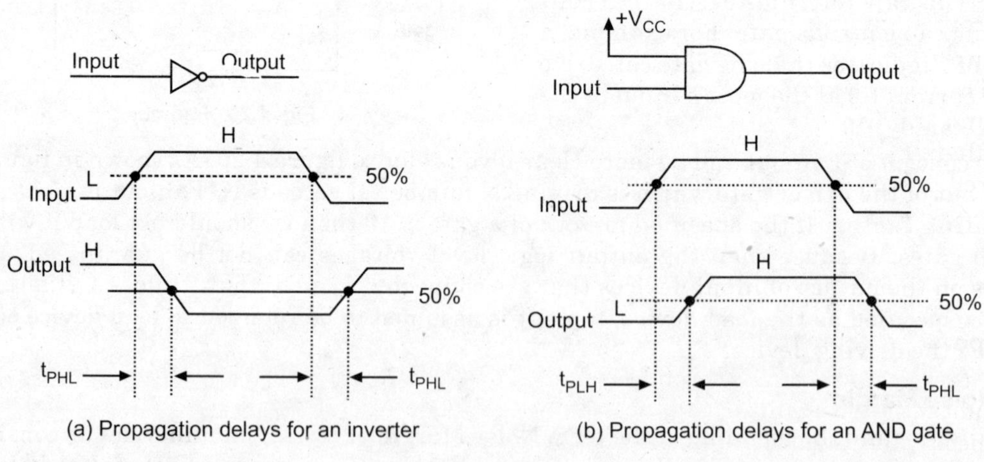

(a) Propagation delays for an inverter (b) Propagation delays for an AND gate

Fig. 4.22.

From figure 4.22, it may be observed that there are following two propagation delays:

1. t_{PHL}: The propagation delay measured when the output is making a transition from HIGH (1) to LOW (0) state.

2. t_{PLH} : The propagation delay measured when the output makes a transition from LOW (0) to HIGH (1) state.
   (a) The values of t_{PHL} and t_{PLH} are not always same. If they are not equal then the one which is higher is considered as the propagation delay time of the gate.
   (b) The propagation delays are measured between the points corresponding to 50% levels as shown in figure 4.22.

Ideally propagation delay should be zero and practically it should be as short as possible. The values of propagation delays are used as a measure of relative speed of logic circuits. For example, a logic circuit with a propagation time of 5 ns will be a faster logic circuit than the one with 10 ns propagation time, under the specified load conditions.

4.3.5. Power Dissipation

As a result of applied voltage and currents flowing through the logic ICs, some power will be dissipated in it, in the form of heat. This power is in milliwatts. Care should be taken to reduce the power dissipation taking place in the logic IC in order to protect the IC against damage due to excessive temperature, to reduce the loading on power supplies etc. Another importance of power dissipation is that the product of power dissipation and propagation time is always constant. Therefore, reduced power dissipation may lead to increase in propagation delay. Usually there is only one power supply terminal on any IC. It is denoted by V_{CC} for the TTL ICs and V_{DD} for the CMOS ICs. The power drawn by an IC from the power supply is given by

$$P = V_{CC} \times I_{CC}$$

Where I_{CC} is the current drawn from the power supply.

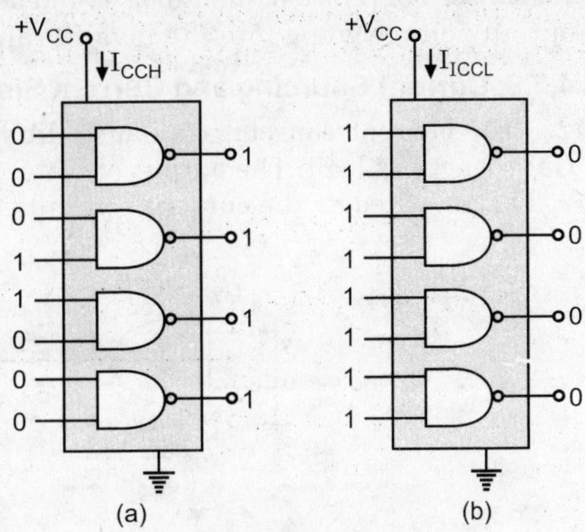

Fig. 4.23. Power dissipation

For many ICs the current drawn from the supply will be dependent on the logic states of the circuits on chip. Figure 4.23 (a) shows a NAND gate IC with all its outputs high. The current drawn from the source under such conditions is denoted by I_{CCH}.

Figure 4.23 (b) shows another extreme condition where the outputs of all the NAND gates are '0'. The current drawn from the source under such conditions is denoted by I_{CCL}. The values of I_{CCH} and I_{CCL} are measured with open circuited outputs (no load), because the load will change these values. I_{CCH} and I_{CCL} are of different values. So, an average value of them is generally considered to calculate the average power dissipation.

$$I_{CC(avg)} = \frac{I_{CCH} + I_{CCL}}{2} \quad \text{...(4.1)}$$

$$P_{D(av)} = V_{CC} \times I_{CC(avg)} \quad \text{...(4.2)}$$

4.3.6. Operating Temperature

The temperature range acceptable for the consumer and industrial applications is 0° to 70° C and that for the military applications is – 55°C to 125°C. The performance of gates will be in the specified limits over these temperature ranges.

4.3.7. Figure of Merit (Speed Power Product SPP)

The figure of merit of a logical family is the product of power dissipation and propagation delay. It is called as the speed power product. The speed is specified in seconds and power is specified in W.

Hence, Figure of merit = Propagation delay time × Power dissipation

In practice, the value of figure of merit should be as low as possible. Figure of merit is always a compromise between speed and power dissipation. That means if we try to reduce the propagation delay then the power dissipation will increase and vice-versa. The speed-power product is used as a common means for measuring and comparing the overall performance of different IC family. Suppose that an IC family has an average propagation delay of 20 ns and an average power dissipation of 5 mW. Then its SPP is given by,

$$SPP = 20 \text{ ns} \times 5 \text{ mW} = 100 \times 10^{-12} \text{ Watt-second} = 100 \text{ picojoules}$$

4.3.8. Invalid Voltage Levels

The operation of a logic circuit will be proper if and only if its input voltage levels are kept outside the invalid voltage range. That means the input voltage should be either lower than $V_{IL(max)}$ or higher than $V_{IH(min)}$. The invalid input voltage will produce an unpredictable output response. Therefore it should be avoided. When the output is overloaded, there is a possibility of output voltage going into the invalid range.

4.3.9. Current Sourcing and Current Sinking

The current sourcing action is illustrated in figure 4.24 (a). Gate-1 acts as a source and Gate-2 acts as load. The output of Gate-1 is high. It supplies a current I_{IH} to the input of Gate-2. This is called as the current sourcing action.

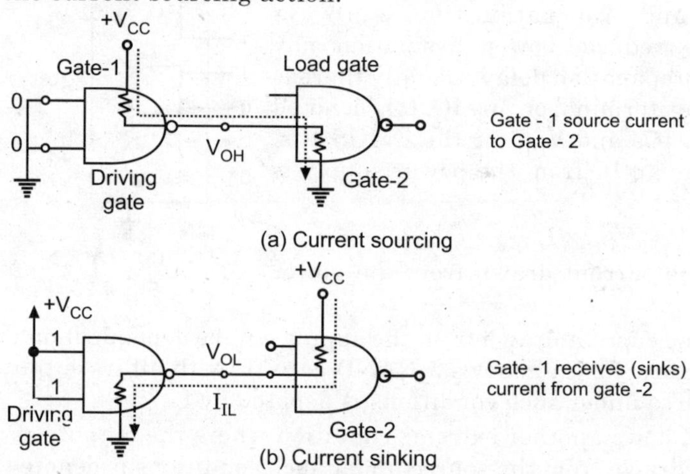

Fig. 4.24.

The current sinking action has been demonstrated in figure 4.24 (b). Gate-2 which is the load gate acts as a load. As soon as the output of Gate-1 goes low, the current starts flowing into the output terminal of Gate-1 as shown in figure 4.24(b). Thus, when gate-1 is accepting the current through its output terminal, it is said that the current sinking is taking place.

4.4 RESISTOR TRANSISTOR LOGIC (RTL)

1. Basic Aspects

RTL logic is the earliest logic family but it has becomes obsolete now. The long form of RTL is resistor transistor, logic which suggests

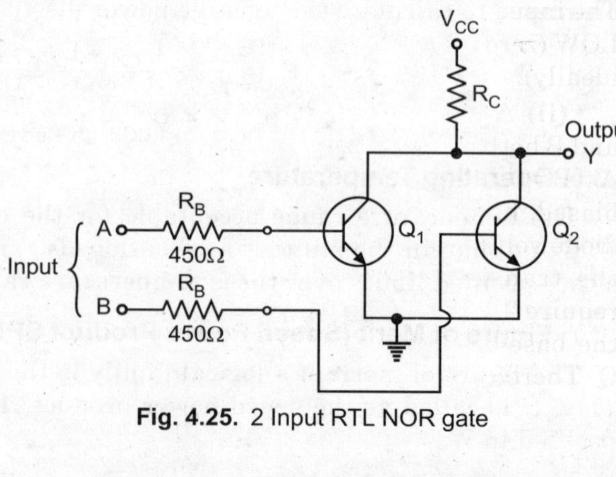

Fig. 4.25. 2 Input RTL NOR gate

that it uses resistors and transistors. The basic RTL gate is a NOR gate as shown in figure 4.25.

2. Working Operation

1. A = B = 0

(i) With A = B = 0, both the transistors remain off. The current through R_C is zero, so drop across it is zero.

(ii) The output voltage is equal to V_{CC} i.e., 1.

Therefore, for A = B = 0. Y = 1

2. A = 0, B = 1

(i) With A = 0 and B = 1, transistor Q_1 is off and Q_2 goes into saturation.

(ii) Hence, the output voltage $V_o = V_{CE2(sat)}$ i.e., 0.

Therefore, for A = 0, B = 1, Y = 0

3. A = 1, B = 0:

(i) With A = 1 and B = 0, transistor Q_1 is saturated and Q_2 is turned off.

(ii) Hence, the output voltage $Y = V_{CE1\,(sat)}$ i.e., 0.

Therefore, for A = 1, B = 0, Y = 0

4. A = B = 1

(i) With A = B = 1, both the transistors saturate and output voltage is equal to saturation voltage

Therefore, A = B = 1, Y = 0

3. Important characteristics of RTL family

(i) Poor noise margin
(ii) Poor fan out
(iii) Low speed
(iv) High power dissipation

4.5 DIODE TRANSISTOR LOGIC (DTL)

1. Basic Aspects

A basic DTL circuit, working as a NAND gate is shown in figure 4.26. It consists of input stage comprising of diodes D_1, D_2 and resistor R_D forming an AND gate. It is followed by a transistorized inverter.

2. Working Operation

(i) A and B are the input terminals. The input voltages A and B can be either LOW (zero volts ideally) or HIGH (+ V_{CC} ideally).

(ii) A and B both are LOW: If A and B both are connected to ground, i.e., A = B = 0, then D_1 and D_2 will get forward biased, hence the potential at M is one diode voltage drop = 0.7 V. But, to drive the transistor Q into conduction we require 2.1 V to forward bias D_3, D_4 and the base-emitter junction of transistor Q. Therefore, Q is cut off and output.

$$Y = V_{CC} = \text{logic } 1$$

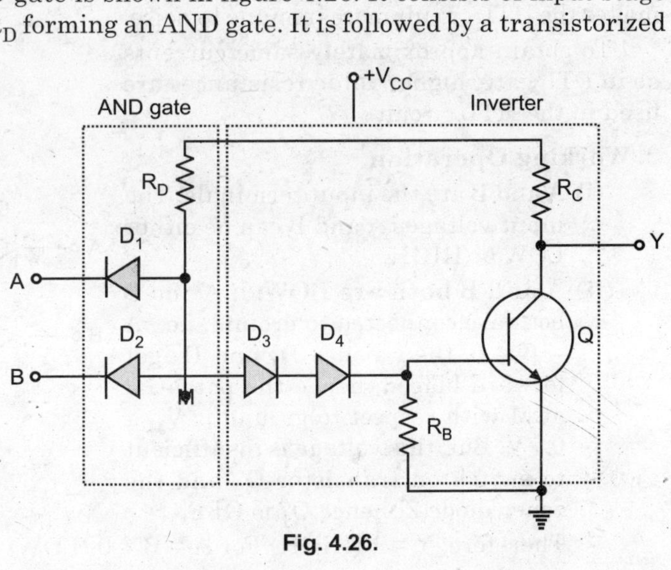

Fig. 4.26.

Therefore, $Y = 1$ (HIGH) for $A = B = 0$ (LOW)

(iii) Either A or B is LOW: If any one of the inputs, A or B is connected to ground with the other terminal connected to + V_{CC}, then the corresponding diode will conduct. Again $V_M \cong 0.7$ V and Q will be cut off giving output Y = Logic 1.

Therefore, $Y = 1 \begin{cases} \text{if } A = 0 \text{ and } B = 1 \text{ or} \\ \text{if } A = 1 \text{ and } B = 0 \end{cases}$

(iv) A and B both are HIGH: If A and B both are connected to + V_{CC}, then both the diodes D_1 and D_2 will be reverse biased and do not conduct. Diodes D_3 and D_4 are forward biased and base current is supplied to transistor Q via R_D, D_3 and D_4. Transistor Q is driven into saturation and the output voltage Y will be pulled down to a Low voltage.

Therefore, $Y = 0$ (LOW) for $A = B = 1$ (HIGH)

This discussion reveals that the circuit operates as a NAND gate.

3. Advantages

(i) Greater fan-out is possible because of high input impedance of subsequent gates in logic 1 state.

(ii) It has greater noise immunity.

(iii) Use of diodes rather than capacitors and resistors make the DTL circuit more economical in IC form.

4.6 HIGH THRESHOLD LOGIC GATE (HTL) (PTU, Jalandhar, Sem. Exam., 2006-07)

1. Basic Aspects

When the industrial environment is noisy due to the presence of motors, high voltage switches, etc. the above logic families will not work satisfactorily. Hence, we require a logic family having very high noise immunity, like the HTL family.

The HTL family uses higher supply voltage like 15 V instead of 5 V. An HTL circuit working as a NAND gate is shown in figure 4.27.

The use of higher supply voltage and a zener diode in place of diodes as in DTL circuit, makes the HTL circuit more immune to noise.

To obtain approximately same currents as in DTL gate, higher value resistances are used in the HTL circuit.

2. Working Operation

(i) A and B are the input terminals. The input voltages A and B can be either LOW or HIGH.

(ii) **A and B both are LOW:** If A and B both are connected to ground, i.e., A = B = 0, then diodes D_1 and D_2 get forward biased, hence the potential at M with respect to ground is $V_M \cong 0.7$ V. But, this voltage is insufficient to switch on transistor Q_2 and the zener diode Z, hence Q_1 is OFF.

Therefore, $Y = 1$ (HIGH) For $A = B = 0$ (LOW)

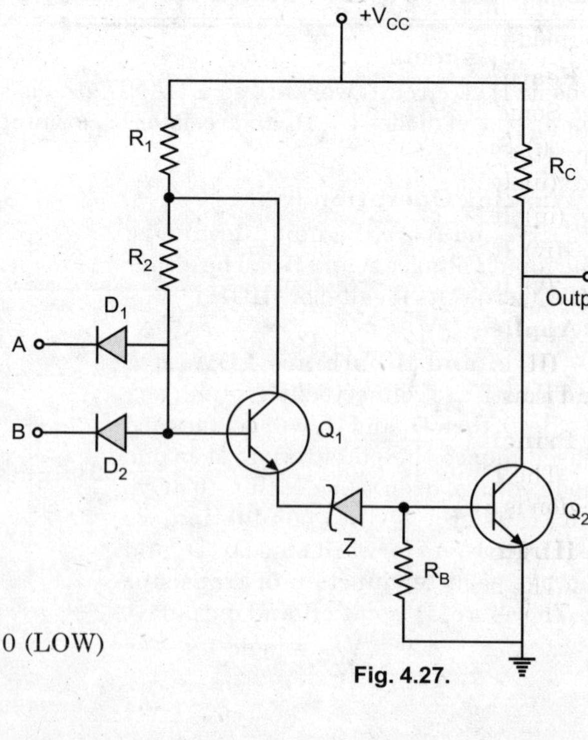

Fig. 4.27.

(iii) Either A or B is LOW: If any one of the inputs, A or B is connected to ground with the other terminal connected to + V_{CC}. Then the corresponding diode conducts. Again $V_M \cong$ 0.7 V and Q_2 will still be OFF giving output Y = logic 1.

Therefore, $$Y = 1 \begin{cases} \text{if } A = 0 \text{ and } B = 1 \text{ or} \\ \text{if } A = 1 \text{ and } B = 0 \end{cases}$$

The voltage at point M required to forward bias Q_1, Q_2 and the zener diode is,

$$V_M = V_{BE2} + V_z + V_{BE1} = 0.7 + 6.8 + 0.7 = 8.2 \text{ V}$$

(Where, V_z = zener breakdown voltage)

Therefore, minimum voltage at A and B input required is,

$$V_{in} = V_M - V_{FD1/D2} = 8.2 - 0.7 = 7.5 \text{ V}$$

Hence, V_{in} required is much higher than the noise levels present in the environment, which make HTL gate more noise immune.

(iv) A and B both are HIGH: If A and B both are connected to + V_{CC}, then both the diodes D_1 and D_2 get reverse biased and do not conduct. Now, Q_2 gates switched ON, conducts the zener diode and switching ON Q_1 and output voltage Y is pulled down to a low voltage.

Therefore, $\quad Y = 0$ (LOW) $\quad$ for A = B = 1 (HIGH)

The above discussion shows that the circuit operates a NAND gate.

3. Advantages

(i) Higher noise immunity

(ii) Input-output characteristics are independent of temperature variable.

4. Disadvantages

Slow speed due to increased supply voltage, resulting in use of high value resistors.

4.6.1. Integrated-Injection Logic (IIL)

The integrated injection logic (IIL or I^2L logic) **is based on** the principle of DCTL i.e., direct coupled transistor logic.

1. Features of IIL

Some of the important features of IIL are as under:

(i) Simplicity

(ii) It is easier and cheaper to fabricate.

(iii) It is a saturated bipolar logic.

(iv) It consumers very little power.

(v) It is used for medium and large scale integration.

2. Application

IIL is used for large scale integration such as SBP 9900 a 16 bit microprocessor developed by Texas Instruments.

3. Principle of IIL

The IIL is based on the concept of merging the components that means one semiconductor region is a part of two or more devices. Hence, it is also called as merged transistor, logic (MTL).

4. IIL Inverter

The basic IIL inverter is shown in figure 4.28.

5. Working Operation

1. $V_i = 0$ (Low)

If $V_i = 0$ i.e., at low logic level, then Q_1 is off, so I_{B1} is equal to zero. The input source acts like a sink for current I_1. Hence, I_2 flows through the base of Q_2 and drives it into saturation. Since, Q_2 is on, output will be low.

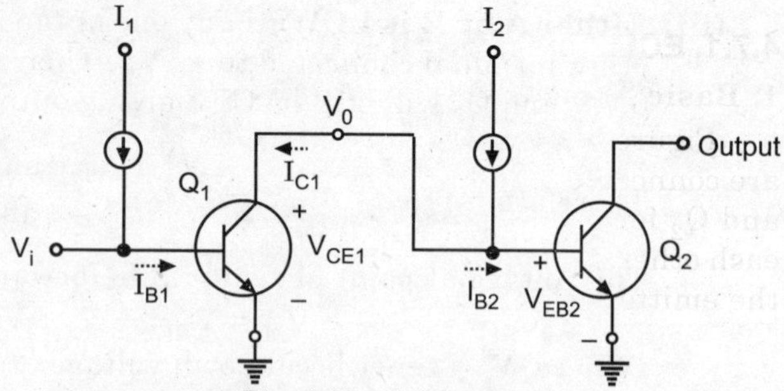

Fig. 4.28. IIL inverter

2. $V_i = 1$ (High)

If V_i is high, then the base current I_{B1} has two components, one is I_1 and the other is from source V_i. So, Q_1 goes into saturation. Hence, $V_{CE1} = V_{CE\ sat} = 0.2$ V. This will drive Q_2 into cutoff and Q_1 acts as sink for I_2. Since, Q_2 is off, output will be high. Thus, the circuit of figure 4.28 acts as in inverter.

6. Characteristics of IIL

(i) Speed of operation depends on the charging current.
(ii) Propagation delay is inversely proportional to charging current.
(iii) Power dissipation is proportional to charging current.
(iv) Figure of merit is in the range of 0.1 to 0.7 PJ.
(v) Very small silicon area is required.
(vi) Packaging density is in the range of 120 to 200 gates per mm^2.

4.7 EMITTER COUPLED LOGIC (ECL)

1. Basic Aspects

As shown in the classification of logic families, TTL is a saturated bipolar family whereas ECL is an unsaturated bipolar family. In TTL gates, the transistors enters into deep saturation. So, the switching speed is limited by the storage delay time. But, since the transistors used in the ECL family do not enter into saturation, it is possible to switch them at much faster speeds. Emitter coupled logic (ECL) is the fastest of all the logic families. Therefore, it is preferred in all the high speed applications.

2. Features, Merits and Demerits of ECL Family

Transistor operate in either cut off or active regions. They are not allowed to operate in saturation. They can switch at very fast speeds. In fact ECL is the fastest logic family. Typically the propagation delay is 1 ns per gate. The logic voltage levels are adjusted to be very close to each other. So, as to avoid the transistor saturation, noise margin is low because the logic are close to each other. Hence, ECL gates are susceptible to noise. Transistors in active region consume more power. Therefore, ECL gates consume more power.

3. Logic Levels in ECL

There are two most widely used ECL families namely 10XXX or 10 k family and 100 XXX or 100 k family. The voltage levels for the same is specified in table 4.2.

Table 4.2. Logic levels for ECL

Series	Supply voltage	V_{OL} Volts	V_{OH} Volts	V_{IL} Volts	V_{IH} Volts
10 XXX	5.2 Volts	–1.7	– 0.9	– 1.4	– 1.2
100 XXX	4.5 Volts	– 1.7	– 0.9	– 1.4	– 1.2

4.7.1. ECL 2-input OR/NOR Gate

1. Basic Aspects

Figure 4.29 shows a two input ECL OR/NOR gate. The emitter or transistor Q_1, Q_2 and Q_3 are connected together so this logic is called as emitter coupled logic. The pair of transistors Q_1 and Q_2 forms an emitter coupled different amplifier Q_1 and Q_3 are connected in parallel with each other. Q_4 and Q_5 are two emitter follower stages. The NOR and OR outputs are obtained at the emitters of Q_4 and Q_5 respectively.

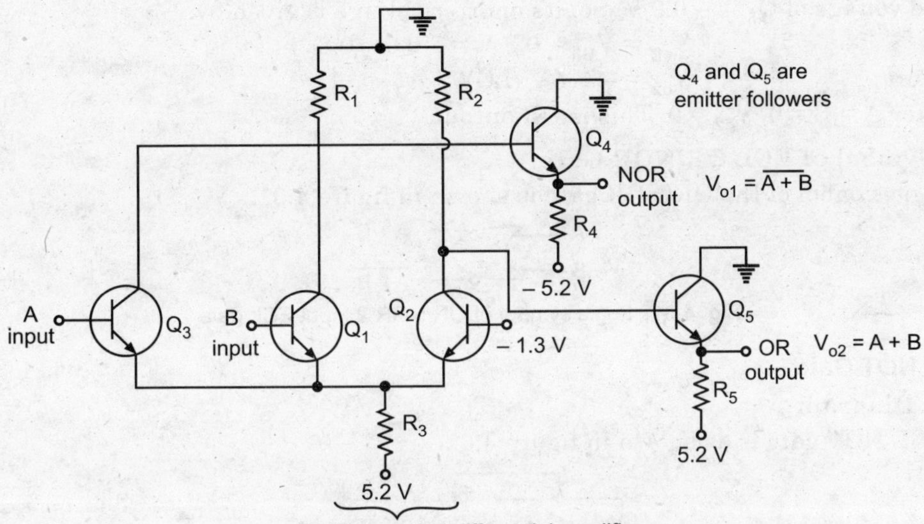

Fig. 4.29. Two input ECL OR/NOR gate

2. Working Operation

When A = B = 0, i.e., when $V_A = V_B = -1.7$ V, the equivalent circuit has been shown in figure 4.30. The base voltages of Q_1, Q_2 and Q_3 are – 1.7 V, – 1.3 V and – 1.7 V respectively. That means the base voltage of Q_2 is most positive. Therefore, Q_2 will conduct heavily and go into saturation. Its base to emitter voltage drop, i.e., $V_{BE2} = 0.8$ V so the emitter voltage of all the transistors will be equal to – 2.1 volts. This voltage will reverse bias the base emitter junctions of Q_1 and Q_2 so, Q_1 and Q_2 both will be off.

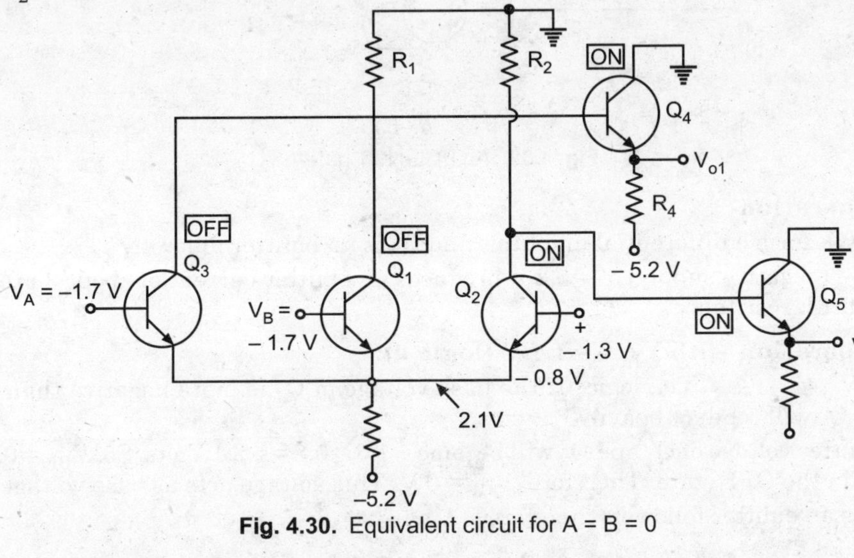

Fig. 4.30. Equivalent circuit for A = B = 0

Since, Q_1 is OFF, the current flowing through R_1 will only be the base current of Q_4. The value of R_1 is selected in such a way that the voltage drop across R_1 is 0.1 Volts. Hence, the base voltage of Q_4 is – 0.1 Volts. So, the emitter of Q_4 (conducting) will be given by,

$$V_{o1} = V_{B4} - V_{BE4} = -0.1 - 0.8$$

Therefore, $$V_{o1} = -0.9 \text{ V (HIGH)}$$

Thus, with A = B = 0, V_{o1} = 1. This is the NOR output of the circuit. Now, consider Q_2. It is already conducting. The voltage drop across R_2 due to I_{C2} is adjusted to be equal to 0.9 Volts. Hence base voltage of Q_5 is – 0.9 V. So, its emitter voltage is given by,

$$V_{o2} = V_{B5} - 0.8 \text{ V} = -0.9 - 0.8$$

Therefore, $$V_{o2} = -1.7 \text{ V (LOW)}$$

Thus, for A = B = 0, V_{o2} = 0. This is OR output.

3. Logic Symbol of ECL OR/NOR Gate

The logic symbol of the OR/NOR gate is shown in figure 4.31.

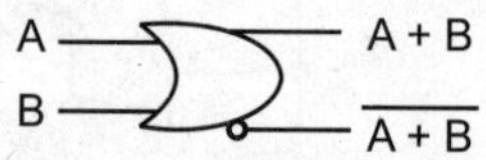

Fig. 4.31. Logic symbol of OR/NOR 2 input ECL gate

4.7.2. ECL NOT Gate

1. Circuit Diagram

The ECL NOT gate is as shown in figure 4.32.

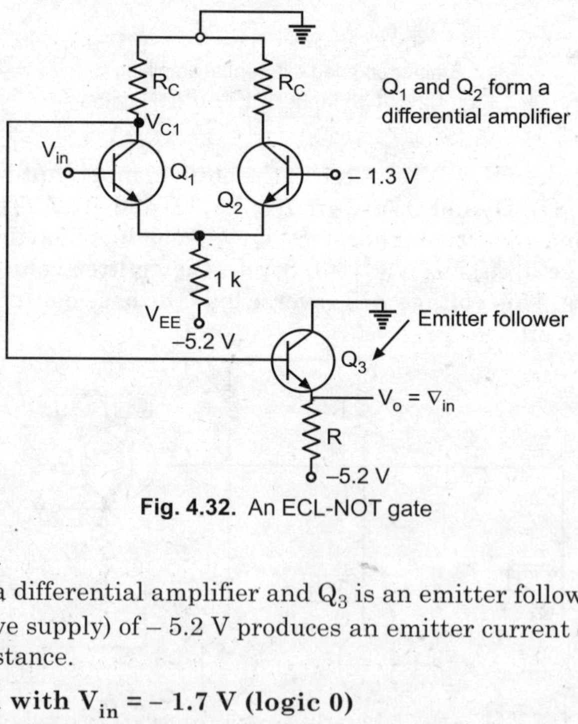

Fig. 4.32. An ECL-NOT gate

2. Working Operation

(i) Q_1 and Q_2 form a differential amplifier and Q_3 is an emitter follower.

(ii) The V_{EE} (negative supply) of – 5.2 V produces an emitter current of about 3 mA that flows through 1 k resistance.

3. Working Operation with V_{in} = – 1.7 V (logic 0)

(i) When V_{in} = – 1.7 V, i.e., logic '0', the base voltage of Q_1 is more negative than that of Q_2. Hence, Q_2 will conduct heavily.

(ii) The emitter voltage of Q_1 and Q_2 will become – 1.3 – 0.8 = – 2.1 Volts. So, V_{BE} = 0.4 V and Q_1 will be in the OFF state. Therefore, V_{C1} = 0 V. This voltage acts as base voltage for Q_3.

(iii) As Q_3 is an emitter follower, therefore,

$$V_o = V_{C1} - V_{BE3} = 0 - 0.8 = -0.8 \text{ V}$$

(iv) But – 0.8 V is logic '1' level for ECL. Thus, $V_o = 1$ when $V_{in} = 0$ and inversion takes place.

4. Working Operation with V_{in} = – 0.8 V (logic 1)

(i) If $V_{in} = -0.8$ V, i.e., logic 1 then base of Q_1 is less negative than that of Q_2. Hence Q_1 conducts heavily.

(ii) The value of R_C is such that a voltage of – 0.9 V is developed across it. So, $V_{C1} = -0.9$ V.

(iii) $V_{C1} = V_{B3} = -0.9$ V. Hence, the output voltage is given by $V_o = V_{C1} - V_{BE3} = -0.9 - 0.8 = -1.7$ V which is logic 0.

(i) Thus, when V_{in} = logic 1 the output V_o = logic 0 and the circuit works as an inverter.

(ii) Table 4.3 summarizes the operation of ECL inverter.

Table 4.3. Summary of operation ECL inverter

V_{in}	Q_1	Q_2	Q_{C1}	V_{C2}	$V_o = V_{C1} - 0.8$ V
– 1.7 V (logic 0)	OFF	ON	0 V	– 0.9 V	– 0.8 V (logic 1)
– 0.8 V (logic 1)	ON	OFF	– 0.9 V	0 V	– 1.7 V (logic 0)

4.7.3. ECL Characteristics

Some of the most important ECL characteristics are as under:

(i) Maximum propagation delay : For the latest ECL gates the propagation delay is very short, of the order of 500 ps.

(ii) Average power dissipation : 5 mW (for advanced ECL gates)

(iii) Speed-power product (SPP): 0.5 pJ.

(iv) Worst case noise margin is only 150 mV. So, ECL device are unreliable to work in noisy environments.

(v) ECL devices generally produce an output and its complement. (e.g. OR/NOR). So, additional inverter is necessary.

(vi) Due to the emitter follower stages, the output impedance is very low.

(vii) Due to low output impedance, the fan out is typically equal to 25.

(viii) Typical power dissipation for the standard ECL gate is 25 mW which is somewhat higher than that of 74 AS series.

(ix) The current flowing in ECL circuits remains almost constant so no current transients are observed and so associated noise also is less.

4.7.4. Advantages and Disadvantages of ECL Family

1. Advantages

(i) It is the fastest logic family.

(ii) High fan out due to high input resistance and low output resistance. Typically fan out = 25.

(iii) Excellent speed power products.

(iv) It can provide two outputs simultaneously e.g. OR and NOR. No additional inverter is needed.

2. Disadvantages

(i) High power dissipation.

(ii) Its high speed generates voltage and current transients.

(iii) Limited logic swing. Hence, more vulnerable to noise.

4.8 TTL FAMILY

The long form of TTL is transistor transistor logic. The digital ICs in the TTL family use only transistors to perform the basic logic operations. TTL ICs were first developed in 1965 and they were known as standard TTL. This version of TTL currents is not available now. The standard TTL has been improved to a great extent over the years. TTL devices are still used as "glue" logic which connects more complex devices in digital systems. The bipolar TTL family as a whole is becoming obsolete, but we can study it in order to understand basic important things about the logic families.

4.8.1. The Multiple Emitter Transistor

In the TTL circuits to follow, we are going to use a very special type of transistor. The normal transistor has only three terminals namely collector, base and emitter. But, this special transistor has more than one emitters, as shown in figure 4.33 (a) and its equivalent circuit is shown in figure 4.33 (b). The multiple emitter input transistor can have upto eight emitters, for an eight input NAND gate.

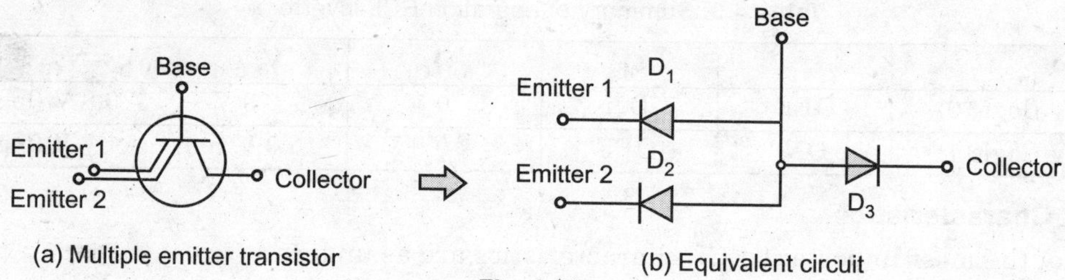

Fig. 4.33.

In the equivalent circuit of figure 4.33 (b), diodes D_1 and D_2 represent the two base to emitter, junctions whereas D_3 represents the collector to base junction. In the forthcoming discussions, we are going to replace the multiple emitter transistor by its equivalent circuit shown in figure 4.33 (b). This transistor can be turned ON by forward biasing either (or both) the diodes D_1 and D_2. This transistor will be in the OFF state if and only if both the base-emitter junctions (D_1 and D_2) are reverse biased.

4.8.2. Two Input TTL-NAND Gate

1. Circuit Diagram

A two input TTL-NAND gate is shown in figure 4.34. A and B are the two inputs while Y is the output terminal of this NAND gate.

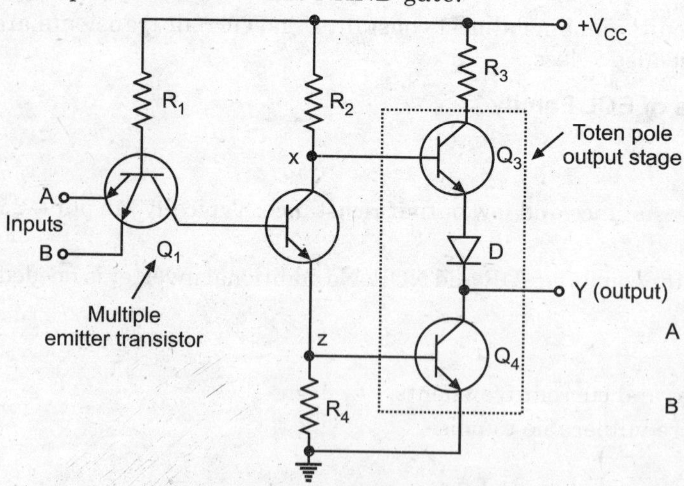

Fig. 4.34. Two input TTL NAND gate

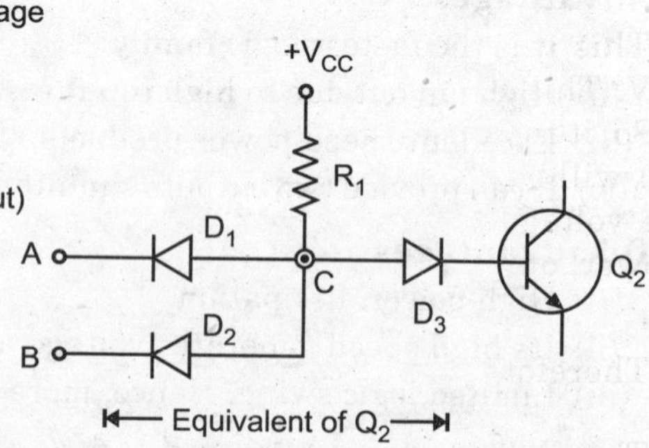

Fig. 4.35. Transistor Q_1 is replaced by its equivalent

2. Working Operation

In order to understand the operation of this circuit, let us replace transistor Q_1 by its equivalent circuit as shown in figure 4.35.

(i) A and B are the input terminals. The input voltages A and B can be either LOW (zero volts ideally) or HIGH (+ V_{CC} ideally).

(ii) **A and B both LOW:** If A and B both are connected to ground, then both the B-E junctions of transistor Q_1 are forward biased.

Hence, diodes D_1 and D_2 in figure 4.35 will conduct to force the voltage at point C in figure 4.35 to 0.7 V. This voltage is insufficient to forward bias base-emitter junction of Q_2. Hence, Q_2 will remain OFF. Therefore, its collector voltage V_x rises to V_{CC}. As transistor Q_3 is operating in the emitter follower mode, output Y will be pulled up to high voltage.

Therefore, $\quad Y = 1$ (HIGH) $\quad$ for A = B = 0 (LOW)

The equivalent circuit for this input condition is shown in figure 4.36 (a).

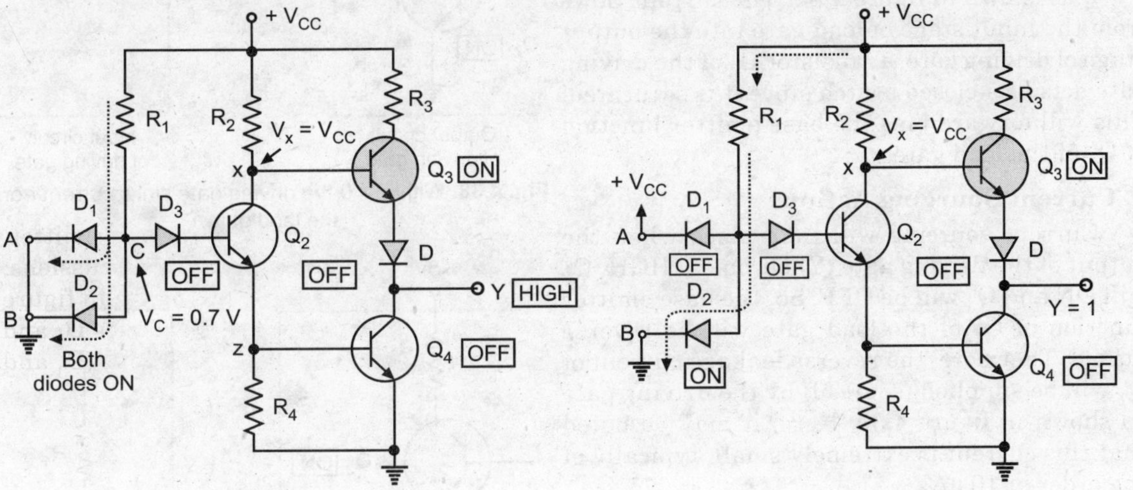

(a) Equivalent circuit for A = B = 0

(b) Equivalent circuit for A = 1, B = 0

Fig. 4.36.

(iii) **Either A or B LOW:** If any one input (A or B) is connected to ground with the other terminal left open or connected to + V_{CC}, then the corresponding diode (D_1 or D_2) will conduct.

This will pull down the voltage at 'C' to 0.7 V. This voltage is insufficient to turn ON Q_2. So, it remains OFF. So, collector voltage V_x of Q_2 will be equal to V_{CC}. This voltage acts as base voltage for Q. As Q_3 acts as an emitter follower, output Y will be pulled to V_{CC}.

$$\text{Therefore, } Y = 1 \begin{cases} \text{if } A = 0 \text{ and } B = 1 \\ \text{if } A = 1 \text{ and } B = 0 \end{cases}$$

The equivalent circuit for this mode is shown in figure 4.36 (b).

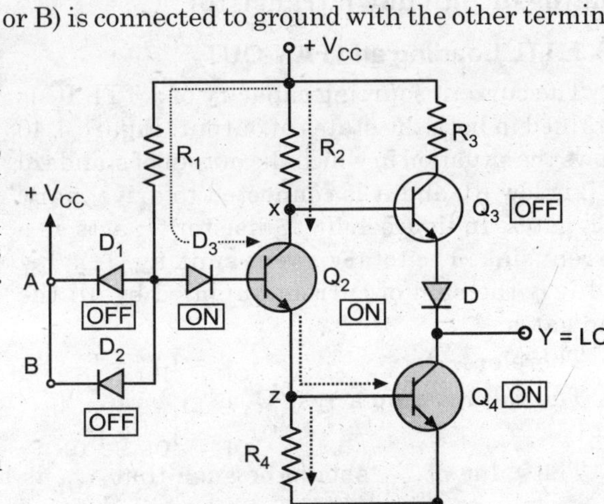

Fig. 4.37. Equivalent circuit for A = B =1

(iv) A and B both HIGH: If A and B both are connected to +V_{CC}, then both the diodes D_1

and D_2 will be reverse biased and do not conduct. Therefore, diode D_3 is forward biased and base current is supplied to transistor Q_2 via R_1 and D_3.

As Q_2 conducts, the voltage at X will drop down and Q_3 will be OFF, whereas voltage at Z (across R_3) will increase to turn ON Q_4. As Q_4 goes into saturation, the output voltage Y will be pulled down to a low voltage. The equivalent circuit for this mode of operation is shown in figure 4.37. This discussion reveals that the circuit operates as a NAND gate.

3. Current Sinking Action

The current sinking will take place for output Y = 0 as shown in figure 4.38. The current flows from the input stage of load gate into the output stage of driving gate. Transistor Q_4 of the driving gate acts as a closed switch since it is saturated. This will forward bias the base emitter junction of Q_1 of the load gate.

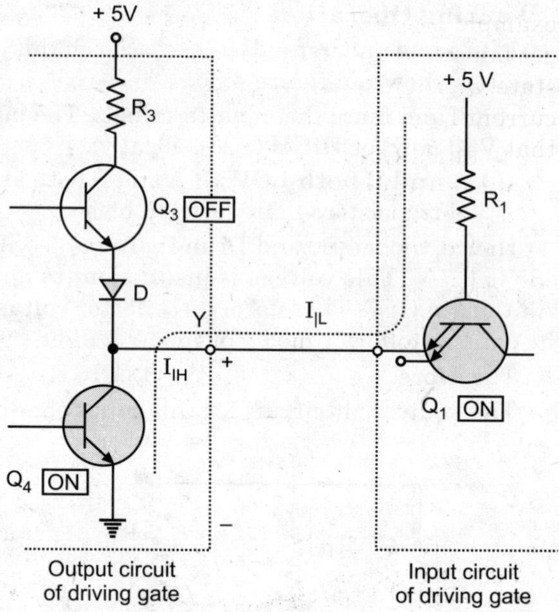

Fig. 4.38. With Y = 0 the driving gate sinks current from the laod gate.

4. Current Sourcing Action

Current sourcing will take place when the output of the driving gate (Y) becomes HIGH. Q_3 will ON and Q_4 will be OFF. So, the base emitter junction of Q_1 of the load gate will be reverse biased. Therefore, the reverse leakage current of Q_1 will be supplied (sourced) by the driving gate as shown in figure 4.39. Here, it may be noted that this current is extremely small, typically of the order of 10 μA.

Transistor Q_3 is called as current sourcing or pull up transistor and Q_4 is called as **current sinking or pull down transistor**.

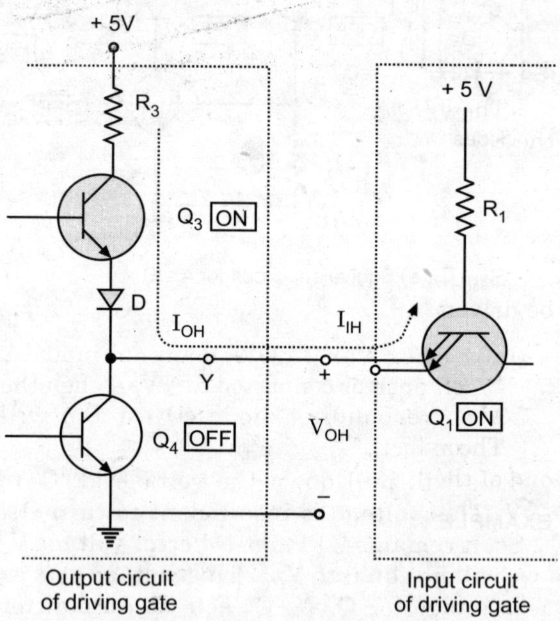

Fig. 4.39. Current sourcing in TTL

4.8.3. TTL Loading and FAN-OUT

The current sourcing capacity of a TTL IC is obtained in both the states of output. Figure 4.40 shows the situation in which the output of standard TTL is low (0) and it is connected to drive many TTL gates. In figure 4.40, transistor Q_4 acts as a current sink. The total current sunk by Q_4 is I_{OL} and it is the sum of current supplied by all the load gates.

Therefore, $$I_{OL} = I_{IL1} + I_{IL2} + \ldots I_{ILn}$$

The voltage drop across Q_4 is given by,

$$V_{OL} = I_{OL} \times R_{ON}$$

The value of I_{OL} should be such that V_{OL} is less than $V_{OL(max)}$ which is 0.4 V for TTL. For

Optimism, is the faith that leads to achievement. Nothing can be done without hope and confidence.

— Helen Keller

example if $I_{IL(max)}$ of each load IC is 1.6 mA and $V_{OL(max)}$ = 0.4 V and $V_{IL(max)}$ = 0.8 V. Then we get fan out of about 10. Figure 4.40 (b) shows the situation in the HIGH state of output. In this state Q_3 is ON, output is V_{OH} and current is being sourced to many TTL gates. The sourced current I_{OH} is equal to the sum of the I_{IH} of all the load ICs. We have to limit I_{OH} to a value such that V_{OH} does not drop to a value below $V_{OH(min)}$.

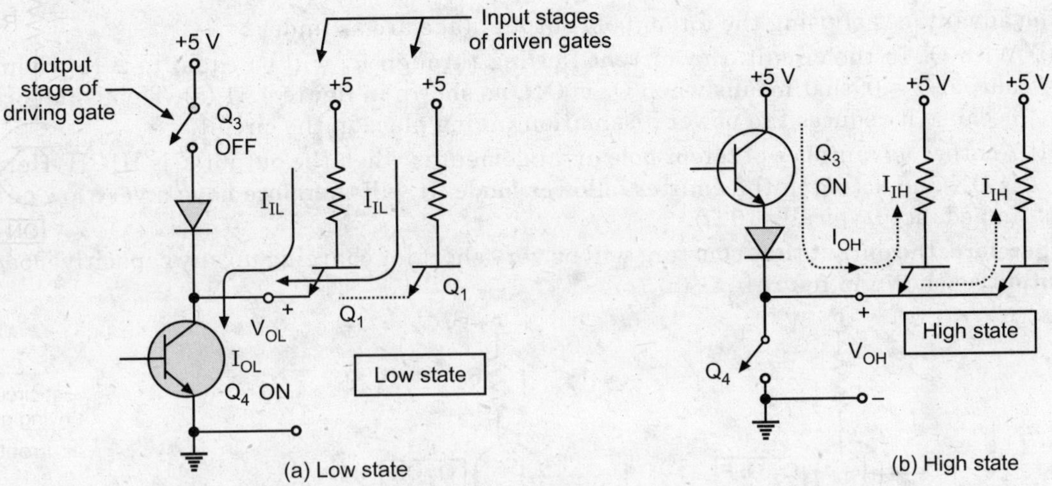

Fig. 4.40. Currents when TTL gate drives several other TTL gates

4.8.4. Determining Fan-out from I_{OH}, I_{OL}, I_{IH} and I_{IL} Values

The Fan-out in the LOW state of output indicates the number of inputs that can be driven in the LOW state is given by,

$$\text{Fan-out (LOW)} = \frac{I_{OL(max)}}{I_{IL(max)}}$$

Similarly, the Fan-out in the HIGH state of output indicates the number of inputs that can be driven in HIGH state is given by,

$$\text{Fan-out (HIGH)} = \frac{I_{OH(max)}}{I_{IH(max)}}$$

The value of Fan-out in the HIGH and LOW states will not always be equal. Then the lower one of them should be chosen as the fan-out of the IC.

EXAMPLE 4.1. The values of currents in different situations are as under:

$$\mathbf{I_{OL(max)} = 20\ mA}$$
$$\mathbf{I_{IL(max)} = 0.3\ mA}$$
$$\mathbf{I_{OH(max)} = 2\ mA}$$
$$\mathbf{I_{IH(max)} = 10\ \mu A}$$

Calculate fan out.

Solution: (1) $\text{Fan-out (LOW)} = \frac{I_{OL(max)}}{I_{IL(max)}} = \frac{20\text{ mA}}{0.3\text{ mA}} = 66.67.$

(2) $\text{Fan-out (HIGH)} = \frac{I_{OH(max)}}{I_{IH(max)}} = \frac{2\text{ mA}}{10\ \mu\text{A}} = 200$

We should choose the overall fan out to be 66.67.

4.8.5. Totem-pole Output Stage

1. Definition

The arrangement of Q_3 and Q_4 on the output side of a TTL NAND gate is called as the totem-pole arrangement.

2. Advantages of totem-pole output stage

The advantages of using the totem-pole output stage are as under :

(i) With Q_3 in the circuit, the current flowing through R_3 will be equal to zero when the output Y = 0, that means when Q_4 is ON, as shown in figure 4.41 (a). This is important because it reduces the power dissipation taking place in the circuit.

(ii) Another advantages of totem-pole arrangement is when the output Y is HIGH. Here Q_3 is ON and acting in the emitter follower mode. It will therefore have a very low output impedance (typically 10 Ω).

Therefore, the output time constant will be very short for charging up any capacitive load on the output as shown in figure 4.41 (b).

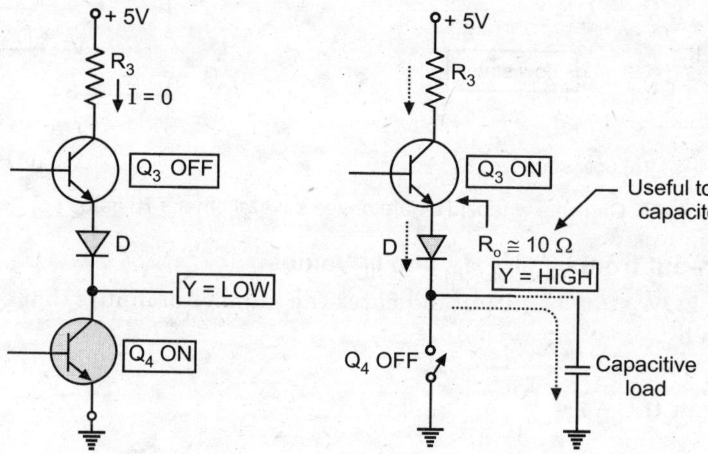

(a) No power dissipation in R_3 (b) Low output resistance

Fig. 4.41.

3. Disadvantages of Totem-pole output

(i) Q_4 in the totem-pole output turns OFF more slowly than Q_3 turns ON.

(ii) So, before Q_4 is completely turned OFF, Q_3 will come into conduction. So, for a very short duration of few nanoseconds, both the transistors will be simultaneously ON.

(iii) This is called as cross conduction and it will draw relatively large current (30 to 40 mA) from the 5 V supply.

4. Function of Diode D

To understand the function of D, let us consider the circuit without D as shown in figure 4.42. We also assume that A = B = 1. So, D_1 and D_2 are OFF, D_3 ON and Q_2

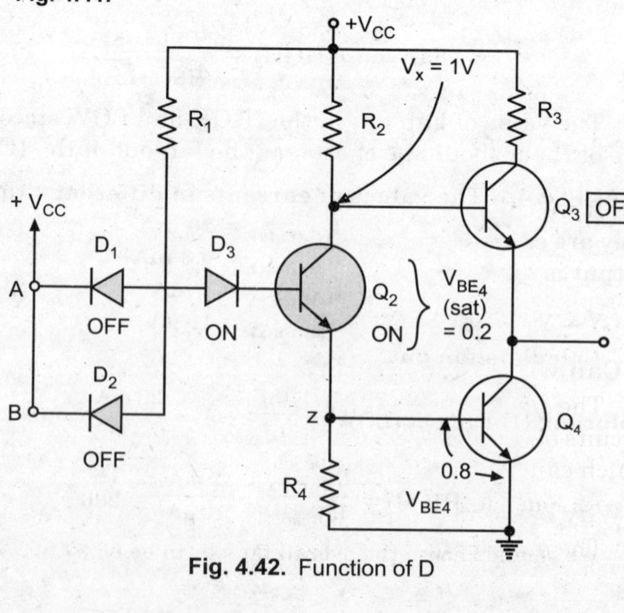

Fig. 4.42. Function of D

is in saturation. Hence, Q_4 is ON (saturation). Therefore, voltage at the collector of Q_2 is given by,

$$V_x = V_{BE4(sat)} + V_{CE2(sat)}$$
$$= 0.8 + 0.2 = 1 \text{ Volts}$$

$V_x = 1$ V will be sufficient to turn ON Q_3 which actually should not happen. So, in order to keep Q_3 in the OFF state diode D is added in the circuit. It is important to avoid simultaneous conduction of Q_3 and Q_4, because it will lead to cross conduction and will increase power dissipation.

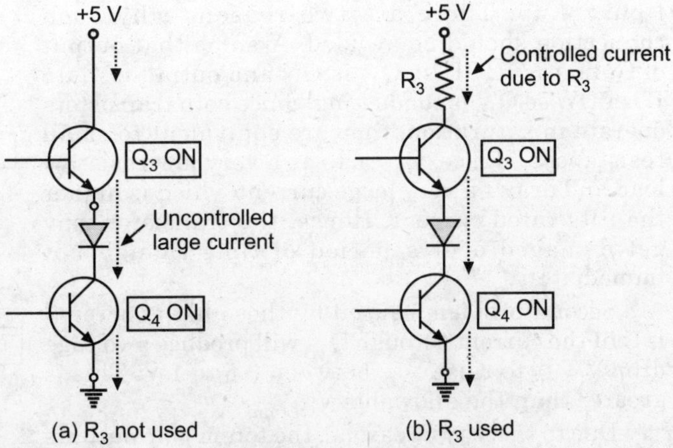

Fig. 4.43. Function of R_3

5. Function of R_3

As discussed earlier, transistor Q_4 in the totem pole arrangement turns OFF more slowly than Q_3 turns ON. So, for few nanoseconds both Q_3 and Q_4 will conduct simultaneously when the output Y is changing state from low to high. If R_3 is not used then there will be no current limiting element in series with Q_3 and Q_4 and a heavy current will be drawn from the source as shown in figure 4.43 (a) which can damage the IC. This can be avoided by limiting the current by inserting resistor R_3 in series with Q_3, a shown in figure 4.43 (b).

6. What is wired AND connection?

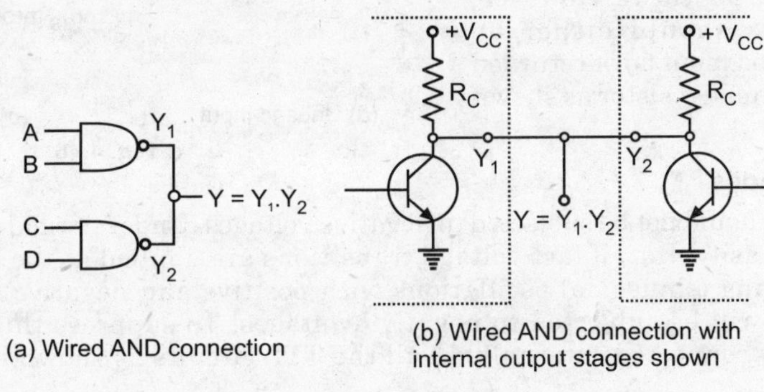

Fig. 4.44.

If we connect the outputs of two gates directly to each other as shown in figure 4.44, then they are said to be wired ANDed with each other. If the outputs are wired ANDed, then the final output is given by,

$$Y = Y_1 \cdot Y_2 = \overline{(A \cdot B)} \cdot \overline{(C \cdot D)} = \overline{AB + CD}$$

7. Can we wired AND the gates having Totem-pole output stage?

The answer is that wired AND connection should not be used for the totem-pole output circuits because of the current spike problem. TTL circuits with open collector outputs are available which can be used for wired AND connection.

8. Why can't we connect the totem pole outputs together?

The outputs of two TTL gates with totem pole output are shown to be connected together in figure 4.45. There are two reasons why such connection should be avoided. Assume that output of Gate A is HIGH so Q_{3A} is ON and output of Gate B is LOW so Q_{4B} is conducting. Since both transistors operate in saturation, they are equivalent to small resistances. Hence, Q_{4B} acts as a very low resistive load and draws a very large current which is higher than its rated current. Hence, the transistor may get damaged over a period of time though not immediately.

Second problem caused by these high currents is that the current through Q_{4B} will produce a voltage drop V_{CE4B} to raise V_{OL} between 0.5 to 1 V. This is greater than the allowable $V_{OL(max)}$

Due to these two reasons, the totem pole outputs should never be directly connected together.

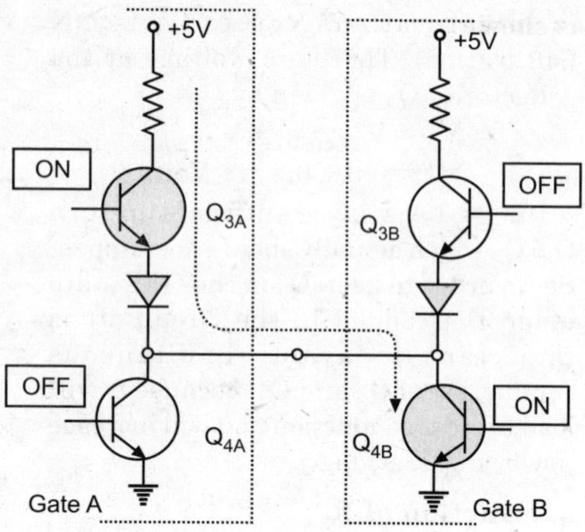

Fig. 4.45. Totem pole outputs tied together

4.8.6. Unconnected Inputs

If any input of a TTL gate is left open, disconnected or floating, then the corresponding base emitter junction of the input transistor Q_1 is not forward biased as shown in figure 4.46 (a). Therefore, the open or floating input acts exactly same way as if a logical 1 is applied to that input. Hence, in TTL ICs all the unconnected inputs are treated as logical 1s. However, the unused inputs should either be connected to some used input (s) or returned to V_{CC} through a suitable resistor as shown in figure 4.46 (b).

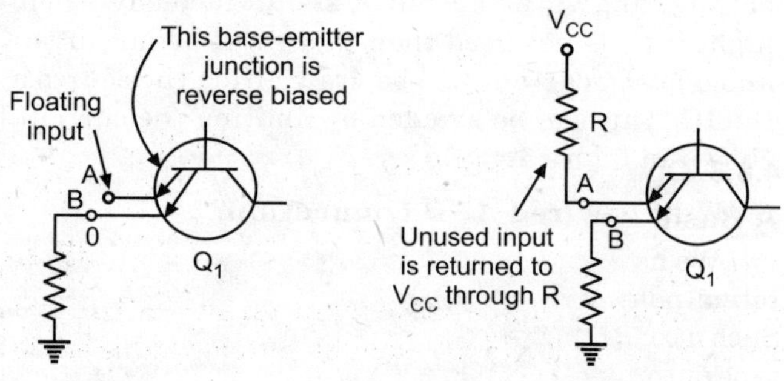

(a) Unused input (b) Standard connection

Fig. 4.46.

4.8.7. Clamping Diodes

The TTL inputs should not be subjected to negative voltages. Under normal operating conditions this gets managed easily. But, if fast voltage transitions are applied at the input, then there is a possibility of ringing (sinusoidal oscillations with positive and negative half cycles). Due to ringing, the inputs will be subjected to negative voltages. To suppress this ringing, clamping diodes are generally connected externally in all the TTL circuits as shown in figure 4.47.

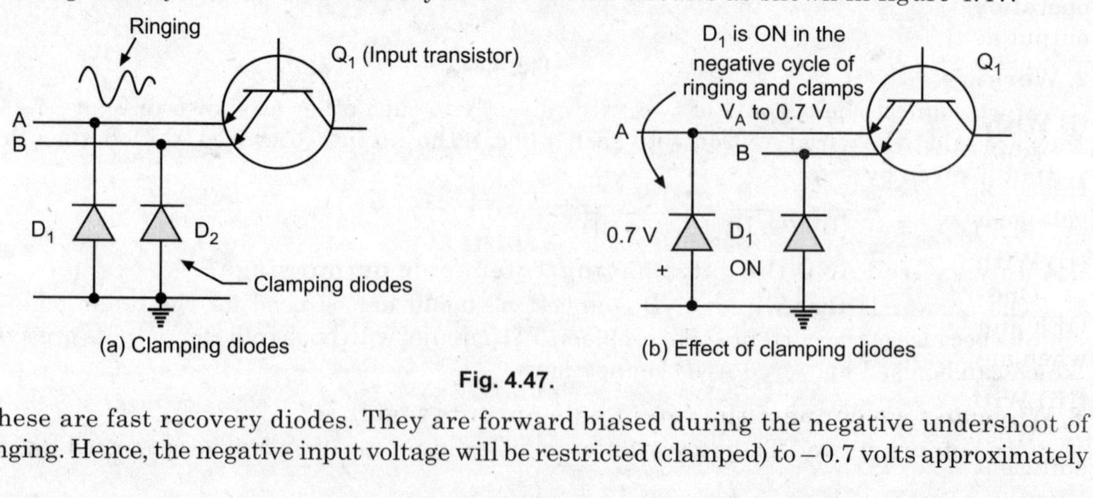

(a) Clamping diodes (b) Effect of clamping diodes

Fig. 4.47.

These are fast recovery diodes. They are forward biased during the negative undershoot of the ringing. Hence, the negative input voltage will be restricted (clamped) to – 0.7 volts approximately as shown in figure 4.47 (b). The clamping action can be better understood by referring to figure 4.48.

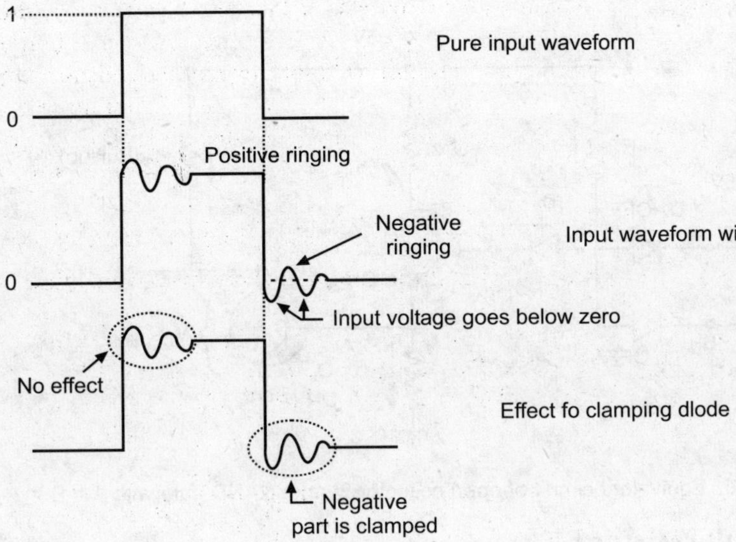

Fig. 4.48. Effect of clamping diodes

4.8.8. Open Collector Outputs

1. Basic Aspects

We have discussed that the gates having totem-pole output can not be wired ANDed. Such a connection becomes possible if another type output stage called open collector output is used. The circuit diagram of a 2-input NAND gate is shown in figure 4.49. We can realize that this is the same TTL NAND gate which we have discussed earlier but with R_3 and Q_3 removed.

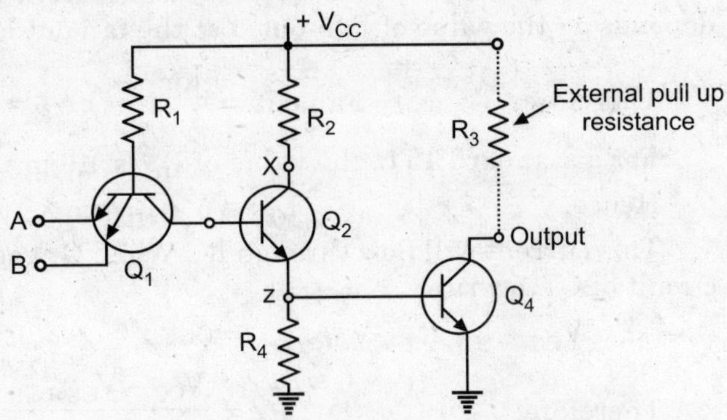

Fig. 4.49. Open collector 2 input NAND gate

The collector point of Q_4 is brought out as output as shown in figure 4.49, therefore, it is called as open collector output. For proper operation it is necessary to connect an external resistance R_3 between V_{CC} and the open collector output as shown in figure 4.49. This resistance is called as pull up resistance.

2. Working Operation

(i) With A = B = 0

With A = B = 0, both the BE junctions of Q_1 are forward biased. Hence, Q_2 remains OFF. Hence, no current flows through R_4. Hence, $V_Z \cong 0$ V. Therefore, Q_4 is OFF and its collector voltage is equal to V_{CC}. Hence, Y = 1 when A = B = 0.

(ii) With A = 0, B = 1 OR A = 1, B = 0

One of the BE junctions is forward biased (the one corresponding to 0 input). Hence, Q_2 is OFF and Q_4 is also OFF. Hence, its collector voltage is equal to V_{CC}. Therefore, output Y = 1 when any one input is low.

(iii) With A = B = 1.

When both the inputs are high, transistor, Q_1 is turned OFF. Hence, Q_2 will be turned ON. sufficient voltage is developed across R_4. Base current is applied to Q_4 and Q_4 goes into saturation.

Hence, the output voltage is equal to $V_{CE(sat)}$ of Q_4. Thus, Y = 0 when A = B = 1. The equivalent circuit for this mode is shown in figure 4.50.

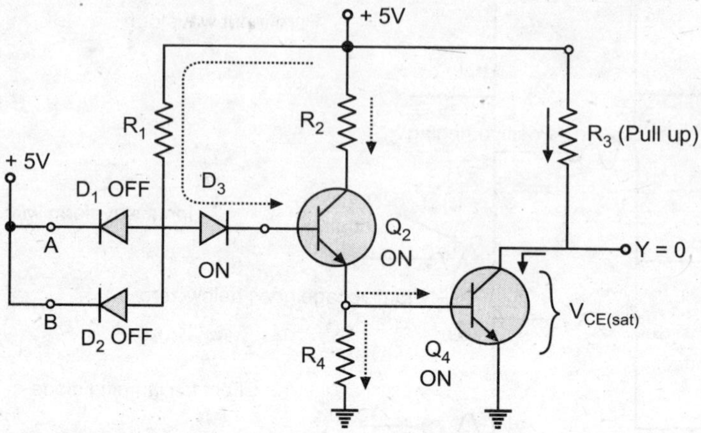

Fig. 4.50. Equivalent circuit of open collector 2-input NAND gate with A = B = 1.

3. Calculation of pull resistor R_3

The value of R_3 depends on how much current the gate is supposed to source. That means R_3 depends on the value of Fan-out. Let the fan out be 5.

Therefore, $\text{Fan out} = \frac{I_{OL}}{I_{IL}}, \quad \therefore \quad 5 = \frac{I_{OL}}{I_{IL}}$

For a standard TTL, the value of I_{IL} is 1.6 mA.

Hence, $I_{OL} = 5 \times I_{IL} = 5 \times 1.6 \text{ mA} = 8 \text{ mA}$

This current will flow through R_3. When Q_4 goes into saturation. Apply KVL to the collector circuit of Q_4 to write.

$$V_{CC} - I_{OL} R_3 - V_{CB4(sat)} = 0$$

Therefore, $$R_3 = \frac{V_{CC} - V_{CE4(sat)}}{I_{OL}} = \frac{5-0.2}{8\times10^3} = 600\ \Omega$$

4. Disadvantages of open collector output

(i) The value of pull up resistance is high (few kΩ). Therefore, if the load capacitance is large then the RC time constant (R_3C) becomes large. This slows down the switching speed of Q_4 and hence the gate.

(ii) Second disadvantage is increased power dissipation. When Q_4 is ON, a large current flows through the pull up R_3. Hence power dissipation is increased. This problem is eliminated if we use totem-pole output arrangement.

5. Advantage

The main advantage of open collector output is that **Wired ANDing** becomes possible.

4.8.9. Wired ANDing

Wired ANDing means tying the outputs of gates together to obtain AND function. It is possible to connect the outputs of two or more gates together as shown in figure 4.51 (a). The wired ANDing is represented schematically as shown in figure 4.51 (b).

Opportunities are like sunrises. If you wait too long you miss them. — **William Arthur Ward.**

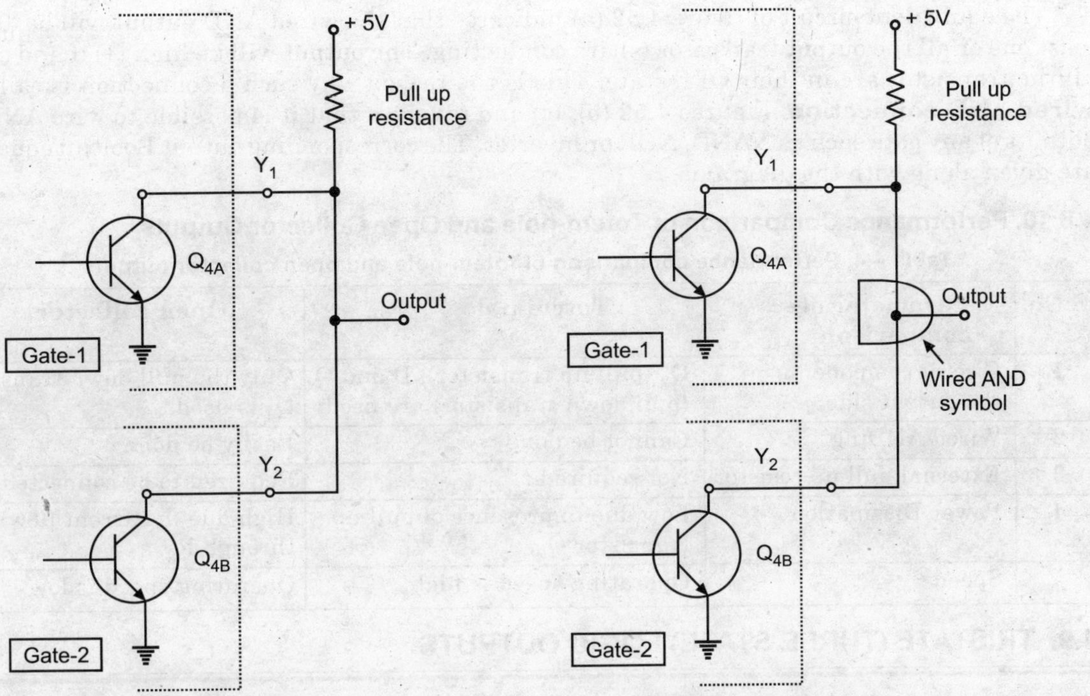

(a) Wired ANDing of two open collector TTL gates

(b) Symbol of wired ANDing

Fig. 4.51. Schematic representation of wired ANDing.

Q_{4A} and Q_{4P} represent the output transistors of Gate-A and Gate-B respectively. A common pull up resistance is used for all the output transistors. The transistors Q_{4A} and Q_{4B} are operated as switches. They operate in either saturation or cut off regions. Hence, the wired ANDing equivalent circuit is as shown in figure 4.52 (a).

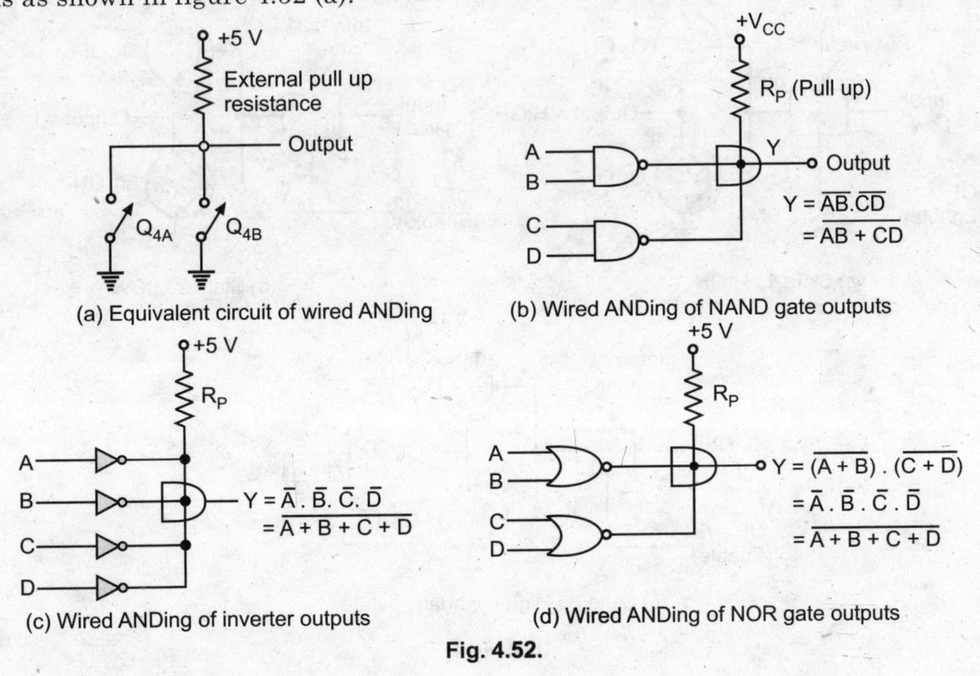

(a) Equivalent circuit of wired ANDing

(b) Wired ANDing of NAND gate outputs

(c) Wired ANDing of inverter outputs

(d) Wired ANDing of NOR gate outputs

Fig. 4.52.

The equivalent circuit of figure 4.52 (a) indicates that the wired AND output will be '0' if at least one or all the output transistor (s) are conducting. The output will be high (1) if and only if all the transistor are in their OFF state. This is the reason why such a connection is called as **wired AND connection**. Figures 4.52 (b), (c) and (d) show that it is possible to wire AND the outputs of any gate such as NAND, NOR or inverter. The corresponding output Boolean equations are given along with the diagrams.

4.8.10. Performance Comparison of Totem-pole and Open Collector Outputs

Table 4.4. Performance comparison of totem-pole and open collector outputs

S.No.	Parameter of comparison	Totem-pole	Open collector
1.	Circuit components on the output side	Q_3 (pull up transistor), D and Q_4 (pull down transistor) are used.	Only the pull down transistor Q_4 is used.
2.	Wired ANDing	Cannot be done	Easily be done
3.	External pull up resistor	Not required	Required to be connected.
4.	Power Dissipation	Low due to presence of pull up transistor Q_3.	High due to current flowing through R_3.
5.	Speed	Operating speed is high.	Operating speed is low

4.9. TRISTATE (THREE STATE) LOGIC OUTPUTS

We have discussed the totem-pole and open collector output configurations. Now, let us study the third type of output circuitry used in TTL and CMOS logic families. It is called as the **tristate configuration**. It is called as tristate because it allows three possible output states namely :

(i) HIGH
(ii) LOW, and,
(iii) High impedance (Hi-Z) state.

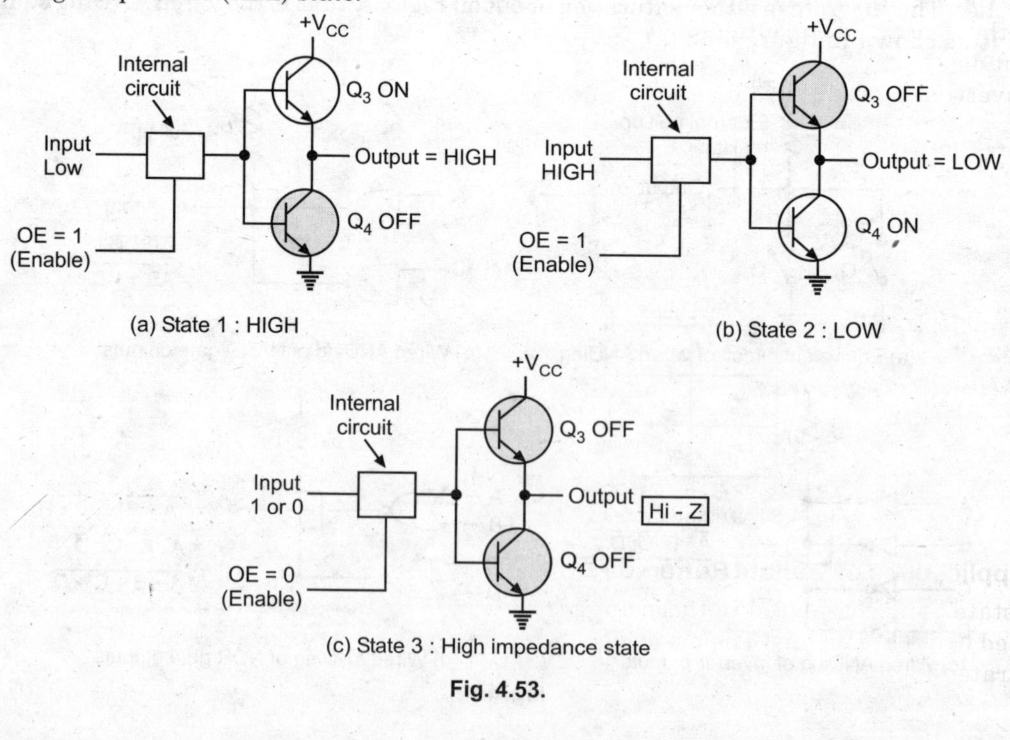

Fig. 4.53.

An additional input called output enable (OE) is introduced. Both the pull up and pull down transistors (Q_3 and Q_4) are being used. The output stage block diagram of a tri-state inverter is shown in figure 4.53 (a). The enable signal OE decides the operation of transistors Q_3 and Q_4. When OE = 1, then the inverter operates as a normal inverter. If input is 0 then output will be 1 (HIGH) as shown in figure 4.53 (a). The pull up transistor Q_3 will be ON.

Similarly, with OE = 1 and input high (1), the output will be low because the pull down transistor Q_4 will be ON as shown in figure 4.53 (b).

High Impedance State

If the enable input OE = 0 (zero), then irrespective of the status of input, both the transistors will remain off as shown in figure 4.53 (c). This state of operation is called as **high impedance (Hi-Z) state**. In this state the output terminal is essentially open circuit i.e., not connected anywhere.

4.9.1. Advantages of Tristate

(i) The outputs of tristate ICs can be connected together, i.e., they can share the use of a common wire. But in doing so there is no reduction in switching speed (this problem is faced for open collector (ICs).

(ii) There is no reduction in speed because the tristate output when enabled, operates as a totem-pole TTL. So, we get all the advantages of totem-pole output such as low output impedance and high speed.

(iii) However, it may be noted that when tristate outputs are connected together, we should enable only one of them at a time.

(iv) Otherwise, two or more active outputs will fight to take control of the common line. This results in very high currents which can damage the ICs and produce invalid output voltage levels.

4.9.2. Tristate Buffers

A tristate buffer is a circuit which connects or isolates its input from its output. In other words it controls the passage of signal from input to output. Tristate buffers are of two types, namely the non-inverting buffers and inverting buffer. The inverting tristate buffers will invert the input data while passing it to the output. Figure 4.54 shows the symbols and truth tables for a non inverting buffer.

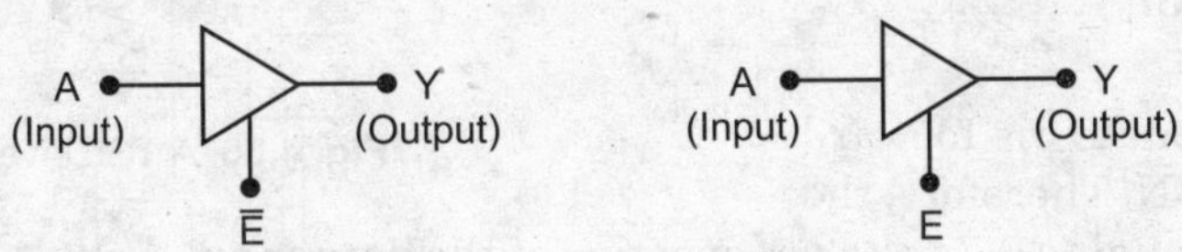

Truth table

$\overline{E}$	Y
0	A
1	Hi – Z

Truth table

$\overline{E}$	Y
0	Hi – Z
1	A

Fig. 4.54. Tristate non-inverting buffers, symbols and truth tables

4.9.3. Applications of Tristate Buffers

Tristate buffers are used in those applications in which several signals are simultaneously connected to a common bus. The basic idea about the common bus and the use of tristate buffers is illustrated in figure 4.55 (a).

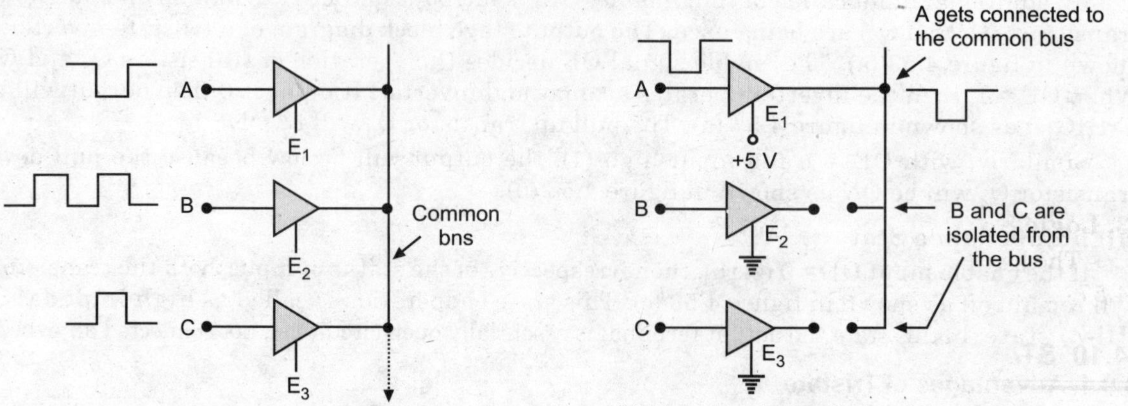

(a) A, B, C are connected to a common bus via tristate buffers

(b) A is getting transmitted on the bus

Fig. 4.55.

Three tristate buffers are used for connecting the inputs A, B and C to the common bus. As shown in figure 4.55 (b), the enable signal E is connected to + 5 V. Therefore, A get connected to the bus. Since, enable terminals E_2 and E_3 are connected to ground, the second and third buffer of figure 4.55 (b) will go into the high impedance state. So, B and C will be isolated from the common bus.

4.9.4. A Tri-State Inverter

1. Basic Aspects

Figure 4.56 shows the circuit diagram of a tri-state logic (TSL) inverter. It has an enable input (OE) and a data input A and one output. The output stage is totem pole type which has been discussed earlier.

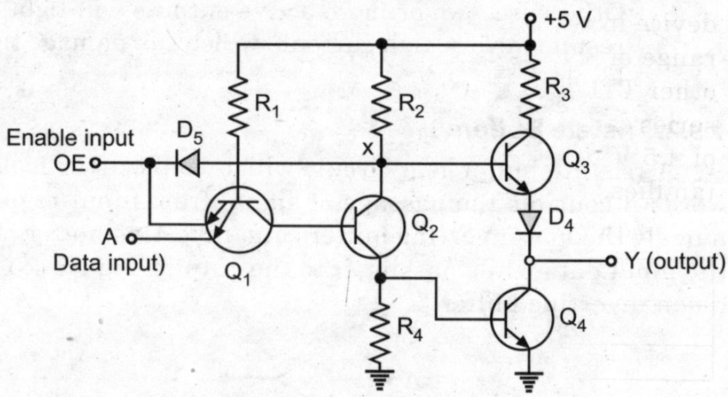

Fig. 4.56. A TSL inverter

2. Working Operation

(i) When OE = LOW

When OE = 0, diode D_5 is forward biased and is turned ON. Therefore, the voltage at point X is equal to 0.7 volts irrespective of the data input A. $V_X = 0.7$ V is not sufficient to turn ON Q_3 so, it remains OFF. OE is also connected to one emitter of transistor Q_1. As OE = '0', transistor Q_1 is ON, hence Q_2 is OFF (as discussed for the NAND gate). As Q_2 is OFF, current through R_4 is zero. So Q_4 is in the OFF state. Thus, with OE = LOW (0), both the output transistors Q_3 and Q_4 will be OFF and the inverter will be in the **High-impedance** state.

(ii) When OE = HIGH

When OE = 1 (HIGH), the diode D_5 will be reverse biased. So, it will be ineffective. Also the corresponding emitter base junction of Q_1 will be reverse biased. So, the circuit of TSL inverter now acts as a TTL NAND gate with only one input A. If A = 0, then Q_1 is ON, so Q_2 is OFF, so $V_X = V_{CC}$, hence Q_3 turns ON and output Y = 1. If A = 1, then Q_1 is OFF, so Q_2 is ON, so V_x is low, but sufficient voltage appears across R_4, so Q_4 goes into saturation and Y = 0. Thus, with OE = 1, the given circuit operates as a normal inverter. Table 4.5 summarizes the operation of the TSL inverter.

Table 4.5. Operation of TSL inverter

Enable input	Data input	output
OE = 1	A = 0	Y = 1
OE = 1	A = 1	Y = 0
OE = 0	A = 0 or 1	Hi-Impedance

3. Logic Symbol

The logic symbol of a TSL inverter is shown in figure 4.57.

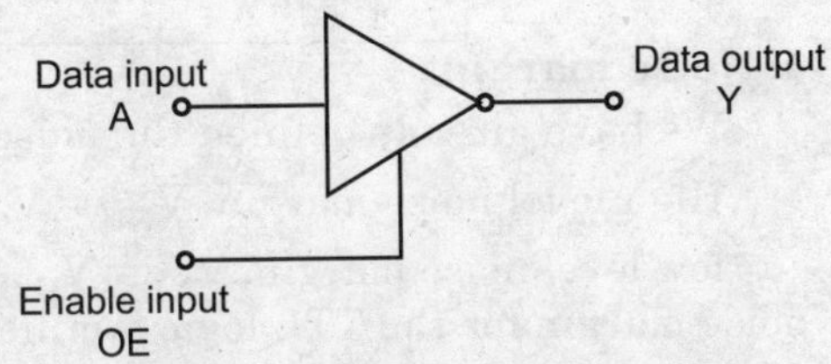

Fig. 4.57. Logic symbol of TSL inverter

4.10 STANDARD TTL CHARACTERISTICS

1. Basic Aspects

In 1964 Texas instruments introduced the transistor transistor logic (TTL) which is a widely used family of digital devices. TTL ICs are fast, inexpensive and easy to use. Earlier, we have learnt the general characteristics for all the logic families. Now, let us learn them for TTL logic family. The Texas instrument first introduced two TTL series namely 54 series and 74 series.

2. Supply voltage and temperature ranges

The most widely used TTL devices series is the 74 series, e.g., 7400, 7402, 7408 etc. Any device in the 74 series works over a temperature range of 0° to 70° C and over a supply voltage range of 4.75 to 5.25 V. This series is used for the consumer and industrial applications. The other TTL series is 54 series which is developed for the military applications. Devices from 54 series can operate over a temperature range of – 55° C to 125° C and over a supply voltage range of 4.5 V to 5.5 V. The table 4.6 gives the supply voltage and temperature ranges for two TTL families.

Table 4.6.

TTL Series	Supply voltage range	Temperature range
74 series	4.75 V to 5.25 V	0° C to 70° C
54 series	4.5 V to 5.5 V	– 55°C to 125° C

3. Voltage levels

Table 4.7 shows the input and output logic voltage levels for the TTL 74 series. The worst case values of various voltage levels for 74 series are indicated by the maximum and minimum values in table 4.7. The worst case values correspond to the worst case conditions of power supply, loading and temperature values.

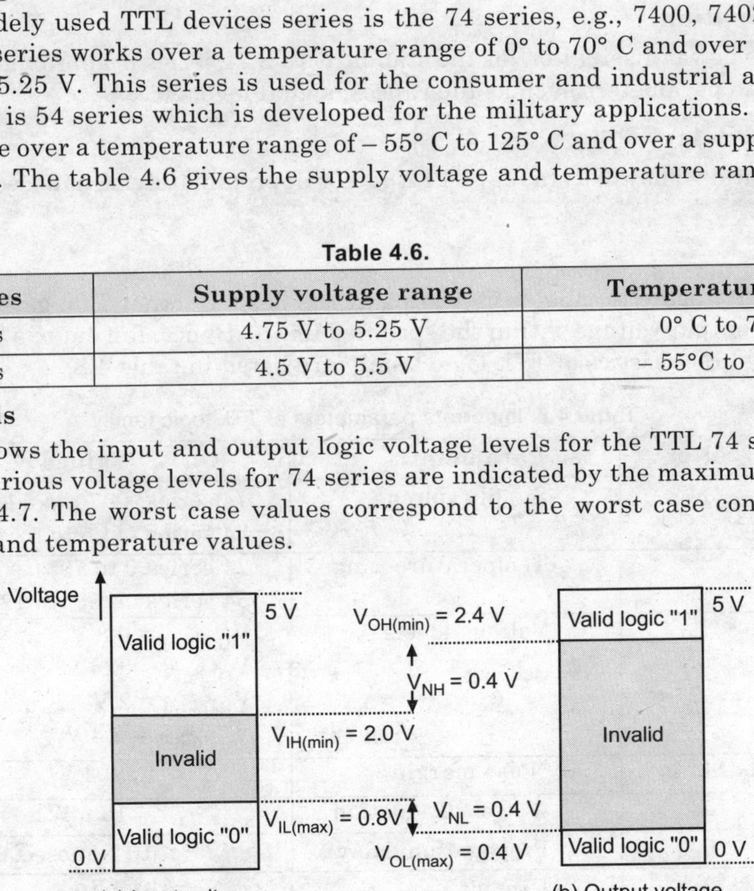

Fig. 4.58. TTL input and output profile

Table 4.7. Voltage levels for TTL 74 series

Voltage	Minimum	Typical	Maximum
V_{OL}	–	0.2 V	0.4 V
V_{OH}	2.4 V	3.4 V	–
V_{IL}	–	–	0.8 V
V_{IH}	2.0 V	–	–

4. Noise margin

We have already defined the noise margins as under:

High level noise margin, $V_{NH} = V_{OH(min)} - V_{IH(min)}$, and,

low level noise margin, $V_{NL} = V_{IL(max)} - V_{OL(max)}$. Substituting the values from table 4.8, the noise margin for the TTL logic families can be calculated.

Therefore, $V_{NH} = 2.4 - 2 = 0.4$ V

$V_{NL} = 0.8 - 0.4 = 0.4$ V

Thus, noise margin for the TTL family is 0.4 V. This means that as long as the induced noise voltage is less than 0.4 V, the operation of the TTL ICs will be unaffected. The noise margins have been shown in figure 4.58.

5. Power Dissipation

The average power dissipation for the standard TTL 74 series is approximately 10 mW. It is dependent on the parameters such as tolerances, signal level etc.

6. Propagation delay

The propagation delay of a standard TTL gate is approximately 10 nanoseconds (nanosecond $= 10^{-9}$ seconds).

7. Fan out

A standard TTL gate is capable of driving at the most 10 other TTL gates simultaneously, maintaining its output voltage within the specified limits. Hence, fan out of a TTL gate is 10. All the important characteristics of TTL logic family are listed in table 4.8.

Table 4.8. Important parameters of TTL logic family

S.No.	Parameter	Values
1.	Supply voltage	74 series: (4.75 to 5.25 V) 54 series : (4.5 to 5.5 V)
2.	Temperature range	74 series 0 to 70° C 54 series : – 55° to 125°C
3.	Voltage levels	$V_{IL(max)} = 0.8$ V $V_{OL(max)} = 0.4$ V $V_{IH(min)} = 2$ V $V_{OH(min)} = 2.4$ V
4.	Noise margin	0.4 V
5.	Power dissipation	10 mW
6.	Propagation delay	10 Nanosec
7.	Fan out	10
8.	Figure of merit	100

4.11 TTL SUB FAMILIES (PTU, Jalandhar, Sem. Exam., 2008-09)

The standard TTL series which we have discussed so far has evolved into many other sub series. They are collectively called as TTL sub families and they provide a wide range of speed and power capabilities.

Some of the sub-family members are as under:

(i) High speed (H) and low power (L) TTL.

(ii) Schottky (S) and low power Schottky (LS) TTL.

(iii) Advanced Schottky (AS) TTL.

(iv) Advanced low power (ALS) Schottky TTL.

(v) Fast (F) TTL.

4.11.1. Standard TTL 74 Series

We have already discussed the standard 74 series of TTL logic in the earlier sections. Even though all these devices are readily available, they have not remained the obvious choice of a designer because the advance versions are available.

4.11.2. Schottky TTL (74 S Series)

7400 series uses saturated BJT switching. That means, many of the internal transistor will go into saturation while conducting. It requires longer time to turn off a saturated transistor due to presence of large storage charge. ***This limits the switching speed of the circuit***. The delay introduced due to storage charge is called as storage-time delay (t_s). The 74 S series reduces this storage time delay. It achieves this by not allowing the transistors to go into deep saturation. Transistors are prohibited from going into deep saturation by connecting a Schottky barrier diode (SBD) between the collector and base of each transistor as shown figure 4.59.

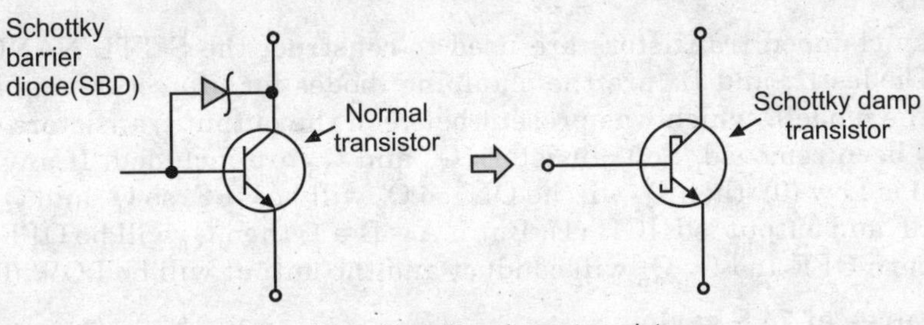

Fig. 4.59. Schottky clamp transistor

4.11.3. Schottky Clamp Transistor

As we connect SBD between the collector and base of a transistor, it gets converted into a schottky clamp transistor (SCT) as shown in figure 4.59.

The forward voltage of SBD is only 0.25 volts. So, if the transistor tries to go into saturation (i.e., if CB junction becomes forward biased) then SBD will be forward biased and starts conducting. So, some of the input current gets diverted through SBD and the base current will reduce (see figure 4.60 (a)). Due to reduced base current, the transistor can not go into saturation (to saturate a transistor we need to supply a large base current). As the transistor is not allowed to saturate, the storage time delay at the time of turn OFF will be automatically reduced.

Little minds are tarned and sub dued by misfortune; but great minds rise about it.

— **Washington Irving**

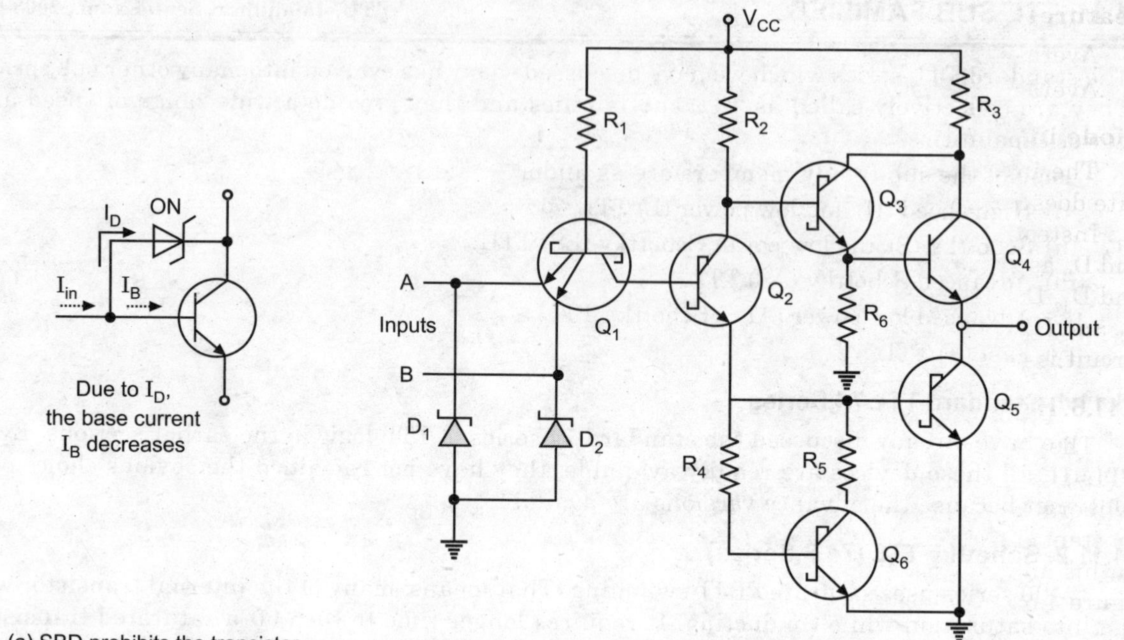

(a) SBD prohibits the transistor from going into saturation

(b) NAND gate in S-TTL

Fig. 4.60.

4.11.4. NAND Gate in S-TTL

1. Basic Aspects

The Schottky clamped transistors are used to construct the S-TTL NAND gate shown in figure 4.60(b). Diodes D_1 and D_2 are the clamping diodes for protection against the negative input voltage. The diode D which was present between the output transistors of standard TTL NAND gate has been removed. So transistors Q_3 and Q_6 are included. If any one or both the inputs (A and B) is Low (0), then Q_1 will be ON, so Q_2 will be OFF, so Q_3 and Q_4 will be ON and Q_5 Q_6 will be OFF and output is HIGH (1). But if A = B = 1, then Q_1 will be OFF so Q_2 turns ON, Q_3 and Q_4 will turn OFF and Q_5 Q_6 will conduct and the output will be LOW (0).

2. Salient Features of 74 S series

(i) The most important features of Schottky TTL are as follows:

Average propagation delay : 3 ns.
Average power dissipation : 20 mW.

(ii) So the propagation delay has reduced from 10 ns for the standard TTL to 3 ms, but the power dissipation has doubled.

(iii) The 74 S NAND gate uses a darlington pair Q_3 Q_4 in order to provide a shorter rise time when switching from ON to OFF state.

(iv) The large power dissipation takes place due to smaller value resistors used for the Schottky TTL circuit.

4.11.5. Low Power Schottky TTL (LS – TTL)

74 LS series is a low power Schottky TTL series. Here, L corresponds to low power and S for Schottky. This series also uses the Schottky clamp transistors but the resistors used are of larger value compared to those for 74 S series. The large resistor values will reduce the power dissipation. But the switching times will increase due to increased resistance values.

Features of 74 LS series

Average power dissipation : 2 mW

Average propagation delay : 9.5 ns

Diode input section

The input section of the low power Schottky gate does not contain the multi emitter transistor Q_1. Instead it uses two Schottky barrier diodes D_1 and D_2 as shown in figure 4.61. The 20 kΩ resistor and D_1, D_2 an AND gate on the input side. Diodes D_3 and D_4 are the clamping diodes. The remaining circuit is same as the standard TTL NAND gate.

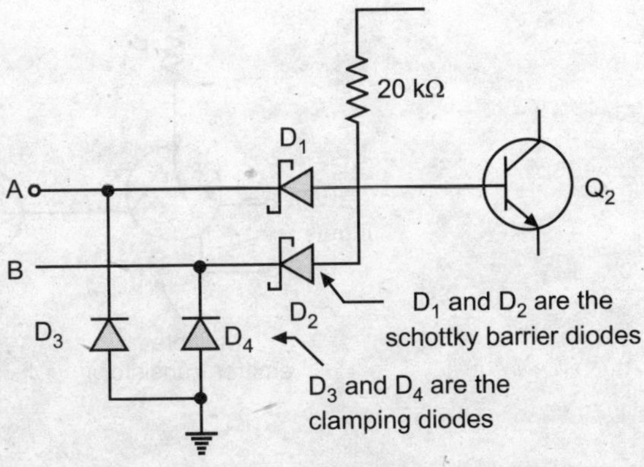

Fig. 4.61. Input section of a low power Schottky TTL gate.

4.11.6. High Speed TTL (74 H series)

The 74 H series corresponding to the high speed TTL ICs. The values of resistors used are deliberately kept small so as to reduce the time constant and increase the switches speed. In addition to this an additional transistor Q_5 is connected to form a darlington pair with Q_3 as shown in figure 4.62.

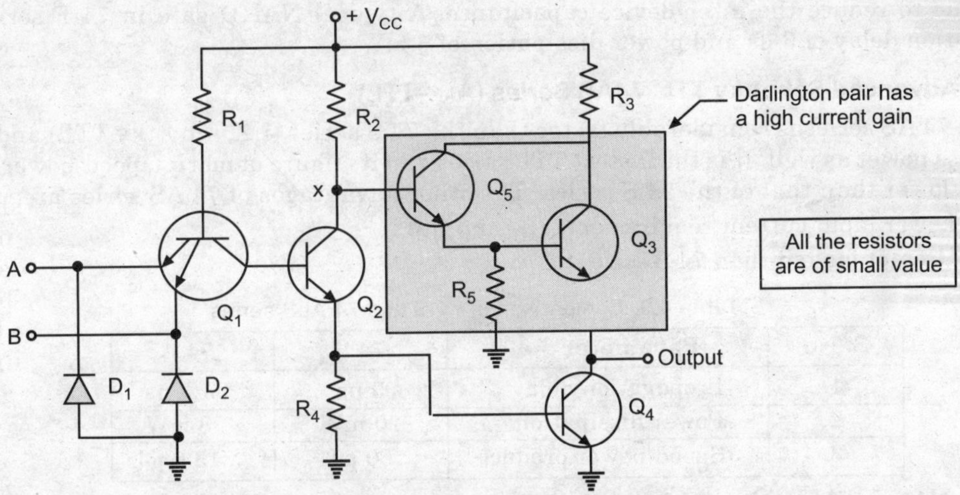

Fig. 4.62. 2-input high speed NAND gate

The switching speed further increases due to the high current gain offered by the Q_3 Q_5 darlington connection. The remaining circuit is same as that of the standard TTL NAND gate discussed earlier. Due to darlington connection of Q_3 and Q_5, the threshold voltage at "X" required to turn on Q_3 wil be 1.4 V. Hence diode D in the standard TTL totem-pole does not appear in this circuit. The propagation delay is 6 ns and average power dissipation is about 30 mW, for the 74 H series.

4.11.7. Low Power TTL (L-TTL)

The 2 input low power TTL NAND gate is shown in figure 4.63. Note that this circuit is exactly same as that of the standard TTL NAND gate except for the use of high value resistors. This will reduce the power dissipation. But the switching speed will reduce, i.e., switching time will increase. The power dissipation is about 1 mW and the propagation delay is around 32 ns for the low power TTL i.e., 74 L series.

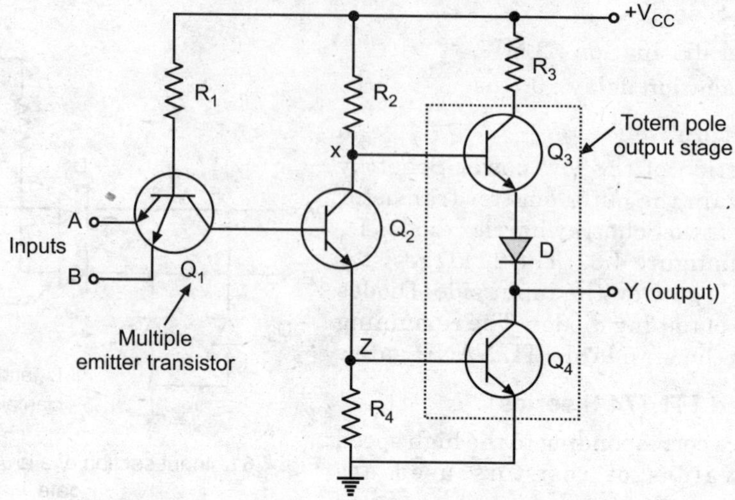

Fig. 4.63. 2-input low power TTL NAND gate

4.11.8. Fast (F) TTL (74 F series)

This series has been produced to achieve reduced propagation delays. It uses a new IC fabrication technique to reduce the inter device capacitance. A typical NAND gate in 74 F series has a propagation delay of 3 ns and power dissipation of 6 mW.

4.11.9. Advanced Schottky TTL, 74AS Series (A_S – TTL)

The 74 AS series is considerably faster than the 74 S series (i.e., schottky TTL) and it needs much less power as well. It is the fastest TTL series and its figure of merit (speed-power product) is much lower than that of the 74 S series. The other advantages of 74 AS series are:

(i) Lower input current requirements (I_{IL} and I_{IH})

(ii) Greater fan out than 74 S series

Table 4.9. Comparison of 74 S and 74 ALS series

S. No.	Parameter	74 S	74 AS
1.	Propagation delay	3 ns	1.7 ns
2.	Power dissipation	20 mW	8 mW
3.	Speed-power product	60 pJ	13.6 pJ

4.11.10. Advanced Low Power Schottky TTL 74 ALS Series

The 74 ALS series requires less power as compared to 74 LS series. In fact there is an improvement in both power as well as speed. 74 ALS has the lowest value of speed-power product of all the TTL. Table 4.10 gives the comparison of 74 LS and 74 ALS series.

Table 4.10. Comparison of 74 LS and 74 ALS series

S.No.	Parameter	74 LS	74 ALS
1.	Propagation delay	9.5 ns	4 ns
2.	Power dissipation	2 mW	1.2 mW
3.	Speed-power product	19 pJ	4.8 pJ

4.12 COMPARISON OF TTL SERIES CHARACTERISTICS

Table 4.11 shows the typical values of various characteristics of different TTL IC series.

Table 4.11. Typical TTL series characteristics

	74	74 S	74 LS	74 AS	74 ALS	74 F
Performance ratings :						
Propagation delay (ns)	9	3	9.5	1.7	4	3
Power dissipation (mW)	10	20	2	8	1.2	6
Speed-power product (pJ)	90	60	19	13.6	4.8	18
Max. clock rate (MHz)	35	125	45	200	70	100
Fan-out (same series)	10	20	20	40	20	33
Voltage parameters						
$V_{OH(min)}$	2.4	2.7	2.7	2.5	2.5	2.5
$V_{OL(max)}$	0.4	0.5	0.5	0.5	0.4	0.5
$V_{IH(min)}$	2.0	2.0	2.0	2.0	2.0	2.0
$V_{IL(max)}$	0.8	0.8	0.8	0.8	0.8	0.8
Current parameters						
$I_{OH(min)}$ mA	– 0.4	– 1	– 0.4	– 2	– 0.4	–1
$I_{OL(min)}$ mA	16	20	8	20	8	20
$I_{IH(max)}$ mA	40	50	20	20	20	20
$I_{IL(max)}$ mA	– 1.6	– 2	– 0.4	– 0.5	– 0.1	– 0.6

4.12.1. Comparison Based on SPP

This is the simplest way of comparing the performance of TTL subfamilies. The speed power product of the IC should be as low as possible. Figure 4.64 shows the graphical comparison of various TTL series.

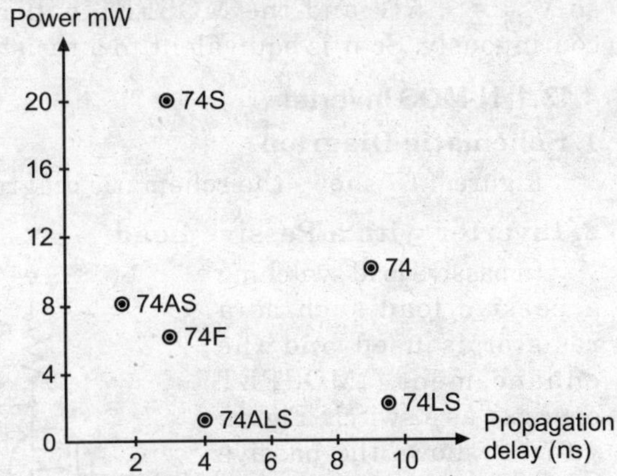

Fig. 4.64. Comparison based on SPP

4.12.2. Comparison Based on Speed

The logic TTL family having the shortest propagation delay is called as the fastest family. Figure 4.64 (a) shows the arrangement of TTL ICs in the descending order of speed. It shows time 74 AS or 74 F are the fastest series while 74 is the slowest one.

Speed	Name of TTL series
Fastest ↑	74 AS/74 F
	74 S
	74 ALS
	74 LS
	74
Slowest	74 L

(a) Comparison based on speed

Power consumption	TTL series
Lowest ↓	74 L
	74 ALS
	74 LS
	74 F
	74 AS
	74
Highest	74 S

(b) Comparison based on power

Fig. 4.65.

4.12.3. Comparison Based on Power

Comparison of various TTL subfamilies based on the power consumption is shown in figure 4.65 (b). It is evident from figures 4.65 (a) and (b) that as speed is increased the power dissipation also increases.

4.13 DIGITAL MOSFET CIRCUITS

1. Classification

Digital circuits which use MOSFETs are classified into following three categories:

(i) PMOS (ii) NMOS and (iii) CMOS

The P-MOS circuits use the P-channel enhancement MOSFETs. PMOS circuits use only N-channel enhancement MOSFETs and CMOS (complementary MOS) uses both P and N channel MOSFETs.

2. N-MOS Circuits

The N-MOS circuits use the N-channel MOSFET as a switch. The most interesting thing is the resistors are replaced by MOSFETs which act as resistors.

3. MOSFET as a resistor

Figure 4.66 shows the connection diagram to operate an N-channel MOSFET as a resistor. The gate and drain of the MOSFET are connected to each other, so V_{GS} = + 5 V and the MOSFET will remain on continuously. So it is equivalent to a resistance.

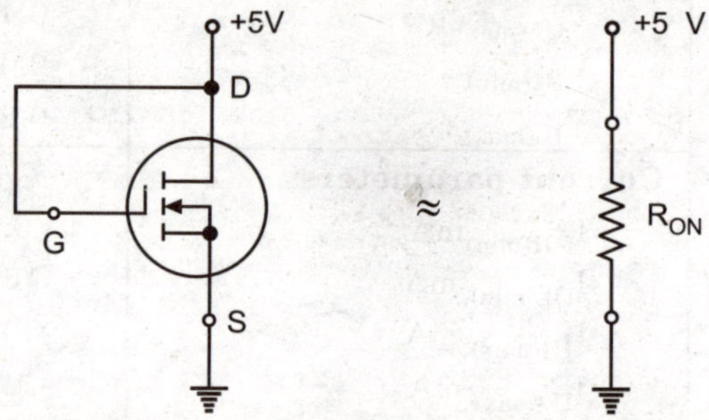

Fig. 4.66. MOSFET as resistance

4.13.1. N-MOS Inverter

1. Schematic Diagram

Figure 4.67 shows the schematic diagram of NMOS inverter V_{in} is the output voltage.

2. Inverter with a Passive Load

In passive load switching a passive load such as a resistor is used and the enhancement MOSFET operates as a switch. Figure 4.67 (a) shows the passive load switching circuit with resistor R_L acting as a passive load.

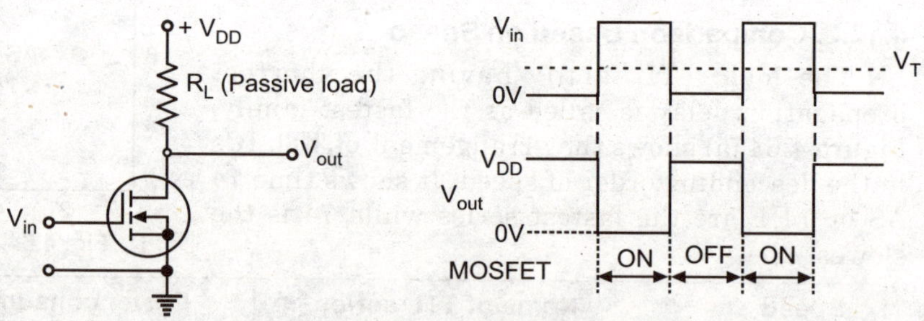

(a) Passive load switching circuit (b) Waveforms

Fig. 4.67.

3. Working Operation

(i) A rectangular waveform V_{in} is applied at the gate of the enhancement mode MOSFET. As shown in figure 4.67 (b) V_{in} will either be less than the threshold voltage V_T or it will be higher than V_T.

(ii) When $V_{in} < V_T$, the MOSFET will be OFF. The current flowing through R_L will be zero and the output voltage will be equal to + V_{DD} as shown in figure 4.67(b).

(iii) When $V_{in} > V_T$, the MOSFET is fully ON and is equivalent to a closed switch. The output voltage will be equal to zero as shown in figure 4.67(b).

(iv) The passive load switching circuit thus acts as an inverter because the output voltage waveform is an inverted version of the input voltage.

(v) But when fabricating in the integrated circuit form, a resistance occupies a lot of space.

Hence, it is replaced by a MOSFET which acts like a resistor. Such a resistance is called as the active load.

4. NMOS inverter

The circuit diagram for an NMOS inverter is shown in figure 4.68. Here, the upper MOSFET Q_2 acts as a load resistor, while the lower MOSFET Q_1 acts as a switch. The input voltage is a rectangular waveform and it is applied at the gate of MOSFET Q_1. The active load switching is preferred in fabrication of integrated circuits because the MOSFET acting as a load resistor needs much smaller space as compared to an actual passive resistor. Therefore, it is possible to accommodate more devices keeping the area of an integrated circuit unchanged. MOSFET Q_1 is biased to operate as a switch. It turns on for a high voltage applied at its gate and turns OFF when the gate voltage is low. So Q_2 will be ON when V_{in} = 5 V and it will be OFF when V_{in} = 0 Volts. Q_2 will be fully on continuously and acts as a resistance.

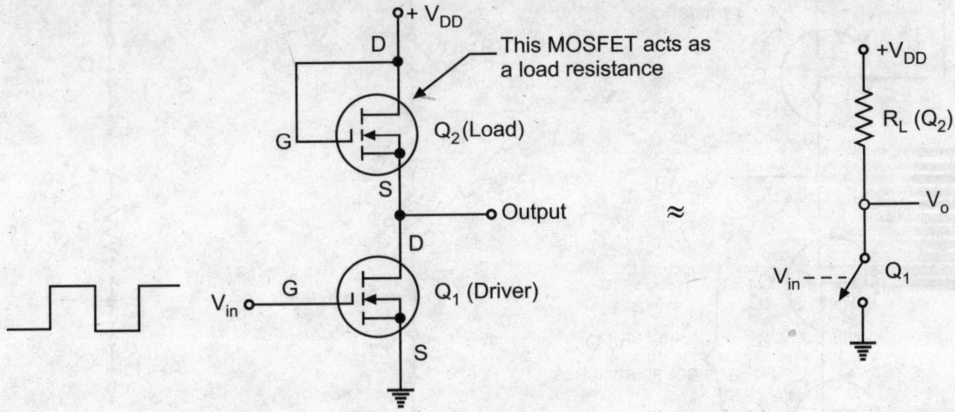

Fig. 4.68. NMOS inverter

The input output waveforms are as shown in figure 4.69.

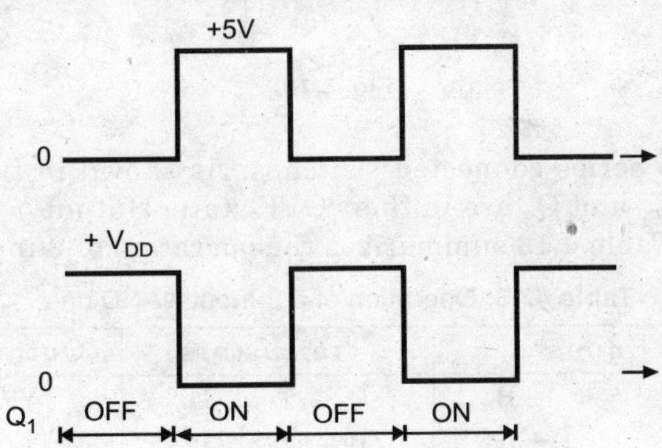

Fig. 4.69. Waveforms for NMOS inverter

The operation of NMOS inverter is summarized in table 4.12.

Table 4.12. Operation of NMOS inverter

$V_{in} = V_{GS}$	Q_1	V_0
– 0 V (logic 0)	OFF	+ 5 V (logic 1)
+ 5 V (logic 1)	ON	0 V (logic 0)

4.13.2. NMOS NAND gate

1. Basic Aspects

Figure 4.70 (a) shows the schematic diagram of a NAND gate using only N-type MOSFETs. MOSFETs Q_1 and Q_2 are operated as switches. The gates of Q_1 and Q_2 are used as two inputs A and B of the NAND gate, whereas Q_3 works as a resistance.

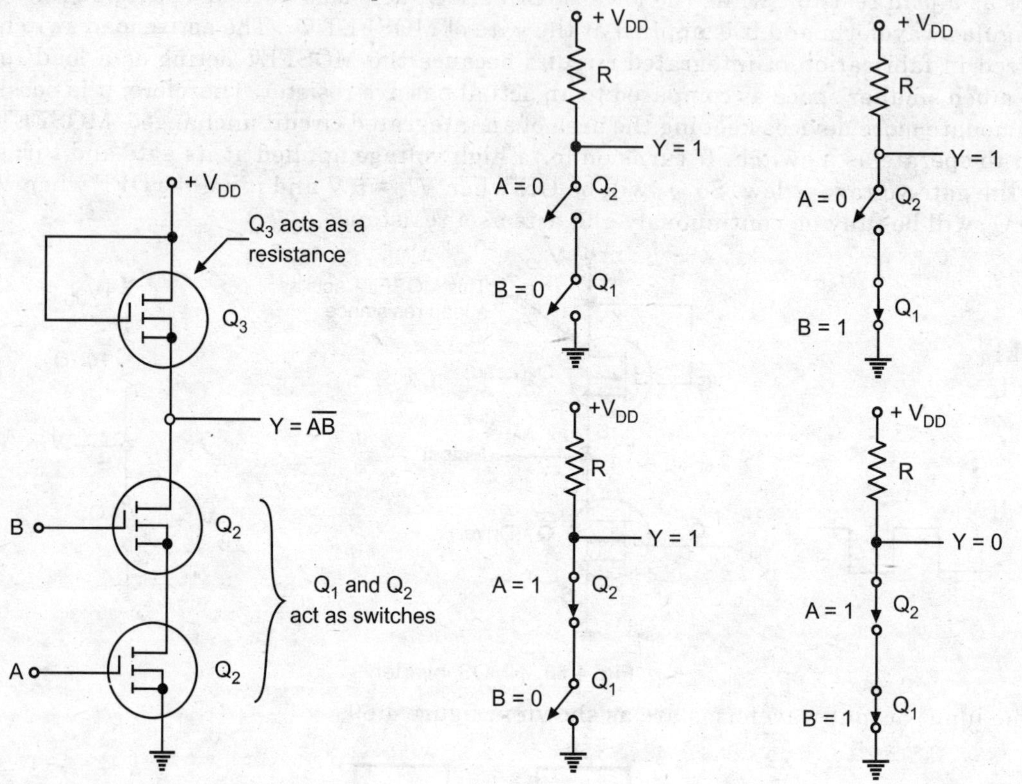

Fig. 4.70.

2. Working Operation

Q_1 and Q_2 act as two series connected switches. As shown in figure 4.70 (b), the output Y = 1 if any one or both Q_1 and Q_2 are in their OFF state. Output Y = 0 only when Q_1 and Q_2 conduct simultaneously. Table 4.13 summarizes the operation of 2-input NAND gate.

Table 4.13. Operation of a 2-input NAND gate

Input		Transistors		Output
A	B	Q_1	Q_2	Y
0	0	OFF	OFF	1
0	1	OFF	ON	1
1	0	ON	OFF	1
1	1	ON	ON	0

4.13.3. 2 input NOR gate using N-MOS

1. Basic Aspects

A two input NOR gate using NMOS is shown in figure 4.71. The N-channel MOSFETs M_1 and M_2 are used as switches and are connected in parallel. The third N-channel MOSFET M_3 acts as an active load i.e., as a resistor.

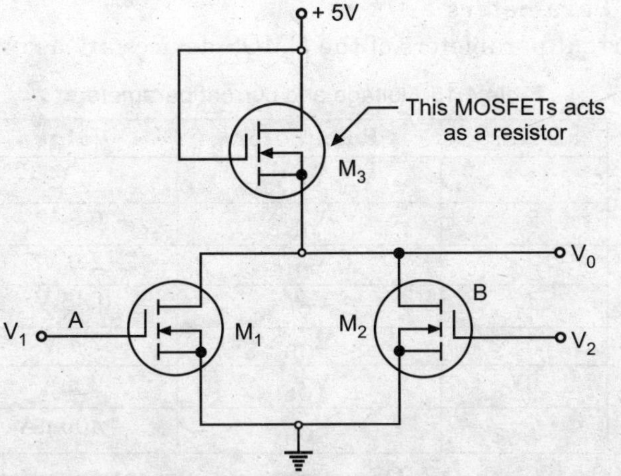

Fig. 4.71. A two input NOR gate using N-type MOSFETs

2. Working Operation

(i) If $V_1 = 0$ volt and $V_2 = 0$ then both the MOSFETs will be OFF and current through R_D is zero. Hence, the output voltage $V_o = +5$ volts i.e., high.

$$\therefore \text{ If } V_1 = V_2 = 0 \text{ then } V_o = 1$$

(ii) If $V_1 = 0$ and $V_2 = +5$ V, then M_1 will be OFF but M_2 will be ON and will act as a closed switch. Hence, the output voltage $V_o = 0$.

$$\therefore \text{ If } V_1 = 0 \text{ and } V_2 = 1 \text{ then } V = 0$$

(iii) If $V_1 = +5$ V and $V_2 = 0$, then M_1 will be ON and acts as a closed switch. Hence the output voltage will be equal to zero. M_2 will be OFF and will act as an open switch.

$$\therefore \text{ If } V_1 = 1 \text{ and } V_2 = 0 \text{ then } V_o = 0$$

(iv) If $V_1 = V_2 = 5$V then both the MOSFETs will be ON. Both will be equivalent to close switches and the output voltage will be zero.

3. Truth table

The truth table of this two input NOR gate has been shown in table 4.14.

Table 4.14. Truth table of two input NOR gate

Inputs		MOSFET		Output
V_1	V_2	M_1	M_2	V_o
0	0	OFF	OFF	+ 5 (logic 1)
0	+ 5 (logic 1)	OFF	ON	0
+ 5 V (logic 1)	0	ON	OFF	0
+ 5 V	+ 5 V	ON	ON	0

4.13.4. Characteristics of NMOS Devices

Important characteristics of N-channel MOSFET:

(i) High input resistance

(ii) High input capacitance

(iii) Small input current

As NMOSFET is used in the NMOS family, the above mentioned characteristics will influence the characteristics of NMOS family.

Voltage and Current Parameters

The voltage and current parameters of the NMOS devices are as under:

Table 4.15. Voltage and current parameters

S.No.	Parameter	Value
1.	V_{CC} or V_{DD}	5 V
2.	V_{IL}	0.8 V
3.	V_{IH}	2.0 V
4.	V_{OL}	0.45 V
5.	V_{OH}	2.4 V
6.	V_{OL}	2 mA
7.	V_{OH}	– 400 μA

As observed from table 4.15, the NMOS voltage and current parameters are compatible with the TTL parameters. Therefore, their interface with TTL gates is easy.

Fan out

Since the input current of MOSFET is very small, one NMOS gate can drive a large number of MOSFET gates. Hence, the fan-out is large (Typically 30).

Speed

But when a large number of NMOS devices are driven the capacitance at the output of the driving gate increases to a great extent. This will reduce the speed of NMOS gate.

Propagation delay time

Due to the presence of large capacitance on the input and output sides of MOS devices, the propagation delay time is very large. Another reason for large propagation delay is that the resistances through which these capacitors are charged are also large. However, the development in MOS technology has increased the operating speed to a level comparable to TTL.

Power dissipation

Power consumption in the MOS gates is very small as compared to that in TTL. This makes them very much suitable for large scale integration.

Typically power dissipation is 0.1 mW per gate.

Unconnected Inpus

All the MOS devices are likely to get damaged due to electrostatic discharge (ESD),

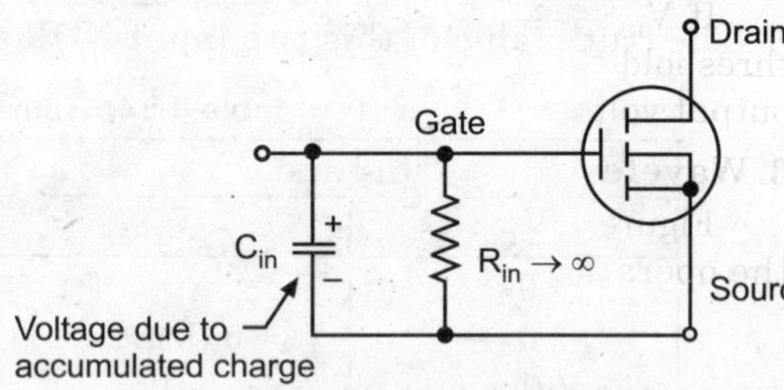

Fig. 4.72. Effect of input capacitance

We know that the gate terminal of a MOSFET is insulated from the channel due to inclusion of SiO_2 layer. This will make the input resistance of the device very large (ideally infinite). The gate leakage current of the MOSFET (I_{GSS}) is of the order of few pico amperes whereas the gate reverse current for a JFET is of the order of few nA. Input capacitance of a MOSFET is large due to its isolated gate structure. This capacitance can accumulate excess static charge. If the voltage across the input capacitance, due to accumulated static charge exceeds the breakdown voltage of gate then the MOSFET will get damaged. To avoid the damage due to ESD, the user has to take certain precautions while handling the MOS devices. They are as under:

(i) Store the MOS devices in conductive foam.

(ii) All the instruments, metal benches used in assembly or testing should be connected to earth.

(iii) The assembler's or handler's wrist should be connected to earth via a long wire and a high value series resistor.

(iv) Do not remove a MOS device from the circuit when the power is ON.

(v) Do not apply signal to a MOS device when the dc voltage applied to it is zero.

4.13.5. A PMOS Inverter

1. Circuit Diagram

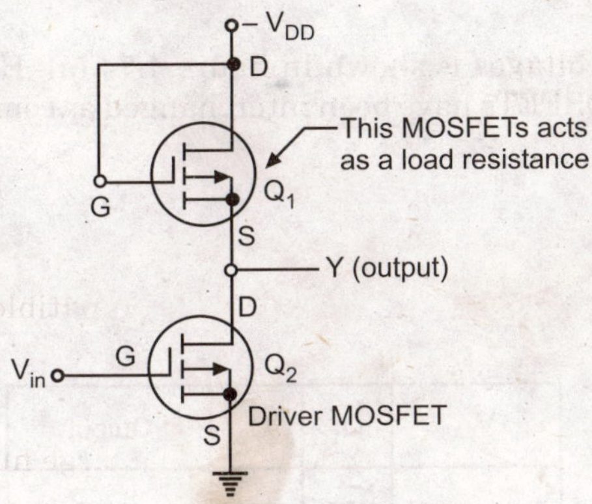

(a) PMOS inverter

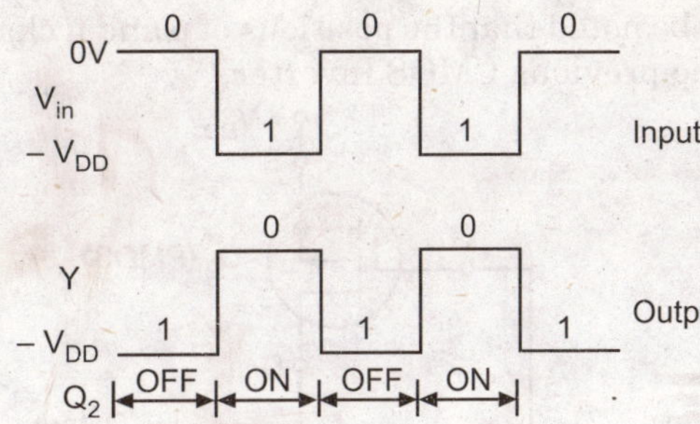

(b) Waveforms

Fig. 4.73.

The PMOS logic uses only P-channel MOSFETs. The schematic diagram of a PMOS inverter is shown in figure 4.73 (a). Both MOSFETs are P channel MOSFETs (enhancement type). Note that Q_1 remains permanently ON and acts as a resistor. But Q_2 acts as a switch depending on the level of input voltage applied at its gate.

2. Working Operation

(i) With V_{in} = 0 Volts. (logic 0)

If $V_{in} = 0$ volts, V_{GS} of Q_2 is also zero. Since it is a p-channel MOSFET, it is turned OFF completely with $V_{GS} = 0$. Therefore Y output will be High ($-V_{DD}$).

(ii) With V_{in} = – 5 V (logic 1)

If $V_{in} = -5$V, then V_{GS} of Q_2 is also – 5 V. This voltage is more negative than the typical threshold voltage ($V_T = -1.5$ V). Hence Q_2 will be turned ON fully and acts as a closed switch. So, output voltage Y is LOW (0). Thus inversion is achieved.

3. Waveforms

Figure 4.73 (b) shows the input output waveforms for the PMOS inverter. Table 4.16. summarizes the operation of PMOS inverter.

Table 4.16. Summary of PMOS Inverter Operation

Input V_{in}	Q_2	Output Y
0 Volts (0)	OFF	$-V_{DD}$ (1)
$-V_{DD}$ (1)	ON	0 (logic 0)

4.14. CMOS LOGIC

(PTU, Jalandhar, Sem. Exam.; 2008-09)

CMOS stands for complementary MOSFET. It is obtained by using a p-channel MOSFET and n-channel MOSFET simultaneously. The P and N channel MOSFETs are connected in series, with their drains connected together and output taken from common drain point. input is applied at the common gate terminal formed by connecting two gates together.

Advantages of CMOS Logic

CMOS logic has several advantages over P-MOS and N-MOS families. Some of them are as under:

(i) CMOS is faster

(ii) It consumes less power than other MOS families.

4.14.1. CMOS Inverter Circuit Diagram

1. Circuit Diagram

CMOS inverter with positive input and output voltages is shown in figure 4.74 (a). Here, it may be noted that the positions of p and n channel MOSFETs have been interchanged as compared to the previous CMOS inverter.

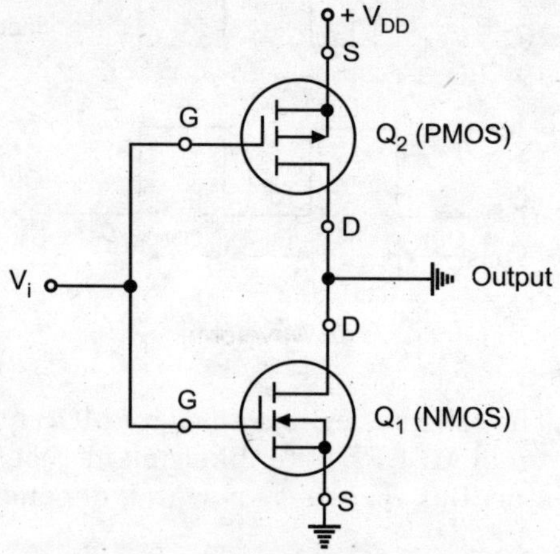

(a) CMOS inverter with positive voltages

V_i	Q_1	Q_2	Output
0V (logic 0)	OFF	ON	V_{DD} (logic 1)
V_{DD}(logic 1)	ON	OFF	0 (logic 0)

(b) Summary of operation

Fig. 4.74.

2. Working Operation

(i) With V_i = 0 Volts (logic 0)

With $V_i = 0$ volts, the gate to source voltage V_{GS} of Q_1 (NMOS) will be 0 volts, hence it will be OFF. But V_{GS} of Q_2 (PMOS) will be equal to $-V_{DD}$. So, Q_2 will be ON. Hence the output voltage will be equal to $+V_{DD}$ i.e., logic 1.

(ii) With $V_i = V_{DD}$ (logic 1)

With $V_i = +V_{DD}$, the gate to source voltages of the two MOSFETs are as under:

Q_1 (NMOS) : $V_{GS1} = V_{DD}$

Q_2 (PMOS) : $V_{GS2} = 0$ Volts

Hence, Q_1 will be turned ON and Q_2 will be OFF. Hence, output gets connected to ground i.e., it will be 0. Thus, the inversion will take place. The operation of this inverter has been summarized in figure 4.74 (b).

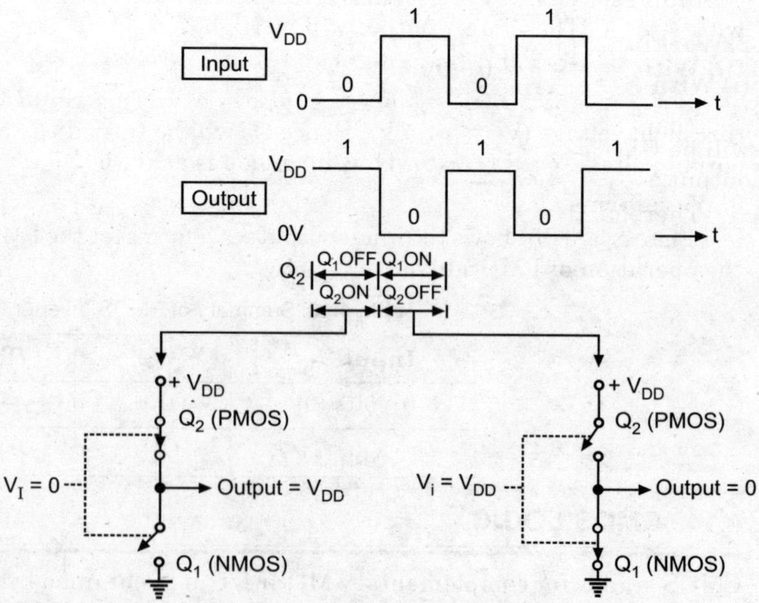

Fig. 4.75. Input output voltage waveforms and equivalent circuits for a CMOS inverter

3. Waveforms and equivalent circuits

The input output voltage waveforms for the inverter are shown in figure 4.75.

4.14.2. CMOS NOR Gate

1. Circuit Diagram and Equivalent Circuit

Figure 4.76 shows the CMOS 2-input NOR gate. It is obtained by modifying the CMOS inverter circuit.

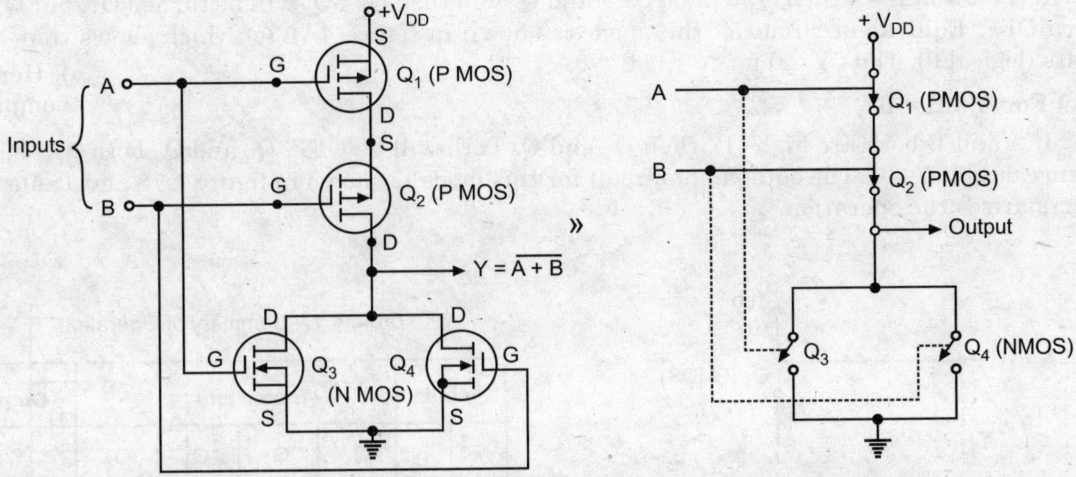

(a) CMOS NOR gate (b) Equivalent circuit

Fig. 4.76.

Q_1 and Q_2 are p-channel MOSFETs connected in series and Q_3 Q_4 are n-channel MOSFETs connected in parallel with each other. Input A is connected to the gates of Q_1 and Q_3 while input B is connected to the gates of Q_2 and Q_4.

> **Important Point :** Output $Y = V_{DD} = 1$ if Q_1 and Q_2 both ON and Q_3 Q_4 both OFF. And output $Y = 0$ if Q_1 or Q_2 or both are OFF and Q_3 or Q_4 both are ON.

2. Working Operation

(i) With A = B = 0

$V_{GS1} = -V_{DD}$, $V_{GS2} = -V_{DD}$, Hence Q_1 and Q_2 will be ON. $V_{GS3} = 0$, $V_{GS4} = 0$, So Q_3 and Q_4 will be OFF. The equivalent circuit for this mode is shown in figure 4.77 (a), which shows that output $Y = V_{DD}$ (logical 1).

Therefore, for A = 0, B = 0, output Y = 1.

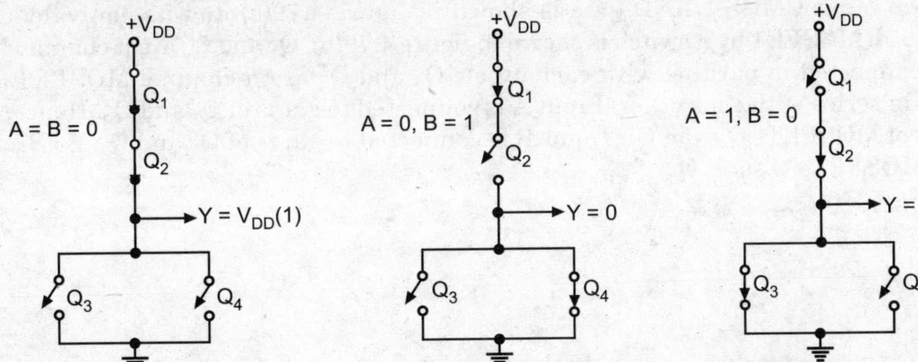

(a) For A = B = 0 (b) For A = 0, B = 1 (c) For A = 1, B = 0

Fig. 4.77. Equivalent circuits

(ii) With A = 0, B = 1

If A = 0 and B = 1, then then Q_1 will remain ON, but Q_2 will turn OFF. Q_3 will remain OFF but Q_4 will be ON. The equivalent circuit for this mode is shown in figure 4.77 (b) which shows that output Y = 0 Volts (logical 0). Thus Y = 0 for A = 0 and B = 1.

(iii) With A = 1, B = 0

If A = 1 and B = 0 then Q_1 will be OFF and Q_2 will turn ON. Q_3 will be turned ON but Q_4 will turn OFF. Equivalent circuit of this mode is shown in figure 4.78 (c) which shows that Y = 0 volts (logical 0). Thus Y = 0 for A = 1, B = 0.

(iv) For A = B = 1

If A and **B** both are high (1), then Q_1 and Q_2 both will be OFF. Q_3 and Q_4 both will be ON. Hence output Y = 0. The equivalent circuit for this mode is shown in figure 4.78 and Table 4.17. summarizes the operation.

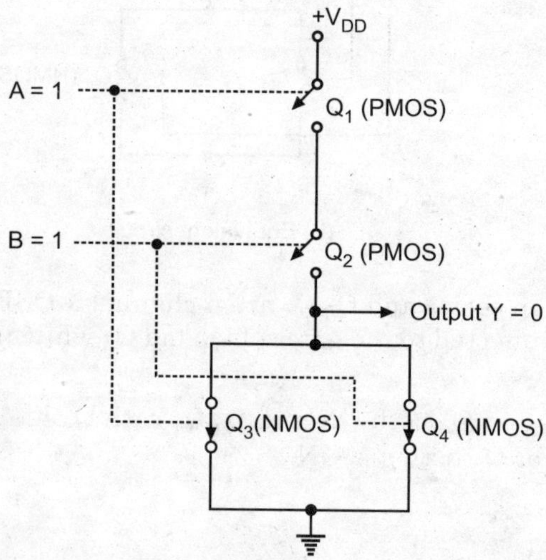

Fig. 4.78. Equivalent circuit for A = B = 1

Table 4.17. Summary of Operation

Inputs		Transistors				Output
A	B	Q_1	Q_2	Q_3	Q_4	Y
0	0	ON	ON	OFF	OFF	V_{DD} (1)
0	1	ON	OFF	OFF	ON	0
1	0	OFF	ON	ON	OFF	0
1	1	OFF	OFF	ON	ON	0

4.14.3. CMOS NAND Gate

1. Basic Aspects

The two input CMOS NAND gate is shown in figure 4.79(a) and its equivalent circuit by replacing each MOSFET by a switch is shown in figure 4.79(b). Q_1 and Q_2 are p-channel MOSFETs. They are connected in parallel with each other. Q_3 and Q_4 are n-channel MOSFETs. They are connected in series with each other. Input A is connected to gates of Q_1 and Q_3. Hence, A controls the status of MOSFETs Q_1 and Q_3. Input B is connected to gates of Q_2 and Q_4. So B controls the status of MOSFETs Q_2 and Q_4.

Progress is impossible without change, and those who cannot change their minds cannot change anything.

— G.B. Shaw

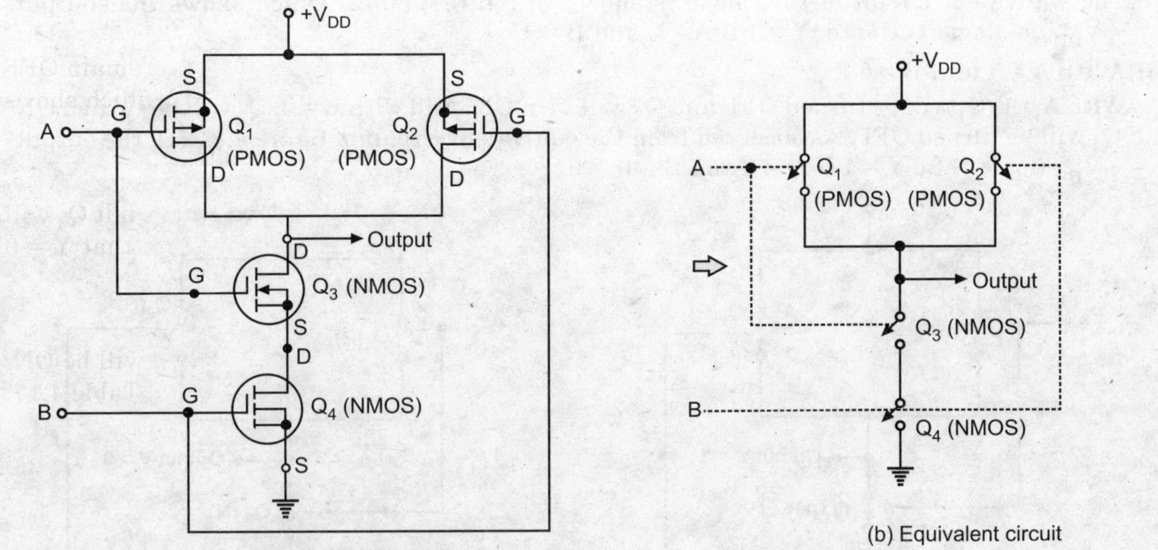

(a) 2-input CMOS NAND gate

(b) Equivalent circuit

(*a*) 2-input CMOS NAND gate

(*b*) Equivalent circuit

Fig. 4.79.

2. Working Operation

(i) A = B = 0

With A = 0 and B = 0, both the P-MOSFETs i.e., Q_1 and Q_2 will be ON. But both the N-MOSFETs i.e., Q_3 and Q_4 will be OFF. As seen from the equivalent circuit of figure 4.80 (a), the output Y = + V_{DD} (logic 1).

So Y = 1 if A = B = 0.

(ii) With A = 0 and B = 1

With A = 0 and B = 1, Q_1 will continue to be ON and Q_3 continues to be OFF. But Q_2 will now turn OFF and Q_4 will be turned ON.

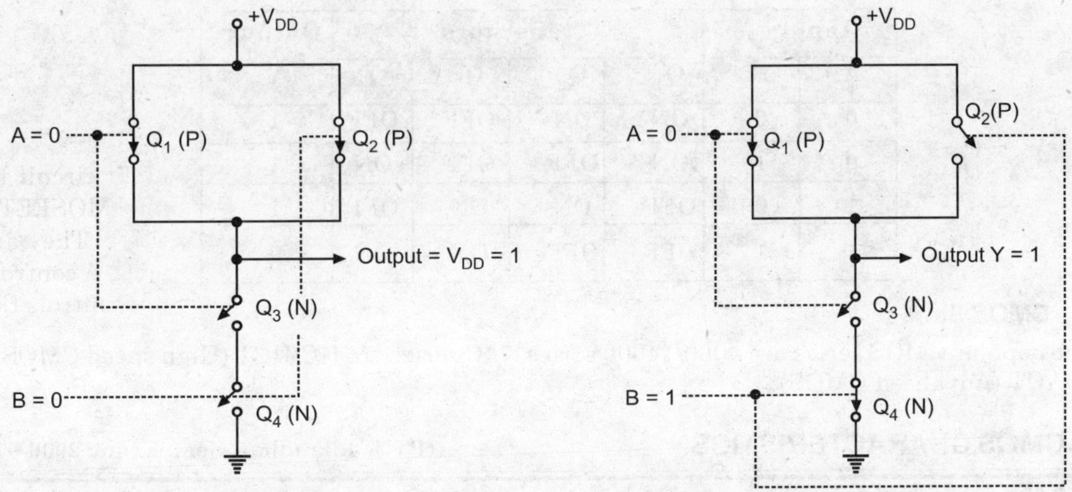

(a) A = B = 0

(b) A = 0, B = 1

Fig. 4.80. Equivalent circuits

The equivalent circuit of this mode is shown in figure 4.80 (b) which shows that output $Y = +V_{DD}$ i.e., logic 1. Hence, Y = 1 if A = 0 and B = 1.

(iii) With A = 1 and B = 0

With A = 1, Q_1 will be turned OFF and Q_3 will turn ON. And with B = 0, Q_2 will be turned ON and Q_4 will be turned OFF. As observed from the equivalent circuit of figure 4.81 (a), the output $Y = +V_{DD}$ (logic 1) So Y = 1 if A = 1 and B = 0.

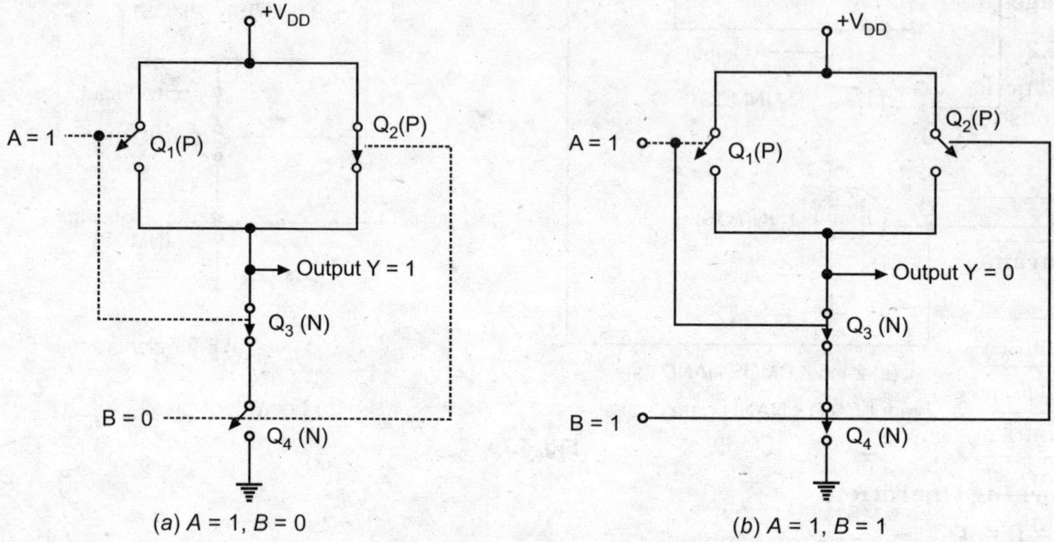

Fig. 4.81. Equivalent circuits

(iv) With A = 1, B = 1

With A = B = 1 both P-MOSFETs i.e., Q_1 and Q_2 will be OFF and both the N-MOSFETs i.e., Q_3 and Q_4 will be ON. The equivalent circuit of this mode is shown in figure 4.81(b). It shows that output Y = 0 (LOW). Hence, Y = 0 if A = B = 1.

The operation of 2-input CMOS NAND gate is summarized in Table 4.18.

Table 4.18. Summary of operation of a CMOS NAND gate

Inputs		Transistors				Output
A	B	Q_1	Q_2	Q_3	Q_4	Y
0	0	ON	ON	OFF	OFF	1
0	1	ON	OFF	OFF	ON	1
1	0	OFF	ON	ON	OFF	1
1	1	OFF	OFF	ON	ON	0

4.14.4. CMOS Series

The popular CMOS series are 4000/14000 series, 74C series, 74 HC/HCT (High speed CMOS), 74 AC/ACT (Advanced CMOS).

4.15. CMOS CHARACTERISTICS

(PTU, Jalandhar, Sem. Exam; 2006-07)

Some of the important CMOS characteristics are as under:

4.15.1. Power Supply Voltage

The 4000/14000 series and 74C series CMOS ICs are capable of operating over a wide range of power supply voltage (typically 3 V to 15 V). So CMOS ICs from these series are extremely versatile. They can operate on 3V batteries are well as 5V TTL compatible power supplies. These ICs can operate on higher power supply voltages, when higher noise margin is required. The other CMOS families such as 74 HC/HCT,74 AC/ACT and 74 AHC/AHCT operate over a voltage range of 2 to 6V.

4.15.2. Logic Voltage Levels

The logic voltage levels for different CMOS series have different values. Table 4.19. lists various logic voltage levels for different CMOS series.

Table 4.19. Logic voltage levels (in volts) with $V_{DD} = V_{CC} = +5$ V

	CMOS							TTL			
Parameter	**4000 B**	**74 HC**	**74 HCT**	**74 AC**	**74 ACT**	**74 AHC**	**74 AHCT**	**74**	**74 LS**	**74 AS**	**74 ALS**
$V_{IH(min)}$	3.5	3.5	2.0	3.5	2.0	3.85	2.0	2.0	2.0	2.0	2.0
$V_{IL(min)}$	1.5	1.0	0.8	1.5	0.8	1.65	0.8	0.8	0.8	0.8	0.8
$V_{OH(min)}$	4.95	4.9	4.9	4.9	4.9	4.4	3.15	2.4	2.7	2.7	2.7
$V_{OL(max)}$	0.05	0.1	0.1	0.1	0.1	0.44	0.1	0.4	0.5	0.5	0.4
V_{NH}	1.45	1.4	2.9	1.4	2.9	0.55	1.15	0.4	0.7	0.7	0.7
V_{NL}	1.45	0.9	0.7	1.4	0.7	1.21	0.7	0.4	0.3	0.3	0.4

Important Points from table 4.19.

(i) All the voltage levels shown in table 4.19 correspond to a supply voltage of +5 volts.

(ii) Here, it may be noted that $V_{OH} \cong 5$ V and $V_{OL} \cong 0$ V.

4.15.3. Noise Margins

Table 4.19 contains the high level and low level noise margins V_{NH} and V_{NL} for each CMOS and TTL series. These may be calculated as under:

$$V_{NH} = V_{OH(min)} - V_{IH(min)}$$
$$V_{NL} = V_{IL(max)} - V_{OL(max)}$$

Hence, CMOS devices will have higher noise margins as compared to those of TTL. Hence, CMOS ICs should be preferred to TTL for operation in the noisy environment. The noise margins in CMOS will increase further if the supply voltages are increased further.

4.15.4. Power Dissipation

The power dissipation of CMOS devices is extremely low when they are in the static (not changing) state. Typically the dc power dissipation is 2.5 nW per gate when $V_{DD} = 5$ V and 10 nW for $V_{DD} = 10$V. This is too small as compared to TTL gates ($P_D = 10$ mW). Hence CMOS devices are preferred for the battery operated systems. But power dissipation is low under the dc operating conditions and at low frequencies. But, power dissipation increases as the circuit switching frequency is increased. The relation between power dissipation and frequency is demonstrated in table 4.20.

The wheel of change moves on, and those who were down go up and those who were up go down.

— **R.N. Tagore**

Table 4.20. Relation Between P_D and frequency

Frequency	0 (dc)	100 kHz	1 MHz
Power dissipation PD	10 nW	0.1 mW	1 mW

4.15.5. Fan Out

The input resistance of CMOS devices is very high (10^{12} Ω). So their input current is very very small almost zero. Therefore, one CMOS gate can drive a large number of other CMOS gates. Hence, fan out of CMOS devices will be large as compared to fan out of TTL. Typically, the fan out is restricted to 50 for operation below 1 MHz. Why is fan out restricted to only 50? This is because the input capacitance of each CMOS gate is about 5 pF. These capacitors act as load on the driving gate as shown in figure 4.82. The charging current for these capacitor is supplied by the driving gate. This will limit the fan-out to 50.

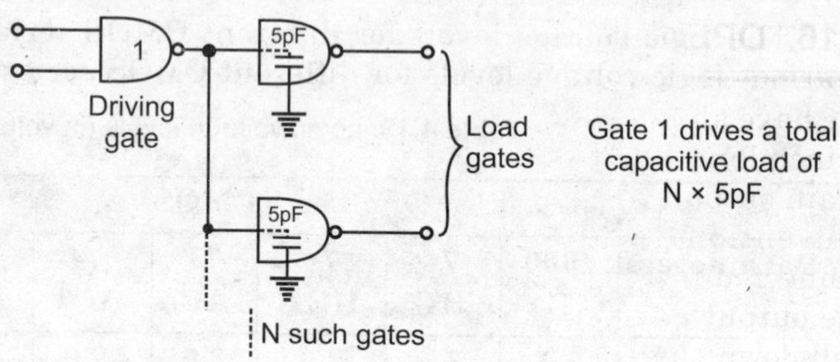

Fig. 4.82. Fan out and switching speed are dependent on the input capacitance of load gates

4.15.6. Switching Speed

The output resistance of CMOS is low in both the states (0 or 1) of output. Hence, even though it has to drive large capacitance loads, the switching speed can be somewhat faster then the NMOS or PMOS devices. The NAND gate of 4000 series has following values of propagation delays,

$$\text{Average } t_{pd} = 50 \text{ ns at } V_{DD} = 5\text{V}$$
$$\text{Average } t_{pd} = 25 \text{ ns at } V_{DD} = 10\text{V}$$

The average time delay for various CMOS ICs are given in table 4.21.

Table 4.21. Switching speeds for various CMOS ICs at V_{DD} = 5 V

Family	4000	74 HC/HCT	74 AC/ACT	74 AHC
Propagation Delay t_{pd}	50 ns	8 ns	4.7 ns	4.3 ns

4.15.7. Unused Inputs

It may be noted that the CMOS inputs should never be left floating. All the CMOS inputs should be either connected to 0 V (ground) or V_{DD}, or to another inputs. This is necessary to avoid permanent damage of CMOS ICs. Sometimes there are some unused gates on a chip. The inputs of such gates also should be connected to ground or + V_{DD}. The CMOS ICs are damaged due to induced voltages at the floating inputs due to noise or static charges. Such voltages can bias the P-MOS and N-MOS in their conduction state that may cause overheating and damage.

4.15.8. Static Sensitivity

The MOS ICs are susceptible to the damages due to static electricity. Such static electricity (voltages upto about 30 kV) gets generated by our simple day to day actions. For example, if a person walks on a carpet, then it will generate a voltage of 30 kV. If he touches the gate of a MOSFET, then the oxide layer will breakdown, due to large voltage and the device will be permanently damaged. Such a damage is called as damage due to electrostatic discharge (ESD), and we have to use a resistor diode protection network for protection.

4.15.9. Latch Ups

Due to the peculiar construction of the MOSFETs, there exist parasitic (unwanted) PNP and

NPN transistors in the MOSFET structures. Under certain operating conditions, the parasitic transistor will turn ON and will latch up i.e., remain ON permanently. This may force a large current through the CMOS device and damage it permanently. Latch ups can take place when the maximum voltage rating of the device is exceeded. Latch ups can also take place due to high voltage spikes or ringing at the input. To avoid such latch ups, we have to use well regulated power supply with minimum spikes on the V_{DD} line. This supply with current limit will help to minimize the device current if latch up takes place. Manufacturers provide an internal protection circuit for avoiding the latch ups.

4.16. OPEN DRAIN OUTPUTS

We have already observed the significance of open collector output for the TTL family. The same principle is extended to the CMOS family, as well. Such CMOS gates are known as the open drain outputs. Figure 4.83 shows a block diagram of an open drain CMOS gate. The outputs of conventional CMOS outputs should never be connected together. In the common drain output, the output is taken at the drain of the N-channel pull-down MOSFET. The drain terminal is left open (not connected to any other circuitry). The P-channel pull-up MOSFET is replaced by a diode as shown in figure 4.83. It is possible to wire AND the CMOS ICs having open drain outputs. An external pull up resistance is required to be connected between the open drain and supply voltage + V_{DD}.

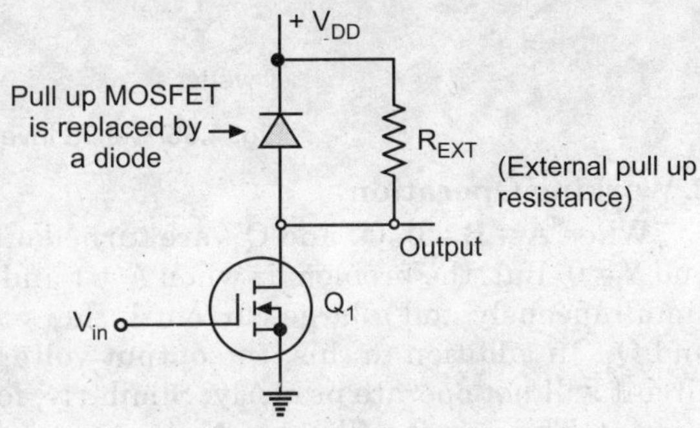

Fig. 4.83. Open drain output

Here, it may be noted that the open collector or open drain circuit can not actively force their output HIGH (1). They are capable of forcing the output LOW (0), The pull up resistance R_{EXT} will decide the charging current and the charging time constant for the capacitive load. As the value of R_{EXT} is in kΩ, the charging time constant will be large. This makes the open drain circuits slow. So the switching speed of open drain CMOS devices is always less then the conventional CMOS devices.

4.16.1. Applications of Open Drain Output

The open drain output CMOS are used as buffer/driver. A buffer or driver is a logic circuit which can supply a larger voltage or current then the ordinary logic gate. So, if we connect a buffer after an ordinary logic gate, then the gate becomes capable of sourcing higher currents.

4.16.2. Application of Open Drain Inverter

(i) The 74 HC 05 is an open-drain hex inverter. It has a current sinking capability of 25 mA.

(ii) Figure 4.84 shows how to drive a control relay using 74 AHC 74 D-FF. This D-FF is interfaced with 74 HC 05 NOR inverter.

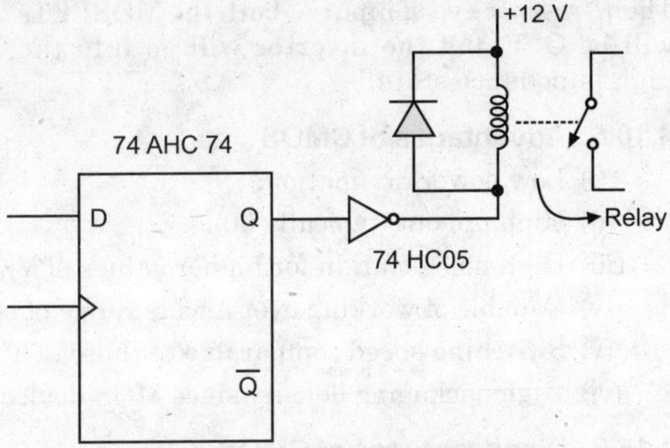

Fig. 4.84. Application of open drain inverter

4.16.3. Wired Logic

1. Circuit Diagram

Figure 4.85 shows the two CMOS inverter with their outputs connected together.

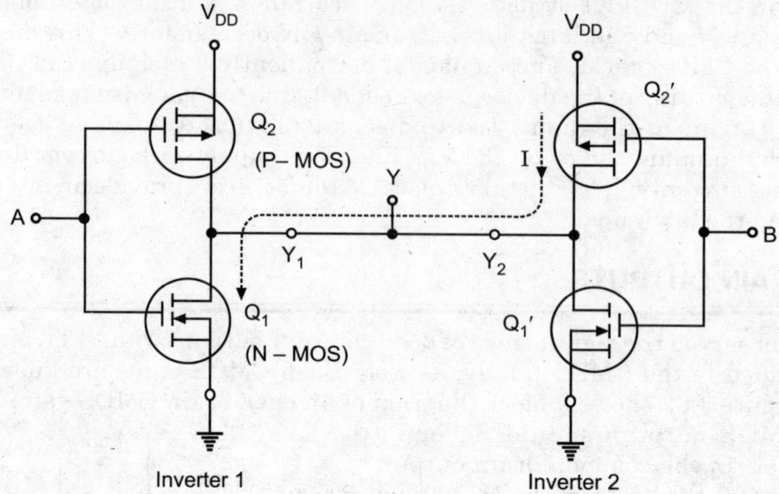

Fig. 4.85. CMOS inverter with outputs connected

2. Working Operation

When A = B = 0. Q_1 and Q_1' are turned off. Hence, Y = 1. When A = B = 1, Q_1 and Q_1' are on and Y = 0. But, the problem is when A = 1 and B = 0. As shown in figure 4.85, Q_1 and Q_2' conduct simultaneously and a large current I, flows through them. This large current will damage Q_2' and Q_1. In addition to this, the output voltage $Y = V_{DD}/2$ which is neither 0 nor 1. Hence, the circuit will not operate properly. Similarly, for A = 0 and B = 1. Q_2 and Q_1' will conduct a large current. The circuit will not operate properly.

3. Conclusion

The wired logic should not be used for CMOS logic circuits.

4.16.4. Tristate Logic Output

Figure 4.86 shows an inverting driver along with logic gates. When both the MOSFETs are in the OFF state, the output is in its high impedance state. When the enable input E = 1, the inverter operates in a normal way. Either Q_1 is ON or Q_2 is ON, depending on the input A.

But, if E = 0, then neither Q_1 nor Q_2 is ON. Then, irrespective of input A, both the MOSFETs will be OFF and the inverter will go into the high impedance state.

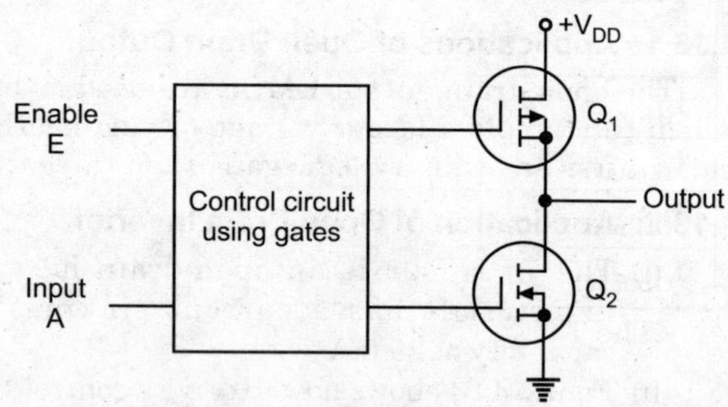

Fig. 4.86. A tristate CMOS inverter

4.16.5. Advantages of CMOS

(i) Low power dissipation.
(ii) High fan out (typically 50).
(iii) High noise margin for higher values of V_{DD}.
(iv) Capable of working over a wide range of supply voltage.
(v) Switching speed comparable to those of TTL.
(vi) High packaging density since MOS devices need less space.

4.16.6. Disadvantages of CMOS

(i) Propagation delays longer than those of TTL (25 to 100 ns).
(ii) Slower than TTL.

(iii) Susceptible to damages due to static charge.
(iv) Latch ups can take place.
(v) Need protection circuitry.

4.17. INTERFACING

(PTU, Jalandhar, Sem. Exam., 2005-06)

The word interfacing means connecting the output (s) of one system to the input (s) of another system or circuits with different electrical characteristics. If the electrical characteristics of the two circuits or systems are different then a direct connection can not be established between them. Hence, an *interface* circuit is required to be inserted between the *driver* circuit and the *load* circuit. The task of the interface circuit is to accept the output of the driver circuit and *condition* it so that it is compatible with the load circuit.

Fig. 4.87. Interfacing between two different systems

ICs from the same family are designed to be connected together without any special considerations. But, when we connect the output of an IC to the input of an IC of different type, we have to be concerned about the voltage and current levels of both the ICs.

4.17.1. TTL Driving CMOS

When we interface different types of ICs, it is necessary to check whether the driving IC is capable of meeting the current and voltage requirements of the load IC or not. For example Table 4.22 reveals that the output current capability of TTL ICs is much higher than the input current values of CMOS ICs. Therefore, there is no problem for TTl to drive CMOS as far as current is concerned. But, there is a problem when we compare the voltage levels of TTL and CMOS. Because $V_{OH(min)}$ of a TTl series is very low as compared with $V_{IH(min)}$ required for the CMOS series like 400B, 74 HC or 74 AC. In such situations something has to done to raise the output voltage to an acceptable voltage level for CMOS.

Table 4.22. Input output currents for standard devices with a supply voltage of 5V

Parameter	CMOS		TTL		
	4000 B	74 HC/HCT	74	74 LS	74 AS
$I_{IH(max)}$	1 μA	1 μA	40 μA	20 μA	20 μA
$I_{IL(max)}$	1 μA	1 μA	1.6 mA	0.4 mA	0.5 mA
$I_{OH(max)}$	0.4 mA	4 mA	0.4 mA	0.4 mA	2 mA
$I_{OL(max)}$	0.4 mA	4 mA	16 mA	8 mA	20 mA

The TTL driving CMOS has been shown in figure 4.88.

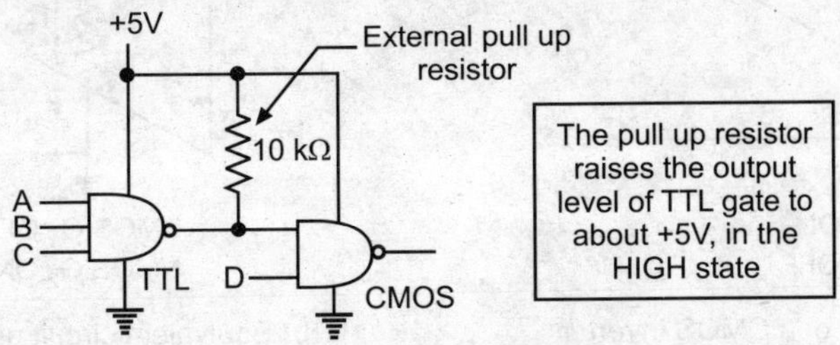

Fig. 4.88. TTL driving CMOS

Due to the presence of pull up resistor will result in rise in the TTL output to approximately

+ 5V in the HIGH state. This will provide the sufficient voltage level at the input of the CMOS gate. Such a pull up resistance is not required if the CMOS gate belongs to 74 HCT or 74 ACT family, since these families can accept the TTL outputs directly.

4.17.2. TTL Driving High Voltage CMOS

But, if the CMOS IC is operated with a supply voltage V_{DD} greater than 5V the interfacing becomes a little bit difficult. For example with V_{DD} = 10 V, the CMOS input requires $V_{IH(min)}$ = 7V. The outputs of many TTL devices cannot be operated at a voltage higher than + 5V. Hence, the technique of using the pull up resistance can not use. The interfacing of TTL and high voltage CMOS is shown in figure 4.89.

When it is not possible to pull up the totem-pole output to + V_{DD}, a buffer is used as interface between TTL totem-pole and high voltage CMOS gate. The 7406 is an inverting CMOS buffer. It has an output voltage rating of 30 V. Hence, the arrangement of buffer along with a pull up resistance of 10 kΩ, as shown in figure 4.89.

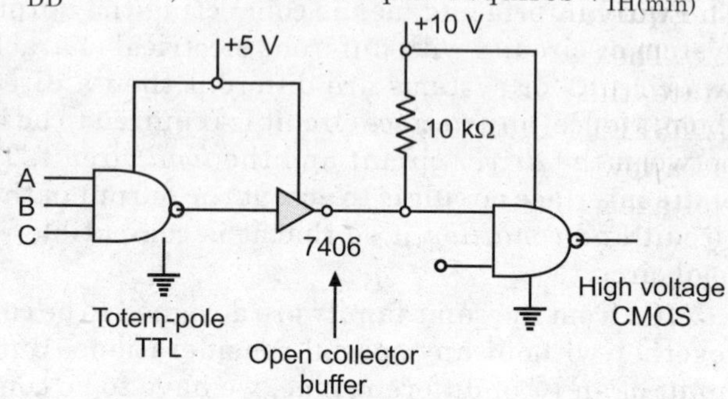

Fig. 4.89. TTL to high voltage CMOS interfacing, using open collector buffer.

4.17.3. Interfacing using Level-Shifter

Another way to interface TTL with high voltage CMOS is to use a voltage level translator as shown in figure 4.90. The voltage translator circuit accepts a low voltage from TTL and translates it into a high voltage output for CMOS. The example of level-shifter is 4504 B.

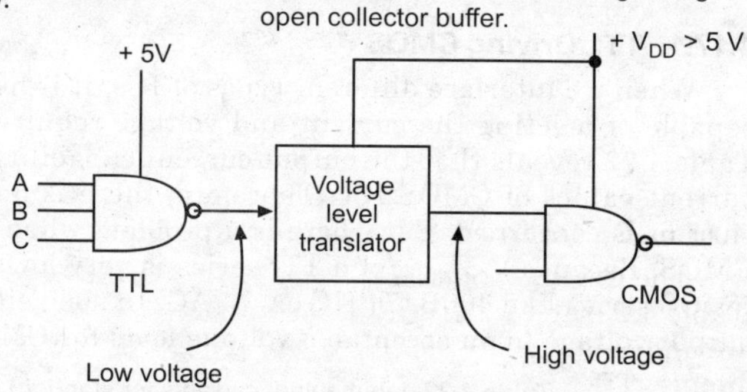

Fig. 4.90. Interfacing using level-shifter

4.17.4 CMOS Driving TTL

1. Basic Aspects

Before we consider the problem of CMOS interfacing with TTL, let us review the output characteristics of CMOS in two different states. Figure 4.91 (a) shows the equivalent output circuit in the HIGH state output, whereas figure 4.91 (b) shows the equivalent output.

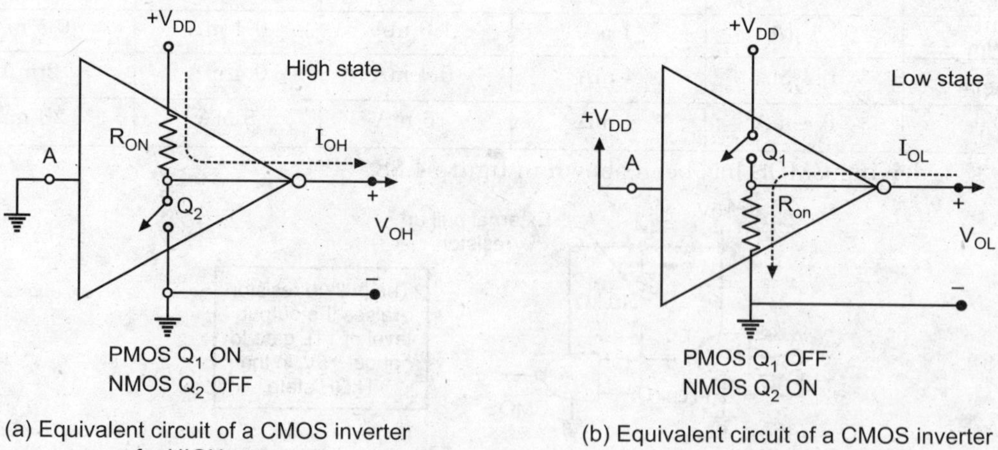

(a) Equivalent circuit of a CMOS inverter for HIGH state

(b) Equivalent circuit of a CMOS inverter for LOW state

Fig. 4.91.

2. Equivalent Circuit in HIGH state

In the HIGH state of output, the P-channel MOSFET Q_1 will be ON, hence, it is replaced by the ON state resistance R_{ON}. Whereas the N-channel MOSFET Q_2 will be OFF. Hence, it is replaced by an open switch as shown in figure 4.91 (a). Therefore, the CMOS output is equivalent to a V_{DD} source with a source resistance of R_{ON}. (R_{ON} ranges from 100 to 1000Ω).

3. Equivalent circuit in LOW state

The equivalent circuit of a CMOS inverter in the low state is shown in figure 4.91 (b). In this state, the P-MOSFET Q_1 is OFF and represented by an open switch. Whereas the N-channel MOSFET is ON. Hence, it is replaced by resistance R_{ON}. Therefore, the CMOS inverter now acts as a low resistance connected to ground and sinks current. Table 4.23. presents the various voltage and current levels for the standard CMOS (4000 B) and TTL (74) series for comparison.

Table 4.23.

S. No.	Parameters for Driving gate (CMOS) of 4000 B series	Parameters for the load gate (TTL) of 74 series
1.	$V_{OH(min)}$ = 4.95 V	$V_{IH(min)}$ = 2 V
2.	$V_{OL(max)}$ = 0.05 V	$V_{IL(max)}$ = 0.8 V
3.	$I_{OH(max)}$ = 0.4 mA	$I_{IH(max)}$ = 40 μA
4.	$I_{OL(max)}$ = 0.4 mA	$I_{IL(max)}$ = 1.6 mA

For low state output (rows 2 and 4); For high state output (rows 1 and 3)

4.17.5. CMOS Driving TTL in the HIGH State

The CMOS output voltage is high (1). Hence, we consider the first and third parameters in Table 4.23. *i.e.,* V_{OH} and I_{OH} for CMOS and V_{IH} and I_{IH} for TTL. Looking at these parameters, we understand that the CMOS output can supply sufficient voltage as well as current to the TTL inputs. Therefore, there is no need to use any special interface for high state.

4.17.6. CMOS Driving TTL in the LOW State

The CMOS output is low (0). Hence, let us consider the second and fourth parameters in table 4.23. i.e., V_{OL}, I_{OL} for CMOS gate and V_{IL} and I_{IL} for TTl gate. Looking at these parameters, we understand that the low state output voltage V_{OL} of CMOS satisfied the V_{IL} requirement of TTL. But the low level output current of a CMOS gate (I_{OL}) is not sufficient to drive even a single TTL gate with $I_{IL(max)}$ = 1.6 mA. Under such circumstances, we should use a non-inverting buffer between the CMOS and TTL gates as shown in figure 4.92, to increase the current sourcing capacity.

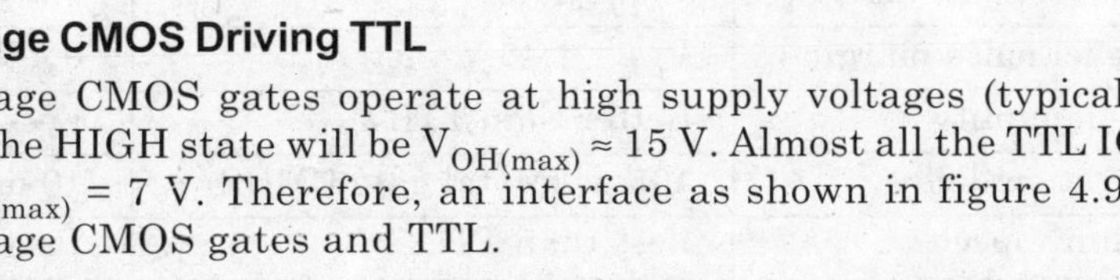

Fig. 4.92. CMOS driving TTL in low state

4.17.7. High Voltage CMOS Driving TTL

The high voltage CMOS gates operate at high supply voltages (typically 15 V). So their output voltage in the HIGH state will be $V_{OH(max)} \approx 15$ V. Almost all the TTL ICs has a maximum input voltage $V_{IH(max)}$ = 7 V. Therefore, an interface as shown in figure 4.93 will be required between high voltage CMOS gates and TTL.

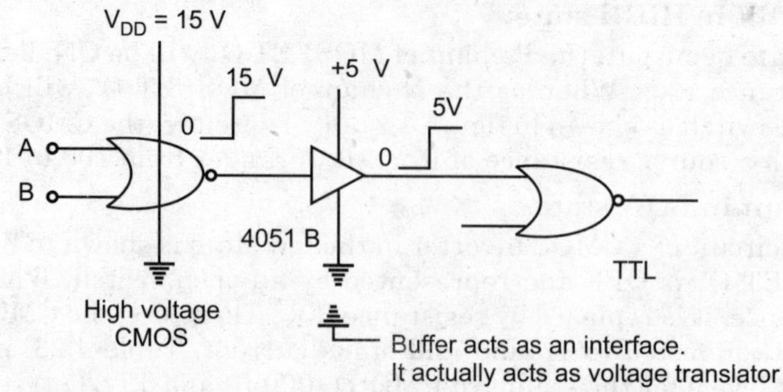

Fig. 4.93. High voltage CMOS driving TTL

The buffer 4051 B operates on +5 V supply. It is a non-inveting buffer. So, it converts a 15 V HIGH output of the CMOS gate into a 5 V input for the TTL gate. Thus the 4051 B buffer acts like a level translator.

4.17.8. Interfacing CMOS with ECl and TTL

The logic voltage levels is TTL are extremely different from those of TTL and CMOS. Therefore to interface ECL gates with the other logic families such as CMOS and TTL we need to use the interface in the form of a voltage translator. Voltage translators are actually the voltage level shifters. They are available in the ECL family. The TTL/CMOS to ECL and ECL to TTL/CMOS interfacing is shown in figure 4.94.

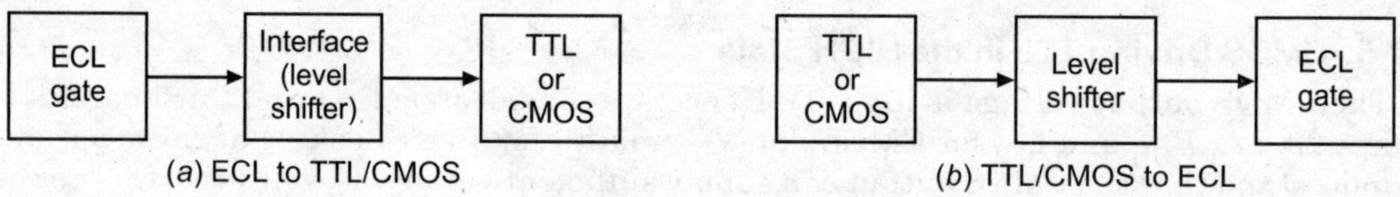

Fig. 4.94. Interfacing

4.18. PERFORMANCE COMPARISONS (PTU, Jalandhar, Sem. Exam., 2004-05)(05 Marks)

4.18.1. Performance Comparison of CMOS and TTL

Table 4.24. Comparison of CMOS and TTL

S.No.	Parameter of comparison	CMOS	TTL
1.	Device used	N-channel MOSFET and P-channel MOSFET	Bipolar junction transistor
2.	$V_{IH(min)}$	3.5 V (V_{DD} = 5 V)	2 V
3.	$V_{IL(max)}$	1.5 V	0.8 V
4.	$V_{OH(min)}$	4.95 V	2.7 V
5.	$V_{OL(max)}$	0.05 V	0.4 V
6.	High level noise margin	V_{NH} = 1.45 V	0.4 V
7.	Low level noise margin	V_{NL} = 1.45 V	0.4 V
8.	Noise immunity	Better than TTL	Less than CMOS
9.	Propagation delay	105 ns (Metal gate CMOS)	10 ns. (Standard TTL)
10.	Switching speed	Less than TTL.	Faster than CMOS

11.	Power dissipation per gate.	P_D = 0.1 mW. Hence used for battery backup applications	10 mW
12.	Speed power product.	10.5 pJ	100 pJ
13.	Dependence of P_D on frequency	P_D increases with increase in	P_D does not depend on frequency.
14.	Fan out	Typically 50	10
15.	Unconnected inputs	Unused inputs should be returned to GND or V_{DD}. They should never be left floating.	Inputs can remain floating The floating inputs are treated as logic 1s.
16.	Component density	More than TTL since MOSFETs need smaller space while fabricating an IC.	Less than CMOS since BJT needs more space.
17.	Operating areas	MOSFETs are operated as switches, i.e., in the ohmic region or cut off region.	Transistors are operated in saturation or cut off regions.
18.	Power supply voltage	Flexible from 3 V to 15 V	Fixed equal to 5 V

4.18.2. Performance Comparison of TTL, CMOS and ECL

S.No.	Parameter of comparison	TTL	CMOS	ECL
1.	Propagation delay	10 ns	70 ns	500 ps
2.	Noise margin	0.4 V	0.45 V_{DD}	150 mV
3.	Power dissipation	10 mW	0.01 mW	5 mW
4.	Fan out	10	50	25
5.	Figure of merit	100 pJ	0.7 pJ	0.5 pJ

4.19 CLASSIFICATION BASED ON CIRCUIT COMPLEXITY

ICs are also classified according to the number of components or gates in the case of digital ICs integrated on the same chip as follows

(i) Small scale integration (SSI) < 10 components

(ii) Medium scale integration LSI < 100 components

(iii) Large scale integration LSI > 100 components

(iv) Very large scale integration VLSI > 1000 Components

SUMMARY

- Various digital ICs available in market belong to various types. These types are known as families.
- Based on the components and devices, internally used, the digital IC families are named as RTL (Resistor transistor logic). TTL (Transistor Transistor Logic), DTL (Diode Transistor Logic). CMOS etc.
- The digital circuits are in the integrated circuit (IC) form the elements used for building such circuits are diodes, resistors, transistors and MOSFETs.
- When a diode is driven from reverse condition to forward condition or in the opposite direction, it does not change its state instantaneously.
- The forward recovery time (t_{rf}) is defined as the time required for the diode current to change from 10 to 90 percent of its final value when the diode is switched from OFF to ON state. In practice, the forward recovery time is very short as compared to the reverse recovery time.

- Commercial switching type diodes are available with time t_{rr} ranging from 1 nanosecond (ns) upto as high as 1 microsecond (μ s). Higher values of t_{rr} correspond to the diodes intended to handle larger currents.
- IN 4153 is a switching diode. It has a reverse recovery time of a few nanoseconds under the test conditions. IN 4153 is high speed switching diode for computer and general purpose applications.
- Transfer characteristics is the graph of controlled quantity versus the controlling quantity. For the transistor to operate as a switch, it is operated in CE configuration. So, let us see the transfer characteristics of a transistor in CE configuration.
- The turn on and turn off times are important in deciding the maximum switching frequency of the transistor. In order to operate the transistor at a high frequency the turn on and turn off times should be as short as possible.
- The MOS-logic family uses MOSFET as the basic device the way TTL uses BJT. The MOSFETs are of two types namely the depletion MOSFETs and the enhancement type MOSFETs. In logic circuits, the enhancement MOSFETs are used. We use the E-MOSFET as a switch.
- The major disadvantage of MOSFETs is that they are susceptible to static charges. Due to accumulated static charges, MOSFETs can get damaged. By using the proper handling procedures, it is possible to minimize the possibility of damage. But still TTL proves to be more reliable and durable for laboratory work.
- The bipolar families of logic circuits construct, especially from components fabricate bipolar transistors on the chip. In the bipolar category there are three basic families called Diode transistor logic (DTL). Transistor transistor logic (TTL) and Emitter Coupled Logic (ECL). DTL uses diodes and transistors, TTL uses transistors almost exclusively.
- The MOS family fabricates MOS field effect transistors (MOSFETs). In the MOS category there are three logic families namely PMOS (p-channel MOSFETs) family, NMOS (n-channel MOSFETs) family and CMOS (complementary MOSFETs) family.
- The output of a logic gate does not change its state instantaneously in response to the change in the state of its input. There is a time delay between these two events, which is called as the propagation delay. Thus, propagation delay is defined as time delay between the instant of application of an input pulse and the instant of occurrence of the corresponding output pulse.
- When the industrial environment is noisy due to the presence of motors, high voltage switches, etc. the above logic families will not work satisfactorily. Hence, we require a logic family having very high noise immunity, like the HTL family.
- Some of the most important ECL characteristics are as follows:
    1. Maximum propagation delay : For the latest ECL gates the propagation delay is very short, of the order of 500 ps.
    2. Average power dissipation : 5 mW (for advanced ECL gates)
    3. Speed-power product (SPP): 0.5 pJ.
    4. Worst case noise margin is only 150 mV. So, ECL device are unreliable to work in noisy environments.
    5. ECL devices generally produce an output and its complement. (e.g. OR/NOR). So, additional inverter is necessary.
    6. Due to the emitter follower stages, the output impedance is very low.
    7. Due to low output impedance, the fan out is typically equal to 25.
    8. Typical power dissipation for the standard ECL gate is 25 mW which is somewhat higher than that of 74 AS series.
    9. The current flowing in ECL circuits remains almost constant so no current transients are observed and so associated noise also is less.
- The long form of TTL is transistor transistor logic. The digital ICs in the TTL family use only transistors to perform the basic logic operations. TTL ICs were first developed in 1965 and they were known as "standard TTL". This version of TTL currents is not available now.
- Wired ANDing means tying the outputs of gates together to obtain AND function. It is possible to connect the outputs of two or more gates together.
- A tristate buffer is a circuit which connects or isolates its input from its output. In other words it controls the passage of signal from input to output. Tristate buffers are of two types, namely the non-inverting

buffers and inverting buffer. The inverting tristate buffers will invert the input data while passing it to the output.

- CMOS stands for complementary MOSFET. It is obtained by using a p-channel MOSFET and n-channel MOSFET simultaneously. The P and N channel MOSFETs are connected in series, with their drains connected together and output taken from common drain point. input is applied at the common gate terminal formed by connecting two gates together.

SHORT QUESTIONS WITH ANSWER

Q.1. Write importance of Diode Switching Times.

Ans. (i) The reverse recovery time decides the time required to switch the diode off. Hence, it decides, how fast or at what maximum rate we can switch the diode on and off.

(ii) Thus, reverse recovery time decides the highest switching frequency of a diode.

(iii) The reverse recovery time depends on the transition capacitance of the diode. So the transition capacitance of switching diodes is small.

(iv) The maximum switching frequency is given by,

$$f_{max} = \frac{1}{10\, t_{rr}}$$

Therefore, as t_{rr} decreases, f_{max} will increase.

Q.2. What are Types of E-MOSFETs?

Ans. The E-MOSFETs are of two types namely

(i) P channel enhancement MOSFETs (PMOS)

(ii) N channel enhancement MOSFETs (NMOS).

Q. 3. What is meant by Fan-in and Fan-out?

Ans. Fan-in : Fan-in is defined as the number of inputs a gate has. For example, a two input gate will have a fan-in equal to 2.

Fan-out : Fan-out is defined as the maximum number of inputs of the same IC family that a gate can drive without falling outside the specified output voltage limits. Higher the fan out higher the current supplying capacity of a gate.

Q.4. What is meant by Noise Margin?

Ans. Noise is an unwanted electrical disturbance which may induce some voltage in the connecting wires used between two gates or from a gate output to load. **Noise immunity** is defined as the ability of a logic circuit to tolerate the noise without causing any unwanted changes in the output. A quantitative measure of noise immunity is called as **noise margin**.

Q. 5. Write salient features of IIL.

Ans. Some of the important features of IIL are:

(i) Simplicity

(ii) It is easier and cheaper to fabricate.

(iii) It is a saturated bipolar logic.

(iv) It consumers very little power.

(v) It is used for medium and large scale integration.

Q. 6. What are characteristics of IIL?

Ans. (i) Speed of operation depends on the charging current.

(ii) Propagation delay is inversely proportional to charging current.

(iii) Power dissipation is proportional to charging current.

(iv) Figure of merit is in the range of 0.1 to 0.7 PJ.

(v) Very small silicon area is required.

(vi) Packaging density is in the range of 120 to 200 gates per mm^2.

Q.7. Write advantages and Disadvantages of ECL Family.

Ans. Advantages

(i) It is the fastest logic family.

(ii) High fan out due to high input resistance and low output resistance. Typically fan out = 25.
(iii) Excellent speed power products.
(iv) It can provide two outputs simultaneously e.g. OR and NOR. No additional inverter is needed.

Disadvantages

(i) High power dissipation.
(ii) Its high speed generates voltage and current transients.
(iii) Limited logic swing. Hence, more vulnerable to noise.

Q. 8. What are advantages of Tristate?

Ans. Advantages of Tristate

(i) The outputs of tristate ICs can be connected together, i.e., they can share the use of a common wire. But in doing so there is no reduction in switching speed (this problem is faced for open collector (ICs).
(ii) There is no reduction in speed because the tristate output when enabled, operates as a totem-pole TTL. So, we get all the advantages of totem-pole output such as low output impedance and high speed.
(iii) However, note that when tristate outputs are connected together, we should enable only one of them at a time.
(iv) Otherwise, two or more active outputs will fight to take control of the common line. This will result in very high currents which can damage the ICs and produce invalid output voltage levels.

Q. 9. Write characteristics of NMOS devices.

Ans. Important characteristics of N-channel MOSFET:

(i) High input resistance
(ii) High input capacitance
(iii) Small input current

As NMOSFET is used in the NMOS family, the above mentioned characteristics will influence the characteristics of NMOS family.

Q. 10. What are the advantages of CMOS?

Ans. (i) Low power dissipation.
(ii) High fan out (typically 50).
(iii) High noise margin for higher values of V_{DD}.
(iv) Capable of working over a wide range of supply voltage.
(v) Switching speed comparable to those of TTL.
(vi) High packaging density since MOS devices need less space.

REVIEW QUESTIONS

1. What are the different logic families? Write their characteristics.
2. What is passive pull-up load and active pull-up load?
3. Give the advantages of active pull-up over passive pull-up.
4. Explain with diagram the working of High-threshold logic (HTL). State the advantage of HTL over DTL?
5. Explain briefly the operation of TTL NAND gate in tri-state output configuration.
6. Explain the use of multi-emitter inputs.
7. Define the following terms regarding a logic family:
   (i) Noise margin (ii) Propagation delay
8. Explain the ECL circuit.
9. Compare the performance of TTL, CMOS and ECL logic.
10. Explain the features of complementary symmetry logic (CMOS)
11. State the application of CMOS logic inverter.
12. What are the precautions taken while using CMOS ICs?
13. Explain 7400 TTL series. Sketch pin-outs of different ICs of 7400 series and state their functions.
14. Mention the advantages and disadvantages of TTL, CMOS and ECL IC families.

5

COMBINATIONAL LOGIC CIRCUITS

Syllabus

Combinational Circuits: Design procedure – Adders, Subtractors, Serial adder/Subtractor, Parallel adder/encoder/decoder, parity checker, code converters, Implementaion of combinational logic using MUX.

5.1 INTRODUCTION

Basically, the digital circuits are one of the following types :

(i) Combinational circuits

(ii) Sequential circuits.

A combinational circuit may be defined as a logic circuit the output of which depends only upon the combination of the inputs. The output does not depend on the past value of inputs or outputs. Therefore, combinational circuits do not need any memory. Figure 5.1 shows the block diagram of a combinational circuit

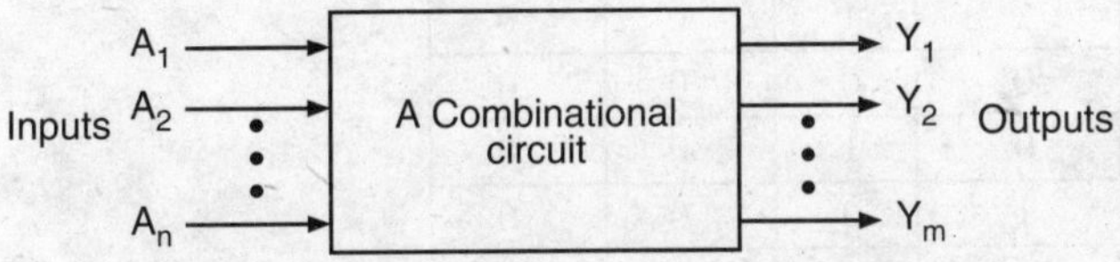

Fig. 5.1. Block diagram of a combinational circuit

A combinational circuit can have a number of inputs and a number of outputs. The circuit of figure 5.1 has n inputs and m outputs. Between the inputs and outputs, logic gates are connected, and hence, combinational circuit basically consists of logic gates. A combinational circuit operates in following three steps :

(i) It accepts n-different inputs.

(ii) The combination of gates operates on the inputs.

(iii) m different outputs are produced as per requirement.

Chapter Outline

5.2 COMBINATIONAL LOGIC DESIGN PROCEDURE

The various steps involved in designing procedure of a combinational logic may be listed as under :

(i) We will be given a problem.

(ii) Then, we determine the number of inputs and outputs and assign letter symbols to input and output variables.

(iii) After that we write a truth table relating the inputs and outputs.

(iv) Then, we write K-map for each output and obtain the simplified Boolean expression for each output.

(v) Lastly, we draw the logic diagram (combinational circuit).

DOUGLAS ENGELBART

What is commonly known today as the computer mouse was devised in 1963 by Douglas Engelbart and was originally known as an X-Y position indicator for a display system. Engelbart's early mouse was an electronic marble that was housed in a wooden box that had a red button and a copper wire tail.

EXAMPLE 5.1. A circuit has four inputs and two outputs. One of the outputs is high when majority of inputs are high. The second output is high only when all inputs are of same type. Design the combinational circuit.

Solution: (i) First, we assign symbols to input and output variables.

Let the four inputs be A, B, C, D and the two outputs be Y_1 and Y_2.

(ii) Then, we write the truth table.

The truth table has been given in Table 5.1.

Table 5.1. Truth table relating the inputs and outputs

	Inputs				Outputs	
Decimal	**A**	**B**	**C**	**D**	Y_1	Y_2
0	0	0	0	0	0	1
1.	0	0	0	1	0	0
2.	0	0	1	0	0	0
3.	0	0	1	1	0	0
4.	0	1	0	0	0	0
5.	0	1	0	1	0	0
6.	0	1	1	0	0	0
7.	0	1	1	1	1	0
8.	1	0	0	0	0	0
9.	1	0	0	1	0	0
10.	1	0	1	0	0	0
11.	1	0	1	1	1	0
12.	1	1	0	0	0	0
13.	1	1	0	1	1	0
14.	1	1	1	0	1	0
15.	1	1	1	1	1	1

Observations from the truth table

From the truth table, we note the following aspects :

(a) $Y_1 = 1$, when number of 1 inputs is higher than the number of 0 inputs.

(b) $Y_2 = 1$, when $A = B = C = D$.

(iii) Then, we write K-map for each output and get simplified expression.

K-maps for the two outputs and the corresponding simplified Boolean expressions have been given in figures 5.2 (a) and (b).

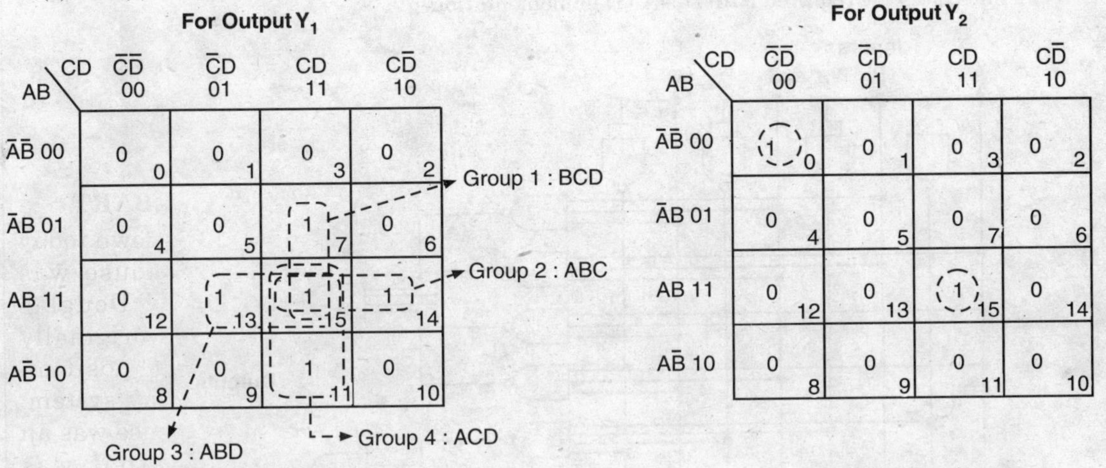

Simplified expression

$Y_1 = BCD + ABC + ABD + ACD$

(a) K-map and simplification for Y_1

Simplified expression

$Y_2 = \bar{A}\,\bar{B}\,\bar{C}\,\bar{D} + A\,B\,C\,D$

(b) K-map and simplification for Y_2

Fig. 5.2.

(iv) Next, let us draw the logic diagram.

The logic diagram has been shown in figure 5.3.

> **DO YOU KNOW?**
>
> It is often necessary to translate between and among various methods of representing quantities with binary numbers.

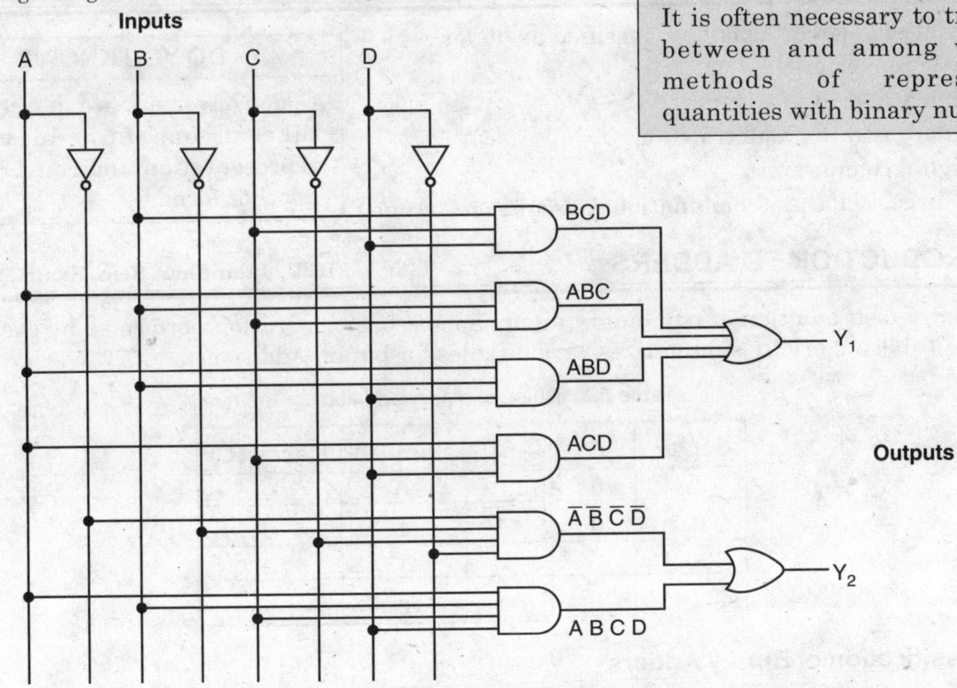

Fig. 5.3. Illustration of Logic diagram

EXAMPLE 5.2. Draw the combinational logic of Problem 5.1 using only NAND gates.

Solution: For drawing the logic diagram of figure 5.3 by using only the NAND gates, we make the following changes.

(i) First, we replace inverters by NAND inverters.
(ii) Then, we replace every AND gate by a NAND and every OR gate by a bubbled NAND.
(iii) Lastly, we draw the circuit using only NAND gates.

Figure 5.4 illustrates the required NAND-NAND implementation.

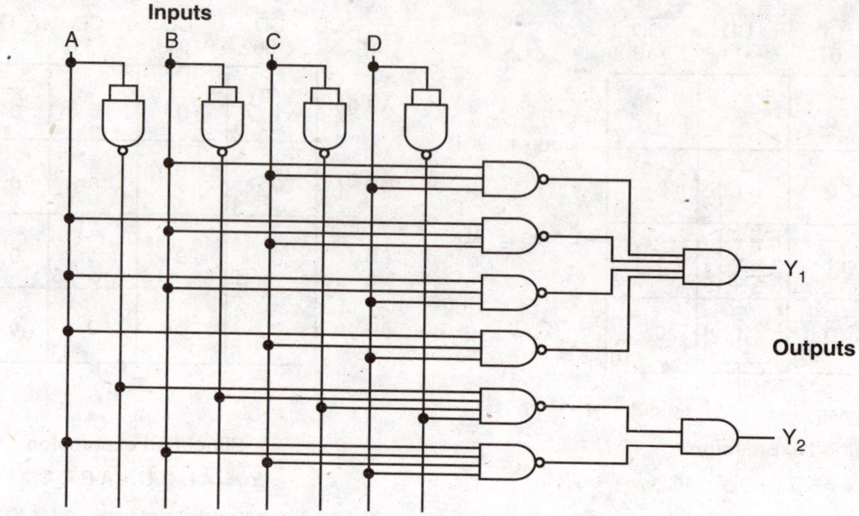

Fig. 5.4. NAND-NAND implementation

5.3. CLASSIFICATION OF COMBINATIONAL CIRCUITS

The combinational circuits may be classified as under :

(i) Code converters.
(ii) Binary and BCD adders.
(iii) Binary and BCD subtractors.
(iv) Digital comparators.

Let us discuss all these combinational circuits one by one.

> **DO YOU KNOW?**
> Code converters are devices that take in one form of binary representation and convert it to another form.

5.4. INTRODUCTION TO ADDERS

(PTU, Jalandhar, Sem. Exam., 2004-05)

As a matter of fact, addition of two binary digits is most basic operation performed by the digital computers. Table 5.2 briefly summarizes various rules for binary addition.

Table 5.2. Rules of binary addition

S. No.	A	B	Sum (S)	Carry (C)
1.	0	0	0	0
2.	0	1	1	0
3.	1	0	1	0
4.	1	1	0	1

5.4.1. Classification of Binary Adders

In this subarticle, let us discuss digital circuits that are used to add two binary numbers. The binary adders may be of following two types :

(i) Half adder, and
(ii) Full adder

5.5. AN HALF ADDER

Half adder is a combinational logic circuit with two inputs and two outputs. It is the basic building block for addition of two single bit numbers. This circuit has two outputs namely carry and sum. Figure 5.5(a) shows the block diagram of half adder. An half adder circuit is designed to add two single bit binary numbers A and B. Hence, the truth table of a half adder will be as shown in figure 5.5(b).

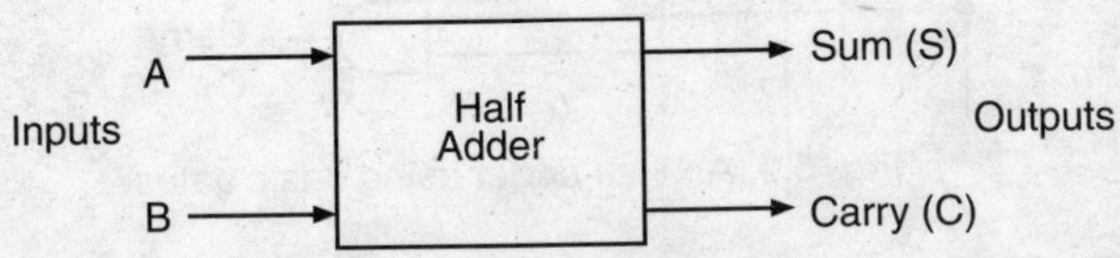

(*a*) Block diagram

S. No.	Inputs		Outputs	
	A	B	Sum	Carry
1.	0	0	0	0
2.	0	1	1	0
3.	1	0	1	0
4.	1	1	0	1

(*b*) Truth table

Fig. 5.5 An Half Adder

5.5.1. Karnaugh-Maps (K-Maps) and Simplified Expressions for Outputs

Figures 5.6 (a) and 5.6 (b) illustrates the K-maps for carry and sum outputs.

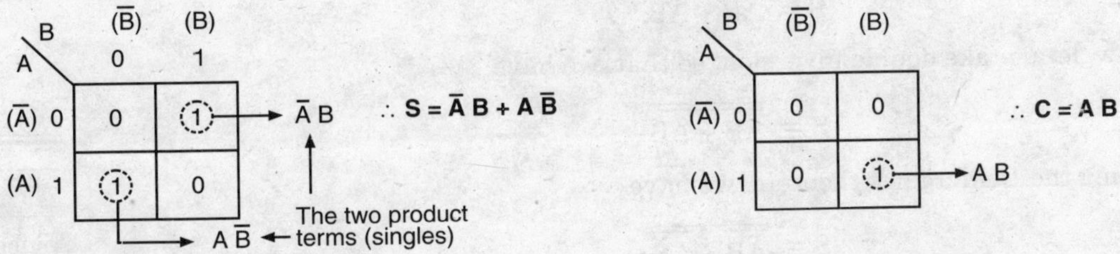

(*a*) K-map for sum output (*b*) K-map for carry output

Fig. 5.6.

Boolean expressions for the sum (S) and carry (C) output may be obtained from the K-maps as under :

$$S = \bar{A}B + A\bar{B} = (A \oplus B) \quad ...(5.1)$$

and $C = AB$

Therefore, figure 5.7 shows the half adder circuit.

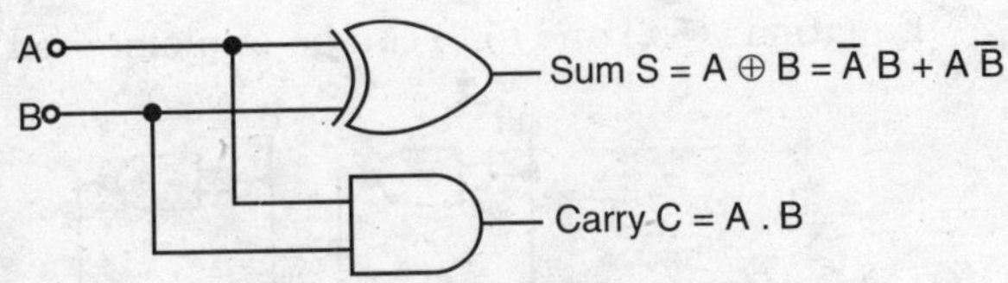

Fig. 5.7 An Half adder circuit

5.5.2. Drawback of Half Adder

Figure 5.8 shows the principle of adding two 2-bit numbers A and B.

Let Number $A = A_1 A_0$

and Number $B = B_1 B_0$

Then, the addition will take place as under

	A_1	A_0	← Number A
+	B_1	B_0	← Number B
+	C_0		← Carry generated from the addtion $(A_0 + B_0)$
C_1	S_1	S_0	

Fig. 5.8. Two bit binary addition

A half adder can add A_0 and B_0 to produce S_0 and C_0. However, the addition of next bits requires the addition of A_1, B_1 and C_0. The addition of three bits is not possible to perform by using an half adder circuit. Therefore, practically we cannot use a half adder.

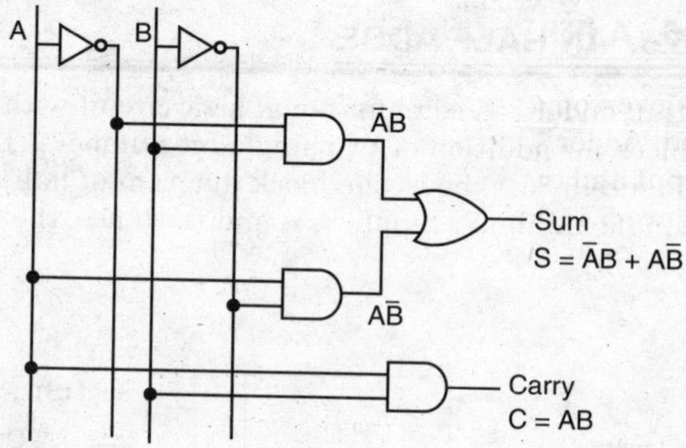

Fig. 5.9. An Half adder using basic gates

5.5.3. Half Adder Using Basic Gates

Let us consider equation (5.1) which states that,

Sum $\quad S = \bar{A}B + A\bar{B}$

and, Carry $\quad C = AB$

These equations can be implemented using the basic gates as depicted in figure 5.9.

5.5.4. Half Adder Using Only NAND Gates

Again, equation (5.1) states that,

Sum, $\quad S = \bar{A}B + A\bar{B}$

and Carry, $\quad C = AB$

We consider the expressions separately, i.e.,

$$S = \bar{A}B + A\bar{B}$$

Now, let us take double inversion, so that, we have

$$S = \overline{\overline{\bar{A}B + A\bar{B}}}$$

Using the DeMorgan's theorem, we have

$$S = \overline{\overline{\bar{A}B} \cdot \overline{A\bar{B}}} \quad ...(5.2)$$

Similarly, we have $\quad C = AB = \overline{\overline{AB}} \quad ...(5.3)$

Equations (5.2) and (5.3) can be implemented using only NAND gates as shown in figure 5.10.

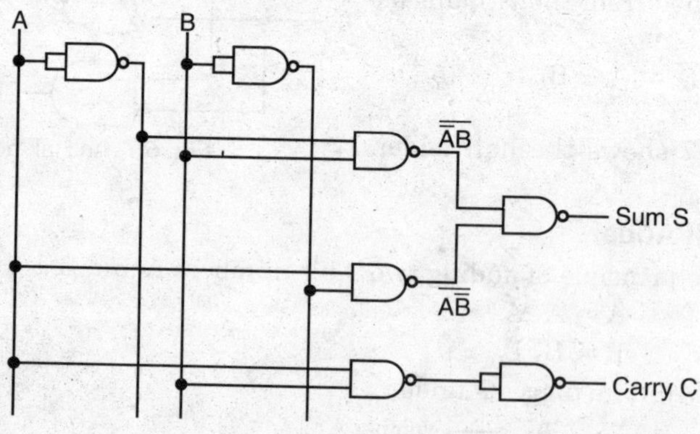

Fig. 5.10. An Half adder using only NAND gates

Natural ability without education has more often raised a man to glory and virtue than education without natural ability.

— **M.T. Cicero**

5.6. A FULL ADDER

(PTU, Jalandhar, Sem. Exam, 2009-10)

To overcome the drawback of an Half Adder circuit, we develop a 3 single bit adder circuit called Full Adder. It can add two one-bit numbers A and B, and carry C_{in}. Basically, a full adder is a three input and two output combinational circuit. Figure 5.11 (a) depicts block diagram of a full adder and figure 5.11(b) illustrates its truth table.

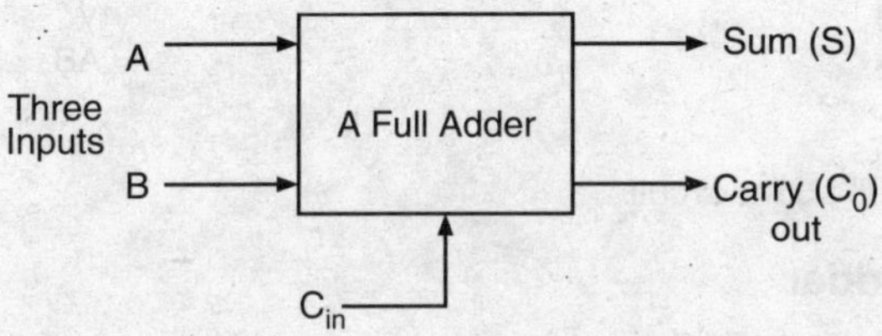

(a) Block diagram

Inputs			Outputs	
A	B	C_{in}	S	C_0
0	0	0	0	0
0	0	1	1	0
0	1	0	1	0
0	1	1	0	1
1	0	0	1	0
1	0	1	0	1
1	1	0	0	1
1	1	1	1	1

(b) Truth table

Fig. 5.11. A Full adder

5.6.1. K-maps for Full Adder

Figure 5.12 and figure 5.13 show the K-maps for the sum (S) and carry out (C_0) outputs and the corresponding Boolean expressions.

K-map for the sum output

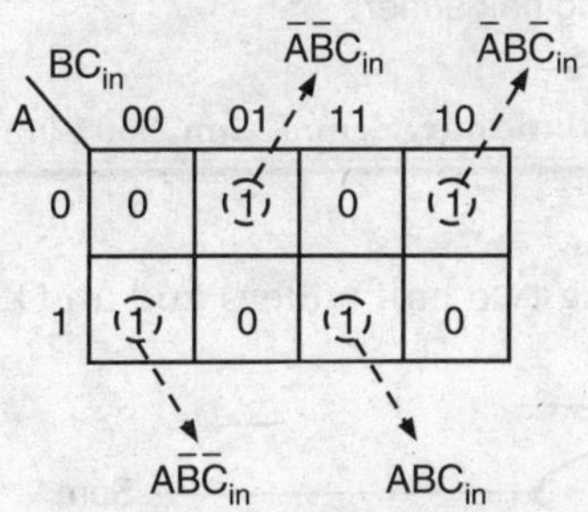

Expression for sum output

$$S = \bar{A}\bar{B}C_{in} + \bar{A}B\bar{C}_{in} + ABC_{in} + A\bar{B}\bar{C}_{in}$$

$$S = C_{in}\underbrace{(\bar{A}\bar{B} + AB)}_{\text{EX-NOR}} + \bar{C}_{in}\underbrace{(\bar{A}B + A\bar{B})}_{\text{EX-OR}}$$

Therefore, $S = C_{in}(\overline{\bar{A}B + A\bar{B}}) + \bar{C}_{in}(\bar{A}B + A\bar{B})$

Let $X = \bar{A}B + A\bar{B}$

Therefore, $S = C_{in}\bar{X} + \bar{C}_{in}X = C_{in} \oplus X$

or $S = C_{in} \oplus (\bar{A}B + A\bar{B})$

But, $\bar{A}B + A\bar{B} = A \oplus B$

Therefore, $S = C_{in} \oplus A \oplus B$

Fig. 5.12. K-map and simplification for sum

K-map for C_0 output

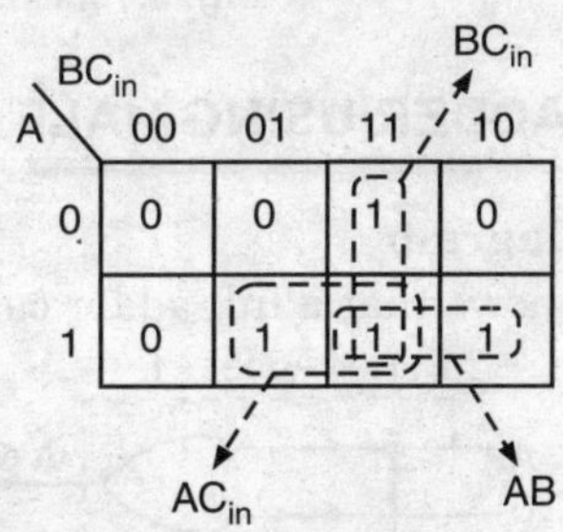

Expression for C_0 output

$$C_0 = AB + AC_{in} + BC_{in}$$

Fig. 5.13. K-map and simplification for Carry out

DO YOU KNOW?

In digital systems, many devices must often share the same data path. This data path is often called a data bus.

To accomplish great things, we must not only act, but also dream: not only plan, but also believe.

— **Anatole France**

5.6.2. A Logic Diagram for Full Adder

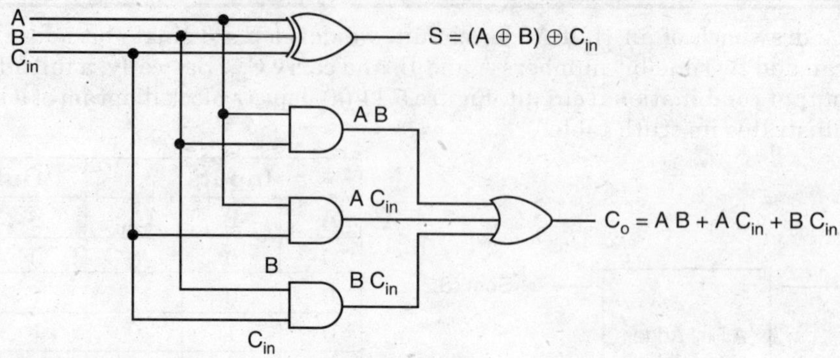

Fig. 5.14. A Full adder circuit

5.6.3. Construction of Full adder using Half Adder

A full adder circuit can also be constructed using two half adders as depicted in figure 5.15.

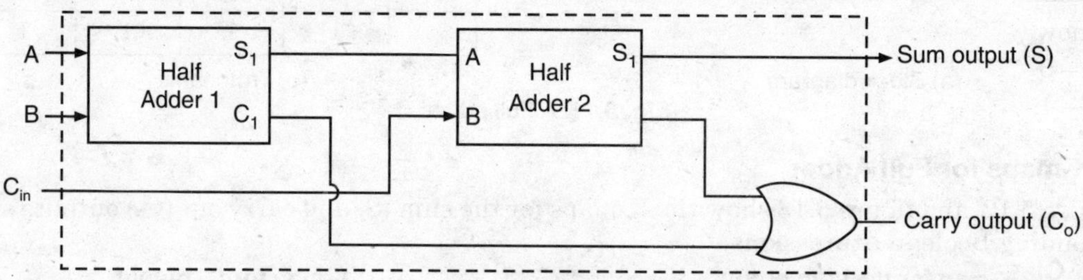

Fig. 5.15. Illustration of a Full adder using half adders

5.7. FULL ADDER USING HALF ADDERS (PTU, Jalandhar, Sem. Exam, 2007-08)(10 marks)

1. Circuit Diagram

Figure 5.16 shows that a full adder can be implemented using two half adders and an OR gate.

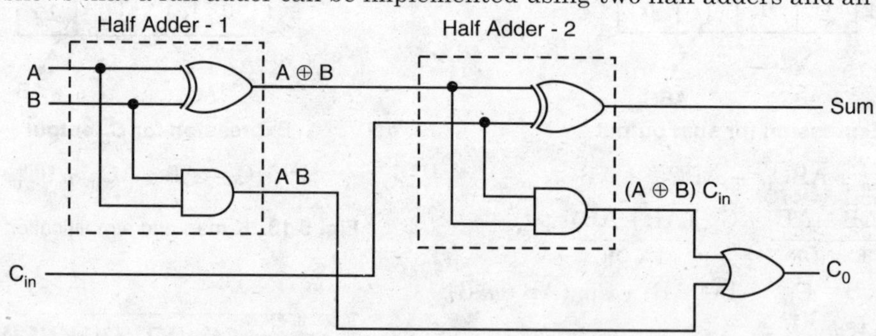

Fig. 5.16. Full adder using two half adders

Now, let us illustrate that the circut in figure 5.16 acts as a full adder.

2. Proof

Let us consider figure 5.16 and write the expression for sum output as under:

$$S = (A \oplus B) \oplus C_{in} = A \oplus B \oplus C_{in}$$

This expression is same as that obtained for a full adder.

Now, we write the expression for carry output C_0 as under:

$$C_0 = (A \oplus B)\, C_{in} + AB = (\bar{A}B + A\bar{B})\, C_{in} + AB$$

$$C_0 = \bar{A}BC_{in} + A\bar{B}C_{in} + AB = \bar{A}BC_{in} + A\bar{B}C_{in} + AB\,(1 + C_{in})$$
$$C_0 = \bar{A}BC_{in} + A\bar{B}C_{in} + AB + ABC_{in}$$
$$C_0 = BC_{in}\,(\bar{A} + A) + A\bar{B}C_{in} + AB = BC_{in} + A\bar{B}C_{in} + AB$$
$$C_0 = BC_{in} + A\bar{B}C_{in} + AB\,(1 + C_{in}) = BC_{in} + A\bar{B}C_{in} + AB + ABC_{in}$$
$$C_0 = BC_{in} + AB + AC_{in}(\bar{B} + B)$$

Therefore, we have $C_0 = BC_{in} + AB + AC_{in}$ **Hence Proved**

The above expression is same that for a full adder. Therefore, we have showed that circuit shown in figure 5.16, really acts like a full adder.

5.7.1. Applications of A Full Adder

A full adder acts as the basic building block of the 4 bit/8 bit binary/ BCD adder ICs such as 7483.

5.8. BINARY SUBTRACTORS

We know that the rules for binary subtraction are as under.

0 – 0 = 0
0 – 1 = 1 with borrow 1
1 – 0 = 1
1 – 1 = 0

Here, it may be noted that in the second case, (0 –1), it is essential to borrow a 1.

5.8.1. Classification of Binary Subtractors

The binary subtractors may be classified as under:

(i) An Half subtractor
(ii) A Full subtractor

5.9. AN HALF SUBTRACTOR

1. Definition

Half subtractor may be defined as a combinational circuit with two inputs and two outputs (i.e., difference and borrow). A half subtractor produces the difference between the two binary bits at the input and also produces an output (i.e., borrow) to indicate if a 1 has been borrowed. In subtraction (A – B), A is called as minuend bit and B is called as subtrahend bit.

2. Truth Table

Table 5.3 illustrates a truth table showing the outputs of a half subtractor for all the possible combinations of input.

Table 5.3. Truth table for an half subtractor

	Inputs		Outputs	
S. No.	A	B	Difference D (A – B)	Borrow B_0
1.	0	0	0	0
2.	0	1	1	1
3.	1	0	1	0
4.	1	1	0	0

3. K-maps for Difference and Borrow Outputs

Figure 5.17 shows the K-maps for the two outputs of a half subtractor.

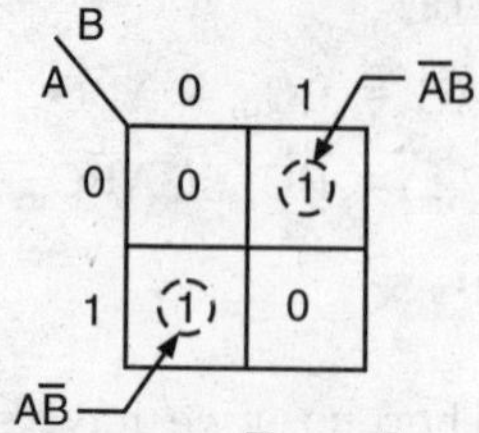

Therefore, difference $D = \bar{A}B + A\bar{B}$

Hence, $D = A \oplus B$

(a) K-map and simplification for difference output

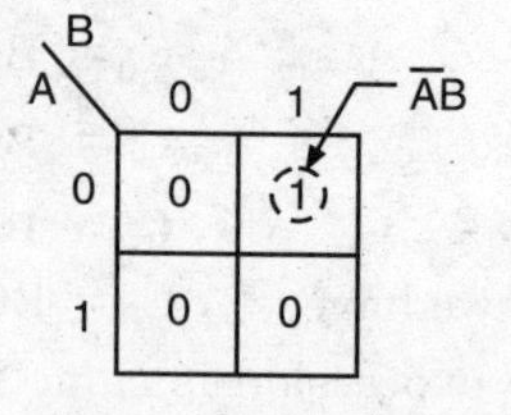

Therefore, borrow $B_0 = \bar{A}B$

(b) K-map and simplification for the borrow output

Fig. 5.17.

4. Logic Diagram

Figure 5.17 (*c*) shows a logic diagram for an half subtractor.

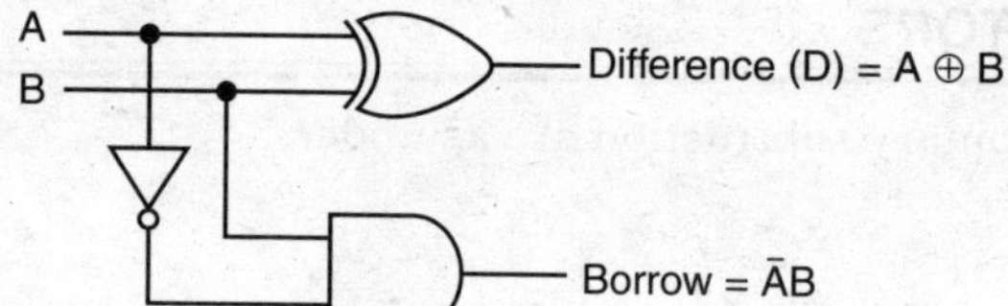

Fig. 5.17. (*c*) An Half subtractor circuit

5. Drawback of An Half subtractor

An half subtractor can only perform the subtraction of two binary bits. However, while performing the subtraction, it does not take into account the borrow of the lower significant stage.

6. An Half subtractor Using Basic Gates

The expressions for the difference and borrow outputs may be listed as under:

$$D = \bar{A}B + A\bar{B} \text{ and } B_0 = \bar{A}B$$

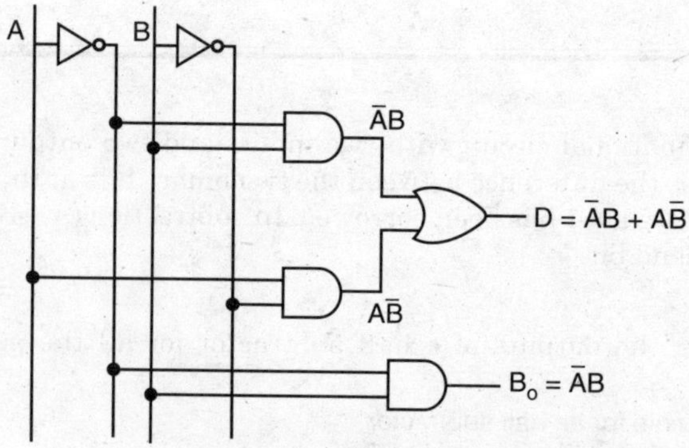

Fig. 5.18. An Half subtractor using basic gates

> **DO YOU KNOW?**
>
> Even though many devices can be *riding* on the bus, there can be only one bus *driver* at any one time. Thus, devices must take turns applying logic signals to the data bus.

7. An Half subtractor Using NAND Gates only

The expression for difference output is given by

$$D = \bar{A}B + A\bar{B}$$

Now, taking the double inversion, we obtain

$$D = \overline{\overline{\bar{A}B + A\bar{B}}}$$

Applying the DeMorgan's theorem, we obtain

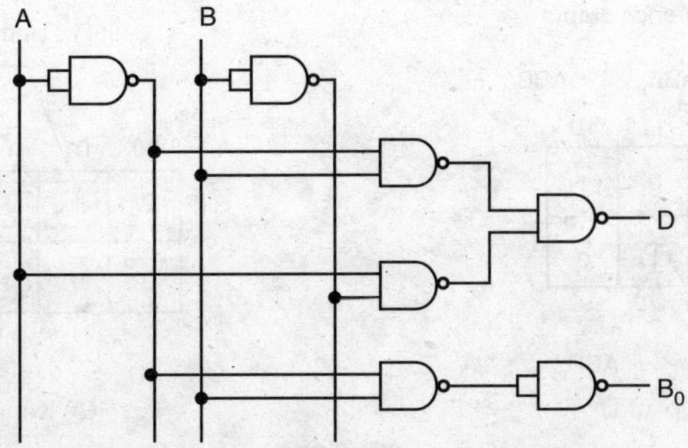

Fig. 5.19. An Half subtractor using NAND gates only

$$D = \overline{\overline{\bar{A}B} . \overline{A\bar{B}}} \qquad ...(5.4)$$

Similarly, the other output B_0 will be given by,

$$B_0 = \bar{A}B = \overline{\overline{\bar{A}B}} \qquad ...(5.5)$$

Equations (5.4) and (5.5) can be implemented as depicted in figure 5.19.

5.10. A FULL SUBTRACTOR

(PTU, Jalandhar, Sem. Exam., 2006-07) (10 marks)

1. Definition

The drawback of an half subtractor can be overcome by a full subtractor. Basically, a full subtractor is a combinational circuit with three inputs A, B and B_{in} and two outputs D and B_0. Here, A is the minuend, B is subtrahend, B_{in} is the borrow produced by the previous stage, D is the difference output and B_0 is the borrow output.

2. Truth Table for a Full subtractor

Table 5.4 represents a truth table for a full subtractor.

Table 5.4. Truth table for a full subtractor

	Inputs			Outputs	
S. No.	**A (Minuend)**	**B (subtrahend)**	**B_{in} (Previous borrow)**	**(A – B – B_{in})**	**B_0**
1.	0	0	0	0	0
2.	0	0	1	1	1
3.	0	1	0	1	1
4.	0	1	1	0	1
5.	1	0	0	1	0
6.	1	0	1	0	0
7.	1	1	0	0	0
8.	1	1	1	1	1

3. K-maps and Simplifications

Figures 5.20 (a) and 5.20 (b) show K-maps for D and B_0 outputs.

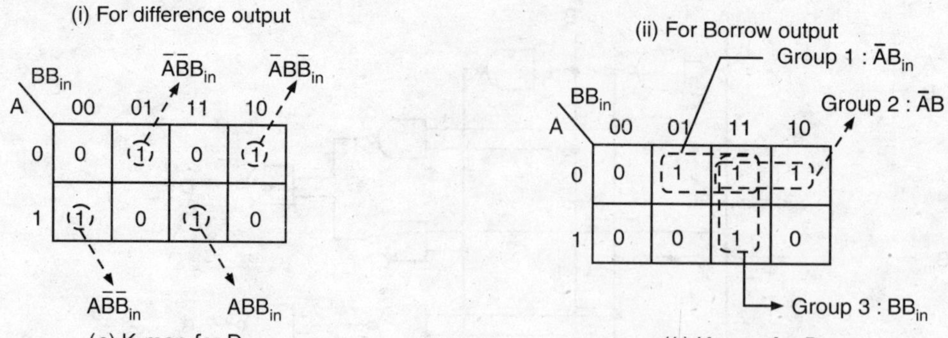

(*a*) K-map for D (*b*) K-map for B_0

Fig. 5.20.

Therefore, $D = \bar{A}\bar{B}B_{in} + \bar{A}B\bar{B}_{in} + A\bar{B}\bar{B}_{in} + ABB_{in}$

Hence, $B_0 = \bar{A}B_{in} + \bar{A}B + BB_{in}$

4. Simplification for difference output

From figure 5.20 (a), we have, $D = \bar{A}\bar{B}B_{in} + \bar{A}B\bar{B}_{in} + A\bar{B}\bar{B}_{in} + AB\,B_{in}$

$$= B_{in}\underbrace{(\bar{A}\bar{B} + AB)}_{\text{EX-NOR}} + \bar{B}_{in}\underbrace{(\bar{A}B + A\bar{B})}_{\text{EX-OR}}$$

Therefore, we have $D = B_{in}\overline{(A \oplus B)} + \bar{B}_{in}(A \oplus B)$

Let $A \oplus B = C$,

Hence, we write $D = B_{in}\,\bar{C} + \bar{B}_{in}C = B_{in} \oplus C$

Therefore, $D = B_{in} \oplus A \oplus B$

5. Simplification for Borrow output

From figure 5.20 (b), we have

$$B_0 = \bar{A}B_{in} + \bar{A}B + BB_{in}$$

Here, no further simplification is possible.

6. Logic diagram for a full subtractor

Figure 5.21 illustrates the logic diagram for a full subtractor.

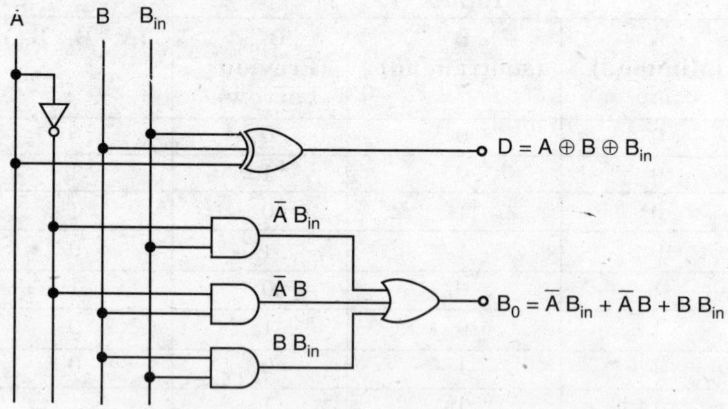

Fig. 5.21. Illustration of Logic diagram for a full subtractor

5.10.1. Construction of a Full Subtractor using Half Subtractors

1. Logic Diagram

Figure 5.22 depicts the implementation of a full subtractor using two half subtractors and an OR gate.

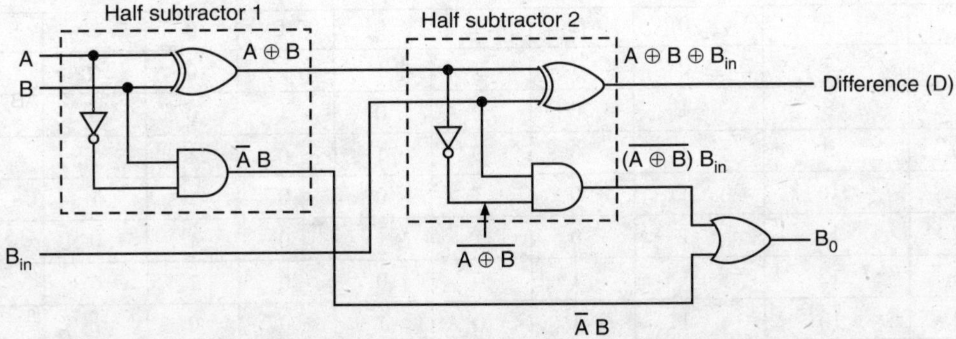

Fig. 5.22. Construction of a full subtractor using half subtractors

Let us illustrate that the circuit shown in figure 5.22 really acts as a full subtractor.

2. Proof

Let us consider figure 5.21 to write the expression for difference output *D* as under :

$$D = (A \oplus B) \oplus B_{in} = A \oplus B \oplus B_{in}$$

This is same as the expression for D output for a full subtractor.

Now, we write the expression for Borrow output B_{out} as under:

$$B_0 = \overline{(A \oplus B)}\, B_{in} + \bar{A}B = (\overline{\bar{A}B + A\bar{B}})\, B_{in} + \bar{A}B$$

$$B_0 = (\bar{A}\bar{B} + AB)\, B_{in} + \bar{A}B = \bar{A}\bar{B}B_{in} + ABB_{in} + \bar{A}B$$

$$B_0 = \bar{A}\bar{B}B_{in} + ABB_{in} + \bar{A}B\,(1 + B_{in}) \qquad [\because (1 + B_{in}) = 1]$$

$$B_0 = \bar{A}\bar{B}B_{in} + ABB_{in} + \bar{A}B + \bar{A}BB_{in}$$

$$B_0 = \bar{A}\bar{B}B_{in} + BB_{in}\,(A + \bar{A}) + \bar{A}B$$

$$B_0 = \bar{A}\bar{B}B_{in} + BB_{in} + \bar{A}B \qquad [\because (A + \bar{A}) = 1]$$

$$B_0 = \bar{A}\bar{B}B_{in} + BB_{in} + \bar{A}B\,(1 + B_{in})$$

$$B_0 = \bar{A}\bar{B}B_{in} + BB_{in} + \bar{A}B + \bar{A}BB_{in}$$

$$B_0 = \bar{A}B_{in}\,(\bar{B} + B) + BB_{in} + \bar{A}B$$

Therefore, we have $B_0 = \bar{A}B_{in} + BB_{in} + \bar{A}B$ **Hence Proved**

Here, it may be noted that this expression is exactly same as that for B_0 of the full subtractor.

EXAMPLE 5.3. Design a combinational circuit whose input is a four bit number and at the output, we obtain the two's complement of the input.

Solution: Let $A_3\, A_2\, A_1\, A_0$ be the four bit input word and $B_3\, B_2\, B_1\, B_0$ represents the two's complement of input.

(i) First, let us write the truth table.

Table 5.5 shows the required truth table.

Table 5.5. Truth table

S. No.	Decimal	Input				Output			
		A_3	A_2	A_1	A_0	B_3	B_2	B_1	B_0
1.	0	0	0	0	0	0	0	0	0
2.	1	0	0	0	1	1	1	1	1
3.	2	0	0	1	0	1	1	1	0
4.	3	0	0	1	1	1	1	0	1
5.	4	0	1	0	0	1	1	0	0
6.	5	0	1	0	1	1	0	1	1

7.	6	0	1	1	0	1	0	1	0
8.	7	0	1	1	1	1	0	0	1
9.	8	1	0	0	0	1	0	0	0
10.	9	1	0	0	1	0	1	1	1
11.	10	1	0	1	0	0	1	1	0
12.	11	1	0	1	1	0	1	0	1
13.	12	1	1	0	0	0	1	0	0
14.	13	1	1	0	1	0	0	1	1
15	14	1	1	1	0	0	0	1	0
16	15	1	1	1	1	0	0	0	1

(*ii*) Next, we shall write K-maps and simplify.

Simplified expression

$$B_3 = \bar{A}_3A_0 + \bar{A}_3A_1 + \bar{A}_3A_2\bar{A}_1\bar{A}_0 + A_3\bar{A}_2\bar{A}_1\bar{A}_0$$

or $$B_3 = \bar{A}_3 (A_0 + A_1) + \bar{A}_1\bar{A}_0 (\bar{A}_3A_2 + A_3\bar{A}_2)$$

or $$B_3 = \bar{A}_3 (A_0 + A_1) + \bar{A}_1\bar{A}_0 (A_3 \oplus A_2)$$

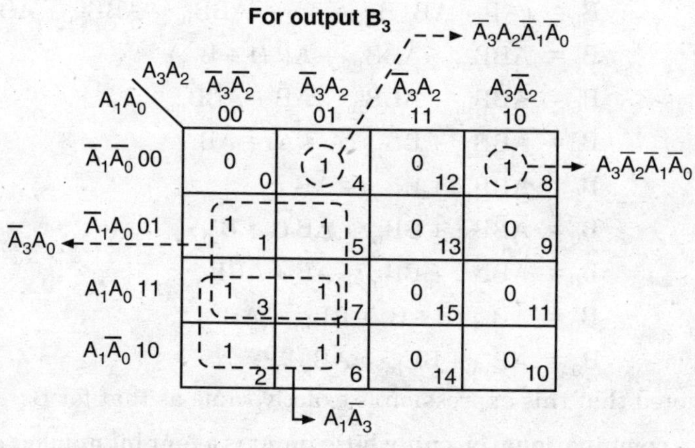

Fig. 5.23. K-map and simplification for output B_3

Simplified expression

$$B_2 = A_2\,\bar{A}_1\bar{A}_0 + \bar{A}_2\,A_0 + \bar{A}_2\,A_1$$

$$= A_2\,\bar{A}_1\bar{A}_0 + \bar{A}_2\,(A_0 + A_1)$$

But, $\bar{A}_1\bar{A}_0 = \overline{(A_1 + A_0)}$

Therefore, $B_2 = A_2\,\overline{(A_1 + A_0)} + \bar{A}_2\,(A_0 + A_1)$

Let $A_0 + A_1 = X$

Therefore, $B_2 = A_2\,\bar{X} + \bar{A}_2\,X = A_2 \oplus X$

Hence, $B_2 = A_2 \oplus (A_0 + A_1)$

To accomplish great things, we must not only act, but also dream: not only plan, but also believe.

— **Anatole France**

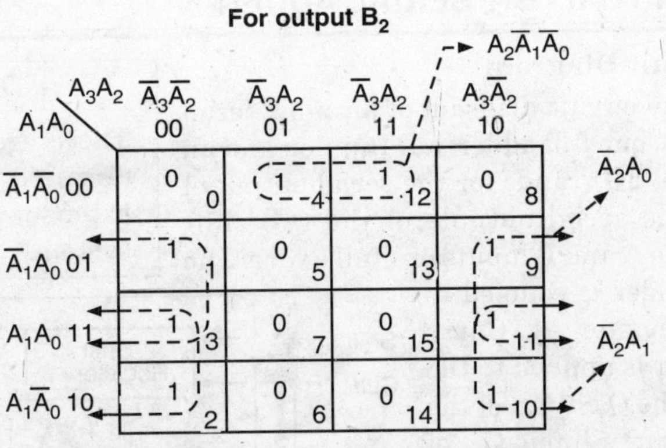

Fig. 5.24. K-map and simplification for output B_2

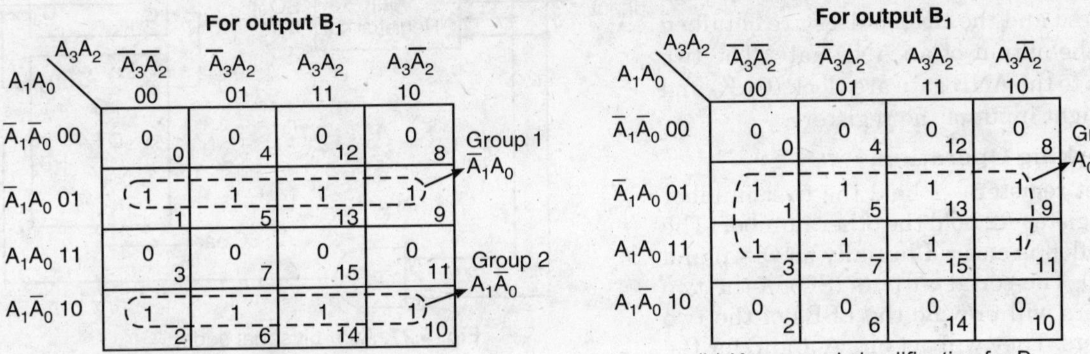

(a) K-map and simplification for B_1

(b) K-map and simplification for B_0

Fig. 5.25.

(iii) Finally, let us realize the combinational circuit using basic gates.

Figure 5.26 shows the required combinational circuit.

In figure 5.26, we have

$$B_3 = \overline{A}_3 (A_0 + A_1) + \overline{A}_1 \overline{A}_0 (A_3 \oplus A_2)$$
$$B_2 = A_2 \oplus (A_0 + A_1)$$
$$B_1 = A_1 \oplus A_0$$
$$B_0 = A_0$$

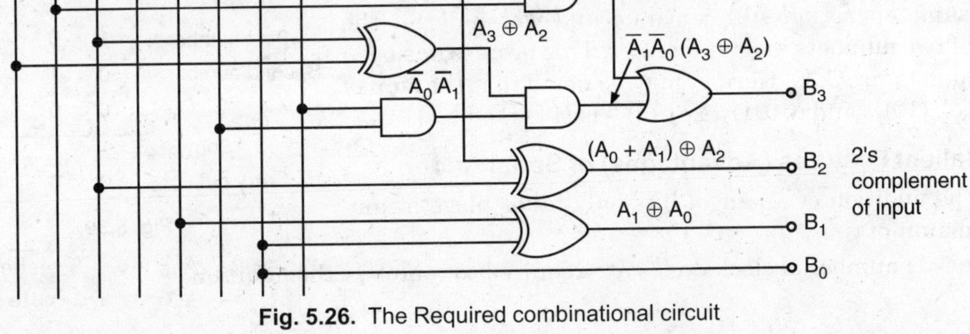

Fig. 5.26. The Required combinational circuit

5.11. INTRODUCTION TO n - BIT SERIAL ADDER

1. Definition and Circuit Diagram

Figure 5.27 shows the circuit diagram of an n-bit serial adder. Two shift registers, one full adder, a D flip-flop and an AND gate are used. Here, SO stands for the serial output of shift register. Hence, the serial outputs of the two shift registers are applied to the A and B inputs of a full adder. The sum output of the full adder is coupled to the serial input SI of register –1. Carry output C_0 of the full adder is applied to the data input of the D F/F. The Q output of the flip-flop is connected to carry input C_{in} of full adder. The D F/F is positive edge triggered and the clock pulses are obtained from the output of an AND gate. The two inputs to the AND gate are clock (CLK) and shift right input of the registers.

> **DO YOU KNOW?**
>
> There are several combinational circuits that are employed extensively in the design of digital system.

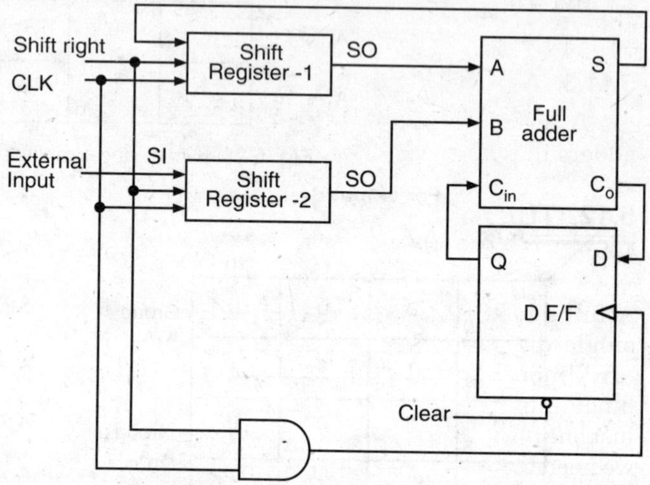

Fig. 5.27. An n-bit serial adder

2. Working Operation

Let register – 1 hold the first number and register – 2 hold the other number. The D flip-flop is cleared initially so Q = 0 and C_{in} = 0. The serial outputs (SO) of the two registers will provide the LSBs of the two numbers. They will act bits A and B for the full adder. The full adder will add these bits and produce sum S and carry out C_0. Thus the addition of LSBs is complete. This has been shown in figure 5.28. Now, a clock pulse is applied to both the shift registers. Hence, the two numbers are right shifted by one bit each. The clock pulse gets applied to the D F/F also and $Q = C_{in} = C_{out} = 0$. The adder adds the two bits available at the SO outputs of the two registers and C_{in} as shown in figure 5.29.

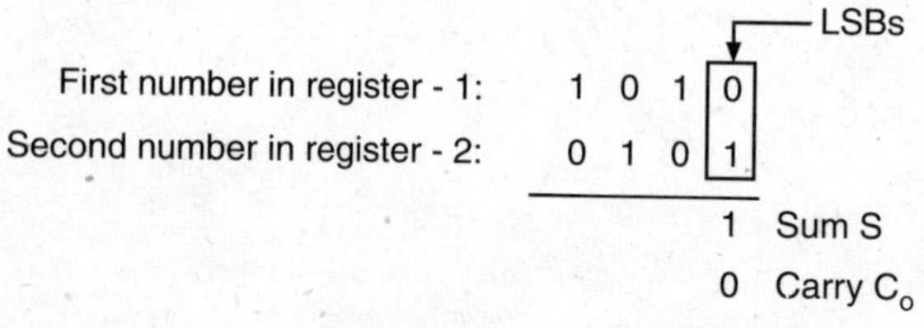

				LSBs
First number in register - 1:	1	0	1	0
Second number in register - 2:	0	1	0	1
Sum S				1
Carry C_o				0

Fig. 5.28. Addition of LSBs

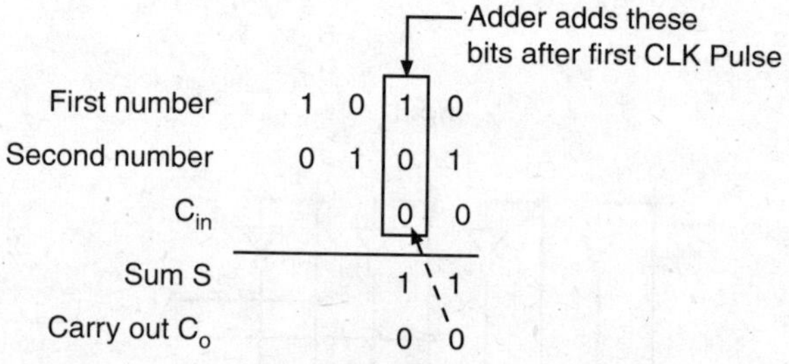

(Adder adds these bits after first CLK Pulse: third column)

First number	1	0	1	0
Second number	0	1	0	1
C_{in}			0	0
Sum S			1	1
Carry out C_o			0	0

Fig. 5.29.

The same operation will continue and we shall get the addition of two numbers in a bit manner. This process has been illustrated in figure 5.30. Here, it may be noted that the sum of $(1010)_2$, i.e., $(10)_{10}$ and $(0101)_2$, i.e., $(5)_{10}$ is $(1111)_2$, i.e., $(15)_{10}$.

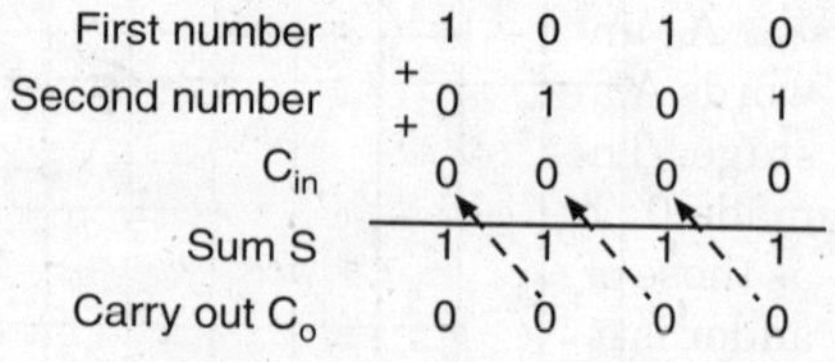

First number	1	0	1	0
+ Second number	0	1	0	1
+ C_{in}	0	0	0	0
Sum S	1	1	1	1
Carry out C_o	0	0	0	0

Fig. 5.30.

5.11.1. Salient Features (Advantages) of Serial Adder

(i) The addition of a pair of bits only takes place at any instant of time.

(ii) (n – 1) number of clock cycles are required to complete the addition.

(iii) The process of addition continues until the shift right control is disabled.

(iv) Sum is available in the serial form.

5.11.2. Drawbacks of Serial Addition

(i) More time is required to complete the addition.

(ii) A complicated circuit is required.

(iii) The circuit contains more number of components.

(iv) The sum and carry are available in the serial form, hence, result of addition can not be observed at once.

5.11.3. A Serial Subtractor

A serial subtractor circuit is same as that of the serial adder except for one change. The full adders in figure 5.27 have been replaced by the full subtractors.

5.12. THE n-BIT PARALLEL ADDER

(PTU, Jalandhar, Sem., Exam., 2003-04)(10 marks)

A full adder is capable of adding only two single digit binary numbers along with a carry input. But, in practice, we need to add binary numbers which are much longer than just one bit. To add two n-bit binary numbers, we need to use the n-bit parallel adder shown in figure 5.31. It makes use of a number of full adders in cascade. The carry output of the previous full adder is connected to the carry input of the next full adder as illustrated in figure 5.31.

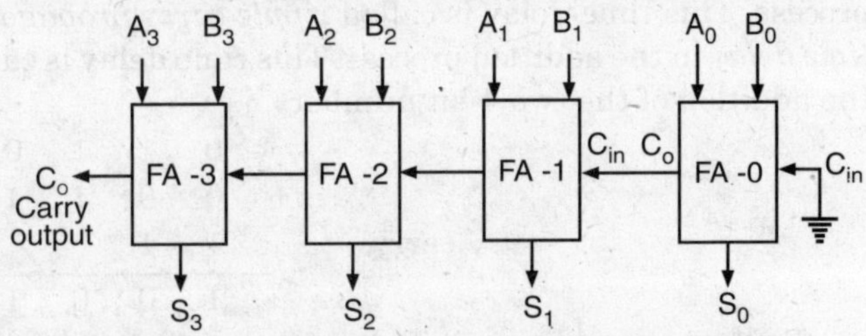

Fig. 5.31. Block diagram of a n-bit parallel adder

5.12.1. A Four Bit Parallel Adder Using Full Adders

Figure 5.32 shows the block diagram of a four bit parallel adder using full adders.

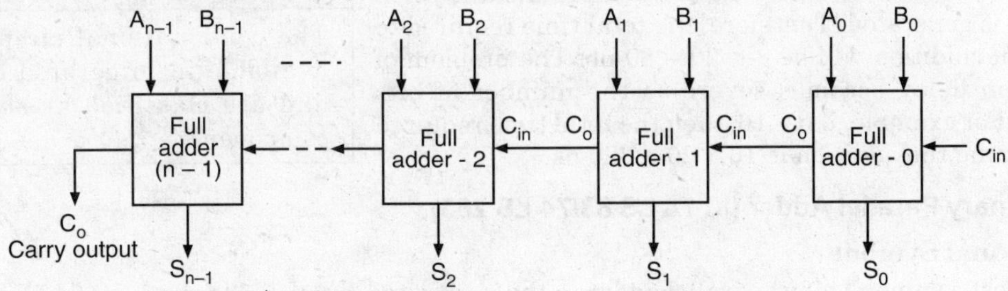

Fig. 5.32. Block diagram of a four-bit parallel adder

A_0 and B_0 represent the LSBs of the four bit words A and B. Hence, full Adder – 0 is the lowest stage. Therefore, its C_{in} has been permanently made 0. The rest of connections are exactly same as those of n-bit parallel adder. The four-bit parallel adder is a very common logic circuit. It is normally shown by a block diagram as illustrated in figure 5.33.

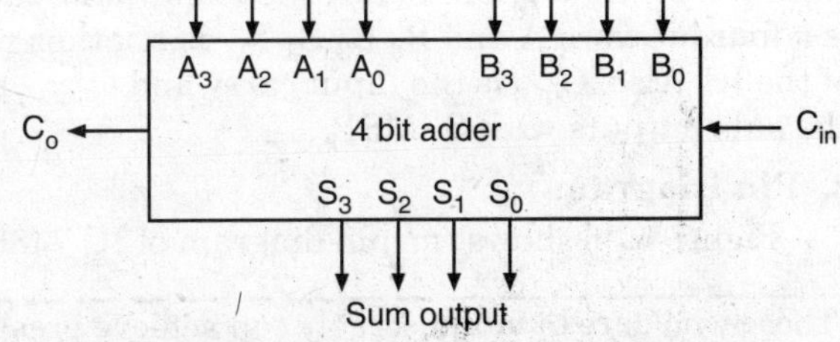

Fig. 5.33. Block diagram of 4-bit parallel adder

EXAMPLE 5.4. Design a 4-bit adder using three full adders and one half adder.

Solution: Figure 5.34 shows the block diagram of the required 4-bit adder. The half adder is used in the least significant position to add A_0 and B_0. The three full adders are used thereafter.

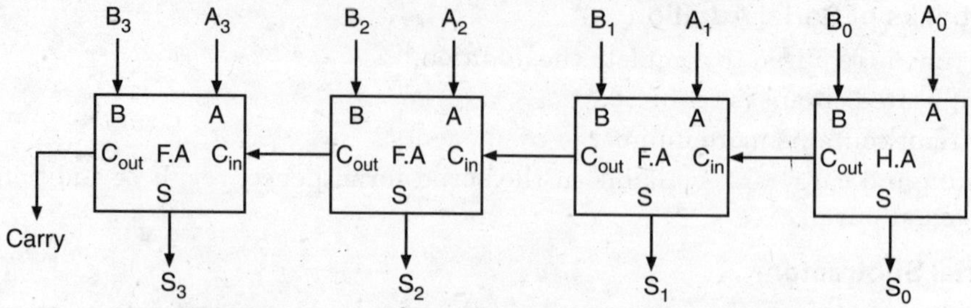

Fig. 5.34. A 4-bit adder using full adders and half adder

5.12.2. Concept of Propagation Delay in a Parallel Adder

In parallel adder, carry out of the previous stage is connected to carry in of the next stage. Therefore, the carry is said to be propagated like ripple from the LSB stage to the MSB stage. This is called ripple carry propagation. The ripple carry propagation results in *time delay* in the addition process. This time-delay is called *ripple carry propagation.* The ripple carry propagation results in *time delay* in the addition process. This time delay is called as the *propagation delay*. Let us consider the addition of the two 4-bit numbers, i.e.,

$$\begin{array}{lrcccc} & & 0 & 1 & 1 & 0 \\ & + & 0 & 0 & 1 & 1 \\ \text{carry} & & 1 & 1 & & \\ \hline & & 1 & 0 & 0 & 1 \end{array}$$

From above addition, it is obvious that the addition of 1's in the second position produces a carry which is added to the bits in the third position and the carry produced in the third position is added to the bits in the fourth position. Hence, the sum bit produced in the MSB position depends on the carry generated due to additions in the preceding stages. Infact, is the real problem. If the propagation delay of each full adder is say 20 ns, then sum S_3 will reach its correct value after $3 \times 20 = 60$ ns from the instant when LSB carry is generated. Therefore, the total time required to perform the addition will be $4 \times 20 = 80$ ns. The problem of propagation delay becomes severe as the number of bits increases. For example, if $n = 16$, then the total time required for performing the addition is $16 \times 20 = 320$ ns.

> **DO YOU KNOW?**
>
> The combinational circuits are available in integrated circuits and are classified as standard components.

5.12.3. Binary Parallel Adder (IC 74 LS 83/74 LS 283)

1. Functional Symbol

The most common binary parallel adder in the integrated circuit (IC) form is IC 74 LS 83/74 LS 283. It is a 4-bit parallel adder, which consists of four interconnected full adders alongwith the lookahead carry circuit. The IC 7483 and 74 283 are TTL MSI (medium scale integration) with the same pin configuration. Figure 5.35 shows the functional symbol of IC 74 LS 283. Here, $A_3\ A_2\ A_1\ A_0$ is a four bit word A and $B_3\ B_2\ B_1\ B_0$ is another word B. Both these words are applied at the inputs of the IC. Also, $C_{in\ 0}$ is the input carry and $C_{out\ 3}$ represents the output carry S_3, S_2, S_1, S_0 represent the sum outputs with S_3 MSB.

2. Pin Diagram

Figure 5.36 shows the pin diagram of IC 7483. This IC adds the two four bit words A and B and

Those who dare to fail miserably can achieve greatly. — **Robert F. Kennedy** the bit at $C_{in\ 0}$ on the bit-by-bit basis to produce a four bit sum output along with carry output at $C_{out\ 3}$.

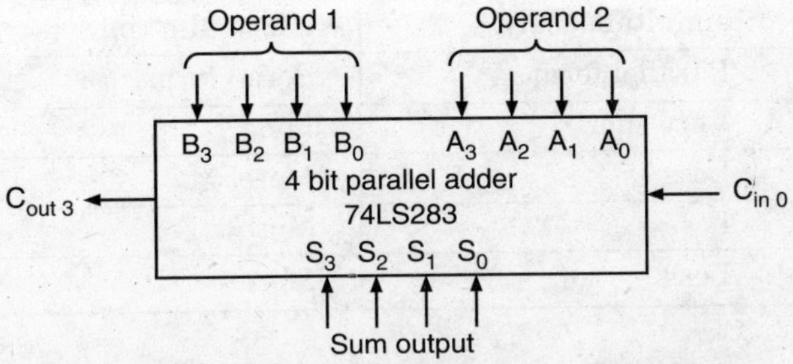

Fig. 5.35. Functional symbol for 74 LS 283

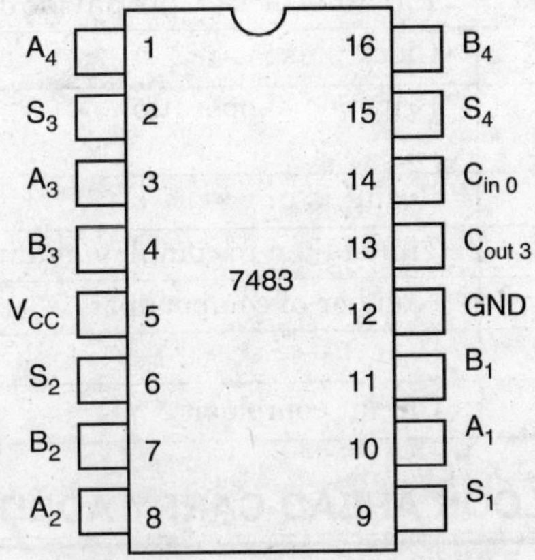

Fig. 5.36. Pin diagram of IC-7483

5.12.4. A Four Bit Binary Adder using IC 7483

Figure 5.37 shows a four bit binary adder. Here, it may be noted that the carry input $C_{in\ 0}$ has been connected to ground.

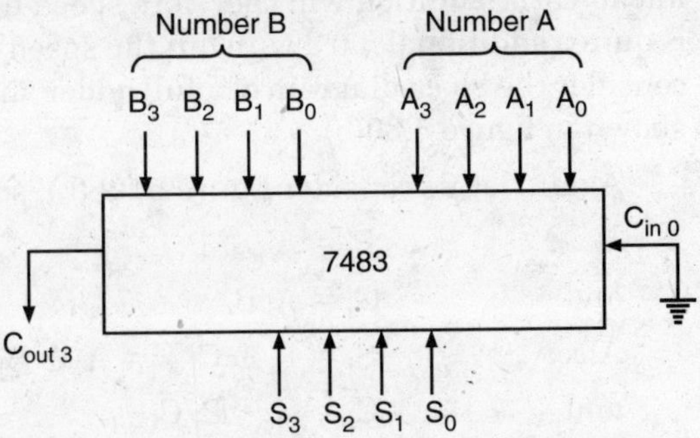

Fig. 5.37. A 4-bit binary adder using IC 7483

5.12.5. Cascading of Adders

If it is required to add two 8-bit numbers using the 4-bit parallel adder 74283, then, we have to cascade the four bit adders. Figue 5.38 illustrates an 8-bit adder using two 4-bit adders, similarly, it is possible to form an adder tree to make an n-bit adder.

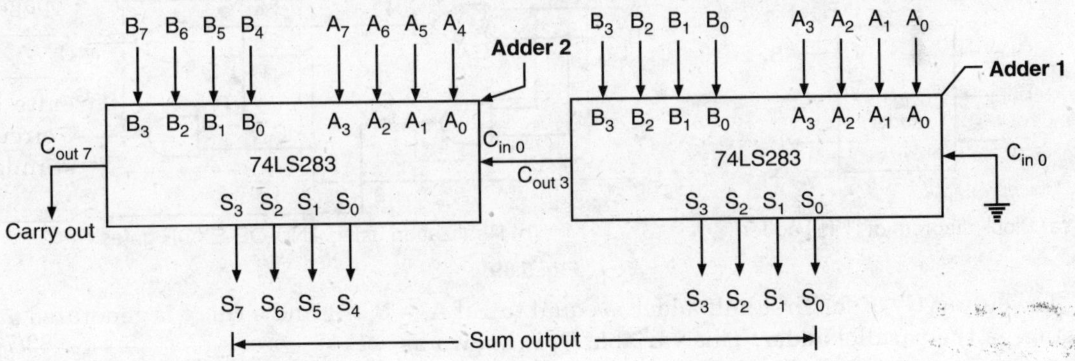

Fig. 5.38. 8-bit addition using 4-bit adders

$A_7 - A_0$ and $B_7 - B_0$ are the two eight bit numbers to be added. Adder-1 in figure 5.38 adds the four LSB bits of the two numbers, i.e., $A_3 - A_0$ and $B_3 - B_0$. The carry input of first adder has been connected to ground (logic 0). The $C_{out\ 3}$ output of adder-1 is connected to C_{in0} input of adder-2. This adder adds this carry and the four MSB bits of the two numbers. Also, $C_{out\ 3}$ of adder-2 acts as the final output carry and the sum output is from S_7 through S_0.

Action may not always bring happiness, but there is no happiness without action. — **Disraeli**

5.12.6. Performance Comparison of Serial and Parallel Addition

S.No.	Parameter of Comparison	Parallel adder	Serial adder
1.	Clock pulse	Not essential	Essential
2.	Principle of operation	All the bits are added simultaneously	One pair of bits is added at a time
3.	Result is in	Parallel form	Serial form
4.	Time taken to complete addition	Very short	Long
5.	Number of components	Less	More
6.	Cost	Less	More
7.	Circuit complexity	Less	More

5.13. LOOK AHEAD-CARRY ADDITION **(PTU, Jalandhar, Sem., Exam, 2006-07)(10 mark)**

This is the addition technique which eliminates the problem due to interstage carry delay. The look-ahead-carry addition will therefore speed up the addition process. The adder with look ahead carry requires additional hardware but the speed of this adder is independent of the number of bits. Let us consider the block diagram of a full adder shown in figure 5.39(a) and its AND-OR-EXOR realization shown in figure 5.39(b).

Again, let us consider figure 5.39 (b), so that, we have

$$P_i = A_i \oplus B_i \quad ...(5.6)$$

and $$G_i = A_i B_i \quad ...(5.7)$$

Also, $$S_i = P_i \oplus C_{i-1} = A_i \oplus B_i \oplus C_{i-1} \quad ...(5.8)$$

and $$C_i = G_i + P_i C_{i-1} \quad ...(5.9)$$

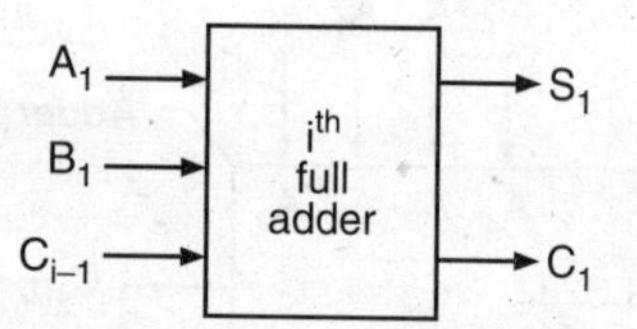

(a) Block diagram of ith full adder

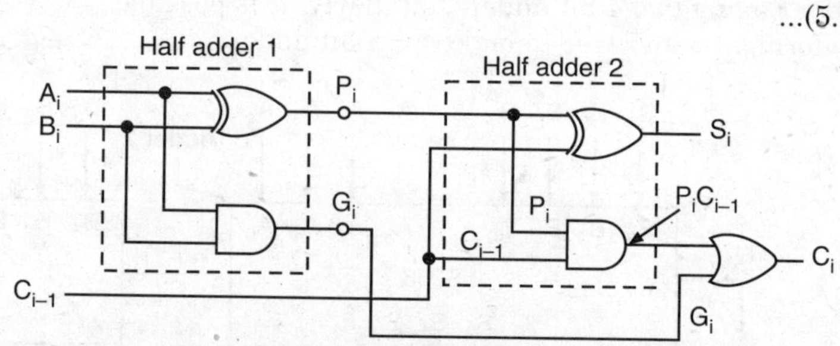

(b) Realization using AND-OR-EXOR gates

Fig. 5.39.

The output G_i of the first half adder is equal to 1 if $A_i = B_i = 1$ and a carry is generated at the ith stage of the parallel adder. This variable G_i is known as Carry Generate and its value does not depend on the input carry, i.e., C_{i-1}. The variable P_i is called as Carry propagate because this term is associated with the propagation of carry from C_{i-1} to C_i. Now, let us consider equation (5.9), i.e., $C_i = G_i + P_i C_{i-1}$. Using this equation, we can write the expression for the carry output of each stage in a 4 bit parallel adder as under :

> **DO YOU KNOW?**
>
> Combinational circuits perform specific digital functions commonly needed in the design of define systems.

$$P_i = A_i \oplus B_i$$

Stage	Expression for carry output	
0	$C_0 = G_0 + P_0\,C_{-1}$	...(5.10)
1.	$C_1 = G_1 + P_1\,C_0 = G_1 + P_1\,(G_0 + P_0\,C_{-1})$	
	Therefore, $C_1 = G_1 + P_1G_0 + P_0P_1C_{-1}$	...(5.11)
2.	$C_2 = G_2 + P_2C_1 = G_2 + P_2\,(G_1 + P_1G_0 + P_0P_1C_{-1})$	
	Therefore, $C_2 = G_2 + P_2G_1 + P_2P_1G_0 + P_2P_1P_0C_{-1}$	...(5.12)
3.	$C_3 = G_3 + P_3G_2 + P_3P_2G_1 + P_3P_2P_1G_0 + P_3P_2P_1P_0C_{-1}$	...(5.13)

In the expressions stated above, the variable involved are as under :

G_0, G_1, G_2, G_3, P_0, P_1, P_2, P_3 and C_{-1}.

Out of them, the G variables are generated from the A and B inputs using AND gates (as illustrated in equation (5.7). And the P variables are obtained again directly from A and B inputs using EX-OR gates (as illustrated in equation 5.6). If the G, P and C_{-1} are a time available, then it is possible to produce the carry outputs C_0, C_1, C_2 and C_3 by using 2-level realization (AND - OR or NAND-NAND) etc. The advantage of generating the carry outputs using this method is that the propagation delay of any two gates will be involved. These carry outputs are then connected to the carry inputs of the succeeding stages. This eliminates the problem of carry rippling. The logic circuit of a look ahead carry generator has been shown in figure 5.40 and it is based upon equations (5.6) through (5.13).

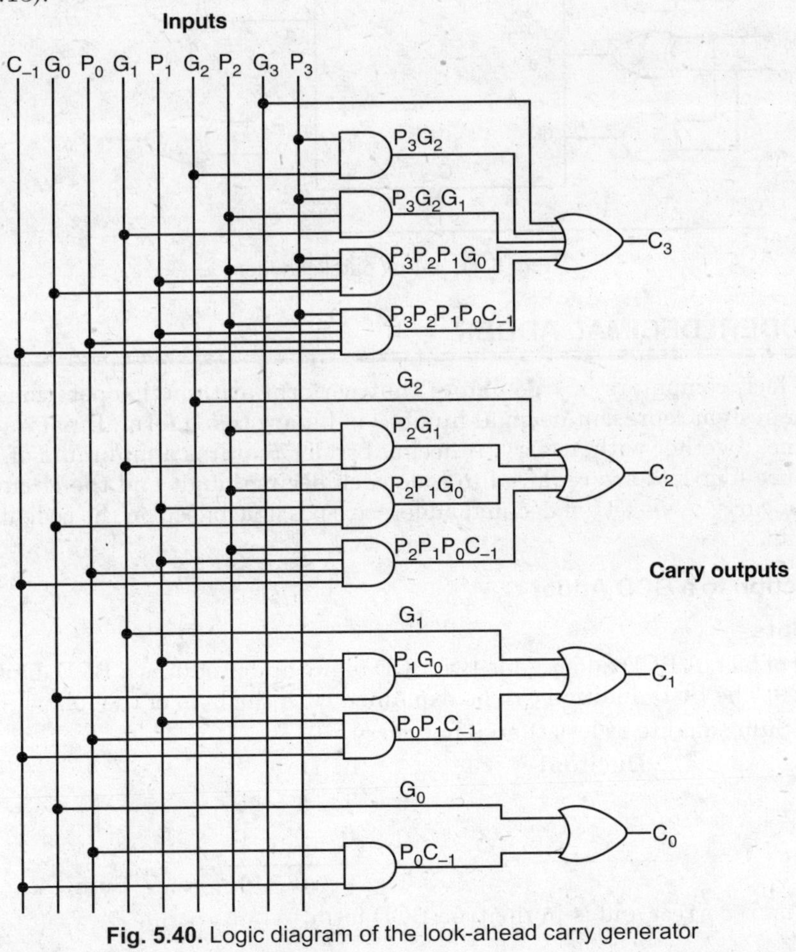

Fig. 5.40. Logic diagram of the look-ahead carry generator

5.13.1. A Four Bit Adder with Look-Ahead Carry

Figure 5.41 shows the block diagram of a four bit parallel adder using the look-ahead carry generator. The P_i and G_i variables are obtained from the inputs A_i and B_i by using the EX-OR and AND gates respectively. The carry outputs C_0 through C_3 are produced simultaneously by the look ahead carry generator. These carry outputs and P_i variables are EX-ORed to produce the sum outputs S_0 through S_3.

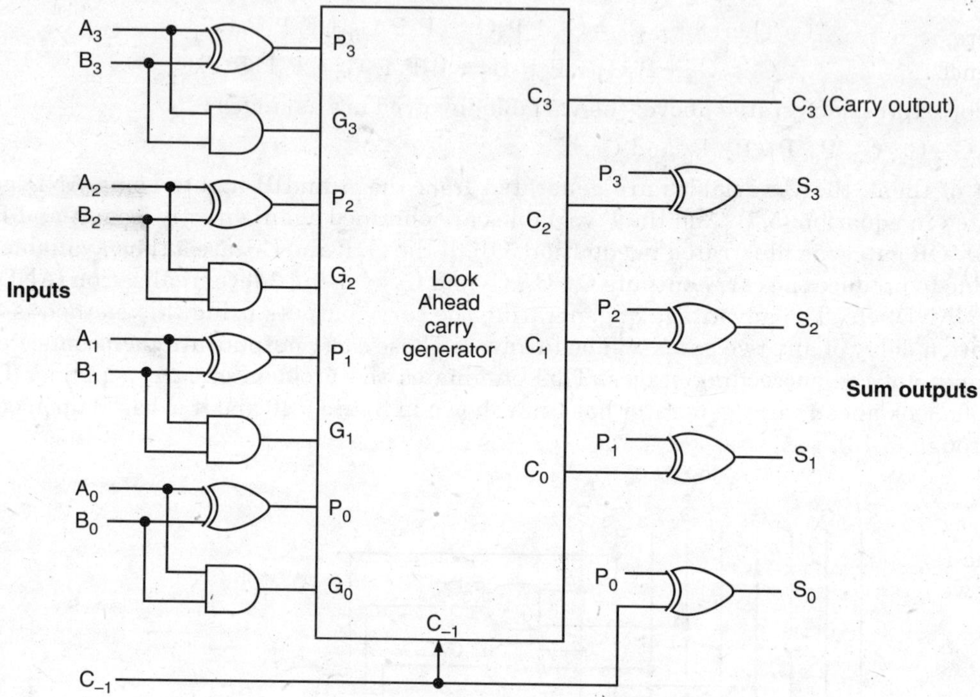

Fig. 5.41. A 4-bit parallel adder

5.14. BCD ADDER: DECIMAL ADDER

As a matter of fact, computers or calculators that perform arithmetic operations directly in the decimal number system represent decimal number in binary coded form. Binary addition requires two bits at a time, together with a carry. A decimal adder requires a minimum of nine inputs and five outputs. Since four bits are required to code each decimal digit and the circuit must have an input and output carry. A variety of decimal adders are possible based on the code used to represent the decimal digits.

5.14.1 Introduction to a BCD Adder

1. Basic concepts

As a matter of fact, A BCD adder adds two BCD digits and produces a BCD digit. A BCD cannot be greater than 9. The BCD addition can be explained with the help of the following three cases.

***(i)* Case I:** Sum equal to or less than 9 but carry = 0

	Decimal	BCD	
	7	0 1 1 1	
+	1	0 0 0 1	
	8	1 0 0 0	← Sum

Here, the sum is correct and is in the true BCD form.

(ii) Case II: Sum greater than 9 but carry = 0

	Decimal	BCD
	7	0 1 1 1
+	4	0 1 0 0
	11	1 0 1 1 (Invalid BCD number)

Here, it may noted that the sum 1011 is invalid BCD number, so, the answer is not correct. Hence, to correct the answer, we add six (0110) to the invalid BCD answer as under :

```
            1 0 1 1   ← Invalid BCD answer
          + 0 1 1 0   ← Add (6)10 for correction
    ---------------
    0 0 0 1 0 0 0 1   ← Correct answer
    \_____/ \_____/
       1       1
```

(iii) Case III: Sum less than or equal to 9 but carry = 1

Let us consider the following addition:

```
 Decimal          BCD
    9           1 0 0 1
  + 8       +   1 0 0 0
 -----      -----------
   17         1 0 0 0 1   ← Wrong result of addition
              |  \_____/
Final carry ←-      └→ Sum is invalid BCD number
```

The result of addition is 0001 0001 = $(11)_{10}$ which is not correct. Hence, to correct the wrong result, we have to add six (0110) as shown below:

```
   0 0 0 1   0 0 0 1   ← Incorrect answer
 + 0 0 0 0   0 1 1 0   ← Add 6 for correction
 ---------   -------
   0 0 0 1   0 1 1 1   ← Correct BCD answer
   \_____/   \_____/
      1         7
```

2. Conclusions

(i) The two given BCD numbers are to be added using the rules of binary addition.

(ii) If sum is less than or equal to 9 and carry = 0, then no correction is necessry. The sum is correct and is in the true BCD form.

(iii) However, if sum is invalid BCD or carry = 1, then the result is wrong and requires some correction.

(iv) The wrong result can be corrected by adding six (0110) to it.

3. The Block Diagram of a BCD Adder

From the above conclusions, we understand that the 4-bit, BCD adder must consist of the following blocks.

(i) A 4-bit binary adder to add the given numbers A and B.

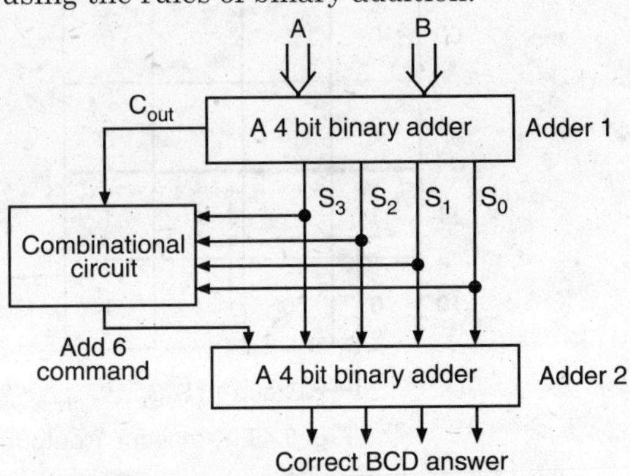

Fig. 5.42. The Block diagram of a BCD adder

(ii) A combinational circuit to check if sum is greater than 9 or carry =1.

(iii) One more 4-bit binary adder to add six (0110) to the incorrect sum if sum > 9 or carry = 1.

The block diagram of such a BCD adder has been shown in figure 5.42. Here, we have to design the combinational circuit which senses if sum is greater than 9 or carry = 1.

4. Design of Combinational Circuit

The output of combinational circuit must be 1 if the sum produced by adder 1 is greater than 9, i.e., 1001. The truth table has been given in Table 5.6.

Table 5.6. Truth table for combinational circuit design

	Inputs				Output	
Sum bits of adder-1 →	S_3	S_2	S_1	S_0	Y	
	0	0	0	0	0	Sum is valid BCD number Therefore, $Y = 0$
	0	0	0	1	0	
	0	0	1	0	0	
	0	0	1	1	0	
	0	1	0	0	0	
	0	1	0	1	0	
	0	1	1	0	0	
	0	1	1	1	0	
	1	0	0	0	0	
	1	0	0	1	0	
	1	0	1	0	1	Sum is invalid BCD number Therefore, $Y = 1$
	1	0	1	1	1	
	1	1	0	0	1	
	1	1	0	1	1	
	1	1	1	0	1	
	1	1	1	1	1	

K-map

Figure 5.43 shows the K-map for Y output of the combinational circuit.

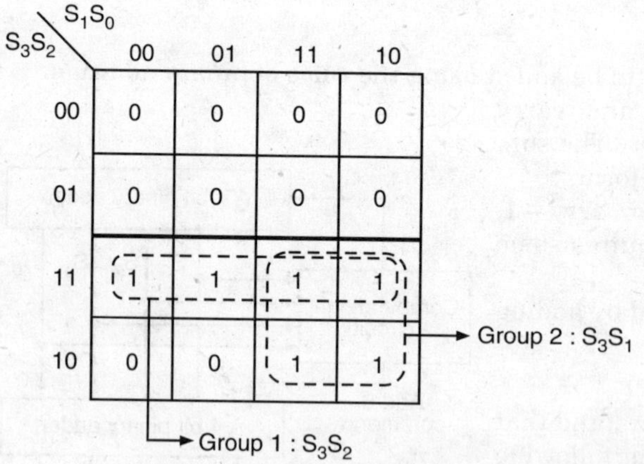

Fig. 5.43. K-map for Y output

DO YOU KNOW?

The two general forms for logic expressions are the sum-of-products form and the product-of-sums form.

Therefore, the simplified Boolean expression will be

$$Y = S_3S_2 + S_3S_1$$

Figure 5.44 shows BCD adder. The output of the combinational circuit must be 1 if C_{out} of adder-1 is high. Therefore, Y is ORed with C_{out} of adder 1 as shown in figure 5.44. The output of combinational circuit is connected to B_1B_2 inputs of adder-2 and $B_3 = B_1 = 0$ as they are connected to ground permanently. This make $B_3\ B_2\ B_1\ B_0 = 0\ 1\ 1\ 0$ if $Y' = 1$. The sum outputs of adder-1 are applied to $A_3\ A_2\ A_1\ A_0$ of adder-2. The output of combinational circuit is to be used as final output carry and the carry output of adder-2 is to be ignored.

5. Working Operation

Case I: Sum ≤ 9 and carry = 0,

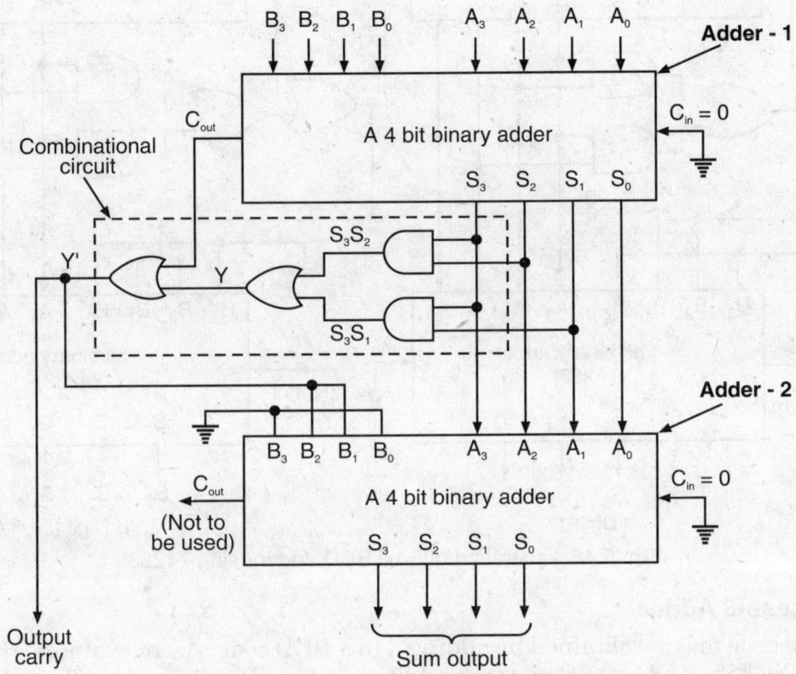

Fig. 5.44. Illustration of a 4-bit BCD adder

The output of combinational circuit $Y' = 0$. Hence, $B_3\ B_2\ B_1\ B_0 = 0000$ for adder-2. Therefore, output of adder-2 is same as that of adder-1.

Case II: Sum > 9 and carry = 0

If $S_3\ S_2\ S_1\ S_0$ of adder-1 is greater than 9, then output Y of combinational circuit becomes 1.

Therefore, $B_3\ B_2\ B_1\ B_0 = 0110$ (of adder-2)

Hence, six (0110) will be added to the sum output of adder-1. We get the corrected BCD result at the output of adder-2.

Case III: Sum ≤ 9 but carry = 1

As carry output of adder-1 is high, we have, $Y' = 1$.

Therefore, $B_3\ B_2\ B_1\ B_0 = 0\ 1\ 1\ 0$ (of adder-2)

Hence, 0 1 1 0 will be added to the sum output of adder-1. We get the corrected BCD result at the output of adder-2. Thus, the four bit BCD addition can be carried out using the binary adder.

> **DO YOU KNOW?**
>
> One approach to the design of a combinational logic circuit is to :
>
> (i) construct its truth table.
> (ii) convert the truth table to a sum-of-products expression,
> (iii) simplify the expression using Boolean algebra or K-mapping,
> (iv) implement the final expression.

5.14.2. 8-Bit BCD Adder Using IC 74283 **(PTU, Jalandhar, Sem. Exam; 2006-07) (10 marks)**

The 8-bit BCD adder using the 4-bit BCD adder 74283 has been shown in figure 5.45. We have to use two 4-bit adders. The carry out of the lower 4 bit adder is applied to the carry input of the next 4-bit adder. The concept of adding 6, i.e., 0 1 1 0 for correction purpose is used in this circuit as well. Figure 5.45 illustrates 8-bit BCD adder using 74283.

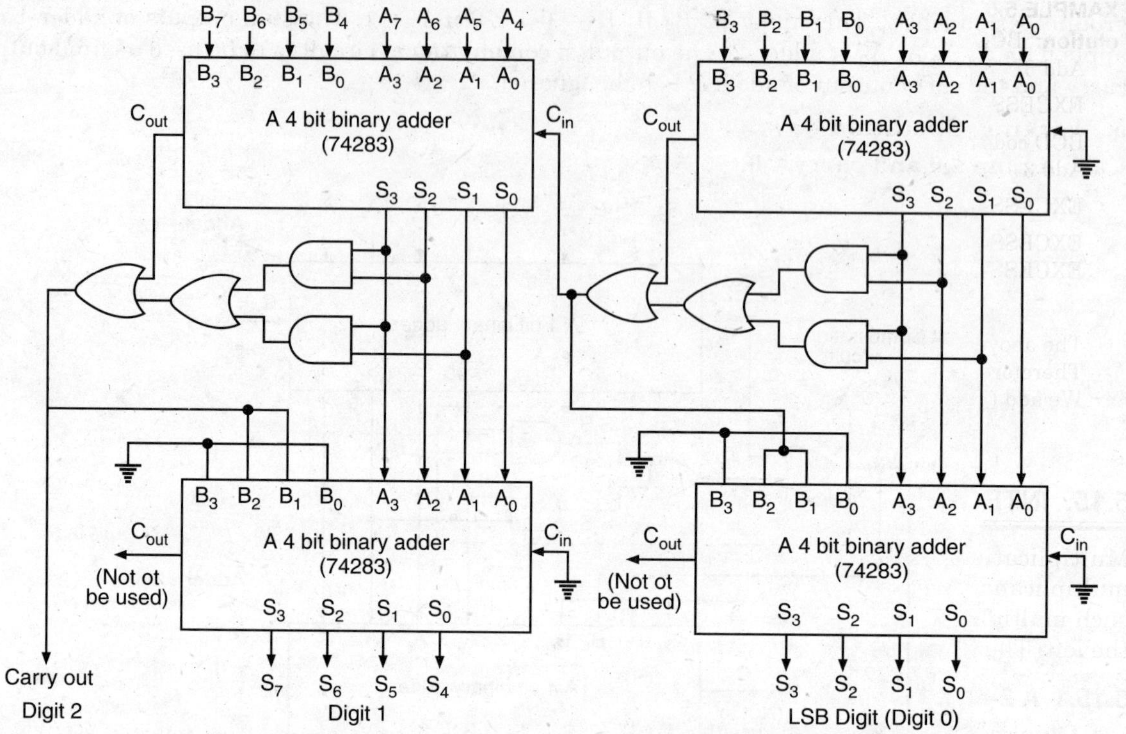

Fig. 5.45. Illustration of 8-bit BCD adder using 74283

5.14.3. An Excess-3 Adder

An Excess-3 code may be obtained by adding 3 to a BCD code. As an example, the EXS-3 code of 3 is 0011. The EXCESS-3 adder performs the addition of two EXS-3 number. The following steps are followed to perform EXCESS-3 code addition. Figure 5.46 shows 4-bit Excess-3 adder.

(i) First, we add two EXCESS-3 number.

(ii) Then, If the above sum produces a carry, we add 3 (0011) to the sum of two digits. If the above sum (step 1) does not produce a carry, we subtract 3 from the above sum (or) add 13 to the sum of digits.

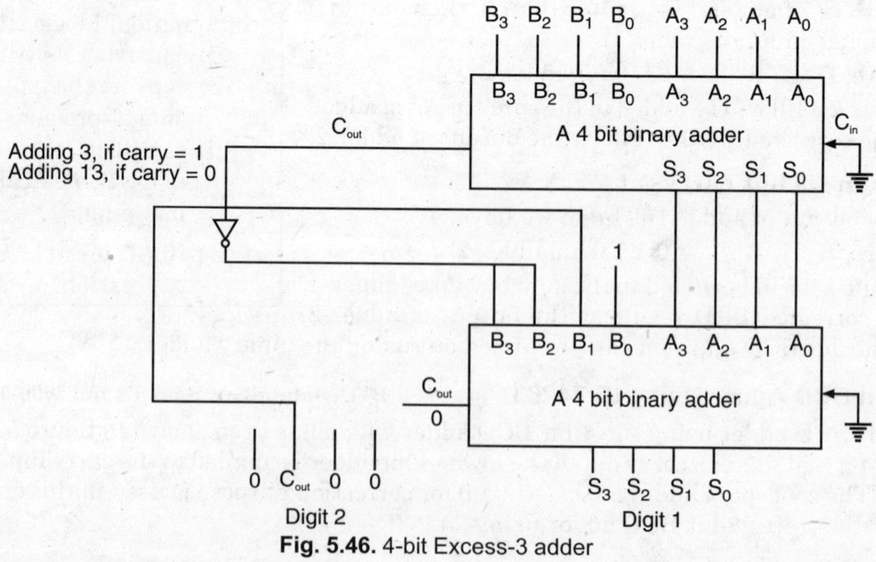

Fig. 5.46. 4-bit Excess-3 adder

EXAMPLE 5.5. Perform $(8)_{10} + (9)_{10}$ using EXCESS-3 addition.

Solution:		
BCD code of 8 is		1000
Add 3		0011
EXCESS-3 code of 8 is		1011
BCD code of 9 is		1001
Add 3		0011
EXCESS-3 code of 9 is		1100
EXCESS-3 code of 8 is		1011
EXCESS-3 code of 9 is		1100
	1	0111

The above sum produces a carry, so, we add 3 for each 4 bit group.

Therefore, Sum is	1	0111
We add (3) for correction	0011	0011
	0100	1010

5.15. INTRODUCTION TO BINARY MULTIPLIERS (PTU, Jalandhar, Sem. Exam., 2004-05)

Multiplication of binary number is performed in the same way as in decimal numbers. The multiplicand is multiplied by each bit of the multiplier starting from the least significant bit. Each such multiplication forms a partial product, successive partial products are shifted one position to the left. The final product is obtained from the sum of the partial products.

5.15.1 A 2-bit by 2-bit Binary Multiplier

Let us consider the multiplication of two 2-bit number as shown in figure 5.47. The multiplicand bits are B_1 and B_0, the multiplier bits are A_1 and A_0 and the products is C_3 C_2 C_1 C_0. The first partial product 1 formed by multiplying A_0 by B_1B_0. The multiplication of two bits such as A_0 and B_0 produces 1 if both bits are 1; otherwise, it produces a 0. This indicated to an AND operation. Therefore, the partial product can be implemented with AND gates as shown in the figure 5.47. The second partial product is formed by multiplying A_1 by B_1B_0 and shifted one position to the left. The two partial products are added with two half adder (HA) circuits. Usually there are more bits in the partial products and it is necessary to use full adders to produce the sum of the partial products. Note that the least significant bit of the product does not have to go through an adder since it is formed by the output of the first AND gate.

		B_1	B_0
		A_1	A_0
		A_0B_1	A_0B_0
	A_1B_1	A_1B_0	
C_3	C_2	C_1	C_0

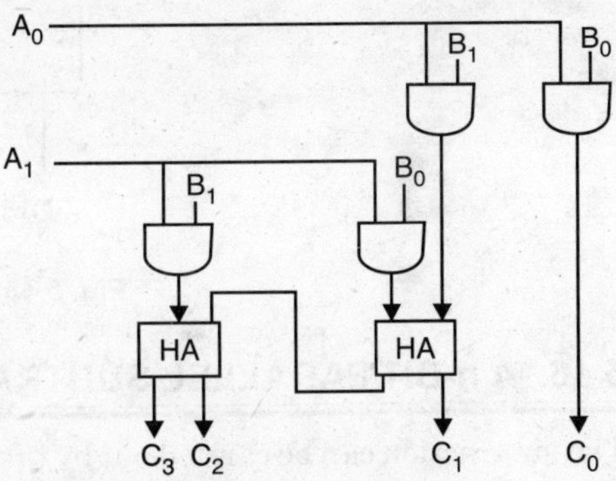

Fig. 5.47. A 2-bit by 2-bit binary multiplier

5.15.2. A 4-bit by 3-bit Multiplier

A combinational circuit binary multiplier with more bits can be constructed in a similar fashion. A bit of the multiplier is ANDed with each bit of the multiplicand in as many levels as there are bits in the multiplier. The binary output of each level of AND gates is added with the partial products of the previous level to form a new partial product. The last level produces the product. For J multiplier bits and K multiplicand bits, we need (J × K) AND gates and (J – 1) K-bit adders to produce a product of J + K bits.

Now, let us consider a multiplier circuit that multiplies a binary number of 4-bits by a number of 3-bits. Let the multiplicand represented by $B_3B_2B_1B_0$ and the multiplier by $A_2A_1A_0$. Since, K = 4 and J = 3, we require 12 AND gates and two 4-bit adders to produce a product of seven bits. The logic diagram of 4-bit by 3-bit multiplier has been shown in figure 5.48.

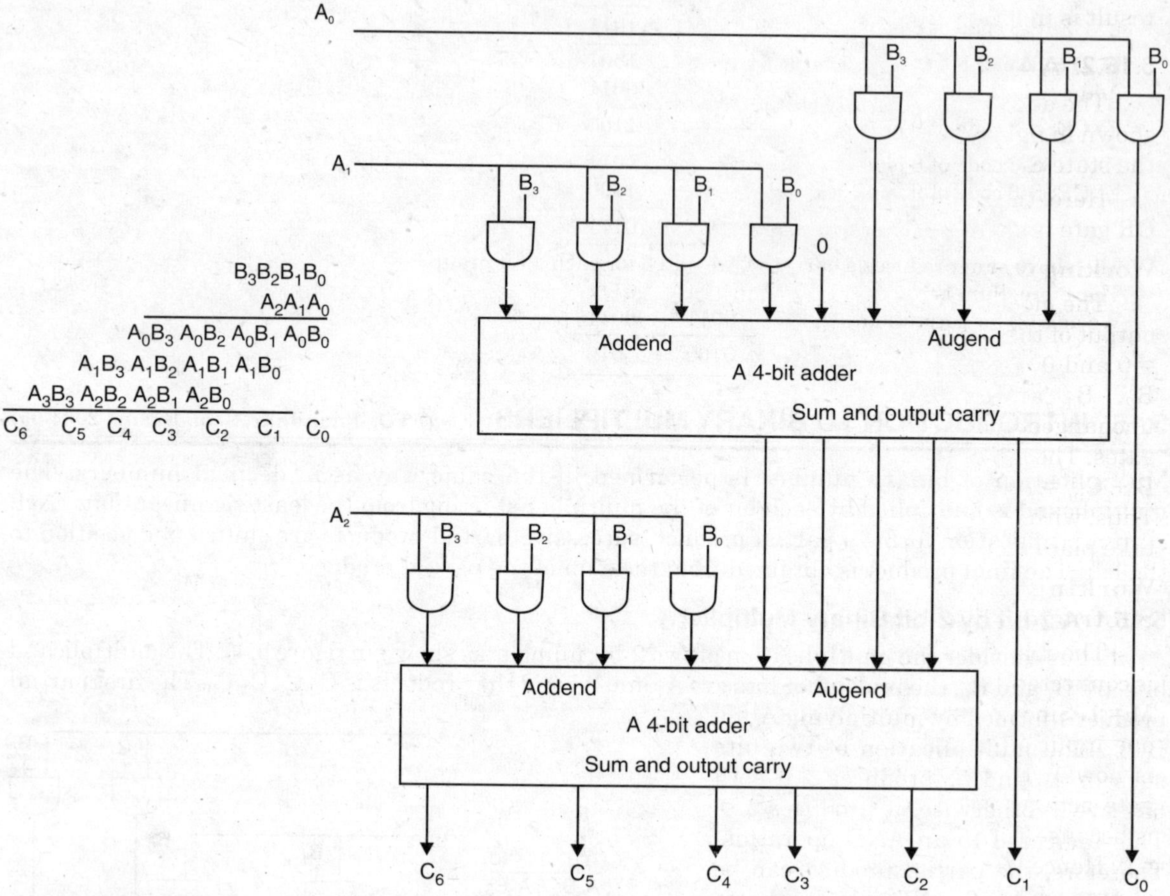

Fig. 5.48. A 4-bit by 3-bit binary multiplier

5.16. A n-BIT PARALLEL SUBTRACTOR (PTU, Jalandhar, Sem. Exam., 2007-08) (10 marks)

The subtraction can be carried out by taking the 1's or 2's complement of the number to be subtracted. For example, we can perform the subtraction (A – B) by adding either 1's complement or 2's complement of B to A. This means that we can use a binary adder to perform the binary subtraction.

5.16.1. 4-Bit Parallel Subtractor using 2's Complement

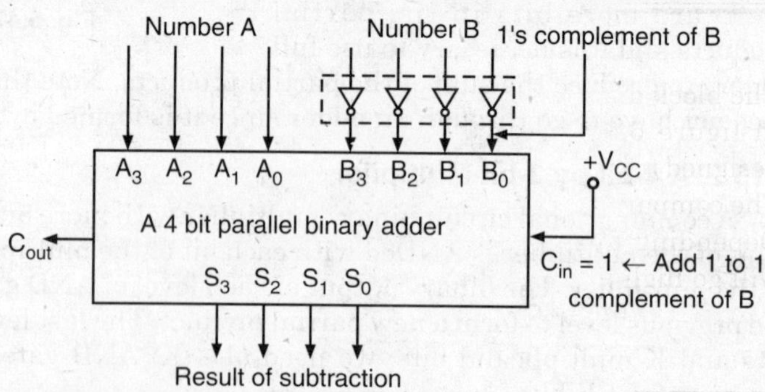

Fig. 5.49 (a). A 4-bit parallel binary subtractor using 2's complement

A 4-bit parallel subtractor using a 4-bit parallel adder has been shown in figure 5.49 (a). The number to be subtracted (B) is first passed through inverters to obtain its 1's complement. Then 1 is added to 1's complement of B, by making C_{in} = 1. Thus we obtain the 2's complement of B. The 4-bit adder then adds A and 2's complement of B to produce the subtraction. Also, $S_3 S_2 S_1 S_0$ represent the result of binary subtraction (A – B) and carry output C_{out} represents the polarity of the result. If A > B, then C_{out} = 0 and the result is in true binary form but if A < B then, C_{out} = 1 and the result is in the 2's complement form.

5.16.2. A 4-Bit Binary Parallel Adder/Subtractor

The addition or subtraction of two 4-bit binary numbers can be obtained using the same circuit shown in figure 5.49 (b). The operation performed by this circuit (Addition or subtraction depends on the state of the mode select input).

Here, the number B is applied to the adder through a set of EX-OR gates. One input of each EX-OR gate is connected to the mode-select input (M).

Working Operation as Adder (M = 0)

The mode select (M) input is connected to ground. Therefore, M = 0. Since M = 0, C_{in} = 0 and the output of the EX-OR gates will be the same number B applied at their inputs. This is because $0 \oplus 0 = 0$ and $0 \oplus 1 = 1$. Hence, B_3, B_2, B_1 and B_0 will pass unchanged through the EX-OR gates. The adder then adds A + B + C_{in} = A + B since C_{in} = 0. Thus, with M = 0, addition will take place.

Working Operation as Subtractor (M = 1)

The mode select (M) input is connected to V_{CC}. Therefore, M = 1. Since M = 1, C_{in} = 1 and one input of each EX-OR gate is now 1. Hence, each EX-OR gate acts as an inverter. This is because $1 \oplus 0 = 1$ and $1 \oplus 1 = 0$. Hence, each bit of word B is inverted by the EX-OR inverters. Thus, we get the 1's complement of number B at the output of EX-OR gates. The inverted number B adds with C_{in} = 1 to give the 2's complement of B. Hence, the adder will add A with the 2's complement of B and the result is actually the subtraction A – B. Hence, with M = 1, this circuit works as a 2's complement subtractor.

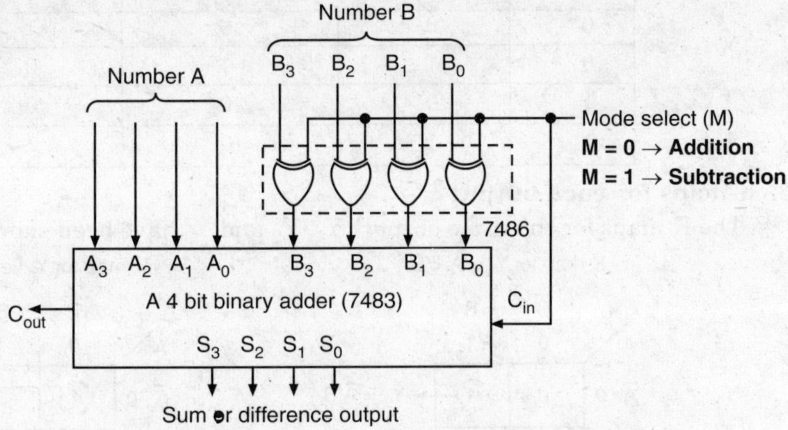

Fig. 5.49 (b). A 4-bit binary parallel adder/subtractor

5.17. MAGNITUDE COMPARATOR : DIGITAL COMPARATOR

(PTU, Jalandhar, Sem. Exam., 2004-05)

The block diagram of an n-bit digital comparator has been shown in figure 5.50. Digital comparator is a combinational circuit, designed to compare the two n-bit binary words applied at its input. The comparator has three outputs namely A > B, A = B and A < B. Depending upon the result of comparison, one of these outputs will go high.

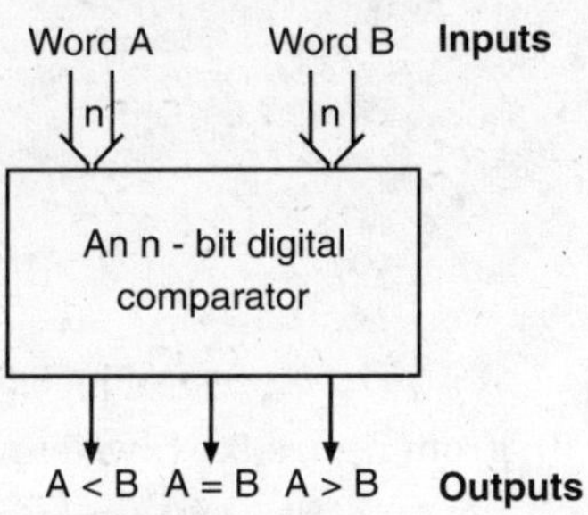

Fig. 5.50. Block diagram of an n-bit comparator

For every minute you are angry, you lose sixty seconds of happiness.

— **Ralph Waldo Emerson**

5.17.1. 1-Bit Magnitude Comparator

1. Definition

The one bit comparator is a combinational logic circuit with two inputs A and B and three outputs namely A < B, A = B and A > B. It compares the two single bit numbers A and B and produces an output that indicates the result of the comparison. This is obvious from the truth table as given in Table 5.7.

> **DO YOU KNOW?**
>
> Magnitude comparators serve as an indicator of the relationship between two binary numbers, with outputs that show >, <, and

2. Truth table of one-bit comparator

Table 5.7. Truth table of a one-bit comparator

Inputs		Outputs		
A	B	$Y_1 = A < B$	$Y_2 = A = B$	$Y_3 = A > B$
0	0	0	1	0
0	1	1	0	0
1	0	0	0	1
1	1	0	1	0

3. K-maps for each output

The K-maps for the three outputs Y_1, Y_2 and Y_3 have been shown in figure 5.51.

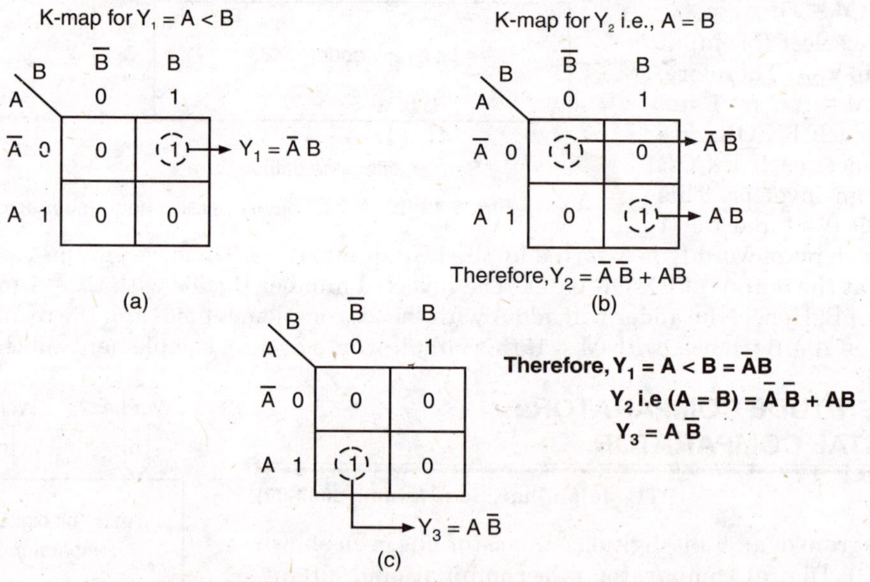

Fig. 5.51. K-maps for the three outputs of a one-bit comparator

From figure 5.51, we can write the expressions for the three outputs as under :

$$Y_1 = (A < B) = \bar{A}B$$

$$Y_2 = (A = B) = \bar{A}\bar{B} + AB = \overline{A \oplus B}$$

$$Y_3 = A\bar{B}$$

The expression for Y_2 is nothing but the expression for an EX-NOR gate. Hence the single bit comparator can be realized as shown in figure 5.52.

> **DO YOU KNOW?**
>
> The K-map is a graphical method for representing a circuit's truth table and generating a simplified expression for the circuit output.

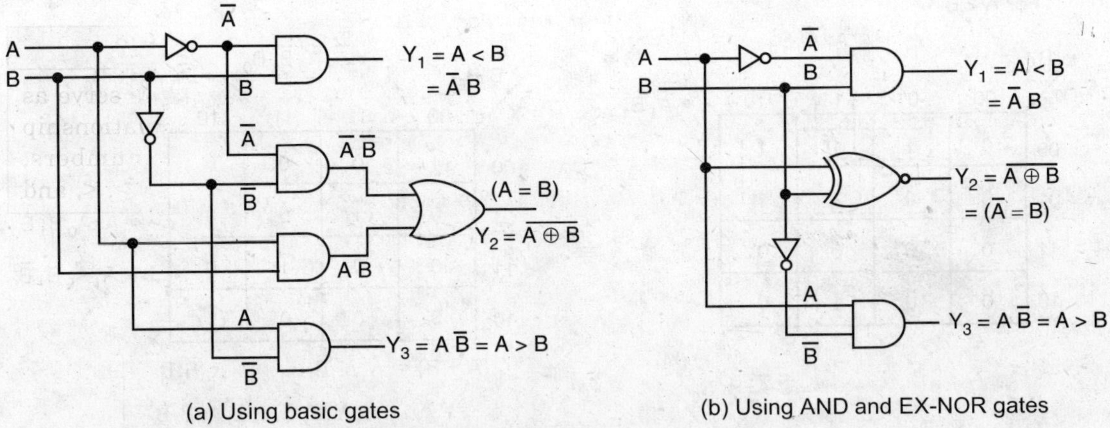

Fig. 5.52. Realization of one bit comparator

5.17.2. A 2-Bit Comparator

For a 2-bit comparator, each input word is 2-bit long. The truth table of a 2-bit comparator has been shown in Table 5.8.

Table 5.8. Truth table for a 2-bit comparator

Inputs				Outputs		
A_1	A_0	B_1	B_0	A < B	A = B	A > B
0	0	0	0	0	1	0
0	0	0	1	1	0	0
0	0	1	0	1	0	0
0	0	1	1	1	0	0
0	1	0	0	0	0	1
0	1	0	1	0	1	0
0	1	1	0	1	0	0
0	1	1	1	1	0	0
1	0	0	0	0	0	1
1	0	0	1	0	0	1
1	0	1	0	0	1	0
1	0	1	1	1	0	0
1	1	0	0	0	0	1
1	1	0	1	0	0	1
1	1	1	0	0	0	1
1	1	1	1	0	1	0

Formation of K-Maps

The K-maps for the three outputs and corresponding simplified expressions have been shown in figures 5.53 (a), 5.53 (b) and 5.54. We have

$$A < B = \overline{A}_1 \overline{A}_0 B_0 + \overline{A}_1 B_1 + \overline{A}_0 B_1 B_0$$

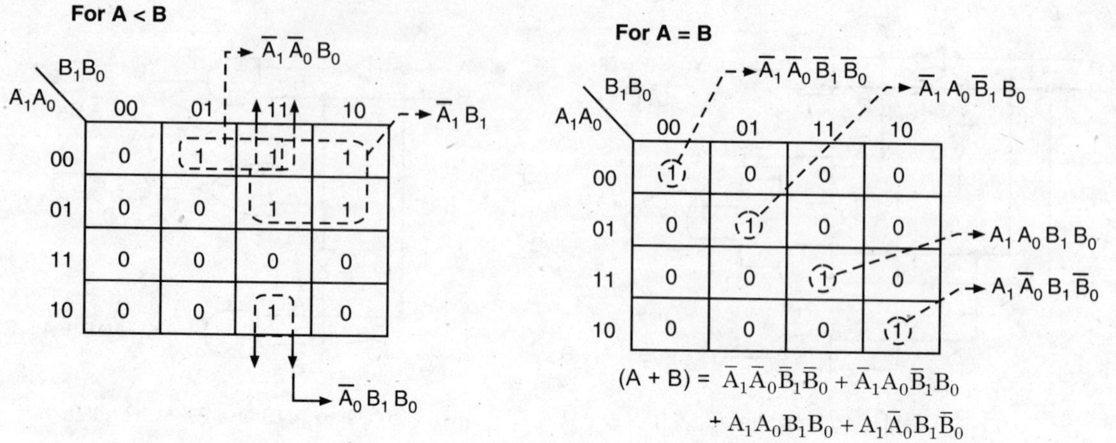

(*a*) K-map for output A < B (*b*) K-map for output A = B

Fig. 5.53

For A > B,

Simplified expression :

$$A > B = A_0 \bar{B}_1 \bar{B}_0 + A_1 A_0 \bar{B}_0 + A_1 \bar{B}_1$$

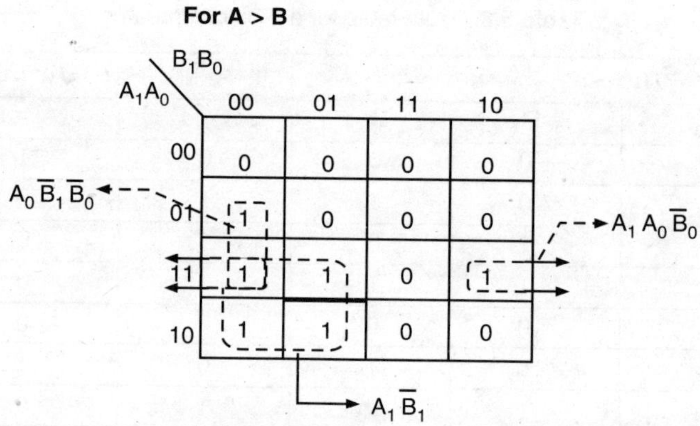

Fig. 5.54. K-map for output *A* > *B*

Simplification for output A = B

Let us consider the K-map for output A = B shown in figure 5.53 (b).

The expression for A = B given by

$$(A = B) = \bar{A}_1 \bar{A}_0 \bar{B}_1 \bar{B}_0 + \bar{A}_1 A_0 \bar{B}_1 B_0 + A_1 A_0 B_1 B_0 + A_1 \bar{A}_0 B_1 \bar{B}_0$$

$$(A = B) = \bar{A}_0 \bar{B}_0 (\bar{A}_1 \bar{B}_1 + A_1 B_1) + A_0 B_0 (\bar{A}_1 \bar{B}_1 + A_1 B_1)$$

$$(A = B) = (\bar{A}_1 \bar{B}_1 + A_1 B_1)(\bar{A}_0 \bar{B}_0 + A_0 B_0)$$

$$(A = B) = (A_1 \odot B_1)(A_0 \odot B_0) \qquad \text{where } \odot = \text{EX-NOR}$$

Logic diagram for a two bit comparator

The logic diagram for the 2-bit digital comparator has been shown in figure 5.55.

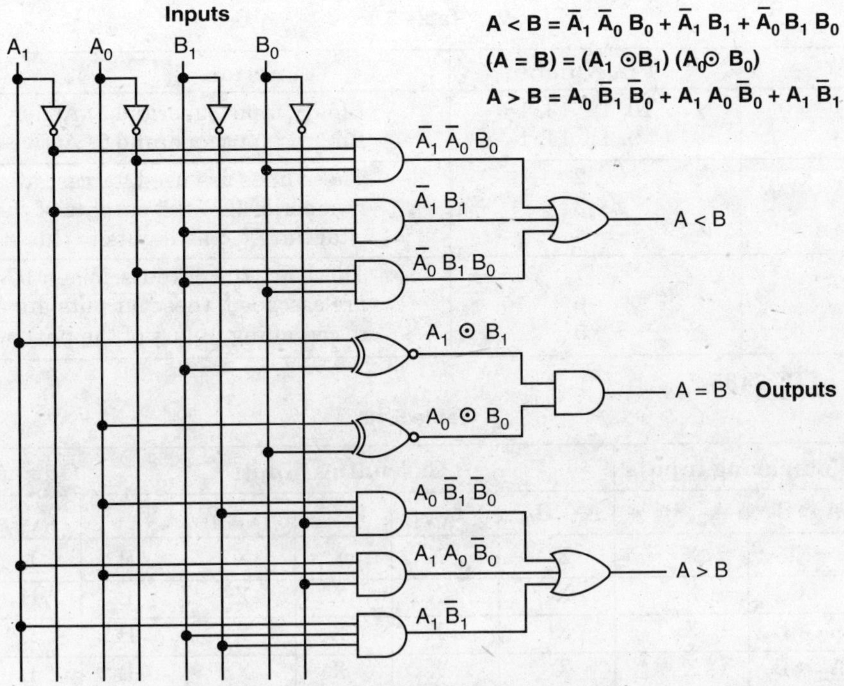

Fig. 5.55. Logic diagram for a 2-bit comparator

5.17.3. A 4-Bit Magnitude Comparator (IC 7485)

As a matter of fact, IC 7485 is a four bit comparator in integrated circuit (IC) form. It compares two 4-bit words A($A_3 - A_0$) and B ($B_3 - B_0$). It is possible to cascade more than one IC 7485 to compare words of almost any length. Figure 5.56 (a) shows the pin configuration and figure 5.56 (b) shows the logic symbol of IC 7485.

> **DO YOU KNOW?**
>
> An exclusive-OR circuit has the expression $x = \overline{A}B + A\overline{B}$. Its output x will be HIGH only when inputs A and B are at opposite logic levels.

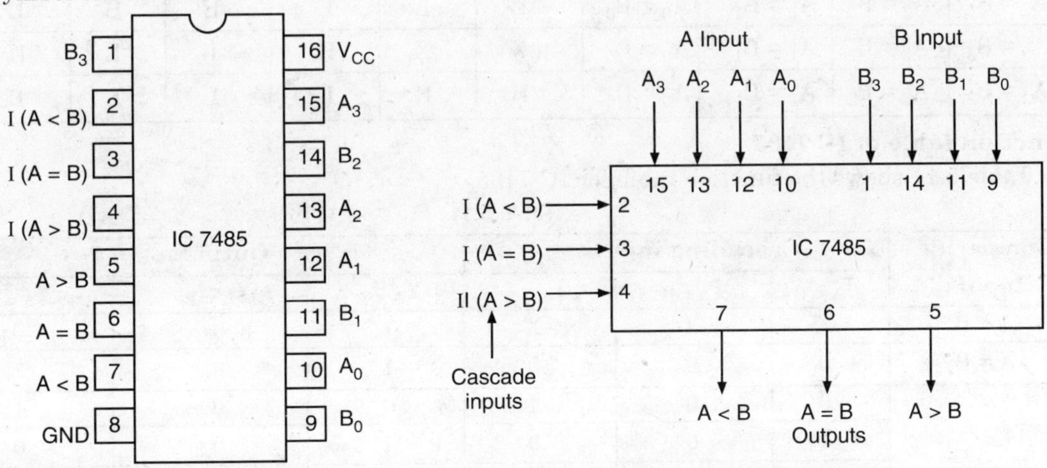

(a) Pin configuration (b) Logic diagram of IC 7485

Fig. 5.56

Pin names and their functions

The pin names and their functions have been listed in Table 5.9.

Table 5.9

Pin Name	Pin Number	Function
A_0 to A_3 B_0 to B_3	10, 12, 13, 15 9, 11, 14, 1	Binary input (operand 1) Active high Binary input (operand 2) Active high
I (A < B) I (A = B) I (A > B)	2 3 4	These lines are used for cascading a number of IC 7485 outputs of the previous stage are fed as inputs to this stage
A < B A = B A > B	7 6 5	These are the outputs. When ICs 7485 are cascaded, these outputs are applied to cascading inputs of the next stage.

Truth table of IC 7485

Table 5.10

Comparing inputs				Cascading inputs			Outputs		
A_3, B_3	A_2, B_2	A_1, B_1	A_0, B_0	$I_{A>B}$	$I_{A<B}$	$I_{A=B}$	$Y_{A>B}$	$Y_{A<B}$	$Y_{A=B}$
$A_3 > B_3$	X	X	X	X	X	X	H	L	L
$A_3 < B_3$	X	X	X	X	X	X	L	H	L
$A_3 = B_3$	$A_2 > B_2$	X	X	X	X	X	H	L	L
$A_3 = B_3$	$A_2 < B_2$	X	X	X	X	X	L	L	L
$A_3 = B_3$	$A_2 = B_2$	$A_1 > B_1$	X	X	X	X	H	L	L
$A_3 = B_3$	$A_2 = B_2$	$A_1 < B_1$	X	X	X	X	L	H	L
$A_3 = B_3$	$A_2 = B_2$	$A_1 = B_1$	$A_0 > B_0$	X	X	X	H	L	L
$A_3 = B_3$	$A_2 = B_2$	$A_1 = B_1$	$A_0 < B_0$	X	X	X	L	H	L
$A_3 = B_3$	$A_2 = B_2$	$A_1 = B_1$	$A_0 = B_0$	H	L	L	H	L	L
$A_1 = B_3$	$A_2 = B_2$	$A_1 = B_1$	$A_0 = B_0$	L	H	L	L	H	L
$A_3 = B_3$	$A_2 = B_2$	$A_1 = B_1$	$A_0 = B_0$	L	L	L	H	H	L
$A_3 = B_3$	$A_2 = B_2$	$A_1 = B_1$	$A_0 > B_0$	X	X	H	L	L	H
$A_3 = B_3$	$A_2 = B_2$	$A_1 = B_1$	$A_0 = B_0$	H	H	L	L	L	L

Function table of IC 7485

Table 5.11 shows the function table for IC 7485.

Table 5.11

Comparing inputs	Cascading inputs			Outputs		
	$I_{A<B}$	$I_{A=B}$	$I_{A>B}$	$Y_{A<B}$	$Y_{A=B}$	$Y_{A>B}$
A < B	X	X	X	1	0	0
A = B	0	0	0	1	0	1
	0	0	1	0	0	1
	1	0	0	1	0	0
	1	0	1	0	0	0
	X	1	X	0	1	0
A > B	X	X	X	0	0	1

Truth table 5.11 tells us that if I (A < B) or I (A > B) is equal to 1 and if the other two cascading inputs are at logic 0 then the corresponding output (A < B or A > B) will be active. But, if A = B, then the corresponding output will be active if and only if I (A = B) is at logic 1 irrespective of the other two inputs.

General description

As shown in truth table (Table 5.11), the MSB are compared first, i.e., A_3 is compared with B_3. Depending upon whether $A_3 > B_3$ or $A_3 < B_3$, the outputs Y (A > B) or Y (A < B) are activated. But, if $A_3 = B_3$, then the next MSB bits B_2 and A_2 are compared. If $A_2 = B_2$, then comparison of A_1 and B_1 is performed and so on. If $A_3\,A_2\,A_1\,A_0 = B_3\,B_2\,B_1\,B_0$ then the IC 7485 will check the cascading inputs. The decision-making will then take place as under :

Case I : If I (A > B) = 1 and I (A = B) = I (A < B) = 0

This indicates that at the previous stage, LSB of A is greater than LSB of B.

Case 2 : If I (A = B) = 1

If I (A = B) = 1, then the chip will not check the status of I (A < B) and I (A > B) inputs and will understand that the LSB of A is equal to LSB of B.

DO YOU KNOW?

An exclusive-NOR circuit has the expression x = $\bar{A}\bar{B}$ + AB. Its output x will be HIGH only when inputs A and B are at the same logic level.

Cascading of IC 7485

As a matter, IC 7485 is a four bit magnitude comparator. If we want to increase the size of the words beyond 4-bit, then the cascading is required to be done. An 8-bit comparator using IC 7485 has been shown in figure 5.57. The outputs of IC 7485-1 are applied to the cascading inputs of IC 7485 – 2. The two 8-bit words to be compared are A ($A_7 - A_0$) and B ($B_7 - B_0$). These are divided in two 4-bit words each and then applied to the two comparator ICs.

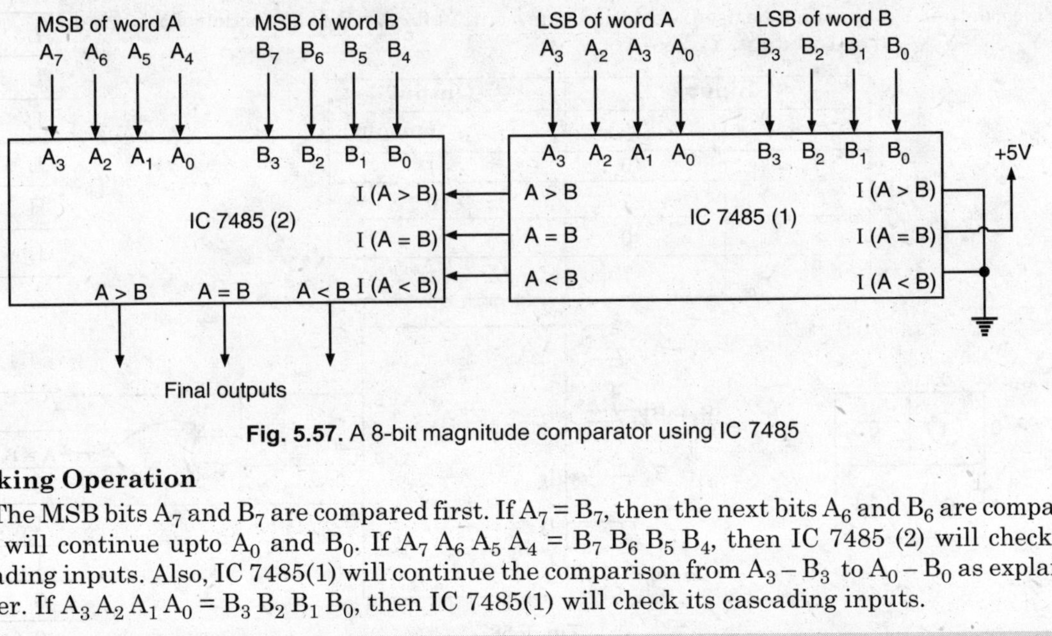

Fig. 5.57. A 8-bit magnitude comparator using IC 7485

Working Operation

The MSB bits A_7 and B_7 are compared first. If $A_7 = B_7$, then the next bits A_6 and B_6 are compared. This will continue upto A_0 and B_0. If $A_7\,A_6\,A_5\,A_4 = B_7\,B_6\,B_5\,B_4$, then IC 7485 (2) will check the cascading inputs. Also, IC 7485(1) will continue the comparison from $A_3 - B_3$ to $A_0 - B_0$ as explained earlier. If $A_3\,A_2\,A_1\,A_0 = B_3\,B_2\,B_1\,B_0$, then IC 7485(1) will check its cascading inputs.

Important Point : Since IC 7485(1) itself is the LSB chip, we have to adjust the status of its cascade inputs, so that they indicate that A = B. Hence, I (A = B) should be connected to + 5 V, i.e., logic 1.

EXAMPLE 5.6. Design a 5 bit magnitude comparator using magnitude comparator chip (7485).

Solution: We know that the output of a comparator is based on the data inputs as well as cascading inputs. Also, comparison always begins from MSB bit. Since, only four bits are available per variable, we accommodate the fifth bit for A_0 and B_0, i.e., LSB's on $I_{A<B}$ and $I_{A>B}$ inputs. This leaves with $I_{A=B}$, whose connection can be found out based on function Table 5.12.

Table 5.12

Comparing inputs		Cascading inputs			Outputs		
		$I_{A<B}$	$I_{A=B}$	$I_{A>B}$	$Y_{A<B}$	$Y_{A=B}$	$Y_{A>B}$
	(a)	0	0	0	1	0	1
	(b)	0	0	1	0	0	1
A = B	(c)	1	0	0	1	0	0
	(d)	1	0	1	0	0	0
	(e)	X	1	X	0	1	0

Condition (e) is ruled out, since it always indicate $Y_{A=B} = 1$.

The other cases imposes a condition that $I_{A=B}$ must be ground.

Cases (b) and (c) gives true output. However, for case (a) and (d), output, $Y_{A=B}$ must be at logic 1. This can be implemented by considering that

	Cascading Input			Output		
(a)	(0)	(0)	(0)	(1)	(0)	(1)
(b)	(1)	(0)	(1)	(0)	(0)	(0)
		Grounded			Not used	

Since, output $Y_{A=B}$ cannot be used, we design a new output for A = B as a function of $Y_{A=B}$ and $Y_{A<B}$, i.e., when $Y_{A>B} = Y_{A<B}$ 0 or 1. Output, $Y_{A=B} = 1$.

Input		Output
$Y_{A>B}$	$Y_{A<B}$	$Y_{A=B}$ (Output)
0	0	1
0	1	0
1	0	0
1	1	1

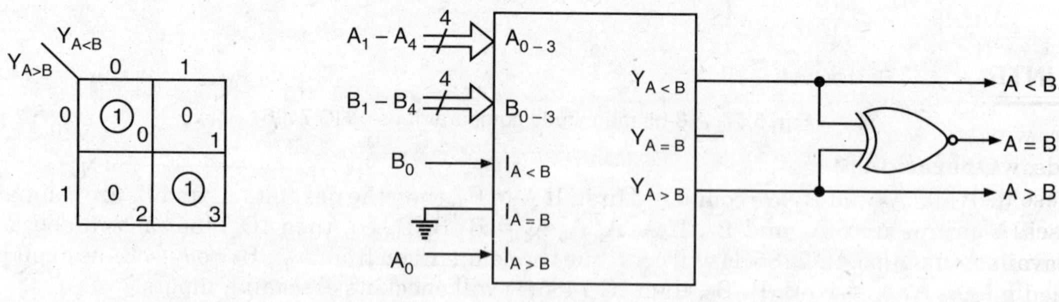

Fig. 5.58.

We have,
$$Y_{A=B} = \bar{Y}_{A>B}\,\bar{Y}_{A<B} + Y_{A<B}\,Y_{A>B} = Y_{A>B} \odot Y_{A<B}$$

The final circuit diagram has been shown in figure 5.58.

EXAMPLE 5.7. Design a 24-bit magnitude comparator using minimum magnitude comparator chips (7485).

Solution: Let us consider figure 5.59.

Here we can cascade six chips. But the problem is of propagation delay, therefore, circuit is bit slower. The 24-bit input is fed to five comparators. Since, there are only 20 inputs per variable, the remaining four are accommodated as explained earlier. This connection is not done for LSB nibble comparator.

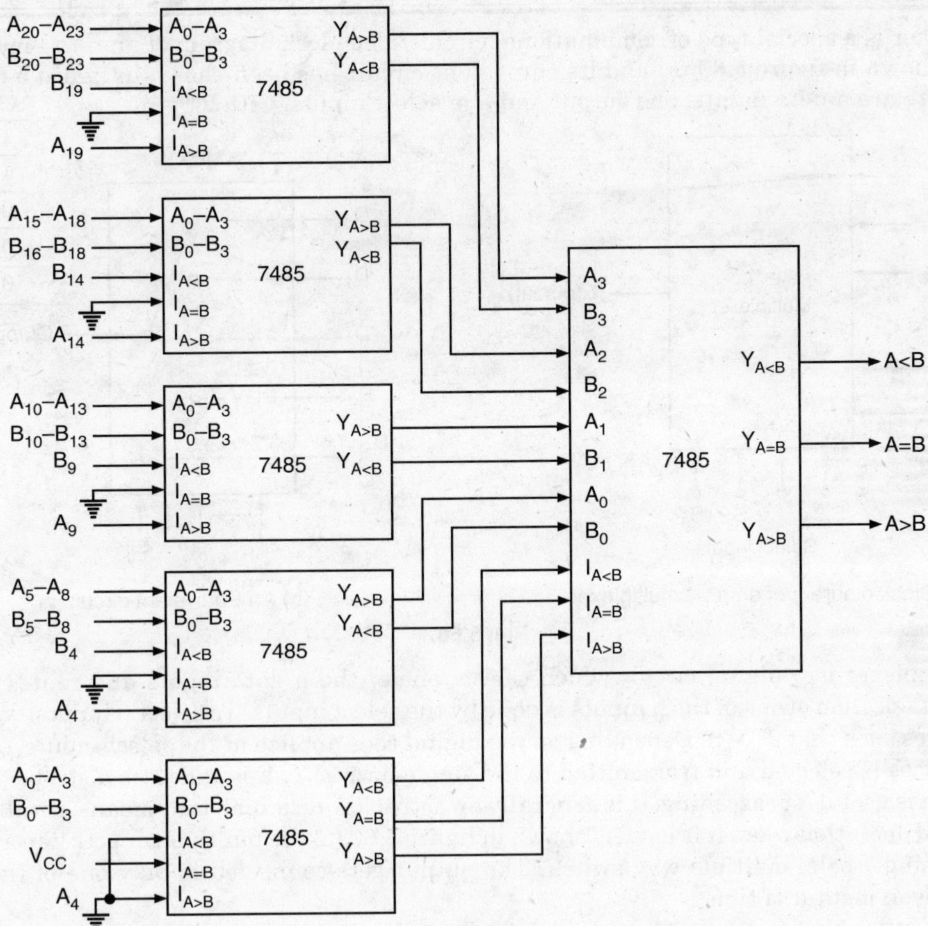

Fig. 5.59.

5.18. INTRODUCTION TO MSI AND LSI CIRCUITS (PTU, Jalandhar, Sem. Exam., 2005-06)

Earlier, we have discussed the traditional methods of designing combinational circuits. In these methods, we used to simplify the expression and realize the simplified expression using logic gates. But these methods are suitable for small circuits. They cannot be used when the circuit complexity increases. Now several combinational circuits are available in the integrated (MSI) form and are easily available in market. Depending upon the level of complexity, the integrated circuits may be classified into four categories as under :

(i) SSI : Small scale integration
(ii) MSI : Medium scale integration
(iii) LSI : Large scale integration
(iv) VLSI : Very large scale integration

DO YOU KNOW?

Each of the basic gates (AND, OR, NAND, NOR) can be used to enable or disable the passage of an input signal to its output.

Because of availability of combinational circuits in the IC form; the design of combinational circuits has been simplified to a great extent. It also improves the reliability of the system because the number of external wired connections. Now, let us discuss the MSI devices such as multiplexer, demultiplexer, adder, parity generators, priority encoders, decoders etc.

5.19. MULTIPLEXER (DATA SELECTOR)

A multiplexer is a special type of combinational circuit. The block diagram of an n-to-l multiplexer has been shown in figure 5.60 (a) and its equivalent circuit has been shown in figure 5.60 (b). As shown, there are n-data inputs, one output and a m select inputs, with $2^m = n$.

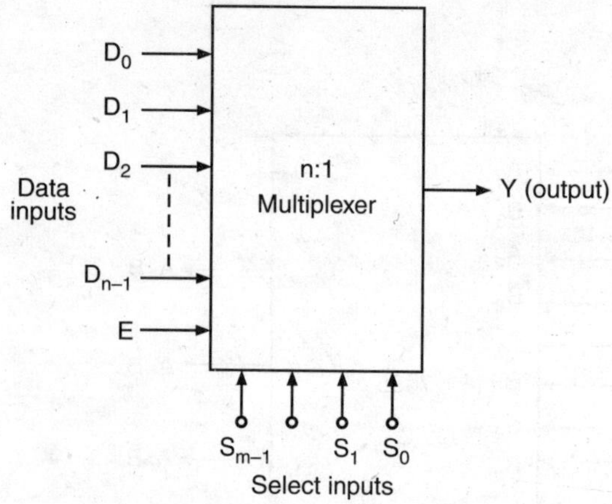

(a) Block diagram of an n : 1 multiplexer

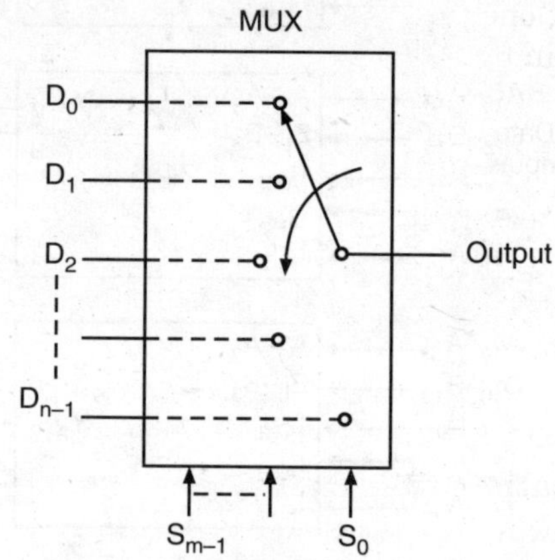

(b) An Equivalent circuit

Fig. 5.60.

A multiplexer is a digital circuit which selects one of the n data inputs and routes it to the output. The selection of one of the n inputs is done by the select inputs. To select n inputs, we require m select lines such that $2^m = n$. Depending on the digital code applied at the select inputs, one out of n data sources is selected and transmitted to the single output Y. E is called as a strobe or enable input which is useful for cascading. It is generally an active low terminal, that means, it will perform the required operation when it is low. As shown in figure 5.60 (b) the multiplexer acts like a digitally controlled single pole, multiple way switch. The output gets connected to only one of the n data inputs at given instant of time.

5.19.1. Necessity of Multiplexers

In most of the electronic systems, the digital data is available on more than one lines. It is essential to route this data over a single line. Under such circumstances, we require a circuit which select one of the many inputs at a time. This circuit is nothing else but a multiplexer which has many inputs, one output and some select inputs. Multiplexer improves the reliability of the digital system because it reduces the number of external wired connections.

5.19.2. Advantages of Multiplexers

(i) It reduces the number of wires.

(ii) Hence, it reduces the circuit complexity and cost.

(iii) We can implement many combinational circuits using MUX.

(iv) It simplifies the logic design.

(v) It does not need the K-maps and simplification.

> **DO YOU KNOW?**
>
> A decoder is a device whose output is activated only when a unique binary combination (code) is presented on its inputs.

5.19.3. Classification of Multiplexers

Multiplexers may be classified as under :

(i) 2 : 1 multiplexer

(ii) 4 : 1 multiplexer

(iii) 8 ; 1 multiplexer

(iv) 16 : 1 multiplexer

5.20. A 2 : 1 MULTIPLEXER

The block schematic of a 2 : 1 multiplexer has been shown in figure 5.61(a). It has two data inputs D_0 and D_1, one select input S, an enable input and one output. The truth table of this MUX has been shown in figure 5.61 (b).

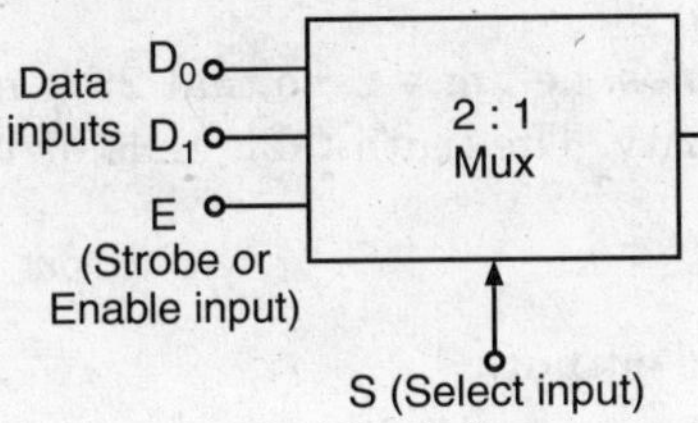

(a) Block diagram

Enable	Select i/p S	Output Y
0	×	0
1	0	D_0
1	1	D_1

(b) Truth table

Fig. 5.61. 2 : 1 Multiplexer

Realization of a 2 : 1 MUX using gates

Now, we write a more elaborate truth table for 2 : 1 MUX as under :

Table 5.13. Truth table of a 2 : 1 MUX

Enable E	Select S	D_1	D_0	Output Y	
0	×	×	×	0	
1	0	×	0	0	
1	0	×	1	1	$\leftarrow Y = E\bar{S}\,D_0$
1	1	0	×	0	
1	1	1	×	1	$\leftarrow Y = ES\,D_1$

From the truth table, it is obvious that Y =1 for only two conditions.

Therefore, we have $Y = E\bar{S}D_0 + ESD_1$ or $Y = E(\bar{S}D_0 + SD_1)$

The realization using gates has been shown in figure 5.62.

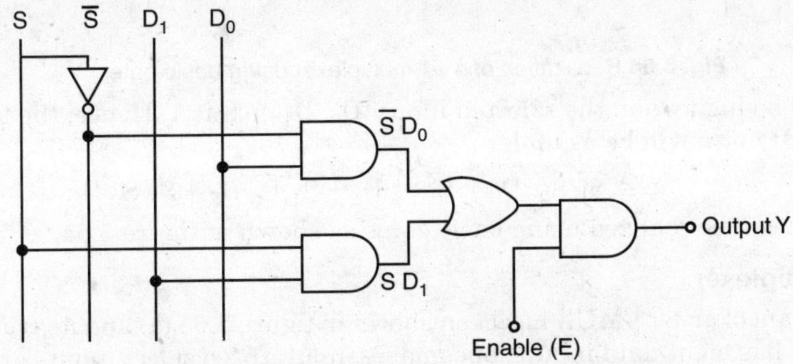

Fig. 5.62. Realization of 2 : 1 MUX using gates

5.20.1. A 4 : 1 Multiplexer

Figure 5.63 shows the block diagram of a 4 : 1 multiplexer and Table 5.14 gives its truth table.

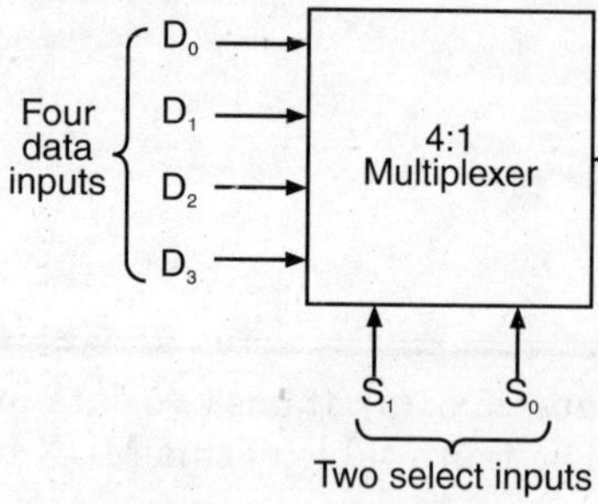

Fig. 5.63. 4 : 1 Multiplexer

Table 5.14

Select inputs		Output
S_1	S_0	Y
0	0	D_0
0	1	D_1
1	0	D_2
1	1	D_3

Here, it may be noted that n = 4, hence, number of select lines, i.e., m = 2, so that $2^m = n$. The strobe terminal (G) has not been included to reduce the complexity. The truth table tells us that if $S_1 S_0 = 00$, the data bit D_0 is selected and routed to output

Therefore, we have $\quad Y = D_0$ when $S_1 S_0 = 00$

Similarly, if $S_1 S_0 = 01$, then D_1 is selected and routed to the output.

Therefore, $\quad Y = D_1$ when $S_1 S_0 = 01$

Also, $Y = D_2$ for $S_1 S_0 = 10$ and $Y = D_3$ for $S_1 S_0 = 11$

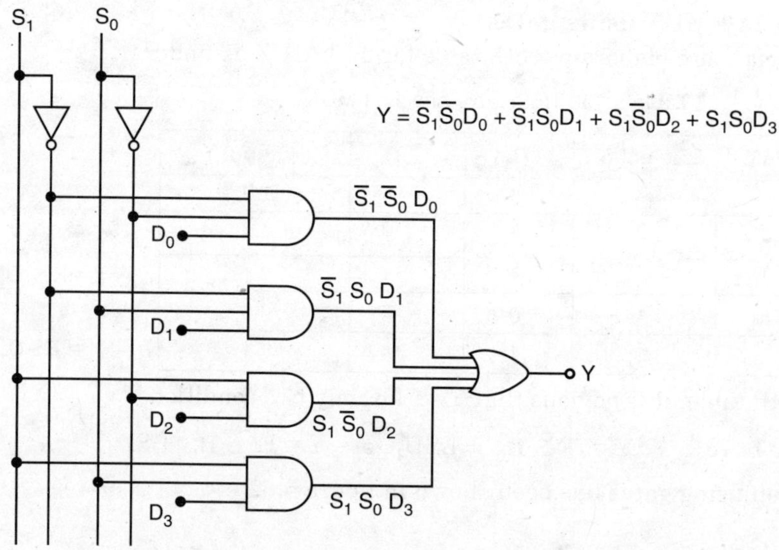

Fig. 5.64 Realization of 4 : 1 multiplexer using basic gates

The output will be high when the selected input (D_0, D_1 etc.) is 1. Hence, the logical expression for output in the SOP form will be as under :

$$Y = \overline{S_1}\,\overline{S_0}\,D_0 + \overline{S_1}\,S_0\,D_1 + S_1\,\overline{S_0}\,D_2 + S_1\,S_0\,D_3$$

This expression can be realized using basic gates as shown in figure 5.64.

5.20.2. A 8 : 1 Multiplexer

The block diagram of an 8 : 1 MUX has been shown in figure 5.65 (a) and its truth table is shown in figure 5.65 (b). It has eight data inputs, one enable input, three select inputs and one output.

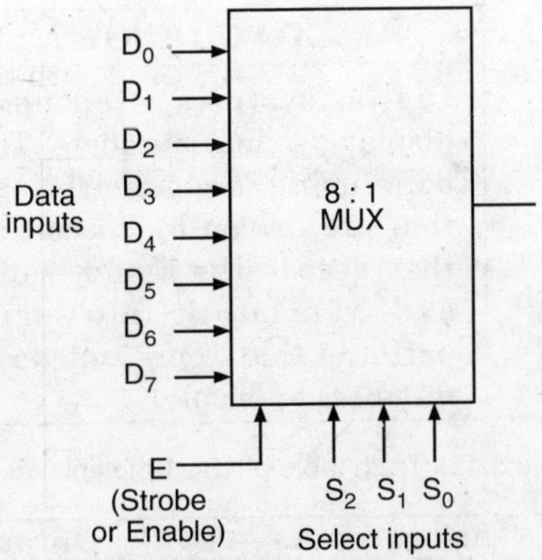

(a) Block diagram

Enable	Select inputs			Output
E	S_2	S_1	S_0	Y
0	×	×	×	0
1	0	0	0	D_0
1	0	0	1	D_1
1	0	1	0	D_2
1	0	1	1	D_3
1	1	0	0	D_4
1	1	0	1	D_5
1	1	1	0	D_6
1	1	1	1	D_7

(b) Truth table

Fig. 5.65. 8 : 1 Multiplexer

Operating principle

When the strobe or enable input is 0, the output will be 0 irrespective of any input. With E = 1, we can select any one of the eight data inputs and connect it to the output. For example, if S_2 S_1 S_0 = 0 1 1, then the data input D_3 is selected and output Y will follow the input D_3.

Realization using gates

The realization of 8 : 1 MUX using gates has been shown in figure 5.66.

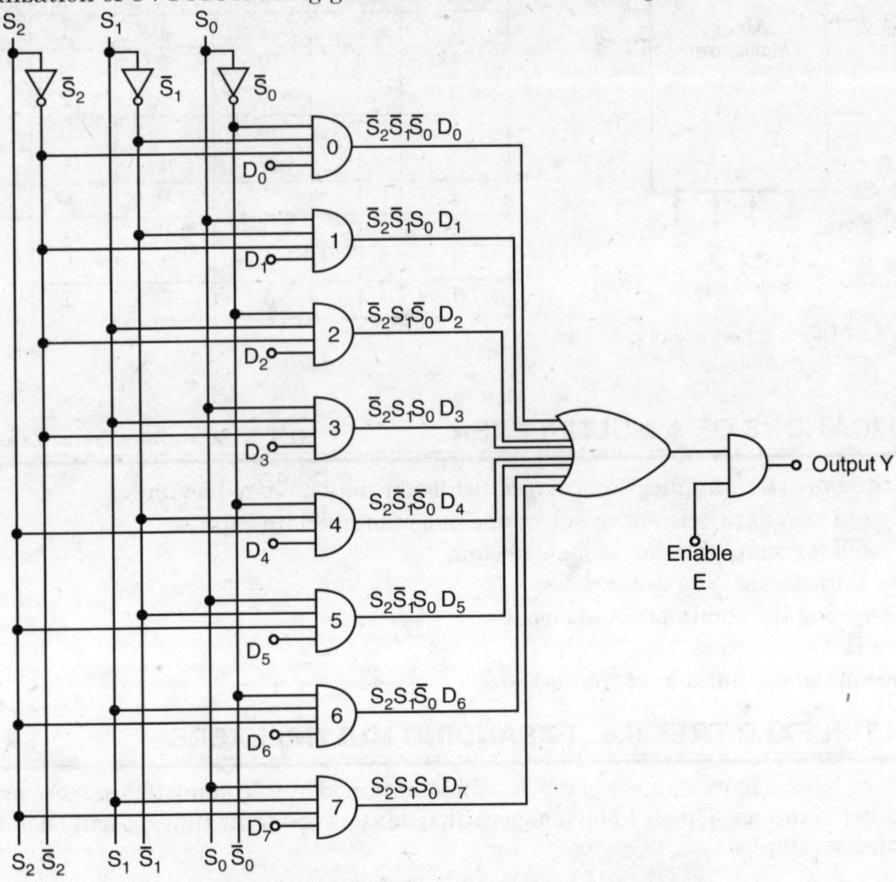

Fig. 5.66. Illustration of a 8 : 1 MUX using gates

5.20.3. A 16 : 1 MUX

The block diagram of a 16 : 1 MUX has been shown in figure 5.67. It has 16 data inputs, 4 select lines, one enable or strobe input and one output.

Operating principle

When the strobe or enable input E = 0, the multiplexer output Y = 0 irrespective of the status of other inputs. With E = 1, we can select one of the 16 inputs with the help of select input lines and connect the selected input to the output. The truth table of a 16 : 1 multiplexer is as under :

> **DO YOU KNOW?**
>
> Digital systems often need to display decimal numbers. This is done using 7-segment displays that are driven by special chips that decode the binary number and translate it into segment patterns that represent decimal numbers to people.

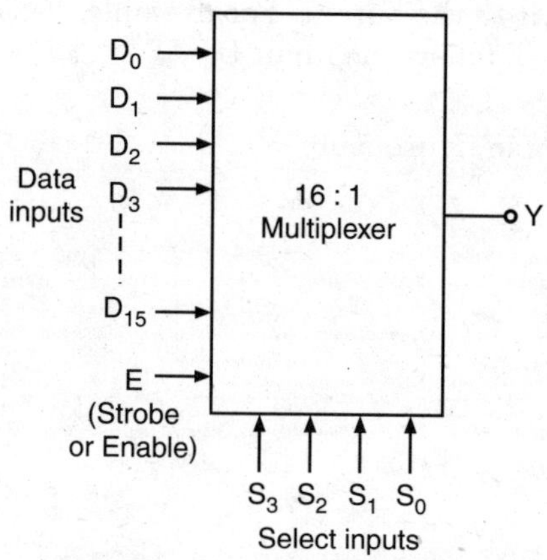

Fig. 5.67 Block schematic of a 16 : 1 multiplexer

Table 5.15. Truth table of 16 : 1 multiplexer

Enable	Select inputs				Output
E	S_3	S_2	S_1	S_0	Y
0	0	0	0	0	D_0
1	0	0	0	1	D_1
1	0	0	1	0	D_2
1	0	0	1	1	D_3
1	0	1	0	0	D_4
1	0	1	0	1	D_5
1	0	1	1	0	D_6
1	0	1	1	1	D_7
1	1	0	0	0	D_8
1	1	0	0	1	D_9
1	1	0	1	0	D_{10}
1	1	0	1	1	D_{11}
1	1	1	0	0	D_{12}
1	1	1	0	1	D_{13}
1	1	1	1	0	D_{14}
1	1	1	1	1	D_{15}

5.21. APPLICATIONS OF A MULTIPLEXER

(PTU, Jalandhar, Sem. Exam., 2006-07)

Some of the important applications of a multiplexer may be listed as under :

(i) It is used as a data selector to select one out of many data inputs.
(ii) It is used for simplification of logic design.
(iii) In the data acquisition system.
(iv) In designing the combinational circuits.
(v) In the D/A converters.
(vi) To minimize the number of connections.

5.22. A MULTIPLEXER TREE (i.e., EXPANDING MULTIPLEXERS)

The multiplexers having more number of inputs can be obtained by cascading two or more multiplexers with less number of inputs. This is known as a **multiplexer tree**. This concept will be more obvious after solving few examples.

EXAMPLE 5.8. Obtain an 8 : 1 multiplexer using two 4 : 1 multiplexers.

Solution: The cascading of two 4 : 1 multiplexer results in 8 : 1 multiplexer as shown in figure 5.68.

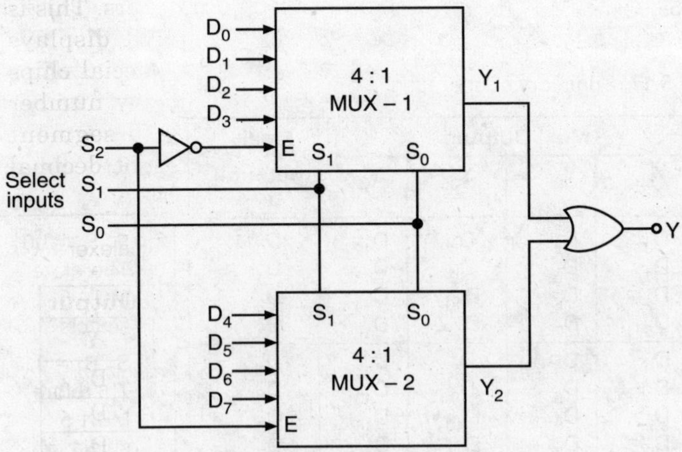

Fig. 5.68. 8 : 1 multiplexer by cascading two 4 : 1 multiplexers

Table 5.16. Truth table

Select inputs			Output	
S_2	S_1	S_0	Y	
0	0	0	D_0	MUX 1 is enable
0	0	1	D_1	
0	1	0	D_2	
0	1	1	D_3	
1	0	0	D_4	MUX 2 is enable
1	0	1	D_5	
1	1	0	D_6	
1	1	1	D_7	

There are in all eight data inputs (D_0 through D_7). The select lines S_1 and S_0 of both 4 : 1 multiplexers are connected in parallel whereas a third select input S_2 is used for enabling one multiplexer at a time. Also, S_2 is connected directly to the enable (E) terminal of MUX-1 whereas $\overline{S}_2$ is connected to the enable terminal of MUX-2. The outputs of the two multiplexers are ORed to obtain the final output Y.

EXAMPLE 5.9. Implement a 16 : 1 multiplexer using 4 : 1 multiplexers.

Solution: Let us consider figure 5.69, for implementation.

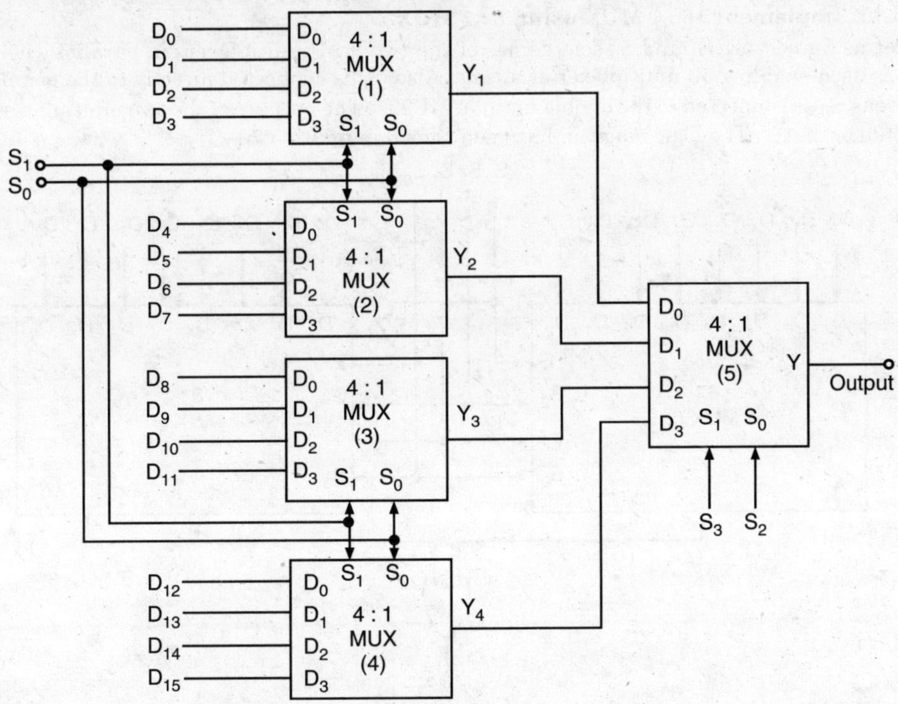

Fig. 5.69. 16 : 1 multiplexer using 4 : 1 multiplexers

Description

The select input S_1 and S_0 of the multiplexers 1, 2, 3 and 4 are connected together. The select inputs S_3 and S_2 are applied to the select input S_1 and S_0 of MUX-5. The output Y_1, Y_2, Y_3, Y_4 are applied to the data inputs D_0, D_1, D_2 and D_3 of MUX-5 as shown in figure 5.69.

The operation can be summarized using Table 5.17.

Table 5.17. Summary of operation

Select inputs				Mux. Outputs				Final output Y	
S_3	S_2	S_1	S_0	Y_1	Y_2	Y_3	Y_4		
0	0	0	0	D_0	D_4	D_8	D_{12}	D_0	$S_3 S_2 = 00$ Therefore, MUX-5 selects Y_1
0	0	0	1	D_1	D_5	D_9	D_{13}	D_1	
0	0	1	0	D_2	D_6	D_{10}	D_{14}	D_2	
0	0	1	1	D_3	D_7	D_{11}	D_{15}	D_3	
0	1	0	0	D_0	D_4	D_8	D_{12}	D_4	$S_3 S_2 = 01$ Therefore, MUX-5 selects Y_2
0	1	0	1	D_1	D_5	D_9	D_{13}	D_5	
0	1	1	0	D_2	D_6	D_{10}	D_{14}	D_6	
0	1	1	1	D_3	D_7	D_{11}	D_{15}	D_7	
1	0	0	0	D_0	D_4	D_8	D_{12}	D_8	$S_3 S_2 = 10$ Therefore, MUX-5 selects Y_3
1	0	0	1	D_1	D_5	D_9	D_{13}	D_9	
1	0	1	0	D_2	D_6	D_{10}	D_{14}	D_{10}	
1	0	1	1	D_3	D_7	D_{11}	D_{15}	D_{11}	
1	1	0	0	D_0	D_4	D_8	D_{12}	D_{12}	$S_3 S_2 = 11$ Therefore, MUX-5 selects Y_4
1	1	0	1	D_1	D_5	D_9	D_{13}	D_{13}	
1	1	1	0	D_2	D_6	D_{10}	D_{14}	D_{14}	
1	1	1	1	D_3	D_7	D_{11}	D_{15}	D_{15}	

EXAMPLE 5.10. Implement 16 : 1 MUX using 8 : 1 MUX.

Solution: Let us connect S_2, S_1 and S_0 select lines of the two 8 : 1 multiplexers in parallel whereas the MSB select input is used enabling on multiplexer at a time. Also, S_3 is connected directly to the enable (E) input of MUX-2 whereas $\bar{S}_3$ is connected to the enable input of MUX-1. The outputs of the two multiplexers are ORed to obtain the final output Y. The logic diagram has been shown in figure 5.70.

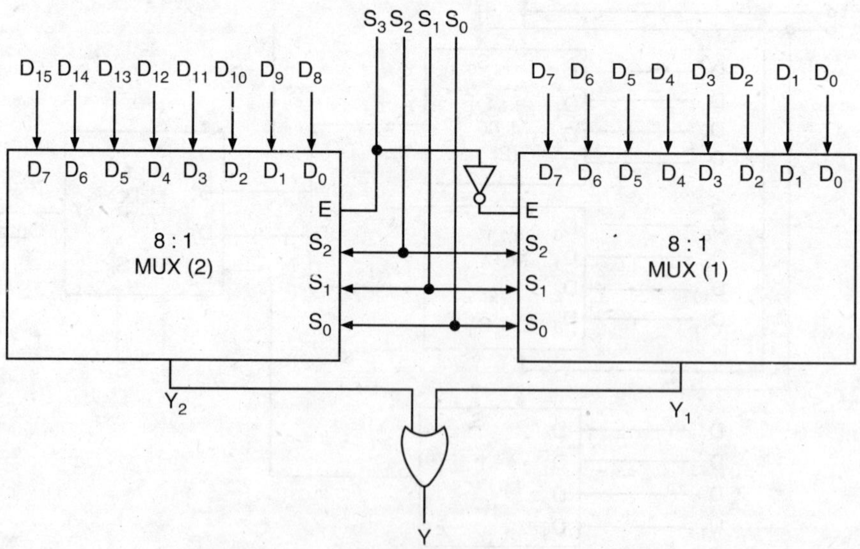

Fig. 5.70. 16 : 1 MUX using two 8 : 1 multiplexers

EXAMPLE 5.11. Implement a 4 : 1 multiplexer using 2 : 1 multiplexers.

Solution: Let us consider figure 5.71 for the implementation.

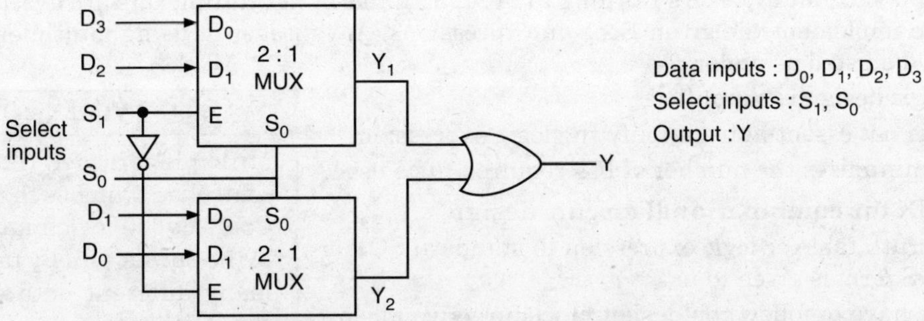

Fig. 5.71. 4 : 1 multiplexer using 2 : 1 multiplexers

EXAMPLE 5.12. Draw a 32 : 1 Multiplexer tree using 4 : 1 Multiplexers.

Solution: A 32 : 1 multiplexer using 4 : 1 multiplexers has been shown in figure 5.72.

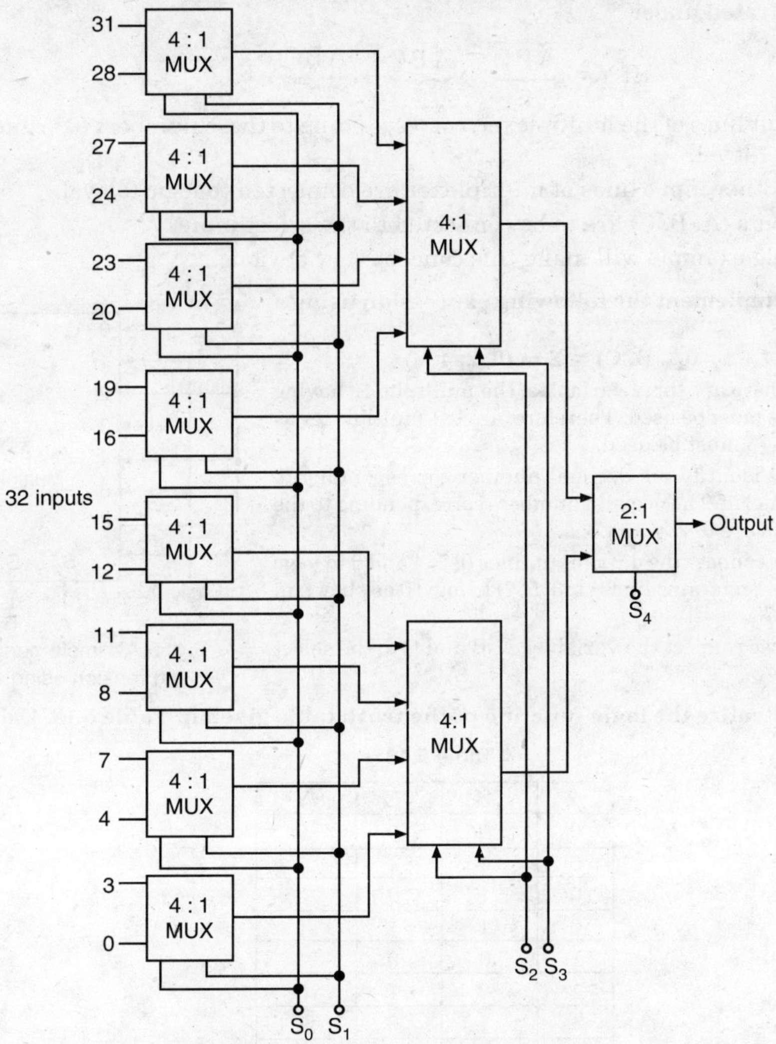

Fig. 5.72. A 32 : 1 MUX using 4 : 1 MUX

5.23. IMPLEMENTATION OF COMBINATIONAL LOGIC USING MULTIPLEXERS

As a matter of fact, multiplexers ICs for 2 : 1, 4 : 1, 8 : 1 and 16 : 1 multiplexers are available. We can use them to implement the given Boolean expressions. Advantages of using multiplexers for logic design may be listed as under :

(i) Logic design is simplified.

(ii) It is not essential to simplify the logic expression.

(iii) It minimizes the number of ICs required to be used.

DO YOU KNOW?

Multiplexers act like digitally controlled switches that select and Connect one logic input at a time to the output pin. By taking turns, many different data signals can share the same data path using multiplexers.

Use of MUX for combinational circuit design

(i) A truth table or logic expression in standard SOP or POS form is given to us.

(ii) We have to follow the design procedure given below to use MUX as a circuit element.

Design Procedure

(i) First, we identify the decimal number corresponding to each minterm in the given expression as illustrated under.

$$\text{If } Y = \underbrace{\overline{A}\,\overline{B}\,\overline{C}}_{0} + \underbrace{A\overline{B}C}_{5} + \underbrace{ABC}_{7}$$

(ii) The input lines of the multiplexer, corresponding to these numbers (0, 5 and 7) are connected to logic 1 level.

(iii) All the other input lines of multiplexer are connected to logic 0 level.

(iv) The inputs (A, B, C) are to be connected to the select inputs.

The following example will make this concept very obvious.

EXAMPLE 5.13. Implement the following expression using a multiplexer :

$$f(A, B, C) = \Sigma m\,(0, 2, 4, 6)$$

Solution: Since there are three variables, the multiplexer having three select inputs must be used. Therefore, an 8 : 1 multiplexer as shown in figure 5.73 must be used.

(i) First, we identify the decimal number corresponding to each minterm. The decimal numbers corresponding to the minterms are 0, 2, 4 and 6.

(ii) Then, we connect the data input lines 0, 2, 4 and 6 to logic 1 and the remaining lines (1, 3, 5, 7) to logic 0 as shown in figure 5.73.

(iii) Finally, we connect the variables A, B and C to the select inputs.

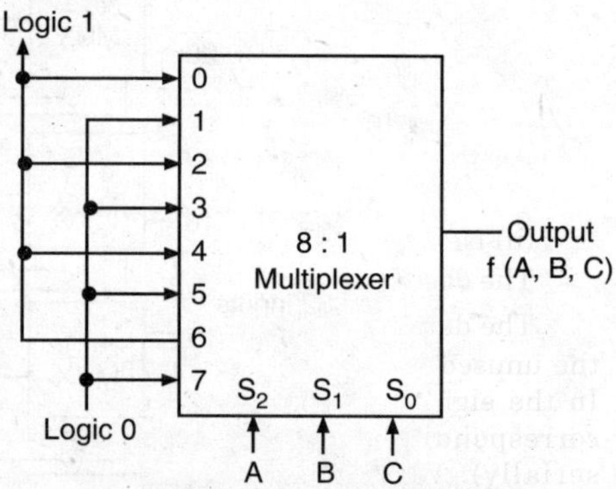

Fig. 5.73. Implementation of a logic expression using a multiplexer

EXAMPLE 5.14. Realize the logic function of the truth table given in Table 5.18. Using a multiplexer.

Table 5.18

A	B	C	Y
0	0	0	0
0	0	1	1
0	1	0	0
0	1	1	1
1	0	0	1
1	0	1	0
1	1	0	1
1	1	1	0

Solution:

(i) First, we express the output Y in the standard SOP form, i.e.,

$$Y = \Sigma\, m\, (1, 3, 4, 6)$$

(ii) Then, we connect the data inputs 1, 3, 4 and 6 to logic 1 and the remaining to logic 0.

(iii) After that we connect the inputs A, B, C to the select lines S_2, S_1 and S_0 respectively.

(iv) Finally, we draw the logic diagram as shown in figure 5.74.

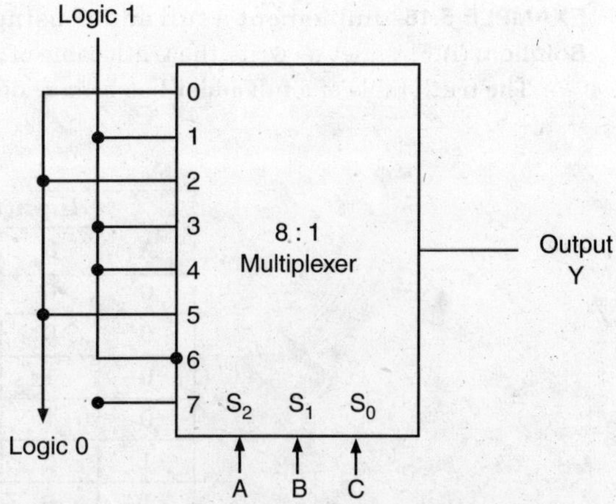

Fig. 5.74. Logic diagram for Example 3.14.

EXAMPLE 5.15. Implement the logic function of the last example using a 4 : 1 multiplexer.

Solution. The logic function to be implemented is given by

$$f(A, B, C) = \Sigma\, m\, (1, 3, 4, 6)$$

(i) First, let us apply two variables B and C to the select inputs.

As shown in figure 5.75, the two input variables B and C are applied to the select lines S_1 and S_0 respectively.

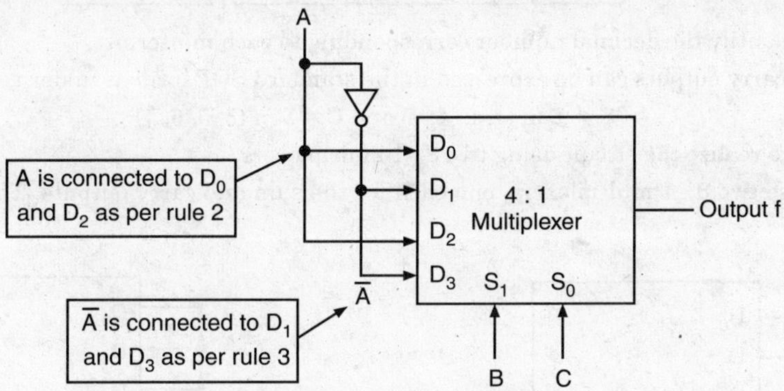

Fig. 5.75. Logic diagram

(ii) Next, let us write the design table.

The design table has been shown in figure 5.76.

The data inputs D_0 to D_4 have been written at the top of the table and the two possible values A and $\overline{A}$ of the unused variable A have been written. In the eight boxes, the decimal numbers corresponding to the minterms (0 to 7 serially). We encircle those minterms corresponding to which the output is 1 (minterms 1, 3, 4, 6).

(iii) Lastly, we check each column in the design table.

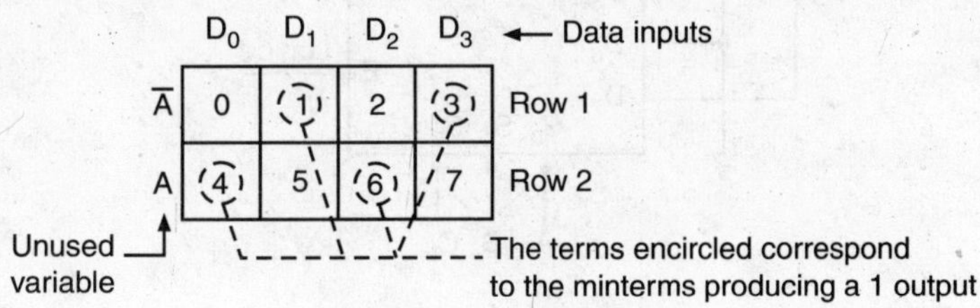

Fig. 5.76. Design table

The columns of design table are inspected using the following rules :

(i) If both the minterms in a column are not circled, then apply logic 0 to the corresponding data input. Note that there is no such column in the implementation table.

(ii) If only the minterm in the second row is encircled (column 1 and 3) then A must be applied to that data input. Hence, we must apply A to the D_0 and D_2 inputs.

(iii) If only the minterm in the first row is encircled, (columns 2 and 4), then $\overline{A}$ must be connected to that data input. Hence, we should apply $\overline{A}$ to the D_1 and D_3 inputs.

(iv) If both the minterms in a column are encircled, then we apply a logic 1 to the corresponding data input. Here, it may be noted that there is no such column in our implementation table.

(v) Finally, we draw the logic diagram.

The logic diagram has been shown in figure 5.76.

EXAMPLE 5.16. Implement a full adder using 8 : 1 multiplexer.

Solution: (i) First, let us write the truth table of a full adder.

The truth table of a full adder has been shown in Table 5.19.

Table 5.19

Inputs			Outputs	
A	B	C_{in}	Sum(S)	Carry (C)
0	0	0	0	0
0	0	1	1	0
0	1	0	1	0
0	1	1	0	1
1	0	0	1	0
1	0	1	0	1
1	1	0	0	1
1	1	1	1	1

(ii) Next, we identify the decimal number corresponding to each minterm.

The sum and carry outputs can be expressed in the standard SOP form as under :

$$S = \Sigma m (1, 2, 4, 7) \text{ and } C = \Sigma m (3, 5, 6, 7)$$

(iii) Finally, we realize the circuit using two 8 : 1 multiplexers.

We have to use two 8 : 1 multiplexers, one each for the sum and carry outputs. The realization has been shown in figure 5.77.

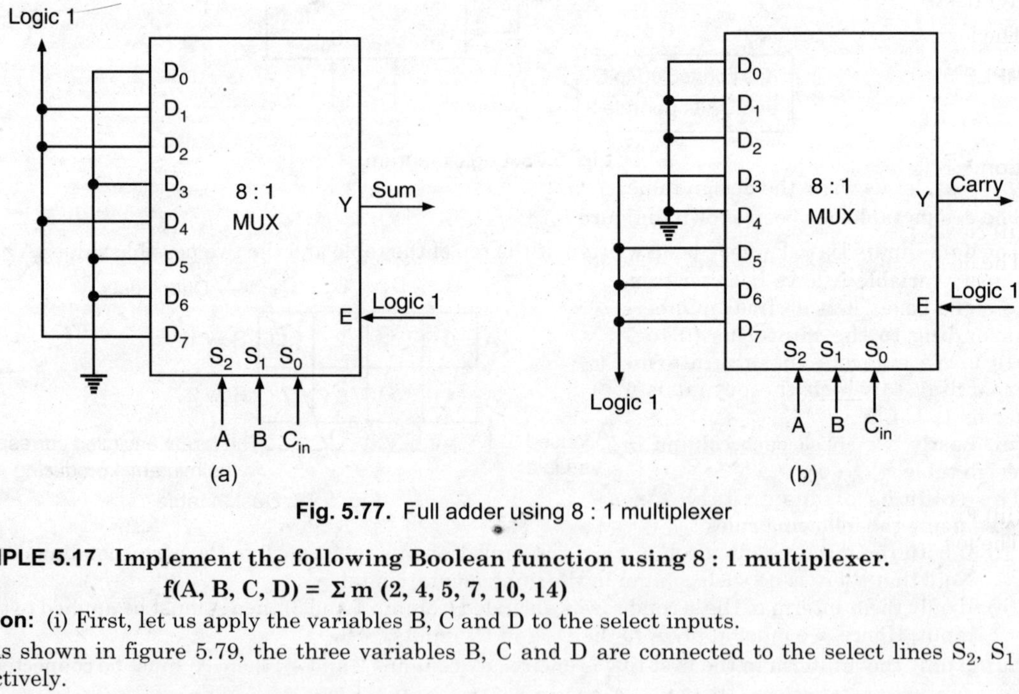

Fig. 5.77. Full adder using 8 : 1 multiplexer

EXAMPLE 5.17. Implement the following Boolean function using 8 : 1 multiplexer.

$$\mathbf{f(A, B, C, D) = \Sigma m (2, 4, 5, 7, 10, 14)}$$

Solution: (i) First, let us apply the variables B, C and D to the select inputs.

As shown in figure 5.79, the three variables B, C and D are connected to the select lines S_2, S_1 and S_0 respectively.

(ii) Next, we write the design table.

The design table has been shown in figure 5.78.

Four steps to achievement: Plan purposefully, prepare prayerfully, proceed positively, pursuspersistently.

— William Arthur Ward

	D_0	D_1	D_2	D_3	D_4	D_5	D_6	D_7	
$\overline{A}$	0	1	(2)	3	(4)	(5)	6	(7)	Row 1
A	8	9	(10)	11	12	13	(14)	15	Row 2
Input to MUX :	0	0	1	0	$\overline{A}$	$\overline{A}$	A	$\overline{A}$	

Fig. 5.78. Design table

In the sixteen boxes, we have entered the decimal numbers corresponding to the minterms 0 to 15 serially use. Encircle those minterms correspond to an output = 1 (minterms, 2, 4, 5, 7, 10, 14).

(iii) Next, we inspect each column in the design table.

(a) If both the minterms in a column are not circled, then we apply logic 0 to the corresponding data input. Here, it may be noted that there is no such column in our implementation table.

(b) If only the minterm in the second row is encircled then A must be applied to that data input. Hence, we must apply A to D_6 input.

(c) If only the minterm in the first row is encircled, then $\overline{A}$ must be connected to that data input. Here, it may noted that there is no such column in our implementation table.

(d) If both the minterms in a column are encircled, then we apply a logic 1 to the corresponding data 1 input. D_2 is the column where both rows are encircled.

(iv) Lastly, we draw the logic diagram.

The logic diagram has been shown in figure 5.79.

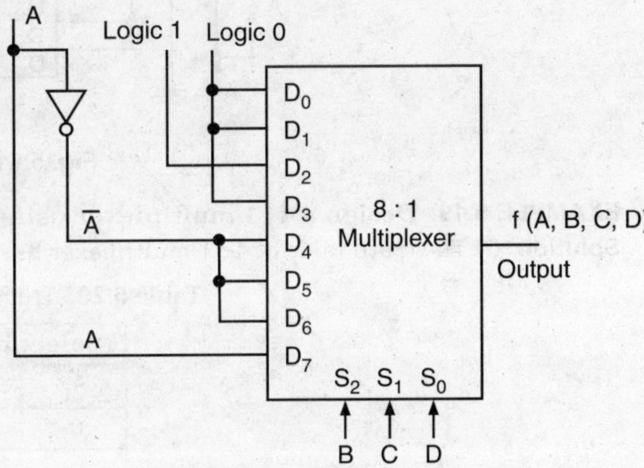

Fig. 5.79. Logic diagram

EXAMPLE 5.18. Implement the following function using 16 : 1 multiplexer :

$$f(A, B, C, D, E) = \Sigma m\ (2, 4, 5, 7, 10, 14, 15, 16, 17, 25, 26, 30, 31)$$

(PTU, Jalandhar, Sem. Exam., 2004-05)

Solution: (i) First, let us apply the variable B, C, D, E to the select inputs.

As shown in figure 5.80, the variable B, C, D, E are applied to the select lines S_3, S_2, S_1, S_0 respectively.

(ii) Next, we write the design table.

The design table has been shown in figure 5.80.

Inputs	D_0	D_1	D_2	D_3	D_4	D_5	D_6	D_7	D_8	D_9	D_{10}	D_{11}	D_{12}	D_{13}	D_{14}	D_{15}
$\overline{A}$	(0)	(1)	(2)	(3)	4	5	(6)	7	(8)	(9)	(10)	11	12	(13)	14	(15)
A	16	(17)	18	19	(20)	21	22	23	(24)	25	26	27	28	29	30	31
Input to MUX	$\overline{A}$	1	$\overline{A}$	$\overline{A}$	A	0	$\overline{A}$	0	1	$\overline{A}$	$\overline{A}$	0	0	$\overline{A}$	0	$\overline{A}$

Fig. 5.80. Design table

(iii) Finally, we realize the logic circuit

The logic circuit has been shown in figure 5.81.

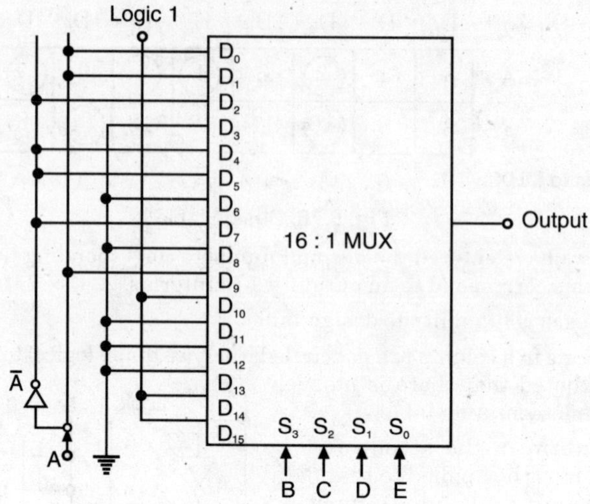

Fig. 5.81. Logic diagram

EXAMPLE 5.19. Design a 4 : 1 multiplexer using only NAND gates.

Solution: (i) The truth table of 4 : 1 multiplexer has been shown in Table 5.20.

Table 5.20. Truth table of 4 : 1 multiplexer

Select inputs		Output
S_1	S_0	Y
0	0	D_0
0	1	D_1
1	0	D_2
1	1	D_3

(ii) The Boolean expression for the output is given by

$$Y = \bar{S}_1 \bar{S}_0 D_0 + \bar{S}_1 S_0 D_1 + S_1 \bar{S}_0 D_2 + S_1 S_0 D_0$$

Let $Y = A + B + C + D$

Taking double bar of RHS, we obtain

$$Y = \overline{\overline{A+B+C+D}} = \overline{\bar{A} \cdot \bar{B} \cdot \bar{C} \cdot \bar{D}} \quad \text{(DeMorgan's theorem)}$$

Substituting the values of A, B, C and D, we obtain

$$Y = \overline{\left(\overline{\bar{S}_1 \bar{S}_0 D_0}\right) \cdot \left(\overline{\bar{S}_1 S_0 D_1}\right) \cdot \left(\overline{S_1 \bar{S}_0 D_2}\right) \cdot \left(\overline{S_1 S_0 D_3}\right)}$$

(iii) Finally, we implement this using only NAND gates as shown in figure 5.82.

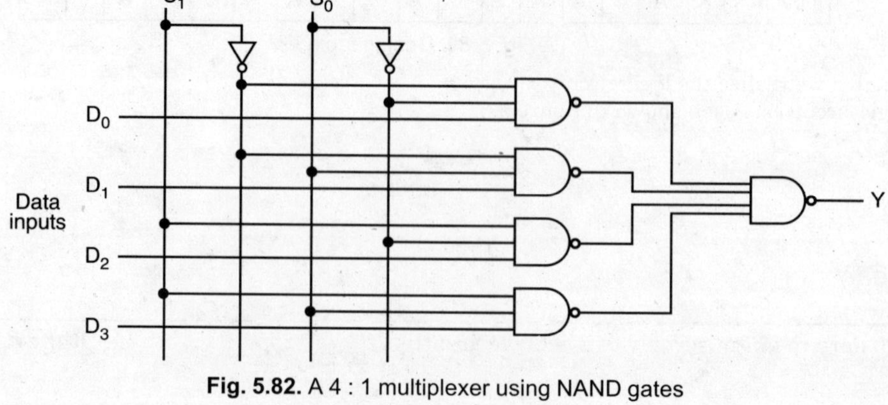

Fig. 5.82. A 4 : 1 multiplexer using NAND gates

EXAMPLE 5.20. Implement the following Boolean function using all 4 : 1 multiplexers

$$f(A, B, C, D, E) = \Sigma m\ (0, 1, 2, 3, 6, 8, 9, 10, 13, 15, 17, 20, 24)$$

Solution: We shall use the 16 : 1 multiplexer implemented in Example 5.18 in order to implement the given expression.

(i) Let us write the design table.

The design table is as shown below.

Table 5.21. Design table

Inputs	D_0	D_1	D_2	D_3	D_4	D_5	D_6	D_7	D_8	D_9	D_{10}	D_{11}	D_{12}	D_{13}	D_{14}	D_{15}
$\overline{A}$	(0)	(1)	(2)	(3)	4	5	(6)	7	(8)	(9)	(10)	11	12	(13)	14	(15)
A	16	(17)	18	19	(20)	21	22	23	(24)	25	26	27	28	29	30	31
Input to MUX	$\overline{A}$	1	$\overline{A}$	$\overline{A}$	A	0	$\overline{A}$	0	1	$\overline{A}$	$\overline{A}$	0	0	$\overline{A}$	0	$\overline{A}$

(ii) Next, we Implement the function.

The implementation using 4 : 1 multiplexers has been shown in figure 5.83.

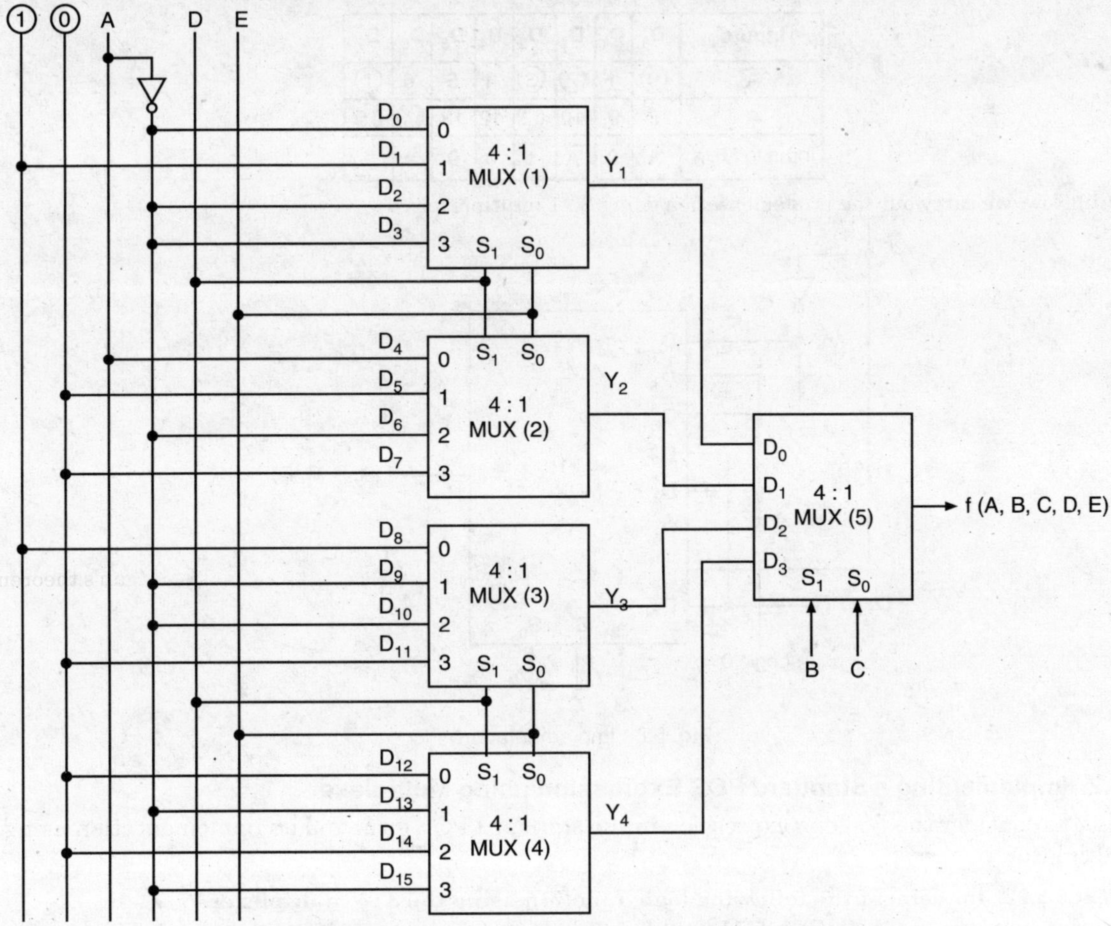

Fig. 5.83. Given function is implemented using all 4 : 1 multiplexers

5.23.1 Implementation of a Logical Expression in the Non-standard Form

Upto now, we have discussed the implementation of expressions in the standard SOP form.

However, if the given expression is not in the standard form, then we have to first bring it to the standard form and then proceed. Example 5.21 illustrates this concept.

EXAMPLE 5.21. Implement the following Boolean expression using 8 : 1 multiplexer :

$$f(A, B, C, D) = \overline{A}\overline{B}\overline{D} + ABC + \overline{B}CD + \overline{A}CD$$

Solution: (i) First, let us convert the given expression to standard SOP form.

we have, $f(A, B, C, D) = \overline{A}\overline{B}\overline{D}(C + \overline{C}) + ABC(D + \overline{D}) + \overline{B}CD(A + \overline{A}) + \overline{A}CD(B + \overline{B})$

$= \overline{A}\overline{B}C\overline{D} + \overline{A}\overline{B}\overline{C}\overline{D} + ABCD + ABC\overline{D} + A\overline{B}CD + \overline{A}\overline{B}CD + \overline{A}BCD + \overline{A}\overline{B}CD$

This is the expression in the standard SOP form.

Therefore, $f(A, B, C, D) = \Sigma m\,(2, 0, 15, 14, 11, 3, 7, 3)$

Since, $X + X = X$, Therefore, $\overline{A}\overline{B}CD + \overline{A}\overline{B}CD = \overline{A}\overline{B}CD$

Therefore, $f(A, B, C, D) = \Sigma m\,(0, 2, 3, 7, 11, 14, 15)$

(ii) Next, we write the design table.

Table 5.22. Design table

Inputs	D_0	D_1	D_2	D_3	D_4	D_5	D_6	D_7
$\overline{A}$	(0)	1	(2)	(3)	4	5	6	(7)
A	8	9	10	(11)	12	13	(14)	(15)
Input to MUX	$\overline{A}$	0	$\overline{A}$	1	0	0	A	1

(iii) Now, we carry out the implementation using 8 : 1 multiplexer.

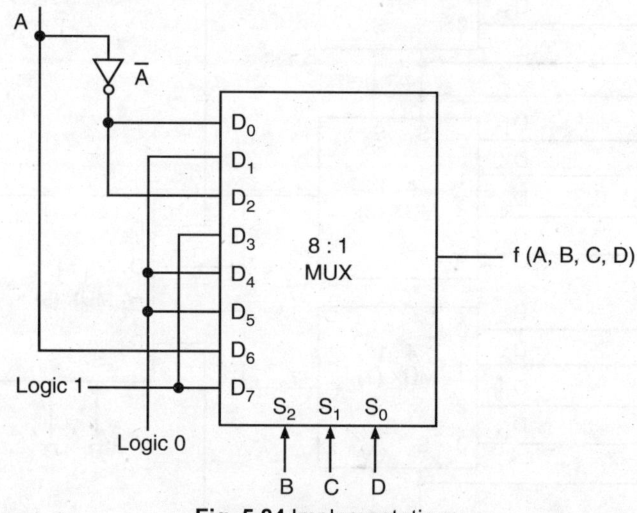

Fig. 5.84 Implementation

5.23.2. Implementing a Standard POS Expression using Multiplexer

Let us consider the Boolean expression in the standard POS form and its implementation using a multiplexer.

EXAMPLE 5.22. Implement the following logic function using the 8 : 1 multiplexer.

$$f(A, B, C) = \Pi M\,(0, 1, 3, 5, 7)$$

Solution: In the given expression, the maxterms have been specified. Hence we must connect the corresponding data input (D_0, D_1, D_3, D_5 and D_7) to logic 0. Also, we connect the remaining data inputs to logic 1. Then, we connect the input variables A, B, C to the select inputs S_2 S_1 and S_0 respectively. The logic diagram has been shown in figure 5.85.

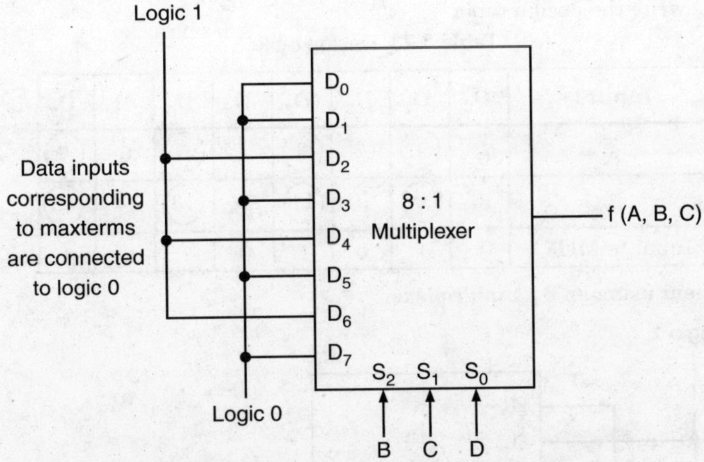

Fig. 5.85. Implementation of standard POS expression

EXAMPLE 5.23. Implement the following logic function using 4 : 1 multiplexer

$$f(A, B, C) = \Pi M\,(0, 1, 3, 5, 7)$$

Solution: Let us connect B and C to the select inputs S_1 and S_0 respectively. For the connections of data inputs (D_0 to D_3) and A, we write down the design table as shown in figure 5.86.

	D_0	D_1	D_2	D_3	
$\bar{A}$	0	1	(2)	3	Row 1
A	(4)	5	(6)	7	Row 2
Input to MUX:	A	0	1	0	

We encircle the maxterms which are not included in the given boolean expression

Fig. 5.86. Design table

Here, it may be noted that as the given expression is in the POS form, we must encircle those maxterms which are not included in the given boolean expression. The rules for deciding the input to the data lines (D_0 to D_3) are, however, the same as those stated earlier for the SOP form. The logic diagram has been shown in figure 5.87.

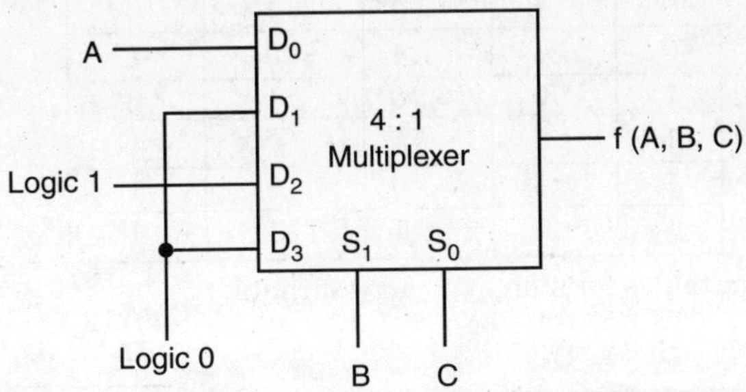

Fig. 5.87. Implementation using 4 : 1 multiplexer

EXAMPLE 5.24. Implement the following function using an 8 : 1 multiplexer.

$$f(A, B, C, D) = \Pi M\,(0, 2, 4, 6, 8, 10, 12, 14)$$

Solution: (i) First, let us write the design table.

Table 5.23. Design table

Inputs	D_0	D_1	D_2	D_3	D_4	D_5	D_6	D_7
$\overline{A}$	0	1	2	3	4	5	6	7
A	8	9	10	11	12	13	14	15
Input to MUX	0	1	0	1	0	1	0	1

(ii) Next, we implement using an 8 : 1 multiplexer.

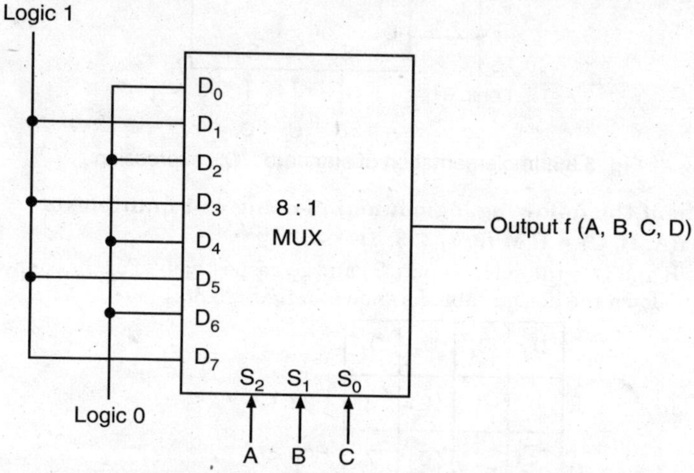

Fig. 5.88. Implementation using 8 : 1 multiplexer

EXAMPLE 5.25. Implement a full adder circuit using two 4 : 1 multiplexer.

Solution: (i) First, let us write the truth table of full adder.

Table 5.24. Truth table of a full adder

Inputs			Output	
A	B	C_{in}	Sum	Carry
0	0	0	0	0
0	0	1	1	0
0	1	0	1	0
0	1	1	0	1
1	0	0	1	0
1	0	1	0	1
1	1	0	0	1
1	1	1	1	1

(ii) Now, we write the design tables for sum and carry outputs.

Sum

	D_0	D_1	D_2	D_3
$\overline{A}$	0	1	2	3
A	4	5	6	7
Input to Mux	A	$\overline{A}$	$\overline{A}$	A

Carry

	D_0	D_1	D_2	D_3
$\overline{A}$	0	1	2	3
A	4	5	6	7
Input to Mux	0	A	A	1

Fig. 5.89. Design tables for sum and carry outputs

(iii) Finally, we draw the logic diagram.

The full adder using 4 : 1 multiplexer has been shown in figure 5.90.

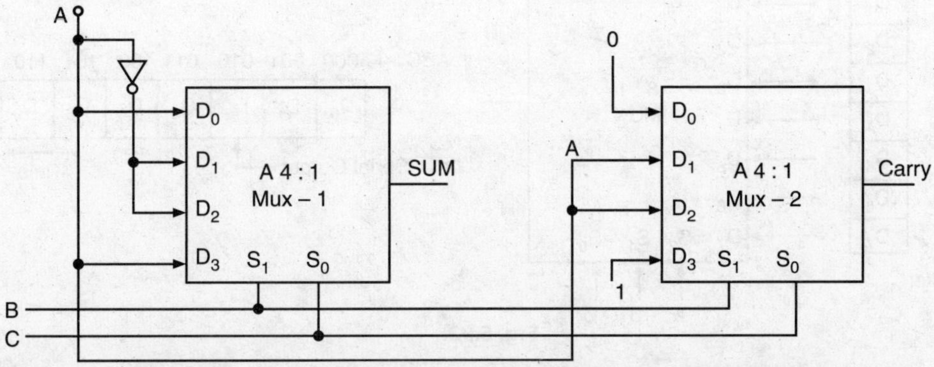

Fig. 5.90. Implementation of a full adder using two 4 : 1 multiplexer

5.23.3. Implementation of Boolean Expression with Don't Care Conditions

Sometimes the Boolean expressions are given with don't care conditions. We know that the don't care conditions can be treated as logic 1's or 0's. Example 5.26 illustrates the use of multiplexer to implement such equations.

EXAMPLE 5.26. Implement the following Boolean expression using an 8 : 1 multiplexer

$$f(A, B, C, D) = \Sigma\, m\,(1, 3, 5, 10, 11, 13, 14) + d(0, 2).$$

Solution: The don't care conditions are assumed to be logic 1's. The design table is shown in figure 5.91(a) and the corresponding logic diagram has been shown in figure 5.91(b).

	D_0	D_1	D_2	D_3	D_4	D_5	D_6	D_7
$\bar{A}$	(0)	(1)	(2)	(3)	4	(5)	6	7
A	8	9	(10)	(11)	12	(13)	(14)	15
Input to Mux	$\bar{A}$	$\bar{A}$	1	1	0	1	A	0

(a) Design table

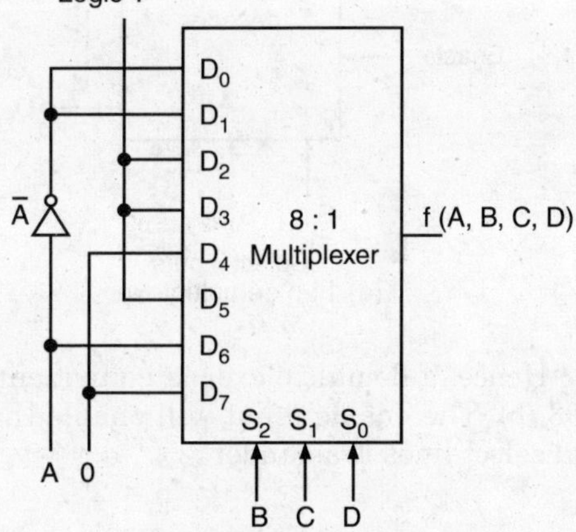

(b) Implementation

Fig. 5.91.

5.23.4. Multiplexer as parallel to Serial Converter

Figure 5.92 shows the use of 8 : 1 MUX as a parallel to serial converter. The select lines are applied to the output of a 3-bit binary counter. Therefore, S_2 S_1 S_0 will change from 000 to 111 continuously. For each combination of select lines, one data input (D_0 through D_7) will be connected to the output. Hence, after one complete cycle of operation, the parallel input word (D_0 D_7) gets converted into a serial word as shown in figure 5.92.

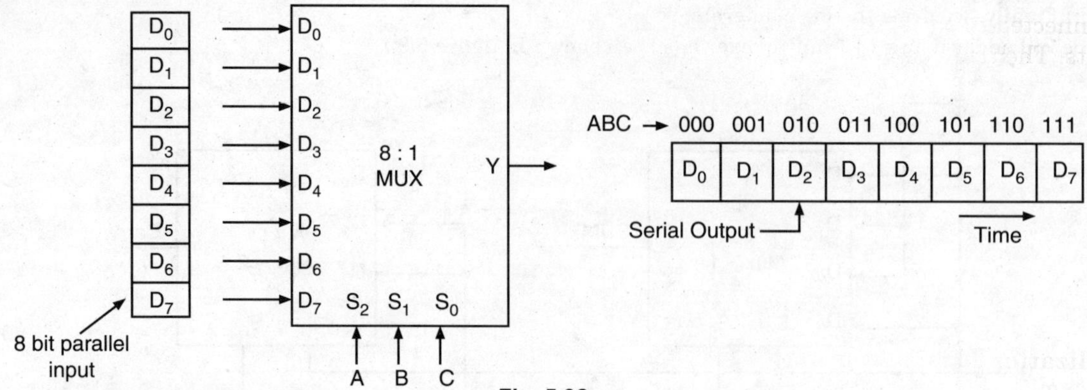

Fig. 5.92.

5.24. DEMULTIPLEXER

The block diagram of a demultiplexer or decoder is shown in figure 5.93 (a). It has only one input, n outputs, and m select inputs. A demultiplexer performs the reverse operation of a multiplexer, i.e., it receives one input and distributes it over several outputs. At a time, only one output line is selected by the select lines and the input is transmitted to the select output line.

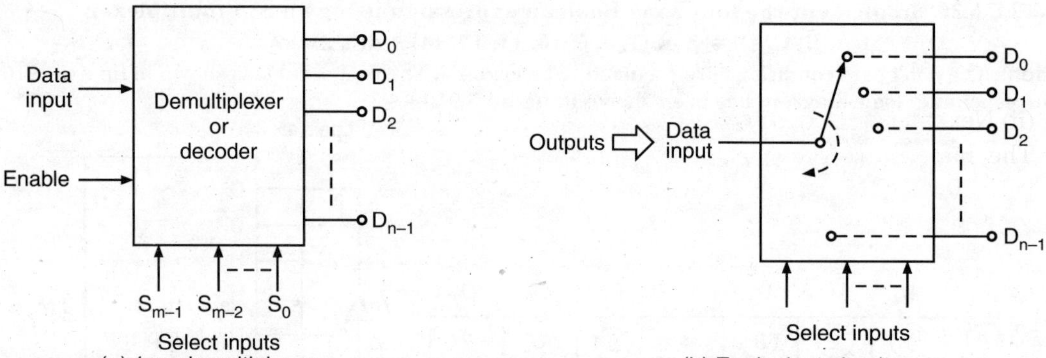

(a) 1 : n demultiplexer (b) Equivalent circuit

Fig. 5.93.

Hence, a demultiplexer is equivalent to a single pole multiple way switch as shown in figure 5.93 (b). The enable input will enable the demultiplexer. The relation between the n output lines and select lines is as under :

$$n = 2^m.$$

5.25. CLASSIFICATION OF DEMULTIPLEXERS **(PTU, Jalandhar, Sem. Exam., 2004-05)**

Similar to the multiplexers, the demultiplexers may also be classified as under :

(i) 1 : 2 demultiplexer
(ii) 1 : 4 demultiplexer
(iii) 1 : 8 demultiplexer
(iv) 1 : 16 demultiplexer

5.25.1. A 1 : 2 Demultiplexer

A 1 : 2 demultiplexer has one data input D_{in}, one select input S_0, one enable (E) input and two outputs Y_0 and Y_1. Also, D_{in} is connected to Y_0 if $S_0 = 0$ and $E = 1$. Similarly, D_{in}

DO YOU KNOW?

Demultiplexers are used at the other end of the data path to separate the signals that are sharing a data path and distribute them to their respective destinations.

is connected to Y_1 if $S_0 = 1$ and $E = 1$. If $E = 0$, then both the outputs will be 0 irrespective of the inputs. The truth table of the 1 : 2 demultiplexer is as under :

Table 5.25. Truth table of demux 1 : 2

Enable	Select	Outputs	
E	S_0	Y_1	Y_0
0	×	0	0
1	0	0	D_{in}
1	1	D_{in}	0

Realization of 1 : 2 Demux

(i) To realize the 1 : 2 demultiplexer using gates, we have to write the truth table as under :

Table 5.26. Truth table of demux 1 : 2

Enable	Data	Select	Outputs	
E	D_{in}	S_0	Y_1	Y_0
0	×	×	0	0
1	0	0	0	0
1	1	0	0	1
1	0	1	0	0
1	1	1	1	0

(rows 2–3) ⟶ Y_0 is connected to D_{in}

(rows 4–5) ⟶ Y_1 is connected to D_{in}

(ii) Next, we write the K-maps.

The maps for the two outputs are as under :

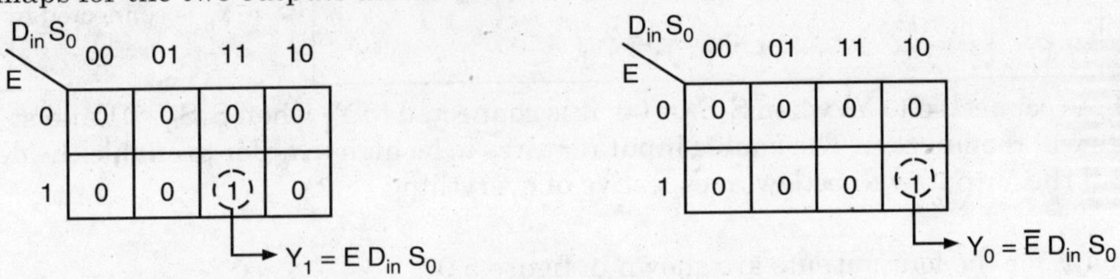

Fig. 5.94. K-maps

Thus, the expressions for Y_0 and Y_1 are as under :

$$Y_0 = ED_{in}\,\bar{S}_0 \quad \text{and} \quad Y_1 = E\,D_{in}\,S_0$$

The 1 : 2 demux using gates has been shown in figure 5.95.

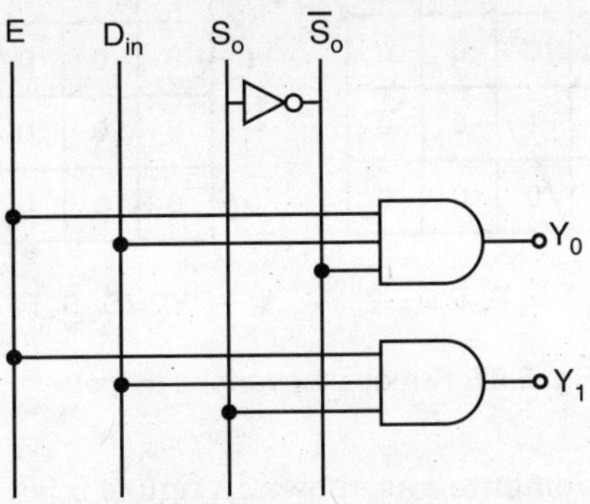

Fig. 5.95. 1 : 2 demux using gates

5.25.2. A 1: 4 Demultiplexer

The 1 : 4 demultiplexer has been shown in figure 5.96 and its truth table is given in Table 5.27.

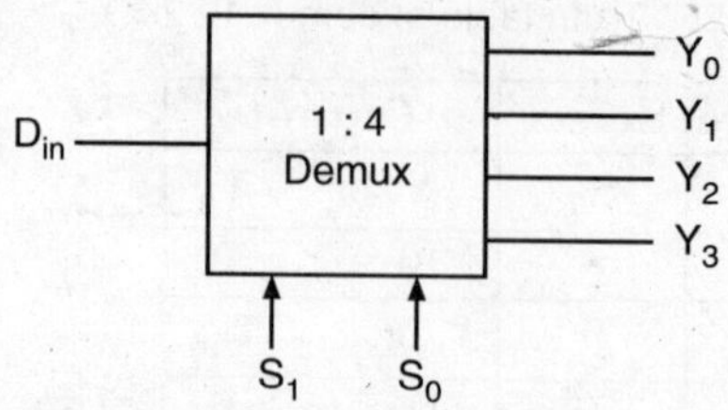

Fig. 5.96. Block diagram of 1 : 4 demultiplexer

Table 5.27. Truth table for 1 : 4 demultiplexer

		Enable		Outputs				
E	D_{in}	S_1	S_0	Y_0	Y_1	Y_2	Y_3	
1	0	0	0	0	0	0	0	Y_0 is connected to D_{in}
1	1	0	0	1	0	0	0	
1	0	0	1	0	0	0	0	Y_1 is connected to D_{in}
1	1	0	1	0	1	0	0	
1	0	1	0	0	0	0	0	Y_2 is connected to D_{in}
1	1	1	0	0	0	1	0	
1	0	1	1	0	0	0	0	Y_3 is connected to D_{in}
1	1	1	1	0	0	0	1	

Here, D_{in} is connected to Y_0 when $S_1S_0 = 00$, it is connected to Y_1 when $S_1S_0 = 01$ and so on. The other outputs will remain zero. The enable input requires to be high in order to enable the demux. If $E = 0$, then all the outputs will be low irrespective of everything.

K-maps

The K-maps for the four outputs are shown in figure 5.97.

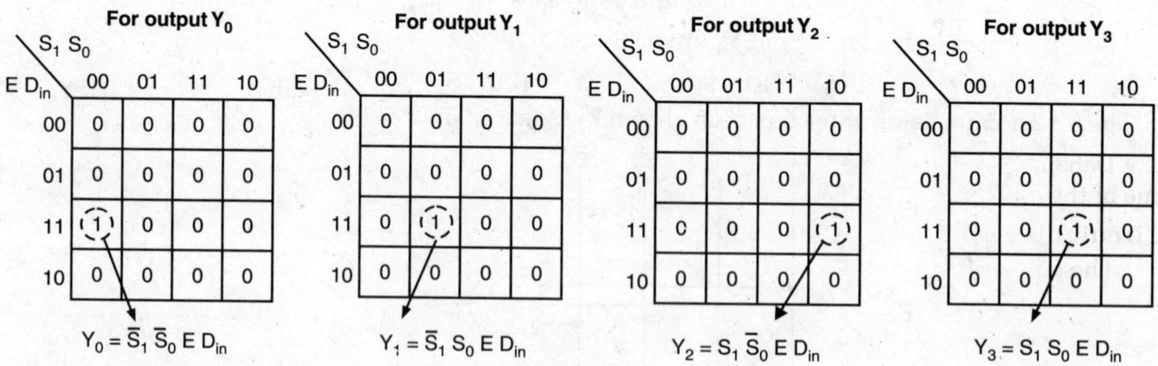

Fig. 5.97. K-maps for various outputs

Implementation

The 1 : 4 Demux has been implemented as shown in figure 5.98.

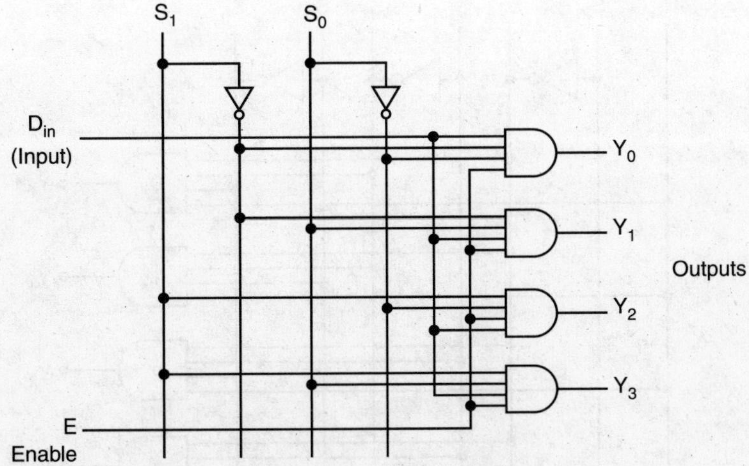

Fig. 5.98. A 1 : 4 demultiplexer

5.25.3. A 1 : 8 Demultiplexer

The 1 : 8 demux is shown in figure 5.99 (a). It has one data input, eight outputs, three select inputs and an enable input E. The truth table is shown in figure 5.99(b).

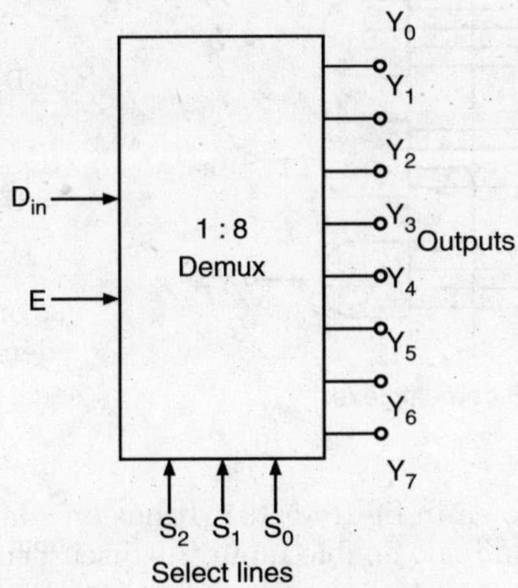

(*a*) 1 : 8 Demux

Enable	Select			Outputs							
E	**S_2**	**S_1**	**S_0**	**Y_7**	**Y_6**	**Y_5**	**Y_4**	**Y_3**	**Y_2**	**Y_1**	**Y_0**
0	×	×	×	0	0	0	0	0	0	0	0
1	0	0	0	0	0	0	0	0	0	0	D_{in}
1	0	0	1	0	0	0	0	0	0	D_{in}	0
1	0	1	0	0	0	0	0	0	D_{in}	0	0
1	0	1	1	0	0	0	0	D_{in}	0	0	0
1	1	0	0	0	0	0	D_{in}	0	0	0	0
1	1	0	1	0	0	D_{in}	0	0	0	0	0
1	1	1	0	0	D_{in}	0	0	0	0	0	0
1	1	1	1	D_{in}	0	0	0	0	0	0	0

(*b*) Truth table for 1 : 8 demux

Fig. 5.99.

Depending upon the combination of the select inputs S_2 S_1 S_0, the data input D_{in} is connected to one of the eight outputs. As an example, if S_2 S_1 S_0 = 1 1 0, then D_{in} is connected to output Y_6.

Circuit diagram

The circuit diagram of 1 : 8 demultiplexer has been shown in figure 5.100.

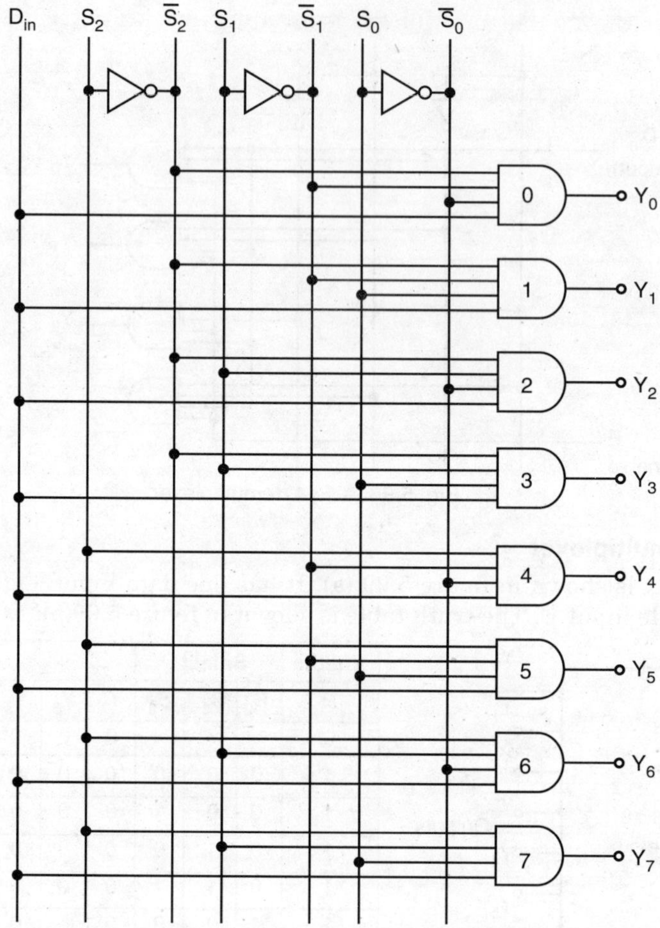

Fig. 5.100. Logic circuit of a 1 : 8 demultiplexer

5.25.4. A 1 : 16 Demultiplexer

The 1 : 16 demultiplexer block diagram has been shown in figure 5.101. It has one data input D_{in}, four select inputs S_3, S_2, S_1 and S_0 sixteen outputs and one enable input. As discussed earlier, the combination of select inputs will select one of the sixteen outputs and connect it to D_{in}.

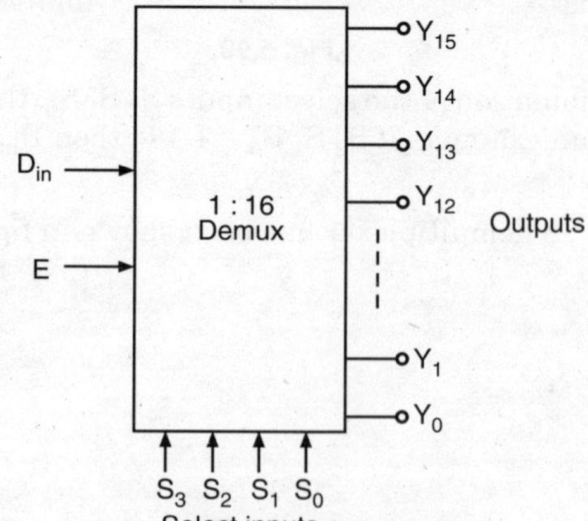

Fig. 5.101. 1 : 16 demultiplexer

The truth table of the 1 : 16 demultiplexer is as under :

Table 5.28. Truth table of 1 : 16 demultiplexer

Select Inputs				Outputs															
S_3	S_2	S_1	S_0	Y_{15}	Y_{14}	Y_{13}	Y_{12}	Y_{11}	Y_{10}	Y_9	Y_8	Y_7	Y_6	Y_5	Y_4	Y_3	Y_2	Y_1	Y_0
0	0	0	0	0	0	0	0	0	0	0	0	0	0	0	0	0	0	0	D_{in}
0	0	0	1	0	0	0	0	0	0	0	0	0	0	0	0	0	0	D_{in}	0
0	0	1	0	0	0	0	0	0	0	0	0	0	0	0	0	0	D_{in}	0	0
0	0	1	1	0	0	0	0	0	0	0	0	0	0	0	0	D_{in}	0	0	0
0	1	0	0	0	0	0	0	0	0	0	0	0	0	0	D_{in}	0	0	0	0
0	1	0	1	0	0	0	0	0	0	0	0	0	0	D_{in}	0	0	0	0	0
0	1	1	0	0	0	0	0	0	0	0	0	0	D_{in}	0	0	0	0	0	0
0	1	1	1	0	0	0	0	0	0	0	0	D_{in}	0	0	0	0	0	0	0
1	0	0	0	0	0	0	0	0	0	0	D_{in}	0	0	0	0	0	0	0	0
1	0	0	1	0	0	0	0	0	0	D_{in}	0	0	0	0	0	0	0	0	0
1	0	1	0	0	0	0	0	0	D_{in}	0	0	0	0	0	0	0	0	0	0
1	0	1	1	0	0	0	0	D_{in}	0	0	0	0	0	0	0	0	0	0	0
1	1	0	0	0	0	0	D_{in}	0	0	0	0	0	0	0	0	0	0	0	0
1	1	0	1	0	0	D_{in}	0	0	0	0	0	0	0	0	0	0	0	0	0
1	1	1	0	0	D_{in}	0	0	0	0	0	0	0	0	0	0	0	0	0	0
1	1	1	1	D_{in}	0	0	0	0	0	0	0	0	0	0	0	0	0	0	0

Circuit diagram

The circuit diagram of 1 : 16 demultiplexer has been shown in figure 5.102.

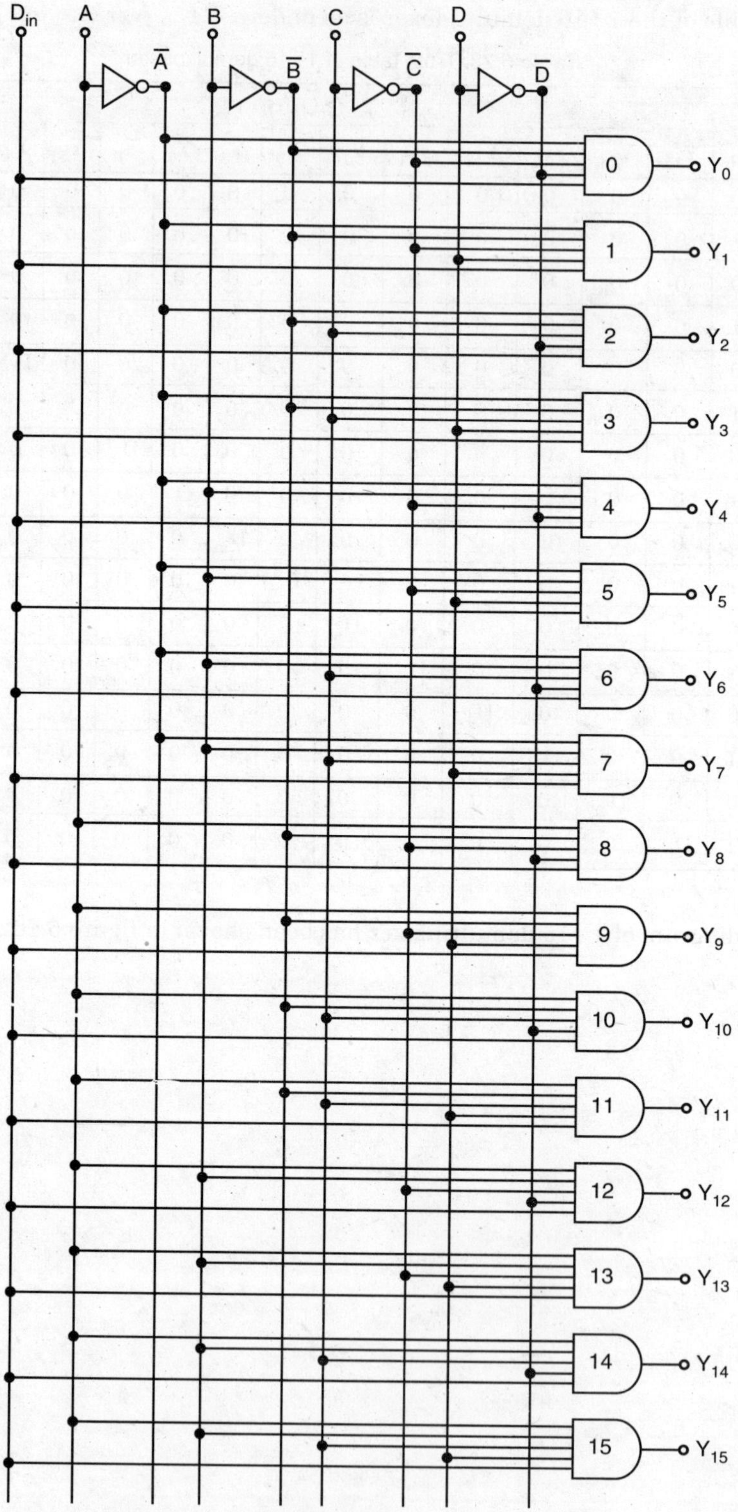

Fig. 5.102. Circuit diagram of 1 : 16 demultiplexer

5.26. CONCEPT OF A DEMULTIPLEXER TREE

Similar to multiplexer, we can construct the demultiplexer with more number of lines using demultiplexers having lower number lines. This is called as demultiplexer tree. It is also called as cascading of demultiplexers. This concept will be very obvious by solving the following examples.

EXAMPLE 5.27. Obtain a 1 : 4 line demux using 1 : 2 demultiplexers.

Solution: Let us consider 5.103 for the solution. The select lines S_0 of both the 1 : 2 demultiplexers are connected together and brought out as S_0 line of the 1 : 4 demux. The second select line S_1 is connected directly to the enable (E_2) input of demux 2 whereas inverted S_1 is connected to the enable input of demux-1.

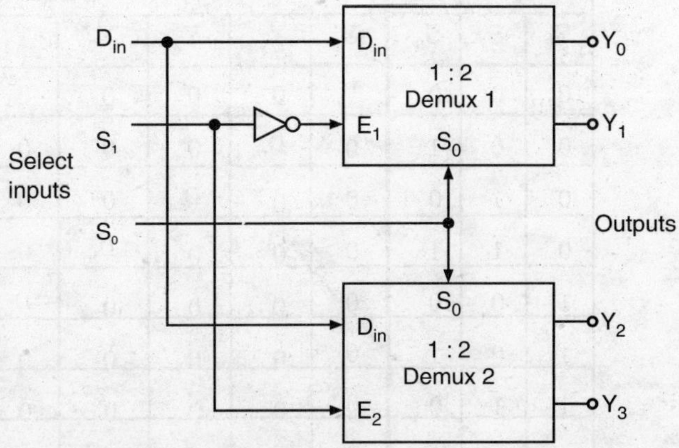

Fig. 5.103 1 : 4 demultiplexer using 1 : 2 demultiplexers

The truth table of the circuit has been shown in the following table.

Table 5.29. Truth table of 1 : 4 demux using two 1 : 2 demux

Select inputs		Outputs				
S_1	S_0	Y_3	Y_2	Y_1	Y_0	
0	0	0	0	0	D_{in}	Demux - 1 enabled
0	1	0	0	D_{in}	0	
1	0	0	D_{in}	0	0	Demux - 2 enabled
1	1	D_{in}	0	0	0	

EXAMPLE 5.28. Obtain a 1 : 8 line demultiplexer using two 1 : 4 line demultiplexers.

Solution: The 1 : 8 demultiplexer using two 1 : 4 demultiplexers has been shown in figure 5.104. This is similar to cascading of multiplexers. The select line S_1 and S_0 of the two 1 : 4 demultiplexers are connected in parallel with each other and S_2 is used for selecting one of the two 1 : 4 demultiplexers. S_2 is connected directly to Enable (E) input of Demux-2 whereas inverted S_2 is connected to the enable input of demux-1. The truth table of this circuit has been shown in Table 5.30.

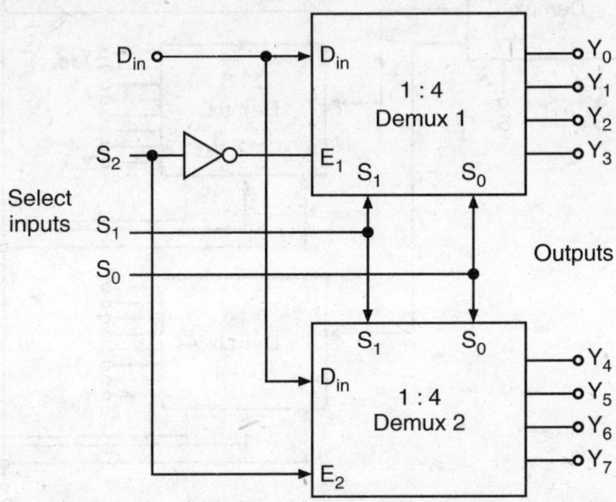

Fig. 5.104. A 1 : 8 demultiplexer using two 1 : 4 demultiplexers

Table 5.30. Truth table of 1 : 8 demultiplexer using two 1 : 4 demultiplexers

Select inputs			Outputs								
S_2	S_1	S_0	Y_0	Y_1	Y_2	Y_3	Y_4	Y_5	Y_6	Y_7	
0	0	0	D_{in}	0	0	0	0	0	0	0	Demux – 1 enabled
0	0	1	0	D_{in}	0	0	0	0	0	0	
0	1	0	0	0	D_{in}	0	0	0	0	0	
0	1	1	0	0	0	D_{in}	0	0	0	0	
1	0	0	0	0	0	0	D_{in}	0	0	0	Demux – 2 enabled
1	0	1	0	0	0	0	0	D_{in}	0	0	
1	1	0	0	0	0	0	0	0	D_{in}	0	
1	1	1	0	0	0	0	0	0	0	D_{in}	

EXAMPLE 5.29. Obtain 1 : 32 demultiplexer using 1 : 8 demultiplexers.

Solution: The required circuit diagram is shown in figure 5.105. Four 1 : 8 demux ICs are being used alongwith one 1 : 4 demux IC. Here, S_4, S_3, S_2, S_1 and S_0 are the five select inputs, D_{in} is the data inputs and Y_0 to Y_{31} are the 32 outputs.

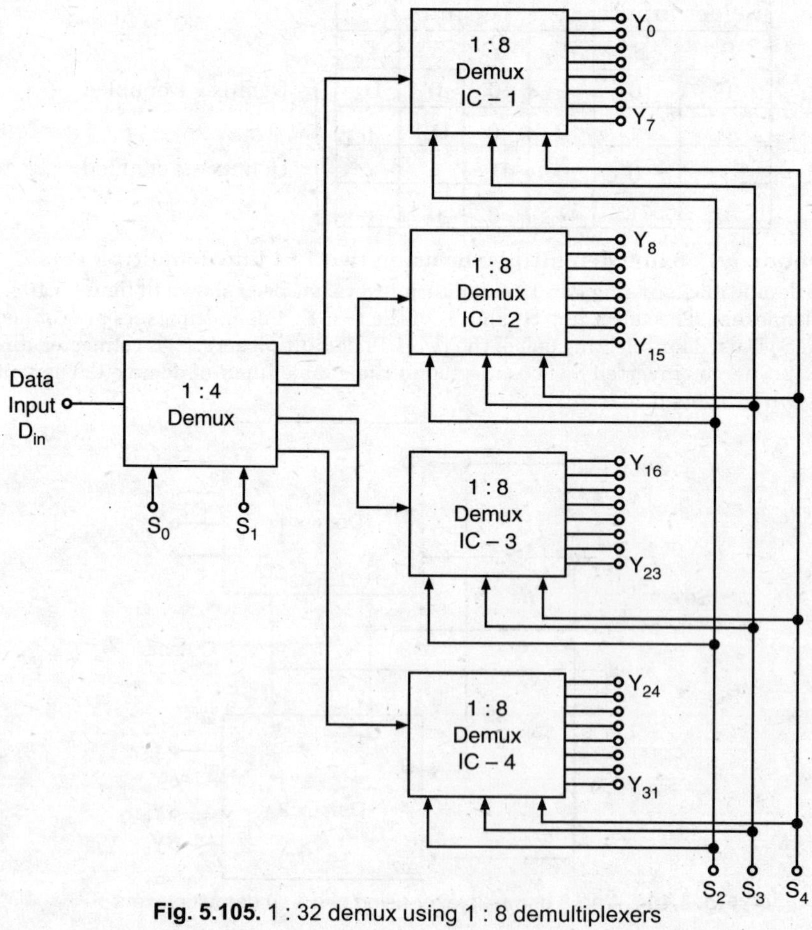

Fig. 5.105. 1 : 32 demux using 1 : 8 demultiplexers

EXAMPLE 5.30. Implement the full subtractor using a 1 : 8 demultiplexer.

Solution: The full subtractor has three inputs A, B and B_{in} and two outputs namely difference D and borrow out (B_{out}). The truth table of a full subtractor is as under :

Table 5.31. Truth table of a full subtractor

Inputs			Outputs	
A	B	B_{in}	D	B_{out}
0	0	0	0	0
0	0	1	1	1
0	1	0	1	1
0	1	1	0	1
1	0	0	1	0
1	0	1	0	0
1	1	0	0	0
1	1	1	1	1

From the truth table, we can express the difference and borrow out outputs in standard SOP forms as under :

$$D = f(A, B, B_{in}) = \Sigma m\ (1, 2, 4, 7)$$

and
$$B_{out} = f(A, B, B_{in}) = \Sigma m\ (1, 2, 3, 7)$$

We connect D_{in} to logic 1 permanently and correct A, B and B_{in} to the select inputs S_2, S_1, S_0 respectively as shown in figure 5.106.

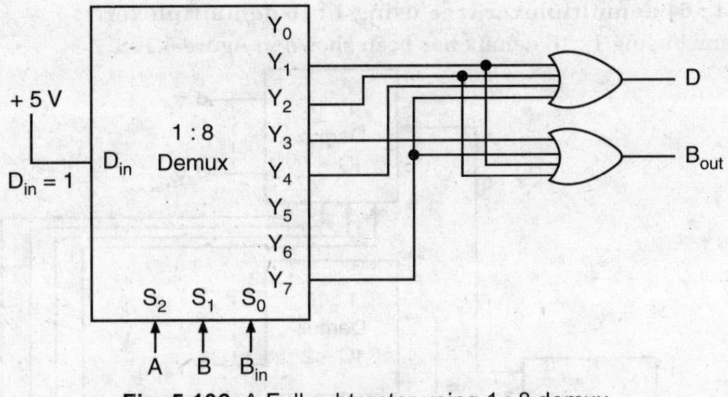

Fig. 5.106. A Full subtractor using 1 : 8 demux

As D_{in} is connected to 1, we get the required minterms at the Demux outputs. We have to OR the required minterms to obtain the D and B_{out} outputs as shown in figure 5.106.

EXAMPLE 5.31. Implement a full adder using demultiplexer

Solution: (i) First, let us write the truth table of the full adder.

Table 5.32. Truth table of a full adder

Inputs			Outputs	
A	B	C_{in}	Sum	Carry
0	0	0	0	0
0	0	1	1	0
0	1	0	1	0
0	1	1	0	1
1	0	0	1	0
1	0	1	0	1
1	1	0	0	1
1	1	1	1	1

(ii) Next, we write expressions for sum and carry.

$$\text{Sum} = \Sigma\, m\,(1, 2, 4, 7)$$

Therefore, outputs Y_1, Y_2, Y_4 and Y_7 should be ORed to get sum output

$$\text{carry} = \Sigma\, m\,(3, 5, 6, 7)$$

Therefore, outputs Y_3, Y_5, Y_6 and Y_7 should be ORed to get the carry output.

Finally, we carry out the implementation using a 1 : 8 demultiplexer.

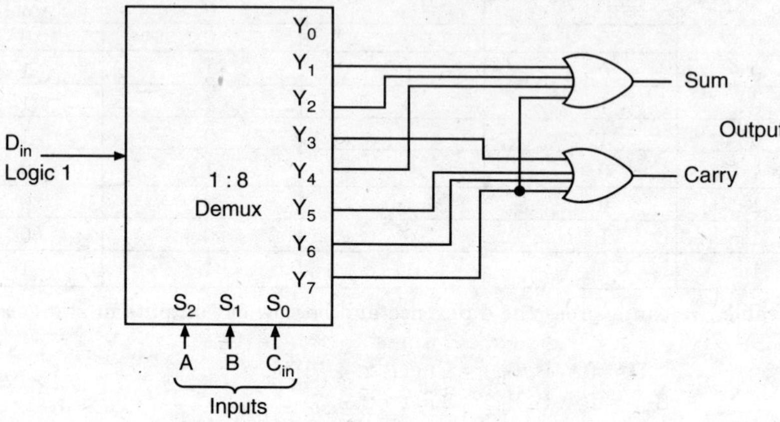

Fig. 5.107. A Full adder using a demultiplexer

EXAMPEL 5.32. Draw 1 : 64 demultiplexer tree using 1 : 16 demultiplexer.

Solution: The 1 : 64 Demux using 1 : 16 demux has been shown in figure 5.108.

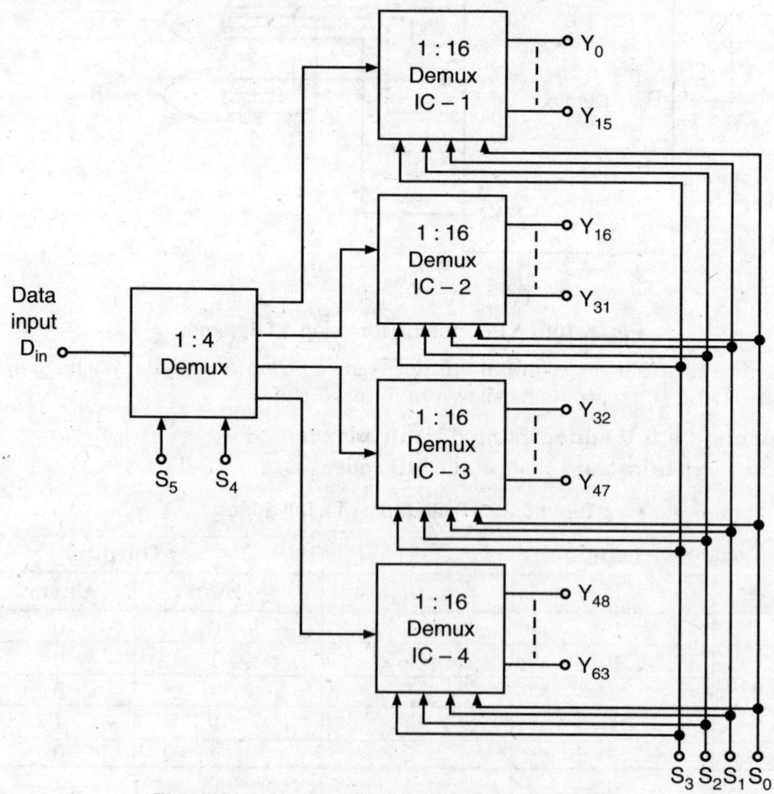

Fig. 5.108. A 1 : 64 Demux using 1 : 16 Demux

S_5 to S_0 are select inputs. Y_0 to Y_{63} outputs. D_{in} is data input.

5.27. PERFORMANCE COMPARISON OF MULTIPLEXER AND DEMULTIPLEXER

Table 5.33. Performance Comparison of MUX and DEMUX

S. No.	Parameter of comparison	Multiplexer	Demultiplexer
1.	Type of logic circuit	Combinational	Combinational
2.	Number of data inputs	n	1
3.	Number of select inputs	m	m
4.	Number of data output	1	n
5.	Relation between input/output lines and select lines	$n = 2^m$	$n = 2^m$
6.	Operation principle	Many to 1 or as data selector	1 to many or data distributor
7.	Application	(i) As a universal logic circuit, we can implement any combinational circuit using a MUX. (ii) In time division multiplexing at the sending end.	(i) We can implement some combinational circuit (ii) In TDM system at receiving end.

5.28. A DECODER

(PTU, Jalandhar, Sem. Exam., 2004-05)

A decoder is a combinational circuit. It converts the n-bit binary information at its into a maximum of 2^n output lines. For example, if n = 2, then, we can design upto 2 : 4 decoder.

5.28.1. A 2 to 4 Line Decoder

The block diagram of a 2 to 4 line decoder has been shown in figure 5.109. A and B are the two inputs whereas D_0 through D_3 are the four outputs. Table 5.34 shows the truth table which explains the operation of the decoder. It shows that each output is 1 for only a specific combination of inputs

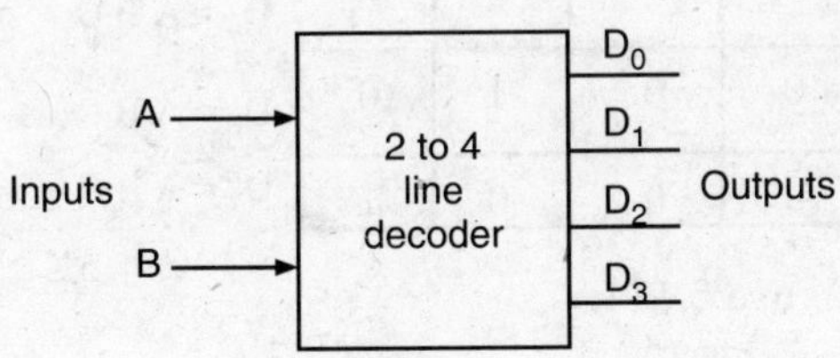

Fig. 5.109. Block diagram of a 2 line to 4 line decoder

Table 5.34. Truth table of a 2 line to 4 line decoder

Inputs		Outputs			
A	**B**	**D_0**	**D_1**	**D_2**	**D_3**
0	0	1	0	0	0
0	1	0	1	0	0
1	0	0	0	1	0
1	1	0	0	0	1

The Boolean expressions for the four outputs are as under :

$$D_0 = \overline{A}\,\overline{B} \qquad D_1 = \overline{A}\,B$$

$$D_2 = A\overline{B} \qquad \text{and} \quad D_3 = A\,B$$

The logic diagram for a 2 to 4 line decoder has been shown in figure 5.110.

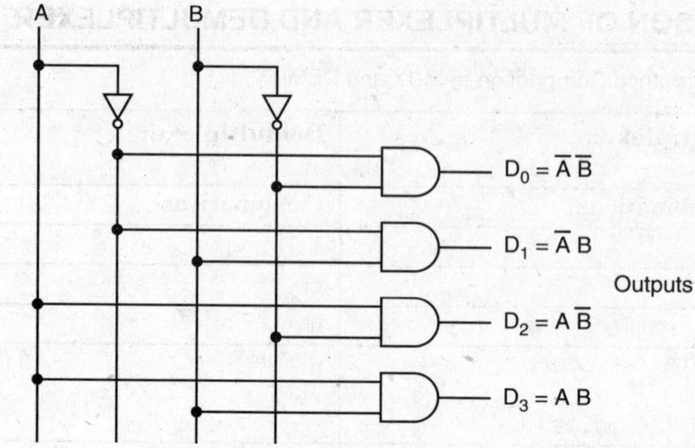

Fig. 5.110. A 2 to 4 line decoder

> **DO YOU KNOW?**
>
> Decoders are widely used in the memory system of a computer where they respond to the address code generated by the central processor to activate a particular memory location.

5.28.2. A 3 to 8 Line Decoder

The truth table of a 3 to 8 line decoder has been shown in Table 5.35.

Table 5.35. Truth table for 3 to 8 line decoder

Inputs			Outputs								
A	**B**	**C**	**D_0**	**D_1**	**D_2**	**D_3**	**D_4**	**D_5**	**D_6**	**D_7**	
0	0	0	1	0	0	0	0	0	0	0	$D_0 = \overline{A}\,\overline{B}\,\overline{C}$
0	0	1	0	1	0	0	0	0	0	0	$D_1 = \overline{A}\,\overline{B}\,C$
0	1	0	0	0	1	0	0	0	0	0	$D_2 = \overline{A}\,B\,\overline{C}$
0	1	1	0	0	0	1	0	0	0	0	$D_3 = \overline{A}\,B\,C$
1	0	0	0	0	0	0	1	0	0	0	$D_4 = A\,\overline{B}\,\overline{C}$
1	0	1	0	0	0	0	0	1	0	0	$D_5 = A\,\overline{B}\,C$
1	1	0	0	0	0	0	0	0	1	0	$D_6 = A\,B\,\overline{C}$
1	1	1	0	0	0	0	0	0	0	1	$D_7 = A\,B\,C$

The logic diagram for 3 to 8 decoder is shown in figure 5.111.

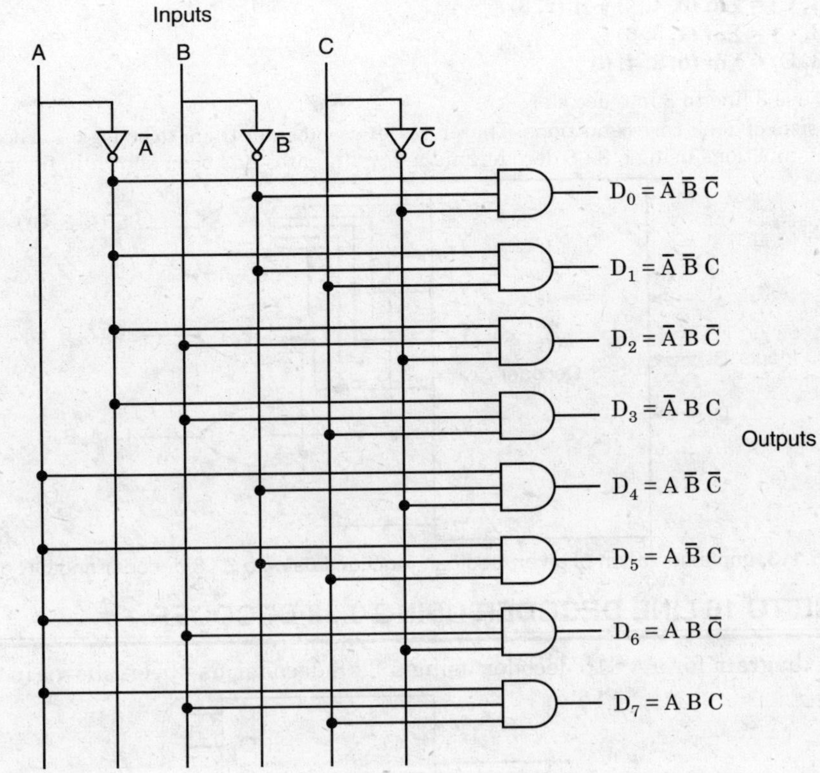

Fig. 5.111. A 3 to 8 line decoder

5.28.3. Implementation of Boolean Functions using Decoder

As a matter of fact, a combination of decoder and gates can be used to implement that Boolean expressions. This has been demonstrated in the next example.

> **DO YOU KNOW?**
>
> Many MSI decoders have several outputs, each one corresponding to only one of the many possible input combinations.

EXAMPLE 5.33. Implement the following Boolean function using a 3 : 8 decoder and external gates.

$$f(A, B, C) = \Sigma\, m\, (2, 4, 5, 7)$$

Solution:

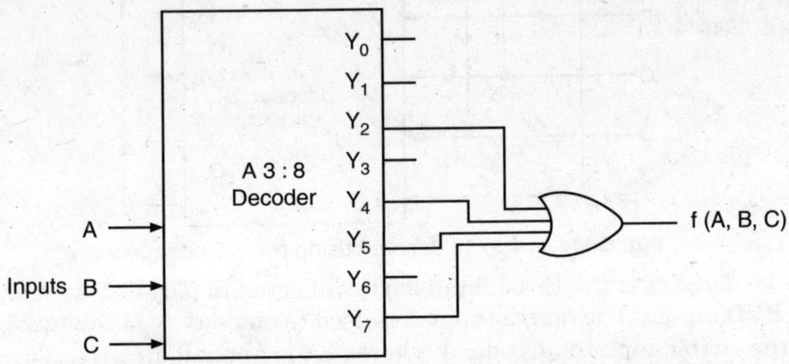

Fig. 5.112. Implementation of Boolean equation using decoder and gate

The decoder produces minterms. The outputs Y_2, Y_4, Y_5 and Y_7 are ORed to produce the required output. The logic implementation of the given logic function has been shown in figure 5.112.

EXAMPLE 5.34. Implement the following output function using a suitable decoder.

f_1 **(A, B, C) = Σm (0, 4, 7) + d (2, 3)**
f_2 **(A, B, C) = Σm (1, 5, 6)**
f_3 **(A, B, C) = Σm (0, 2, 4, 6)**

Solution: Let us use 3 line to 8 line decoder.

Here, f_1 consists of don't care conditions. Therefore, let us consider them to be logic 1. The implementation of the given three functions using a 3 : 8 decoder and a few OR gates has been shown in figure 5.113.

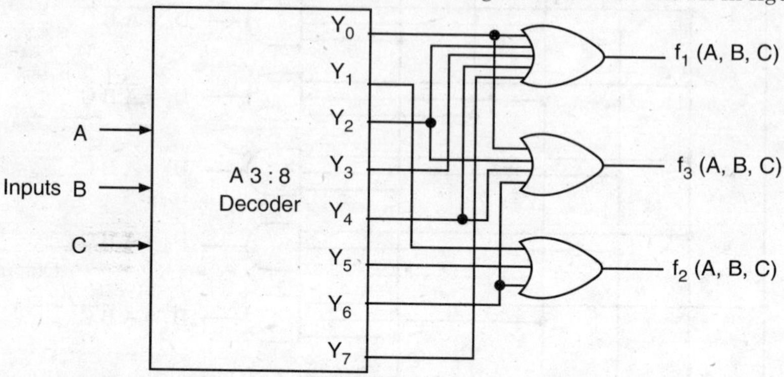

Fig. 5.113. Implementation of given Boolean functions using a 3 : 8 decoder and OR gates

5.29. A 4 LINE TO 16 LINE DECODER USING 3 : 8 DECODER

The connection diagram for a 4 : 16 decoder using a 3 : 8 decoder has been shown in figure 5.114.

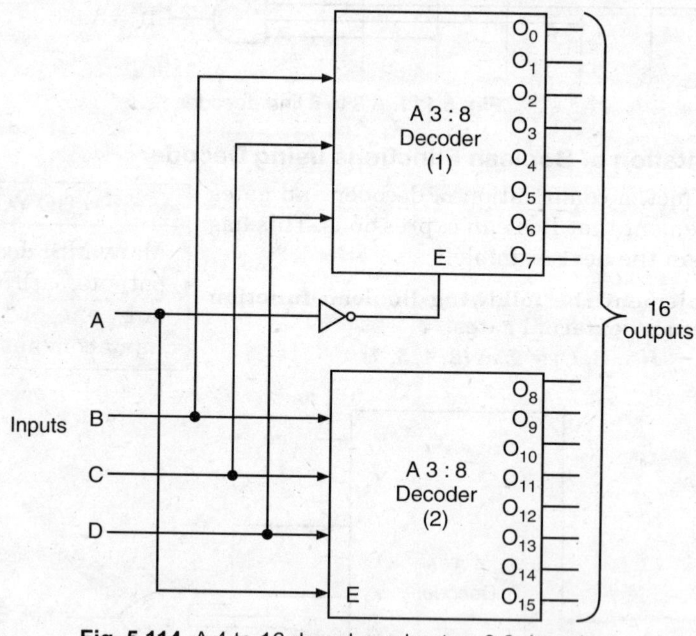

Fig. 5.114. A 4 to 16 decoder using two 3:8 decoders

Here, it may be noted that the three inputs are connected in parallel, with each other and then connected to the BCD inputs. The fourth input A is used to enable one of the two 3 : 8 decoders. Also, $\bar{A}$ is applied to the enable input of decoder 1 whereas A is applied directly to decoder 2.

5.29.1. A BCD to 7 Segment Decoder using a Common Cathode Display

If the common cathode type display is used, then the output of the decoder must be high (1) to turn on the display segment. The truth table for such a decoder is given by Table 5.36.

Table 5.36. Truth table for BCD to seven segment decoder with a common cathode display

Decimal	Inputs				Outputs						
	B_3	B_2	B_1	B_0	a	b	c	d	e	f	g
0	0	0	0	0	1	1	1	1	1	1	0
1	0	0	0	1	0	1	1	0	0	0	0
2	0	0	1	0	1	1	0	1	0	0	1
3	0	0	1	1	1	1	1	1	0	0	1
4	0	1	0	0	0	1	1	0	1	1	1
5	0	1	0	1	1	0	1	1	1	1	1
6	0	1	1	0	1	0	1	1	1	1	1
7	0	1	1	1	1	1	1	0	0	0	0
8	1	0	0	0	1	1	1	1	1	1	1
9	1	0	0	1	1	1	1	1	0	1	1

K map and simplifications

The K-maps and simplified equations for all the seven outputs have been shown in figure 5.115.

Output a

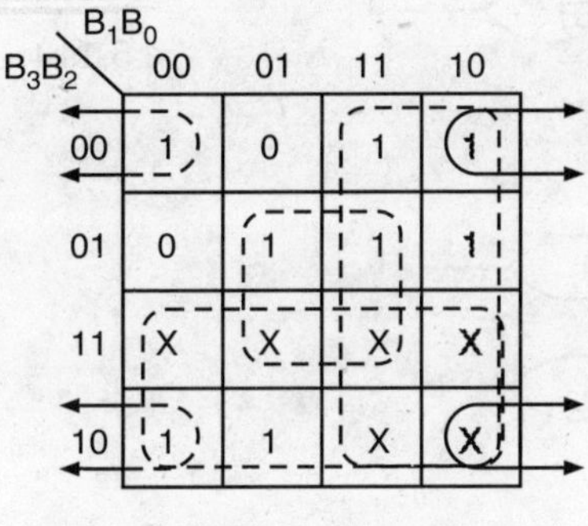

$a = B_3 + B_1 + B_2B_0 + \overline{B}_2\overline{B}_0$

Output b

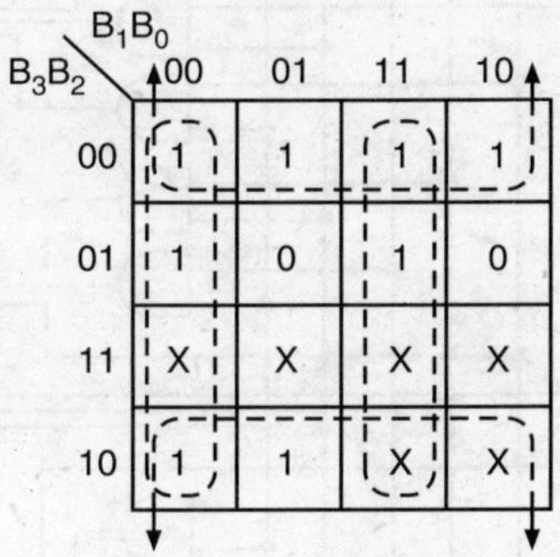

$b = \overline{B}_2 + \overline{B}_1\overline{B}_0 + B_1B_0$

Output c

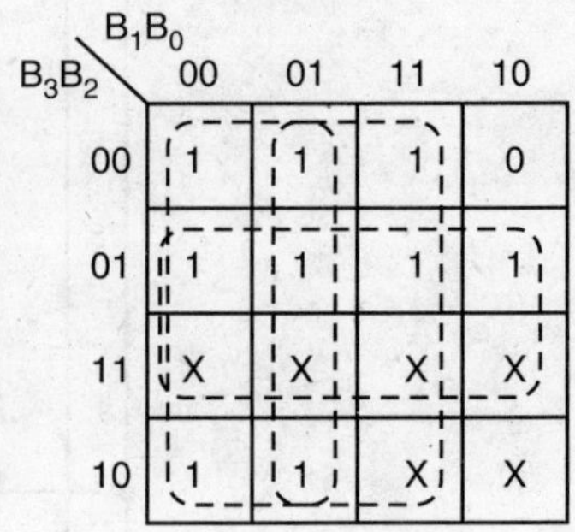

$c = B_2 + \overline{B}_1 + B_0$

Output d

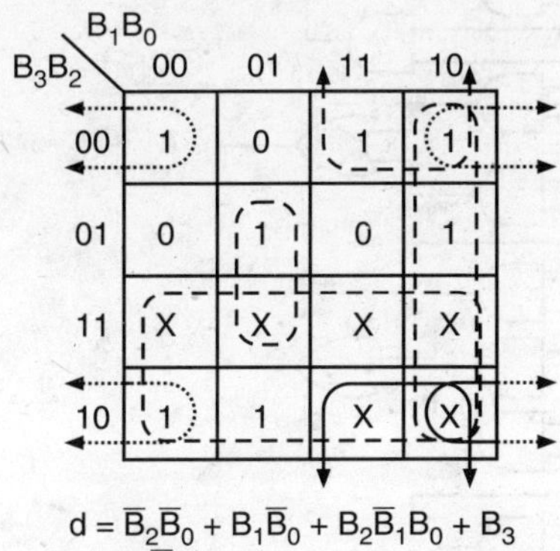

$d = \overline{B}_2\overline{B}_0 + B_1\overline{B}_0 + B_2\overline{B}_1B_0 + B_3 + \overline{B}_2B_1$

Output e

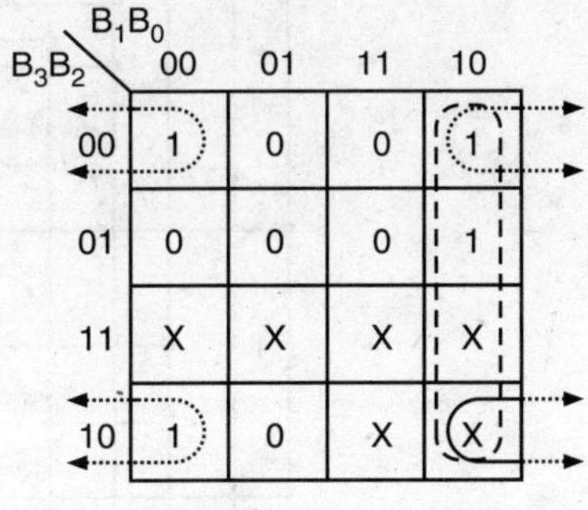

$e = \overline{B}_2\overline{B}_0 + B_1\overline{B}_0$

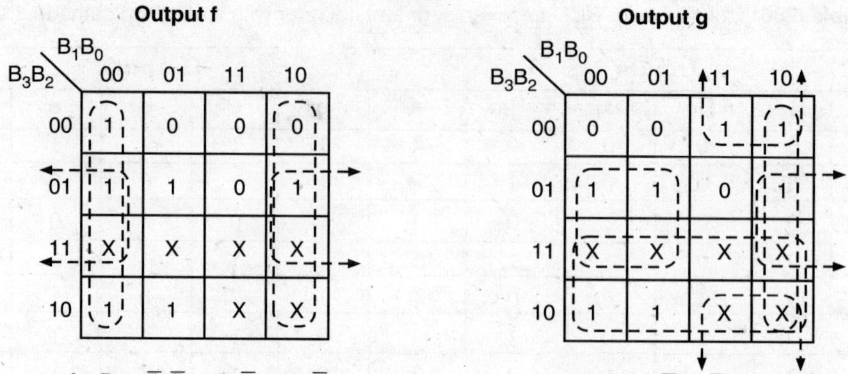

$f = B_3 + \overline{B}_1\overline{B}_0 + B_2\overline{B}_1 + B_2\overline{B}_0$ $\qquad$ $g = B_3 + B_2\overline{B}_1 + \overline{B}_2B_1 + B_1\overline{B}_0$

Fig. 5.115. K maps for various outputs

Logic diagram

The logic diagram for BCD to seven segment decoder has been shown in figure 5.116.

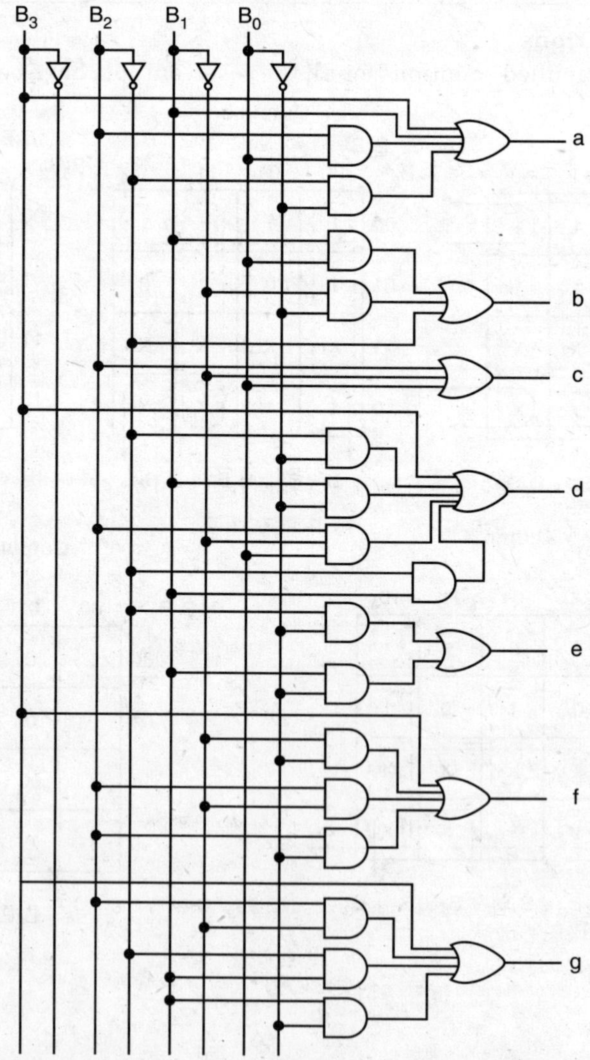

Fig. 5.116. BCD to seven segment decoder for a common cathode type display

EXAMPLE 5.35. Implement the BCD to seven segment decoder for a common cathode display using 4:16 decoder.

Solution : Considering the truth table (Table 5.36), we can write the logic expressions for a, b... f in terms of the decoder outputs as under :

> **DO YOU KNOW?**
>
> The segment elements can be light-emitting diodes, liquid crystals, or glowing electrodes surrounded by neon gas.

$$a = O_0 + O_2 + O_3 + O_5 + O_6 + O_7 + O_8 + O_9$$
$$b = O_0 + O_1 + O_2 + O_3 + O_4 + O_7 + O_8 + O_9$$
$$c = O_0 + O_1 + O_3 + O_4 + O_5 + O_6 + O_7 + O_8 + O_9$$
$$d = O_0 + O_2 + O_3 + O_5 + O_6 + O_8 + O_9$$
$$e = O_0 + O_2 + O_6 + O_8$$
$$f = O_0 + O_4 + O_5 + O_6 + O_8 + O_9$$
$$g = O_2 + O_3 + O_4 + O_5 + O_6 + O_8 + O_9$$

The implementation using 4 line to 16 line decoder has been shown in figure 5.117.

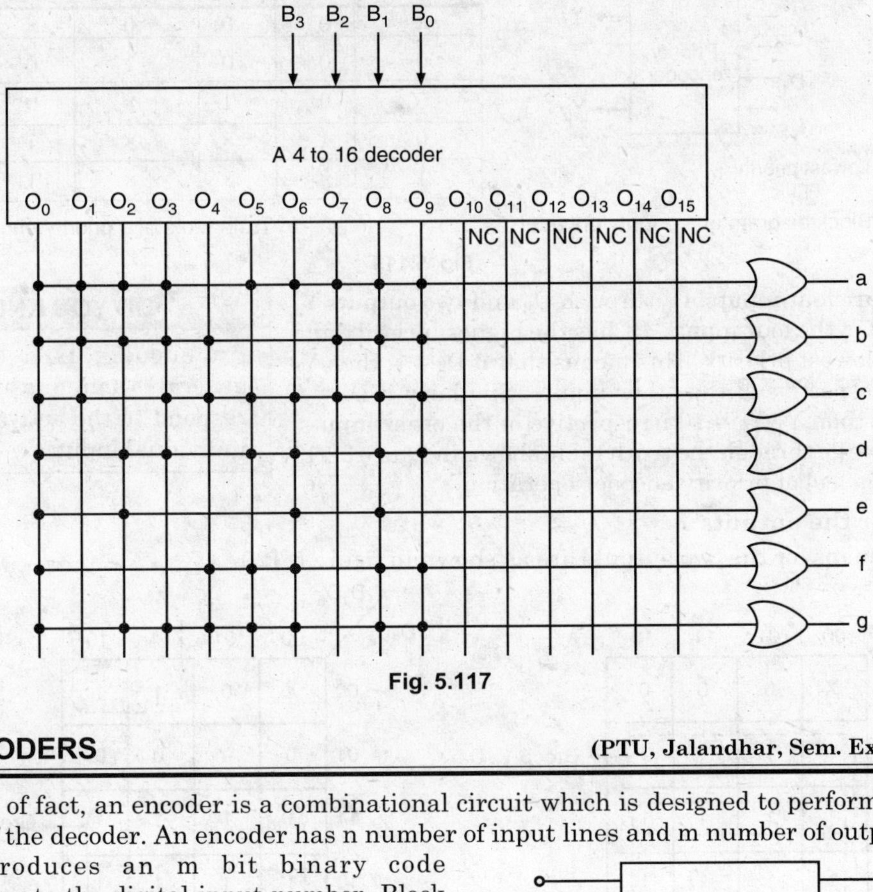

Fig. 5.117

5.30 ENCODERS

(PTU, Jalandhar, Sem. Exam., 2004-05)

As a matter of fact, an encoder is a combinational circuit which is designed to perform the inverse operation of the decoder. An encoder has n number of input lines and m number of output lines. An encoder produces an m bit binary code corresponding to the digital input number. Block diagram of an encoder is shown in figure 5.118. The encoder accepts an n input digital word and converts it into an m bit another digital word. The internal combinational circuit of the encoder is designed accordingly.

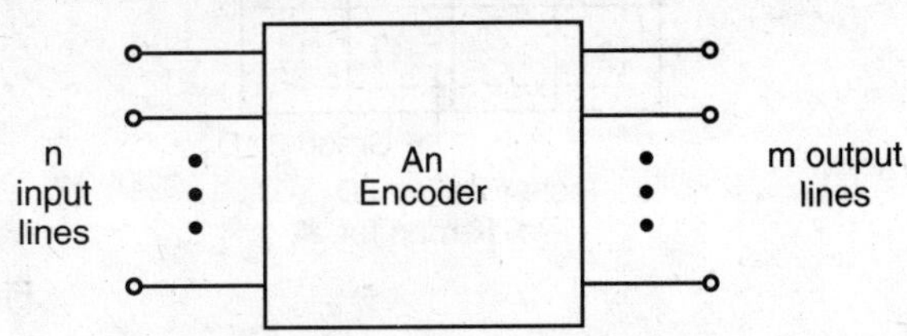

Fig. 5.118. Block diagram of an encoder

5.30.1. Classification of Encoders

The encoders may be classified as under :

(*i*) Priority encoders.

(*ii*) Decimal to BCD encoder.
(*iii*) Octal to binary encoder.
(*iv*) Hexadecimal to binary encoder.

5.31 A PRIORITY ENCODER

This is a special type of encoder. In this encoder, priorities are given to the input lines. If two or more input lines are 1 at the same time, then the input line with highest priority shall be considered. The block diagram of a priority encoder is shown in figure 5.119 (*a*) and its truth table is shown in figure 5.119(*b*).

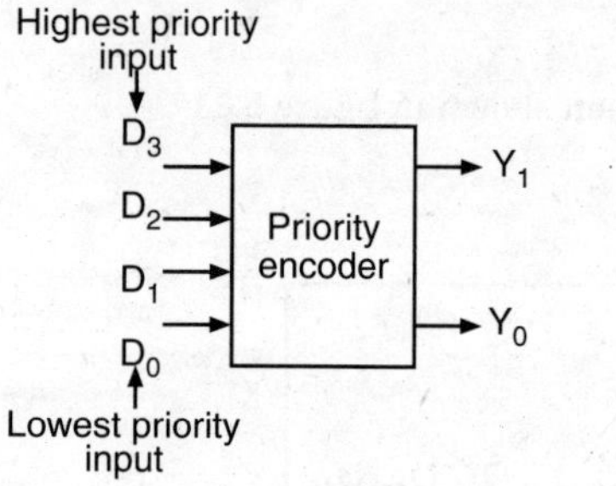

(*a*) Block diagram of a priority encoder

Highest	Inputs		Lowest	Outputs	
D_3	D_2	D_1	D_0	Y_1	Y_0
0	0	0	0	×	×
0	0	0	1	0	0
0	0	1	×	0	1
0	1	×	×	1	0
1	×	×	×	1	1

(*b*) Truth table of a priority encoder

Fig. 5.119

There are four inputs, D_0 through D_3 and two outputs Y_1 and Y_0. Out of the four inputs, D_3 has the highest priority and D_0 has the lowest priority. This means that if $D_3 = 1$, then $Y_1\,Y_0 = 11$ irrespective of the other inputs. Similarly if $D_3 = 0$ and $D_2 = 1$ then $Y_1\,Y_0 = 10$ irrespective of the other inputs. Carefully, we go through the truth table shown in figure 5.119 (b) to get the feel of priority encoder operation.

> **DO YOU KNOW?**
>
> An encoder is a device that generates a unique binary code in respond to the activation of each individual input.

K-maps for the outputs

The K maps for the two outputs are as shown in figure 5.120

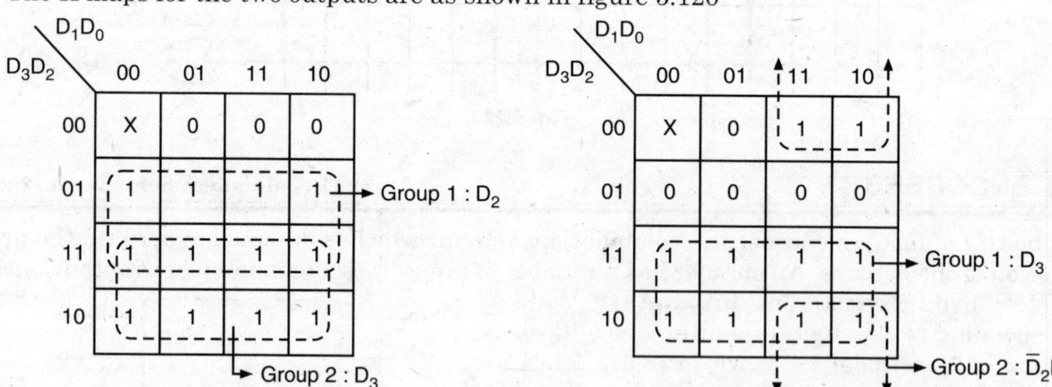

There fore, $Y_1 = D_3 + D_2$
(*a*) K-map for Y_1

There fore, $Y_0 = D_3 + \overline{D}_2\,D_1$
(*b*) K-map for Y_0

Fig. 5.120

Implementation

The logic diagram for the priority encoder is shown in figure 5.121.

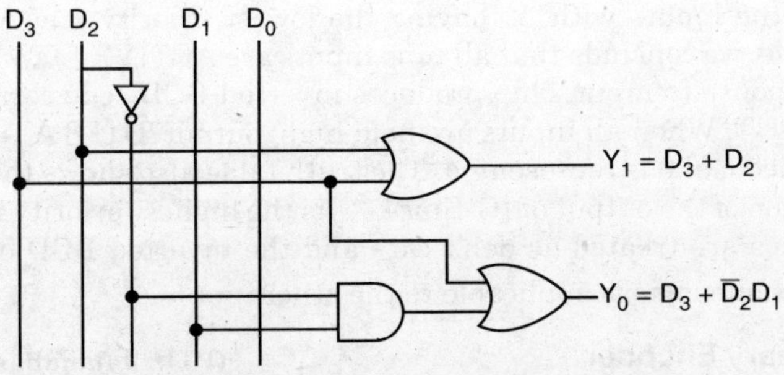

Fig. 5.121. Logic diagram of a priority encoder

Pin configuration and logic symbol

Figure 5.122 (b) shows the pin configuration and figure 5.122 (a) shows the logic symbol of IC 74147.

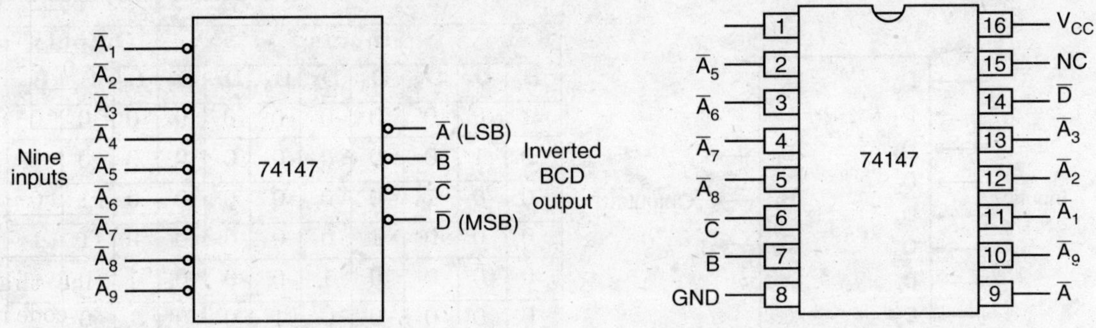

(a) Logic symbol of IC 74147 (b) Pin configuration of 74147

Fig. 5.122

Here, A_1 to A_9 are the active low inputs and A, B, C, D are the active low outputs.

Truth table

Figure 5.123 shows the truth table of decimal to BCD encoder IC 74147.

Inputs									**Outputs (Inverted BCD)**				**Normal BCD**			
$\bar{A}_1$	$\bar{A}_2$	$\bar{A}_3$	$\bar{A}_4$	$\bar{A}_5$	$\bar{A}_6$	$\bar{A}_7$	$\bar{A}_8$	$\bar{A}_9$	**D**	**C**	**B**	**A**	**D**	**C**	**B**	**A**
1	1	1	1	1	1	1	1	1	1	1	1	1	0	0	0	0
×	×	×	×	×	×	×	×	0	0	1	1	0	1	0	0	1
×	×	×	×	×	×	×	0	1	0	1	1	1	1	0	0	0
×	×	×	×	×	×	0	1	1	1	0	0	0	0	1	1	1
×	×	×	×	×	0	1	1	1	1	0	0	1	0	1	1	0
×	×	×	×	0	1	1	1	1	1	0	1	0	0	1	0	1
×	×	×	0	1	1	1	1	1	1	0	1	1	0	1	0	0
×	×	0	1	1	1	1	1	1	1	1	0	0	0	0	1	1
×	0	1	1	1	1	1	1	1	1	1	0	1	0	0	1	0
0	1	1	1	1	1	1	1	1	1	1	1	0	0	0	0	1

Fig. 5.123. Truth table for 74147

Here, A_1 to A_9 are the inputs with A_1 having the lowest priority and A_9 having the highest priority. From truth table, we conclude that all nine inputs are ACTIVE LOW representing decimal digit from 1 to 9. In response to input, chip produces inverted BCD code corresponding to highest numbered ACTIVE INPUT. When all inputs are held high, output $\bar{D}\bar{C}\bar{B}\bar{A} = 1\,1\,1\,1$, i.e., DCBA = $(0000)_2 = (0)_{10}$. Thus, a decimal 0 is represented. The truth table also shows the normal BCD output which is actually inversion of the output of IC. Since, A_9 is the highest priority input, if $\bar{A}_9 = 0$, then, the remaining input lines are treated as don't care and the inverted BCD output is produced as $\bar{D}CB\bar{A} = 0\,1\,1\,0$. The same logic is applicable to the other inputs.

5.31.1. An Octal to Binary Encoder

(PTU, Jalandhar, Sem. Exam., 2007-08)

The octal to binary encoder has 8-input lines and 3-output lines. Corresponding to the eight input octal numbers we get three bit binary output. Here, it may be noted that in encoders, only one input will have a one value at any given time. Figure 5.124 shows the block diagram of an octal to binary encoder and Table 5.37 gives its truth table.

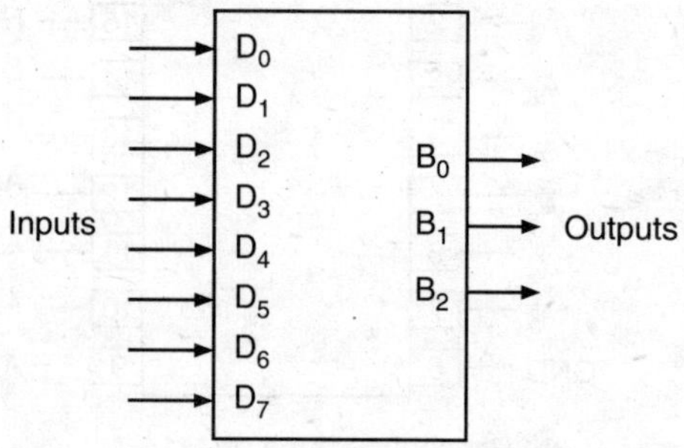

Fig. 5.124. Block diagram of an octal to binary encoder

Table 5.37. Truth table of an octal to binary encoder

Inputs								Outputs		
D_0	D_1	D_2	D_3	D_4	D_5	D_6	D_7	B_0	B_1	B_2
1	0	0	0	0	0	0	0	0	0	0
0	1	0	0	0	0	0	0	0	0	1
0	0	1	0	0	0	0	0	0	1	0
0	0	0	1	0	0	0	0	0	1	1
0	0	0	0	1	0	0	0	1	0	0
0	0	0	0	0	1	0	0	1	0	1
0	0	0	0	0	0	1	0	1	1	0
0	0	0	0	0	0	0	1	1	1	1

Logical expressions for the outputs

Considering the truth table, we can write the logical expressions for the outputs as under :

$$B_0 = D_4 + D_5 + D_6 + D_7$$
$$B_1 = D_2 + D_3 + D_6 + D_7$$
$$B_2 = D_1 + D_3 + D_5 + D_7$$

Implementation using basic gates

An octal to binary encoder using the basic gates has been shown in figure 5.125.

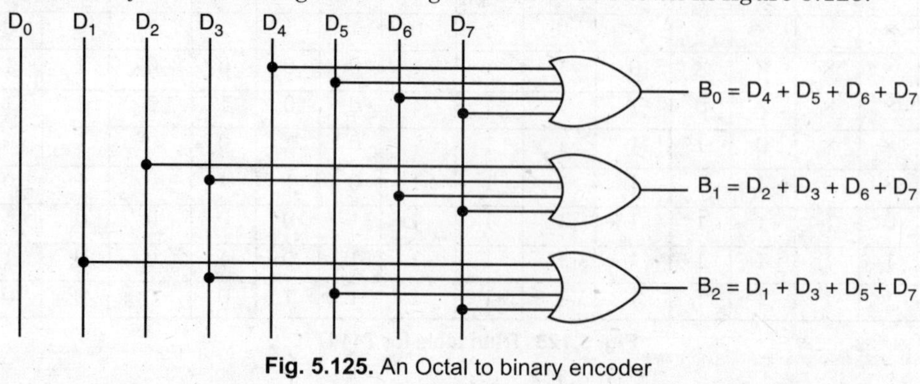

Fig. 5.125. An Octal to binary encoder

5.32 PARITY GENERATORS/CHECKERS

A parity bit is an additional bit which is added to a binary word in order to make the number of 1's in the new word formed, even (even parity) or odd (odd parity).

The even and odd parity has been illustrated in figure 5.126.

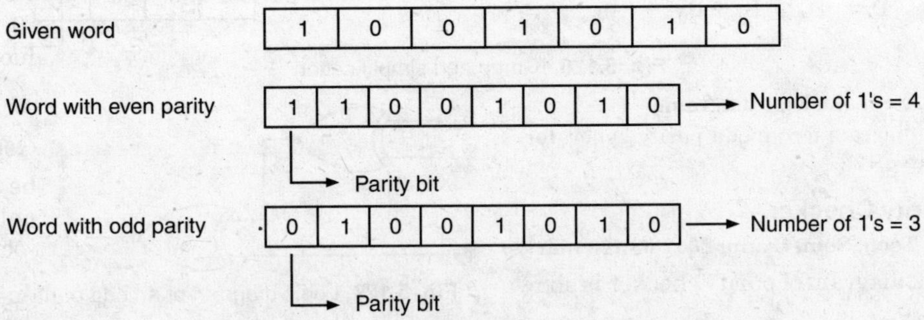

Fig. 5.126. Illustration of even and odd parity

Parity bits are added in order to detect any error occurred in the process of transmission. The words with parity bits are transmitted by the transmitter. At the receiver, these words are received and checked for errors. The receiver checks parity of the received word and compares it with the parity of the transmitted word. If they are same then there is no error. But, if the parity of received word is different from that of transmitted word, then the receiver understands that an error is present. In the even parity system, the added parity bit will make the total number of 1's an even number. In the odd parity system, the added parity bit will make the total number of 1's an odd number.

5.32.1. A Parity Generator

A Parity generator is a logic circuit which generates the parity bits for even parity or odd parity. Block diagram of the parity generator is shown in figure 5.127. The digital word is applied at the input and a parity bit for either even parity or odd parity.

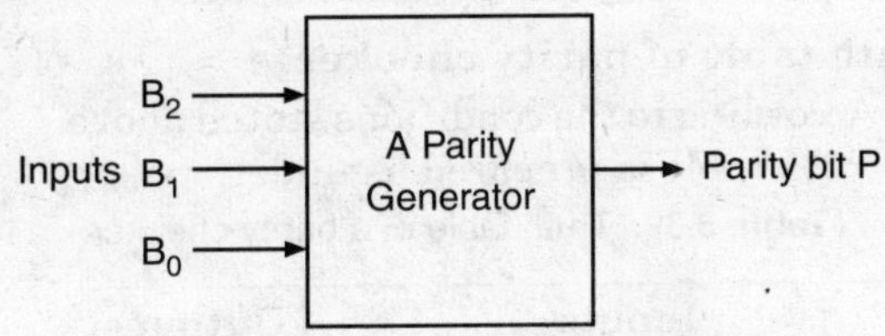

Fig. 5.127. Block diagram of a parity generator

EXAMPLE 5.36. Design a parity generator using the basic gates to produce digital words with odd parity. Assume the input to be three bit binary words.

Solution: (i) First, we write truth table.

Table 5.38 shows the required truth table.

Table 5.38

Inputs			Output
B_2	B_1	B_0	Odd parity bit P
0	0	0	1
0	0	1	0
0	1	0	0
0	1	1	1
1	0	0	0
1	0	1	1
1	1	0	1
1	1	1	0

These parity bits make the total number of 1's odd

(ii) Now, we obtain K-map and simplify.

$$P' = \bar{B}_2\,\bar{B}_1\,\bar{B}_0 + \bar{B}_2\,B_1\,B_0 + B_2\,\bar{B}_1\,B_0 + B_2\,B_1\,\bar{B}_0$$

$$P = \bar{B}_2\,(\bar{B}_1\,\bar{B}_0 + B_1\,B_0) + B_2\,(\bar{B}_1\,B_0 + B_1\,\bar{B}_0)$$

$$= \bar{B}_2\,(\overline{B_1 \oplus B_0}) + B_2\,(B_1 \oplus B_0)$$

Therefore, $P = B_2 \odot (B_1 \oplus B_0)$

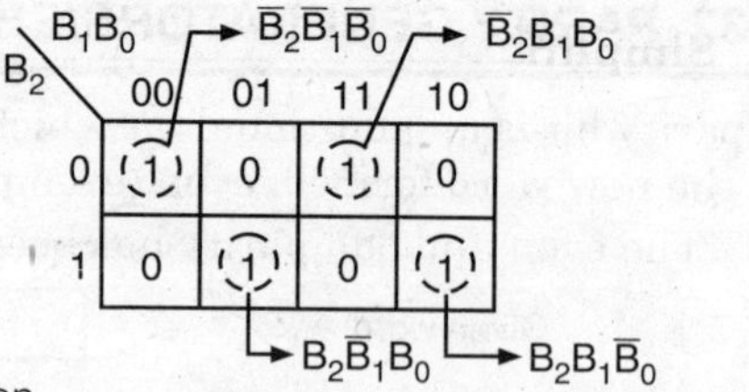

Fig. 5.128. K-map and simplification

(iii) Next, we make logic diagram.

The logic diagram for an odd parity generator is shown in figure 5.129.

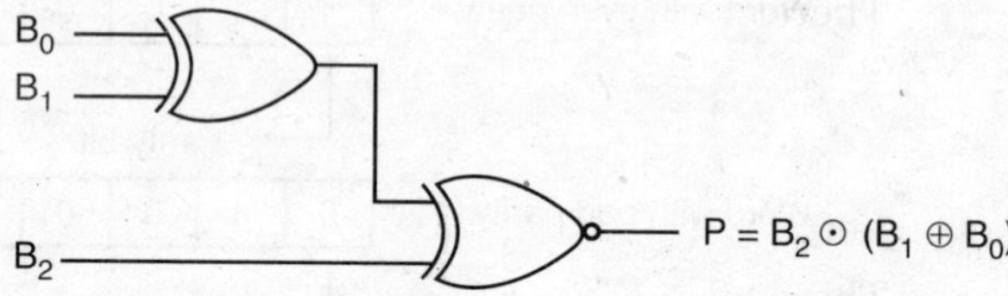

Fig. 5.129. Logic diagram of an odd parity generator

5.32.2. A Parity Checker

(U.P. Tech, Sem. Exam; 2007-08)(10 marks)

The block diagram of parity checker is shown in figure 5.130.

The parity checker circuit is at the receiver. The received word with a parity bit is applied to the parity checker. The parity checker will check the parity of the received word and produce its output. Since the parity of transmitted words is odd, the parity checker will treat the received words with odd parity as the correct words and those with even parity as the incorrect words. The output of parity checker circuit is denoted by parity error output PEO.

PEO = 1 .... If error is present, i.e., if the received four bit word has an even parity.

PEO = 0 .... If there is no error, i.e., if the received four bit word has an odd parity.

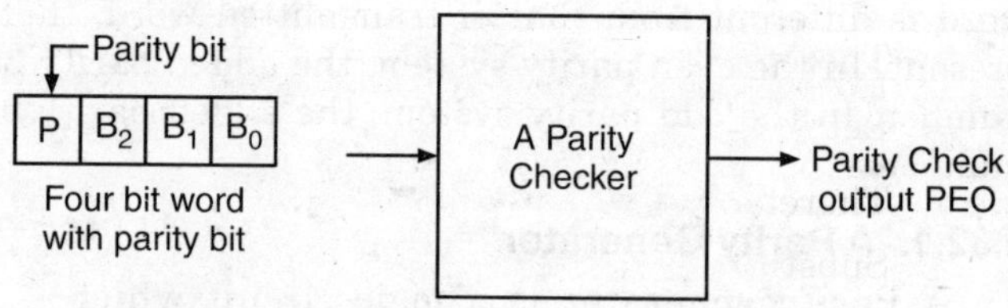

Fig. 5.130. Block diagram of a parity checker

Truth table of parity checker

According to the conditions stated above, the truth table is prepared as under:

Table 5.39. Truth table of a parity checker

Inputs				Output
P	B_2	B_1	B_0	PEO
0	0	0	0	1
0	0	0	1	0
0	0	1	0	1
0	0	1	1	1
0	1	0	0	0
0	1	0	1	1
0	1	1	0	1
0	1	1	1	0
1	0	0	0	0
1	0	0	1	1
1	0	1	0	1
1	0	1	1	0
1	1	0	0	1
1	1	0	1	0
1	1	1	0	0
1	1	1	1	1

1 ≡ error (even parity)

0 ≡ No error (odd parity)

K - maps and simplifications

For PEO

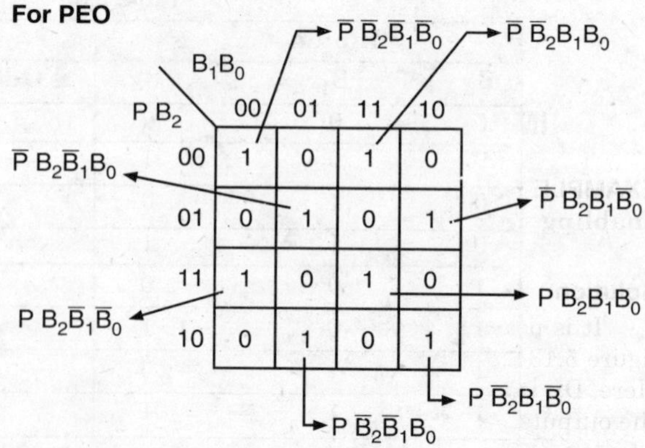

Fig. 5.131. K-map for a parity checker output

Simplification:

$$\text{PEO} = \bar{P}\,\bar{B}_2\,\bar{B}_1\,\bar{B}_0 + \bar{P}\,\bar{B}_2\,B_1\,B_0 + \bar{P}\,\bar{B}_2\,\bar{B}_1\,B_0 + \bar{P}\,B_2\,B_1\,\bar{B}_0$$
$$+ P\,B_2\,\bar{B}_1\,\bar{B}_0 + P\,B_2\,B_1\,B_0 + P\,\bar{B}_2\,\bar{B}_1\,B_0 + P\,\bar{B}_2\,B_1\,\bar{B}_0$$
$$= \bar{B}_1\,\bar{B}_0\,(\bar{P}\,\bar{B}_2 + P\,B_2) + B_1\,B_0\,(\bar{P}\,\bar{B}_2 + P\,B_2)$$
$$+ \bar{B}_1\,B_0\,(\bar{P}\,B_2 + P\,\bar{B}_2) + B_1\,\bar{B}_0\,(\bar{P}\,B_2 + P\,\bar{B}_2)$$

Therefore, $\text{PEO} = (\bar{P}\,\bar{B}_2 + P\,B_2)\,(\bar{B}_1\,\bar{B}_0 + B_1\,B_0) + (\bar{P}\,B_2 + P\,\bar{B}_2)\,(\bar{B}_1\,B_0 + B_1\,\bar{B}_0)$

$$= (P \odot B_2)\,(B_1 \odot B_0) + (P \oplus B_2)\,(B_1 \oplus B_0)$$
$$= \overline{(P \oplus B_2)}\,\overline{(B_1 \oplus B_0)} + (P \oplus B_2) + (B_1 \oplus B_0)$$

Let $X = P \oplus B_2$ and $Y = (B_1 \oplus B_0)$

Therefore, $\text{PEO} = \bar{X}\,\bar{Y} + X\,Y = X \odot Y = \overline{X \oplus Y}$

Substituting the values, we obtain

$$\text{PEO} = \overline{(P \oplus B_2) \oplus (B_1 \oplus B_0)}$$

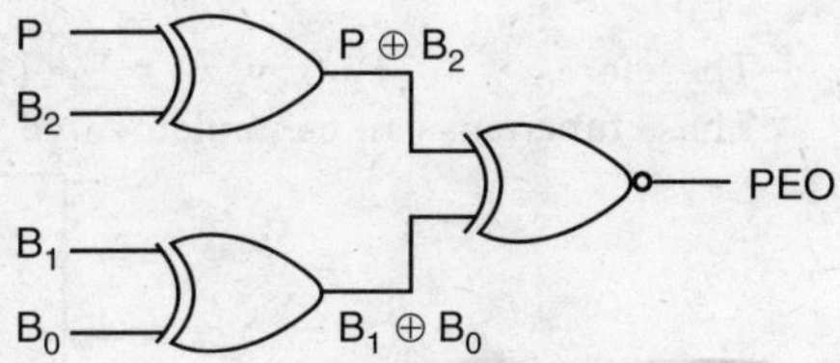

Fig. 5.132.

Logic diagram

Logic diagram is as shown in figure 5.132.

We have, $B_0 = (G_3 \oplus G_2)\,(\bar{G}_1\,\bar{G}_0 + G_1\,G_0) + \overline{(G_3 \oplus G_2)}\,(\bar{G}_1\,G_0 + G_1\,\bar{G}_0)$

Therefore, $B_0 = (G_3 \oplus G_2)\,\overline{(G_1 \oplus G_0)} + \overline{(G_3 \oplus G_2)}\,(G_1 \oplus G_0)$

Let $X = G_3 \oplus G_2$ and $Y = (G_1 \oplus G_0)$

Therefore, $B_0 = X\bar{Y} + \bar{X}Y = X \oplus Y$

Substituting for X and Y, we obtain,

$$B_0 = G_3 \oplus G_2 \oplus G_1 \oplus G_0$$

The Gray to Binary code converter is as shown in figure 5.133.

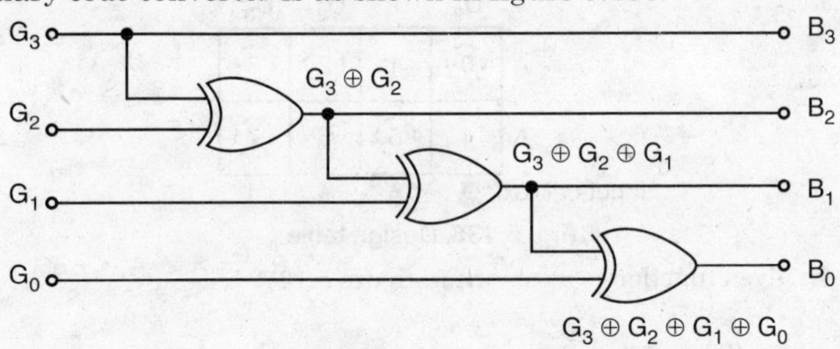

Fig. 5. 133. A Gray to Binary code converter

IMPORTANT SOLVED EXAMPLES

EXAMPLE 5.37. Demultiplexer is decoder circuit with an additional enabling inputs. Do you agree with the above statement?

(PTU, Jalandhar, Sem. Exam, 2006-07)

Solution : Demultiplexer as Decoder

It is possible to operate a demultiplexer as a decoder. Let us consider figure 5.134 which shows a 1 : 4 demultiplexer operating as a 2 : 4 decoder. Here, D_{in} is the data input, S_1 S_0 are the select lines and Y_3 through Y_0 are the outputs.

In order to operate it as a 2 : 4 decoder, we have to use S_1 S_0 as inputs, keep D_{in} open and use Y_3 to Y_0 as outputs as shown in figure 5.134.

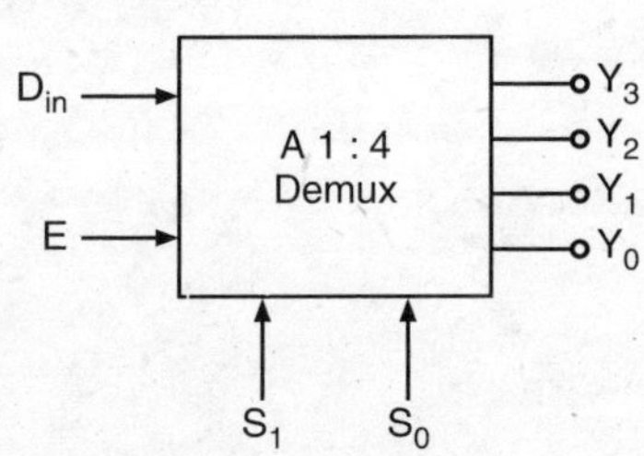

Fig. 5.134. A 1 : 4 Demux

EXAMPLE 5.38. A combinational circuit is defined by the following Boolean functions. Design circuit with a decoder and external gates:

$F_1(x, y, z) = x'y'z' + xz$

$F_2(x, y, z) = xyz' + x'z$

Solution : Function F_1 can be written in standard SOP form as under :

$$F_1(x, y, z) = \bar{x}\,\bar{y}\,\bar{z} + (y + \bar{y})\,x\,z$$

$$F_1(x, y, z) = \bar{x}\,\bar{y}\,\bar{z} + x\,y\,z + x\,\bar{y}\,z$$

This function can be expressed in terms of minterms as under :

$$F_1(x, y, z) = \Sigma m\,(0, 5, 7)$$

Similarly, function F_2 can be expressed in terms of minterms as under :

$$F_1(x, y, z) = x\,y\,\bar{z} + \bar{x}\,z = x\,y\,\bar{z} + \bar{x}\,z\,(y + \bar{y})$$

$$= x\,y\,\bar{z} + \bar{x}\,z\,y + \bar{x}\,\bar{y}\,z$$

Therefore, $F_2(x, y, z) = \Sigma m\,(1, 3, 6)$

These functions can be implemented using a 3 : 8 decoder and logic gates, as shown in figure 5.135.

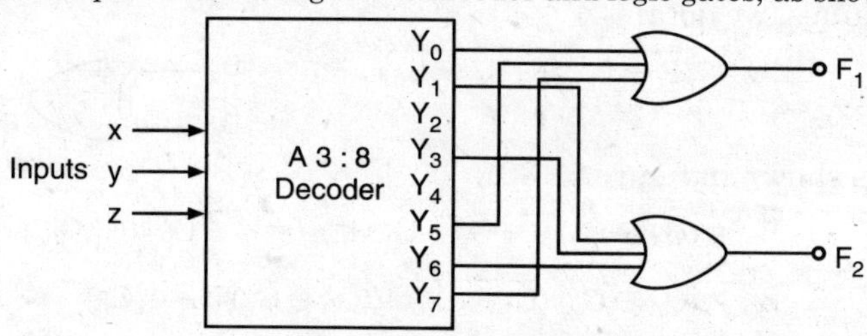

Fig. 5.135.

EXAMPLE 5.39. Using four-input multiplexers, implement the following function

$F(A, B, C) = \Sigma m\,(0, 2, 3, 5, 7)$ **(PTU, Jalandhar, Sem. Exam, 2006-07)**

Solution : Given that $F(A, B, C) = \Sigma m\,(0, 2, 3, 5, 7)$

Inputs B and C are applied as select line inputs of the 4 : 1 MUX, S_1 and S_0 respectively.

	D_0	D_1	D_2	D_3
$\bar{A}$	(0)	1	(2)	(3)
A	4	(5)	6	(7)
Input ot Mux :	$\bar{A}$	A	$\bar{A}$	1

Fig. 5.136. Design table

The logic diagram for given function is as shown in figure 5.137.

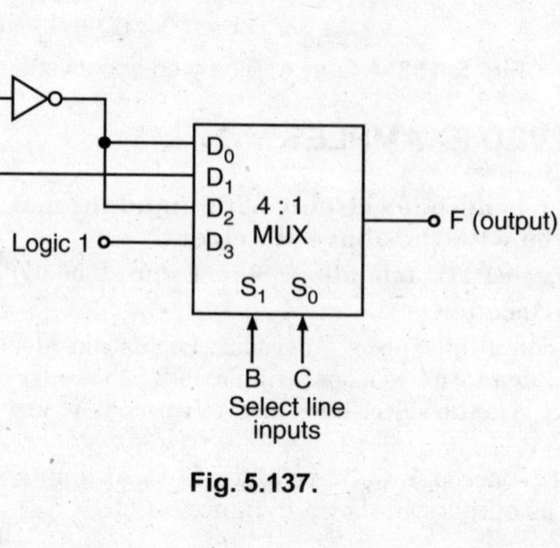

Fig. 5.137.

EXAMPLE 5.40. Implement 4-variables switching function using MUX F = Σ (0, 2, 4, 6, 8, 10, 12) for representing negative function. **(PTU, Jalandhar, Sem. Exam, 2004-2005) (7 Marks)**

Solution: The given function is,

$$F = \Sigma\,(0, 2, 4, 6, 8, 10, 12)$$

For negative logic, the output of the MUX must be low when the given inputs are high, or the output is high when the given inputs are low. The required logic diagram is as shown in figure 5.138.

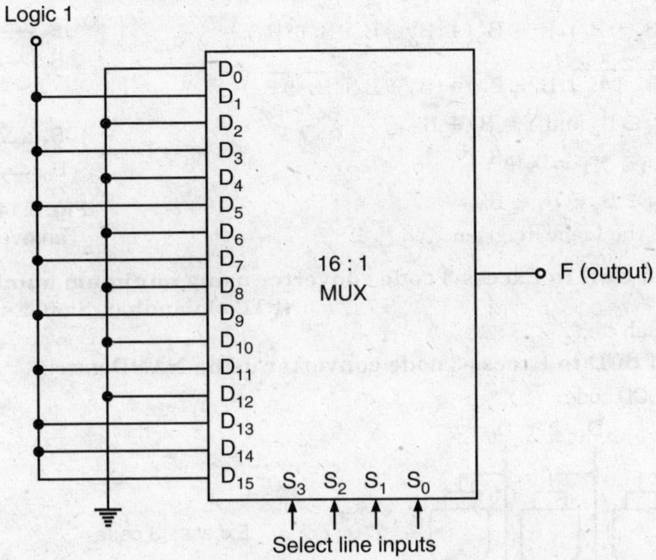

Fig. 5.138.

EXAMPLE 5.41. Design a parity generator to generate even parity bit for a 4 bit word. Use EX-OR and $\overline{\text{EX-OR}}$ gate. **(PTU, Jalandhar, Sem. Exam., 2008-09)**

Solution. (i) First, we write truth table.

Truth table

Inputs				Output even parity bit P
B_3	B_2	B_1	B_0	
0	0	0	0	0
0	0	0	1	1
0	0	1	0	1
0	1	0	0	0
0	1	0	0	1
0	1	0	1	0
0	1	1	0	0
0	1	1	1	1
1	0	0	0	1
1	1	0	1	0
1	0	1	0	0
1	0	1	1	1
1	1	0	0	0
1	1	0	1	1
1	1	1	0	1
1	1	1	1	0

(ii) K-map and simplification:

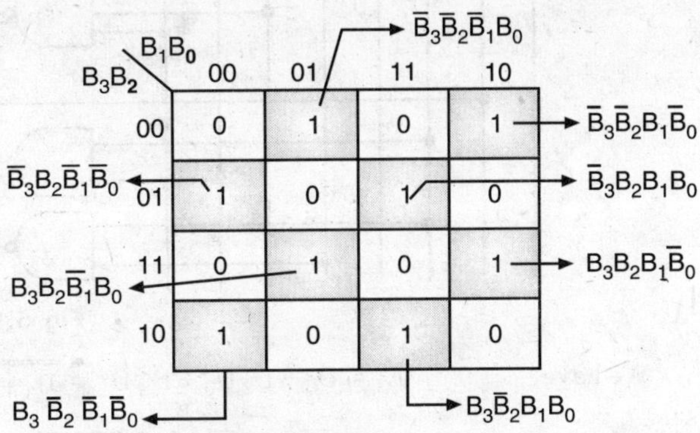

Fig. 5.139.

$$P = \bar{B}_3\bar{B}_2\bar{B}_1B_0 + \bar{B}_3\bar{B}_2B_1\bar{B}_0 + \bar{B}_3B_2\bar{B}_1\bar{B}_0 + \bar{B}_3B_2B_1B_0 + B_3B_2B_1\bar{B}_0 + B_3\bar{B}_2\bar{B}_1\bar{B}_0 + B_3B_2\bar{B}_1B_0 + B_3\bar{B}_2B_1B_0$$

$$P = \bar{B}_3\bar{B}_2(\bar{B}_1B_0 + B_1\bar{B}_0) + B_3B_2(B_1\bar{B}_0 + \bar{B}_1B_0) + \bar{B}_3B_2(\bar{B}_1\bar{B}_0 + B_1B_0) + B_3\bar{B}_2(\bar{B}_1\bar{B}_0 + B_1B_0)$$

$$= (\bar{B}_3\bar{B}_2 + B_3B_2)(\bar{B}_1B_0 + B_1\bar{B}_0) + (\bar{B}_3B_2 + B_3\bar{B}_2)(\bar{B}_1\bar{B}_0 + B_1B_0)$$

$$= (B_3 \odot B_2)(B_1 \oplus B_0) + (B_3 \oplus B_2)(B_1 \odot B_0)$$

$$= \left(\overline{B_3 \oplus B_2}\right)(B_1 \oplus B_0) + (B_3 \oplus B_2)\left(\overline{B_1 \oplus B_0}\right)$$

Let $X = B_3 \oplus B_2$ and $Y = B_1 \oplus B_0$

Therefore, $P = \bar{X}Y + X\bar{Y} = X \oplus Y$

or $P = B_3 \oplus B_2 \oplus B_1 \oplus B_0$

(iii) Lastly, we make the Logic diagram.

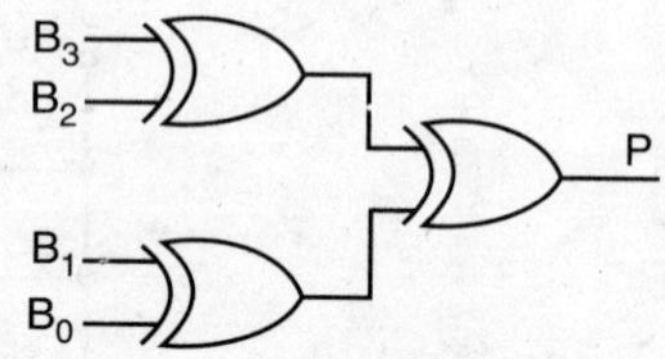

Fig. 5.140. Logic diagram of an even parity generator

EXAMPLE 5.42. Design a BCD to Excess-3 code converter using minimum number of NAND gates.
(PTU, Jalandhar, Sem. Exam., 2008-09)(5 marks)

Solution: Please go through text.

Implementation of BCD to Excess-3 code converter using NAND gates:

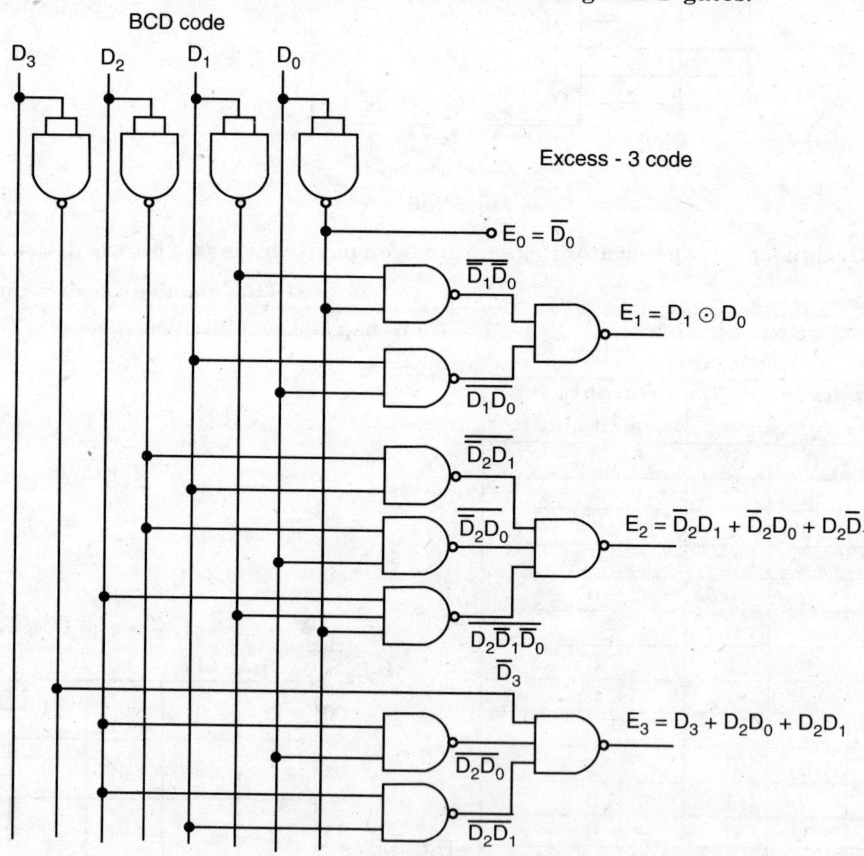

Fig. 5.141.

We have $E_3 = D_3 + D_2D_0 + D_2D_1 = \overline{\overline{D_3 + D_2D_0 + D_2D_1}} = \overline{\bar{D}_3 \cdot \overline{D_2D_0} \cdot \overline{D_2D_1}}$

$$E_2 = \bar{D}_2D_1 + \bar{D}_2D_0 + D_2\bar{D}_1\bar{D}_0 = \overline{\overline{\bar{D}_2D_1 + \bar{D}_2D_0 + D_2\bar{D}_1\bar{D}_0}} = \overline{\overline{\bar{D}_2D_1} \cdot \overline{\bar{D}_2D_0} \cdot \overline{D_2\bar{D}_1\bar{D}_0}}$$

$$E_1 = \bar{D}_1\bar{D}_0 + D_1D_0 = \overline{\overline{\bar{D}_1\,\bar{D}_0 + D_1D_0}} = \overline{\overline{\bar{D}_1\,\bar{D}_0}\cdot\overline{D_1D_0}}$$

$$E_0 = \bar{D}_0$$

EXAMPLE 5.43. Implement the following function using a 4-to-16 line decoder:
$f = \Sigma m\ (1, 2, 4, 7, 8, 11, 12, 13)$

Solution: We have $f = \Sigma m\ (1, 2, 4, 7, 8, 11, 12, 13)$

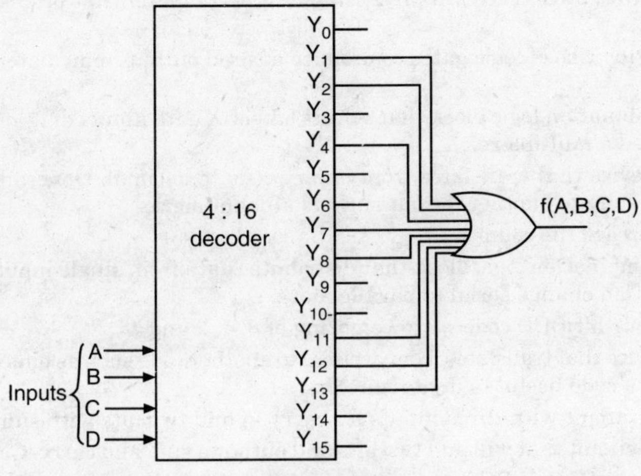

Fig. 5.142.

EXAMPLE 5.44. Implement the following expression using a multiplexer:
$f(A,B,C,D) = \Sigma m\ (0, 2, 3, 6, 8, 9, 11, 12, 14)$.

(PTU, Jalandhar, Sem. Exam., 2006-07)

Solution: (i) First, we apply the variables B, C and D to the select inputs.

The three variables B, C and D are connected to the select lines S_2, S_1 and S_0 respectively.

(ii) Then, we write the design table:

	D_0	D_1	D_2	D_3	D_4	D_5	D_6	D_7
$\bar{A}$	(0)	1	(2)	(3)	4	5	(6)	7
A	(8)	(9)	10	(11)	(12)	13	(14)	15
Input to MUX:	1	A	$\bar{A}$	1	A	0	1	0

(iii) Lastly, we draw the logic diagram

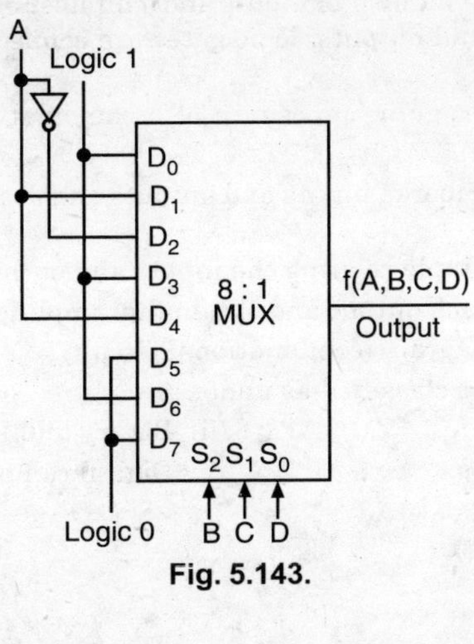

Fig. 5.143.

GLOSSARY

1. **Active-matrix display :** A high-quality expensive color LCD using active-matrix technology which involves the use of thin-film transistors.
2. **Adder :** Combinational logic circuit generates sum and carry outputs from any set of binary inputs. Half-adders and full-adders are two fundamental adder circuits.
3. **Anode :** Positive section of a device such as a diode or LED.
4. **Arithmetic logic unit :** Part of central processing unit of computer that processes data using arithmetic and logic operations.
5. **Combinational logic :** Use of logic gates to produce desired output immediately. No memory or latching characteristics.
6. **Data selector :** Combination logic block that selects one-of-*X* data inputs and connects that information to the output. Also called a multiplexer.
7. **Decoder :** A logic device that translates from binary code to decimal. Generally, it translates processed data into a digital system to another format such as alphanumeric.
8. **Decrement :** To decrease the count by 1.
9. **Demultiplexer :** Combination logic block that distributes data from single input to one-of-X outputs. Also called a distributor. Can change serial to parallel data.
10. **8421 BCD code :** Four-bit BCD code with weighting of 8, 4, 2 and 1.
11. **Encoder :** A logic device that translates from decimal to another code such as binary. Generally, it translates input information to a code useful to digital circuits.
12. **Full-adder :** Digital circuit with three inputs for carry in and two bits with sum and carry out outputs.
13. **Half-adder :** Digital circuit that will add two bits and output a sum and carry. Cannot handle carry inputs.
14. **Light-emitting diode :** Special PN junction that gives off light when current flows through it. Has lens to focus the light.
15. **Multiplex :** In driving displays, to turn on/off one of several displays each for a short time in turn at a high enough frequency so they appear to be lit continuously. In general, transmitting several signals over common lines.
16. **Multiplexer :** Combinational logic block selects one-of-*X* inputs and directs the information to a single output. Also called a data selector. Can change parallel to serial data.

SUMMARY

- Basically, the digital circuits are of one of the following types :
  (i) Combinational circuits
  (ii) Sequential circuits.

  A combinational circuit may be defined as a logic circuit the output of which depends only upon the combination of the inputs. The output does not depend on the past value of inputs or outputs.
- A combinational circuit can have a number of inputs and a number of outputs. This circuit has n inputs and m outputs. Between the inputs and outputs, logic gates are connected, and hence, combinational circuit basically consists of logic gates.
- The various steps involved in designing procedure of a combinational logic may be listed as under :
  (i) We will be given a problem.
  (ii) Then, we determine the number of inputs and outputs and assign letter symbols to input and output variables.
  (iii) After that we write a truth table relating the inputs and outputs.
  (iv) Then, we write K-map for each output and obtain the simplified Boolean expression for each output.
  (v) Lastly, we draw the logic diagram (combinational circuit).
- The combinational circuits may be classified as under :
  (i) Code converters.
  (ii) Binary and BCD adders.
  (iii) Binary and BCD subtractors.
  (iv) Digital comparators.
- The binary adders may be of following two types :
  (i) Half adder, (ii) Full adder

- Half adder is a combinational logic circuit with two inputs and two outputs. It is the basic building block for addition of two single bit numbers. This circuit has two outputs namely carry and sum.
- To overcome the drawback of an Half Adder circuit, we develop a 3 single bit adder circuit called Full Adder. It can add two one-bit numbers A and B, and carry C_{in}. Basically, a full adder is a three input and two output combinational circuit.
- Half subtractor may be defined as a combinational circuit with two inputs and two outputs (i.e., difference and borrow). A half subtractor produces the difference between the two binary bits at the input and also produces an output (i.e., borrow) to indicate if a 1 has been borrowed. In subtraction (A – B), A is called as minuend bit and B is called as subtrahend bit.
- A full adder is capable of adding only two single digit binary numbers along with a carry input. But, in practice, we need to add binary numbers which are much longer than just one bit. To add two n-bit binary numbers, we need to use the n-bit parallel adder
- An Excess-3 code may be obtained by adding 3 to a BCD code. As an example, the EXS-3 code of 0 is 0011. The EXCESS-3 adder performs the addition of two EXS-3 number.
- Because of availability of combinational circuits in the IC form; the design of combinational circuits has been simplified to a great extent. It also improves the reliability of the system because the number of external wired connections.
- A multiplexer is a digital circuit which selects one of the n data inputs and routes it to the output. The selection of one of the n inputs is done by the select inputs. To select n inputs, we require m select lines such that $2^m = n$. Depending on the digital code applied at the select inputs, one out of n data sources is selected and transmitted to the single output Y. E is called as a strobe or enable input which is useful for cascading.
- In most of the electronic systems, the digital data is available on more than one lines. It is essential to route this data over a single line. Under such circumstances, we require a circuit which select one of the many inputs at a time. This circuit is nothing else but a multiplexer which has many inputs, one output and some select inputs. Multiplexer improves the reliability of the digital system because it reduces the number of external wired connections.
- The advantages of multiplexers :
  - (i) It reduces the number of wires.
  - (ii) It reduces the circuit complexity and cost.
  - (iii) We can implement many combinational circuits using MUX.
  - (iv) It simplifies the logic design.
  - (v) It does not need the K-maps and simplification.
- Some of the important applications of a multiplexer may be listed as under :
  - (i) It is used as a data selector to select one out of many data inputs.
  - (ii) It is used for simplification of logic design.
  - (iii) In the data acquisition system.
  - (iv) In designing the combinational circuits.
  - (v) In the D/A converters.
  - (vi) To minimize the number of connections.
- The multiplexers having more number of inputs can be obtained by cascading two or more multiplexers with less number of inputs. This is known as a **multiplexer tree**.
- Similar to multiplexer, we can construct the demultiplexer with more number of lines using demultiplexers having lower number lines. This is called as demultiplexer tree. It is also called as cascading of demultiplexers.

SHORT QUESTIONS WITH ANSWER

Q.1. How can you classify combinational circuits?

Ans. The combinational circuits may be classified as under :

(i) Code converters. (ii) Binary and BCD adders.

(iii) Binary and BCD subtractors. (iv) Digital comparators.

Q.2. What is meant by an Half adder?

Ans. Half adder is a combinational logic circuit with two inputs and two outputs. It is the basic building block for addition of two single bit numbers. This circuit has two outputs namely carry and sum.

Q.3. Explain about Binary multiplier.

Ans. Multiplication of binary number is performed in the same way as in decimal numbers. The multiplicand is multiplied by each bit of the multiplier starting from the least significant bit. Each such multiplication forms a partial product, successive partial products are shifted one position to the left. The final product is obtained from the sum of the partial products.

Q.4. What is necessity of Multiplexers?

Ans. In most of the electronic systems, the digital data is available on more than one lines. It is essential to route this data over a single line. Under such circumstances, we require a circuit which select one of the many inputs at a time. This circuit is nothing else but a multiplexer which has many inputs, one output and some select inputs. Multiplexer improves the reliability of the digital system because it reduces the number of external wired connections.

Q.5. Write the advantages of Multiplexers.

Ans. The advantages of multiplexers are :

(i) It reduces the number of wires.

(ii) It reduces the circuit complexity and cost.

(iii) We can implement many combinational circuits using MUX.

(iv) It simplifies the logic design.

(v) It does not need the K-maps and simplification.

Q.6. What is a combinational circuit?

Ans. In a combinational circuit, the output depends upon present input(s) only i.e., not dependent on the previous input(s). The combinational circuit has no memory element. It consists of logic gates only.

Q.7. Write two characteristics of combinational circuits.

Ans. The two characteristics of combinational circuits are :

(i) In combinational circuits, the output exists as long as the input exists.

(ii) A combinational circuit will always respond in the same fashion to input function, when we apply signal to the input terminal of the combinational logic circuit.

Q.8. What is a half-adder?

Ans. A logic circuit, that can add two 1-bit numbers and produce outputs for sum and carry, is called a half-adder.

Q9. What is a full-adder?

Ans. A binary adder, which can add two 1-bit binary numbers along with a carry bit and produces outputs for sum and carry is called a full-adder.

Q.10. What is the logic signal required to inhibit a NAND gate and what is its output when inhibited.

Ans. Logic 0 is the logic signal required to inhibit a NAND gate and 1 is its output, irrespective of all other inputs, when inhibited.

Q.11. What is multiplexing?

Ans. Multiplexing is an operation performed with digital logic circuits called multiplexers and allows information to the switched from several lines into a single line is a specified sequence.

Q.12. What is an encoder?

Ans. An encoder is a logic circuit that produces coded binary inputs from uncoded inputs.

Q.13. What is a demultiplexer?

Ans. Demultiplexer is a logic circuit that accepts one data input and distributes it over several outputs.

Q.14. What is a decoder?

Ans. Decoder is a logic circuit that decodes from binary to octal, decimal hexadecimal or any other code such as 7-segment circuit etc.

Q.15. What are the applications of multiplexer?

Ans. Multiplexer have the applications of :

(i) data selection and routing
(ii) operation sequencing,
(iii) waveform generation,
(iv) logic function generation,
(v) parallel to serial conversion.

REVIEW QUESTIONS

1. Realize the full adder circuit using NAND gates only.
2. Realize the full subtractor circuit using NAND gates only.
3. Design a four bit BCD adder using IC 7483 and NAND gates only.
4. An combinational circuit has four inputs and one output. The output is equal to 1 when :
   (i) All inputs are equal to 1.
   (ii) None of the inputs are equal to 1.
   (iii) An odd number of inputs are equal to 1.
      (a) Obtain the truth table.
      (b) Obtain the simplified output in sum of product form.
      (c) Find the simplified output in product of sum form.
      (d) Draw the logic diagram for SOP equation of part (b).
5. A staircase light is controlled by two switches, one at the top of stairs and other at bottom of stairs.
   (i) Prepare a truth table for this system.
   (ii) Write minimized logic equation in SOP form.
   (iii) Realize the circuit using minimum number of
      (a) NAND gates and (b) NOR gates
6. Design the combinational logic circuit which accepts three bit input multiplies the input by two and subtracts three from it while giving the output. Implement the circuit using basic gates.
7. Explain the functional diagram and truth table of comparator IC 7485. Implement 3 bit comparator using the same.
8. Write notes on the Look-ahead carry adder.
9. Explain the working of a half adder? Draw its logic diagram.
10. Draw the logic diagram of full-adder and its truth table.
11. Give the circuit diagram of a full-adder using NAND gate and explain. Give its truth table.
12. What is similar between a half-adder and a half subtractor?
13. With suitable block diagram explain the operation of n-bit serial adder.
14. Write a note on a 4-bit parallel binary adder.
15. Draw pin diagram of IC 7483 and explain its operation as 4-bit binary adder.
16. What is half subtractor? Draw the logic diagram and truth table of half subtractor.
17. What is meant by a full subtractor? Draw a full subtractor circuit.
18. Draw the circuit diagram of single digit BCD adder using IC 7483.
19. Implement a full subtractor circuit using only NAND gates.
20. In detail explain the working of 4-bit binary adder/subtractor using IC 7483.
21. Draw half subtractor circuit. Use NAND gates only. Explain its working.
22. What is meant by a multiplexer? Explain with block diagram the principle of multiplexing.
23. Explain with diagram and truth table the operation of 4 : 1 MUX.
24. Explain briefly with pin-diagram the following ICs IC-74157, IC-74151, IC-74150.
25. What is 'Multiplexer tree'?
26. Explain with diagram 64 to 1 multiplexer using four 16 to 1 multiplexer.
27. Explain with diagram 32 : 1 multiplexer using three of 8 : 1 multiplexer.
28. Explain with diagram 32 : 1 multiplexer using two 16 : 1 multiplexer.

29. Draw the circuit diagram of 1 to 4 line demultiplexer and explain the working.
30. Draw circuit diagrams for the following :
    (i) 2 : 1 multiplexer
    (ii) 1 : 2 demultiplexer

NUMERICAL PROBLEMS

1. Implement the following functions using 8 : 1 MUX.
   (i) f (A, B, C, D) = Σm (2, 4, 5, 7, 10, 14)
   (ii) f (A, B, C, D) = Σm (0, 2, 4, 6, 8, 10, 12, 14)
2. Design a full adder using 74138 IC.
3. Design an odd parity generator for a 3-bit message code.
4. Implement the following functions with 8 : 1 MUX.
   (i) F (w, x, y, z) = Σm (0, 1, 3, 5, 8, 9, 15)
   (ii) F (A, B, C) = πM (2, 3, 4, 7)
5. Draw and explain the design of 32:1 MUX using 8 : 1 MUX and 4:1 MUX.
6. Design a full adder using 3 : 8 decoder.
7. Implement 16 : 1 multiplexer using 4 : 1 MUX ICs.
8. Design and implement a BCD to seven segment decoder using truth table, K-maps and logic gates.
9. Using 1 : 8 DEMUX, implement the following function:
   F_1 (A, B, C) = Σm (0, 1, 3, 5, 7)
   F_2 (A, B, C) = Σm (0, 1, 2, 4)
   F_2 (A, B, C) = Σm (1, 2, 3, 4, 6)
10. What is multiplexer? Draw the logic diagram of 4 : 1 multiplexer using gates and explain its working.
11. Implement the following:
    1. 4 bit prime number detector using 8 : 1 multiplexer IC.
    2. 2-bit comparator using 4 : 16 line decoder IC (generate only EQUALITY output).
12. Implement the following functions using 4 : 16 decoder IC:
    (i) f_1(A, B, C, D) = Σm(0, 2, 8, 10) + d (4, 7, 6, 11, 15)
    (ii) f_2(A, B, C, D) = $\bar{C}D + B\bar{C}\bar{D} + BC\bar{D} + A\bar{B}C\bar{D}$
13. Present an algorithm for performing subtraction using adder.
14. Design a parity generator to generate even parity bit for a 4-bit word. Use EX-OR and $\overline{\text{EX-OR}}$ gate.
15. Design a BCD to Excess-3 code converter using minimum number of NAND gates.
16. Implement the following function using a 4-to-16 line decoder:
    f = Σm (1, 2, 4, 7, 8, 11, 12, 13).
17. Implement the expression using a multiplexer
    f(A,B,C,D) = Σm (0, 2, 3, 6, 8, 9, 11, 12, 14)

PART—B

5. **Sequential Circuits:** Flip flips, SR, JK, T, D, and Master slave, Excitation table, Edge trigerring, Level Trigerring, Realization of one flip-flop using other flip flops. Asynchronous/ Ripple counters, Synchronous counters, Modulo-n counter, Ring counters, Classification of sequential circuits-Moore and Mealy, Design of Synchronous counters; state diagram, Circuit implementation, Shift registers.
6. **Memory Devices:** Classification of memories, RAM organization, Write operation, Read operation, Memory cycle, Static RAM Cell-Bipolar, RAM cell, MOSFET RAM cell, Dynamic RAM cell, ROM organization, PROM, EPROM, EEPROM, Field Programmable Gate Arrays (FPGA).
7. **Signal Conversions:** Analog and Digital signals, A/D and D/A conversion techniques (Weighted type, R-2R Ladder type, Counter Type, Dual Slope type, Successive Approximation type).

Above syllabus has been covered in following chapters :

Chapter 6 : Synchronous Circuits : Flip-flops

Chapter 7 : Synchronous Circuits : Shift Registers

Chapter 8 : Synchronous Circuits: Counters

Chapter 9 : Memory Devices

Chapter 10: Signal Conversions

6

SYNCHRONOUS CIRCUITS: FLIP-FLOPS

Syllabus

Synchronous Circuits: Flip flips, SR, JK, T, D, and Master slave, Excitation table, Edge trigerring, Level Trigerring, Realization of one flip-flop using other flip flops. Asynchronous/Ripple counters, Synchronous counters, Modulo-n counter, Ring counters, Classification of sequential circuits-Moore and Mealy, Design of Synchronous counters; state diagram, Circuit implementation, Shift registers.

6.1 INTRODUCTION

6.1.1 Combinational Circuits : Definition

Upto now, we have discussed only the combinational circuits. We know that the output of a combinational circuit at any instant of time, depends only upon the levels present at input terminals. Further, the combinational circuits do not use any memory. This means that the previous state of input does not have any effect on the present state of the circuit. The sequence in which the inputs are being applied has no effect on the output of a combinational circuit.

6.1.2 Sequential Circuits : Definition

In case of a sequential circuit, the *timing* parameter comes into effect. Infact, the output of a sequential circuit depends upon the present time inputs, the previous output and the sequence in which the inputs are applied.

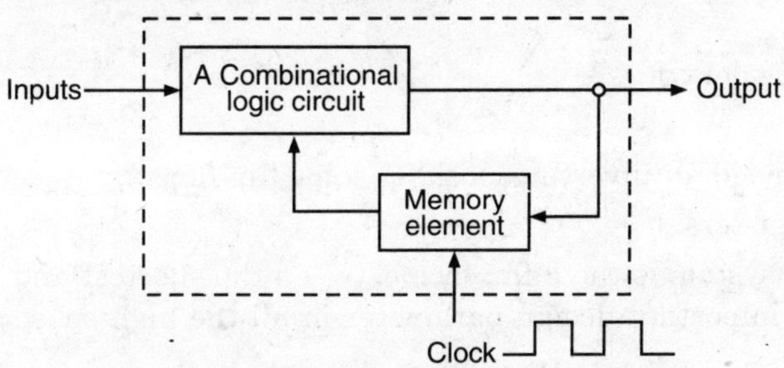

Fig. 6.1. Block diagram of a sequential circuit

Chapter Outline

- Introduction
- One-Bit Memory Cell (Basic Bistable Element)
- S-R Latch using NOR gates
- NAND Latch (S-R) Latch using NAND Gates)
- The Gated S-R Latch i.e., (level triggered S-R Flip-flop)
- The Gated D latch (Level triggered D Flip-flop)
- Clocked S-R Flip-flop (Edge triggered)
- Clocked D Flip-flop (edge triggered)
- JK latch (level triggered JK Flip-flop)
- Clocked J-K Flip-flop (i.e., edge triggered)
- Toggle Flip flop (T Flip-flop)
- Master Slave SR Flip-flop
- Master Slave JK Flip-flop
- Preset and Clear Inputs
- Excitation Table of Flip-flop
- Conversion of One Type of FF to the other
- Applications of Flip-flops

In order to provide the previous input or output, we require a memory element. Therefore, a sequential circuit requires a memory element. Figure 6.1 illustrates the block diagram of a sequential circuit which includes the memory element in feedback path.

1. Present State of Sequential Circuit

The data stored by the memory element at any given instant of time is known as the present state of the sequential circuit.

2. Next State

The combinational circuit operates on the external inputs and the present state to produce new outputs. Some of these new outputs are stored in memory element and known as the next state of the sequential circuit. The most important part of the sequential circuit is the memory element. The memory element of figure 6.1 is known as flip-flop (FF). It is the most basic memory element.

JEFFREY M. WILKINS

Jeffrey M. Wilkins founded the on-line computer service CompuServe in 1969 when he was 25 years old. The company served businesses at first and then evolved when Wilkins realized that the growing popularity of personal computers presented a new market. In 1985 Wilkins founded another communications industry business, Metatec Corporation.

6.1.3 Clock Signal : Definition and Importance

The clock signal depicted in figure 6.2 is a timing signal. Every sequential signal has this timing signal applied. Infact, clock is a rectangular signal as shown in figure 6.2, with a duty cycle equal to 50%. The clock signal repeats itself after every T seconds. Therefore, the clock frequency is F = 1/T.

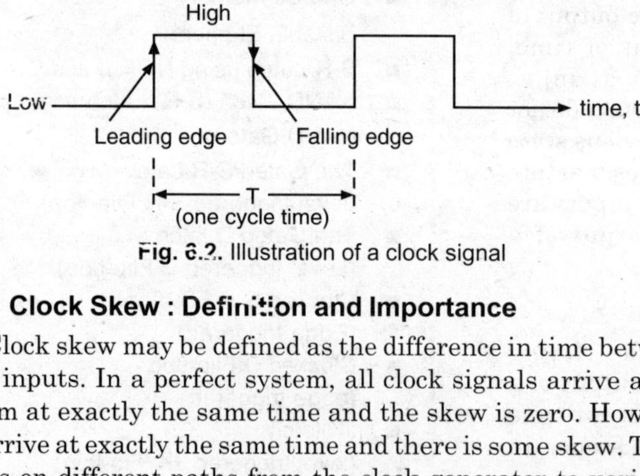

Fig. 6.2. Illustration of a clock signal

DO YOU KNOW?

The flip-flop is known by other names, including latch and bistable miltivibrator.

6.1.4 Clock Skew : Definition and Importance

Clock skew may be defined as the difference in time between the clock edges arriving at a pair of clock inputs. In a perfect system, all clock signals arrive at all the various clock input pins of the system at exactly the same time and the skew is zero. However, in real time systems, the edges do not arrive at exactly the same time and there is some skew. The clock skew occurs because of different delays on different paths from the clock generator to various circuits. The major reasons for this may be described as under :

(i) Different length of wires (wires introduce delay).

(ii) Gates (buffers) on the paths.

(iii) Flip-flops that clock on different edges (need for inverting clock for some flip-flops).

(iv) Gating the clock to control loading of registers.

The maximum allowable clock skew for the system is the difference between the shortest and longest path delays. Actually, clock skew is an important design parameter in all the high-speed clock systems.

6.1.5 Performance Comparison of Combinational and Sequential Circuits

S. No.	Parameter of comparison	Combinational Circuits	Sequential Circuits
1.	Output depend on	Inputs present at that instant of time.	Present inputs and past inputs/ outputs.
2.	Memory	Not necessary	Necessary
3.	Clock input	Not essential	Essential
4.	Important Examples	Adders, subtractors, code converters	Flip-flops, shift registers, counters.

6.2 ONE-BIT MEMORY CELL (BASIC BISTABLE ELEMENT)

1. Definition

A flip-flop is popularly known as a basic digital memory circuit. It has two stable states namely logic 1 state and logic 0 state. It can be designed either using NOR gates or NAND gates. A flip-flop can be designed by using the fundamental circuit shown in figure 6.3. NAND gates 1 and 2 are basically acting as inverters. Therefore, this circuit is known as a cross coupled inverter. Here, output of gate 1 is connected to the input of gate-2 and output of gate 2 is connected to input of gate-1 as shown in figure 6.3.

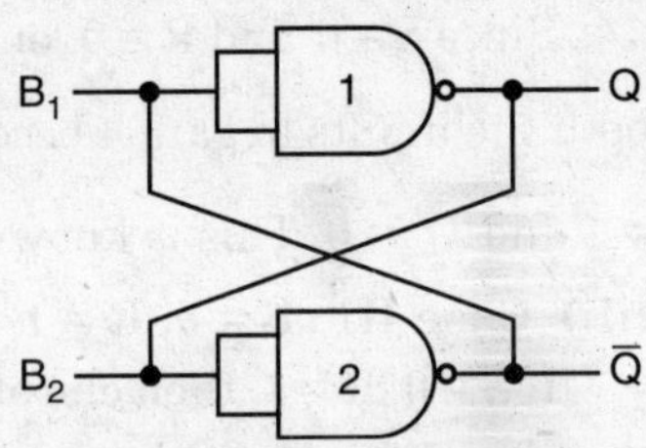

Fig. 6.3. A cross coupled inverter as memory element

2. Working Operation

We assume that output of gate-1, i.e. Q = 1. Therefore, $B_2 = 1$.

Since $B_2 = 1$, output of gate-2, i.e., $\overline{Q} = 0$. This makes $B_1 = 0$. Therefore, Q continues to be equal to 1. In a similar manner, we can demonstrate that if we start with Q = 0, then we shall get Q = 0 and $\overline{Q} = 1$.

3. Conclusions

From the above discussion, we can draw the following conclusions :

(i) The outputs of circuit (Q and $\overline{Q}$) will always be complementary. This means that if Q = 0 then $\overline{Q} = 1$ and vice versa. They will never be equal $Q = \overline{Q} = 0$ or 1 is an invalid state.

(ii) This circuit has two stable states. One stable state corresponds to Q = 1, $\overline{Q} = 0$ and it is called as 1 state or **set state.** On the other hand, another stable state corresponds to Q = 0, $\overline{Q} = 1$ and it is called as 0 state or **reset state.**

(iii) If the circuit is in the reset state (Q = 0, $\overline{Q} = 1$), then it continues to be in the reset state and if it is in the set state (Q = 1, $\overline{Q} = 0$), then it continues to remain in the set state.

(iv) This characteristics of the circuit illustrates that it can store 1 bit of digital information. Therefore, it is called as a 1-bit memory cell.

6.2.1 Latch

1. Definition

The crosscoupled inverter of figure 6.3 is capable of locking or latching the information. Therefore, this circuit is also known as a latch. The drawback of this circuit is that we cannot enter the desired digital data into it. This drawback can be overcome by modifying the circuit as shown in figure 6.4. This modification allows us to enter the desired digital data into the circuit.

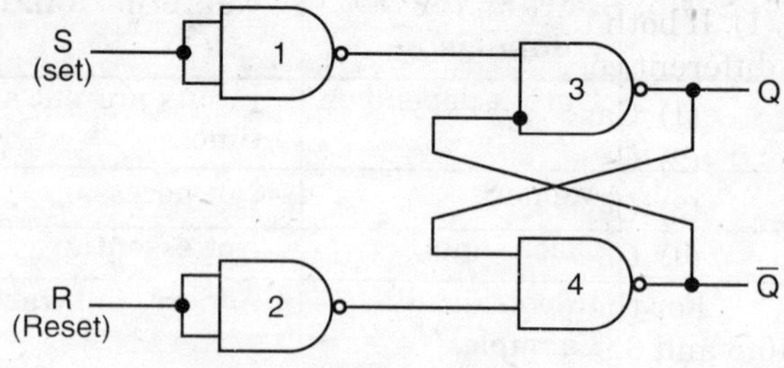

Fig. 6.4. Modified memory cell

2. Working Operation

(i) Case I : S = R = 0

The outputs of gates 1 and 2 become 1. Let Q = 0 and $\overline{Q}$ = 1 initially. Therefore, both the inputs to gate-3 are 1 and the inputs to gate-4 are (01). Therefore, gate-3 output, i.e., Q = 0 and gate-4 output $\overline{Q}$ = 1. Thus, with S = R = 0, there is no change in the state of outputs.

(ii) Case II : S = 1, R = 0

Since S = 1 and R = 0, one of the inputs to gate-3 will be 0. This forces Q output to 1. Therefore, both the inputs to gate-4 become 1. This forces $\overline{Q}$ to 0. Therefore, for S = 1, R = 0, the outputs are Q = 1 and $\overline{Q}$ = 0. This is known as set state.

(iii) Case III : S = 0, R = 1

If S = 0, R = 1, then one of the inputs to gate-4 becomes 0. This forces the $\overline{Q}$ output to 1. Therefore, both the inputs to gate-3 will be 1. This forces Q to 0. Hence, for S = 0, R = 1, the outputs are Q = 0, $\overline{Q}$ = 1. This is called as reset state or clear state.

> **DO YOU KNOW?**
>
> The term latch is used for certain types of flip-flops.

(iv) Case IV : S = R = 1

If S = R = 1, then outputs of gates 1 and 2 become zero. Hence, one of the inputs to gates 3 and 4 will be 0. Therefore, outputs Q and $\overline{Q}$ both become 1. It is not allowed as Q and $\overline{Q}$ should be complementary. Hence, S = R = 1 condition is prohibited.

6.3 S-R LATCH USING NOR GATES

(PTU, Jalandhar, Sem. Exam., 2009-10)

As a matter of fact, latch is a bistable element with two stable states. It has two outputs Q and $\overline{Q}$ which are complementary to each other. Latch is a sequential logic circuit which checks all its inputs continuously and will change its output as soon as the input changes without waiting for the clock signal. Usually, an enable signal is provided for a latch. When the enable signal is active, the output changes as soon as there is a change in input. S-R (set – reset) latch is the simplest type of latch and figure 6.5 shows S-R latch constructed using the NOR gates. Two NOR gates are cross coupled. This means that the output of NOR gate 1 is connected to one of the inputs of the NOR gate 2. And output of NOR gate 2 is connected to one of the inputs of NOR gate 1. The outputs Q and $\overline{Q}$ can be expressed using Boolean equations as under :

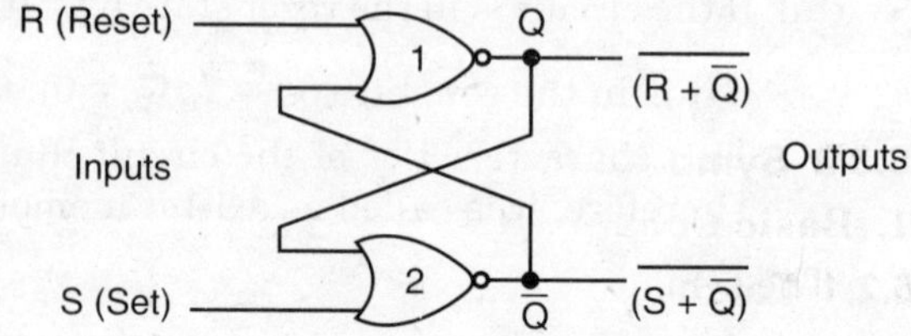

Fig. 6.5. S-R latch using NOR gates

$$Q = \overline{R + \overline{Q}} \text{ and } \overline{Q} = \overline{S + Q}$$

6.3.1 Working Operation of S-R Latch

Recall from chapter 3 that the NOR gate output goes to 0 when any one input becomes HIGH (1). If both the inputs are high (1), then the output is 0. Let us understand the operation under four different conditions.

(1) Case I : S = 1, R = 0

(2) Case II : S = 0, R = 1

(3) Case III : S = 0, R = 0

(4) Case IV : S = 1, R = 1.

For the working operation, we consider figures 6.6, 6.7, 6.8 and 6.9.

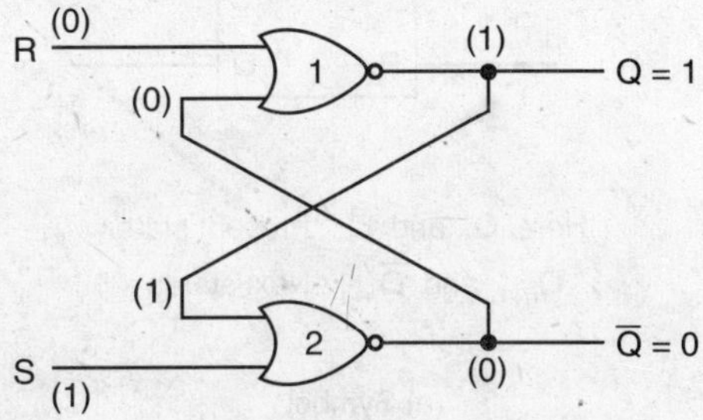

Fig. 6.6.

1. Case I : S = 1, R = 0 : As S = 1, output of NOR gate 2, i.e., $\overline{Q}$ = 0. Hence, both inputs of NOR gate (1) are 0. Therefore, Q = 1. Hence, both inputs of NOR gate (2) are 1. Hence, $\overline{Q}$ = 0. Therefore, For S = 1, R = 0, Q = 1 and $\overline{Q}$ = 0.

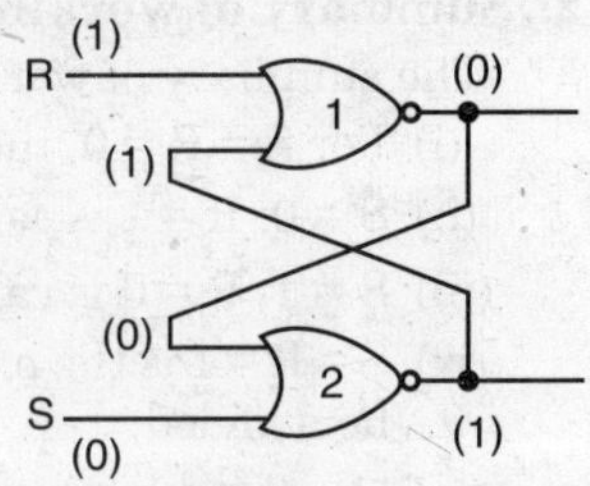

Fig. 6.7.

2. Case II : S = 0, R = 1 : As R = 1, output of NOR-gate (1), i.e., Q = 0. Hence, both inputs to NOR gate (2) are 0. Hence, its output of NOR gate (2), i.e., $\overline{Q}$ = 1. Therefore, for S = 0 and R = 1, Q = 0 and $\overline{Q}$ = 1.

3. Case III : S = 0, R = 0: We know that $\overline{Q} = \overline{S+Q}$ and $Q = \overline{R+\overline{Q}}$. Let us substitute S = R = 0 Therefore, $\overline{Q} = \overline{0+Q}$ and $Q = \overline{0+\overline{Q}}$.

Therefore, $\overline{Q} = \overline{0}.\overline{Q}$ and $Q = \overline{0}\,.\,Q$ according to DeMorgans theorems.

Therefore, if S = R = 0, then Q and $\overline{Q}$ do not change their states.

4. Case IV : S = 1, R = 1 : If S = R = 1, then the outputs Q and $\overline{Q}$ both are forced to 0. Actually, this is an indeterminate state which must be avoided. This state must not be used because it violates the basic requirement of a latch, i.e., Q and $\overline{Q}$ must be complements of each other.

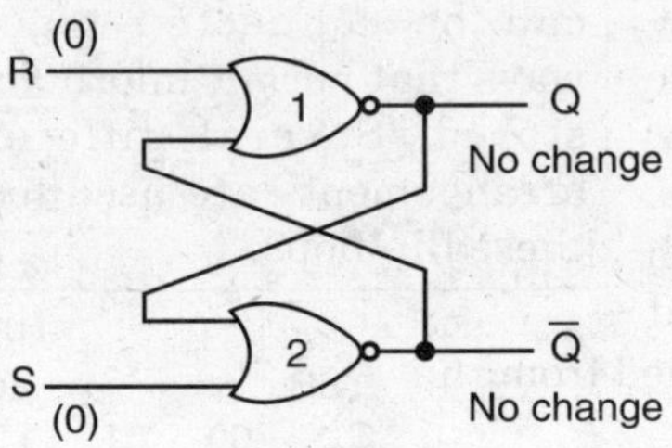

Fig. 6.8.

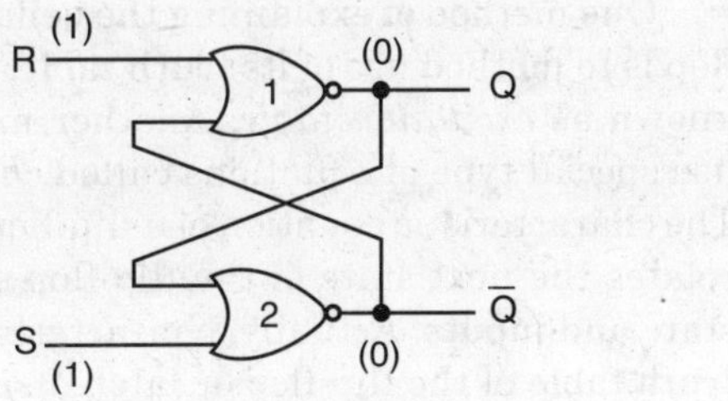

Fig. 6.9.

6.3.2 Symbol and Truth Table of S-R Latch

1. Basic Concepts

The symbol and truth table of S-R latch using NOR gates are as shown in figure 6.10(a) and (b) respectively. In truth table, Q_n and $\overline{Q}_n$ represent the present states of outputs, i.e., the outputs before applying a new set of inputs. Q_{n+1} and $\overline{Q}_{n+1}$ represent the next states of outputs, i.e., the outputs after applying a new set of inputs.

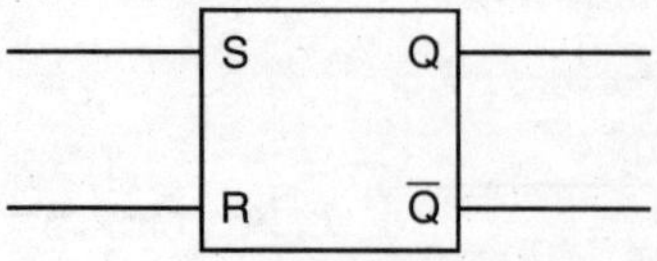

Here, Q_n and $\overline{Q}_n$: Present states

Q_{n+1} and $\overline{Q}_{n+1}$: Next states

(a) Symbol

Inputs		Outputs				Comment
S	R	Q_n	$\overline{Q}_n$	Q_{n+1}	$\overline{Q}_{n+1}$	
0	0	0	1	0	1	No change (NC)
0	0	1	0	1	0	
0	1	0	1	0	1	Reset
0	1	1	0	0	1	
1	0	0	1	1	0	Set
1	0	1	0	1	0	
1	1	0	1	×	×	Prohibited state
1	1	1	0	×	×	

(b) Truth table of S-R latch

Fig. 6.10.

2. Summary of working operation of S-R latch

The summary of working operation of an S-R latch is as under :

(i) For S = R = 0, the latch output does not change.

(ii) S = 0, R = 1 is called as the Reset condition as Q = 0 and $\overline{Q}$ = 1.

(iii) S = 1, R = 0 is called as the Set condition as Q = 1 and $\overline{Q}$ = 0.

(iv) S = R = 1 is the prohibited state. The output is unpredictable. Therefore, this condition must be avoided.

3. Race Condition

The condition S = R = 1 is called as Race condition. When any one input to a NOR gate is 1, its output becomes 0. Thus, both the outputs try to become 0. This is known as the *RACE* condition.

DO YOU KNOW?

The most important memory element is the flip-flop, which is made up of an assembly of logic gates. Even though a logic gate, by itself, has no storage capability, several can be connected together in ways that permit information to be stored. Several different gate arrangements are used to produce these flip-flops.

6.3.3 Characteristic Equation

(PTU, Jalandhar, Sem. Exam; 2006-07) (10 marks)

One method of explaining the behaviour of a latch or flip-flop is to method use of its truth table. The truth table is also known as *excitation table.* Another method of doing it is to use special type of equations called *characteristic equations.* The characteristic equation of a flip-flop is the equation which relates the next state of the flip-flop or latch to the current state and inputs. Actually, characteristic equation is obtained from the truth table of the flip-flop or latch, using the *K*-map.

Characteristic Equation of SR Latch

(PTU, Jalandhar, Sem. Exam., 2006-07)

Let us consider the truth table of SR latch and write the K-map for Q_{n+1} as shown in figure 6.11. After simplification, the characteristic equation of SR latch is expressed as,

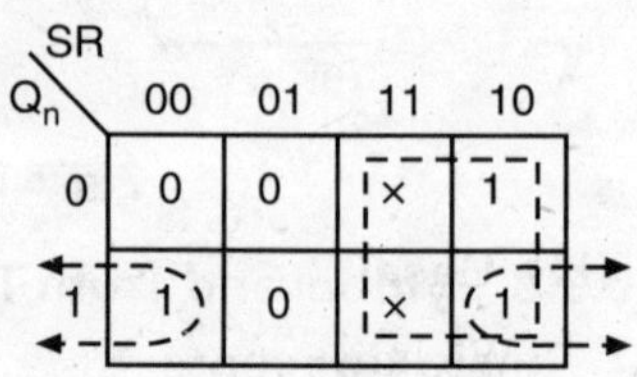

Fig. 6.11. K-map for S-R latch

$$Q_{n+1} = S + \overline{R}\,Q_n \quad ...(6.1)$$

6.4. NAND LATCH (S-R LATCH USING NAND GATES)

1. Basic Concepts

We can also construct a S-R latch with NAND gates as shown in figure 6.12(a). Here, it may be noted that the outputs of two NAND gates are cross connected, in an identical manner as that in the NOR latch.

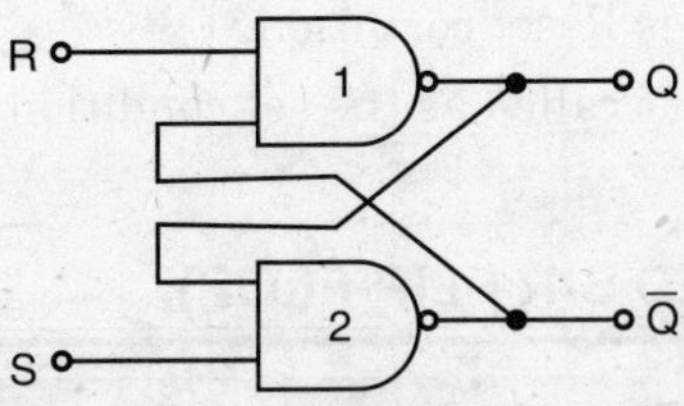

(a) NAND latch

S	R	Q_{n+1}	$\overline{Q}_{n+1}$	
0	0	Race	Race	
0	1	0	1	→ Reset
1	0	1	0	→ Set
1	1	NC (Q_n)	NC($\overline{Q_n}$)	→ Inactive

(b) Truth table

Fig. 6.12.

Figure 6.12 (a) shows the NAND latch and figure 6.12 (b) shows its truth table. The working operation of S-R NAND latch has been summarized in figure 6.13.

(*i*) Case I : S = 0, R = 0 : Race

When any one input of a NAND gate becomes 0, its output is forced to 1. Here, S = R = 0. Therefore, Q and $\overline{Q}$ both are forced to be equal to 1. This is an undeterminate state and hence must be avoided. This is also called as *Race condition*.

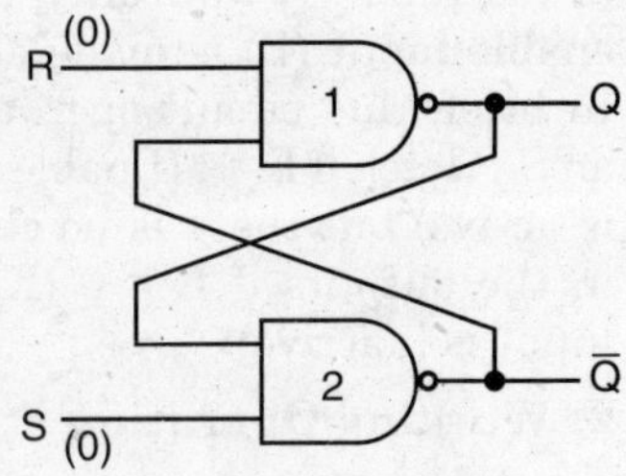

Fig. 6.13. *S* = 0, *R* = 0.

(*ii*) Case II : S = 0, R = 1 : Reset

Since S = 0, it forces $\overline{Q}$ to be 1. Hence, both inputs to NAND gate-1 are 1. Hence, Q = 0. Thus, with S = 0 and R = 1, the outputs are Q = 0 and $\overline{Q}$ = 1. This is called as reset condition.

(iii) S = 1, R = 0 : Set

Since R = 0, Q is forced to 1. Hence, both inputs to NAND gate-2 are 1. Hence, $\overline{Q}$ = 0. Thus, with S = 1 and R = 0, the outputs are Q = 1 and $\overline{Q}$ = 0. This is called as set condition.

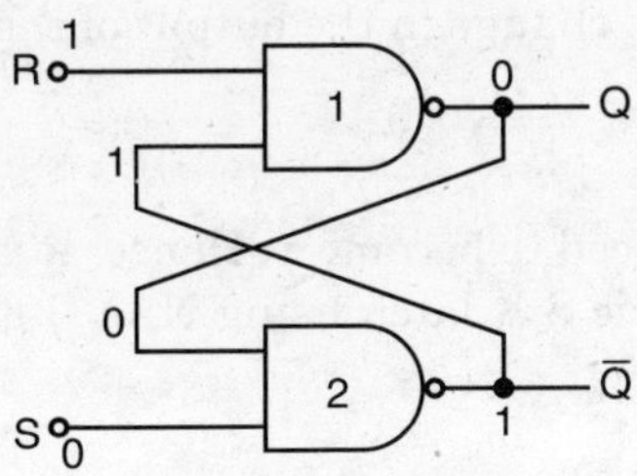

Fig. 6.14.

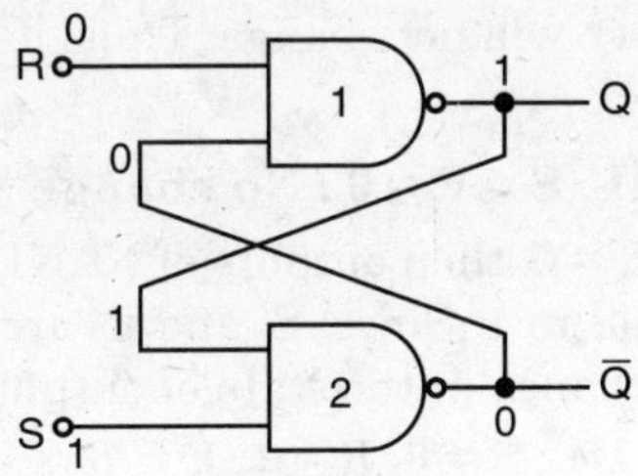

Fig. 6.15.

(iv) Case IV : S = 1, R = 1 : No change

We have $Q_{n+1} = \overline{R \cdot \overline{Q}_n}$ and $\overline{Q}_{n+1} = \overline{S \cdot Q_n}$

Using DeMorgan's theorem,

$$Q_{n+1} = \overline{R} + Q_n \text{ and } \overline{Q}_{n+1} = \overline{S} + \overline{Q_n}.$$

Let us substitute $\overline{R}$ = 0 and $\overline{S}$ = 0 to get, $Q_{n+1} = 0 + Q_n = Q_n$ and $\overline{Q}_{n+1} = 0 + \overline{Q}_n = \overline{Q}_n$.

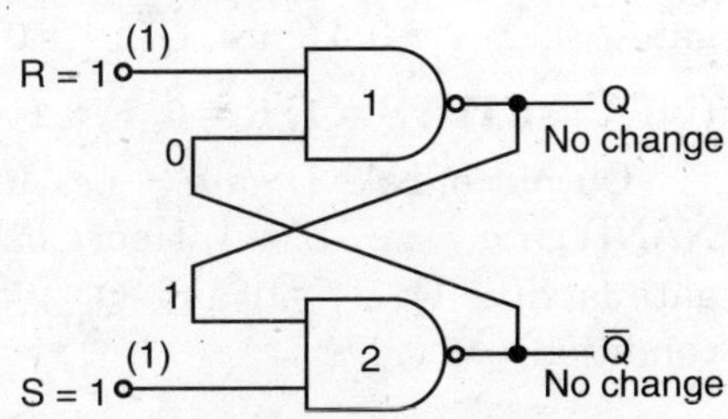

Fig. 6.16.

Thus, there is no change in the output if S = R = 1. The symbol for S-R latch using NAND gates is shown in figure 6.17 along with the summary of operation.

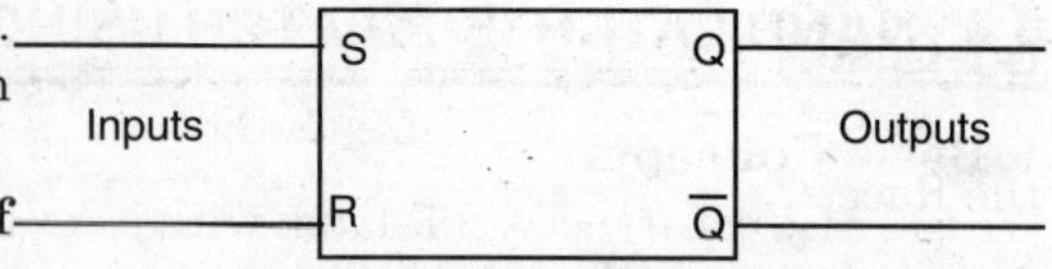

Fig. 6.17. Symbol of S-R latch.

2. Circuit Symbol and Summary of Operation of S-R Latch

(i) For S = 0, R = 0, both the outputs will be forced to become 1. This is RACE condition and must be avoided.

(ii) For S = 0, R = 1, the outputs are Q = 0, $\overline{Q}$ = 1 and it is Reset condition.

(iii) For S = 1, R = 0, the outputs are Q = 1, $\overline{Q}$ = 0 and it is called as the set condition.

(iv) For S = R = 1, there is no change in the output state.

6.5 THE GATED S-R LATCH (i.e., LEVEL TRIGGERED S-R FLIP-FLOP)

1. Basic Concepts

Figure 6.18 shows the gated S-R latch. It is basically the S-R latch using NAND gates with an additional enable input. It is also known as level triggered SR-FF. The outputs of simple S-R latch used to change instantly in response to any change made at the input. But, this does not happen with the gated S-R latch. For this circuit, the change in output takes place if and only if the enable input (E) is made active. In brief, this circuit operates as an S-R latch if E = 1 (Enable input is active) but there is no change in the outputs if E = 0 (Enable input is inactive).

> **DO YOU KNOW?**
>
> The term bistable multivibrator is the more technical name for a flip-flop, but, it is too much of a mouthful to be used regularly.

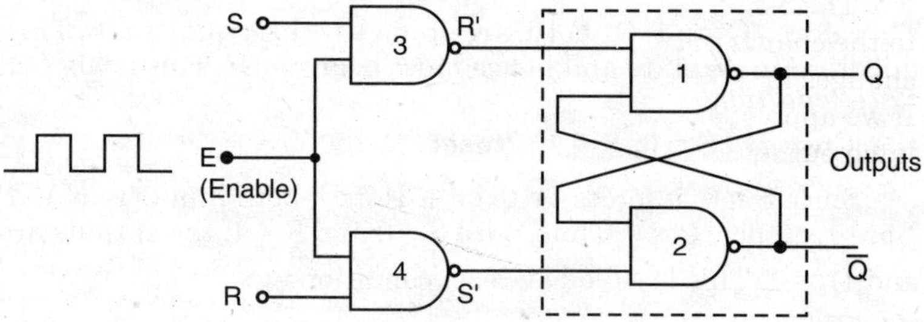

Fig. 6.18. Gated S-R Latch

2. Working Operation

(i) Case I : S = X, R = X, E = 0

Since enable E = 0, the outputs of NAND gates 3 and 4 will be forced to be 1 irrespective of the values of S and R. This means that R′ = S′ = 1. These are the inputs of the S-R latch. Hence, the outputs of NAND latch, i.e., Q and $\overline{Q}$ will not change. Thus, if E = 0, then there is no change in the output of the gated S-R latch.

(ii) Case II : S = R = 0 : No change

If S = R = 0, then outputs of NAND gates 3 and 4 are forced to become 1. Hence, R′ and S′ both will be equal to 1. Since, S′ and R′ are the inputs of the basic S-R latch using NAND gates, there will be no change in the state of outputs.

(iii) Case III : S = 0, R = 1, E = 1

Since S = 0, output of NAND gate-3, i.e., R′ = 1. And, as R = 1 and E = 1, the output of NAND gate-4, i.e., S′ = 0. Hence, $Q_{n+1} = 0$ and $\overline{Q}_{n+1} = 1$. This is the reset condition.

(iv) Case IV : S = 1, R = 0, E = 1

Output of NAND gate 3, i.e., R′ = 0 and output of NAND gate 4, i.e., S′ = 1. Hence output of S-R NAND gate latch is $Q_{n+1} = 1$ and $\overline{Q}_{n+1} = 0$. This is the set condition.

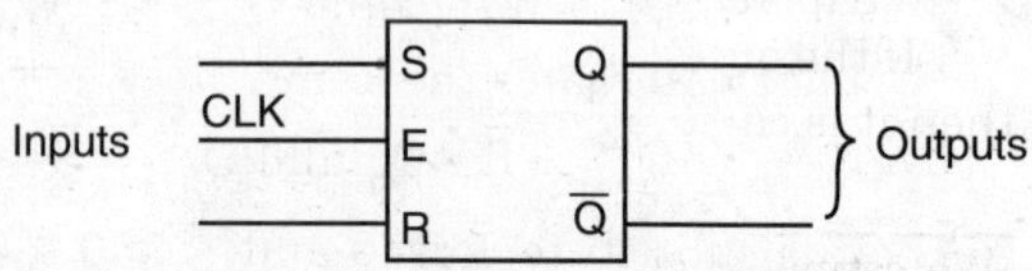

Fig. 6.19. Symbol for S-R latch

(v) Case V : S = 1, R = 1, E = 1

As S = 1, R = 1 and E = 1, the outputs of NAND gates 3 and 4 both are 0, i.e., S′ = R′ = 0. Hence, the Race condition will occur in the basic NAND latch.

3. Symbol and Truth Table

The symbol and truth table of the gates S-R latch are as shown in figures 6.19 and 6.20 respectively.

Inputs				Outputs		Comments
Case	**Enable E**	**S**	**R**	Q_{n+1}	$\overline{Q}_{n+1}$	
I	0	×	×	$\overline{Q}_n$	$\overline{Q}_n$	No change as E = 0
II	1	0	0	$\overline{Q}_n$	$\overline{Q}_n$	No change (NC)
III	1	0	1	0	1	Reset condition
IV	1	1	0	1	0	Set condition
V	1	1	1	Indeterminate		Avoid this condition

Fig. 6.20. Truth table of gated S-R latch

6.5.1 Concept of Level Triggering

The S-R gated latch is said to be a **positive level triggered** circuit, because the circuit responds to the changes in input only when the enable input (E) is HIGH. The output will change as per the changes in input as long as the active level (HIGH for this circuit) is maintained at the enable input. If we apply a rectangular or square waveform at the enable input as shown in figure 6.21, then the latch outputs will change in response to the changes in input, only when the enable input is HIGH.

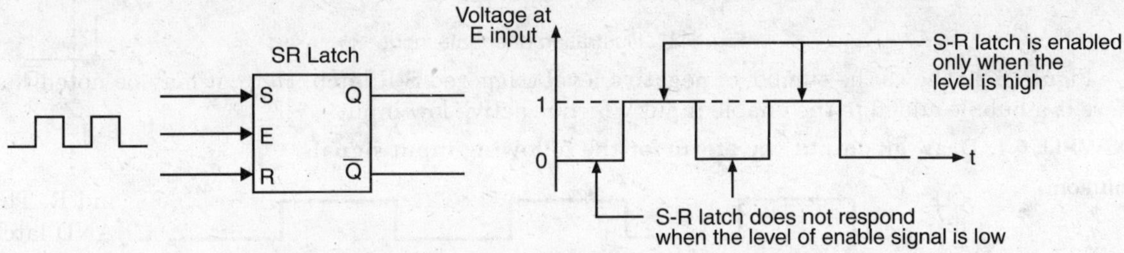

Fig. 6.21. Illustration of level triggered S-R latch

6.5.2 Types of Level Triggered Latches

There are following two types of level triggered latches :

(i) Positive level triggered.

(ii) Negative level triggered.

(i) Positive Level Triggered

If the outputs of S-R latch respond to the input changes, for its enable input at HIGH (1) level, then it is called as the positive level triggered S-R latch.

(ii) Negative Level Triggered

If the outputs of an S-R latch respond to the input changes, for its enable input at LOW (0) level, then it is called as the negative level triggered S-R latch. Such a latch is shown in figure 6.22.

> We cannot control the negative atmosphere of the world, but we can control the atmosphere of our minds.
> — **John C. Maxwell**

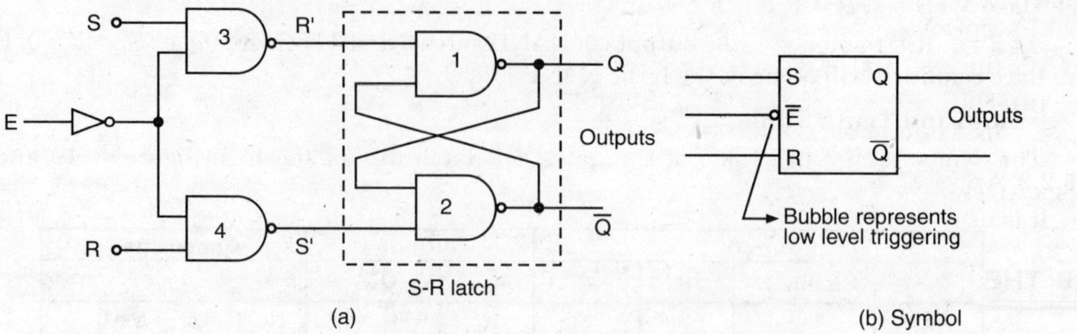

Fig. 6.22. Negative level triggered S-R latch

Due to the additional inverter connected to the enable terminal, this circuit is sensitive to the low level (0) applied to the enable input. If a square wave is applied to the enable input as shown in figure 6.23, then the latch output will respond to input changes only when the square input is low.

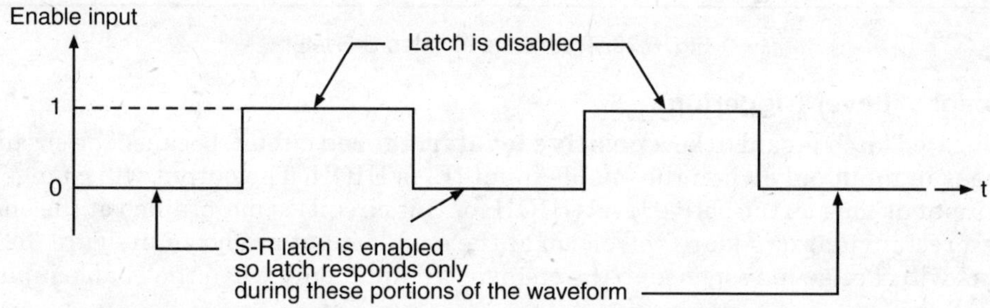

Fig. 6.23. Illustration of enable input

Figure 6.22 shows the symbol of negative level triggered S-R latch. Here, it may be noted that there is a bubble added to the enable input which is active low input.

EXAMPLE 6.1. Draw an output waveform for the following input signals.

Solution:

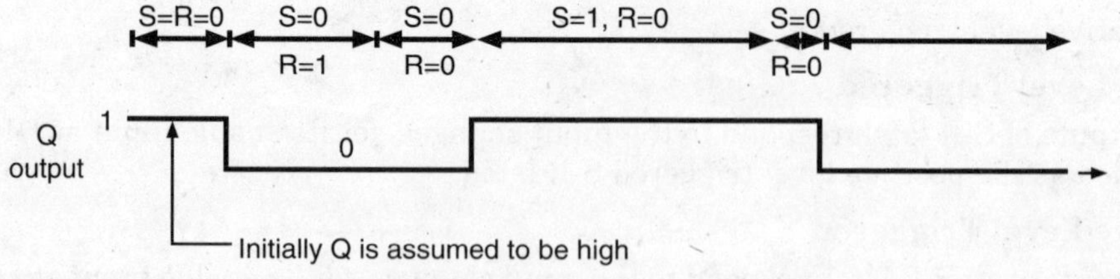

Fig. 6.24. Waveforms for gated S-R latch

6.5.2.1 Drawbacks of S-R Latch

From the truth tables of the S-R latch using NOR and NAND gates, we can draw the following important conclusions:

(i) When S = R = 0 or S = R = 1, the outputs Q and $\overline{Q}$ either don't change (NC) or they are indeterminate (invalid) due to race condition.

(ii) This disadvantage of S-R latch can be overcome by using the gated D latch discussed in the next article.

DO YOU KNOW?

A flip-flop is a logic circuit with a memory characteristic such that its Q and $\overline{Q}$ outputs will go to a new state in response to an input pulse and will remain in that new state after the input pulse is terminated.

6.5.2.2 Application of S-R Latch

It is used as electronic timer.

6.6 THE GATED D LATCH (LEVEL TRIGGERED D FLIP-FLOP)

1. Basic Concept and Circuit Diagram

The circuit diagram of the gated D latch is shown in figure 6.25(a) and its logic symbol is shown in figure 6.25(b). This is also called as level triggered D Flip-flop. Here, it may be noted that D latch is the simple gated S-R latch with a NAND inverter connected between its S and R inputs. This latch has only one input denoted by D. Due to the NAND inverter, S and R will always be the complements of each other. Hence S = R = 0 or S = R = 1, these input conditions will never appear. This will avoid the problems associated with SR = 00 and SR = 11 conditions.

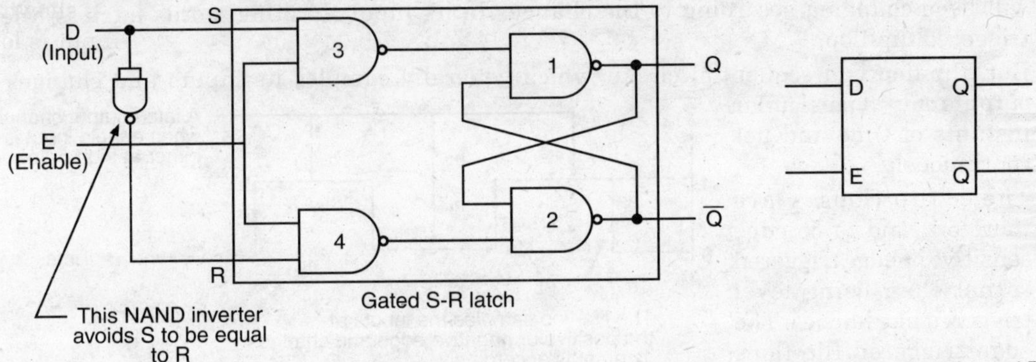

(a) Gated D Latch

(b) Logic symbol of gated D latch

Fig. 6.25.

2. Truth Table

Truth table for the gated D latch is given in Table 6.1.

Table 6.1. Truth table for the gated D latch

S.No.	Inputs		Outputs		Comment
	E	D	Q_{n+1}	$\overline{Q}_{n+1}$	
1.	0	×	Q_n	$\overline{Q}_n$	No change (NC)
2.	1	0	0	1	Reset condition
3.	1	1	1	0	Set condition

3. Working Operation

If E = 0, then the latch is disabled. Hence, there is no change in output. If E = 1 and D = 0, then S = 0 and R = 1. Hence, irrespective of the present state, the next state is $Q_{n+1} = 0$ and $\overline{Q}_{n+1} = 1$. This is the reset condition. On the other side, if E = 1 and D = 1, then S = 1 and R = 0. This will set the latch and $Q_{n+1} = 1$, $\overline{Q}_{n+1} = 0$ irrespective of the present state.

Important Point : From the truth table, it is evident that Q output will follow the D input. If D = 0 then Q = 0 and if D = 1 then Q = 1.

4. Characteristic Equation for D Latch

Let us consider the truth table of D latch and write the K-map for Q_{n+1} as shown in figure 6.26. After simplification we get the characteristic equation of D latch as under :

$$Q_{n+1} = E \cdot D + \overline{E}$$

E \ DQ_n	00	01	11	10
0	Q_n	Q_n	Q_n	Q_n
1	0	0	1	1

Fig. 6.26. K-map for D latch

6.7 CLOCKED S-R FLIP-FLOP (EDGE TRIGGERED) (PTU, Jalandhar, Sem. Exam., 2008-09)

Difference between a latch and flip-flop

(i) Latches and flip-flops both are basically the bistable elements.

(ii) As discussed earlier, a latch has got an enable input. As long as it is active, the latch output will keep changing according to the changes in its input. In other words latch is a level triggered flip-flop.

(iii) But, flip-flop is a sequential circuit which generally samples its inputs and changes its outputs only at particu-lar instants of time and not continuously.

(iv) The flip-flops are therefore said to be edge sensitive or edge triggered rather than being level triggered like latches. The edge triggered flip-flops are also called as clocked flip-flops.

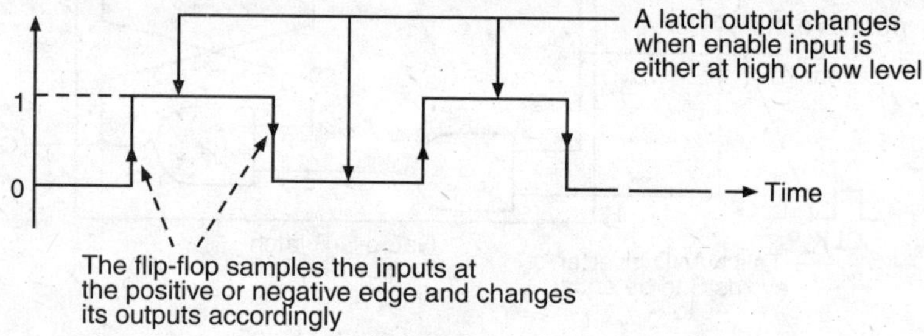

Fig. 6.27.

(v) The rectangular signal applied to the enable (E) input of a latch is shown in figure 6.27. If the same signal is applied at the clock signal to the flip-flop, then its outputs will change only at either rising (positive) edge or at the falling (negative) edge of the clock.

6.7.1 Edge Triggering

For the edge triggered flip-flops, it is essential to apply the clock signal (timing signal) in the form of sharp positive and negative spikes instead of in the form of a pulse train. Such sharp spikes are shown in figure 6.28. These spikes can be derived from the rectangular clock pulses with the help of a passive differentiator circuit shown in figure 6.28.

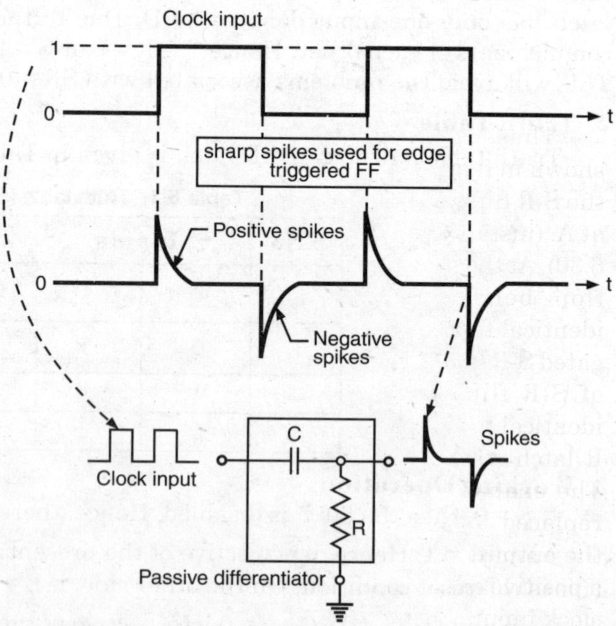

Fig. 6.28. Use of differentiator to obtain sharp edges

6.7.2 Types of Edge Triggered Flip-flops (Clocked FFs)

There are following two types of edge triggered flip-flops :

(i) Positive edge triggered flip-flops.

(ii) Negative edge triggered flip-flops.

Positive edge triggered flip-flops will allow its outputs to change only at the instants corresponding to the rising edges of clock (or positive spikes). Its outputs will not respond to change in inputs at any other instant of time. Negative edge triggered flip-flops will respond only to the negative going edges (or spikes) of the clock.

DO YOU KNOW?

A NAND latch and a NOR latch are simple FFs that respond to logic levels on their SET and RESET inputs.

6.7.3 Positive Edge Triggered S-R Flip-flop

(PTU, Jalandhar, Sem. Exam; 2008-09)

1. Circuit Diagram

Figure 6.29(a) shows the circuit diagram of a positive edge triggered S-R flip-flop and figure 6.29(b) shows the logic symbol for it.

2. Working Operation

Note that the S-R flip-flop of figure 6.29 (a) consists of a differentiator circuit and a gated S-R latch which has already been discussed. C-R_1 acts as a differentiator and converts the rectangular clock pulses into positive and negative spikes. The diode acts as rectifying diode and allows only the positive spikes to pass through, blocking the negative spikes.

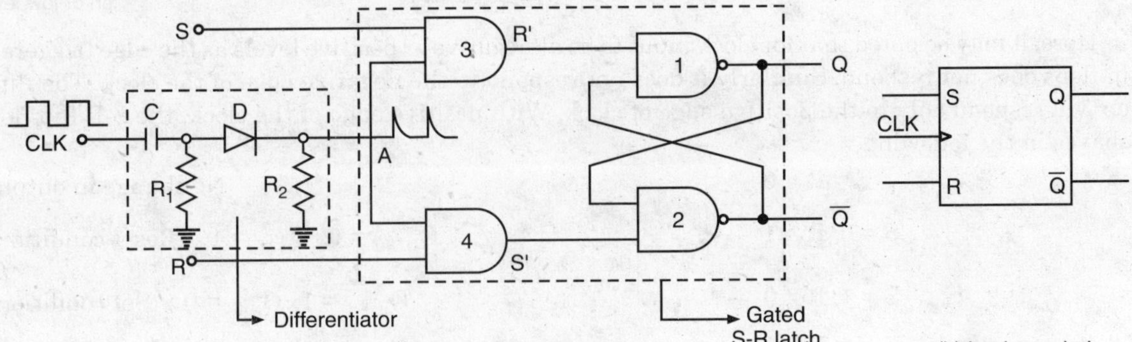

(a) Positive edge triggered S-R flip-flop (b) Logic symbol

Fig. 6.29.

Thus, we get only the positive spikes, corresponding to the leading edges of the clock signal, as shown in figure 6.30 at point A which is enable input of the S-R latch. Hence, the gated S-R latch in the S-R flip-flop will be enabled only for a short duration of time when the positive spikes is present at A. (instants t_1, t_2 .... in figure 6.30). At these instants, the flip-flop behaviour is exactly identical to that of the enabled gated S-R latch. The truth table of S-R flip-flop will also be identical to that of the gated S-R latch with only one change. The enable input will now be replaced by clock input. Also, the outputs will change only if a positive edge is applied to the clock input.

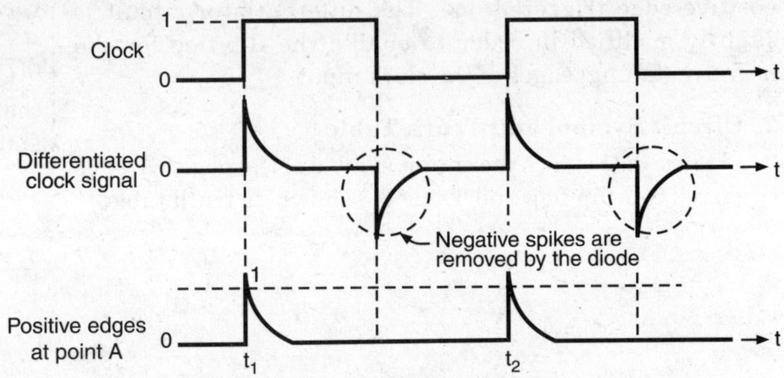

Fig. 6.30. Generation of positive triggering pulses

3. Truth Table

The truth table of a positive edge triggered S-R flip-flop is shown in Table 6.2.

Table 6.2. Truth table of a positive edge triggered SR flip-flop

	Inputs			Outputs		Remark
	CLK	S	R	Q_{n+1}	$\overline{Q}_{n+1}$	
FF is disabled	0	×	×	Q_n	$\overline{Q}_n$	No change (NC)
	1	×	×	Q_n	$\overline{Q}_n$	No change (NC)
	↓	×	×	Q_n	$\overline{Q}_n$	No change (NC)
FF responds only to the positive edges	↑	0	0	Q_n	$\overline{Q}_n$	No change (NC)
	↑	0	1	0	1	Reset
	↑	1	0	1	0	Set
	↑	1	1	Race	Race	Avoid

↓ = Negative edge of clock, ↑ = Positive edge of clock

Here, it may be noted that for clock input to be at negative or positive levels as the edge triggered flip-flops does not respond. Similarly it does not respond to the negative edge of the clock. The flip-flop will respond only to the positive edge of clock. With positive edge of the clock, the S-R flip-flop behaves in the following way.

$S = R = 0 \rightarrow$ No change in output

$S = 0, R = 1 \rightarrow$ $Q_{n+1} = 0, \overline{Q}_{n+1} = 1$ (Reset condition)

$S = 1, R = 0 \rightarrow$ $Q_{n+1} = 1, \overline{Q}_{n+1} = 0$ (Set condition)

$S = R = 1 \rightarrow$ Race condition

6.7.4 Negative Edge Triggered S-R Flip-flop

1. Basic Concepts

The internal circuit (with NAND gates) of the negative edge triggered S-R flip-flop is exactly same as that for the positive edge triggered one. The differentiator circuit is slightly modified in order to enable the flip-flop for the negative (falling) edges of the clock input.

DO YOU KNOW?

Clearing (resetting) a FF means that its output ends up in the $Q = 0/\overline{Q} = 1$ state. Setting a FF means that it ends up in the $Q = 1/\overline{Q} = 0$ state.

2. Circuit Symbol and Truth Table

Figure 6.31 shows the circuit symbol of the negative edge triggered S-R flip-flop and Table 6.3 shows its truth table.

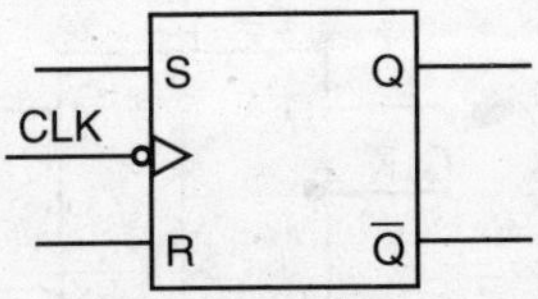

Fig. 4.31. Circuit symbol of negative edge triggered SR FF

Here ↑ = Positive edge of clock
↓ = Negative edge of clock

Table 6.3. Truth table of a negative edge triggered S-R FF

Inputs			Outputs		Remark	
CLK	S	R	Q_{n+1}	$\overline{Q}_{n+1}$		
0	×	×	Q_n	$\overline{Q}_n$	No change (NC)	FF is disabled
1	×	×	Q_n	$\overline{Q}_n$	No change (NC)	
↑	×	×	Q_n	$\overline{Q}_n$	No change (NC)	
↓	0	0	Q_n	$\overline{Q}_n$	No change (NC)	FF responds only to negative edges
↓	0	1	0	1	Reset	
↓	1	0	1	0	Set	
↓	1	1	Race	Race	Avoid	

EXAMPLE 6.2. Draw the waveform for the Q output for the input waveforms shown in figure 6.32.

Solution:

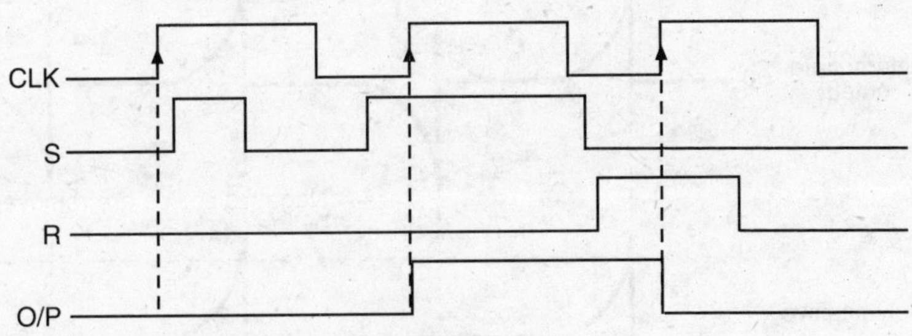

Fig. 6.32.

6.8 CLOCKED D FLIP-FLOP (EDGE TRIGGERED)

(PTU, Jalandhar, Sem. Exam; 2006-07) (10 marks)

The edge triggered D flip-flops can be of following two types :

(i) Positive edge triggered D flip-flop.

(ii) Negative edge triggered D flip-flop.

These flip-flops can be derived from the level triggered D latch which we have discussed earlier.

6.8.1 Positive Edge Triggered D Flip-flop

1. Circuit Diagram and Circuit Symbol

Figure 6.33(a) shows the positive edge triggered D flip-flop. It consists of a gated D latch and a differentiator circuit. The symbol is shown in figure 6.33(b). The clock pulses are applied to the circuit through a differentiator formed by R_1C and a rectifier circuit consisting of diode D and R_2. The NAND gates 1 through 5 form a D latch. The differentiator converts the clock pulses into positive and negative spikes as shown in figure 6.33(c) and the combination of D and R_2 will allow only the positive spikes to pass through, blocking the negative spikes.

Only a life lived for others is the life worthwhile. — **Albert Einstein**

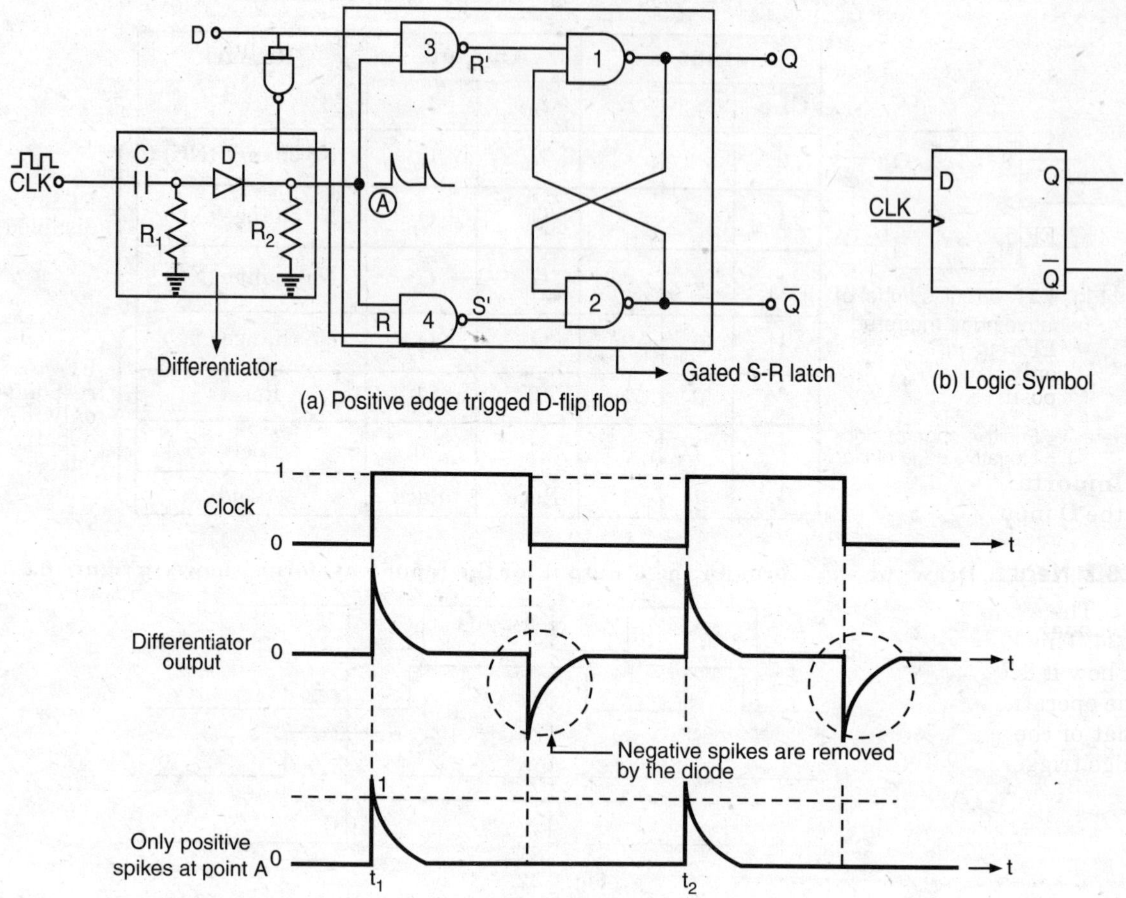

(a) Positive edge trigged D-flip flop

(b) Logic Symbol

(c) Waveform

Fig. 6.33.

The operation of D F/F is exactly same as that of the gated D latch discussed with only one difference. The D latch had an enable terminal whereas the D flip-flop has a clock input. At instants t_1 and t_2, ... the gated D latch is enabled by the positive spikes obtained at point A. Hence the outputs will change based on the inputs at these particular instants only. At any other instants of time, the D flip-flop will not respond to the changes in input.

> **DO YOU KNOW?**
>
> Most cloked FFs also have asycnhronous inputs that can set or clear the FF independently of the clock input.

2. Truth Table

Table 6.4 shows the truth table of a positive edge triggered D flip-flop.

Progress is impossible without change, and those who cannot change their minds cannot change anything.

— **G.B. Shaw**

Table 6.4. Truth table of a positive edge triggered D flip-flop

	Inputs		Outputs		State
	CLK	D	Q_{n+1}	$\overline{Q}_{n+1}$	
FF disabled	0	×	Q_n	$\overline{Q}_n$	No change
	1	×	Q_n	$\overline{Q}_n$	No change
	↓	×	Q_n	$\overline{Q}_n$	No change
FF responds only to the positive edges	↑	0	0	1	Q follows D input
	↑	1	1	0	

Important Point : From the truth table, it is obvious that the Q output of the flip-flop follows the D input.

6.8.2 Negative Edge Triggered D Flip-flop

The symbol for negative edge triggered D flip-flop is shown in figure 6.34. This F/F responds only to the negative edges of the clock pulses. This is how it is different from the positive edge triggered flip-flop. Otherwise, the operation of the negative edge triggered D flip-flop is exactly same as that of the positive edge trigger D flip-flop. The truth table of negative edge triggered D flip-flop is shown in Table 6.5.

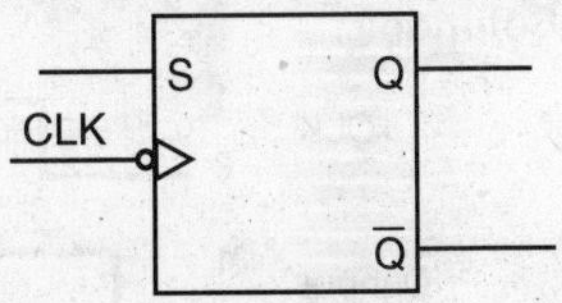

Fig. 6.34. Symbol of negative edge triggered D flip-flop.

Table 6.5. Truth table of a negative edge triggered D flip-flop

	Inputs		Outputs		Comment
	CLK	D	Q_{n+1}	$\overline{Q}_{n+1}$	
FF disabled	0	×	Q_n	Q_n	No change
	1	×	Q_n	$\overline{Q}_n$	No change
	↑	×	Q_n	$\overline{Q}_n$	No change
FF responds only to the positive edges	↓	0	0	1	Q output follows D input
	↑	1	1	0	

EXAMPLE 6.3. Draw the output waveform for a positive edge triggered D flip-flop, if the clock and D input waveforms are shown in figure 6.35.

> The wheel of change moves on, and those who were down go up and those who were up go down.
>
> — **R.N. Tagore**

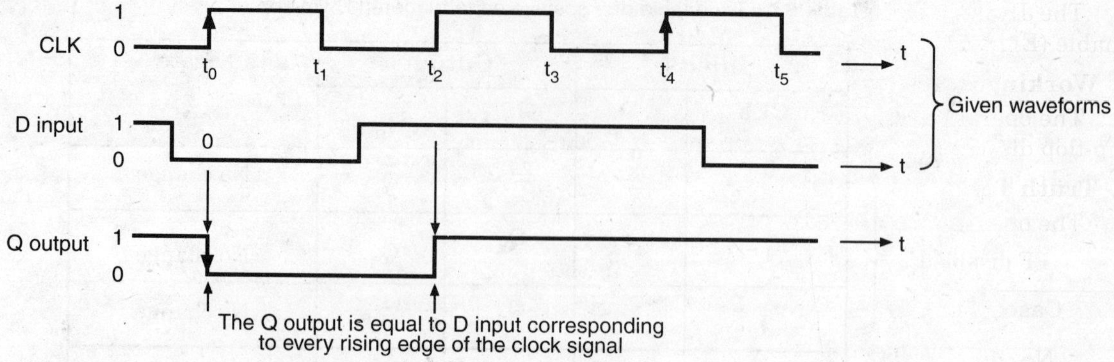

Fig. 6.35.

Solution: At instant $t = t_0$, the first positive edge of the clock arises, and Q output follows the D input to become zero. Similarly at $t = t_2$, Q output follows the D input at that instant. Note that Q output does not respond to any change in D input at any other instant of time.

EXAMPLE 6.4. For the given waveforms in figure 6.36, draw the output (Q) waveform for the negative edge triggered D flip-flop.

Solution:

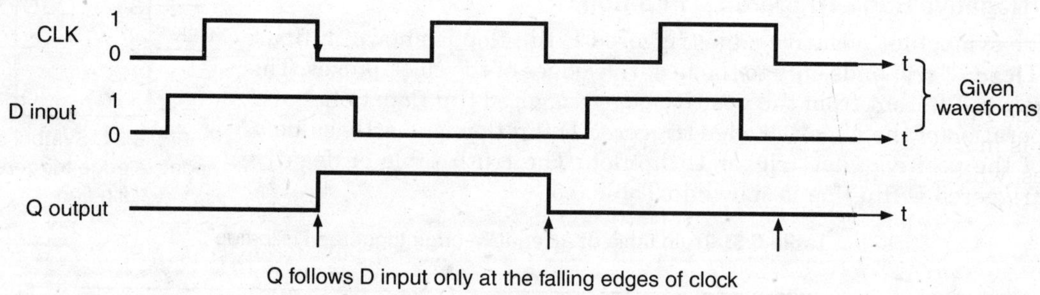

Fig. 6.36.

6.9 J-K LATCH (LEVEL TRIGGERED JK FLIP-FLOP)

1. Circuit Diagram and Circuit Symbol

The JK latch using NAND gate is shown in figure 6.37. It consists of the basic SR latch and an enable input. Note that the output Q and $\overline{Q}$ are returned back and connected to the inputs of NAND gates 4 and 3 respectively.

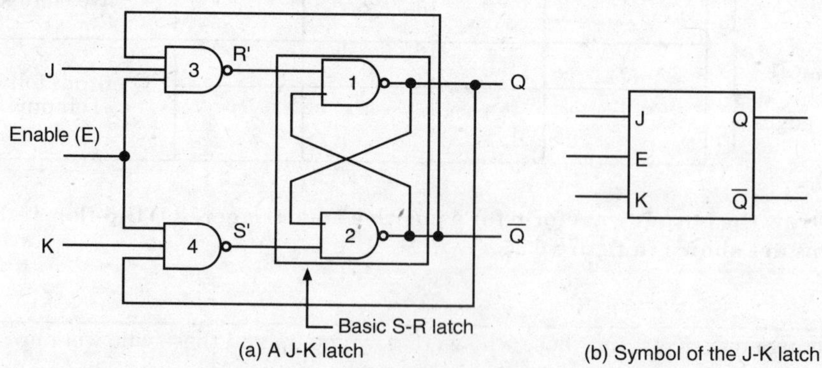

Fig. 6.37.

The JK latch of figure 6.37(a) responds, to the input changes if a positive level is applied at the enable (E) input. Hence it is a positive level triggered latch.

2. Working Operation

The operation of NAND J-K latch is exactly identical to that of the positive edge triggered J-K flip-flop discussed earlier. The only difference between them is that, this circuit is level triggered.

3. Truth Table

The operation of NAND J-K latch has been summarized in Table 6.6.

Table 6.6. Truth table of positive level triggered J-K latch

Case	Inputs			Outputs		State
No.	E	J	K	Q_{n+1}	$\overline{Q}_{n+1}$	
Case I	0	×	×	Q_n	$\overline{Q}_n$	No change
Case II	1	0	0	Q_n	$\overline{Q}_n$	No change
Case III	1	0	1	0	1	Reset
Case IV	1	1	0	1	0	Set
Case V	1	1	1	$\overline{Q}_n$	Q_n	Toggle

6.9.1 Race Around Condition in J-K Latch (PTU, Jalandhar, Sem. Exam., 2006-07)

The *Race Around Condition* that we are going to explain occurs when J = K = 1, i.e., when the latch is in the toggle mode. Let us consider figure 6.38 which shows the waveforms for the various modes, when a rectangular waveform is applied to the *Enable* input.

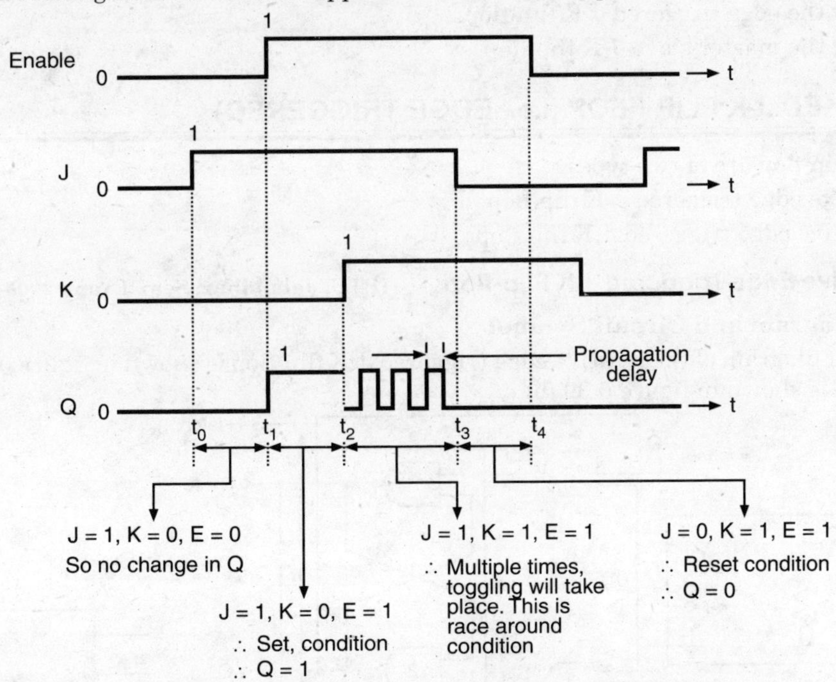

Fig. 6.38. Waveforms for various modes of a J-K latch.

The happiness of every country depends upon the character of its people rather than the form of its government. — **Thomas C. Haliburton**

Interval $t_0 - t_1$

During this interval, J = 1, K = 0 and E = 0. Hence, the latch is disabled and there is no change in Q.

Interval $t_1 - t_2$

During this interval, J = 1, K = 0 and E = 1. Hence, this is a set condition and Q becomes 1.

> **DO YOU KNOW?**
>
> The D-latch is a modified NAND latch that operates like a D flip-flop except that it is not edge triggered.

Interval t_2 - t_3 : Race Around

At instant t_2, J = K = 1 and E = 1. Hence, the J-K latch is in the toggle mode and Q becomes low (0) and $\overline{Q}$ = 1. These changed outputs get applied at the inputs of NAND gates 3 and 4 of the J-K latch. Thus, the new inputs to Gates 3 and 4 are :

NAND - 3 : J = 1, E = 1, $\overline{Q}$ = 1

NAND - 4 : K = 1, E = 1, Q = 0.

Hence R′ will become 0 and S′ will become 1. Therefore, after a time period corresponding to the propagation delay, the Q and $\overline{Q}$ outputs will change to,

$$Q = 1 \text{ and } \overline{Q} = 0.$$

These changed output again get applied to the inputs of NAND gates-3 and 4 and the outputs will toggle again.

Thus, as long as J = K = 1 and E = 1, the outputs will keep toggling indefinitely. This multiple, toggling in the J-K latch is called as Race Around condition. It must be avoided.

Interval $t_3 - t_4$

During this interval, J = 0, K = 1 and E = 1. Hence, it is the reset condition. So, Q becomes zero.

How to Avoid Race Around Condition ?

The race around condition in J-K latch can be avoided by,

(i) Using the edge triggered J-K flip-flop.

(ii) Using the master slave J-K flip-flop.

6.10 CLOCKED J-K FLIP-FLOP (i.e., EDGE TRIGGERED)

Clocked J-K flip-flop are of two types :

(i) Positive edge triggered J-K flip-flop

(ii) Negative edge triggered J-K flip-flop

6.10.1. Positive Edge Triggered J-K Flip-flop **(PTU, Jalandhar, Sem. Exam; 2006-07) (10 marks)**

1. Circuit Diagram and Circuit Symbol

The circuit diagram of the positive edge triggered J-K flip-flop is shown in figure 6.39 (a) and its circuit symbol is shown in figure 6.39 (b).

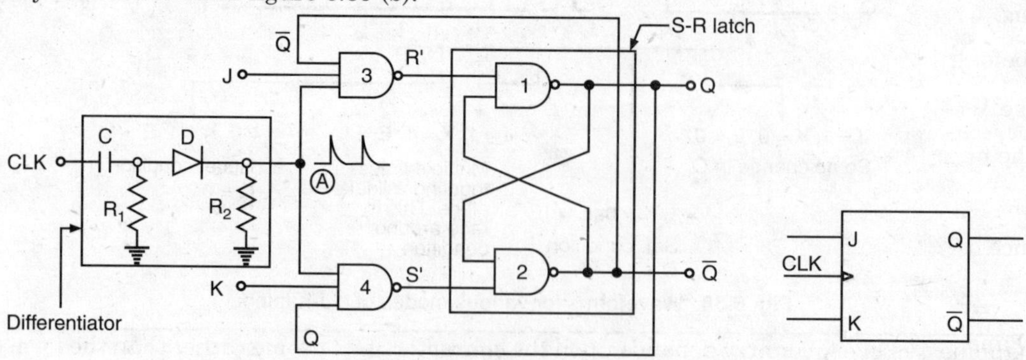

(a) Positive edge triggered J-K flip-flop (b) Logic Symbol

Fig. 6.39.

2. Working Operation

The clock signal is a train of rectangular or square waves. It is passed through a differentiator ($C - R_1$) and a rectifier (D and R_2), to obtain the positive spikes at point A as already discussed. NAND gates 1 and 2 form an S-R latch. The other two NAND gates have three inputs each. The outputs Q and $\overline{Q}$ are fed back to the inputs of gates 4 and 3. According to figure 6.39 (a), we can write the mathematical expressions for S′ and R′ as under :

$$S' = \overline{K . Q . CLK} \text{ and } R' = \overline{J . \overline{Q} . CLK}$$

3. Working Operation

Let us now understand the operation step-by-step.

(i) Case I : CLK = 0 or 1, i.e., level

Since CLK = 0 or 1, i.e., level, this flip-flop is disabled, because the output of differentiator is zero for any level input. Therefore Q and $\overline{Q}$ output do not change their state.

(ii) Case II : If clock edge is falling edge

For the falling edge of the clock, the rectifier ($D - R_2$) will block the negative spikes. Hence, voltage at point *A* is logic 0. So, one input to both the NAND gate 3 and 4 is 0. Hence, both S′ and R′ will be forced to 1. For an S-R latch using NAND gates, the outputs will remain unchanged if S = R = 1.

Therefore, corresponding to the falling edge of clock, the outputs Q and $\overline{Q}$ will remain unchanged.

(iii) Case III : CLK = ↑, J = 0, K = 0

Since J = 0, R′ = 1 and as K = 0, S′ = 1. Since, S′ = R′ = 1, the outputs Q and $\overline{Q}$ will not change their state even though a positive edge of the clock pulse is being applied.

(iv) Case IV : CLK = ↑, J = 0, K = 1

Recalling the expressions for S′ and R′, we have

$$S' = \overline{K . Q . CLK} \text{ and } R' = \overline{J . \overline{Q} . CLK}$$

If the previous state of Q and $\overline{Q}$ is Q = 1 and $\overline{Q}$ = 0, then,

$$S' = \overline{1 . 1 . 1} = 0 \text{ and } R' = \overline{0 . 0 . 1} = 1$$

Hence according to the operation of S-R latch, if S′ = 0, R′ = 1, then Q = 0, and $\overline{Q}$ = 1.

Therefore, **Q = 0 and $\overline{Q}$ = 1**

Thus, with J = 0, K = 1 and positive going clock, the J-K flip-flop will **reset.** If Q = 0 and $\overline{Q}$ = 1 before the application of clock pulse, then there will not be any change in their state.

(v) Case V : CLK = ↑, J = 1, K = 0

If the previous state of Q and $\overline{Q}$ is Q = 0 and $\overline{Q}$ = 1, then, we have

$$S' = \overline{0 . 0 . 0} = 1 \text{ and } R' = \overline{1 . 1 . 1} = 0$$

Hence according to the operation of S-R latch, if S′ = 1 and R′ = 0 then, we have

Q = 1 and $\overline{Q}$ = 0 i.e., the flip-flop is **set.**

If Q and $\overline{Q}$ are already 1, 0, then, there will not be any change in the output state.

* An idealist is one who helps other people to be properous. — **Henry Ford**

(vi) Case VI : CLK = ↑, J = 1, K = 1 (Toggle Mode)

We shall discuss this case for two different previous cases.

- Previously let Q = 1 and $\overline{Q}$ = 0

↓

- $S' = \overline{K \cdot Q \cdot CLK}$ and $R' = \overline{J \cdot \overline{Q} \cdot CLK} = 1$

↓

∴ $S' = \overline{1 \cdot 1 \cdot 1} = 0$ and $R' = \overline{1 \cdot 0 \cdot 1} = 1$

↓

- Hence, the SR latch will reset

↓

- So, $Q_{n+1} = 0$ and $\overline{Q}_{n+1} = 1$

- Previously let Q = 0 and $\overline{Q}$ = 1

↓

- $S' = \overline{1 \cdot 0 \cdot 1}$ and $R' = \overline{1 \cdot 1 \cdot 1}$

↓

- So, S′ = 1 and R′ = 0

↓

- Hence, the SR latch will set

↓

- So, $Q_{n+1} = 1$ and $\overline{Q}_{n+1} = 0$

From the operation discussed above, we conclude that when J = K = 1 and clock hits, then Q and $\overline{Q}$ outputs are inverted, i.e., $Q_{n+1} = \overline{Q}_n$ and $\overline{Q}_{n+1} = Q_n$. This is called as the TOGGLING mode. This is a very important mode of operation.

Table 6.7. Truth table for positive edge triggered J-K FF

Case No.	Inputs			Outputs		State
	CLK	J	K	Q_{n+1}	$\overline{Q}_{n+1}$	
Case I	0 or 1	×	×	No change	No change	Flip-flop is
Case II	↓	×	×	No change	No change	disaled
Case III	↑	0	0	No change	No change	
Case IV	↑	0	1	0	1	Reset
Case V	↑	1	0	1	0	Set
Case VI	↑	1	1	$\overline{Q}_n$	Q_n	Toggle

6.10.2 Characteristic Equation of J-K Flip-flop **(PTU, Jalandhar, Sem. Exam; 2006-07)(10 marks)**

Let us consider the truth table of J-K flip-flop (Table 6.8) and write the K-map for Q_{n+1} as shown in figure 6.40.

Table 6.8. Truth table

J	K	Q_n	Q_{n+1}
0	0	0	0
0	0	1	1
0	1	0	0
0	1	1	0
1	0	0	1
1	0	1	1
1	1	0	1
1	1	1	0

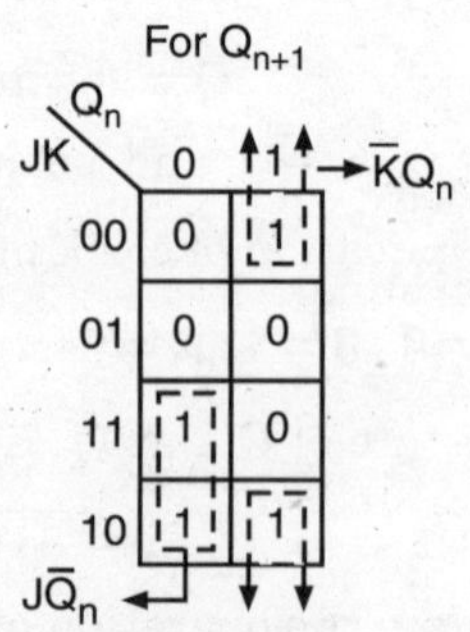

Therefore, $Q_{n+1} = J\overline{Q}_n + \overline{K}Q_n$

Fig. 6.40. K-map for $Q_{n+1} = J\overline{Q}_n + \overline{K}Q_n$.

Gratitude is the memory of the heart. — **J.B. Massies**

Hence, the characteristic equation is $Q_{n+1} = J\overline{Q}_n + \overline{K}Q_n$.

6.10.3 How does an Edge Triggered J-K FF Avoid Race Around Condition ?

For the racing around to take place, it is essential to have the enable input high along with J = K = 1. As the enable input remains high for a long time in a J-K latch, the problem of multiple toggling arises. But, in edge triggered J-K flip-flop, the positive clock pulse is present only for a very short time. Hence, by the time the changed outputs return back to the inputs of NAND gates 3 and 4, the clock pulse has died down to zero. Hence, the multiple toggling can not take place. Thus, the edge triggering avoids the race around condition.

6.11 TOGGLE FLIP-FLOP (T FLIP-FLOP) (PTU, Jalandhar, Sem. Exam., 2008-09)

1. Circuit Diagram and Logic Symbol

Toggle flip-flop is basically a J-K flip-flop with *J* and *K* terminals permanently connected together. It has only one input denoted by T, as shown in figure 6.41(a). The symbol for positive edge triggered T flip-flop is shown in figure 6.41(b) and Table 6.9 shows its truth table.

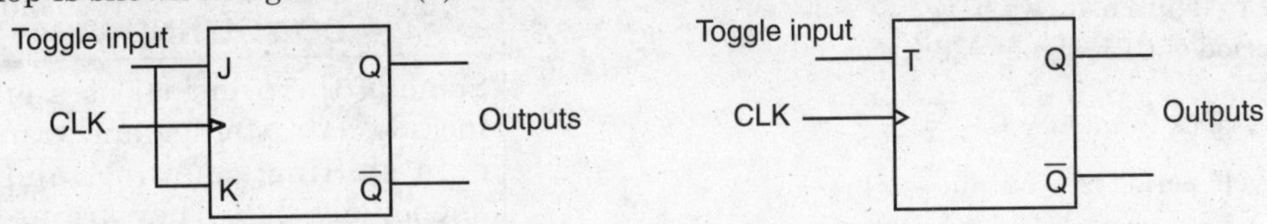

(a) JK FF is converted into T flip-flop (b) Logic symbol of positive edge triggered T flip-flop

Fig. 6.41.

Table 6.9. Truth table of a toggle FF (positive edge triggered)

Inputs		Outputs		State
CLK	**T**	Q_{n+1}	$\overline{Q}_{n+1}$	
↑	0	Q_n	$\overline{Q}_n$	No change
↑	×	Q_n	$\overline{Q}_n$	No change
1	×	Q_n	$\overline{Q}_n$	No change
0	×	Q_n	$\overline{Q}_n$	No change
↑	1	$\overline{Q}_n$	Q_n	Toggle

When T = 0, J = K = 0. Hence, the outputs Q and $\overline{Q}$ will not change. But, if T = 1, then J = K = 1, and the outputs will toggle corresponding to every leading edge of clock signal.

EXAMPLE 6.5. Determine the output *Q* for the set up shown in figure 6.42. What is the relation between the frequency of clock and that of *Q* output ?

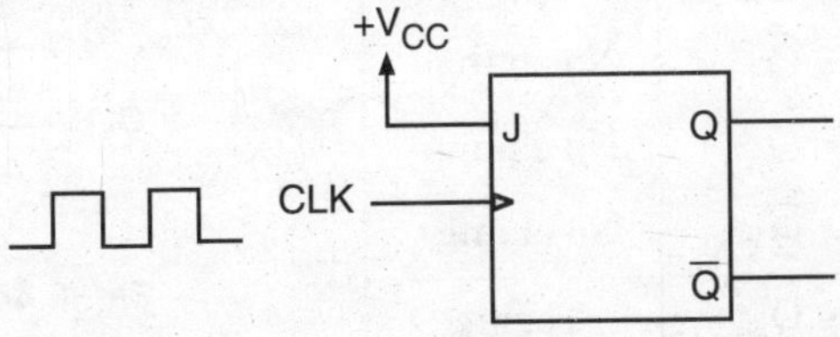

Fig. 6.42. The given set up.

Solution: The flip-flop of figure 6.42 is a toggle flip-flop with *T* = 1. It is a positive edge triggered flip-flop.

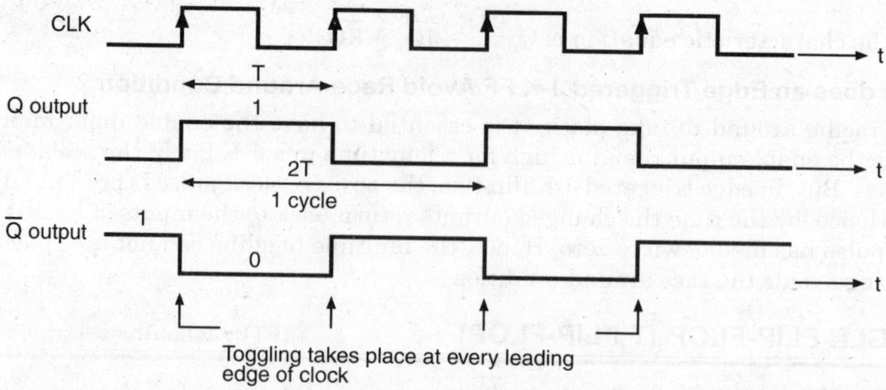

Fig. 6.43. Waveforms for a T F/F.

Hence, the outputs will toggle in response to each rising edge of the clock, as shown in figure 6.43.

Relation between Input and Output Frequency

Referring to figure 6.43, we have

1 cycle period of CLK signal = T

Therefore, clock frequency $f_{CLK} = \frac{1}{T}$

1 cycle period of Q output $= 2T$

Hence, output frequency $f_0 = \frac{1}{2T}$

But, $\left(\frac{1}{T}\right) = f_{CLK}$

Therefore, output frequency $f_0 = \frac{f_{CLK}}{2}$ ...**Ans**

> **DO YOU KNOW?**
>
> Some of the principal uses of FFs include data storage and transfer, data shifting, counting, and frequency division. They are used in sequential circuits that follow a predetermined sequence of states.

Important Point. The T flip-flop divides the clock frequency by 2. Hence, a T flip-flop can be used as a frequency divider.

6.11.1 Negative Edge Triggered T Flip-flop

1. Logic symbol and Truth Table

Figure 6.44 shows the logic symbol of a negative edge triggered toggle flip-flop and Table 6.10 gives its truth table.

Table 6.10. Truth Table

Inputs		Outputs		State
T	CLK	Q_{n+1}	$\overline{Q}_{n+1}$	
0	×	Q_n	$\overline{Q}_n$	No change
1	↑	Q_n	$\overline{Q}_n$	No change
1	0	Q_n	$\overline{Q}_n$	No change
1	1	Q_n	$\overline{Q}_n$	No change
1	↓	$\overline{Q}_n$	Q_n	Toggle

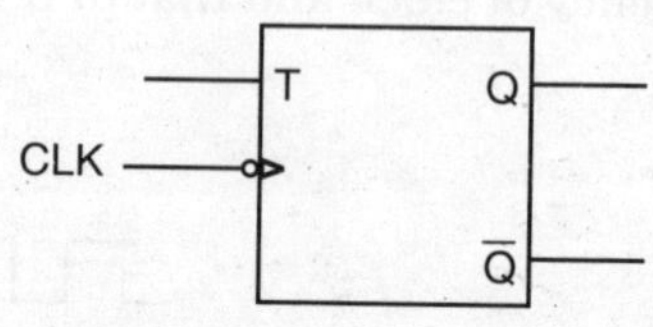

Fig. 6.44. Logic symbol of a negative edge triggered toggle flip-flop.

* Imagination is more important than knowledge. —**Albert Einstein**

2. Waveforms

The waveforms for negative edge triggered toggle F/F have been shown in figure 6.45.

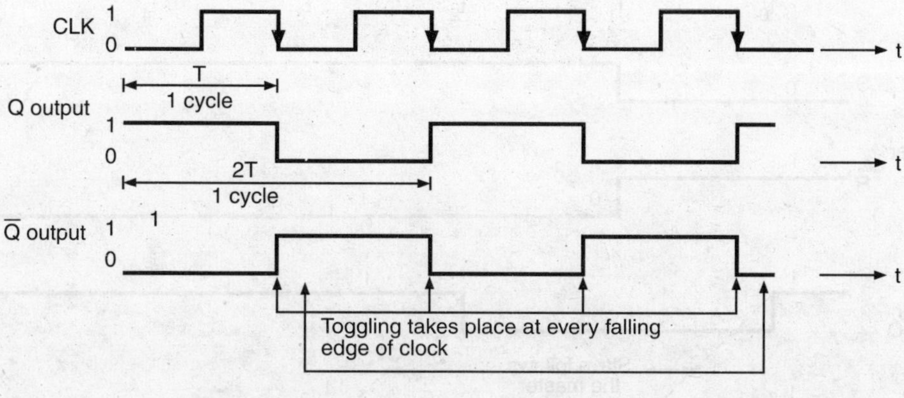

Fig. 6.45. Waveforms for negative edge triggered T F/F.

The toggle input T = 1. The output toggle corresponding to every falling edge of clock. This F/F also acts as a frequency divider. The frequency of Q and $\overline{Q}$ outputs is half the frequency of clock input.

6.11.2 Application of T F/F

The T F/F acts as the basic building block of a ripple counter.

6.12 MASTER SLAVE S-R FLIP-FLOP

(PTU, Jalandhar, Sem. Exam., 2004-05)

1. Basic Concepts

A master slave S-R flip-flop consists of two level triggered flip-flops, and an inverter as shown in figure 6.46. The first flip-flop is called as master. It receives the S-R inputs directly. The clock signal (that acts as enable signal) is applied directly to the master. Outputs (Q and $\overline{Q}$) of the master are applied to the S-R inputs of the slave. The clock is inverted and applied to the slave flip-flop. This will force the slave flip-flop to respond to the low level of the clock.

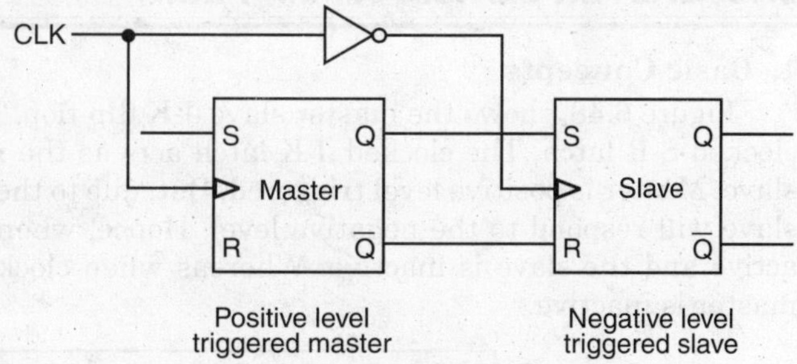

Fig. 6.46. Master slave S-R flip-flop.

2. Waveforms

The waveforms for the master slave flip-flop are as shown in figure 6.47.

3. Working Operation

Let us consider the waveforms shown in figure 6.47. Before instant t_0, the S and R inputs to the master F/F are S = 0 and R = 1. At $t = t_0$, the clock signal becomes 1. This will enable the master and disable the slave. Hence, Q output of master becomes low, i.e., at $t = t_0$ the master resets but there is no change in the slave output. At $t = t_1$, the clock signal becomes low (0). This will enable the slave and disable the master. At instant t_1, the inputs to the slave are S = 0 and R = 1, hence, the slave will reset. Thus, the slave has followed the master. In other words, the output state of master has been transferred to the output of slave.

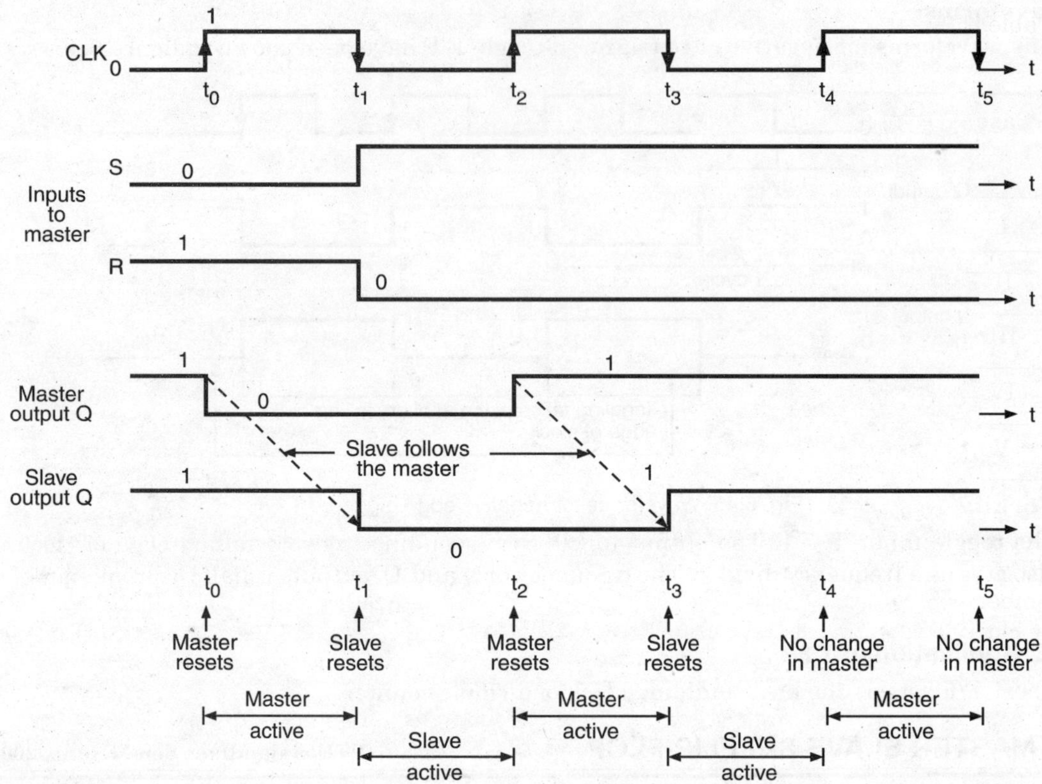

Fig. 6.47. Waveforms for the master slave S R flip-flop.

6.13 MASTER SLAVE J-K FLIP-FLOP

(PTU, Jalandhar, Sem. Exam., 2005-06)

1. Basic Concepts

Figure 6.48, shows the master slave J-K flip-flop. It is a combination of a clocked J-K latch and clocked S-R latch. The clocked J-K latch acts as the master and the clocked SR latch acts as the slave. Master is positive level triggered. But, due to the presence of the inverter in the clock line, the slave will respond to the negative level. Hence, when the clock = 1 (positive level), the master is active and the slave is inactive. Whereas when clock = 0 (low level), the slave is active and the master is inactive.

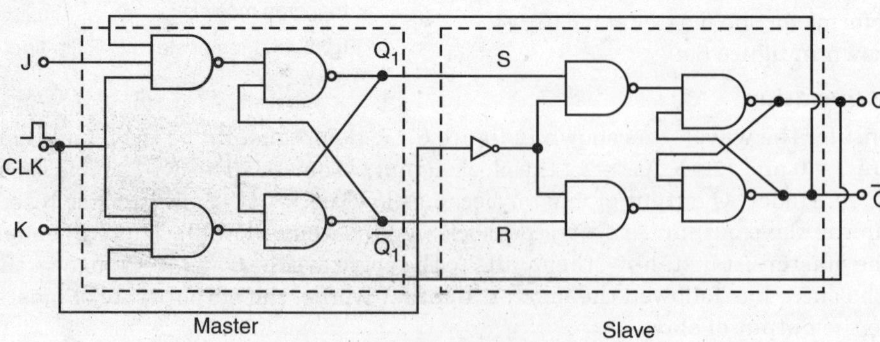

Fig. 6.48. Master slave J-K FF.

2. Truth Table

Table 6.11 gives truth table of master slave J-K flip-flop.

Table 6.11. Truth table of master slave J-K FF

Case	Inputs			Outputs		Remark
	CLK	J	K	Q_{n+1}	$\overline{Q}_{n+1}$	
I	×	0	0	Q_n	$\overline{Q}_n$	No change
II	$\sqcap$ (1)	0	0	Q_n	$\overline{Q}_n$	No change
III	$\sqcap$ (1)	0	1	0	1	Reset
IV	$\sqcap$ (1)	1	0	1	0	Set
V	$\sqcap$ (1)	1	1	$\overline{Q}_n$	Q_n	Toggle

3. Working Operation

Let us discuss the working operation of the master slave J-K FF with reference to its truth table. We must always remember one important fact that in the positive half cycle of the clock, the master is active and in the negative half cycle, the slave is active. This has been shown in figure 6.49.

> **DO YOU KNOW?**
>
> Edge-triggered (clocked) FFs can be triggered to a new state by the active edge of the clock input according to the state of the FF's synchronous control inputs (S, R or J, K or D)

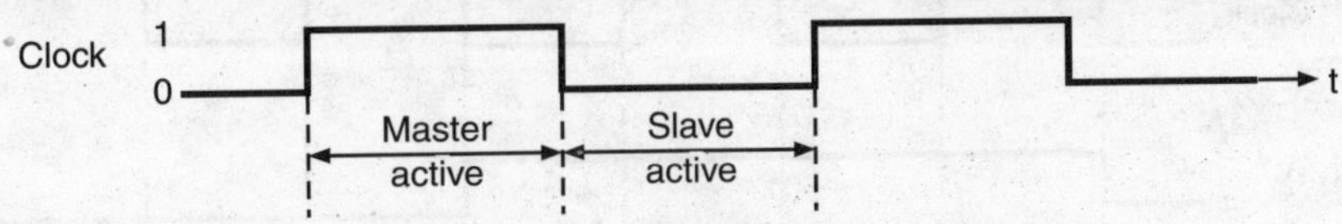

Fig. 6.49.

Case I : Clock = ×, J = K = 0

(i) For clock = 1, the master is active, slave inactive. As J = K = 0, therefore, outputs of master, i.e., Q_1 and $\overline{Q}_1$ will not change. Hence, the S and R inputs to the slave will remain unchanged.

(ii) As soon as clock = 0, the slave becomes active and master is inactive. But, since, the S and R inputs have not changed, the slave outputs will also remain unchanged.

Therefore, outputs will not change if J = K = 0.

Case II : Clock = $\sqcap$, J = K = 0

This condition has been already discussed in case I.

Case III : Clock = $\sqcap$, J = 0 and K = 1

Clock = 1 : Master active, slave inactive.

Therefore, outputs of the master become $Q_1 = 0$ and $\overline{Q}_1 = 1$. This means that S = 0 and R = 1.

Clock = 0 : Slave active, master inactive.

Therefore, outputs of the slave become Q = 0 and $\overline{Q} = 1$.

Again, if clock = 1 : Master active, slave inactive.

Therefore, even with the changed outputs Q = 0 and $\overline{Q} = 1$ fed back to master, its outputs will $Q_1 = 0$ and $\overline{Q}_1 = 1$. This means that S = 0 and R = 1.

Hence, with clock = 0 and slave becoming active, the outputs of slave will remain Q = 0 and $\overline{Q}$ = 1.

Thus, we get a stable output from the Master slave.

Case IV : CLK = ⊓ , J = 1, K = 0

Clock = 1 : Master active, slave inactive.

Therefore, outputs of master become Q_1 = 1 and $\overline{Q}_1$ = 0, i.e., S = 1, R = 0.

Clock = 0 : Master inactive, slave active.

Therefore, Outputs of slave become Q = 1 and $\overline{Q}$ = 0.

Again, if clock = 1, then it can be shown that the outputs of the slave are stabilized to Q = 1 and $\overline{Q}$ = 0.

Case V : CLK = ⊓ , J = 1, K = 1

Clock =1 : Master active, slave inactive.

Therefore, Outputs of master will toggle. So, S and R also will be inverted.

Clock = 0 : Master inactive, slave active.

Therefore, Outputs of the slave will toggle.

These changed outputs are returned back to the master inputs.

But since clock = 0, the master is still inactive. So, it does not respond to these changed outputs.

This avoids the multiple toggling which leads to the race around condition. Thus, the master slave flip-flop will avoid the race around condition.

4. Waveforms

The waveforms for the master slave flip-flop have been shown in figure 6.50.

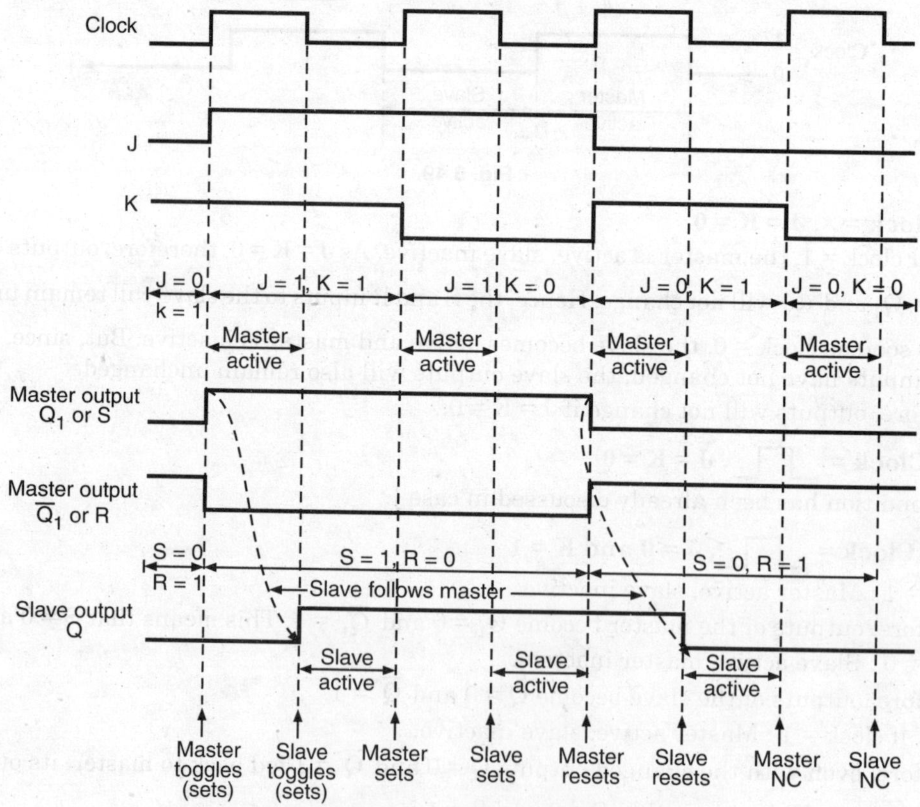

Fig. 6.50. Waveforms of master slave J-K flip-flop.

5. Observations from the Waveforms

We can make the following important observations from the waveforms of the master slave J-K Flip-flop.

(i) The slave always follows the master, after a delay of half clock cycle period.

(ii) The multiple toggling or the race around condition is successfully avoided.

6.14 PRESET AND CLEAR INPUTS

1. Basic Concepts

In the flip-flops discussed so far, when the power switch is turned on, the state of outputs Q and $\overline{Q}$ is uncertain. The outputs can be Q = 0. $\overline{Q}$ = 1 or the other way around. But this uncertainty cannot be tolerated in certain applications. In some applications, it is necessary to initially set or reset the flip-flops. This can be practically achieved by adding two more inputs to a flip-flop, called **preset (PR)** and **clear (CLR)** inputs. These inputs are called as **direct** or **asynchronous inputs** because we can apply them any time between clock pulses without thinking about their synchronization with the clock.

2. Circuit Diagram and Circuit Symbol

The S-R flip-flop with preset and clear is shown in figure 6.51(a) and its logic symbol is shown in figure 6.51(b).

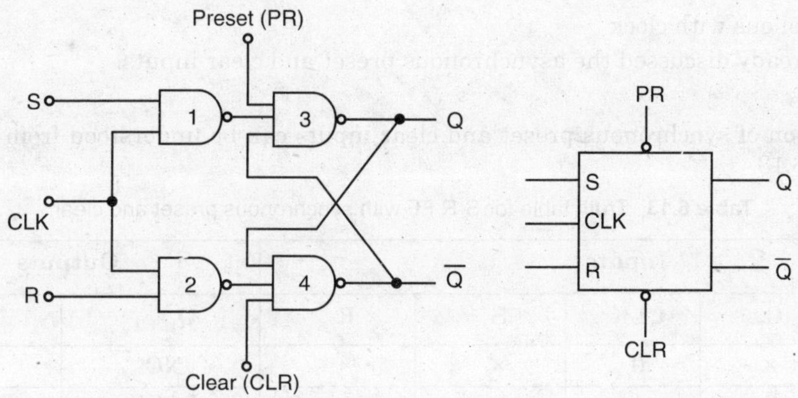

(a) S-R FF with preset and clear (b) Logic Symbol

Fig. 6.51.

Here, it may be noted that both these inputs are active low inputs. This has been indicated by the bubbles on these inputs in figure 6.51.

3. Working Operation

(i) Case I : PR = CLR = 1

If PR = CLR = 1 then both are inactive and the S-R FF operates in accordance with the truth table of the conventional S-R flip-flop.

(ii) Case II : PR = 0 and CLR = 1

As PR = 0, the output of gate 3 of figure 6.51(a) will be 1. This means that Q = 1. Therefore, all the three inputs to the gate 4 will be 1 and $\overline{Q}$ will become 0. Thus, making PR = 0 will set the flip-flop.

(iii) Case III : PR = 1, CLR = 0

If PR = 1 and CLR = 0, then the output of gate-4, i.e., $\overline{Q}$ = 1. Therefore all the three inputs to gate-3 will be 1 and hence Q = 0. Thus, making CLR = 0 will reset the flip-flop.

(iv) Case IV : PR = 0, CLR = 0

This condition should never be used, because it leads to an uncertain state.

4. Truth Table

The operation of SR flip-flop with preset and clear inputs has been summarized in Table 6.12.

Table 6.12. Summary of operation of S-R FF

Inputs			Output	Operation performed
CLK	PR	CLR		
1	1	1	Q_{n+1}	Normal S-R flip-flop
×	0	1	1	FF is set
×	1	0	0	FF is reset

The don't care conditions (×) marked in the CLK column indicate that PR and CLR inputs have higher priority than clock input.

6.14.1 Synchronous Preset and Clear Inputs

1. Basic Concepts

Preset and clear inputs can be of following two types :

(i) Asynchronous with clock

(ii) Synchronous with clock

We have already discussed the asynchronous preset and clear inputs.

2. Truth Table

The operation of synchronous preset and clear inputs can be understood from the truth table given in Table 6.13.

Table 6.13. Truth table for S-R FF with synchronous preset and clear.

Inputs					Outputs	
P_r	C_r	CLK	S	R	Q_{n+1}	$\overline{Q}_{n+1}$
×	×	0	×	×	NC	NC
0	0	1	×	×	RACE	RACE
0	1	1	×	×	0	1
1	0	1	×	×	1	0
1	1	1	0	0	NC	NC
1	1	1	0	1	0	1
1	1	1	1	0	1	0
1	1	1	1	1	RACE	RACE

6.14.2 J-K Flip-flop with Preset and Clear Inputs

1. Basic Concepts and Symbol

A J-K flip-flop with preset and clear inputs is shown in figure 6.52. These are the synchronous preset and clear terminals. Both these inputs are active low and have higher priority than all the other inputs.

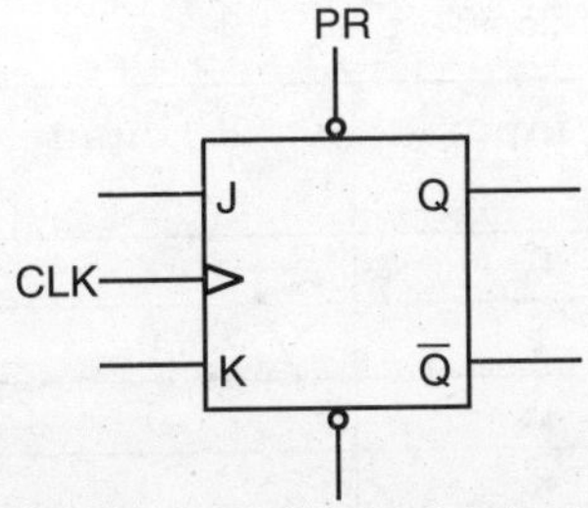

(a) Symbol of a J-K FF with preset and clear inputs

Inputs			Output	Operation performed
CLK	PR	CLR		
1	1	1	Q_{n+1}	Normal J-K FF
×	0	1	1	FF is set
×	1	0	0	FF is reset

(b) Summary of operation of S-R FF

Fig. 6.52.

2. Applications of J-K Flip-flop

(i) Shift register

(ii) Counters

6.15 EXCITATION TABLE OF FLIP-FLOP

1. Basic Concepts

Till now we have written the truth tables of various flip-flops. The truth tables are also known as the **characteristic tables.** But, while designing the sequential circuits, sometimes the present and next state of a circuit is given and we are expected to find the corresponding input condition. Present state is the state prior to application of clock pulse and the next state means the state after application of clock pulse. For example, the outputs of an S-R FF before clock pulse are $Q_n = 0$ and $\overline{Q}_n = 1$ and it is expected that these outputs should remain unchanged after application of clock. Then what must be the values of inputs S_n and R_n to achieve this ? Let us consider truth table of S-R FF to answer this question. The answer is, for the following two conditions the outputs remain unchanged at $Q = 0$ and $\overline{Q} = 1$.

Condition 1 : S_n and $R_n = 0 \rightarrow$ Refer first row of Table 6.14.

Condition 2 : $S_n = 0$ $R_n = 1 \rightarrow$ Refer third row of Table 6.14.

Table 6.14. Truth table of S-R flip-flop

	CLK	S_n	R_n	Q_{n+1}	$\overline{Q}_{n+1}$
Row 1	↑	0	0	Q_n	$\overline{Q}_n$
	↑	0	1	0	1
Row 2	↑	1	0	1	0
	↑	1	1	Race	Race

From the two conditions mentioned above we conclude that S_n input should be equal to 0 and R_n input can be 0 or 1 (don't care) in order to achieve our objective of maintaining $Q = 0$ and $\overline{Q} = 1$ before and after clock. Similarly, we can find the input conditions for all the possible situations that may exist on the output side. The table containing all these output situations and the corresponding input condition is called as the excitation table of a flip-flop. The excitation table of S-R flip-flop has been shown in Table 6.15.

Table 6.15. Excitation table of SR flip-flop

	Present state of Q output	Next state of Q output	S_n input	R_n input
Case I	0	0	0	×
Case II	0	1	1	0
Case III	1	0	0	1
Case IV	1	1	×	0

2. Description of Excitation Table of S-R FF

We have already discussed case I.

(i) Case II : Q should change from 0 to 1

This is nothing but the set condition. Hence, $S_n = 1$ and $R_n = 0$ should be the inputs.

(ii) Case III : Q should change from 1 to 0

This is nothing but the reset condition. Hence, S_n should be 0 and R_n should be 1.

(iii) Case IV : Q should be 1. No change

To satisfy this requirement, we have following two possible input conditions :

Condition 1 : $S_n = R_n = 0$ i.e. No change in output.

Condition 2 : $S_n = 1$ and $R_n = 0$.

From these conditions, we conclude that S_n can be either 0 or 1, i.e., don't care and $R_n = 0$. Hence the inputs corresponding to this case is

$$S_n = \times \quad \text{and} \quad R_n = 0$$

Similarly, we can write the excitation tables for the other flip-flops.

6.15.1 Excitation Table of D Flip-flop

Excitation table of a D flip-flop is given in Table 6.16.

Table 6.16. Excitation table of a D flip-flop

Q output		Input
Present state	Next state	D_n
0	0	0
0	1	1
1	0	0
1	1	1

6.15.2 Excitation Table of J-K Flip-flop

(PTU, Jalandhar, Sem. Exam; 2006-07) (05 marks)

Excitation table of a J-K flip-flop is given in Table 6.17.

Table 6.17.

	Q output		Inputs	
	Present state	Next state	J_n	K_n
Case I	0	0	0	×
Case II	0	1	1	×
Case III	1	0	×	1
Case IV	1	1	×	0

Don't care conditions result due to toggling

Let us consider case II of Table 6.17. For the change in Q output from 0 to 1, we have the following two input conditions :

Condition I : $J_n = 1$ and $K_n = 0$, i.e., set condition.

Condition II : $J_n = 1$ and $K_n = 1$, i.e., toggle.

Hence, $J_n = 1$ and $K_n = \times$ (0 or 1) corresponding to case II. Similarly, for case III also, the don't care condition corresponds to toggle mode.

6.15.3 Excitation Table of T Flip-flop

Excitation table of T flip-flop has been shown in Table 6.18.

Table 6.18. Excitation table of T flip-flop

Q output		Input
Present state	**Next state**	T_n
0	0	0
0	1	1
1	0	1
1	1	0

6.16 CONVERSION OF ONE TYPE OF FF TO THE OTHER TYPE

The conversion from one type of flip-flop to the other requires a systematic approach using the excitation tables and K map simplifications. Figure 6.53 shows a generalized model for conversion from one flip-flop to the other. As shown in figure 6.53, the required flip-flop is actually a combination of given flip-flop and a combinational logic circuit using gates. (Such a combinational circuit is called as the flip-flop conversion logic). The inputs to FF conversion logic are, the flip-flop data inputs and the outputs of given flip-flop.

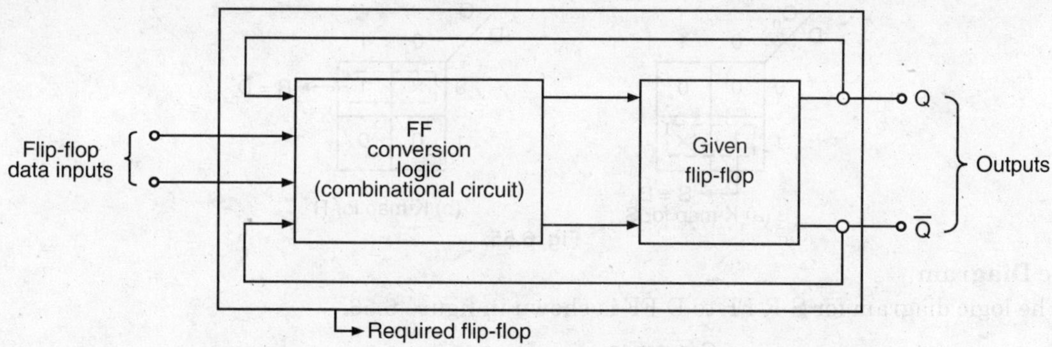

Fig. 6.53. General model used to convert one type of FF to the other type

The conversion logic is designed by combining the excitation tables of both the flip-flops. The truth table of the conversion logic has data inputs and Q and $\overline{Q}$ outputs of the given FF as inputs and the inputs of the given FF are the outputs of the truth table. Then, we draw the K map for each output and obtain the simplified expressions. The conversion logic is then implemented using gates.

6.16.1 Conversion from S-R Flip-flop to D Flip-flop

Let us consider figure 6.54. Here, the given FF is S-R FF and the required FF is D FF. The truth table for the conversion logic is shown in Table 6.19. The inputs are D and Q whereas outputs are S and R.

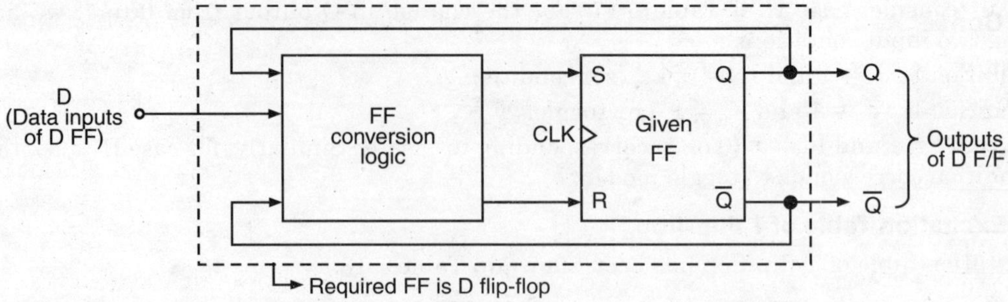

Fig. 6.54.

The truth table is prepared by combining the excitation tables of D FF and S-R FF.

Table 6.19. Truth table for S-R to D FF conversion

Inputs			Outputs	
D	**Present state Q_n**	**Next state Q_{n+1}**	**S**	**R**
0	0	0	0	×
1	0	1	1	0
0	1	0	0	1
1	1	1	×	0

← Entries from excitation table of D FF →

← Entries from excitation table of S-R FF →

Now, let us write the K maps for the S and R outputs as shown in figures 6.55 (*a*) and (*b*).

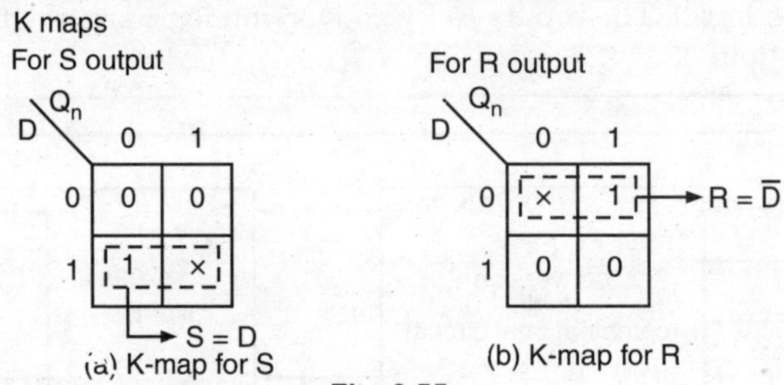

(a) K-map for S (b) K-map for R

Fig. 6.55.

Logic Diagram

The logic diagram for S-R FF to D FF is shown in figure 6.56.

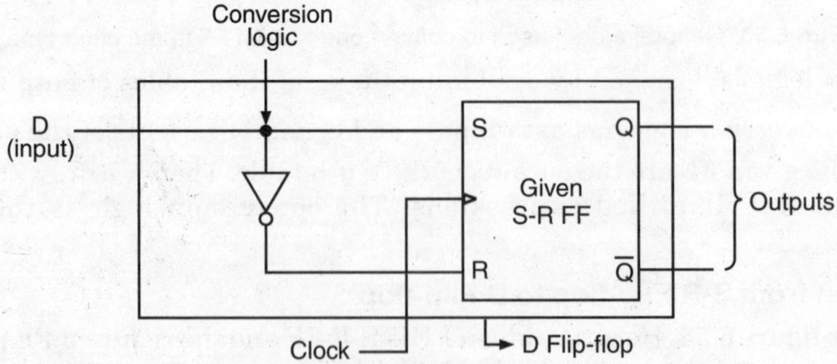

Fig. 6.56. SR flip-flop to D flip-flop.

6.16.2 Conversion of J-K FF to T FF (PTU, Jalandhar, Hyderabad, Sem. Exam; 2003-04)

(i) First, we write the truth table for conversion

The required truth table is obtained from the excitation tables of JK and T flip-flops.

Table 6.20. Truth table for conversion from J-K to T FF

Inputs			Outputs	
T	**Present state Q**	**Next state Q**	**J**	**K**
0	0	0	0	×
1	0	1	1	×
1	1	0	×	1
0	1	1	×	0
← Excitation table of T FF →				
← Excitation table of J-K FF →				

(ii) K maps and simplification

The K maps for outputs J and K are shown in figure 6.57.

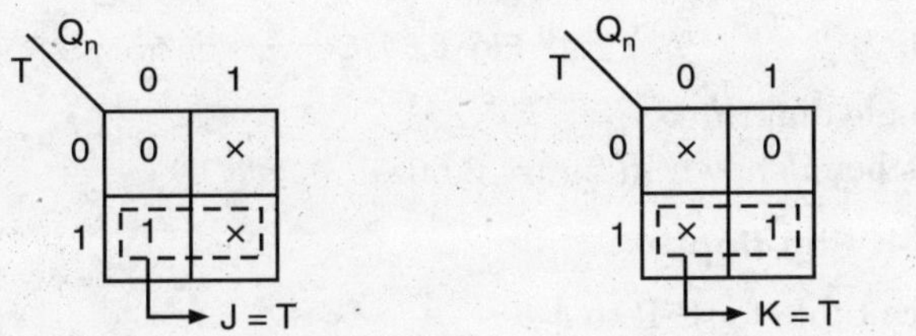

(a) K map for J output (b) K map for K output

Fig. 6.57.

(iii) Lastly, we draw the logic diagram.

The logic diagram has been shown in figure 6.58.

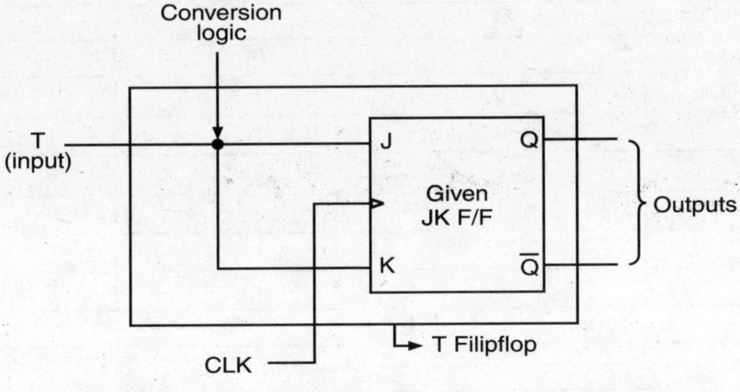

Fig. 6.58. Logic diagram for conversion of JK FF to T FF.

6.16.3 S-R Flip-flop to T Flip-flop

(i) First, we write the truth table.

Table 6.21. Truth table

Inputs			Outputs	
T	**Present state Q_n**	**Next state Q_{n+1}**	**S**	**R**
0	0	0	0	×
1	0	1	1	0
1	1	0	0	1
0	1	1	×	0

(ii) Next, we write the K maps and obtain the expressions for S and R.

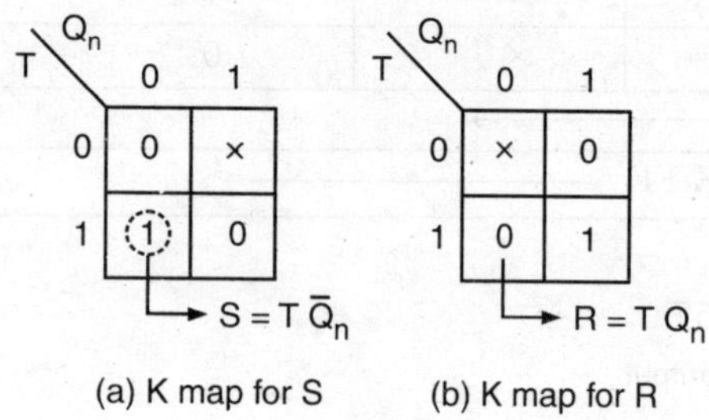

Fig. 6.59.

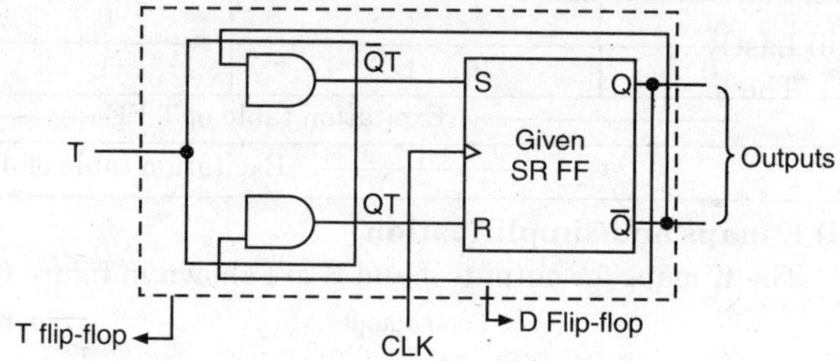

Fig. 6.60. Conversion from S-R flip-flop to T flip-flop

(iii) Lastly, we draw the logic diagram.

The logic diagram has been shown in figure 6.60.

6.16.4 S-R Flip-flop to J-K Flip-flop

(*i*) First, we write the truth table for S-R to J-K.

The truth table for S-R to J-K flip-flop conversion has been shown in Table 6.22.

Table 6.22. Truth table for S-R to J-K FF conversion

Inputs				Outputs	
J	**K**	**Present state Q_n**	**Next state Q_{n+1}**	**S**	**R**
0	0	0	0	0	×
0	1	0	0	0	×
1	0	0	1	1	0
1	1	0	1	1	0
0	1	1	0	0	1
1	1	1	0	0	1
0	0	1	1	×	0
1	0	1	1	×	0

← Excitation table of J-K FF →

← Excitation table of S-R FF →

(*ii*) Next, we write K-maps and simplification.

K-maps for S and R outputs are shown in figure 6.61.

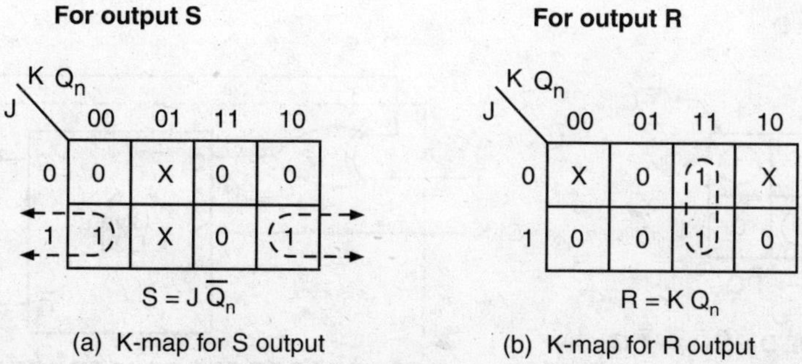

Fig. 6.61.

(iii) Lastly, let us draw the logic diagram.

The logic diagram of SR to JK flip-flop is given in figure 6.62.

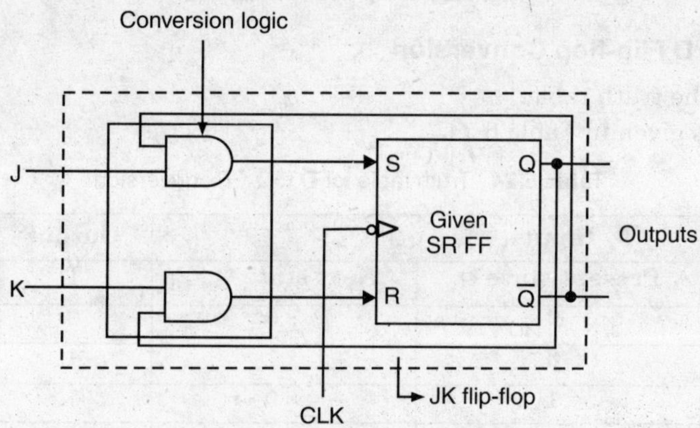

Fig. 6.62. SR to JK flip-flop conversion.

6.16.5 Conversion of D Flip-flop to T Flip-flop

(i) First, we write the truth table for D to T FF conversion

The truth table is as follows :

Table 6.23. Truth table for D to T FF conversion

Inputs			Output
T	Present state Q_n	Next state Q_{n+1}	D
0	0	0	0
1	0	1	1
1	1	0	0
0	1	1	1

(ii) Next, we make K-map and simplify and then make a logic diagram

K-map is shown in figure 6.63 (a) and logic diagram is shown in figure 6.63 (b).

For output D

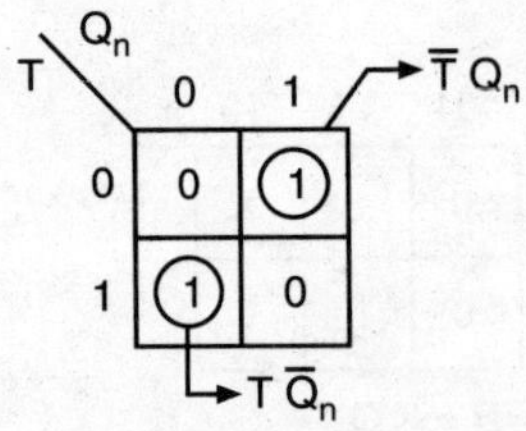

Therefore, $D = T\overline{Q}_n + \overline{T}Q_n$

Therefore, $D = T \oplus Q_n$

(a) K map and simplification

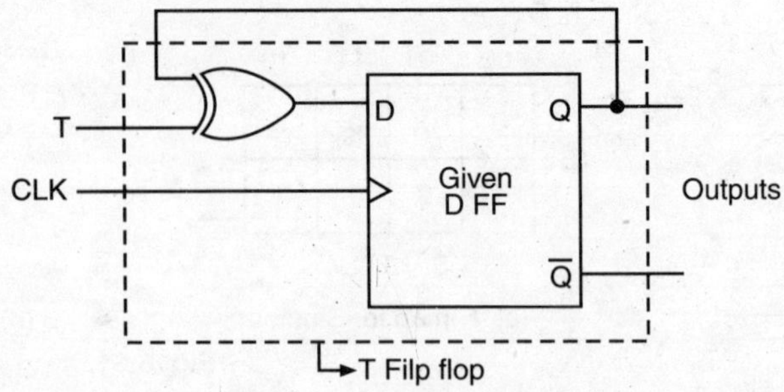

(b) Logic diagram

Fig. 6.63. D flip-flop to T flip-flop.

6.16.6 T Flip-flop to D Flip-flop Conversion

(i) Let us first write the truth table.

The truth table is given in Table 6.24.

Table 6.24. Truth table for D to T FF conversion

Inputs			Output
D	**Present state Q_n**	**Next state Q_{n+1}**	**T**
0	0	0	0
1	0	1	1
0	1	0	1
1	1	1	0

(ii) Next, we make K-maps, simplify and make a logic diagram

The K-map is shown in figure 6.64 (a) and the logic diagram is given in figure 6.64 (b).

For output T

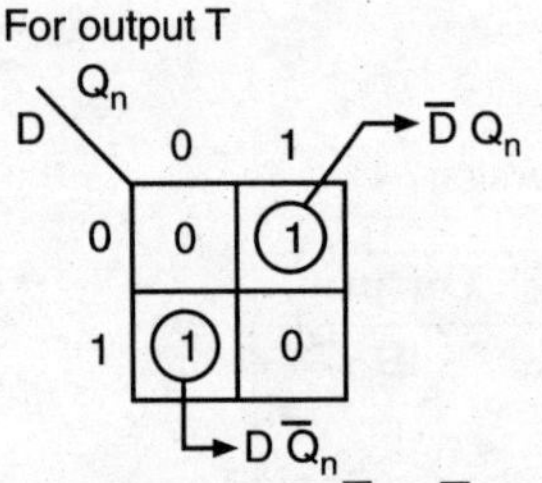

Therefore, $T = D\overline{Q}_n + \overline{D}Q_n$

Therefore, $T = D \oplus Q_n$

(a) K map for T output

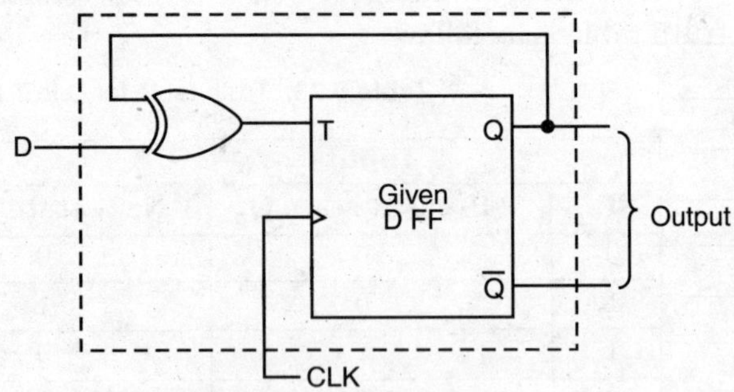

(b) Logic diagram of T to D flip flop conversion

Fig. 6.64.

6.16.7 JK Flip-flop to D Flip-flop Conversion

(i) First, we write the truth table for JK to D conversion.

The truth table is as follows :

Table 6.25. Truth table for J-K to D flip-flop conversion

Inputs			Outputs	
D	**Present state Q_n**	**Next state Q_{n+1}**	**J**	**K**
0	0	0	0	×
1	0	1	1	×
0	1	0	×	1
0	1	1	×	0
← Excitation table of D FF →				
	← Excitation table of JK FF →			

(ii) Next, we make K-maps and simplify.

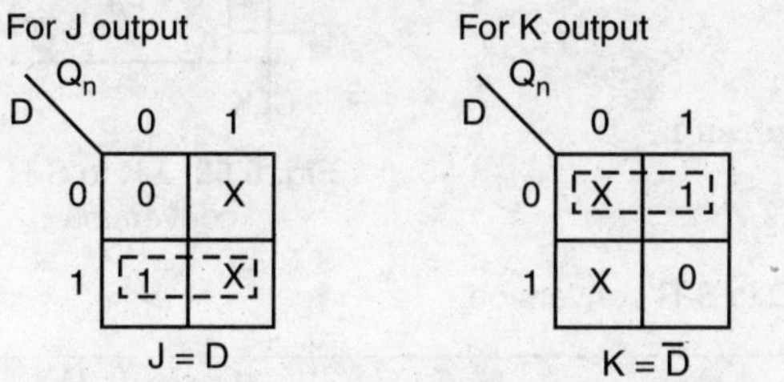

(a) K-map for J output (b) K-map for K output

Fig. 6.65.

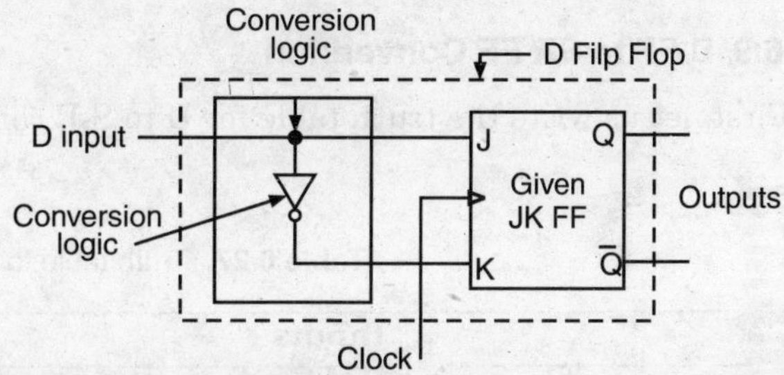

Fig. 6.66. Logic diagram for conversion from JK FF to D FF.

(iii) Lastly, we draw the logic diagram.

The logic diagram for JK flip-flop to D flip-flop is shown in figure 6.66.

6.16.8 J-K Flip-flop to S-R Flip-flop Conversion

(*i*) First, let us write the truth table for J-K to SR.

The truth table is shown in Table 6.26.

Table 6.26. Truth table for J-K to S-R conversion

Inputs				Outputs	
S	**R**	**Present state Q_n**	**Next state Q_{n+1}**	**J**	**K**
0	0	0	0	0	×
0	1	0	0	0	×
1	0	0	1	1	×
1	1	0	1	×	×
0	0	1	0	×	0
0	1	1	0	×	1
1	0	1	1	×	0
1	1	1	1	×	×

(ii) Next, we make K-maps and simplify.

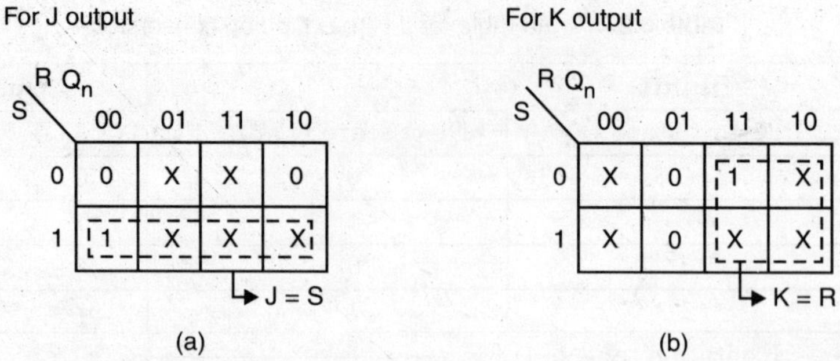

Fig. 6.67.

(iii) Lastly, let us make logic diagram.

The logic diagram for J-K to S-R FF is shown in figure 6.68.

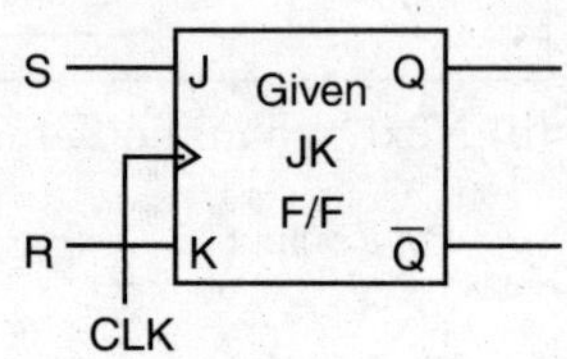

Fig. 6.68. J-K to S-R FF conversion.

6.16.9 D FF to SR FF Conversion

(i) First, let us write the truth table for D to S-R conversion.

Table 6.27. Truth table for D to S-R conversion

Inputs				Output D
S	**R**	**Present state Q_n**	**Next state Q_{n+1}**	
0	0	0	0	0
0	1	0	0	0
1	0	0	1	1
0	1	1	0	0
0	0	1	1	1
1	0	1	1	1

(ii) Next, let us make K-maps and simplify.

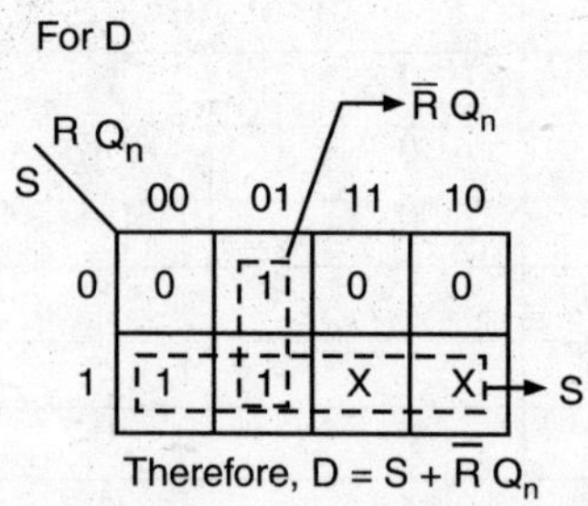

Fig. 6.69. K-map and simplification.

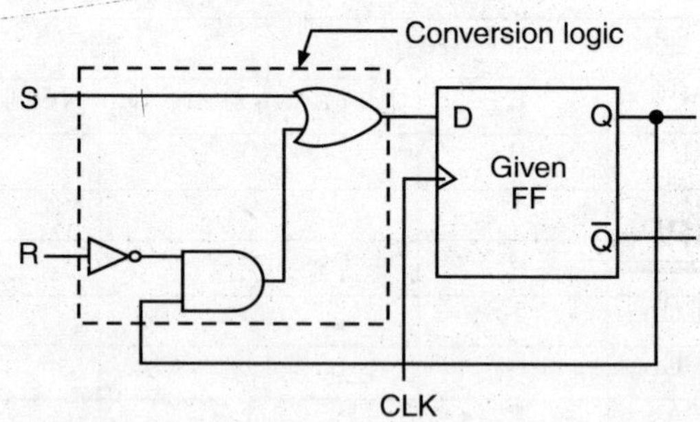

Fig. 6.70. Logic diagram for D to S-R FF conversion.

(iii) Lastly, we make logic diagram.

6.16.10 T FF to SR FF Conversion

(i) First, we write the truth table for T to S-R FF Conversion.

Table 6.28. Truth table for T to S-R FF conversion

Inputs				Output T
S	**R**	**Present state Q_n**	**Next state Q_{n+1}**	
0	×	0	0	0
1	0	0	1	1
0	1	1	0	1
×	0	1	1	0

(ii) Next, we write K-maps and simplify for T output.

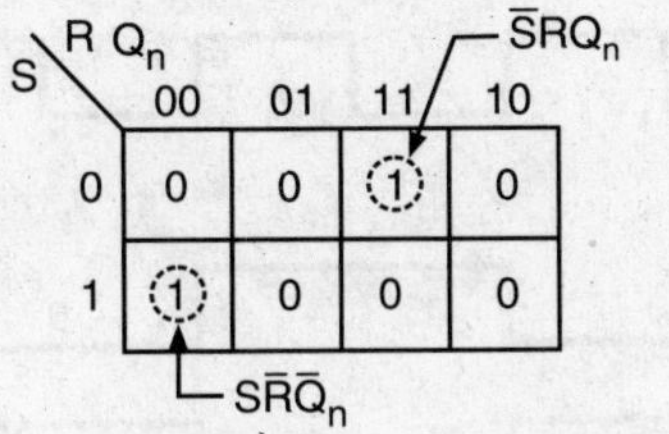

Therefore, $T = S\bar{R}\bar{Q}_n + \bar{S}RQ_n$

Fig. 6.71. K-map and simplification.

(iii) Finally, we make logic diagram.

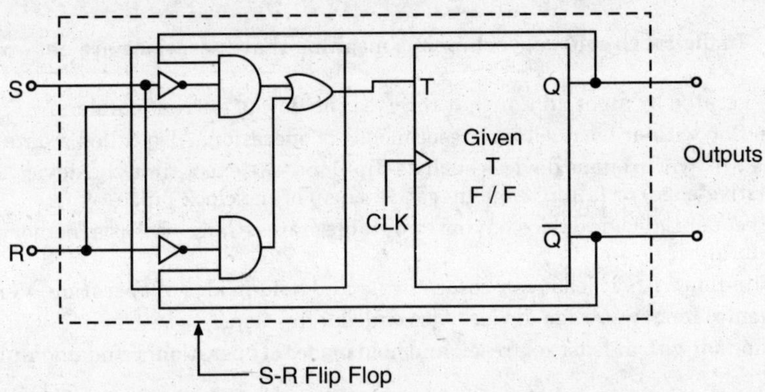

Fig. 6.72. Logic diagram for D to S-R FF conversion.

6.17 APPLICATIONS OF FLIP-FLOPS

(PTU, Jalandhar, Sem. Exam., 2004-05)

Some of the important applications of the flip-flops may be listed as under :

(i) Elimination of keyboard debounce

(ii) As a memory element

(iii) In various types of registers

(iv) In counters/timers

(v) As a delay element.

EXAMPLE 6.6. Draw the waveform at Q for 4 clock pulses. Assume Q = 0 initially as shown in figure 6.73.

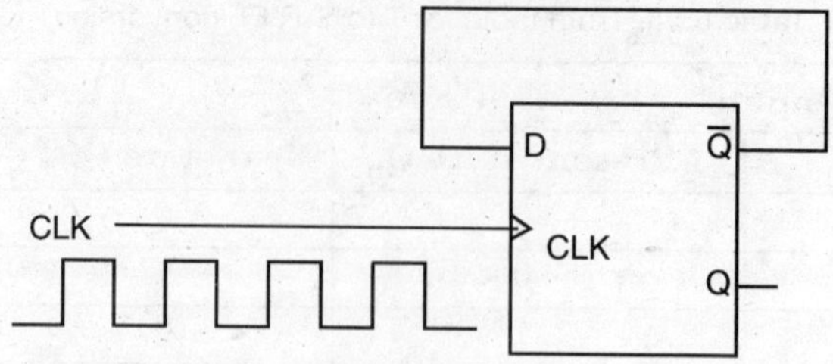

Fig. 6.73.

Solution: The waveforms have been shown in figure 6.74.

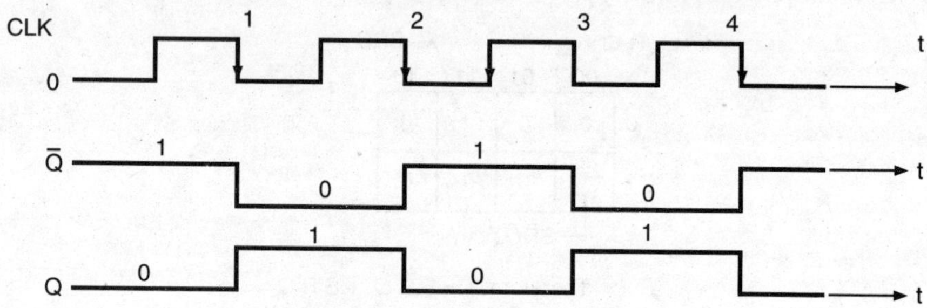

Fig. 6.74. Illustration of waveforms.

GLOSSARY

1. **Asynchronous :** In digital circuits, asynchronous meaning that operations are not executed in step with the clock.
2. **Clock :** Signal generated by an oscillator that receives the flow of current carriers.
3. **D flip-flop :** Flip-flop with at least set and reset modes of operation. Also called a data or delay flip-flop.
4. **Edge triggering :** In synchronous devices such as flip-flops, the exact time the device is activated such as on the rising (positive-edge) or falling edge (negative-edge) of the clock pulse.
5. **Flip-flop :** Basic sequential logic device having two stable states. Can serve as a memory device. Also called a bistable multivibrator.
6. **J-K flip-flop :** Flip-flop with at least set, reset, toggle and hold modes of operation. Very adaptable.
7. **Latch :** Fundamental binary storage device. Also called a flip-flop.
8. **R-S flip-flop :** Flip-flop with at least set, reset, and hold modes of operation. Fundamental latching (memory) circuit.
9. **T flip-flop :** Short for toggle flip-flop. The output toggles to the opposite logic state on repeated clock pulses. Very useful in digital counter circuits.
10. **Trigger :** A pulse that causes a logic device to be activated or change states.

SUMMARY

- We know that the output of a combinational circuit at any instant of time, depends only upon the levels present at input terminals. Further, the combinational circuits do not use any memory. This means that the previous state of input does not have any effect on the present state of the circuit. The sequence in which the inputs are being applied has no effect on the output of a combinational circuit.
- In case of a sequential circuit, the timing parameter comes into effect. Infact, the output of a sequential circuit depends upon the present time inputs, the previous output and the sequence in which the inputs are applied.
- In order to provide the previous input or output, we require a memory element. Therefore, a sequential circuit requires a memory element.
- Clock skew may be defined as the difference in time between the clock edges arriving at a pair of clock inputs. In a perfect system, all clock signals arrive at all the various clock input pins of the system at

exactly the same time and the skew is zero. However, in real time systems, the edges do not arrive at exactly the same time and there is some skew. The clock skew occurs because of different delays on different paths from the clock generator to various circuits.

- A flip-flop is popularly known as a basic digital memory circuit. It has two stable states namely logic 1 state and logic 0 state. It can be designed either using NOR gates or NAND gates.
- As a matter of fact, latch is a bistable element with two stable states. It has two outputs Q and $\overline{Q}$ which are complements of each other. Latch is a sequential logic circuit which checks all its inputs continuously and will change its output as soon as the input changes without waiting for the clock signal. Usually, an enable signal is provided for a latch.
- The S-R gated latch is said to be a **positive level triggered** circuit, because the circuit responds to the changes in input only when the enable input (E) is HIGH. The output will change as per the changes in input as long as the active level (HIGH for this circuit) is maintained at the enable input.
- **Positive Level Triggered :** If the outputs of S-R latch respond to the input changes, for its enable input at HIGH (1) level, then it is called as the positive level triggered S-R latch.
- **Negative Level Triggered :** If the outputs of an S-R latch respond to the input changes, for its enable input at LOW (0) level, then it is called as the negative level triggered S-R latch.
- There are following two types of edge triggered flip-flops :
  (i) Positive edge triggered flip-flops.
  (ii) Negative edge triggered flip-flops.
- Positive edge triggered flip-flops will allow its outputs to change only at the instants corresponding to the rising edges of clock (or positive spikes). Its outputs will not respond to change in inputs at any other instant of time. Negative edge triggered flip-flops will respond only to the negative going edges (or spikes) of the clock.
- The edge triggered D flip-flops can be of following two types :
  (i) Positive edge triggered D flip-flop.
  (ii) Negative edge triggered D flip-flop.
  These flip-flops can be derived from the level triggered D latch.
- For the racing around to take place, it is essential to have the enable input high along with J = K = 1. As the enable input remains high for a long time in a J-K latch, the problem of multiple toggling arises. But, in edge triggered J-K flip-flop, the positive clock pulse is present only for a very short time. Hence, by the time the changed outputs return back to the inputs of NAND gates 3 and 4, the clock pulse has died down to zero. Hence, the multiple toggling can not take place. Thus, the edge triggering avoids the race around condition.
- Some of the important applications of the flip-flops may be listed as under :
  (i) Elimination of keyboard debounce
  (ii) As a memory element
  (iii) In various types of registers
  (iv) In counters/timers
  (v) As a delay element.

SHORT QUESTIONS WITH ANSWER

Q.1. What is meant by sequential circuits ?

Ans. In case of a sequential circuit, the *timing* parameter comes into effect. Infact, the output of a sequential circuit depends upon the present time inputs, the previous output and the sequence in which the inputs are applied.

Q.2. What is One-bit Memory Cell (Basic Bistable Element).

Ans. A flip-flop is popularly known as a basic digital memory circuit. It has two stable states namely logic 1 state and logic 0 state. It can be designed either using NOR gates or NAND gates. NAND gates-1 and 2 are basically acting as inverters. Therefore, this circuit is known as a cross coupled inverter. Here, output of gate-1 is connected to the input of gate-2 and output of gate-2 is connected to input of gate-1

Q.3. Define a Clock signal.

Ans. The clock signal depicted in figure 6.75 is a timing signal. Every sequential signal has this timing signal applied. Infact, clock is a rectangular signal as shown in figure 6.75, with a duty cycle equal to 50%. The clock signal repeats itself after every T seconds. Therefore, the clock frequency is F = 1/T.

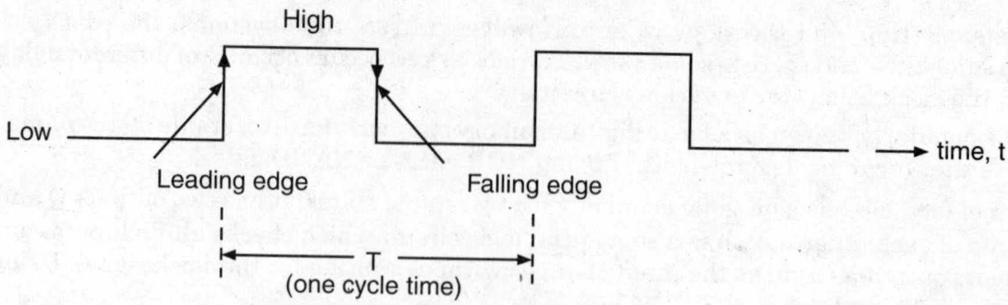

Fig. 6.75. Illustration of a clock signal

Q.4. Carry out the performance comparison of Combinational and sequential circuits.

Ans.

S. No.	Parameter of comparison	Combinational Circuits	Sequential Circuits
1.	Output depend on	Inputs present at that instant of time.	Present inputs and past inputs/ outputs.
2.	Memory	Not necessary	Necessary
3.	Clock input	Not essential	Essential
4.	Important Examples	Adders, subtractors, code converters	Flip-flops, shift registers, counters.

Q.5. What is concept of Level Triggering

Ans. The S-R gated latch is said to be a **positive level triggered** circuit, because the circuit responds to the changes in input only when the enable input (E) is HIGH. The output will change as per the changes in input as long as the active level (HIGH for this circuit) is maintained at the enable input.

Q.6. Write the types of Level Triggered Latches.

Ans. There are following two types of level triggered latches :

(i) Positive level triggered.

(ii) Negative level triggered.

(i) Positive Level Triggered

If the outputs of S-R latch respond to the input changes, for its enable input at HIGH (1) level, then it is called as the positive level triggered S-R latch.

(ii) Negative Level Triggered

If the outputs of an S-R latch respond to the input changes, for its enable input at LOW (0) level, then it is called as the negative level triggered S-R latch.

Q.7. Write the types of Clocked Flip-flop.

Ans. The edge triggered D flip-flops can be of following two types :

(i) Positive edge triggered D flip-flop.

(ii) Negative edge triggered D flip-flop.

These flip-flops can be derived from the level triggered D latch.

Q.8. What is sequential circuit?

Ans. Sequential logic circuit is that circuit in which outputs of the circuit remain in their state, until instructed to change it. In other words sequential logic circuit may be defined as that logic circuit whose outputs are determined by the sequence in which input signals are applied.

Q.9. What is a flip-flop?

And. A flip-flop is a basic memory element that is made of an assembly of logic gates and is used to store 1-bit of information.

Q.10. What is a latch?

Ans. It is a D-type of flip-flop and stores one bit of data.

Q.11. What is an excitation table ?

Ans. Excitation table gives an information about what should be the flip-flop inputs if the outputs are specified before and after the clock pulses.

REVIEW QUESTIONS

1. Mention the types of digital systems ? Explain with the block diagram and also with their operating principle.
2. What is a flip-flop ?
3. State and explain the triggering methods used for flip-flops.
4. What is the function of preset and clear inputs in flip-flop ?
5. Explain with truth table the working of clocked RS flip-flop.
6. State the disadvantages of RS flip-flop. How can they be avoided ?
7. Explain with diagram the working of D type flip-flop. Give its truth table.
8. Give reason why D flip-flop is called as data latch ?
9. Draw the circuit using logic gates of a T-type flip-flop. Draw its symbol and write its truth table.
10. Explain S-R flip-flop using NOR gates.
11. Describe how two cross-coupled NAND gates form a R-S flip-flop. Write its truth table.
12. What are the various types of flip-flops ?
13. Draw the circuit of S-R flip-flop using NAND gate.
14. Draw the schematic diagram of J-K flip-flop and describe its working. Write down its truth table.
15. What is race around condition ?
16. Draw the circuit of J-K flip-flop using NAND gate.
17. Draw a neat diagram of master slave J-K flip-flop. Explain how race around condition is avoided using master slave J-K flip-flop ?
18. Explain the working of the master slave J-K flip-flop.
19. The D input and a single clock pulse are shown in figure 6.76. Compare the resulting Q outputs for : Positive edge triggered flip-flop, negative edge triggered flip-flop and pulse triggered Master-slave flip-flops. The flip-flops are initially RESET.

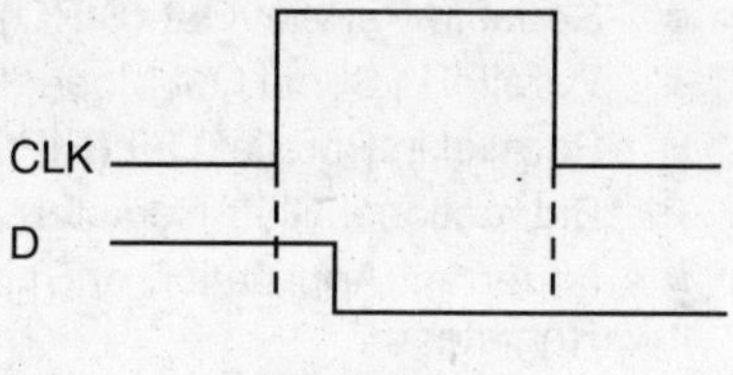

Fig. 6.76.

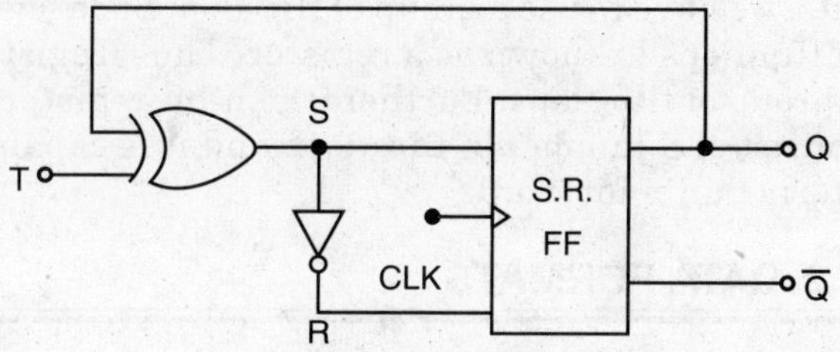

Fig. 6.77.

20. Using a T-FF design a R-S FF.
21. Using NAND gates, sketch a clocked-RS flip-flop. Using this FF, sketch MSJK flip-flop and using this master slave J-K FF, sketch Toggle and Delay flip-flops.

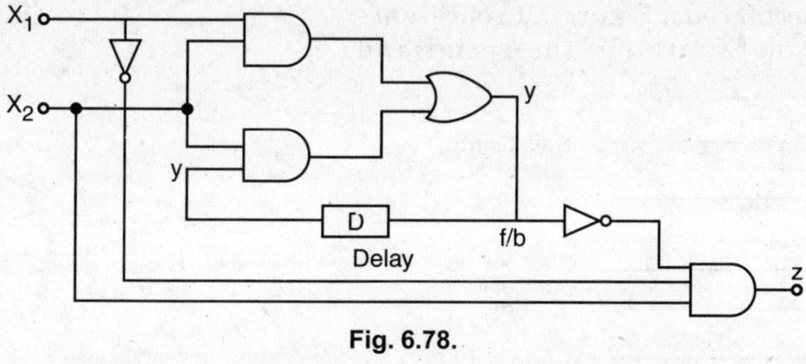

Fig. 6.78.

Implement the state transition diagram shown in figure 6.78, using T-flip-flops.

7

SEQUENTIAL CIRCUITS: SHIFT REGISTERS

Syllabus

Shift Registers: Classification and details.

Chapter Outline

- Introduction
- Data Formats
- Registers Classification
- Buffer Registers
- Shift Registers
- Serial in Parallel Out (SIPO)
- Parallel in Serial Out Mode (PISO)
- Parallel in Parallel Out (PIPO)
- Bidirectional Shift Register
- Important Application of Shift Registers
- Universal Shift Register
- Performance Comparison of Counters and Registers

7.1 INTRODUCTION

In last chapter, we have discussed various applications of flip-flops. One of them was *registers*. As discussed earlier, a flip-flop is a 1 bit memory cell which may be used to store digital data. To increase the storage capacity in terms of number of bits, we have to use a group of flip-flop. Such a group of flip-flops is known as a register. Thus, register is a group of flip-flops. Further, the n-bit register will consist of n number of flip-flops and it is capable of storing an n bit word.

7.2 DATA FORMATS

We can enter the data either in serial (one bit at a time) or in the parallel form (all the bits at a time). Also, the stored data can be retrieved in serial or *parallel* form. Data in the serial form is also called as a *temporal code*, whereas that in the parallel form is called as *special code*. Figure 7.1 represents a four bit digital data 1011 in the serial and parallel manner.

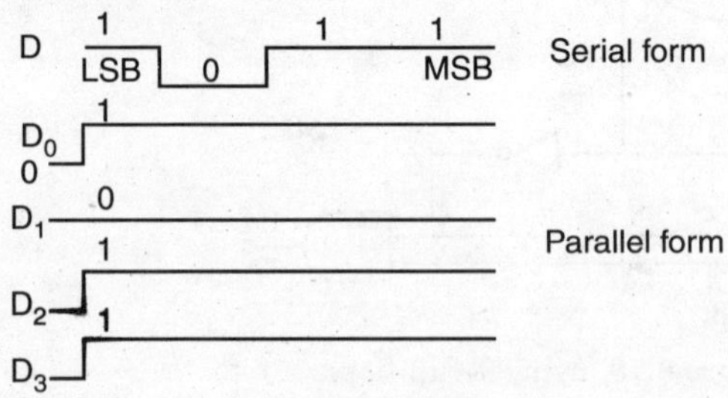

Fig. 7.1. Data representation in serial and parallel form

7.3 REGISTERS CLASSIFICATION

Registers may be classified based on the way in which data are entered and taken out from a register. There may be following four possible modes :

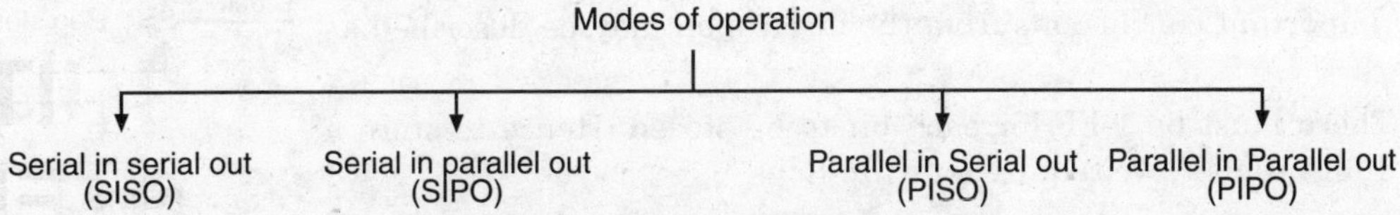

Fig. 7.2. Modes of operation of registers.

Also, it is possible to design registers using discrete flip-flops such as SR, JK or D flip-flops. However, the registers are also available in MSI devices. They are available in 74/54 TTL series as listed in Table 7.1.

Table 7.1. Shift registers available in 74/54 series

S.No.	IC Number	Description of IC
1.	7491, 7491 *A*	8 bit serial in, serial out
2.	7494	4 bit parallel in, serial out
3.	7495	4 bit serial/parallel in, parallel out (shift right, shift left)
4.	7496	5 bit parallel in/parallel out, serial in/serial out

7.4 BUFFER REGISTERS

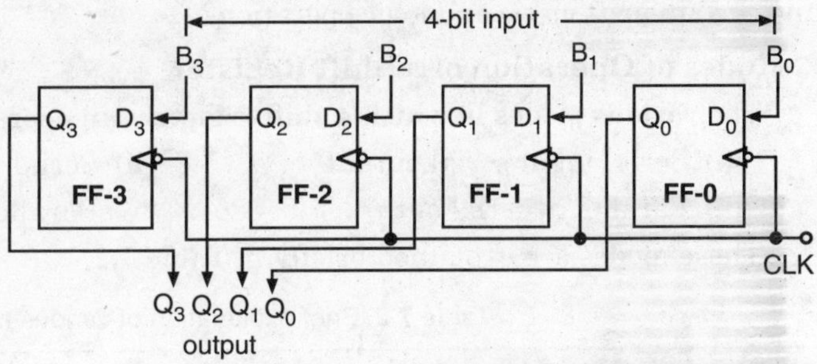

Fig. 7.3. A four-bit buffer register using D flip-flops

1. Definition

Figure 7.3 shows the simplest type of register constructed using four D flip-flops. This is a 4-bit register, but we can construct an *n*-bit register by following the same construction principle.

This register is also known as the *buffer register*. Here, D-flip-flop is negative edge triggered and all the flip-flops are connected to a common clock signal. Therefore, all of them are triggered at the same instant of time. Further, buffer registers are used for temporary storage of digital words.

2. Working Operation

Let us consider that the word to be stored is B_3 B_2 B_1 B_0 = 1 0 1 0. These bits are connected to the D inputs of the four D flip-flops as shown in figure 7.3. Then, we apply the clock pulse. Corresponding to the first negative edge of the clock pulse, the outputs of all the D flip-flops shall follow their respective inputs.

Therefore, we write $Q_3\ Q_2\ Q_1\ Q_0 = B_3\ B_2\ B_1\ B_0 = 1\ 0\ 1\ 0$

Even if the inputs are now changed, the output data remains latched to 1 0 1 0 till the next negative edge of the clock. This means that the buffer register is capable of storing the digital data.

DENNIS C. HAYES

In 1978, 28-year-old Dennis C. Hayes formed what became Hayes Microcomputer Products, Inc. When he began, Hayes hand-assembled and soldered products on a borrowed dining room table in his home. Then, in 1981 the Hayes™ Smartmodem™, which was easily integrated into the computer environment, started a communications revolution.

3. Schematic Diagram

Figure 7.4 shows the schematic diagram of a buffer register.

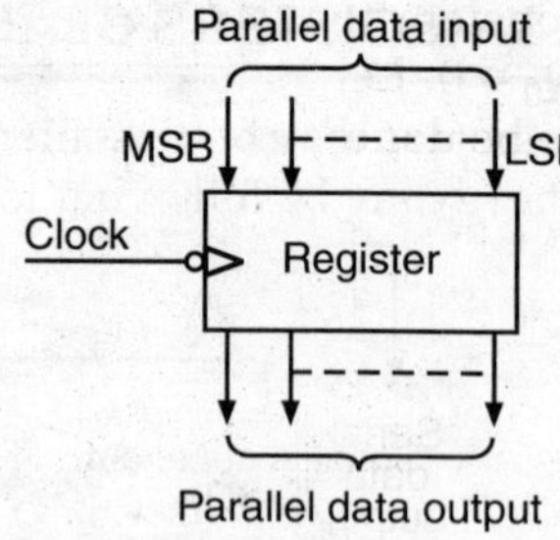

Fig. 7.4. Schematic diagram of buffer register

4. Important conclusions

Few important conclusions from the discussion may be described as under :

(i) There must be 1-FF for each bit to be stored. Hence to store a 4-bit number, we require four flip-flops.

(ii) It is interesting to note that all the four input bits B_3 B_2 B_1 B_0 are loaded into the register simultaneously, i.e., at the same instant of time.

(iii) Hence, this way of applying the input and taking the output is known as parallel input parallel output and the mode of operation is called as *parallel shifting.*

7.5 SHIFT REGISTERS

(PTU, Jalandhar, Sem. Exam., 2006-07)

1. Definition

As a matter of fact, the binary data in a register can be moved within the register from one flip-flop to the other or outside it with application of clock pulses. Those registers which allow such data transfers are known as *shift registers.* Shift registers find applications for data storage, data transfer and certain arithmetic and logic operations.

2. Modes of Operation of a Shift Register

The various modes in which a shift register can operate may be listed as under :

(i) Serial input serial output
(ii) Serial input parallel output
(iii) Parallel in serial out
(iv) Parallel in parallel out.

These modes are explained briefly in Table 7.2.

Table 7.2. Brief explanation of various modes of shift register

S. No.	Mode of operation	Illustration	Comment
1.	Serial input serial output (serial shift right)	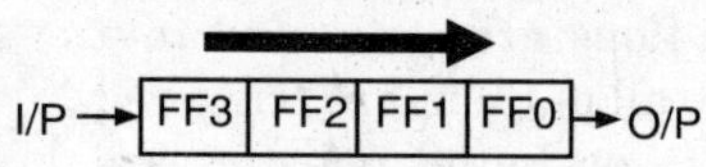	Data bits shift from left to right by 1 position per clock cycle.
2.	Serial input serial output (serial shift left)		Data bits shift from right to left by 1 position per clock.
3.	Serial input parallel output	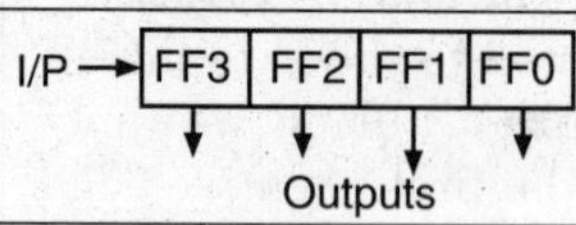	All ouput bits are made available simultaneously after 4-clock pulses.
4.	Parallel input serial output	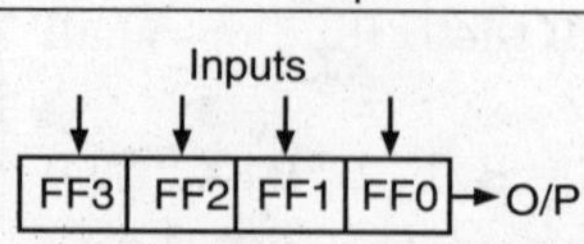	All inputs are loaded simultaneously but output bit-by-bit.

7.5.1 Serial Input Serial Output (Shift Left Mode)

1. Definition

Figure 7.5 shows the serial input serial output type shift register with shift left mode. Let us consider that all the flip-flops initially are in the rest condition. This means that, i.e., $Q_3 = Q_2 = Q_1 = Q_0 = 0$. Let us illustrate the entry of a four bit binary number 1 1 1 1 into the register. When this is to be done, this number must be applied to D_{in} bit-by-bit with the MSB bit applied first.

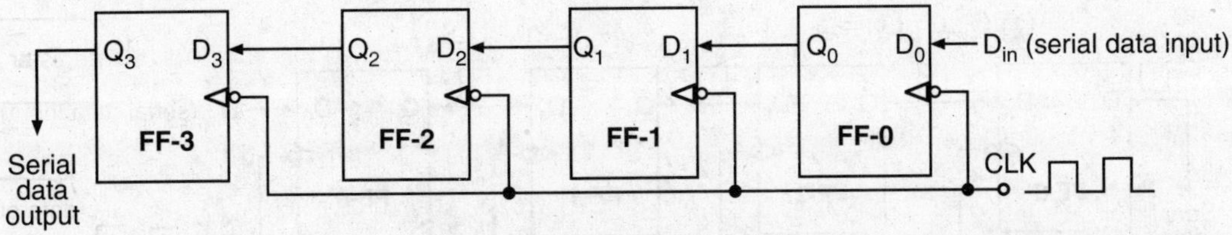

Fig. 7.5. Serial shift left register

This means that the D input of FF-0, i.e., D_0 is connected to serial data input (D_{in}). Also, output of FF-0, i.e., Q_0 is connected to the input of the next flip-flop, i.e., D_1 and so on.

> **DO YOU KNOW?**
>
> A sequential logic system uses FFs, counters and registers, along with logic gates. It outputs and sequence of operatons depend on present and past inputs.

2. Working Operation

Before application of clock signal, let us consider that $Q_3\,Q_2\,Q_1\,Q_0 = 0\,0\,0\,0$ and we apply MSB bit of the number to be entered to D_{in}. Therefore, $D_{in} = D_0 = 1$. Now, we apply the clock. On the first falling edge of clock, the FF-0 is set and the stored word in the register will be

$$Q_3\,Q_2\,Q_1\,Q_0 = 0\,0\,0\,1$$

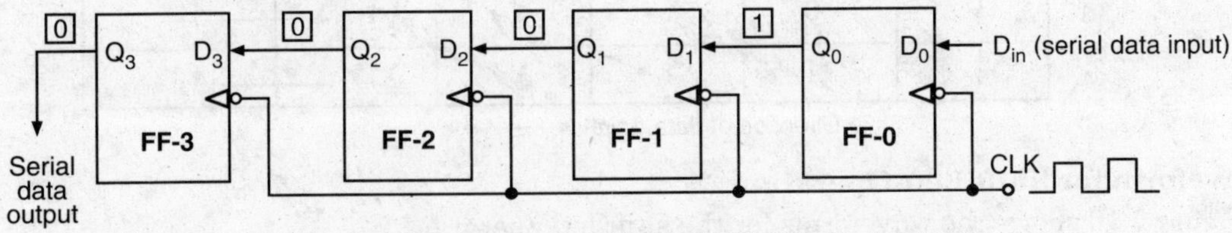

Fig. 7.6. Shift register status after first falling clock edge

Then, we apply the next bit to D_{in}. Hence, $D_{in} = 1$. As soon as the next negative edge of the clock hits, FF-1 will set and the stored word will change to,

$$Q_3\,Q_2\,Q_1\,Q_0 = 0\,0\,1\,1$$

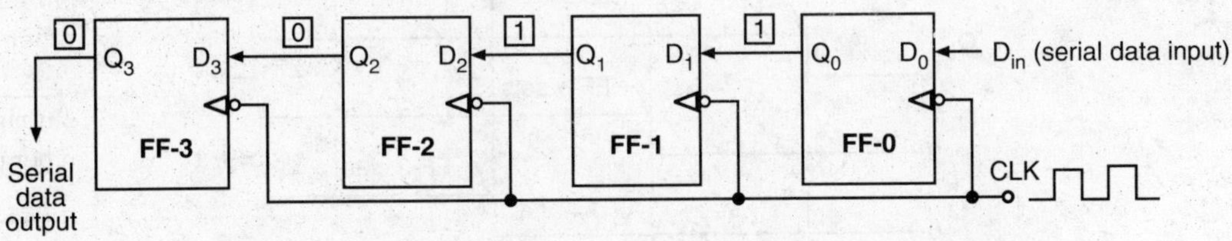

Fig. 7.7. Shift register status after the second falling edge of clock

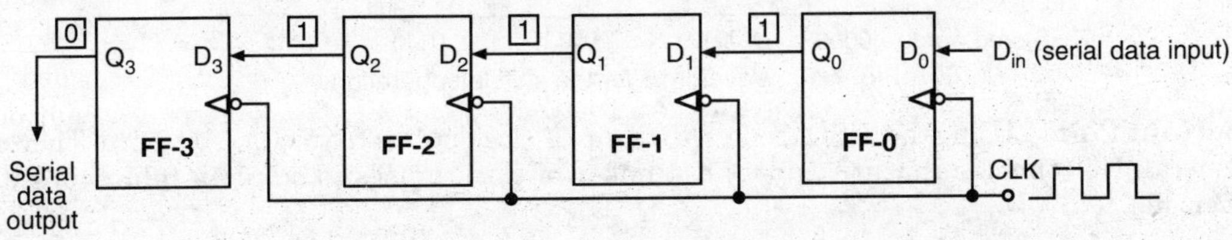

Fig. 7.8. Shift register status after the third falling edge of clock

Next, we apply the next bit to be stored, i.e., 1 to D_{in}. Then, we apply the clock pulse. As soon as the third negative clock edge hits, FF-2 sets and the output will get modified to

$$Q_3\,Q_2\,Q_1\,Q_0 = 0\,1\,1\,1$$

Similarly, with $D_{in} = 1$, and with the fourth negative clock edge arriving, the second word in the register will be given by

$$Q_3\,Q_2\,Q_1\,Q_0 = 1\,1\,1\,1$$

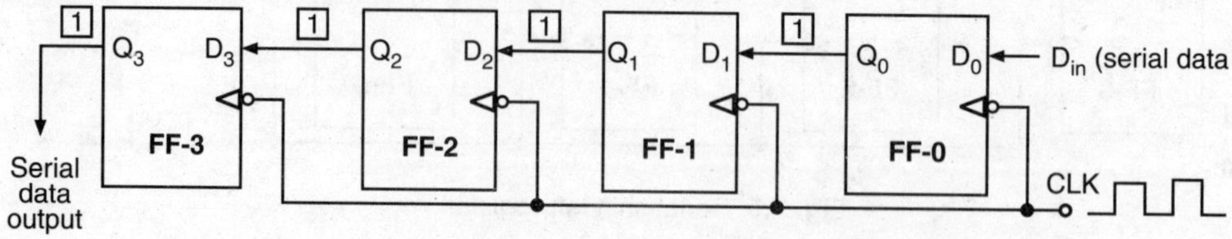

Fig. 7.9. Shift register status after the fourth falling edge of clock

Table 7.3. Brief Explanation of shift left operation

	CLK	Q_3	$Q_2=D_3$	$Q_1=D_2$	$Q_0=D_1$	Serial input $D_{in}=D_0$
Initially		0	0	0	0	
1st	↓	0	0	0	1	1
2nd	↓	0	0	1	1	1
3rd	↓	0	1	1	1	1
4th	↓	1	1	1	1	1

Direction of data travel ←

3. Waveform for Shift Left Operation

Figure 7.10 shows the waveforms for the shift left operation.

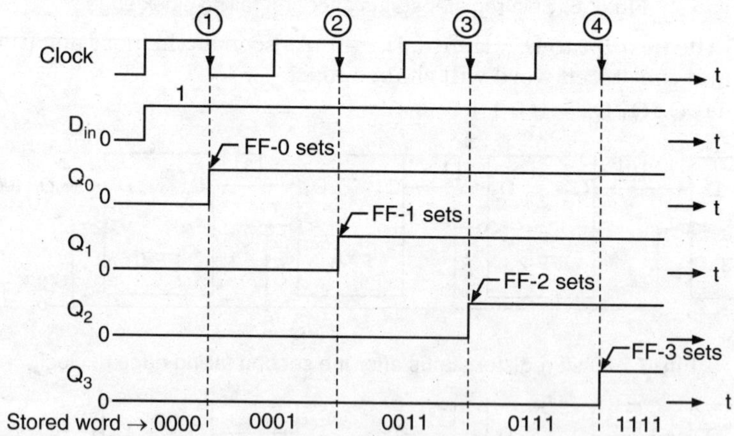

Fig. 7.10. Waveforms for the shift left operation

Important Point : Using the SISO mode, we require 4-clock pulses to store a 4-bit word. Therefore, in general, we conclude that it requires n number of clock pulses to store an n-bit word using SISO mode.

7.5.2 Serial in Serial Out (Shift Right Mode)

1. Definition

Figure 7.11 shows the serial input serial output type shift register with shift right mode. Let us consider that all the flip-flops initially are in the reset condition. This means that $Q_3 = Q_2 = Q_1 = Q_0 = 0$. Let us illustrate the entry of a four bit binary number 1 1 1 1 into the register. When this is to be done, this number must be applied to D_{in} bit-by-bit with the LSB bit applied first.

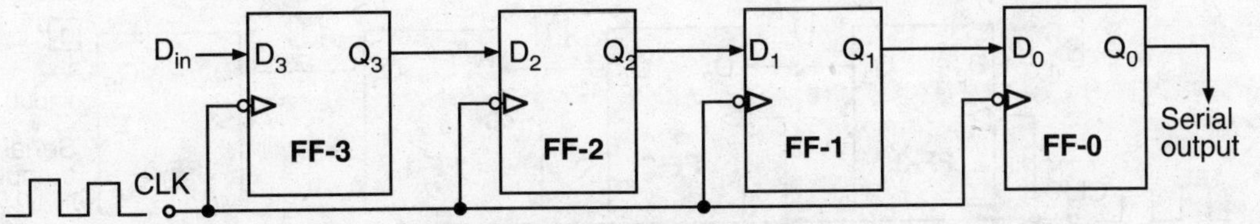

Fig. 7.11. Serial shift right register

This means that the D input of FF-3, i.e., D_3 is connected to serial data input (D_{in}). Output of FF-3, i.e., Q_3 is connected to the input of the next flip-flop, i.e., D_2 and so on.

2. Working Operation

Before application of clock signal, let $Q_3\ Q_2\ Q_1\ Q_0 = 0\ 0\ 0\ 0$ and we apply LSB bit of the number to be entered to D_{in}. Hence, $D_{in} = D_3 = 1$. Then, we apply the clock. On the first falling edge of clock, the FF-3 is set, and the stored word in the register will be given by

$$Q_3\ Q_2\ Q_1\ Q_0 = 1\ 0\ 0\ 0$$

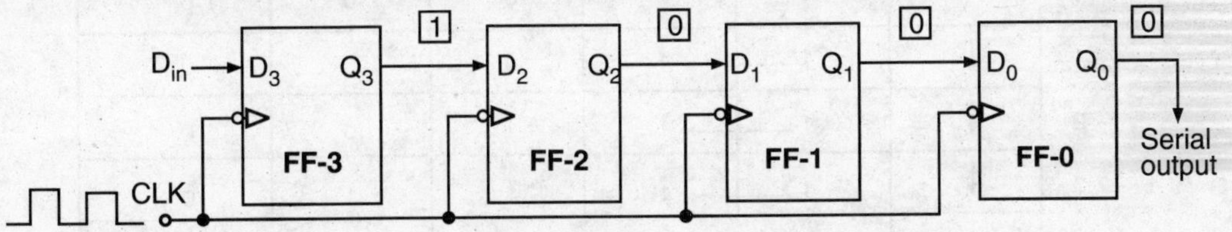

Fig. 7.12. Shift register status after first falling clock edge

Next, we apply the next bit to D_{in}. Hence, $D_{in} = 1$. As soon as the next negative edge of the clock hits, FF-2 sets and the stored word will change to,

$$Q_3\ Q_2\ Q_1\ Q_0 = 1\ 1\ 0\ 0$$

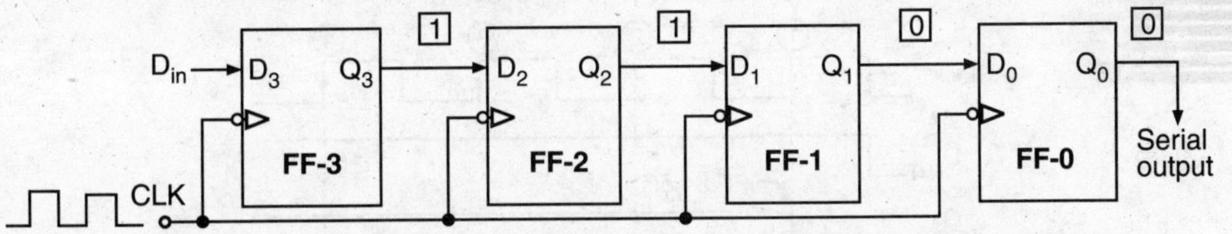

Fig. 7.13. Shift register status after the second falling edge of clock

Further, we apply the next bit to be stored, i.e., 1 to D_{in}. Then, we apply the clock pulse. As soon as the third negative clock edge hits, FF-1 will sets and the output will get modified to

$$Q_3\ Q_2\ Q_1\ Q_0 = 1\ 1\ 1\ 0$$

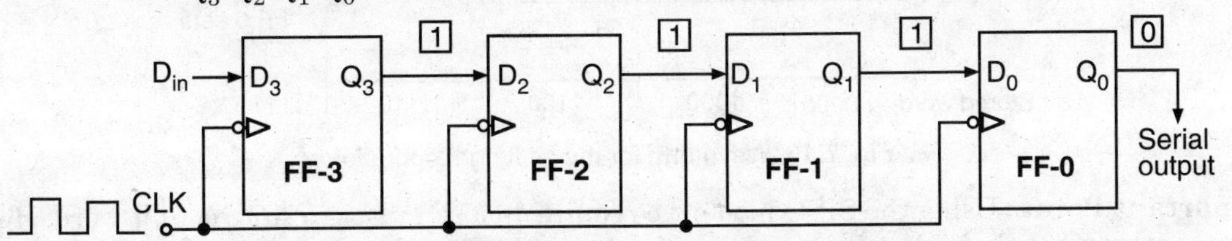

Fig. 7.14. Shift register status after the third falling edge of clock

The very first step towards success in any occupation is to become interested in it.

—**Sir William Osler**

Similarly, with $D_{in} = 1$ and with the fourth negative clock edge arriving, the stored word in the register will be given by

$$Q_3\ Q_2\ Q_1\ Q_0 = 1\ 1\ 1\ 1$$

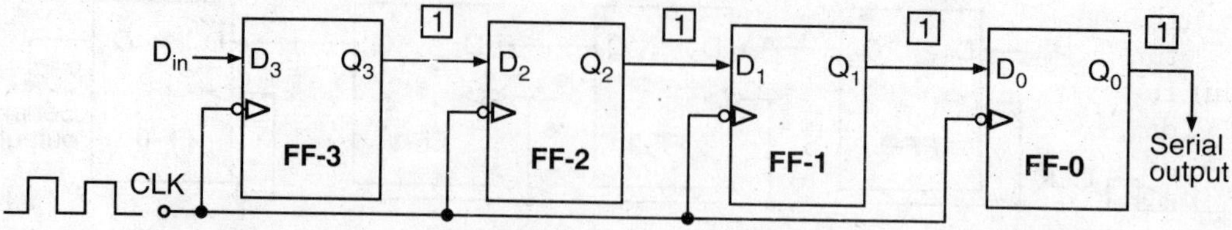

Fig. 7.15. Shift register status after the fourth falling edge of clock

3. Brief Explanation

Table 7.4 summarizes the Shift Right Operation

Table 7.4. Brief Explanation of Shift Right Operation

	CLK	$D_{in} = Q_3$	$Q_3=D_2$	$Q_2=D_1$	$Q_1=D_0$	Serial input $Q_0=D_1$
Initially			0	0	0	0
1st	↓	1	1	0	0	0
2nd	↓	1	1	1	0	0
3rd	↓	1	1	1	1	0
4th	↓	1	1	1	1	1

→ Direction of data travel

4. Waveforms for Shift Right Operation

Figure 7.16 shows the waveforms for shift right operation.

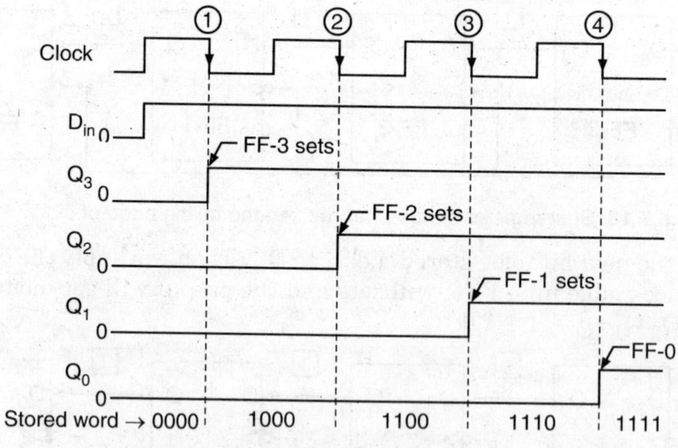

Fig. 7.16. Waveform for the shift right operation

Important Point : Using the SISO mode, we require 4-clock pulses to store a 4-bit word. Hence, in general, we conclude that it requires n number of clock pulses to store an n-bit word using SISO mode.

7.5.3 Applications of Serial Operation

Few important applications of serial operation may be described as under :

(i) The transmission of data from one place to the other takes place in serial manner as shown in figure 7.17.

(ii) It takes a longer time for serial transmission, since the time required to transmit one bit is equal to the time corresponding to one clock cycle.

(iii) However, for long distance communication, where the distances are in kilometres, serial communication has an added advantage that only one conductor is required to be used.

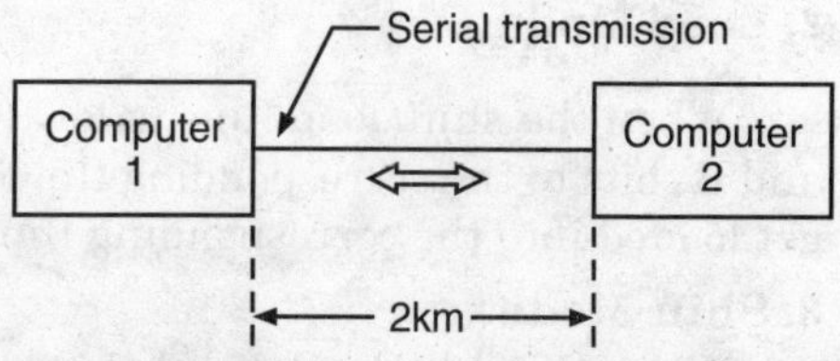

Fig. 7.17. Application of serial operation

7.6 SERIAL IN PARALLEL OUT (SIPO)

(PTU, Jalandhar, Sem. Exam., 2008-09)

In this particular operation, the data is entered serially and then taken out in parallel. This means that first the data is loaded bit-by-bit. The outputs are disabled as long as the loading is taking place. As soon as the loading is complete, and all the flip-flops consist of their required data, the outputs are enabled so that all the loaded data is made available over all the output lines simultaneously. Also, number of clock cycles required to load a four bit word is 4. Therefore, the speed of operation of SIPO mode will remain same as that of SISO mode.

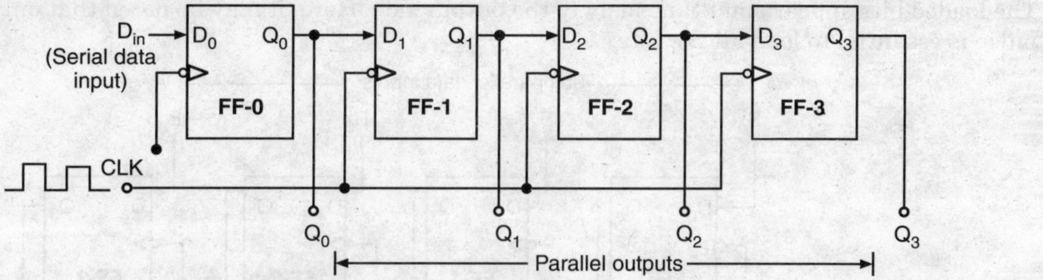

Fig. 7.18. Illustration of serial input parallel output mode

7.7 PARALLEL IN SERIAL OUT MODE (PISO)

1. Definition and Circuit Diagram

In this particular mode, the bits are centered in parallel, i.e., simultaneously as shown in figure 7.19. The circuit shown in figure 7.19 is a four bit parallel input serial output register. Output of previous FF is connected to the input of the next one with the help of a combinational circuit. Then, the binary input word B_0, B_1, B_2, B_3 is applied through the same combinational circuit. There are two modes in which this circuit can work. These modes are shift mode and load mode.

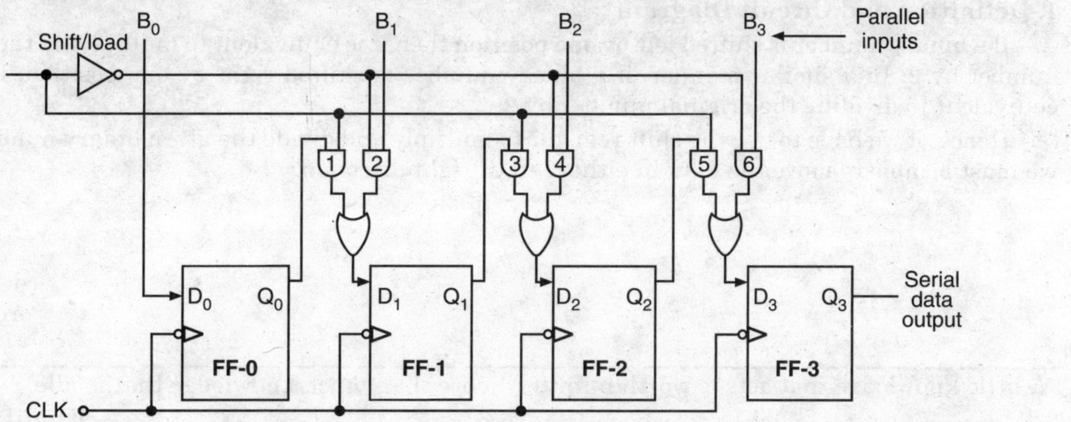

Fig. 7.19. Parallel in serial out shift register

2. Load Mode

When the shift/$\overline{\text{load}}$ line is low (0), the AND gates 2, 4 and 6 become active. They pass B_1, B_2 and B_3 bits to the corresponding flip-flops. On low going edge of clock, the binary inputs B_0 B_1 B_2 B_3 get loaded into the corresponding flip-flops. Therefore, parallel loading takes place.

3. Shift Mode

When the shift/$\overline{\text{load}}$ line is high (1), the AND gates 2, 4, 6 become inactive. Hence, the parallel loading of the data becomes impossible. But, the AND gates 1, 3 and 5 become active. Therefore, the shifting of data from left to right bit-by-bit on application of clock pulses. Thus, the parallel in serial out operation takes place.

7.8 PARALLEL IN PARALLEL OUT (PIPO)

> **DO YOU KNOW?**
>
> A shift register is often used as a way to delay a digital signal by an integral number of clock cycles.

Figure 7.20 demonstrates the parallel in parallel out mode of operation. The 4-bit binary input B_0, B_1, B_2, B_3 is applied to the data inputs D_0, D_1, D_2 and D_3 respectively of the four flip-flops. As soon as a negative clock edge is applied, the input binary bits shall be loaded into the flip-flops simultaneously. The loaded bits appear simultaneously to the output side. Here, it may be noted that only one clock pulse is essential to load all the bits.

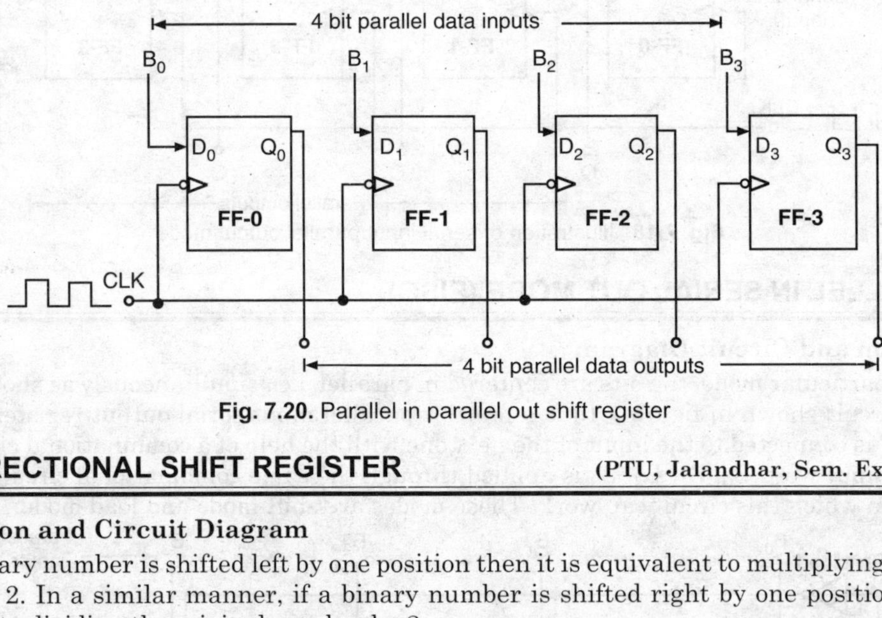

Fig. 7.20. Parallel in parallel out shift register

7.9 BIDIRECTIONAL SHIFT REGISTER

(PTU, Jalandhar, Sem. Exam., 2007-08)

1. Definition and Circuit Diagram

If a binary number is shifted left by one position then it is equivalent to multiplying the original number by 2. In a similar manner, if a binary number is shifted right by one position, then it is equivalent to dividing the original number by 2.

Hence, if we have to use the shift register to multiply and divide the given binary number, then we must be able to move the data in either left or right direction.

A little knowledge that acts is worth infinitely more than much knowledge that is idle.

— **Khalil Gibran**

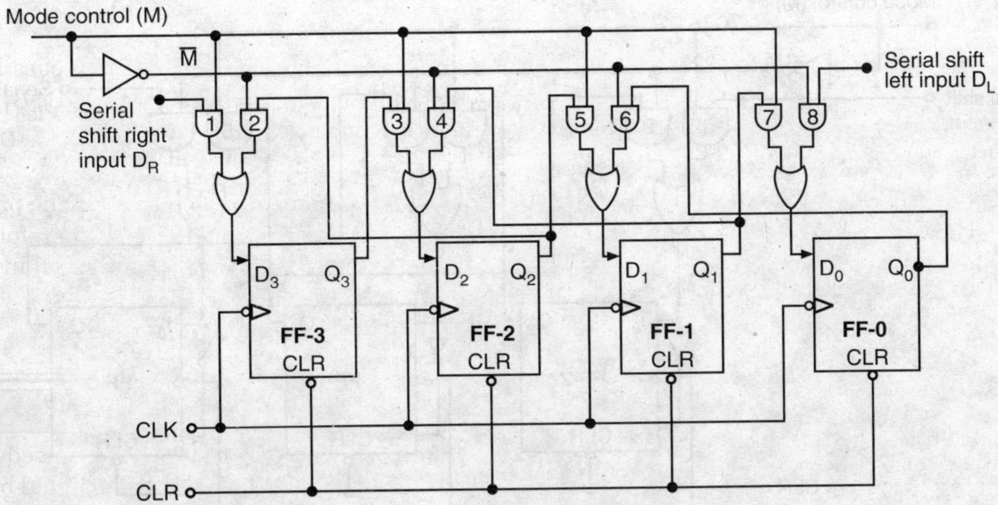

Fig. 7.21. A 4-bit bidirectional shift register

Such a register is known as a bidirectional register. Figure 7.21 shows four-bit bidirectional shift register. There are two serial inputs. These are the serial right shift data input D_R and the serial left shift data input D_L alongwith a Mode select input (M).

2. Working Operation

(i) With M = 1 : Shift right operation

If M = 1, then the AND gates 1, 3, 5 and 7 are enabled whereas the remaining AND gates 2, 4, 6 and 8 will be disabled. Hence, the data at D_R (shift right input) is shifted to right bit-by-bit from FF-3 to FF-0 upon the application of clock pulses. Thus, with M = 1, we get the serial right shift operation.

(ii) With M = 0 : Shift left operation

When the mode control M is connected to 0 then the AND gates 2, 4, 6 and 8 are enabled while 1, 3, 5, 7 are disabled. Hence, the data at D_L (shift left input is shifted left bit by bit from FF-0 to FF-3 on application of the clock pulses. Thus, with M = 0 we get the serial left shift operation. Here, it may be noted that M must be changed only when CLK = 0, otherwise, the data stored in the register may be altered.

7.9.1 A 3-bit Bidirectional Register using the J-K Flip-Flops

Figure 7.22 illustrates a 3-bit bi-directional shift register using J-K flip-flops.

This circuit is same as the 4 bit bi-directional shift register discussed earlier. The only modification is the inclusion of an inverter between the J and K inputs of each flip-flops. With M = 1, the shift right operation takes place as discussed in the previous article and with M = 0, the shift left operation will take place.

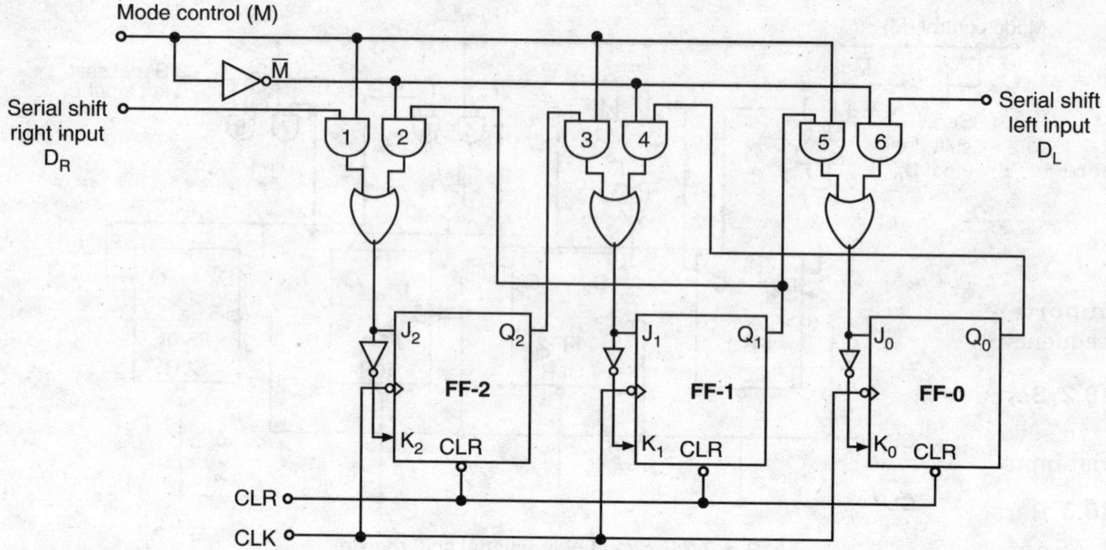

Fig. 7.22. A 3-bit bi-directional shift register using J-K flip-flops

7.10 IMPORTANT APPLICATIONS OF SHIFT REGISTERS

Few common and important applications of a shift register may be described as under :

(i) For temporary data storage
(ii) For multiplication and division
(iii) As a delay line
(iv) As serial to parallel converter
(v) As parallel to serial converter
(vi) As ring counter
(vii) As twisted ring counter or Johnson counter

We have already discussed the use of shift register for temporary storage of data and for multiplication or division. Let us now discuss the remaining applications one-by-one.

7.10.1 Delay Line

(PTU, Jalandhar, Sem. Exam., 2004-05)

1. Definition

A delay line may be defined as a circuit that is used to delay a signal by some fixed time-interval.

Also, a serial input can be used to introduce a time delay of Δt seconds in the digital signal applied at its serial input.

2. Working Operation

The signal which has to be delayed is applied first at the serial input terminal of SISO register. Also, clock signal of frequency, f_{CLK} is applied to all the flip-flops simultaneously. A bit applied at the serial input with required n number of right shifts to reach the serial output. As each shift takes a time equal to the period of 1 clock cycle, the time delay introduced shall be given by.

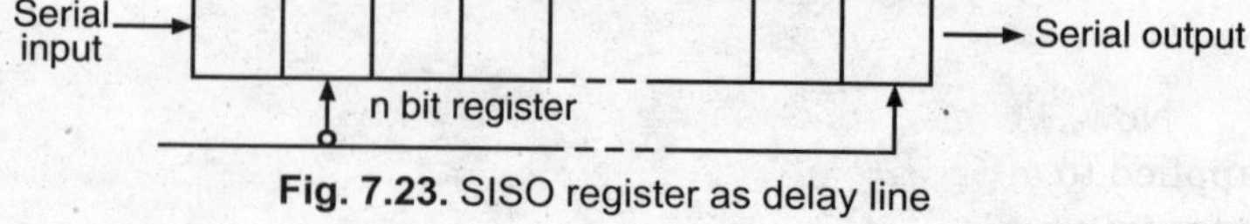

Fig. 7.23. SISO register as delay line

$$\text{Time delay } \Delta t = \text{Number of shifts} \times \text{Time period of 1 clock cycle}$$

$$\text{Time delay } \Delta t = n \times \frac{1}{f_{CLK}}$$

where n = number of shifts

and $\frac{1}{f_{CLK}}$ = time period corresponding to one cycle of clock

Important Point. For an n-bit register, the time delay can be applied by adjusting the clock frequency.

7.10.2 Serial to Parallel Converter

In several applications, the data in serial form required to be converted in parallel form. The serial input parallel output mode of the shift register is used for such applications.

7.10.3 Parallel to Serial Converter

As a matter of fact, the data communication between two computers takes place in the serial transmission form. But internally the data processing takes place in the parallel form. Hence, we require to use a parallel to serial converter to convert the internal parallel data into a serial data suitable for transmission. Further, for this operation, we can use the parallel to serial mode of the shift register.

7.10.4 Ring Counter

1. Definition

Figure 7.24 illustrates a typical application of shift registers which is popularly known as Ring Counter. Here, the connections reveal that they are similar to the connections for shift right operation, except for one change. Actually, output of FF-3 is connected to data input D_0 of FF-0. It is interesting to note that ring counter is a special type of shift register.

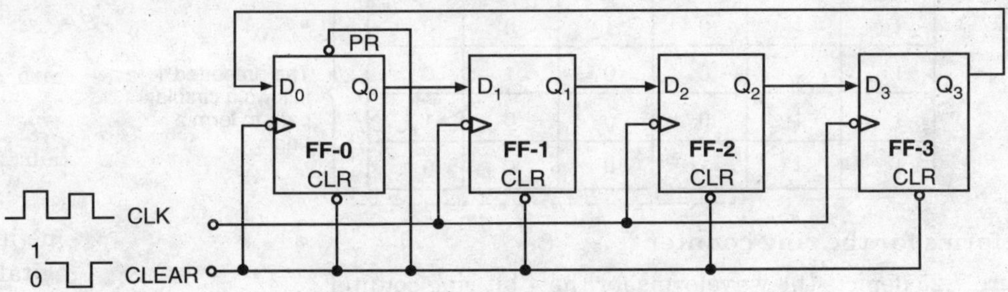

Fig. 7.24. A four bit ring counter

2. Working Operation

Initially, a low clear (CLR) pulse is applied to all the flip-flops. Therefore, FF-3, FF-2 and FF-1 reset but FF-0 shall be present. Hence, the outputs of the shift register will be :

$$Q_3\,Q_2\,Q_1\,Q_0 = 0\;0\;0\;1.$$

Now, we make the clear terminal inactive by applying a high level to it. Next, the clock signal is applied to all the flip-flops simultaneously. Here, it may be noted that all the flip-flops are negative edge triggered.

1. On the first negative edge CLK edge

As soon as the first falling edge of the clock hits, only FF-1 sets since $Q_0 = D_1 = 1$. The FF-0 resets because $D_0 = Q_3 = 0$ and there shall be no change in the status of FF-2 and FF-3. Hence, after the first clock pulse the outputs will be :

$$Q_3\, Q_2\, Q_1\, Q_0 = 0\,0\,1\,0$$

2. On the second falling edge of clock

At the second falling edge of the clock, only FF-2 sets since $D_2 = Q_1 = 1$. FF-1 resets since $D_1 = Q_0 = 0$. However, there is no change in status of FF-3 and FF-0. Therefore, after the second clock pulse, the outputs will be,

$$Q_3\, Q_2\, Q_1\, Q_0 = 0\,1\,0\,0.$$

In a similar manner, after the third clock pulse, the outputs will be,

$$Q_3\, Q_2\, Q_1\, Q_0 = 1\,0\,0\,0$$

Also, after the fourth one the outputs will be,

$$Q_3\, Q_2\, Q_1\, Q_0 = 0\,0\,0\,1.$$

These are the outputs from where we started. Therefore, the operation repeats hereafter.

3. Number of outputs

It is interesting to know that the number of output states for a ring counter will always be equal to the number of flip-flops. Therefore, for a 4-bit ring counter, the number of states will be equal to 4.

Table 7.5 summarizes the working operation of a four bit ring counter.

Table 7.5. Summary of working operation of a ring counter

CLR	CLK	Q_0	Q_1	Q_2	Q_3	
⊔	X	1	0	0	0	← FF-0 preset, others cleared
1	↓	0	1	0	0	The presetted 1 follows a circular path to form a ring
1	↓	0	0	1	0	
1	↓	0	0	0	1	
1	↓	1	0	0	0	

4. Waveforms for the ring counter

Figure 7.25 depicts the waveforms for the 4-bit ring counter.

These waveforms obviously illustrate that the presetted 1 shifts one bit per clock cycle and forms a ring. Therefore, it is named as ring counter.

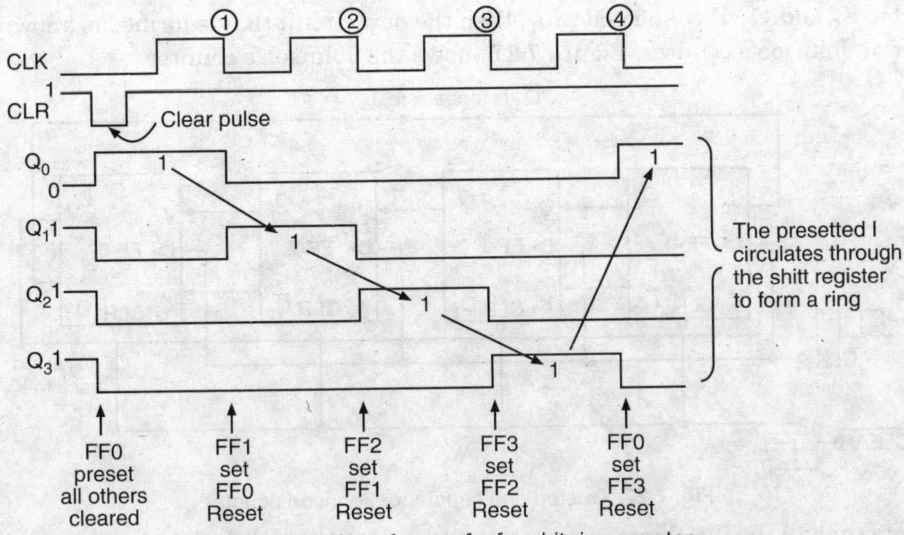

Fig. 7.25. Waveforms of a four bit ring counter

5. Applications of ring counter

Ring counters find applications in those fields in which several operations are to be sequentially controlled. As an example, in resistance welding, the operations called squeeze, bold, weld and off are to be performed sequentially. Thus, we can use a ring counter to initiate all such operations.

6. Ring counter using J-K Flip-Flop

Figure 7.26 clearly illustrates that a ring counter may also be constructed using the J-K Flip-flops.

DO YOU KNOW?

Numerous IC registers are available and may be classified according to whether their inputs are parallel (all bits entered simultaneously), serial (one bit at a time), or both. Likewise registers can have outputs that are parallel (all bits available simultaneously) or serial (one bit available at a time).

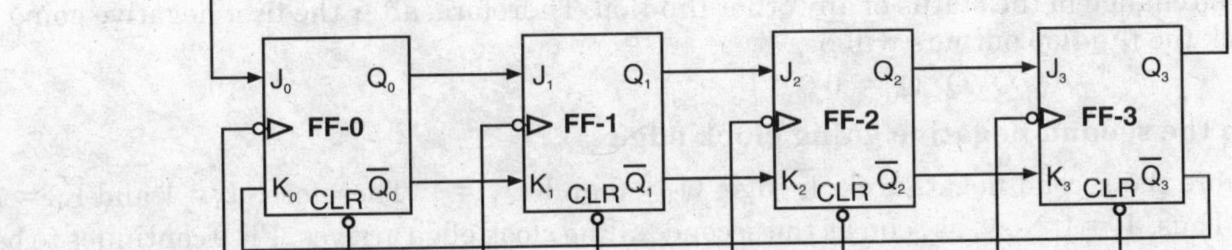

Fig. 7.26. Illustration of ring counter using J-K flip-flops

7.10.5 Johnson's Counter (Twisted/Switch Tail Ring Counter)

1. Definition

As discussed earlier, in a ring counter, the outputs of FF-3 were connected directly to the input of FF-0, i.e., $\overline{Q}_3$ to J_0, Q_3 to K_0. In place of it, if the outputs are crosscoupled to the inputs, i.e., if Q_3

is connected to K_0 and $\overline{Q}_3$ is connected to J_0, then the new circuit that is formed is known as twisted ring counter or Johnson's counter. Figure 7.27 shows the Johnson's counter.

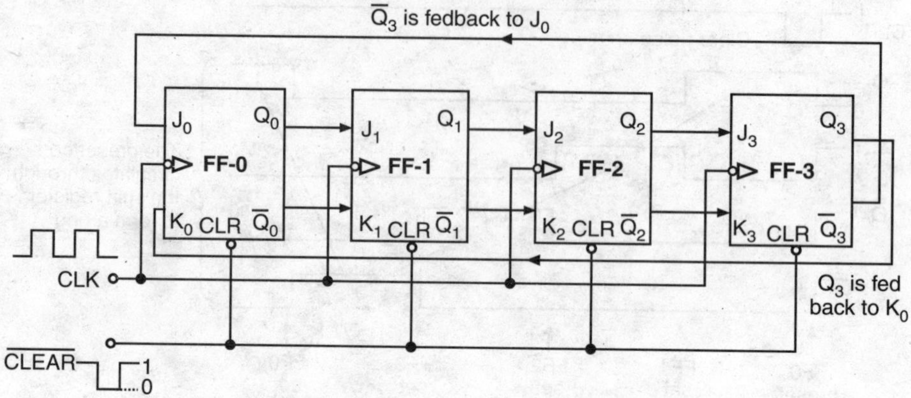

Fig. 7.27. Twisted ring counter or Johnson counter

Here, note that all the flip-flops are negative edge triggered and clock pulses are applied to all of them simultaneously. Further, the clear inputs of all the flip-flops are connected together and connected to an external clear signal. Further, it is important that all these clear inputs are active low inputs.

2. Working Operation

Initially, we apply a short negative going pulse to the $\overline{\text{clear}}$ input of all flip-flops. This resets all the flip-flops. Therefore, initially the outputs are,

$$Q_3\, Q_2\, Q_1\, Q_0 = 0\,0\,0\,0$$

But, $\overline{Q}_3 = 1$ and since, it is coupled to J_0, it is also equal to 1.

Therefore, we write $J_0 = 1$ and $K_0 = 0$ (Initially)

(i) On the first falling edge of clock pulse

As soon as the first negative edge of clock arrives, FF-0 sets. Hence, Q_0 becomes 1. But, there will be no change in the status of any other flip-flop. Therefore, after the first negative going edge of the clock the flip-flop outputs will be,

$$Q_3\, Q_2\, Q_1\, Q_0 = 0\,0\,0\,1$$

(ii) On the second negative going clock edge

Before the second negative clock edge, $Q_3 = 0$ and $\overline{Q}_3 = 1$. Therefore, $J_0 = 1$ and $K_0 = 1$. Also, $Q_0 = 1$. Thus, $J_1 = 1$. Now, as soon as the second falling clock edge arrives, FF-0 continues to be in the set mode and FF-1 now sets. Hence, Q_1 becomes 1 and $\overline{Q}_1 = 0$. There will be no change in the status of any other flip-flop. Therefore, after the second clock edge the outputs will be,

$$Q_3\, Q_2\, Q_1\, Q_0 = 0\,0\,1\,1$$

In a similar manner, after the third clock pulse, the outputs will be,

$$Q_3\, Q_2\, Q_1\, Q_0 = 0\,1\,1\,1$$

Also, after the fourth clock pulse, the outputs will be,

$$Q_3\, Q_2\, Q_1\, Q_0 = 1\,1\,1\,1$$

Here, it may be noted that now $\overline{Q}_3 = 0$, i.e., $J_0 = 0$ and $K_0 = 1$.

Therefore, as soon as the fifth negative going clock pulse strikes, FF-0 resets. But, the outputs of the other flip-flops will remain unchanged. This means that after the fifth clock pulse, the outputs will be,

$Q_3\ Q_2\ Q_1\ Q_0 = 1\ 1\ 1\ 0$ (after the 5th clock pulse)

This operation continues till we reach the all zero output state. This means that $Q_3\ Q_2\ Q_1\ Q_0$ = 0 0 0 0. Table 7.6 briefly summarizes the operation of Johnson's counter.

Table 7.6. Summary of operation of Johnson's counter

$\overline{\text{CLEAR}}$	CLK	Q_3	Q_2	Q_1	Q_0	State number	Decimal equivalent
⊔	Initially	0	0	0	0	1	0
1	↓	0	0	0	1	2	1
1	↓	0	0	1	1	3	3
1	↓	0	1	1	1	4	7
1	↓	1	1	1	1	5	15
1	↓	1	1	1	0	6	14
1	↓	1	1	0	0	7	12
1	↓	1	0	0	0	8	8
1	↓	0	0	0	0	1	0

Here, it may be noted that there are 8 distinct states of output.

Basically, it can be said that the number of states of a Johnson's counter is twice the number of flip-flops used.

3. Waveforms for Johnson's Counter

Figure 7.28 depicts the waveforms for a 4-bit Johnson counter.

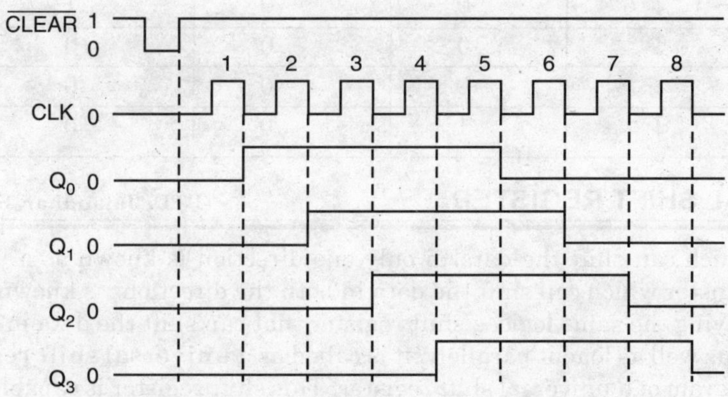

Fig. 7.28. Waveforms of Johnson counter

7.10.6 Johnson's Counter Using D Flip-Flops

As a matter of fact, it is possible to construct the Johnson counter using D flip-flops as shown in figure 7.29.

* Man is the highest animal. Indeed, it is intelligence that separates man from the other animals.
—**J. L. Nehru**

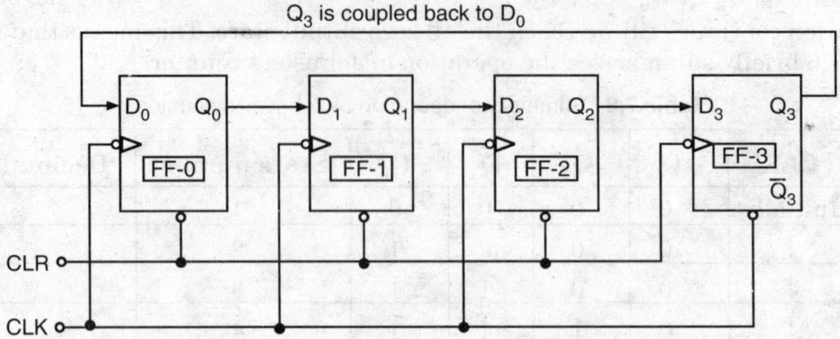

Fig. 7.29. Johnson counter using D flip-flops

Table 7.7 briefly summarizes the working operation of this circuit.

Table 7.7. Summary of operation of Johnson counter

$\overline{CLEAR}$	CLK	Q_0	Q_1	Q_2	Q_3
⊔		1	0	0	0
1	↓	1	1	0	0
1	↓	1	1	1	0
1	↓	1	1	1	1
1	↓	0	1	1	1
1	↓	0	0	1	1
1	↓	0	0	0	1
1	↓	0	0	0	0
1	↓	1	0	0	0

7.11 UNIVERSAL SHIFT REGISTER **(PTU, Jalandhar, Sem. Exam., 2007-08)**

A shift register which can shift the data in only one direction is known as a unidirectional shift register. A shift register which can shift the data in both the directions is known as a bi-directional shift register. Applying the same logic, a shift register that can shift the data in both the directions (shift right or left) as well as load it parallely, it is called as a **universal shift register.** Figure 7.32 shows the logic diagram of a universal shift register. This shift register is capable of performing the following operations :

(i) Parallel loading (parallel input parallel output)

(ii) Left shifting

(iii) Right shifting

The Mode control input is connected to Logic 1 for parallel loading operation whereas it is connected to 0 for serial shifting. With mode control pin connected to ground, the universal shift register acts as a bi-directional register. For serial left operation, the input is applied to the serial output which goes to AND gate-1 in figure 7.30. Whereas for the shift right operation, the serial input is applied to D input (input of AND gate 8). The well known example of universal shift register in the IC form is IC7495.

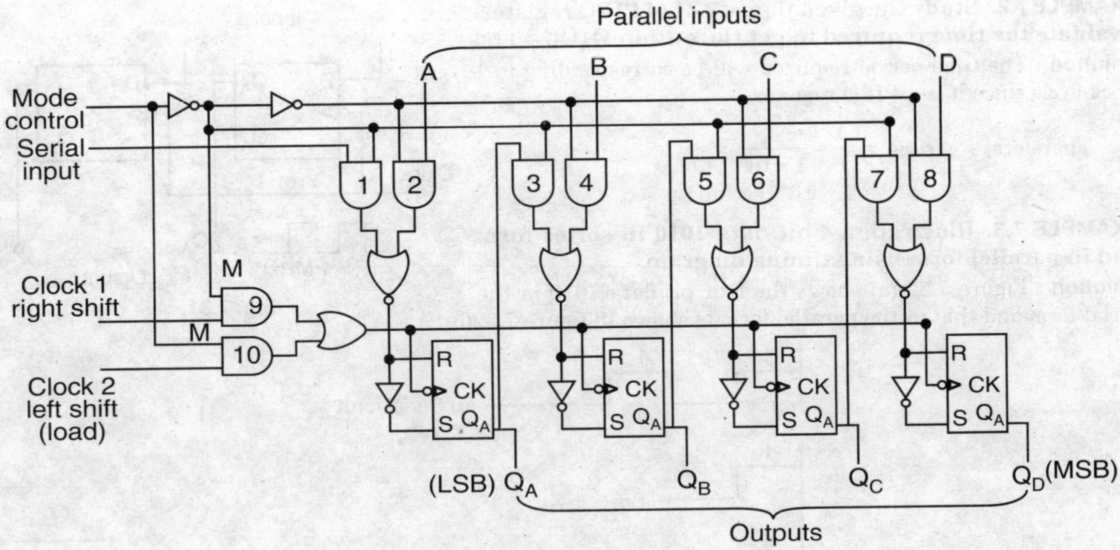

Fig. 7.30. Logic diagram of a universal shift register

7.12 PERFORMANCE COMPARISON OF COUNTERS AND REGISTERS

S. No.	Parameter of comparison	Registers	Counters
1.	Basic building block	R-S, J-K flip-flops	T flip-flops.
2.	Modes of operation	Serial, parallel	Serial up or down.
3.	Change in output state	Output need not follow a sequence	Output will always follow a sequence either in the upward or downward direction.
4.	Types	SISO, SIPO, PIPO and PISO	Up counters, down counters, up/down counters.
5.	Direction of data transfer	Bidirectional	Bidirectional
6.	Number of flip-flops per bit	One	One
7.	Few applications	Data storage, data shifting, multiplication or division	Time and frequency measurement, clock, *A* to *D* converter.

FEW IMPORTANT SOLVED EXAMPLES

EXAMPLE 7.1. To shift hexadecimal number A into 4 flip-flops serial shift register, evaluate time if clock frequency is (i) 5 MHz, (ii) 2 MHz.

Solution: The hex number A = $(1010)_2$. To shift this number into a 4 flip-flop shift register, we require four clock pulses.

(i) **For f = 5 MHz :** The time period of this clock is $T = \frac{1}{f} = \frac{1}{5 \times 10^6} = 0.2\ \mu s$. This is one clock cycle period. Therefore, to shift $(A)_{16}$, we need $4 \times 0.2\ \mu s = 0.8\ \mu$ sec.

(ii) **For f = 2 MHz :** One cycle period $T = \frac{1}{f} = \frac{1}{2 \times 10^6} = 0.5\ \mu s$. Therefore, to shift $(A)_{16}$, we need $4 \times 0.5\ \mu s = 2\ \mu$ sec.

To make mistakes is human; to stumble is commonplace, to be able to laugh at yourself is maturity.
—**William Arthur Ward**

EXAMPLE 7.2. Study the given figure 7.31 of PIPO register. Evaluate the time required to get the output $Q_AQ_B = 11$.

Solution : The time period required will be corresponding to 1 clock cycle since it is a PIPO register

$$\text{Therefore,} \quad \text{time} = \frac{1}{F} = \frac{1}{1\times 10^6} = 1 \times 10^{-6} \text{ sec or } 1\ \mu \text{ sec.}$$

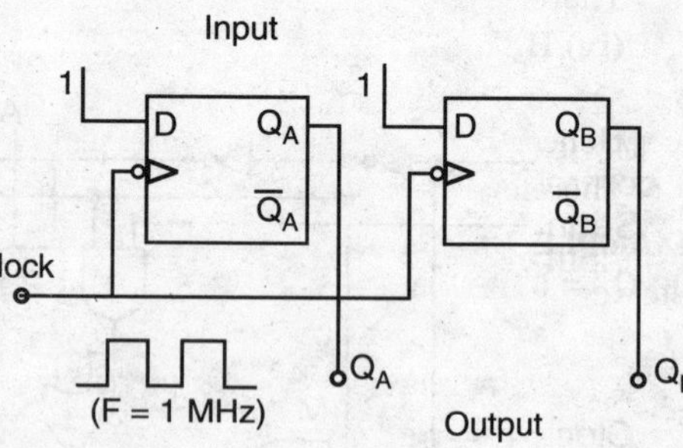

Fig. 7.31.

EXAMPLE 7.3. Illustrate a 4-bit data 1010 in series form and in parallel form using timing diagram.

Solution : Figure 7.32 (a) shows the four bit data 1010 in the serial form and that in the parallel form is shown in figure 7.32(b).

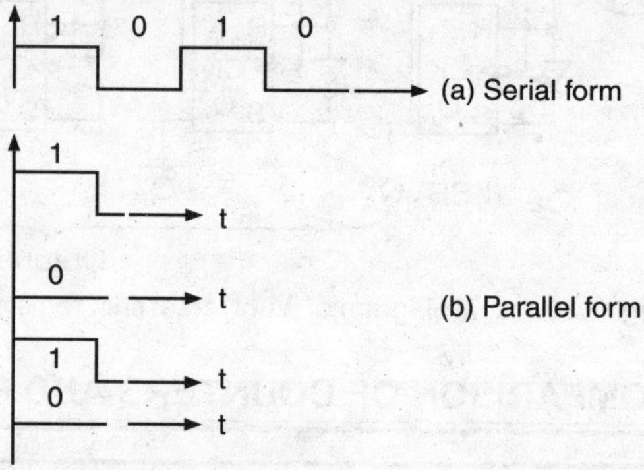

Fig. 7.32. 1010 in serial and parallel form

EXAMPLE 7.4. Define an universal shift register. Draw a 3-bit SISO left shift register. Use D flip-flop. Assume data word 110. Also, write truth table.

Solution : Table 7.8 shows the required truth table.

Table 7.8.

CLK	Q_2	$Q_1 = D_2$	$Q_0 = D_1$	Serial input $D_{in} = D_0$
Initially	0	0	0	
1st ↓	0	0	1	← 1
2nd ↓	0	1	1	← 1
3rd ↓	1	1	1	← 0

← Direction of data travel

EXAMPLE 7.5. How many flip-flops are required to build a shift register to store following numbers.

(i) Decimal 28 **(ii) Binary 6 bits**

(iii) Octal 17 **(iv) Hexadecimal A.**

Solution : In general, we know that *n* number of flip-flops will be required to build a shift register to store 'n' bits.

(i) Decimal 28

$$(28)_{10} = (11100)_2$$

Hence, five bits will be required to represent $(28)_{10}$ in binary format.

Therefore, the number of flip-flops required to store $(28)_{10} = 5$. **Ans.**

(ii) Binary 6 bits

The number of flip-flops required to store binary 6 bits = 6. **Ans.**

(iii) Octal 17

$$(17)_8 = (1111)_2$$

Hence, 4-bits will be required to represent $(17)_8$ in binary format.

Therefore, the number of flip-flops required to store $(17)_8 = 4$. **Ans.**

(iv) Hexadecimal A

$$(A)_{16} = (1010)_2$$

Hence, 4 bits will be required to represents $(A)_{16}$ in binary format.

Therefore, the number of flip-flops required to store $(A)_{16} = 4$. **Ans.**

EXAMPLE 7.6. Consider the given circuit shown in figure 7.33. Here, initial output condition is Q_A Q_B Q_C = 010. Write truth table of output Q_A Q_B Q_C for four clock pulses.

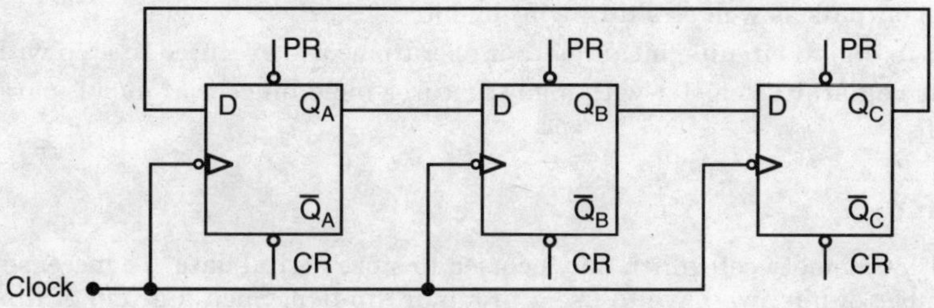

Fig. 7.33.

Solution:

CLK	Q_C	Q_B	Q_A
Initially	0	1	0
1	1	0	0
2	0	0	1
3	0	1	0
4	1	0	0

EXAMPLE 7.7. Determine the output state of a 4-bit SIPO shift register after 3 clock pulses if the serial in terminal is held HIGH and the initial contents of the shift register are :

(i) 1 0 0 1 (ii) 0 0 0 1.

Solution: (i) Let us consider table 7.9 for the answer.

Table 7.9.

CLK	Q_D	Q_C	Q_B	Q_A	Serial input
Initially	1	0	0	1	1
1st	0	0	1	1	1
2nd	0	1	1	1	1
Answers → 3rd	1	1	1	1	1

Therefore, output state after 3rd clock pulse = 1 1 1 1.

(ii) Let us consider Table 7.10 for answer.

Table 7.10.

CLK	Q_D	Q_C	Q_B	Q_A	Serial input
Initially	1	0	0	1	1
1st	0	0	1	1	1
2nd	0	1	1	1	1
3rd	1	1	1	1	1

Therefore, output state after 3rd clock pulse = 1 1 1 1.

GLOSSARY

1. **Register :** A group of temporary memory cells (such as flip-flops) for temporary storages that have a common purpose. For instance, a register might have a name (such as DIRS in a popular microcontroller) and have a specific width (such as 8-16 or 16-bits).
2. **Reset condition :** In a flip-flop, the normal output (Q) has been reset or cleared to 0.
3. **Set condition :** In a flip-flop, the normal output (Q) has been set to 1.
4. **Shift register :** A sequential logic block made up of flip-flops that allows parallel or serial loading and serial or parallel outputs as well as shifting bit-by-bit.
5. **Synchronous :** In digital circuits, meaning that operations are executed in step with the clock.
6. **Universal shift register :** Register with many features including serial-in/out, parallel-in/out, hold, and shift right or left.

SUMMARY

- A flip-flop is a 1 bit memory cell which may be used to store digital data. To increase the storage capacity in terms of number of bits, we have to use a group of flip-flop. Such a group of flip-flops is known as a register.
- Thus, register is a group of flip-flops. Further, the n-bit register will consist of n number of flip-flops and it is capable of storing an n-bit word.
- We can enter the data either in serial (one bit at a time) or in the parallel form (all the bits at a time). Also, the stored data can be retrieved in serial or **parallel** form. Data in the serial form is also called as a **temporal code,** whereas that in the parallel form is called as **special code.**
- Registers may be classified based on the way in which data are entered and taken out from a register. There may be following four possible modes. It is possible to design registers using discrete flip-flops such as S-R, J-K or D flip-flops. However, the registers are also available in MSI devices.
- As a matter, the binary data in a register can be moved within the register from one flip-flop to the other or outside it with application of clock pulses. Those registers that allow such data transfers are known as shift registers. Shift registers find applications for data storage, data transfer and certain arithmetic and logic operations.
- The various modes in which a shift register can operate may be listed as under :
  (i) Serial input serial output
  (ii) Serial input parallel output
  (iii) Parallel in serial out
  (iv) Parallel in parallel out.
- Using the SISO mode, we require 4-clock pulses to store a 4-bit word. Therefore, in general, we conclude that it requires n number of clock pulses to store an n-bit word using SISO mode.
- Few important applications of serial operation may be described as under :
  (i) The transmission of data from one place to the other takes place in serial manner.
  (ii) It takes a longer time for serial transmission, since the time required to transmit one bit is equal to the time corresponding to one clock cycle.
  (iii) However, for long distance communication, where the distances are in kilometres, serial communication has an added advantage that only one conductor is required to be used.
- Few common and important applications of a shift register may be described as under :
  (i) For temporary data storage (ii) For multiplication and division
  (iii) As a delay line (iv) As serial to parallel converter
  (v) As parallel to serial converter (vi) As ring counter
  (vii) As twisted ring counter or Johnson counter
- A delay line may be defined as a circuit that is used to delay a signal by some fixed time-interval.
  Also serial input serial output can be used to introduce a time delay of Δt seconds in the digital signal applied at its serial input.
- In several applications, the data in serial form requires to be converted into parallel form. The serial input parallel output mode of the shift register is used for such applications.

- As a matter of fact, the data communication between two computers takes place in the serial transmission form. But internally the data processing takes place in the parallel form. Hence, we require to use a parallel to serial converter to convert the internal parallel data into a serial data suitable for transmission. Further, for this operation, we can use the parallel to serial mode of the shift register.
- A shift register which can shift the data in only one direction is known as a unidirectional shift register. A shift register which can shift the data in both the directions is known as a bidirectional shift register. Applying the same logic, a shift register that can shift the data in both the directions (shift right or left) as well as load it parallely, it is called as a **universal shift register.**

SHORT QUESTIONS WITH ANSWER

Q. 1. What is meant by Data Formats ? Explain.

Ans. We can enter the data either in serial (one bit at a time) or in the parallel form (all the bits at a time). Also, the stored data can be retrieved in serial or **parallel** form. Data in the serial form is also called as a **temporal code,** whereas that in the parallel form is called as **special code.** Figure 7.34 represents a four bit digital data 1011 in the serial and parallel manner.

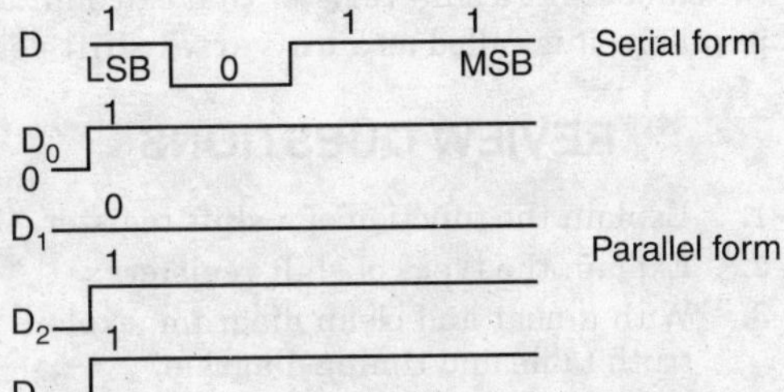

Fig. 7.34. Data representation in serial and parallel form

Q.2. How can you classify register ? Explain.

Ans. Registers may be classified based on the way in which data are entered and taken out from a register. There may be following four possible modes.

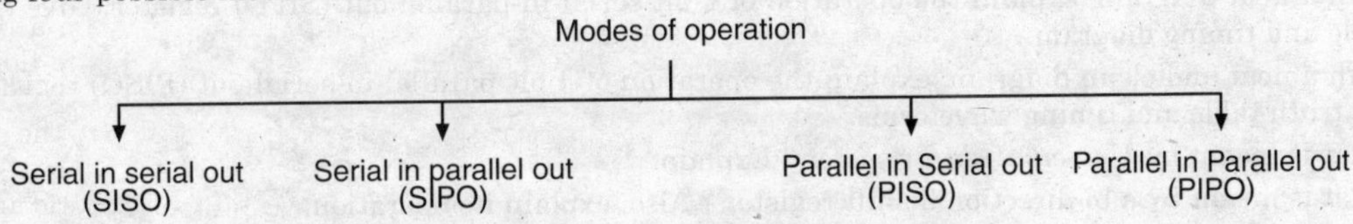

Fig. 7.35. Modes of operation of registers

It is possible to design registers using discrete flip-flops such as S-R, J-K or D flip-flops. However, the registers are also available in MSI devices.

Q.3. What do you mean by shift register ?

Ans. As a matter, the binary data in a register can be moved within the register from one flip-flop to the other or outside it with application of clock pulses. Those registers that allow such data transfers are known as shift registers. Shift registers find applications for data storage, data transfer and certain arithmetic and logic operations.

Q. 4. What are the various modes of operation of a shift Register ?

Ans. The various modes in which a shift register can operate may be listed as under :

(i) Serial input serial output
(ii) Serial input parallel output
(iii) Parallel in serial out
(iv) Parallel in parallel out.

Q. 5. Write the applications of serial operation.

Ans. Few important applications of serial operation may be described as under :

(i) The transmission of data from one place to the other takes place in serial manner.
(ii) It takes a longer time for serial transmission, since the time required to transmit one bit is equal to the time corresponding to one clock cycle.
(iii) However, for long distance communication, where the distances are in kilometres, serial communication has an added advantage that only one conductor is required to be used.

Q.6. Write applications of shift registers.

Ans. Few common and important applications of a shift register may be described as under :

(i) For temporary data storage
(ii) For multiplication and division
(iii) As a delay line
(iv) As serial to parallel converter

(v) As parallel to serial converter (vi) As ring counter
(vii) As twisted ring counter or Johnson counter

Q.7. What is a delay line ?

Ans. A delay line may be defined as a circuit that is used to delay a signal by some fixed time-interval.

Also serial input serial output can be used to introduce a time delay of Δt seconds in the digital signal applied at its serial input.

Q.8. What is meant by universal shift register ? Explain.

Ans. A shift register which can shift the data in only one direction is known as a unidirectional shift register. A shift register which can shift the data in both the directions is known as a bi-directional shift register. Applying the same logic, a shift register that can shift the data in both the directions (shift right or left) as well as load it parallely, it is called as a **universal shift register.**

REVIEW QUESTIONS

1. Explain the function of a shift register. Give its applications.
2. Explain the types of shift registers.
3. With a neat and clean diagram, explain the working operation of a 4-bit left shift register. Also, give its truth table and timing diagram.
4. With a neat and clean diagram, explain the working operation of 4-bit SISO (Serial-in-serial-out) register. Draw the timing diagram and give its truth table.
5. With a neat diagram, explain the operation of 4-bit serial-in-parallel out (SIPO) register. Give the truth table and timing diagram.
6. With a neat and clean diagram, explain the operation of 4-bit parallel-in-serial-out (PISO) register. Give the truth table and timing waveforms.
7. What is meant by *Universal shift register* ? Explain.
8. What is meant by a bi-directional shift register ? Also, explain its operation.
9. Describe any one shift register IC. Sketch its schematic diagram and pin configuration. Give its specifications.
10. Draw circuit diagram of a 3-bit SIPO shift register. Use D flip-flops. Also, explain its working.
11. Define bidirectional shift register. Draw 3-bit bi-directional shift register using D flip-flop.

NUMERICAL PROBLEMS

1. For logic circuit shown in figure 7.36,
   (i) Draw the waveforms at Q_0 and Q_1 for first five clock cycles. Assume initial states of Q_0 and Q_1 as 11.
   (ii) What be the logic states of Q_0 and Q_1 after 7th clock pulse ? Assume initial states of Q_1 and Q_0 as 00.

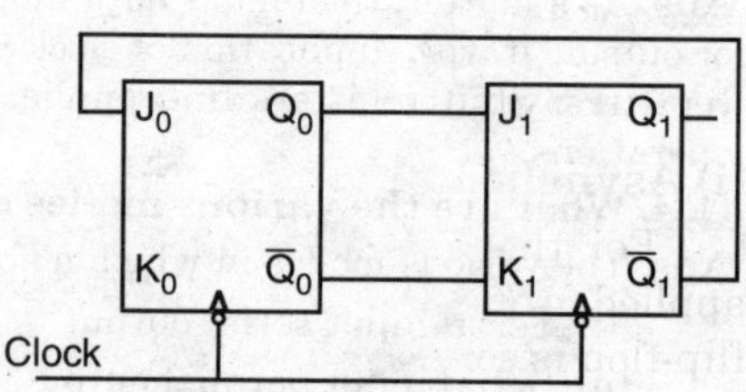

Fig. 7.36. Given circuit

2. Determine the output states of a 4-bit SIPO shift register after four clock pulses if the input terminal is hold high. Also, draw the waveforms.
3. Using shift register, implement a PRBS of 15 states.
4. Draw a 3-bit Johnson counter. Also, draw the releavant output waveforms.
5. For the circuit diagram shown in figure 7.37 assuming initial reset condition, draw the timing diagrams for the outputs Q_1 and Q_2.
6. Draw neat and diagram of a 3-bit universal shift register using J-K flip-flops having right and left shift inputs and mode control M such that. M = 0 .... Left shift, M = 1.... right shift.
7. Explain with the help of neat and clean diagram, the working operation of a 3-bit universal shift register.

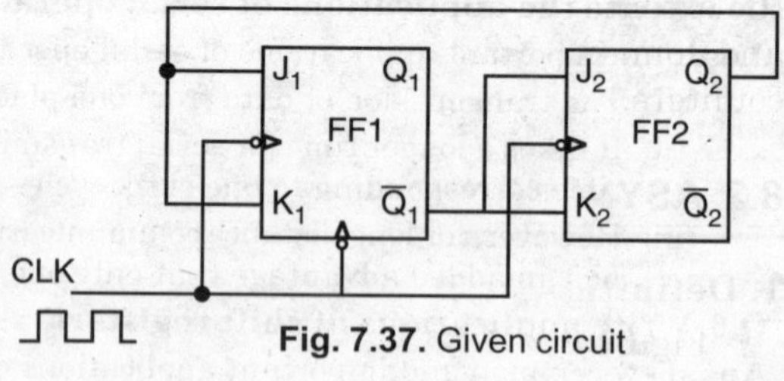

Fig. 7.37. Given circuit

8

SEQUENTIAL CIRCUITS: COUNTERS

Syllabus

Counters: Asynchronous/Ripple counters, Synchronous counters, Modulo-n counter, Ring counters, Classification of sequential circuits-Moore and Mealy, Design of Synchronous counters; State diagram, Circuit implementation.

Chapter Outline

- Introduction to Counters
- Asynchronous/ripple up counters
- Introduction to up/down counters
- Modulus of the counter
- Problems Associated with Ripple Counters
- Synchronous Counters
- Counter Applications
- Modulo – N Synchronous Counters
- Up/down synchronous counter
- Sequential Circuit Design
- Moore Circuit
- Mealy Circuit
- Analysis of Clocked Sequential Circuits
- State Reduction and Assignments
- State Assignment
- Design Procedure for Clocked Sequential Circuits
- Design using Unused States
- Lock out Condition
- Bush Diagram
- Sequence Generator

8.1 INTRODUCTION TO COUNTERS

As a matter of fact, the digital circuit used for counting pulses is known as counter. It is a sequential circuit. Counter is the widest application of flip-flops. It is a group of flip-flops with a clock signal applied. Basically, counters count the number of clock pulses. Hence, with some modifications, it can be used for measuring frequency or time period.

8.1.1. Classification of Counters

Counters are basically of following two types :

(i) Asynchronous or ripple counters.

(ii) Synchronous counters.

(i) Asynchronous or Ripple Counters

For these counters, the external clock signal is applied to one flip-flop and then the output of preceding flip-flop is connected to the clock of next flip-flop.

(ii) Synchronous Counters

In synchronous counters, all the flip-flops receive the external clock pulse simultaneously. Ring counter and Johnson counter are the examples of synchronous counters.

8.2 ASYNCHRONOUS/RIPPLE UP COUNTERS

1. Definition

Figure 8.1 shows the logic diagram of a 2-bit ripple up counter. The number of flip-flops used is 2. A 4-bit counter will use four flip-flops. The toggle (T) flip-flops are being used. But, we can use the *JK* flip-flops also with J and K connected permanently to logic 1. External clock is applied to the clock input of flip-flop A and Q_A output is applied to the clock input of the next flip-flop i.e. FF-B.

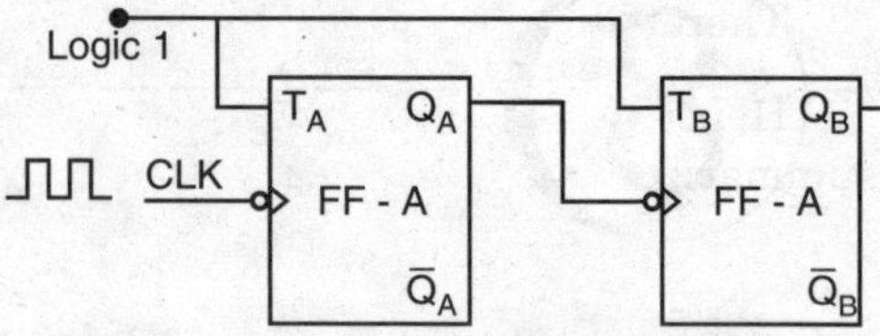

Fig. 8.1. A two bit asynchronous binary up counter

2. Working Operation of Counter

Initially, let both the flip-flops be in reset condition.

Therefore, $Q_B\,Q_A = 00$

(*i*) On the first negative going clock edge

As soon as the first falling edge of the clock hits FF-A, it will toggle as $T_A = 1$. Hence, Q_A will be equal to 1. Also, Q_A is connected to clock input of FF-B. Since Q_A has changed from 0 to 1, it is treated as the positive clock edge by FF-B. There is no change in Q_B because FF-B is a negative edge triggered FF. Hence after the first clock pulse the counter outputs are

$$Q_B\,Q_A = 01$$

(After the first CLK pulse)

(ii) At the Second Falling Edge of Clock

On the arrival of second falling clock edge, FF-A toggles again, to make $Q_A = 0$. This change in Q_A (from 1 to 0) acts as a negative clock edge for FF-B. Hence, it will also toggle and Q_B will become 1. Therefore, after the second clock pulse, the counter outputs are

$$Q_B\,Q_A = 10$$

(After the second CLK pulse)

Here, it may be noted that both the outputs are changing their state. But both the changes do not take place simultaneously Q_A will change first from 1 to 0 and then Q_B will change from 0 to 1. This is due to the propagation delay of FF-A. Hence, both flip-flops will never triggered at the same instant. Therefore, counter is called as an asynchronous counter. This has been shown in figure 8.2.

WILLIAM (BILL) H. GATES III

The chairman and Chief Software Architect of Microsoft Corporation, Gates began programming computers at the age of 13. In 1974, while he was a college student, Gates developed a version of BASIC, the programming language for the first microcomputer. Believing that the personal computer would eventually be found on every office desktop and in every home, Gates and Paul Allen formed Microsoft in 1975.

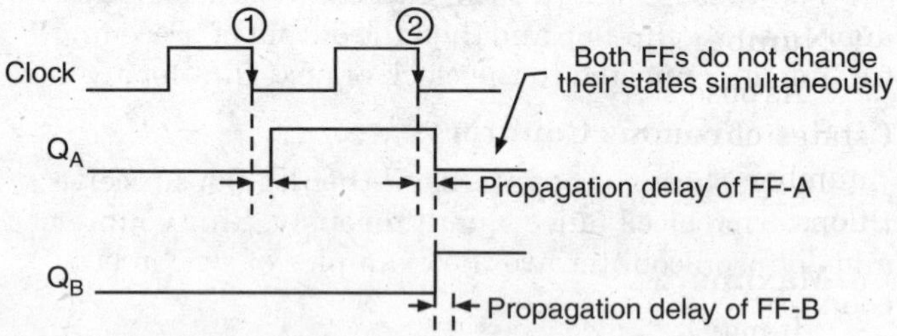

Fig. 8.2. FFs do not change state simultaneously.

(iii) At the Third Falling Edge of Clock

On arrival of the third falling edge, FF-A toggles again and Q_A becomes 1 from 0. Since this is a positive going change, FF-B does not respond to it and remains inactive. Hence, Q_B does not change and continues to be equal to 1.

$$Q_B\,Q_A = 11$$

(After the third CLK pulse)

(iv) At the fourth Negative Clock Edge

On the fourth falling clock edge, FF-A toggles and Q_A changes from 1 to 0. This negative change in Q_A acts as clock pulse for FF-B. Hence, it toggles to change Q_B from 1 to 0.

Therefore, $Q_B\,Q_A = 00$ (After the fourth CLK pulse)

Hence, the counter has reached the original state. The operation will now repeat. Table 8.1 summarizes the working operation of the counter and figure 8.3 shows the timing waveforms.

Table 8.1. Summary of operation of a 2-bit binary ripple counter

Clock	Counter outputs		State number	Decimal equivalent of counter output
	Q_B (MSB)	Q_A (LSB)		
Initially	0	0	—	0
1st (↓)	0	1	1	1
2nd (↓)	1	0	2	2
3rd (↓)	1	1	3	3
4th (↓)	0	0	4	0

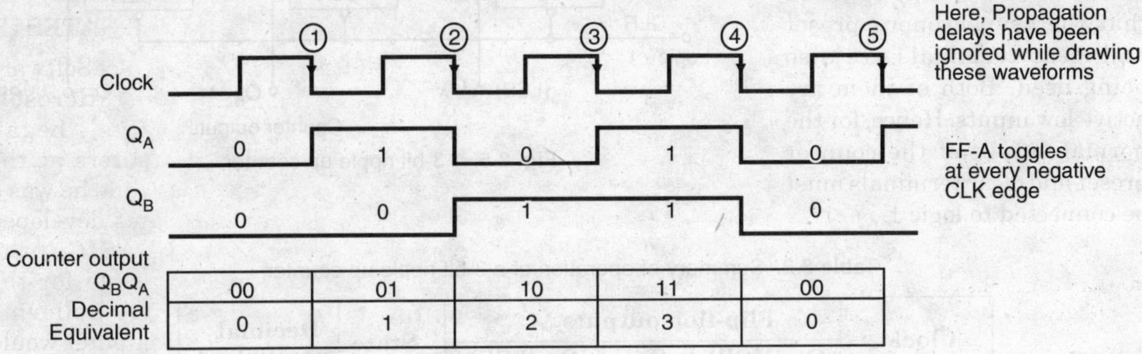

Fig. 8.3. Timing diagram for a 2-bit ripple up counter

3. Why is it called Counter ?

Let us consider figure 8.3. Here, the decimal count corresponds to the number of clock pulses, which counter has received. Thus, this circuit counts the clock pulses. Therefore, it is called as counter.

4. Number of States

As observed from Table 8.1, this counter has four distinct states of output namely 00, 01, 10 and 11. In general, the number of states = 2^n, where n is equal to the number of flip-flops.

> **DO YOU KNOW?**
>
> An up/down counter can be commanded to count up or count down.

5. Maximum Count

It may be observed from Table 8.1, the maximum count is 3 (decimal), i.e., 11 binary. In general, the maximum count = $(2^n - 1)$.

8.2.1 A Two Bit Asynchronous Up Counter Using JK Flip-Flops

A 2-bit asynchronous up counter using JK flip-flops has been shown in figure 8.4. Here, it may be noted that the *J* and *K* inputs of both the flip-flops are connected to logic 1. So, actually, they are converted into *T* flip-flops. The working operation of this circuit is exactly same as that of the counter using the *T* flip-flops.

8.2.2 A 3 Bit Asynchronous Up Counter

(PTU, Jalandhar, Sem. Exam, 2002-03)

1. Definition

We can apply all the basic concepts which were introduced for the 2-bit ripple up counter to the 3-bit ripple up counter. Figure 8.5 shows the logic diagram of a 3-bit ripple up counter. Since it is a 3-bit counter, we require to use 3-flip-flops.

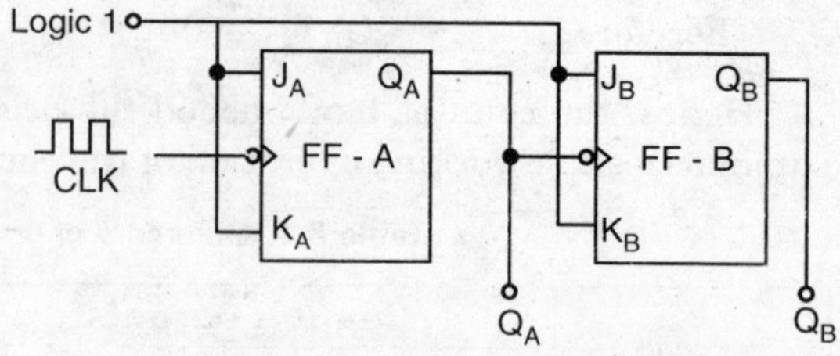

Fig. 8.4. A two bit ripple up counter using JK flip-flops

The working operation of a 3-bit ripple up counter takes place in exactly similar manner as that of a 2-bit counter.

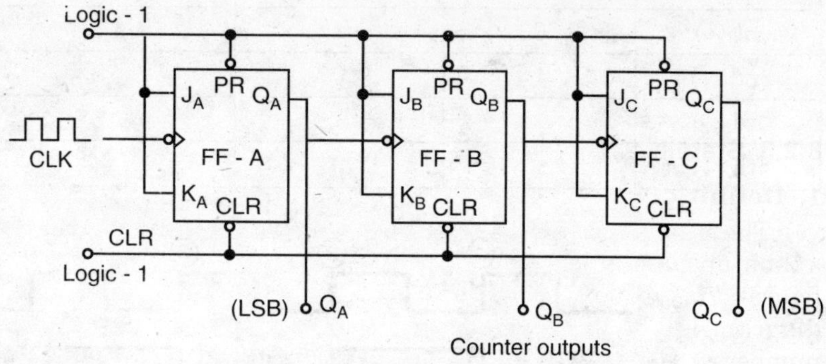

Fig. 8.5. A 3-bit ripple up counter

2. Working Operation

Table 8.2 summarizes the working operation of this counter. Here, it may be noted that the asynchronous preset and clear terminals are also being used. Both of them are active low inputs. Hence, for the normal output of the counter preset and clear terminals must be connected to logic 1.

Table 8.2. Summary of operation of a 3-bit ripple up counter

Clock	Flip-flop outputs			State	Decimal equivalent
	Q_C (MSB)	Q_B	Q_A (LSB)		
Initially	0	0	0	1	0
1st (↓)	0	0	1	2	1
2nd (↓)	0	1	0	3	2
3rd (↓)	0	1	1	4	3
4th (↓)	1	0	0	5	4
5th (↓)	1	0	1	6	5
6th (↓)	1	1	0	7	6
7th (↓)	1	1	1	8	7
8th (↓)	0	0	0	1	0

3. Number of States

Number of states = $2^n = 2^3 = 8$.

4. Maximum Count

Maximum count = $2^n - 1 = 8 - 1 = 7$. Let us consider Table 8.2. The maximum count is Q_C Q_B Q_A = 1 1 1, i.e., decimal 7. It is important to note that Q_C is treated as MSB and Q_A as LSB.

5. Timing Diagram

The timing diagram of a 3-bit ripple up counter have been shown in figure 8.6.

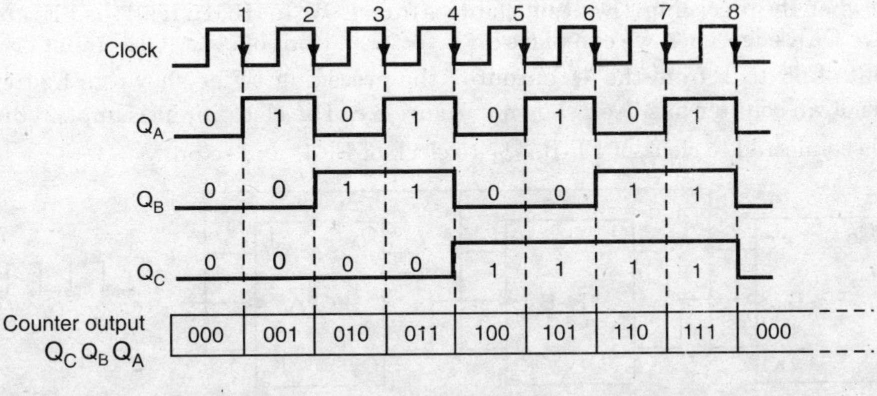

Fig. 8.6. Illustration of timing diagram for a 3-bit ripple up counter

8.2.3 State Diagram of a Counter

1. Definition

The state diagram of a counter represents the states of a counter graphically. As an example, for a 2-bit up counter the state diagram is shown in figure 8.7(a) and for a 2-bit down counter, the state diagram is shown in figure 8.7(b). Here, the number written inside a circle represents the state number, whereas the arrow show the direction of counter (up or down).

Here, it is interesting to note that in the counters only the state is important. Therefore, in the state diagram, we have not shown any input or output conditions.

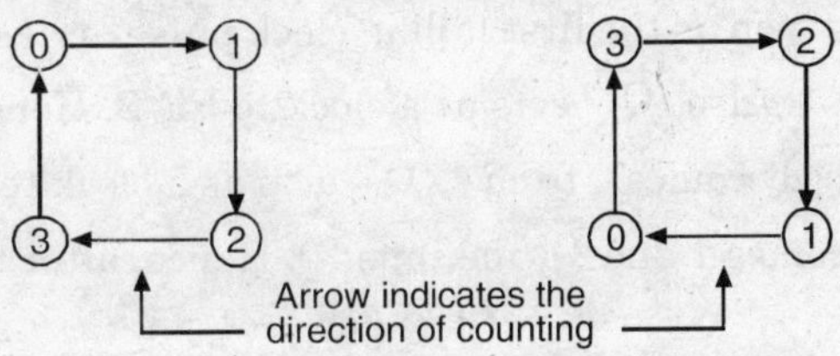

(a) For 2-bit up counter (b) For a 2-bit down counter

Fig. 8.7. State diagram

8.2.4 Asynchronous Down Counter

1. Definition

All the counters discussed so far have counted upwards from zero. Hence, they can be called as **up counters.** However, the counters which can count in the **downward** direction, i.e., from the maximum count to zero are called **down counters.** The count down sequence for a 3-bit asynchronous down counter is as under :

	Q_C	Q_B	Q_A
(7)	1	1	1
(6)	1	1	0
(5)	1	0	1
(4)	1	0	0
(3)	0	1	1
(2)	0	1	0
(1)	0	0	1
(0)	0	0	0

← Flip-flop outputs

Recycle

From this sequence, it is evident that FF-A should toggle at every negative going clock edge but FF-B should change its state each time when Q_A changes from LOW (0) to HIGH (1) and Q_C changes when Q_B changes from LOW to HIGH. Thus in a down counter, each FF except the first one (A) should toggle when the preceding FF output changes from LOW to HIGH. If all the FFs are responding to the negative CLK edge, then we can place an inverter in front of every CLK input or we can drive the CLK input of next FF from the $\overline{Q}$ output of the preceding FF as shown in figure 8.8. A 3-bit asynchronous down counter has been shown in figure 8.8. The clock input is applied directly to FF-A. But, $\overline{Q_A}$ is connected to clock of FF-B, $\overline{Q_B}$ to clock of FF-C and so on.

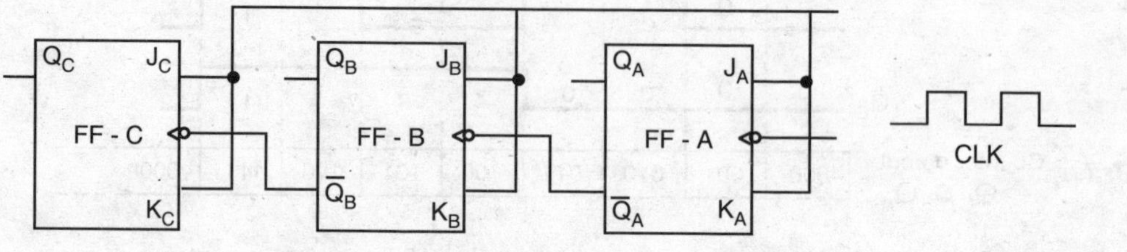

Fig. 8.8. A 3-bit asynchronous down counter

2. Working Operation

Let initially all the flip-flops be in the reset condition.

Therefore, $Q_C\,Q_B\,Q_A = 0\,0\,0$

As soon as the first falling clock pulse arrives, FF-A toggles. So, Q_A becomes 1 and $\overline{Q_A}$ becomes 0 from 1. Also, $\overline{Q_A}$ acts as a clock to FF-B. Hence, FF-B will change its state. Hence, Q_B becomes 1 and $\overline{Q_B}$ becomes 0 from 1. $\overline{Q_B}$ acts as a clock to FF-C. Hence FF-C will change its state. Therefore, Q_C becomes 1 and $\overline{Q_B}$ becomes 0. Hence, after the first clock pulse, the output of counter are,

$$Q_C\,Q_B\,Q_A = 1\,1\,1 \qquad \text{(After the 1st CLK pulse)}$$

Corresponding to the second falling clock edge, FF-A toggles, Q_A becomes 0 and $\overline{Q_A}$ becomes 1.

This positive going change in $\overline{Q_A}$ does not alter the state of FF-B. So, Q_B remains 1 and $\overline{Q_B}$ remains 0. Hence, there is no change in the state of FF-C. Hence, after the second clock pulse, the counter outputs as under:

$$Q_C\,Q_B\,Q_A = 1\,1\,0 \qquad \text{(After the 2nd CLK pulse)}$$

The down counting will thus take place. The timing diagram for the down counter is shown in figure 8.9.

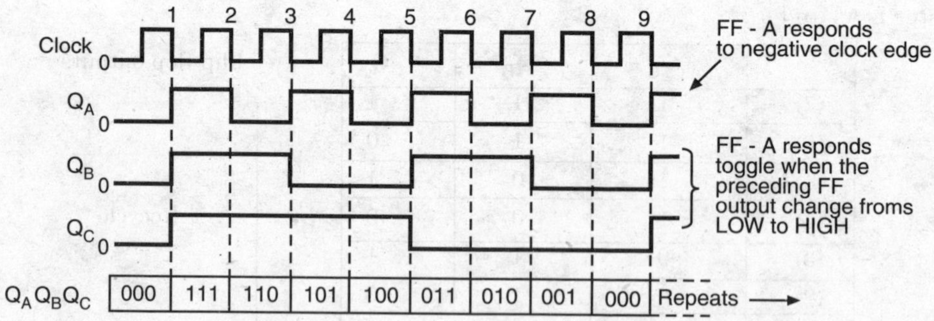

Fig. 8.9. Timing diagram of a 3-bit down counter.

3. State Diagram

The state diagram of a 3-bit down counter has been shown in figure 8.10.

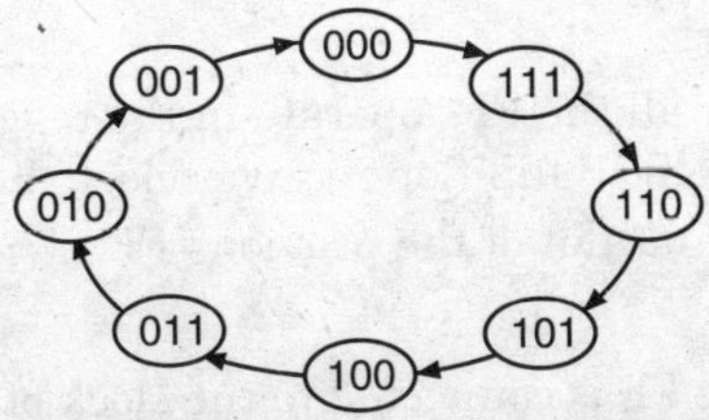

Fig. 8.10. State diagram of a 3-bit ripple down counter.

Important Point : The numbers inside the circles correspond to the state of $Q_C\ Q_B\ Q_A$.

8.2.5 Drawbacks of Ripple Counters

(i) Every flip-flop has its own propagation delay. In ripple counter, the output of the previous FF is used as clock for the next FF.

(ii) Hence, the propagation delay goes on accumulating. For a 3-bit ripple counter, the propagation delay of the first FF gets added to that of the second FF to decide the transition time for the third stage.

(iii) This accumulated time delay is the main problem with the ripple counters, because the propagation delay goes on increasing with increase in number of flip-flops.

(iv) This will put a limitation on the maximum clock frequency.

(v) The frequency f of a clock pulse for reliable working operation of the counter is given by,

$$f \leq \frac{1}{n(t_d) + T_s}$$

where n = Number of flip-flops

T_s = Width of strobe pulse (if used)

and t_d = Propagation delay of one flip-flop.

> **DO YOU KNOW?**
>
> Logic gates can be used to decode (detect) any or all states of **a** counter.

EXAMPLE 8.1. For a 4-bit asynchronous counter, evaluate the maximum clock frequency at which the counter can operate reliably. Assume the propagation delay of each FF to be 40 ns and width of strobe pulse to be 20 ns.

Solution: We know that the maximum clock frequency for reliable operation is expressed as under :

$$f = \frac{1}{n(t_d) + T_s} = \frac{1}{(4 \times 40 \times 10^{-9}) + (20 \times 10^{-9})}$$

Therefore, we get $f = 5.56$ MHz. **Ans.**

8.3 INTRODUCTION TO UP/DOWN COUNTERS **(PTU, Jalandhar, Sem. Exam., 2008-09)**

We have designed the up counters and the down counters separately. However, in practice, both these modes are generally combined together to obtain an UP/DOWN counter. A mode control (M) input is also provided to select either up or down mode. A combinational circuit is required to be designed and used between each pair of flip-flops in order to achieve the up/down operation.

8.3.1 Classification of Up/Down Counters

The up/down counters may be of following two types :

(i) UP/DOWN ripple counters

(ii) UP/DOWN synchronous counters.

8.3.2 UP/DOWN Ripple Counters

In the up/down ripple counter all the FFs operate in the toggle mode. Hence, either T flip-flops or JK flip-flops are to be used. The LSB flip-flop receives clock directly. However, the clock to every other FF is obtained from Q or $\overline{Q}$ output of the previous FF.

(i) UP counting mode (M = 0)

The Q output of the preceding FF is connected to the clock of the next stage if up counting is to be achieved. For this mode, the mode select input M is at logic 0 (M = 0).

(ii) DOWN counting mode (M = 1)

If M = 1, then the $\overline{Q}$ output of the preceding FF is connected to the clock of the next FF. This will operate the counter in the down counting mode.

8.3.3 3-bit and 4-bit up/down Ripple Counters

The design of 3-bit and 4-bit up/down ripple counters has been illustrated in the following examples.

EXAMPLE 8.2. Design a 3-bit binary up/down ripple counter. Draw the timing diagram.

Solution: The requirements of counter are :

(i) 3-bit : Hence, three FFs are required.

(ii) UP/DOWN : Thus, a mode control input is essential.

We know that for a ripple up counter, the Q output of preceding FF is connected to the clock input of the next one. Also, for a ripple down counter, the $\overline{Q}$ output of the preceding FF is connected to the clock input of the next one. Let the selection of Q or $\overline{Q}$ output of the preceding FF be controlled by the mode control input M such that,

If M = 0, UP counting. Hence, we connect Q to CLK

If M = 1, DOWN counting. Hence, we connect $\overline{Q}$ to CLK.

Let us design a combinational logic to satisfy all the requirements stated above.

1. Truth Table of Combinational Circuit

The truth table of such a combinational circuit has been shown in Table 8.3.

Table 8.3. Truth table

Inputs			Output
M	Q	$\overline{Q}$	M
0	0	0	0
0	0	1	0
0	1	0	1
0	1	1	1
1	0	0	0
1	0	1	1
1	1	0	0
1	1	1	1

Y = Q for up counting (M = 0 rows)

Y = $\overline{Q}$ for down counting (M = 1 rows)

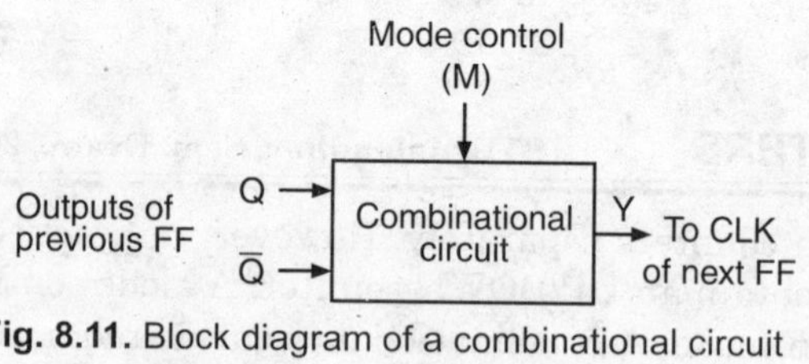

Fig. 8.11. Block diagram of a combinational circuit

2. K-map and Simplified Expression for Output

Figure 8.12 (a) shows the K-map and simplified expression for Y.

3. Logic Diagram for Combinational Circuit

The logic diagram for the combinational circuit has been shown in figure 8.12 (b).

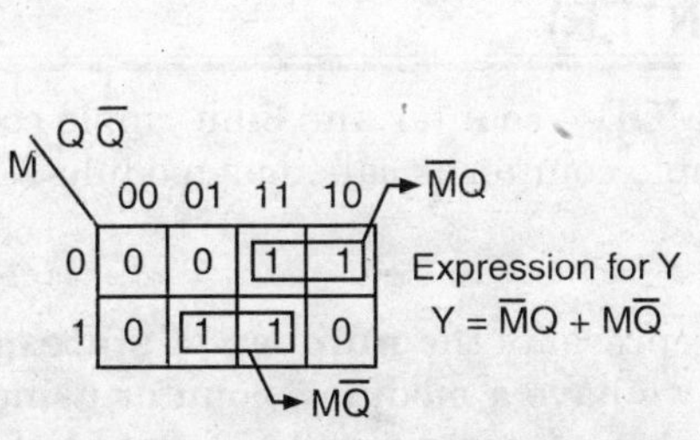

Fig. 8.12 (a) K-map for Y

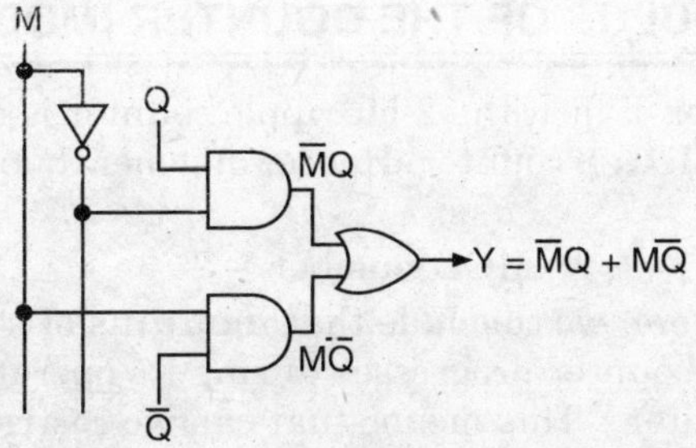

Fig. 8.12 (b)

4. Logic Diagram of a 3-bit Up/Down Counter

The combinational circuit is connected between every pair of flip-flops to obtain the 3-bit up/down counter as shown in figure 8.13.

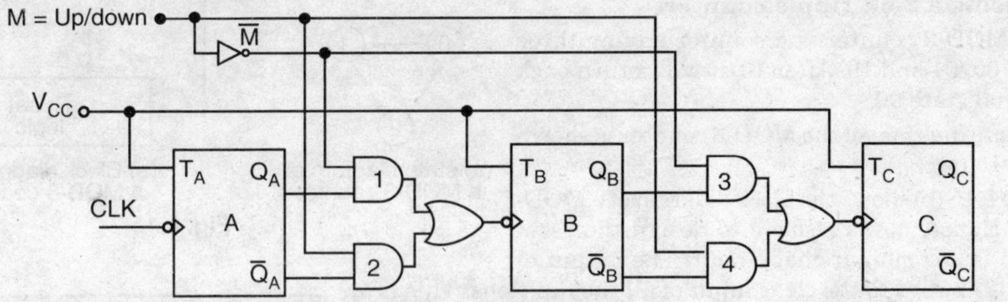

Fig. 8.13. A 3-bit up/down ripple counter

5. Working Operation of a 3-bit Up/Down Ripple Counter

(i) With M = 0 (Up Counting Mode)

If M = 0 and $\overline{M}$ = 1, the AND gates 1 and 3 in figure 8.13 will be enabled whereas the AND gates 2 and 4 will be disabled. Hence, Q_A gets connected to the clock input of FF-B and Q_B gets connected to the clock input of FF-C. These connections are same as those for the normal up counter. Thus, with M = 0, the circuit works as an up counter.

(ii) With M = 1 (Down counting mode)

If M = 1, then AND gates 2 and 4 in figure 8.13 are enabled whereas the AND gates 1 and 3 are disabled. Hence, $\overline{Q_A}$ gets connected to the clock input of FF-B and $\overline{Q_B}$ gets connected to the clock input of FF-C. As discussed earlier, these connections will produce a down counter. Hence, with M = 1, the circuit works as a down counter.

EXAMPLE 8.3. Design a 4-bit up/down ripple counter.

Solution: Let us consider Example 8.2. The combinational circuit is going to the same. However, we have to use 4 flip-flops. The 4-bit up/down counter has been shown in figure 8.14.

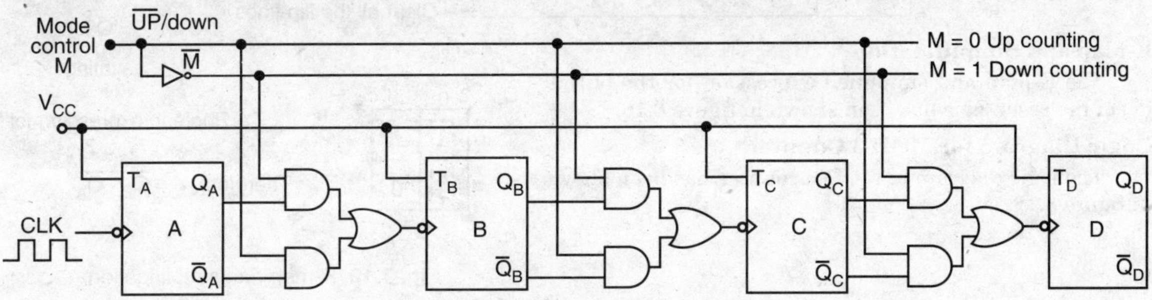

Fig. 8.14. A 4-bit up/down ripple counter.

The circuit works as an up counter with M = 0 whereas it operates as a down counter when M = 1.

8.4 MODULUS OF THE COUNTER (MOD-n COUNTER)

As a matter of fact, the 2-bit ripple counter is called as MOD-4 counter and 3-bit ripple counter is called as MOD-8 counter. Hence, in general, an n-bit ripple counter is called as modulo-N counter where

$$\text{MOD number} = 2^n$$

Therefore, we conclude that **modulus** of a counter represents the **number of states** through which the counter progresses during its operation. Can we have a modulo-5 counter using a 3-bit ripple counter ? This means that can we restrict the number of states to only 5 instead of 8 under normal conditions ? The answer is yes. We can design such modulo counters with the help of the basic ripple counter structure and a combinational logic called reset logic. This concept has been illustrated in Example 8.4 and Example 8.5.

EXAMPLE 8.4. Design a MOD-3 asynchronous counter using a 2-bit ripple counter.

Solution: MOD-3 counter is a counter having three states, i.e., 00, 01 and 10. After 10, it will return back to its original state 00.

The state diagram of the MOD-3 counter is shown in figure 8.15 (a).

Figure 8.15 (b) shows the block diagram of a MOD-3 counter. Hence, now, we have to design the reset logic which is a combinational circuit. The output of reset logic is applied to the clear input of all the flip-flops. This is an active low input. As soon as flip-flop reaches state 2, the reset logic must produce a logic 0 at its output and clear all the FFs to jump back to state 0.

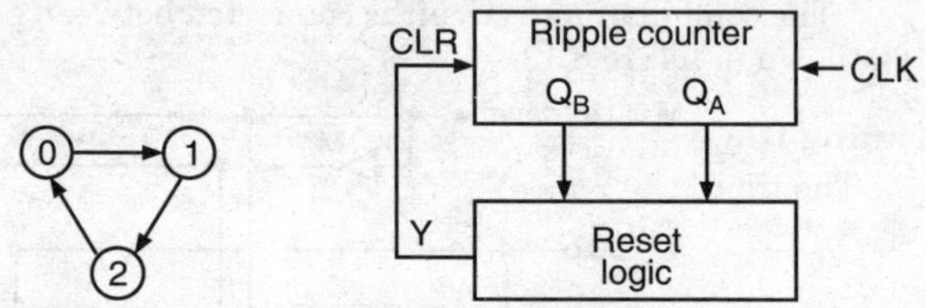

(a) state diagram of a MOD-3 counter (b) Block diagram of a MOD-3 coutner

Fig. 8.15.

DO YOU KNOW?

Synchronous counters with arbitrary counting sequences can be implemented by following a standard design procedure.

Truth Table of Reset Logic

Table 8.4 shows the truth table for the reset logic. Here, it may be noted that for the states 0, 1 and 2 i.e. for outputs 00, 01 and 10, output Y = 1. So CLR is in active. However, as soon as counter enters into the state 3, i.e., output 11, output Y = 0 so CLR = 0, both the flip-flops will be reset and counter is brought to state –0.

Table 8.4. Truth table of reset logic

Flip-flop outputs		Output of reset logic Y	
Q_B	Q_A		
0	0	1	Valid states (3)
0	1	1	
1	0	1	
1	1	0	→ Clear all the flip-flops

K-map and Simplification

The K-map and simplified expression for the output (Y) of the reset logic has been shown in figure 8.16.

Logic Diagram for MOD-3 Counter

The logic diagram of MOD-3 counter has been shown in figure 8.17.

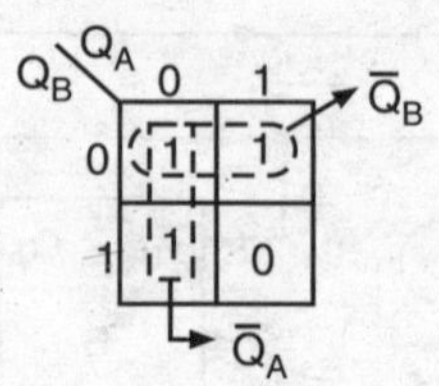

Boolean expression for Y

$$Y = \bar{Q}_B + \bar{Q}_A$$

$$\text{Therefore, } Y = \overline{Q_B \cdot Q_A}$$

Fig. 8.16. K-map and simplification

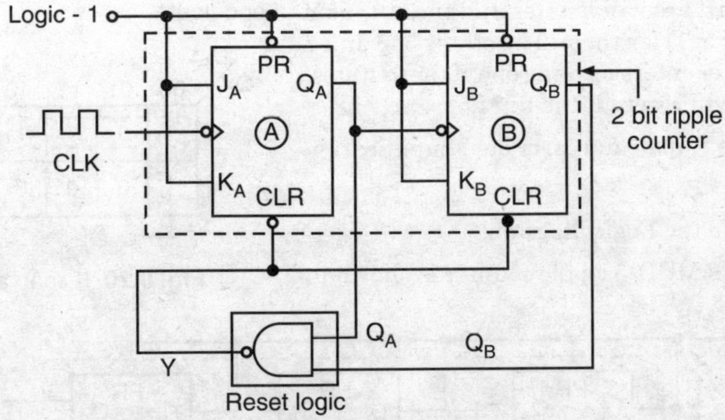

Fig. 8.17. MOD-3 counter using a 2-bit ripple counter

Timing Diagram of MOD-3 Counter

The timing diagram for a MOD-3 counter has been shown in figure 8.18.

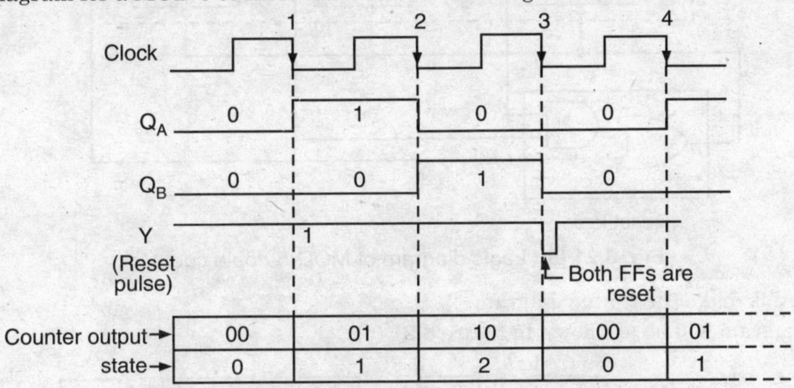

Fig. 8.18. Timing diagram for MOD-3 counter

EXAMPLE 8.5. Design a Modulo-5 ripple counter using a 3-bit ripple counter.

Solution: (i) Let us draw the state diagram.

The state diagram of MOD-5 ripple counter has been shown in figure 8.19.

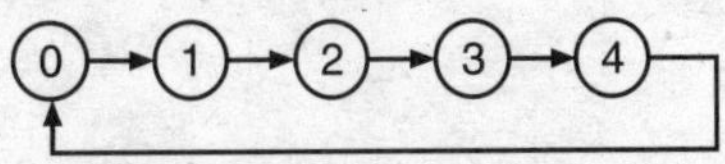

Fig. 8.19. State diagram of a MOD-5 ripple counter.

(ii) Next, let us write truth table for the reset logic.

Table 8.5 shows the truth table for the reset logic.

Table 8.5. Truth table for the reset logic

State	Flip-flop outputs			Output Y of reset logic	
	Q_C	Q_B	Q_A		
0	0	0	0	1	Valid states
1	0	0	1	1	
2	0	1	0	1	
3	0	1	1	1	
4	1	0	0	1	
5	1	0	1	0	Invalid states
6	1	1	0	0	
7	1	1	1	0	

DO YOU KNOW?

Asynchronous counters can produce glitches in decoding gates due to the counter propagation delays. Synchronous counters are less likely to cause decoding glitches. Strobing is a technique for eliminating the effects of decoding glitches.

The states 0 through 4 are valid states and the output Y of reset logic (Y) is inactive (1) for them. The states 5, 6 and 7 are invalid states. If counter enters into any one of these states, then, Y = 0 (active) and will reset all the flip-flops.

(iii) Now, let us make K-map and carry out simplification for Y output.

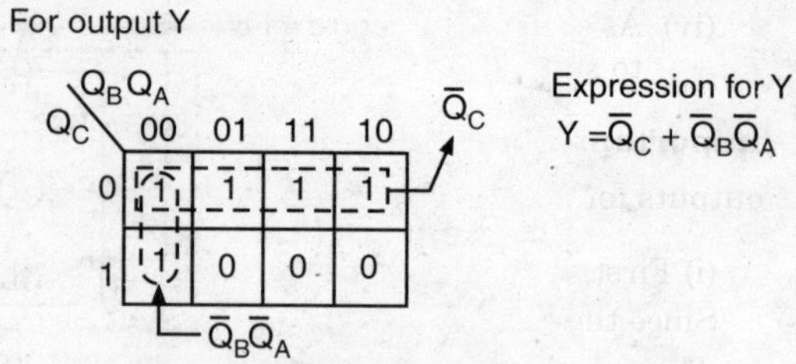

Fig. 8.20. K-map and simplification

(iv) Next, let us make the Logic diagram.

The logic diagram of a MOD-5 ripple counter is shown in figure 8.21 (a).

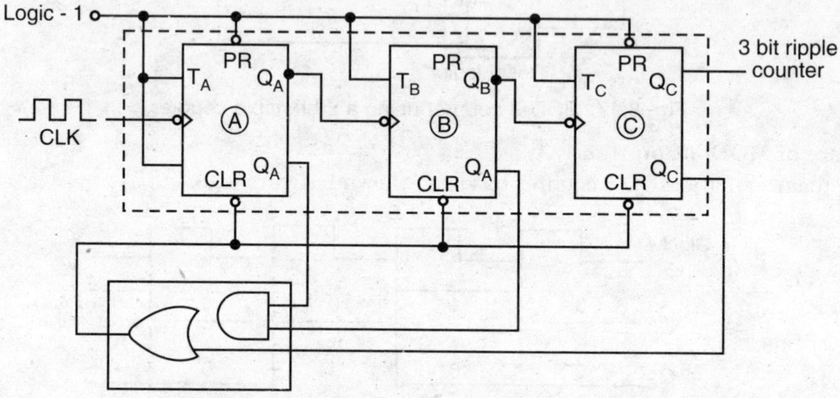

Fig. 8.21 (a). Logic diagram of MOD-5 ripple counter

(v) Finally, let us make the timing diagram.

The timing diagram will be as shown in figure 8.21 (b).

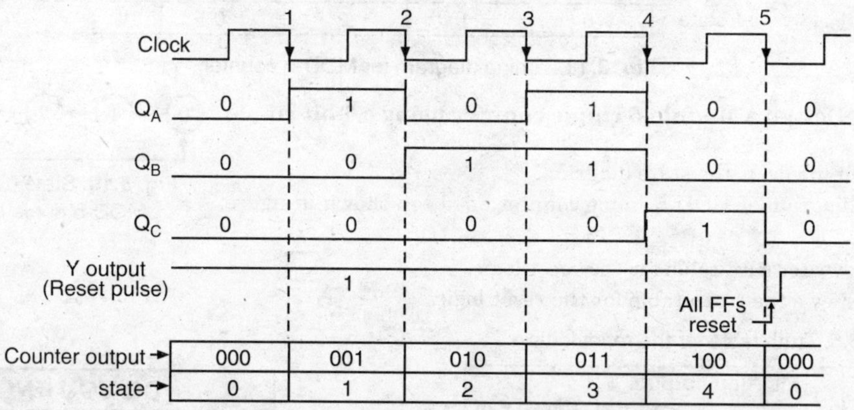

Fig. 8.21 (b). Timing diagram of MOD-5 ripple counter

EXAMPLE 8.6. Design a ripple counter for the state diagram shown in figure 8.22.

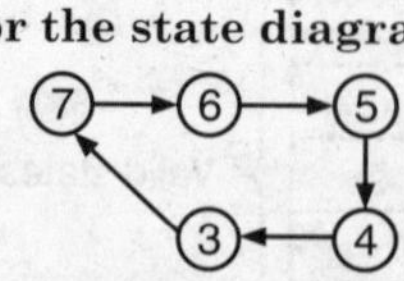

Fig. 8.22. Given state diagram

Solution: The observations from the given state diagram may be listed as under :

(i) The highest state is 7 and the counter is a down counter.

(ii) 7, 6, 5, 4, 3 are the valid states.

(iii) 0, 1, 2 are the invalid states.

(iv) As soon as the count reaches 3, i.e., 011, all the flip-flops should be set and counter should return back to state 7.

Important Point : In order to make the counter to DOWN COUNT, we must treat $\overline{Q_C}$, $\overline{Q_B}$ and $\overline{Q_A}$ as outputs for deciding the state.

(i) First, let us decide the number of flip-flops (n).

Since the highest state is 7, the number of flip-flops (n) is given as under:

$$\text{Highest state} = 2^n - 1$$

or $$8 \leq 2^n$$

or $$n = 3$$

Therefore, 3 flip-flops must be used.

(ii) Next, let us make truth table for the reset logic.

Truth table for the reset logic has been shown in Table 8.6.

Table 8.6. Truth table of reset logic

	State	Flip-flop outputs $\overline{Q_C}$	$\overline{Q_B}$	$\overline{Q_A}$	Output of reset logic Y
Invalid states	0	0	0	0	0
	1	0	0	1	0
	2	0	1	0	0
Valid states	3	0	1	1	1
	4	1	0	0	1
	5	1	0	1	1
	6	1	1	0	1
	7	1	1	1	1

Here, $\overline{Q_C}$, $\overline{Q_B}$, $\overline{Q_A}$ outputs are used because the counter is a DOWN counter

(iii) Next, let us make K-map carry out and simplification for output Y.

(iv) Finally, let us draw the logic diagram.

The logic diagram of the required counter has been shown in figure 8.24.

For output Y

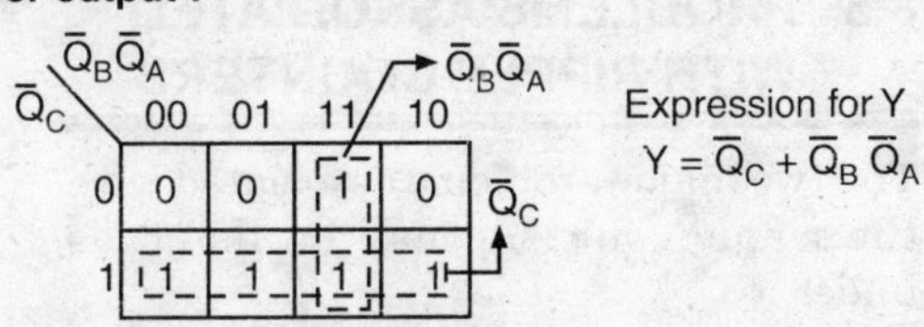

Fig. 8.23. K-map and simplification for Y output

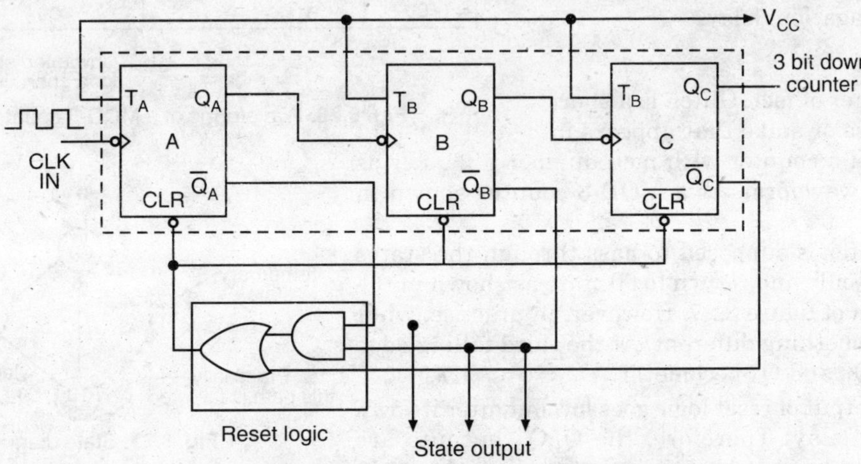

Fig. 8.24. Logic diagram of the required counter

8.4.1 Frequency Division Taking Place in Asynchronous Counters

While discussing flip-flops, we have observed that a flip-flop in toggle mode divides the clock frequency by 2. This means that the frequency of Q or $\overline{Q}$ output of a toggle flip-flop is exactly half of the clock frequency. The concept of frequency division is applicable to the counters as well because we use flip-flops in the toggle mode for counters. Let us consider the timing waveforms of a 3-bit asynchronous up counter.

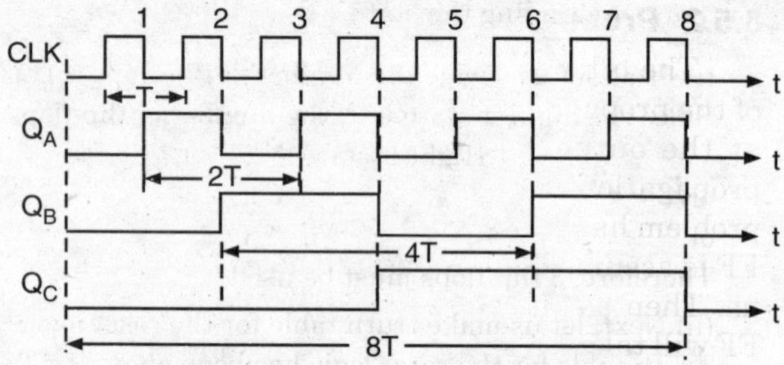

Fig. 8.25. Timing waveforms of a 3-bit asynchronous up counter

Conclusions

Let one cycle period of the clock signal be T sec. Hence the clock frequency $f_{CLK} = (1/T)$ Hz. Output of the least significant flip-flop (i.e., FF-A) has a one cycle period of (2T) as shown in figure 8.25. Hence, we have

$$\text{Frequency of } Q_A \text{ output} = f_A = \frac{1}{2T} = \frac{f_{CLK}}{2}$$

The one cycle period of Q_B is 4T. Hence, the frequency of Q_B is given by

$$f_B = \frac{1}{4T} = \frac{f_{CLK}}{4} \qquad \left(\because \frac{1}{T} = f_{CLK}\right)$$

Similarly, the frequency of Q_C output is given by

$$f_C = \frac{1}{8T} = \frac{f_{CLK}}{8}$$

Important Point : In any counter, the signal at the output of last FF (i.e., MSB) will have a frequency equal to the input clock frequency divided by the MOD number of the counter.

8.5 PROBLEMS ASSOCIATED WITH RIPPLE COUNTERS

The two major problems associated with the ripple counters may be listed as under :

(i) Generation of unwanted short duration pulses called glitch.

(ii) Propagation delay.

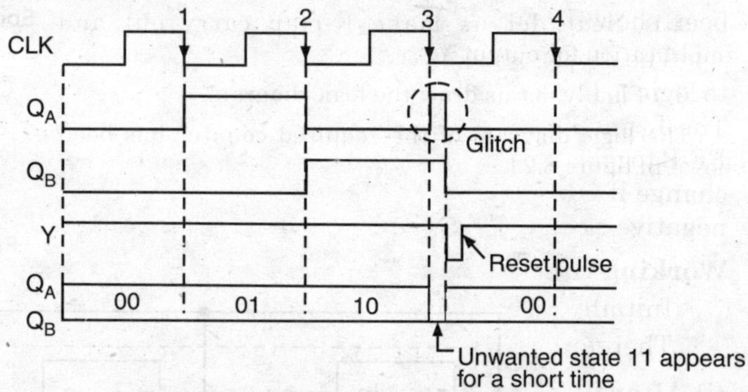

Fig. 8.26. Waveforms of a MOD-3 counter

8.5.1 Glitch

As a matter of fact, Glitch is a short duration pulse or spike that appears in the outputs of a counter with mod number < 2^n. Let us consider the waveforms of a MOD-3 counter shown in figure 8.26.

The counter is supposed to pass through the states 00, 01 and 10 only and return to 00 state as shown in the state diagram of figure 8.27. However, in practice, what happens is something different. At the third falling edge of CLK, the Q_B and Q_A become 11.

Hence, output of reset logic goes low but after its own propagation delay. Therefore, the Q_BQ_A outputs are allowed to be 11 for a short period of time as shown in

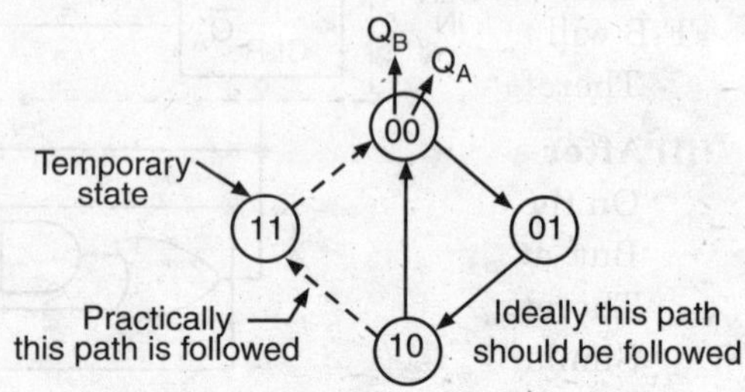

Fig. 8.27. State diagram

figure 8.27. Here, it may be noted that 11 is an unwanted state. In Q_A output waveform, a short pulse called Glitch appears as shown in figure 8.26.

> **DO YOU KNOW?**
> The term state machine refers to a circuit that sequences through a set of predetermined states controlled by a clock and other input signals.

8.5.2 Problem of Propagation Delay

The other major drawback of the ripple counters is that of the propagation delay. Each FF is clocked by the transition at the output of the preceding FF. Due to the inherent propagation delay (t_{pd}) of each flip-flop, the outputs of FFs do not change simultaneously. This problem has been illustrated in figure 8.28. Each FF is assumed to have a propagation delay of 50 ns. Then as observed from figure 8.28, the last FF will take 3 × 50, i.e., 150 ns to change its state after the application of the 4th negative going edge of the clock. In figure 8.28, the clock period is assumed to be 1000 ns. But, its clock frequency is raised so that clock period reduces to 100 ns, then the Q_C Q_B Q_A = 011 state will never appear as the propagation delay of 150 ns is longer than 1 cycle period of clock.

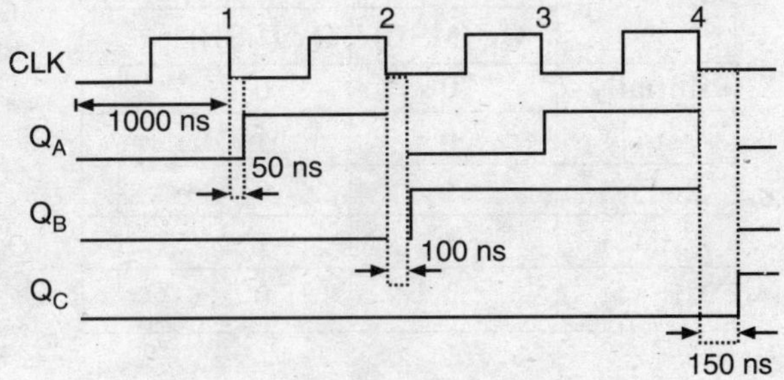

Fig. 8.28. Waveforms showing the effect of propagation delay

8.6 SYNCHRONOUS COUNTERS

(PTU, Jaldanhar, Sem. Exam., 2008-09)

If the *clock* pulses are applied to all the flip-flops in a counter simultaneously, then such a counter is called as **synchronous counter**.

8.6.1 A 2-Bit Synchronous UP Counter

A 2-bit or MOD-4 synchronous counter has been shown in figure 8.29.

The J_A and K_A inputs of FF-A are connected to logic 1. Hence, FF-A works as a toggle flip-flop. The J_B and K_B inputs are connected to Q_A. Hence, FF-B toggles if $Q_A = 1$ and there won't be any state change if $Q_A = 0$, at the instant of application of negative clock edge.

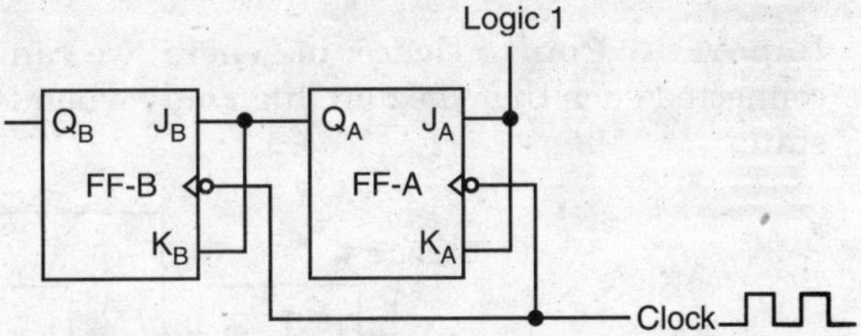

Fig. 8.29. A 2-bit (MOD-4) synchronous counter

Working Operation

Initially, let both the FFs be in the reset state.

Therefore, $Q_B\,Q_A = 0\,0$ (initially)

(i) After first negative clock edge

As soon as the first negative clock edge is applied, FF-A toggles and Q_A will change from 0 to 1. But, at the instant of application of negative clock edge, $Q_A = 0$, therefore, $J_B = K_B = 0$. Hence, FF-B will not change its state. Thus, Q_B will remain 0.

Therefore, $Q_B\,Q_A = 0\,1$ (After the first clock pulse)

(ii) After second negative clock edge

On the arrival of second negative clock edge, FF-A toggles again and Q_A changes from 1 to 0. But, at this instant, Q_A was 1. Hence, $K_B = 1$ and FF-B toggles. Hence, Q_B changes from 0 to 1.

Therefore, $Q_B\,Q_A = 1\,0$ (After the second clock pulse)

Similarly, on application of the third falling clock edge, FF-A toggles from 0 to 1 but there is no change of state for FF-B

Therefore, $Q_B\, Q_A = 11$ (After the third clock pulse)

On application of the next clock pulse, Q_A changes from 1 to 0 as Q_B will also change from 1 to 0. Hence, we have

$Q_B\, Q_A = 0\,0$ (After the fourth clock pulse)

This is the original state. The working operation of counter will repeat after this. The working operation is summarized in Table 8.7 and the timing diagram has been shown in figure 8.30.

Table 8.7. Summary of operation of a 2-bit synchronous counter

Clock	Counter Outputs	
	Q_B (MSB)	Q_A (LSB)
Initially	0	0
1st (↓)	0	1
2nd (↓)	1	0
3rd (↓)	1	1
4th (↓)	0	0

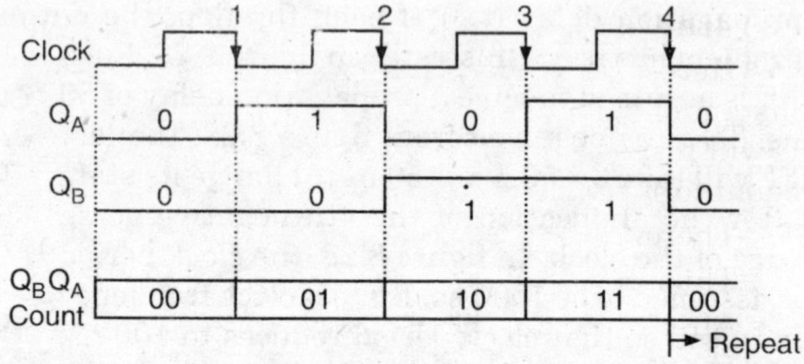

Fig. 8.30. Timing diagram for a 2-bit synchronous counter

8.6.2 A 3-Bit Synchronous Binary UP Counter

(PTU, Jalandhar, Sem. Exam; 2002-03, 2007-08)

We can extend the principle of operation of a 2-bit synchronous counter to a 3-bit counter shown in figure 8.31. Here, FF-A acts as a toggle FF since $J_A = K_A = 1$. Also, Q_A output of FF-A is applied to J_B as well as K_B. Hence, if $Q_A = 1$ at the instant of triggering, then FF-B will toggle, but, if $Q_A = 0$ then FF-B will not change its state. Therefore, Q_A and Q_B are ANDed and the output of AND gate is applied to J_C and K_C. Hence when Q_A and Q_B both are simultaneously high, then, $J_C = K_C = 1$ and FF-C will toggle. Otherwise there is no change in the state of FF-3.

Important Point : Hence, in general we can say that each FF must have its J and K inputs connected such that they are high only when the outputs of all lower order FFs are in the high state.

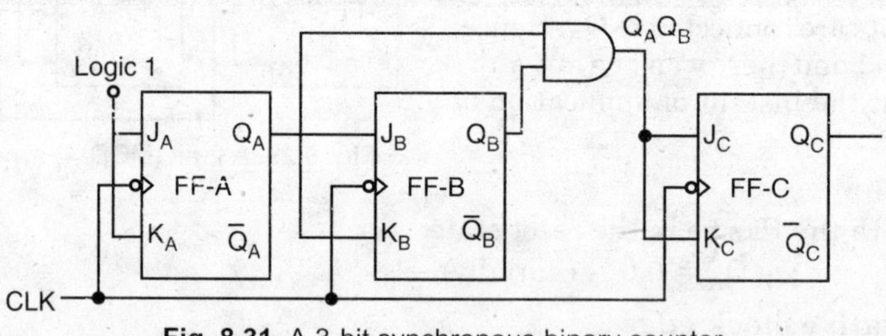

Fig. 8.31. A 3-bit synchronous binary counter

Working Operation

Initially, all the FFs are in their reset state.

Therefore, $Q_C\, Q_B\, Q_A = 0\,0\,0$

(i) 1st Clock Pulse

FF-A toggles and Q_A changes to 1 from 0. But, since $Q_A = 0$ at the instant of application of 1st falling clock edge, $J_B = K_B = 0$ and Q_B does not change state.

Therefore, Q_B remains 0.

Similarly, Q_C also does not change state.

Therefore, $Q_C = 0$

Hence, $Q_C\, Q_B\, Q_A = 0\,0\,1$ (After 1st clock pulse)

(ii) 2nd Clock Pulse

FF-A toggles and Q_A becomes 0. But at the instant of application of 2nd falling clock edge Q_A was equal to 1. Hence $J_B = K_B = 1$. Hence, FF-B will toggle and Q_B becomes 1. Output of AND gate is 0 at the instant of negative clock edge. So $J_C = K_C = 0$. Hence Q_C remains 0.

Therefore, $Q_C\ Q_B\ Q_A = 0\ 1\ 0$ (After the 2nd clock pulse)

(iii) 3rd Clock Pulse

After the 3rd clock pulse, the outputs are $Q_C\ Q_B\ Q_A = 0\ 1\ 1$.

(iv) 4th Clock Pulse

Here, it may be noted that $Q_B = Q_A = 1$. Hence, output of AND gate = 1 and $J_C = K_C = 1$, at the instant of application of 4th negative edge of the clock. Hence, on application of this clock pulse, FF-C will toggle and Q_C changes from 0 to 1. FF-A toggles as usual and Q_A becomes 0. Since Q_C was equal to 1 earlier, FF-B will also toggle to make $Q_B = 0$

Therefore, $Q_C\ Q_B\ Q_A = 1\ 0\ 0$ (after the 4th clock pulse)

Thus, the counting progress.

After the 7th clock pulse, the output is 111 and after the 8th clock pulse, all the flip-flops toggle and change their outputs to 0. Hence, $Q_C\ Q_B\ Q_A = 0\ 0\ 0$ after the 8th pulse and the operation repeats.

Table 8.8. Summary of operation of a three bit synchronous counter

Clock	Q_C	Q_B	Q_A
0	0	0	0
1st (↓)	0	0	1
2nd (↓)	0	1	0
3rd (↓)	0	1	1
4th (↓)	1	0	0
5th (↓)	1	0	1
6th (↓)	1	1	0
7th (↓)	1	1	1

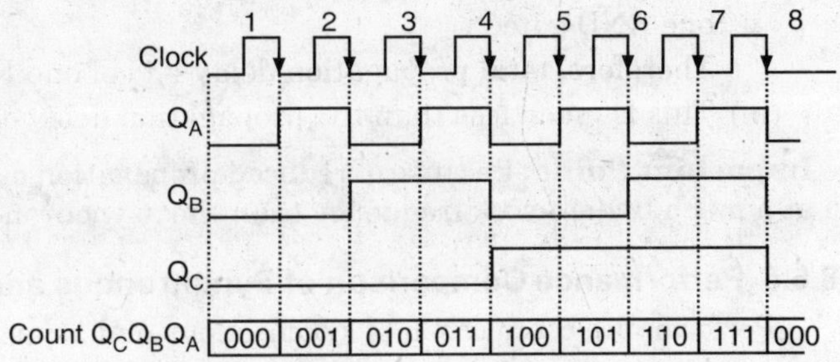

Fig. 8.32. Timing diagram for a 3-bit synchronous counter

Timing Diagram

Timing diagram for a 3-bit synchronous counter has been shown in figure 8.32. Here, it may be noted that the waveforms of synchronous counter are exactly same as those of an asynchronous counter.

8.6.3. A Four Bit Synchronous UP Counter

All the concepts of a synchronous counter are extended to design a 4-bit synchronous counter shown in figure 8.33. Here, it may be noted that J_D and K_D of the FF-D are connected to the output of an AND gate, the inputs of which come from the outputs of the three preceding FFs. Operating principle of this counter is same as that of the 3-bit counter discussed earlier.

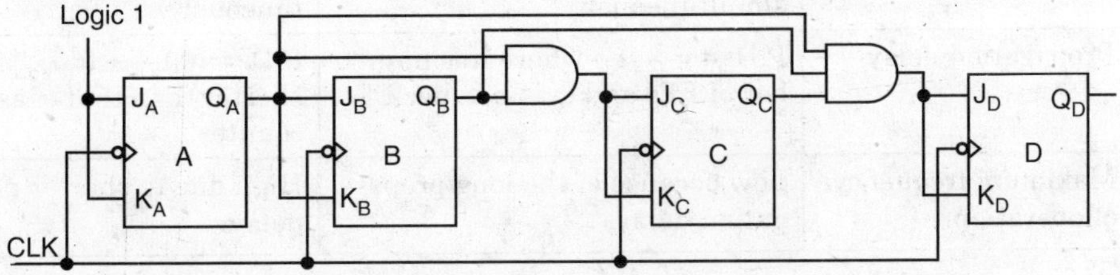

Fig. 8.33. A four-bit synchronous counter

Timing Diagram

The timing diagram of a four-bit synchronous counter has been shown in figure 8.34. Here, it may be noted that the principle of frequency division is applicable to the synchronous counters as well.

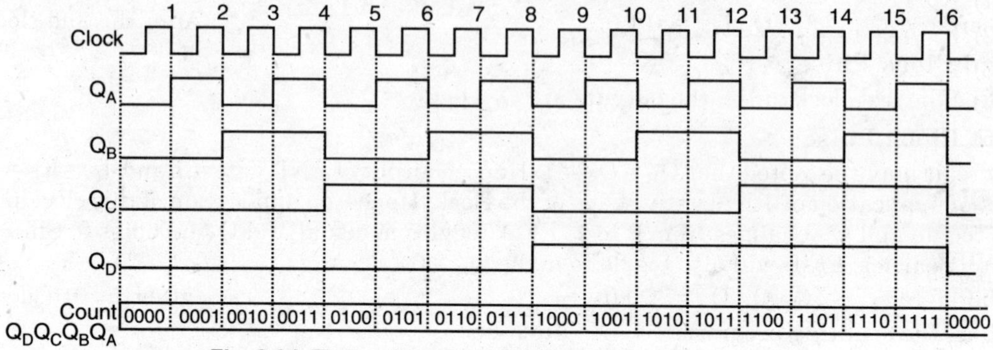

Fig. 8.34. Timing diagram of a 4-bit synchronous counter

8.6.4 Salient Features of a Synchronous Counter

(i) As all the FFs are clocked at the same instant, the propagation delay problem is reduced to a great extent.

(ii) The total propagation delay between the instant of application of clock edge and the instant at which the MSB output changes is equal to sum of propagation delay of one FF and that of one AND gate.

Therefore, total propagation delay = t_d of one FF + t_d of one gate.

(iii) This is much less than the propagation delay of an asynchronous counter.

Important Point : Because of reduced propagation delay, the synchronous counters can operate at a much higher clock frequency than the asynchronous counters.

8.6.5 Performance Comparison of Synchronous and Asynchronous

Performance comparison of synchronous and asynchronous counters is given in Table 8.9.

Table 8.9. Performance comparison of synchronous and asynchronous counters

Sr. No.	Parameter of Comparison	Asynchronous Counter	Synchronous Counter
1.	Circuit complexity	Logic circuit is simple.	With increase in number of states, the logic circuit becomes complicated.
2.	Connection pattern	Output of the preceding FF is connected to clock of the next FF.	There is no connection between output of preceding FF and CLK of next one.
3.	Clock input	All the FFs are not clocked simultaneously.	All FFs receive clock signal simultaneously.
4.	Propagation delay	P.D. = n × (t_d) where n is number of FFs and t_d is p.d. per FF.	P.D. = $(t_d)_{FF} + (t_d)_{gate}$. It is much shorter than that of asynchronous counter.
5.	Maximum frequency of operation.	Low because of the long propagation delay.	High due to shorter propagation delay.

8.6.6 Typical Counter ICs

This section is dedicated to the typical counter ICs and their applications. Let us discuss the following ICs in this article.

(1) IC 7490 : Decade binary counter
(2) IC 7492/93 : 4-bit ripple counter
(3) IC 74161 : 4-bit synchronous binary counter
(4) IC 74191 : Synchronous up/down counter
(5) IC 74192 : UP/DOWN/BCD/Binary counter

8.7 COUNTER APPLICATIONS

1. Definition

There are number of applications for different types of counters which we have discussed so far. In this article, let us discuss one of the important applications namely the frequency counter.

2. Basic Frequency Counter

A frequency counter is a circuit which measures and displays the frequency of the signal applied at its input. Block diagram of the most basic frequency counter is shown in figure 8.35 (a). It consists of a counter, a decoder display and an AND gate. The signal of unknown frequency f_u is applied to one of the inputs of the AND gate whereas the gating or sampling pulses are applied to its other input.

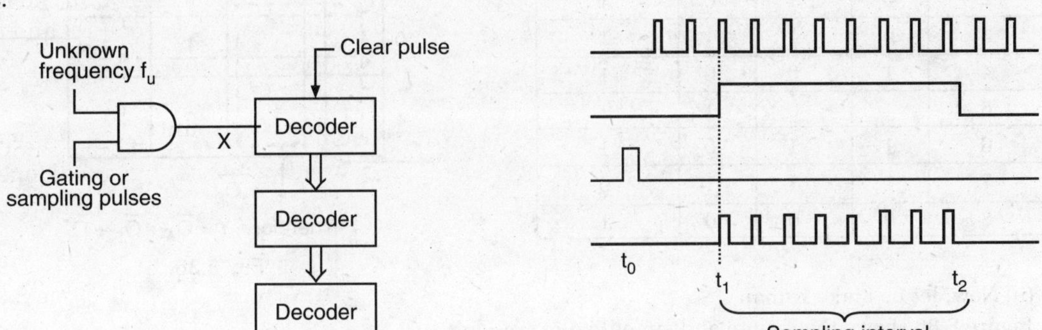

(a) Basic frequency counter (b) Waveforms

Fig. 8.35.

3. Working Operation

The gating or sampling pulses will decide the time duration for which the unknown frequency pulses are allowed to pass through to the counter. As shown in figure 8.35 (b), the counter is cleared at $t = t_0$. The pulses of unknown frequency (f_u are allowed to pass through to the counter for the time duration t_1 to t_2. This is the time duration for which the sampling pulse enables the AND gate. This time duration is called as the **sampling interval**. The counter counts the pulses only during the sampling interval. After the instant t_2, the sampling pulse goes to 0 and the AND gate is disabled. The counting stops at t_2. The number of pulses counted by the counter during the sampling interval is directly proportional to the unknown frequency f_u. The decoder converts counter output to seven segment output suitable for driving a seven segment display.

DO YOU KNOW?

The term state machine is more often used to describe other kinds of sequential circuits. They may have an irregular counting pattern.

4. Drawback

The main drawback of this technique is that the accuracy of this technique depends almost entirely on the duration of the sampling interval. Hence, we have to control the sampling interval very accurately.

8.7.1 Other Applications of Counters

Some of the applications of counters may be listed as under :

1. In digital clock
2. In the frequency counters
3. In time measurement
4. In digital voltmeters
5. In the counter type *A* to *D* converter
6. In the digital triangular wave generator
7. In the frequency divider circuits.

EXAMPLE 8.7. Draw the divide by 7 asynchronous up counter using T flip-flop. Write truth table draw the timing diagram.

Solution: (i) First, let us make the truth table.

The truth table of divide by 7 asynchronous up counter has been shown in Table 8.10. Here, it may be noted that Y represents the output of the combinational circuit used for resetting the flip- flops.

Table 8.10.

CLK	Q_C	Q_B	Q_A	Y Output
Initially	0	0	0	1
1	0	0	1	1
2	0	1	0	1
3	0	1	1	1
4	1	0	0	1
5	1	0	1	1
6	1	1	0	1
7	1	1	1	0
8	0	0	0	1

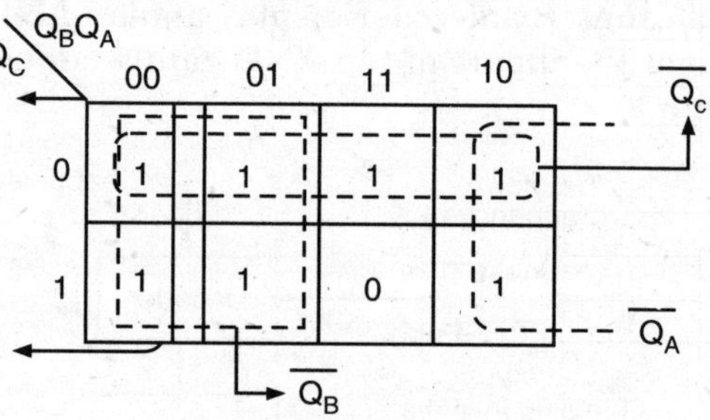

Therefore, $Y = \overline{Q}_A + \overline{Q}_B + \overline{Q}_C$

Fig. 8.36.

(ii) Now, let us make K-map.

Figure 8.36 shows the K-map and simplified expression.

(iii) Next, let us draw the circuit.

Divide by 7 counter has been shown in figure 8.37.

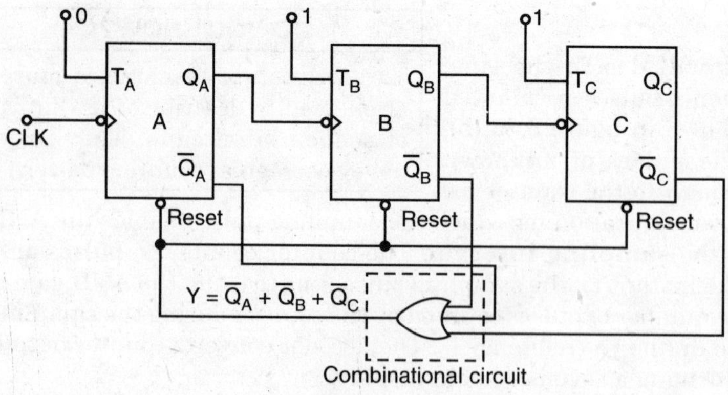

Fig. 8.37.

> **DO YOU KNOW?**
>
> The general distinction between the two terms 'counter' and 'state machine' is that a counter is commonly used to count events, while a state machine is commonly used to control events. The correct descriptive term depends on how we wish to use the sequential circuit.

(iv) Next, we make timing diagrams.

The timing diagram has been shown in figure 8.38.

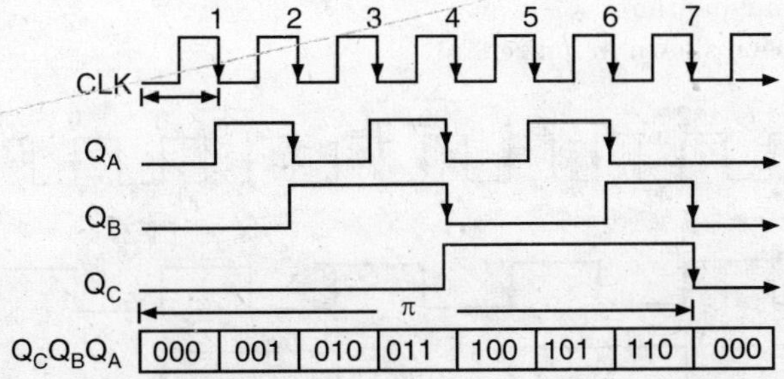

Fig. 8.38. Timing diagram of divide by 7 counter

EXAMPLE 8.8. For MOD-11 asynchronous up counter.

(i) Draw circuit diagram. Use T flip-flop.

(ii) Write truth table.

(iii) Draw timing diagram.

(iv) If the output frequency is 11 kHz what will be the clock input ?

Solution: (i) First let us write the truth table.

Table 8.11.

Q_D	Q_C	Q_B	Q_A	Y Output
0	0	0	0	1
0	0	0	1	1
0	0	1	0	1
0	0	1	1	1
0	1	0	0	1
0	1	0	1	1
0	1	1	0	1
0	1	1	1	1
1	0	0	0	1
1	0	0	1	1
1	0	1	0	1
1	0	1	1	0
1	1	0	0	0
1	1	0	1	0
1	1	1	0	0
1	1	1	1	0

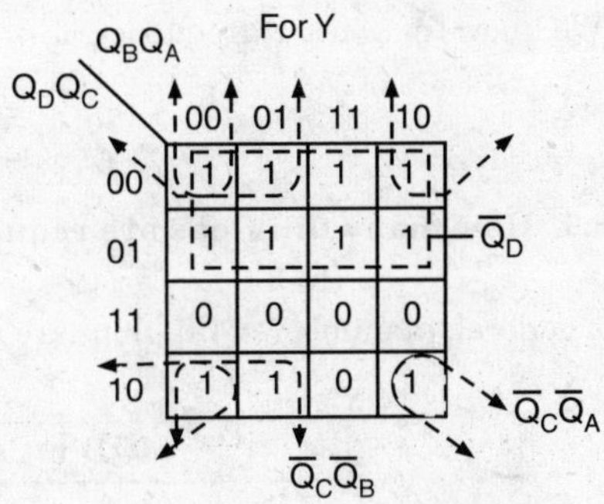

Therefore, $Y = \overline{Q}_D + \overline{Q}_C\overline{Q}_A + \overline{Q}_C\overline{Q}_B$

or $Y = \overline{Q}_D + \overline{Q}_C(\overline{Q}_A + \overline{Q}_B)$

Fig. 8.39. K-map.

(ii) Next, let us draw the K-map.

The K-map has been shown in figure 8.39. The simplified expression for Y output of the combinational circuit is as under :

$$Y = \overline{Q_D} + \overline{Q_C}\left(\overline{Q_A} + \overline{Q_B}\right)$$

(iii) Then, we draw the circuit diagram.

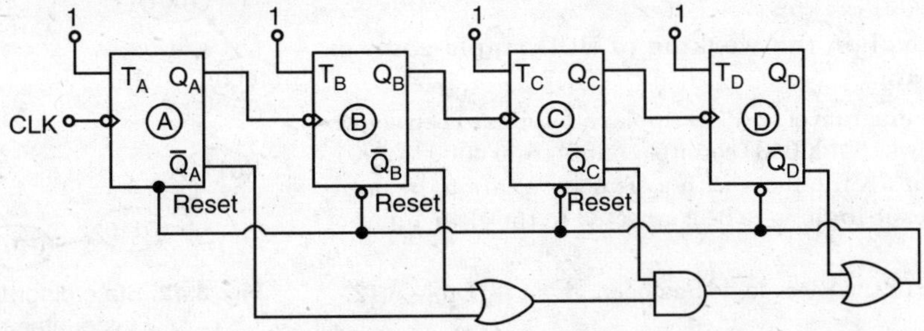

Fig. 8.40. MOD-11 counter

(iv) Finally, we draw timing diagram.

The timing diagram has been shown in figure 8.41.

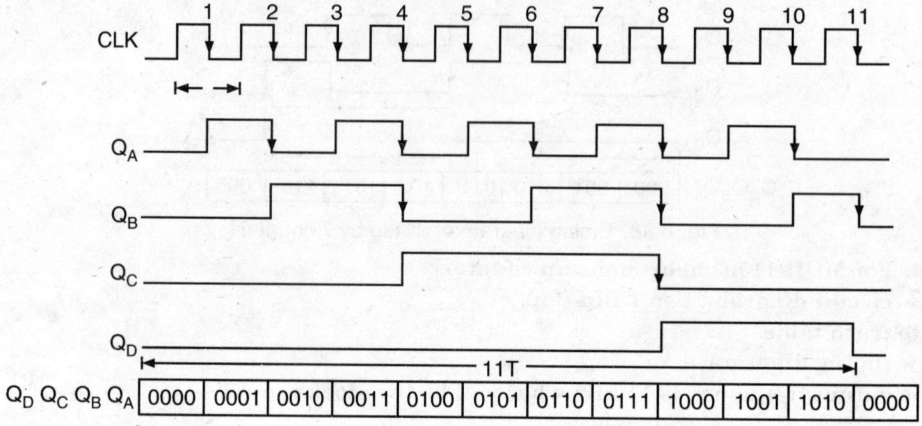

Fig. 8.41. Timing diagram

We know that output frequency $f_0 = \frac{f_{CLK}}{11}$

Therefore, $f_{CLK} = 11\, f_0 = 11 \times 11$ kHz

or $f_{CLK} = 121$ kHz **Ans.**

EXAMPLE 8.9. How many flip-flops are required to construct the following modulus counter? Why?

(a) 5 **(b) 83** **(c) 99** **(d) 10.**

Solution: In general, m number of flip-flops are required to construct MOD-n counter, where

$$n \leq 2^m$$

	MOD n Counter	Number of flip-flops (m)
A	5	3
B	83	7
C	99	7
D	10	4

EXAMPLE 8.10. How many flip-flops are required to count 16 clock pulses ? Why ?

Solution: Basically, to count n clock pulses, m flip-flops are required, where, we have

$$n \leq 2^m$$

Therefore, to count 16 clock pulses, 4 flip-flops are required, since $16 = 2^4$.

EXAMPLE 8.11. Explain the working of BCD ripple counter with timing diagram.

Solution. The state diagram of an BCD ripple counter has been shown in figure 8.42. It shows that a BCD counter counts from 0000 to 1001, i.e., from 0 to 9. Four JK flip-flops with a clear input are to be used. The output of the **reset logic** is to be connected to the clear input of all the FFs.

The truth table of the reset logic has been given in Table 8.12.

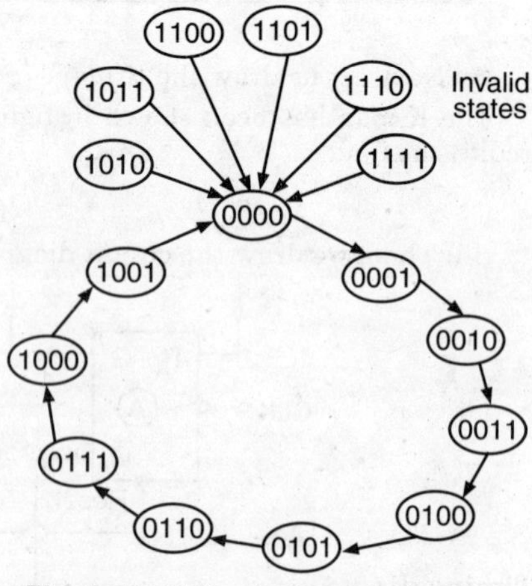

Fig. 8.42. State diagram of a BCD counter

Table 8.12. Truth table of a BCD counter

CLK	Q_D	Q_C	Q_B	Q_A	Output of reset logic	
0	0	0	0	0	1	Valid states
1	0	0	0	1	1	
2	0	0	1	0	1	
3	0	0	1	1	1	
4	0	1	0	0	1	
5	0	1	0	1	1	
6	0	1	1	0	1	
7	0	1	1	1	1	
8	1	0	0	0	1	
9	1	0	0	1	1	
–	1	0	1	0	0	Invalid states
–	1	0	1	1	0	
–	1	1	0	0	0	
–	1	1	0	1	0	
–	1	1	1	0	0	
–	1	1	1	1	0	

K-map and simplification

The K-map for Y output has been shown in figure 8.43.

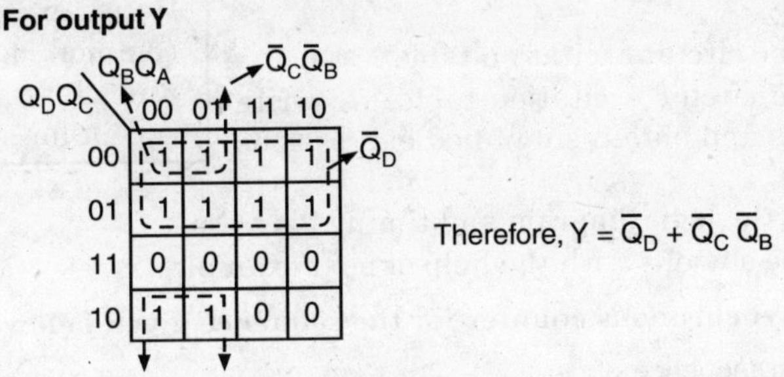

Therefore, $Y = \bar{Q}_D + \bar{Q}_C\,\bar{Q}_B$

Fig. 8.43.

Logic diagram : The logic diagram of a BCD counter has been shown in figure 8.44.

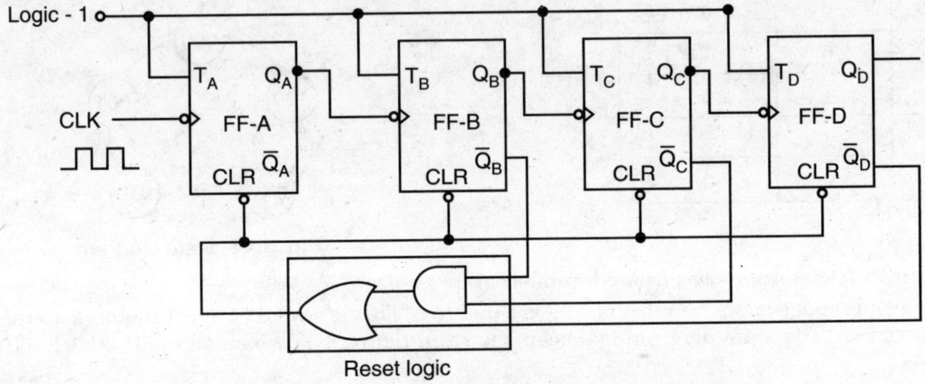

Fig. 8.44. Logic diagram of a BCD counter

Timing diagram

The timing diagram of a BCD counter has been shown in figure 8.45.

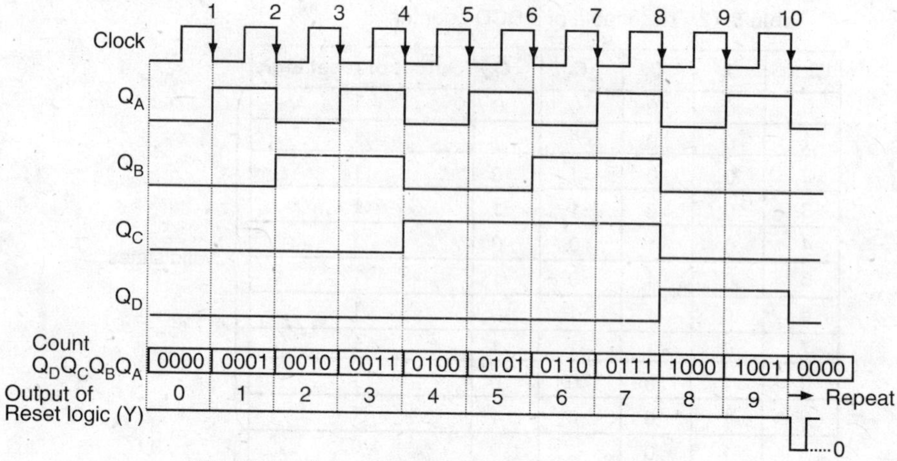

Fig. 8.45. Timing diagram of a BCD ripple counter

8.8 MODULO – N SYNCHRONOUS COUNTERS **(PTU, Jalandhar, Sem. Exam., 2008-09)**

We have already discussed the MOD-N asynchronous counters. Now, let us discuss the Modulo-N synchronous counters.

> **DO YOU KNOW?**
>
> In a synchronous system, all flip-flops are clocked by the same common clock signal, and preset and clear inputs are not used, except for system initialization.

Procedural Steps

(i) First, we decide the number of FFs and type of FF to be used.

(ii) Then, we write the circuit excitation table.

(iii) Finally, from the circuit excitation table, we write down the K-maps and obtain simplified expressions for the outputs.

(iv) Finally, we draw the logic diagram and timing diagram.

This procedure will be obvious with the help of next example.

EXAMPLE 8.12. Design a synchronous counter for the sequence given below.

Table 8.13. Desired sequence

Q_C	Q_B	Q_A
0	0	0
0	0	1
0	1	0
0	1	1
1	0	0

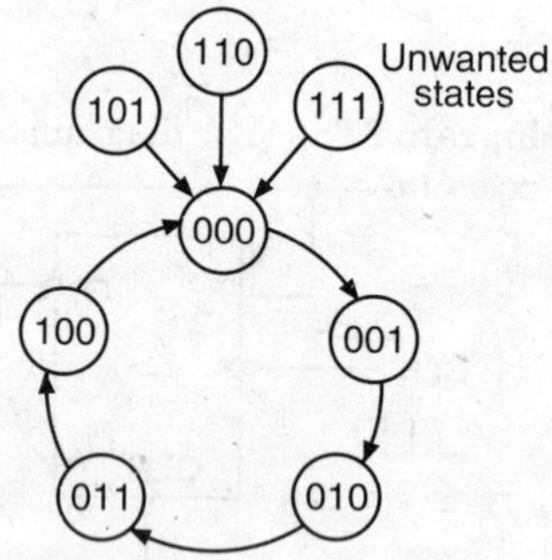

Fig. 8.46. State diagram

Solution: (i) First, we determine the desired number of FFs.

From the given sequence, the number of FFs is equal to 3. This is a MOD-5 synchronous counter since the number of states is 5. The state diagram has been shown in figure 8.53 which shows that 101, 110, 111 are unwanted states.

(ii) Next, we write the excitation table and state table.

The type of FF used is JK flip-flop. The excitation table for a JK FF has been shown in Table 8.14.

Excitation Table of J-K FF

Table 8.14 depicts the excitation table of a J-K FF.

Table 8.14. Excitation table of J-K FF

Present state Q_n	Next state Q_{n+1}	J	K
0	0	0	×
0	1	1	×
1	0	×	1
1	1	×	0

Circuit Excitation Table

The circuit excitation table has been shown in Table 8.15.

Table 8.15. Circuit excitation table

Present state			Next state			Flip-flop inputs					
Q_C	Q_B	Q_A	Q_{C+1}	Q_{B+1}	Q_{A+1}	J_C	K_C	J_B	K_B	J_A	K_A
0	0	0	0	0	1	0	×	0	×	1	×
0	0	1	0	1	0	0	×	1	×	×	1
0	1	0	0	1	1	0	×	×	0	1	×
0	1	1	1	0	0	1	×	×	1	×	1
1	0	0	0	0	0	×	1	0	×	0	×
1	0	1	0	0	0	×	1	0	×	×	1
1	1	0	0	0	0	×	1	×	1	0	×
1	1	1	0	0	0	×	1	×	1	×	1

(iii) K-maps Simplification

K-maps for the J and K inputs of all the FFs and the corresponding simplified expressions have been shown in figure 8.47. Here, it may be noted that the inputs to this combinational circuit are Q_A, Q_B, Q_C and outputs are J_A, K_A through J_C, K_C.

> **DO YOU KNOW?**
>
> In asynchronous (ripple) counters, the clock signal is applied to the LSB FF, and all other FFs are clocked by the output of the preceding FF.

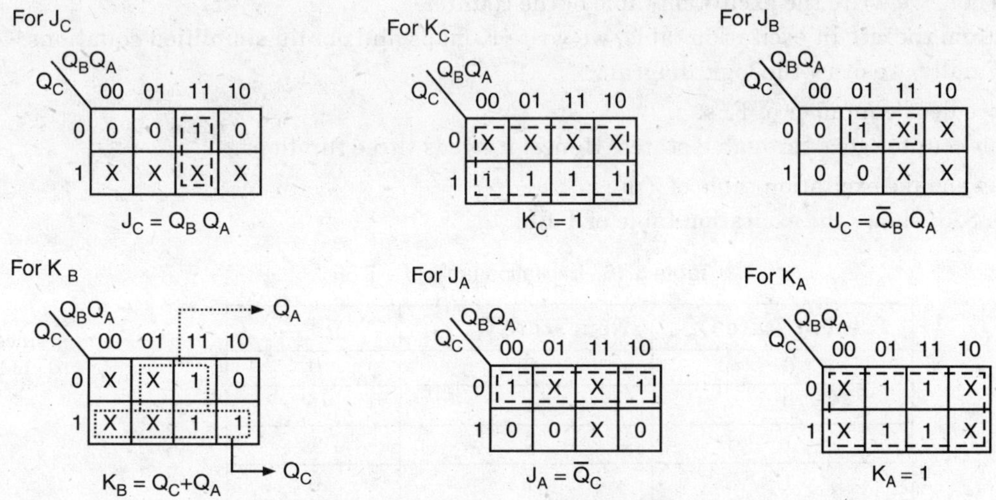

Fig. 8.47. K-maps and simplifications

(iv) Next, we draw the logic diagram.

The logic diagram of MOD-5 synchronous counter has been shown in figure 8.48.

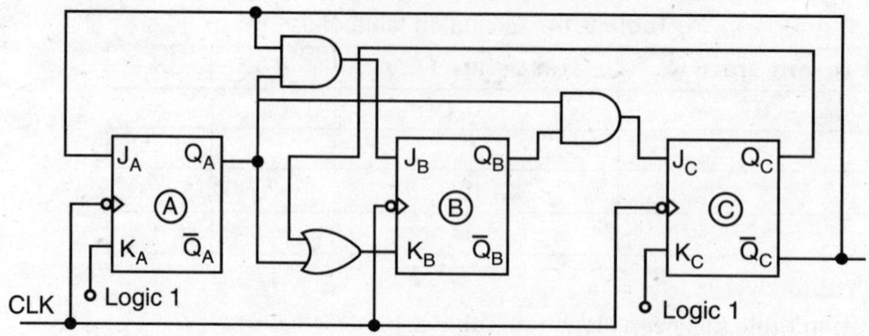

Fig. 8.48. Logic diagram of MOD-5 synchronous counter

(v) Finally, we draw the timing diagram.

The timing diagram of MOD-5 synchronous counter has been shown in figure 8.49.

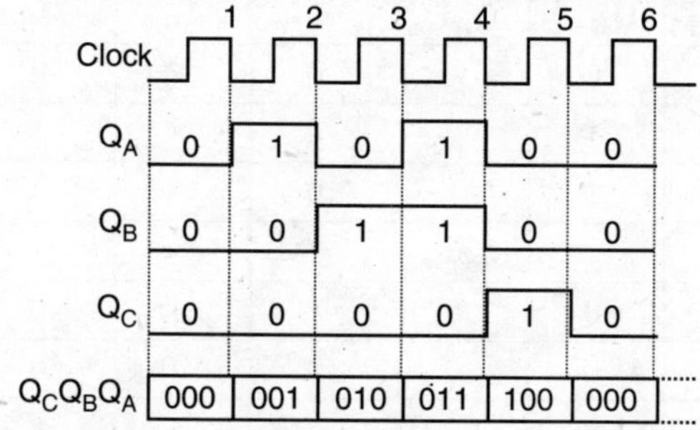

Fig. 8.49. Timing diagram of MOD-5 counter

8.8.1 Design of a 3-Bit Synchronous Counter

Now, let us design a 3-bit synchronous counter first using the T flip-flops and then using J-K flip-flops.

8.8.1.1. Design Using T Flip-Flops

(i) First, we decide the number of flip-flops.

(ii) Next, we write the excitation table of T flip-flop.

(iii) Then, we write the excitation table of the counter.

(iv) From the circuit excitation table, we write K-maps and obtain simplified equations.

(v) Finally, we draw the logic diagram.

(i) First, we decide number of FFs.

A 3-bit counter goes through 8 states. Hence, it needs three flip-flops.

(ii) Next, we make excitation table of T FFs.

Table 8.16 shows the excitation table of T FF.

Table 8.16. Excitation table of a T FF

Present state Q_n	Next state Q_{n+1}	T
0	0	0
0	1	1
1	0	1
1	1	0

(iii) Next, we make state diagram and circuit excitation table.

Count sequence for a 3-bit up counter has been given in Table 8.17 and figure 8.50 shows the corresponding state diagram.

Table 8.17.

Q_C	Q_B	Q_A
0	0	0
0	0	1
0	1	0
0	1	1
1	0	0
1	0	1
1	1	0
1	1	1

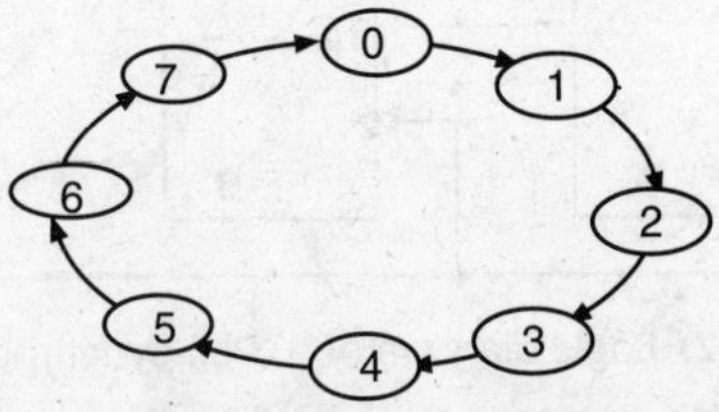

Fig. 8.50. State diagram.

(iv) Table 8.18 shows the circuit excitation table.

Table 8.18. Circuit excitation table

Present state			Next state			Flip-flop inputs		
Q_C	Q_B	Q_A	Q_{C+1}	Q_{B+1}	Q_{A+1}	T_C	T_B	T_A
0	0	0	0	0	1	0	0	1
0	0	1	0	1	0	0	1	1
0	1	0	0	1	1	0	0	1
0	1	1	1	0	0	1	1	1
1	0	0	1	0	1	0	0	1
1	0	1	1	1	0	0	1	1
1	1	0	1	1	1	0	0	1
1	1	1	0	0	0	1	1	1

Table 8.18 has been written by considering the present and next state of an output and the required value to input is written as per the excitation table of T FF. As per the excitation table of a T FF, T_C must be 0. Similarly, all other entries are made.

(v) Then, we write K-maps and obtain simplified expressions.

> **DO YOU KNOW?**
>
> A counter's MOD number is the number of stable states in its counting cycle; it is also the maximum frequency-division ratio.

Let us, consider figure 8.51 for the K-maps corresponding to all the FF inputs. The simplified equations for T_A, T_B and T_C also are shown.

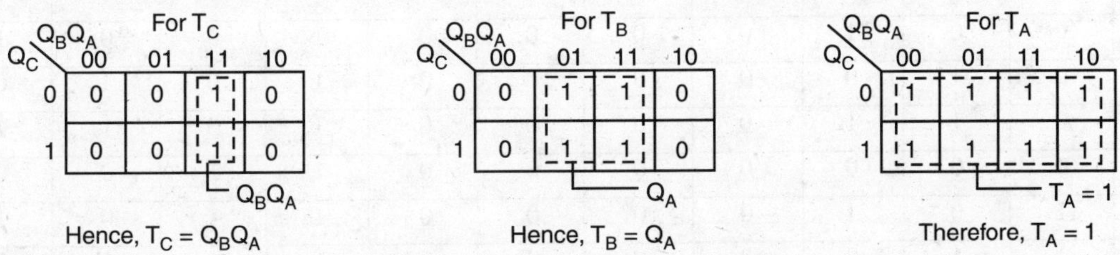

Fig. 8.51. K-maps for different FF inputs

(vi) Finally, we draw the logic diagram.

By using the simplified equations for T_A, T_B and T_C, we can draw the logic diagram for the 3-bit synchronous counter as shown in figure 8.52.

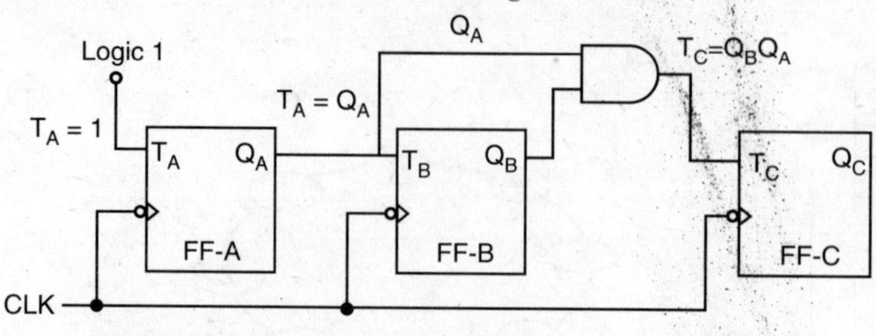

Fig. 8.52. Logic diagram of a 3-bit synchronous counter

DO YOU KNOW?

The normal (maximum) MOD number of a counter is 2^N. One way to modify a counter's MOD number is to add circuitry that will cause it to recycle before it reaches its normal last count.

8.8.2 Synchronous Decade Counter

For the design of a synchronous decade counter, we follow the following steps :

(i) First, we write the excitation table for T FF and circuit excitation table.

Excitation table for T FF is shown in following Table 8.19.

Table 8.19. Excitation table for T flip-flop

Present state Q_n	Next state Q_{n+1}	T
0	0	0
0	1	1
1	0	1
1	1	0

The circuit excitation table has been shown in Table 8.20.

Table 8.20. The circuit excitation table

Present state				Next state				Flip-flop inputs			
Q_D	Q_C	Q_B	Q_A	Q_{D+1}	Q_{C+1}	Q_{B+1}	Q_{A+1}	T_D	T_C	T_B	T_A
0	0	0	0	0	0	0	1	0	0	0	1
0	0	0	1	0	0	1	0	0	0	1	1
0	0	1	0	0	0	1	1	0	0	0	1
0	0	1	1	0	1	0	0	0	1	1	1
0	1	0	0	0	1	0	1	0	0	0	1
0	1	0	1	0	1	1	0	0	0	1	1
0	1	1	0	0	1	1	1	0	0	0	1
0	1	1	1	1	0	0	0	1	1	1	1
1	0	0	0	1	0	0	1	0	0	0	1
1	0	0	1	0	0	0	0	1	0	0	1
1	0	1	0	0	0	0	0	1	0	1	0
1	0	1	1	0	0	0	0	1	0	1	1
1	1	0	0	0	0	0	0	1	1	0	0
1	1	0	1	0	0	0	0	1	1	0	1
1	1	1	0	0	0	0	0	1	1	1	0
1	1	1	1	0	0	0	0	1	1	1	1

(ii) Next, let us make K-maps and carryout simplifications.

The K-maps for T_D, T_C, T_B, T_A and their simplified expressions are given in figures 8.53 (a), (b), (c) and (d).

For T_D

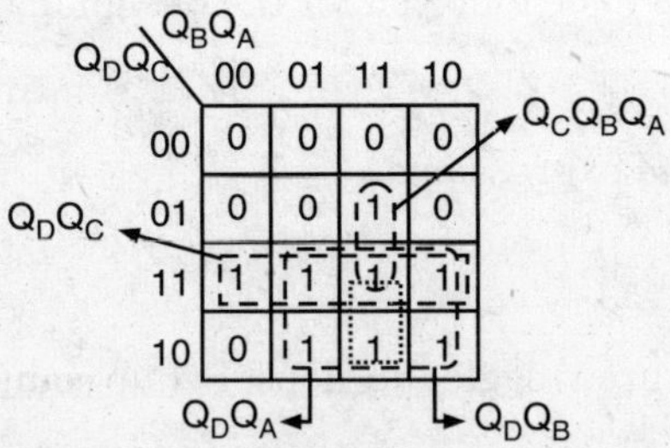

$T_D = Q_C\, Q_B\, Q_A + Q_D\, Q_C + Q_D\, Q_B + Q_D\, Q_A$

Therefore, $T_D = Q_C\, Q_B\, Q_A + Q_D(Q_C + Q_B + Q_A)$

(a) K map for T_D

For T_C

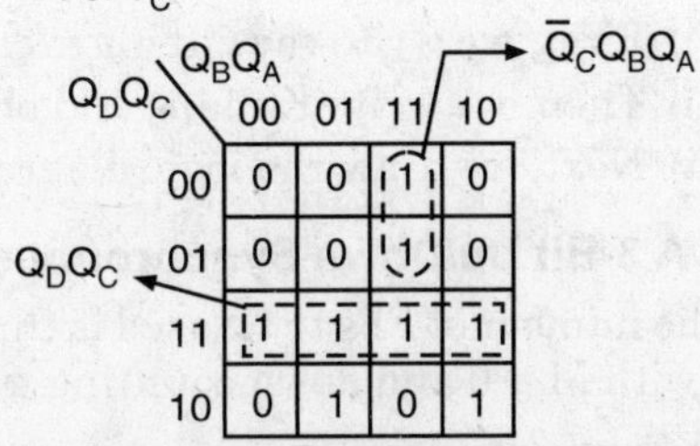

Hence, $T_C = \bar{Q}_D\, Q_B\, Q_A + Q_D\, Q_C$

(b) K map for T_C

For T_B

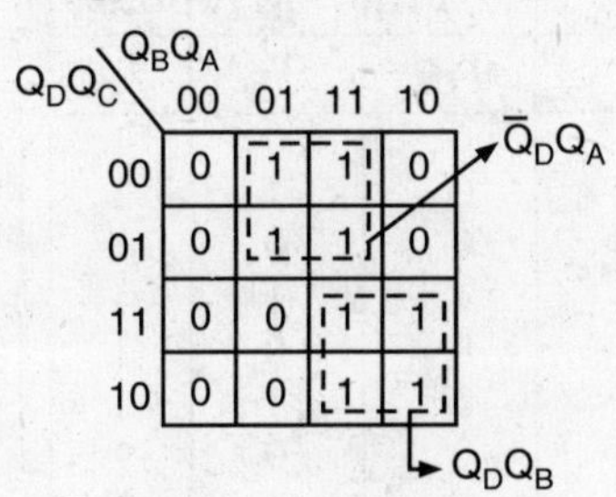

Therefore, $T_B = Q_DQ_B + \bar{Q}_DQ_A$

(c) K map for T_B

For T_A

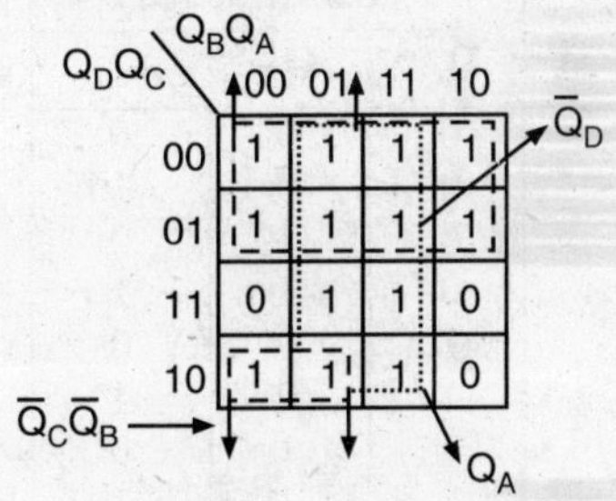

Therefore, $T_A = \bar{Q}_D + \bar{Q}_C\bar{Q}_BQ_A$

(d) K map for T_A

Fig. 8.53.

Figure 8.54 shows the logic diagram of a synchronous decade counter.

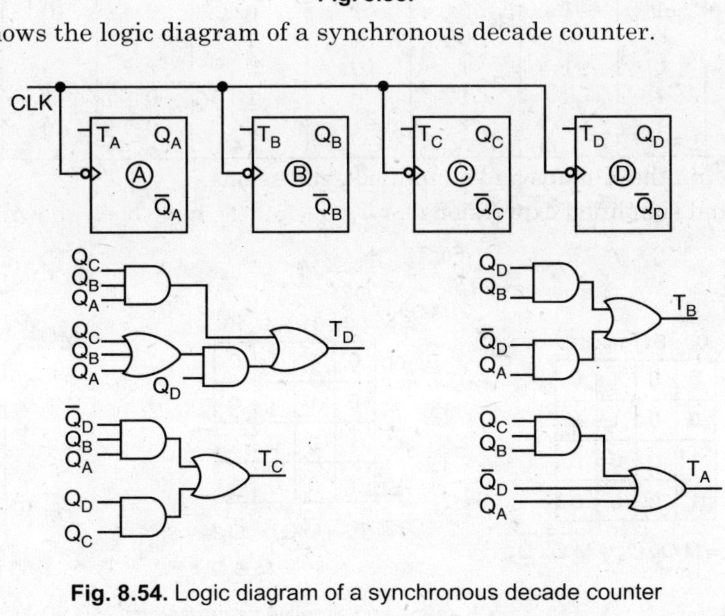

Fig. 8.54. Logic diagram of a synchronous decade counter

8.9 UP/DOWN SYNCHRONOUS COUNTER

(PTU, Punjab, Sem. Exam., 2006-07)

The operating principle of a synchronous up/down counter is same as that of the up/down ripple counters. However, the design procedure and realization are different. The mode control (M) is used for selecting the mode of operation. The procedural steps to be followed for the design may be listed as under :

(i) First, we write the circuit excitation table.

(ii) Then, we write K-maps and obtain the simplified expressions.

(iii) Next, we draw the logic diagram.

8.9.1 A 3-Bit Up/Down Synchronous Counter

The number of FFs to be used is three. We shall use three toggle flip-flops. Let up counting takes place with M = 0 and down counting take place for M = 1.

(i) First, we write the circuit excitation table.

The circuit excitation table has been shown in Table 8.21.

Table 8.21. Excitation table for a 3-bit up/down synchronous counter

Mode	Present state			Next state			Flip-flop inputs			
control M	Q_C	Q_B	Q_A	Q_{C+1}	Q_{B+1}	Q_{A+1}	T_C	T_B	T_A	
0	0	0	0	0	0	1	0	0	1	Up counting
0	0	0	1	0	1	0	0	1	1	
0	0	1	0	0	1	1	0	0	1	
0	0	1	1	1	0	0	1	1	1	
0	1	0	0	1	0	1	0	0	1	
0	1	0	1	1	1	0	0	1	1	
0	1	1	0	1	1	1	0	0	1	
0	1	1	1	0	0	0	1	1	1	
1	0	0	0	1	1	1	1	1	1	Down counting
1	0	0	1	0	0	0	0	0	1	
1	0	1	0	0	0	1	0	1	1	
1	0	1	1	0	1	0	0	0	1	
1	1	0	0	0	1	1	1	1	1	
1	1	0	1	1	0	0	0	0	1	
1	1	1	0	1	0	1	0	1	1	
1	1	1	1	1	1	0	0	0	1	

(ii) Next, let us obtain the K-maps and simplified expressions.

The K-maps and simplified expressions for T_C, T_B and T_A have been shown in figure 8.55.

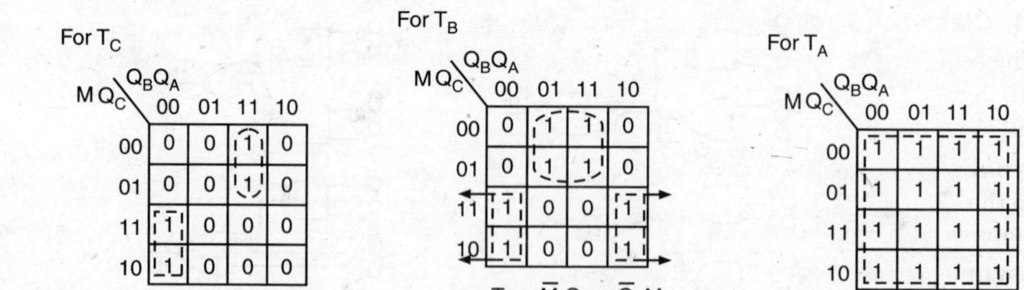

Thererfore, $T_C = \overline{M}\,Q_B\,Q_A + M\,\overline{Q}_B\,\overline{Q}_A$

$T_B = \overline{M}\,Q_A + \overline{Q}_A M$
$= M \oplus Q_A$

Therefore, $T_A = 1$

Fig. 8.55. K-maps and simplified expressions

(iii) Finally, let us draw the logic diagram.

The logic diagram for a 3-bit synchronous up/down counter has been shown in figure 8.56.

> **DO YOU KNOW?**
>
> Counters can be cascaded (chained together) to produce greater counting ranges and frequency-division ratios.

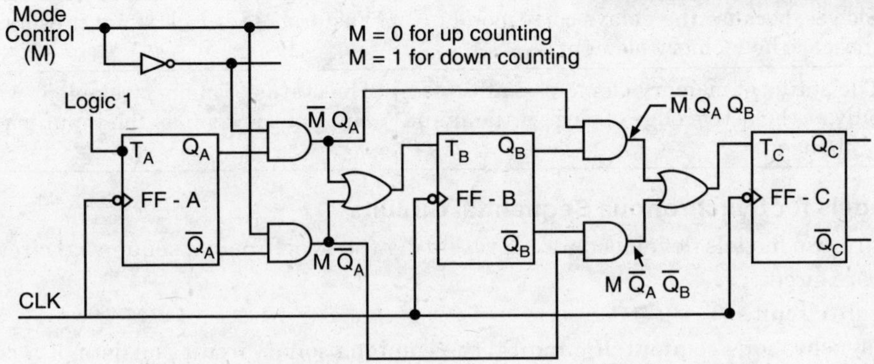

Fig. 8.56. Logic diagram of a 3-bit synchronous up/down counter

8.10 SEQUENTIAL CIRCUIT DESIGN

Figure 8.57 shows the general model of a sequential circut. It shows that a sequential circuit consists of a combinational circuit along with the memory elements. The memory elements will always appear in the feedback path. The output of a sequential circuit depends on the present state as well as the past state.

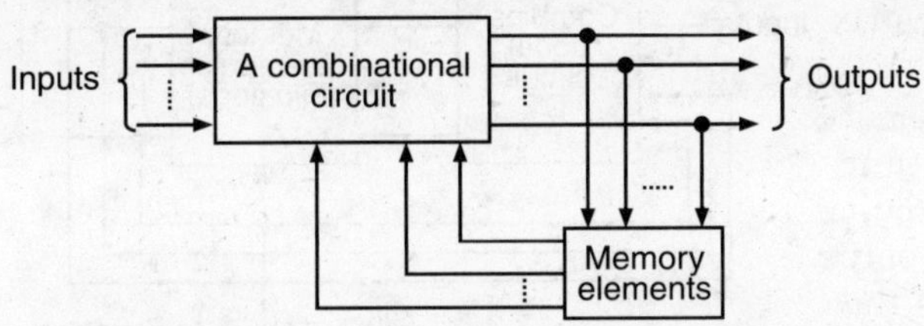

Fig. 8.57. General model of a sequential circuit

The present state is stored in the memory elements. These present states are applied to the combinational circuit alongwith the external inputs. Therefore, the outputs and the next state of the sequential circuit are decided by the external inputs and the present state.

8.10.1 Classification of Sequential Circuits

The sequential circuits may be classified into following two categories :

(i) Synchronous sequential circuits

(ii) Asynchronous sequential circuits

(i) Synchronous Sequential Circuits

In the synchronous sequential circuits, the contents of memory elements can be changed only at the rising or falling edges of clock signal.

(ii) Asynchronous Sequential Circuits

In the asynchronous sequential circuit, the contents of memory elements can be changed at any instant of time.

Table 8.22. Performance comparison of synchronous and asynchronous sequential circuits

S.No.	Synchronous sequential circuits	Asynchronous sequential circuits
1.	These circuits are easy to design.	Difficult to design.
2.	A clocked FF acts as a memory element.	An unlocked FF or time delay element is used as memory element.
3.	Slower, because the delays correspond to those of the memory element.	Faster as the clock is not present.
4.	The status of memory element is affected only at the active edge of clock if the input is changed.	The status of memory element will change any time as soon as the input is changed.

8.10.2 Models for Synchronous Sequential Circuits

There are two models developed for representing the synchronous sequential circuits as under :

(i) Moore circuit

(ii) Mealy circuit.

In the synchronous sequential circuits, the combinational circuit consists of a combination of logic gates whereas the flip-flops are used as memory elements. The synchronous sequential circuit is called as a **Moore circuit** if the output depends only on the present state of flip-flops. The circuit is called as a **Mealy circuit** if the output is dependent on the present state of flip-flops and the external inputs.

8.11 MOORE CIRCUIT

Figure 8.58 shows the general form of Moore circuit.

Here, it may be noted that the most important thing for a Moore circuit that its output depends only on the present state of the memory elements. The external inputs do not influence the outputs. The combinational circuit C_1 on the input side is some times called as the next state decoder. The external inputs and the feedback signals from the memory elements are applied to this combinational circuit. The outputs of the memory elements are applied to another combinational circuit C_2 which is also called as output decoder.

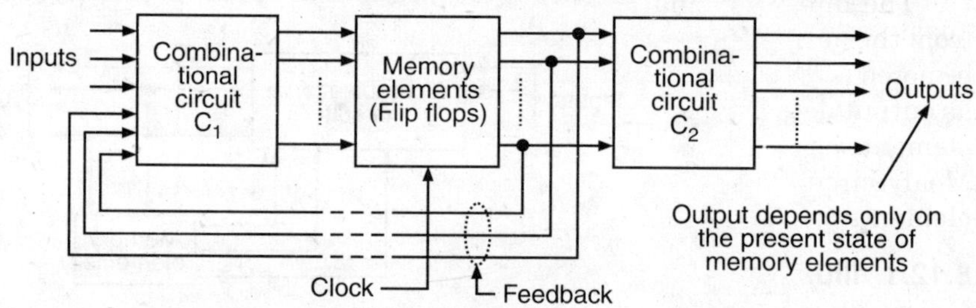

Fig. 8.58. A general form of Moore circuit

8.11.1 Illustrative Example of Moore Circuit

Figure 8.59 shows the example of a Moore circuit. Here, flip-flops A and B act as the memory elements for the circuit. Gates 1 and 2 form the input combinational circuit C_1, because the external inputs A and $\overline{A}$ are applied to one of their inputs, whereas the other inputs are coming from the Q_B and $\overline{Q}_A$ outputs of the memory elements.

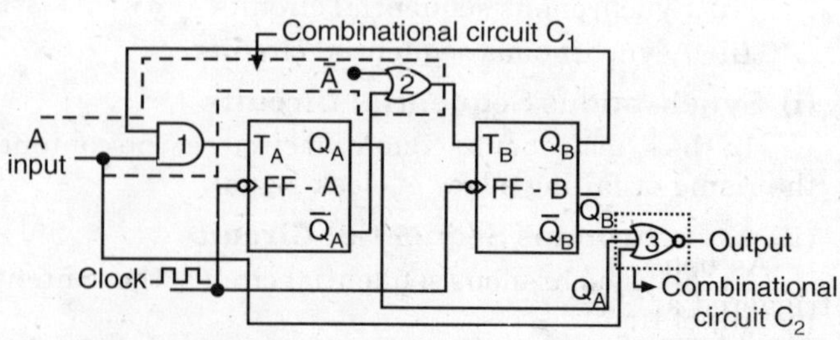

Fig. 8.59. Illustrative example of a Moore circuit

Gate-3 acts as the output decoder or

combinational circuit C_2. The inputs to this gate are from the outputs of the two memory elements FF-A and FF-B.

Therefore, $$Y = \overline{Q_A + Q_B}$$

Both the flip-flops are negative edge triggered and the clock input is applied to both the flip-flops simultaneously. Therefore, the output Y changes only after the application of clock pulse.

8.12 MEALY CIRCUIT

The generalised block diagram of a Mealy circuit has been shown in figure 8.60.

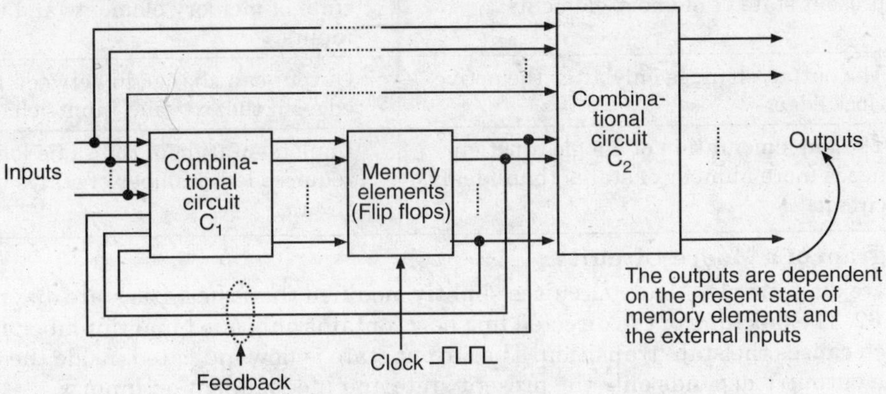

Fig. 8.60. A general form of Mealy circuit

The input combinational circuit which is also called as the next state decoder receives inputs from the outputs of memory element and from the external inputs. But, the most important point to be noted is that the output combinational circuit C_2 (also called as output decoder) receives signal from the output of memory elements as well as the external inputs. Hence, the outputs of Mealy circuit is dependent on the present state of memory elements as well as the external inputs.

> **DO YOU KNOW?**
>
> In a synchronous (parallel) counter, all of the FFs are simultaneously clocked from the input clock signal.

8.12.1 Illustrative Example of Mealy Circuit

Circuit shown in figure 8.61 is an example of Mealy circuit.

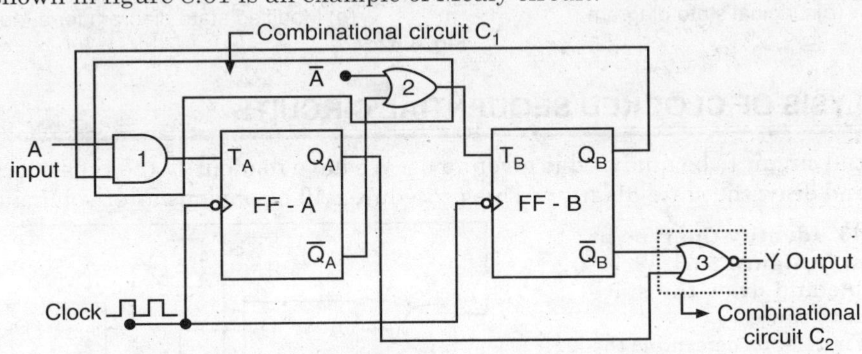

Fig. 8.61. Illustrative example of Mealy circuit

As usual, the flip-flops A and B form the memory element. Both the FFs are negative edge triggered and simultaneously clocked. Gates 1 and 2 form the input combinational circuit C_1 and gate-3 forms the output combinational circuit. Here, it may be noted that the inputs to gate-3 are Q_A, $\overline{Q_B}$ and A. This means that the output is now dependent on the output of memory elements as well as the external input A. Also note that the change in A in between the successive negative clock edges will not affect the outputs of the flip-flops. But, it will definitely affect the final output. This may produce a false output. To eliminate this problem, the external inputs must be allowed to change their state in synchronization with the active edge of clock.

8.12.2 Performance Comparison of Moore and Mealy Models

Table 8.23. Performance Comparison of Moore and Mealy models

S. No.	Moore model	Mealy model
1.	The final output depends only on the present state of memory elements.	The final output depends on the present state of memory elements and the external inputs.
2.	The output changes only after the active clock edge.	Output can change in between the clock edges if the external inputs change.
3.	The implementation of a logic function needs more number of states than Mealy circuit.	Implementation of the same logic function requires less number of states than Moore circuit.

State Diagram of a Moore Circuit

The state diagram of a Moore circuit is slightly modified than the basic state diagram as shown in figure 8.62. The labelling of the directed line now contains only one binary number corresponding to x in which causes the state transition. The output state is now indicated inside the circle. This is because the output Y depends only the present state and independent on input x.

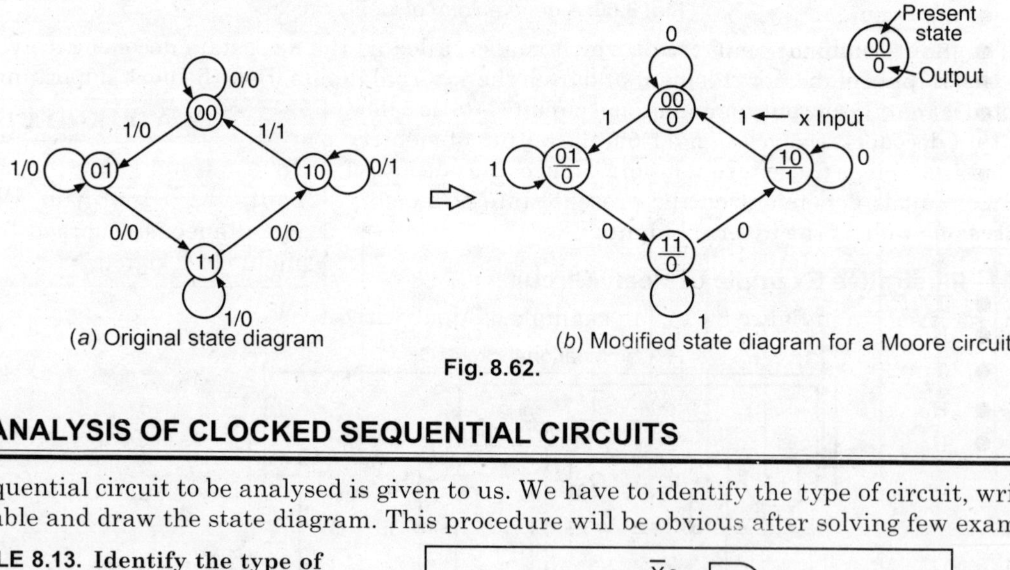

(*a*) Original state diagram (*b*) Modified state diagram for a Moore circuit

Fig. 8.62.

8.13 ANALYSIS OF CLOCKED SEQUENTIAL CIRCUITS

The sequential circuit to be analysed is given to us. We have to identify the type of circuit, write its state table and draw the state diagram. This procedure will be obvious after solving few examples.

EXAMPLE 8.13. Identify the type of circuit given in figure 8.63. Write the state table and draw the state diagram for the same.

Solution: (i) First, let us determine the type of circuit.

Since output $Y = \overline{A + \overline{B}}$, it is dependent only on the present state of the memory elements. Hence, this is a Moore circuit.

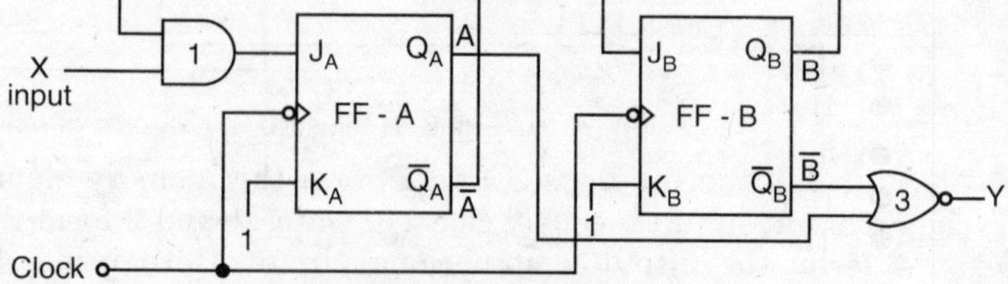

Fig. 8.63. The given circuit

(ii) Then, we write the state table.

The state table has been shown in Table 8.24.

Table 8.24. State table

Present state	Next state		Output
A B	X = 0	X = 1	Y
0 0	0 1	0 0	0
0 1	0 0	1 0	1
1 0	0 0	0 0	0
1 1	0 0	0 0	0

The present state column in the state table represents the FF outputs before application of clock pulse. The next state column represents the outputs of FF outputs after the application of clock pulse. It is dependent on the present state as well as the external input X. Output Y column values are obtained from the present state column of the state table.

Derivation of State Table

The derivation of state table is as under. Let us consider each present state and obtain the corresponding next state and Y output.

Present state AB = 00

Next state with X = 0	Next state with X = 1
● $J_A = X \,.\, B = 0 \,.\, 0 = 0$ and $K_A = 1$	● $J_A = X \,.\, B = 1.0 = 0$ and $K_A = 1$
● Hence $A_{n+1} = 0$	● Hence $A_{n+1} = 0$
● $J_B = \overline{X} \,.\, \overline{A} = 1 \,.\, 1 = 1$ and $K_B = 1$	● $J_B = \overline{X} \,.\, \overline{A} = 0 \,.\, 1 = 0$ and $K_B = 1$
● So FF-B will toggle	
● Hence $B_{n+1} = 1$	● Hence $B_{n+1} = 0$
● Hence next state = 01	● Hence next state = 00

Output Y $= \overline{A + \overline{B}} = \overline{0 + 1}$

Therefore, $Y = 0$

Present state AB = 01

Next state with X = 0	Next state with X = 1
● $J_A = X \,.\, B = 0 \,.\, 1 = 0$ and $K_A = 1$	● $J_A = X \,.\, B = 1 \,.\, 1 = 1$ and $K_A = 1$
● Hence, $A_{n+1} = 0$	● Hence, FF-A toggles
● $J_B = \overline{X} \,.\, \overline{A} = 1 \,.\, 1 = 1$ and $K_B = 1$	● Therefore, $A_{n+1} = 1$
● Hence, $B_{n+1} = 0$ as FF-B toggles	● $J_B = \overline{X} \,.\, \overline{A} = 0 \,.\, 1 = 0$ and $K_B = 1$
● Hence, next state = 00	● Hence, $B_{n+1} = 0$
	● Hence, next state = 00

Therefore, Output Y $= \overline{A + \overline{B}} = \overline{0 + 0} = 1$

Present state AB = 10

Next state with X = 0	Next state with X = 1
● $J_A = X \,.\, B = 0 \,.\, 0 = 0$ and $K_A = 1$	● $J_A = X \,.\, B = 1 \,.\, 0 = 0$ and $K_A = 1$
● Hence, $A_{n+1} = 0$	● Hence, $A_{n+1} = 0$
● $J_B = \overline{X} \,.\, \overline{A} = 1 \,.\, 0 = 0$ and $K_B = 1$	● $J_B = \overline{X} \,.\, \overline{A} = 0 \,.\, 0 = 0$ and $K_B = 1$
● Hence, $B_{n+1} = 0$	● Hence, $B_{n+1} = 0$
● Therefore, next state = 00	● Therefore, the next state is 00

Therefore, Output Y $= \overline{A + \overline{B}} = \overline{1 + 1} = 0$

Similarly, we can obtain the next state and output for the present state of 1 1 as under :

Present state : AB = 11

Next state (X = 0) : 00 next state (X = 1) = 00

Output Y = 0

(iii) Next, let us draw the state diagram.

The state diagram is drawn with the help of state table. In order to illustrate this point, the state table has been repeated in figure 8.64(a) and the corresponding state diagram is shown in figure 8.64(b).

Present state	Next state A_{n+1} B_{n+1}		Output
A B	X = 0	X = 1	Y
0 0	0 1	0 0	0
0 1	0 0	1 0	1
1 0	0 0	0 0	0
1 1	0 0	0 0	0

(a) State table

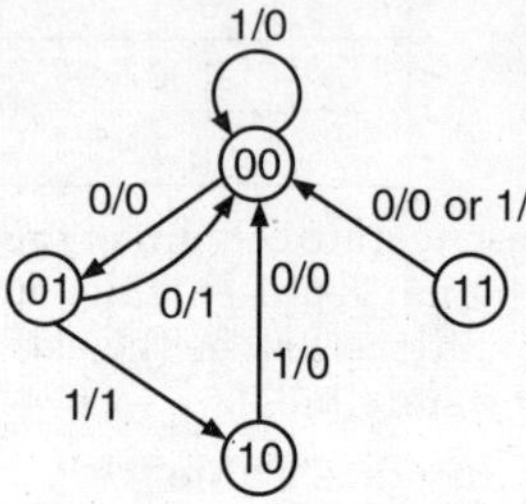

(b) State diagram

Fig. 8.64.

(iv) Finally, let us obtain the modified state diagram.

Figure 8.65 shows the modified state diagram for Moore circuit. Here, it may be noted that the binary number corresponding to Y output has been included inside the circle itself.

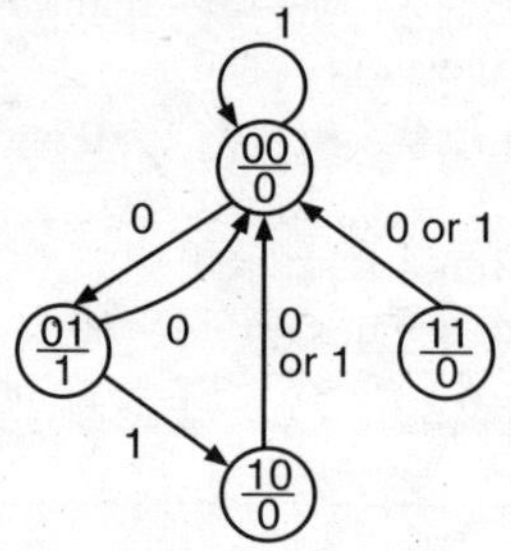

Fig. 8.65. Modified state diagram for the given Moore circuit

> **DO YOU KNOW?**
>
> The maximum clock frequency for an asynchronous counter, f_{max}, decreases as the number of bits increases. For a synchronous counter, f_{max} remains the same, regardless of the number of bits.

EXAMPLE 8.14. Identify the circuit given in figure 8.66. Write the state table and draw state diagram for the same.

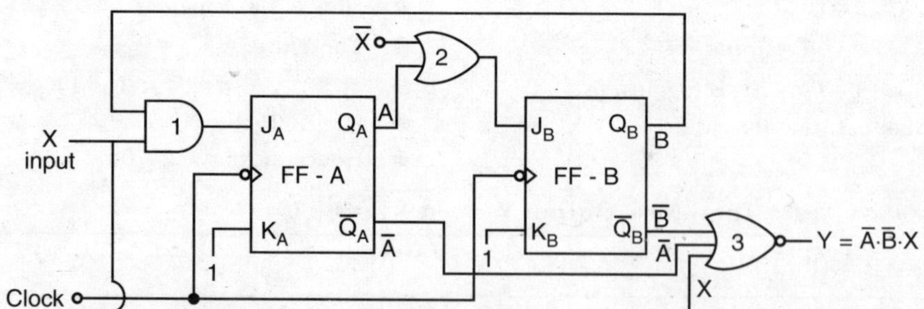

Fig. 8.66. Given circuit

Solution: (i) First, let us identify the type of circuit.

The given circuit is a Mealy circuit since the Y output depends on the present state as well as the X input.

(ii) Then, we write the state table.

> **DO YOU KNOW?**
>
> A decade counter is any MOD-10 counter. ABCD counter is a decade counter that sequences through the 10 BCD codes (0-9).

The state table for given circuit is given in Table 8.25. Note that the output Y contains two columns one each for X = 0 and X = 1. This is because in Mealy circuit, the output depends on external input as well as the present state.

$$Y = \bar{A}.\bar{B}.X \quad ...(i)$$

where $\bar{A}\,\bar{B}$ represents the present state of flip-flops.

Table 8.25. State table

Present state	Next state $A_{n+1}\,B_{n+1}$		Output $Y = \bar{A}.\bar{B}.X$	
A B	**X = 0**	**X = 1**	**X = 0**	**X = 1**
0 0	0 1	0 0	0	1
0 1	0 0	1 0	0	0
1 0	0 1	0 1	0	0
1 1	0 0	0 0	0	0

We have, $J_A = X.B$ and $J_B = \bar{X} + A$

Derivation of State Table

Let us demonstrate the procedure to obtain the next state and output corresponding to first two present states AB = 00 and 01. We can follow the same procedure for the remaining states.

Present state AB = 00

Next state for X = 0	Next state for X = 1
● $J_A = X.B = 0.0 = 0$ and $K_A = 1$ ● Hence, $A_{n+1} = 0$ ● $J_B = \bar{X} + A = 1 + 0 = 1$ and $K_A = 1$ ● Hence, FF-B will toggle ● Therefore, $B_{n+1} = 1$ ● Hence, next state = 01 **Output Y** ● For X = 0, $Y = \bar{A}\bar{B}X = 1.1.0 = 0$ ● For X = 1, $Y = \bar{A}\bar{B}X = 1.1.1 = 0$	● $J_A = X.B = 1.0 = 0$ and $K_A = 1$ ● Hence, $A_{n+1} = 0$ ● $J_B = \bar{X} + A = 0 + 0 = 0$ and $K_A = 1$ ● Hence, $B_{n+1} = 0$ ● Therefore, the next state = 00

Present state AB = 01

Next state for X = 0	Next state for X = 1
● $J_A = X.B = 0.1 = 0$ and $K_A = 1$ ● Hence, $A_{n+1} = 0$ ● $J_B = \bar{X} + A = 1 + 0 = 1$ and $K_A = 1$ ● So FF-B will toggle ● Hence, $B_{n+1} = 0$ ● Therefore, next state = 00 **Output Y** ● For X = 0, $Y = \bar{A}\bar{B}X = 1.0.0 = 0$ ● For X = 1, $Y = \bar{A}\bar{B}X = 1.0.1 = 0$	● $J_A = X.B = 1.1 = 1$ and $K_A = 1$ ● So, FF-A will toggle ● Hence, $A_{n+1} = 0$ ● $J_B = \bar{X} + A = 0 + 0 = 0$ and $K_B = 1$ ● Hence, $B_{n+1} = 0$ ● Therefore, next state = 10

Similarly, we can obtain the next state and output Y for the remaining two present states to complete the state table.

(iii) Finally, let us draw the state diagram.

The state diagram for the given Mealy circuit is shown in figure 8.67 (b). The state table has been repeated in figure 8.67 (a) for reference.

Present state	Next state A_{n+1} B_{n+1}		Output Y	
A B	X = 0	X = 1	X = 0	X = 1
0 0	0 1	0 0	0	1
0 1	0 0	1 0	0	0
1 0	0 1	0 1	0	0
1 1	0 0	0 0	0	0

(a) State table repeated

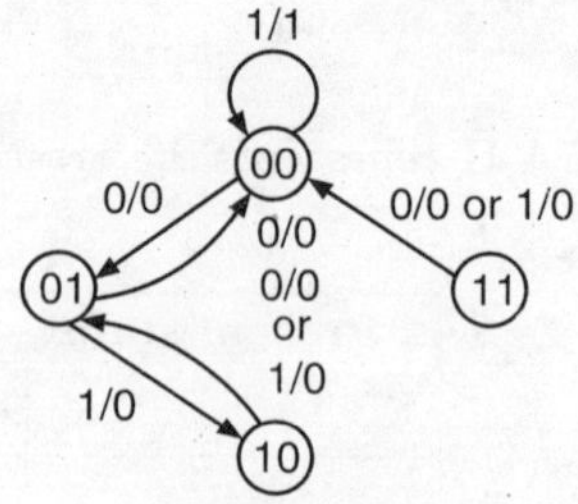

(b) State diagram

Fig. 8.67.

8.14 STATE REDUCTION AND ASSIGNMENTS

The analysis of a sequential circuit always starts from a circuit diagram and ends at writing the state table or drawing the state diagram as we did in last two examples. The design of a sequential circuit starts from a set of specifications and ends at drawing the logic diagram. In this article, let us discuss some properties of sequential circuits which are useful in reducing the number of gates and flip-flops while designing the circuit.

8.14.1 State Reduction

In the state diagram, some states can be redundant states. These redundant states are avoided by the technique called state reduction technique. The state reduction results in elimination of redundant states which ultimately results in reducing the number of flip-flops and gates to be used in a sequential circuit.

How to Identify a Redundant State ?

If two states in a state diagram are equivalent, then one of them is treated as redundant and can be eliminated. The two states are said to be equivalent if every same output and same next state is produced for every possible set of inputs. Reducing the number of states will not alter the input output conditions of the circuit.

8.14.2 Illustrative Example of State Reduction

The specifications of a sequential circuit have been given in the state diagram of figure 8.68.

The steps to be followed for state reduction are as under :

(i) First, we write the state table from the given state diagram.

(ii) Then, we identify the equivalent states.

(iii) Next, we eliminate one of the two equivalent states and draw a new state table.

(iv) In the new state table, we check for equivalent states and continue till all the redundant states are eliminated.

In the example of state reduction, only the input output sequences are important. Therefore, the present states have been denoted by a, b, c, ... g instead of their binary values.

(i) First, let us write the state table.

Table 8.26 represents the state table derived from the given state diagram.

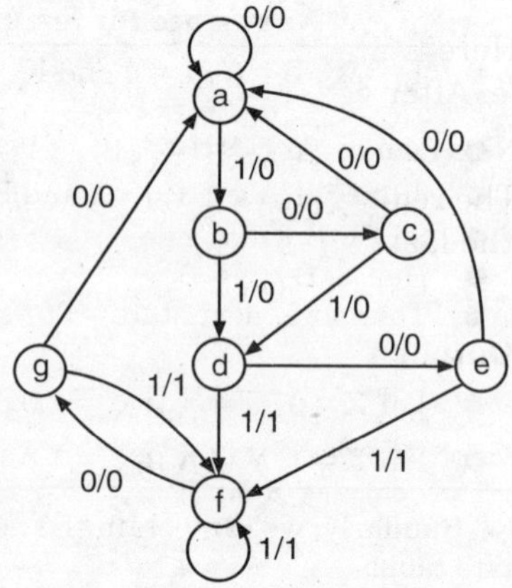

Fig. 8.68. Given state diagram

Table 8.26. State table

Present state	Next state		Output Y	
	X = 0	X = 1	X = 0	X = 1
a	a	b	0	0
b	c	d	0	0
c	a	d	0	0
d	e	f	0	1
e	a	f	0	1
f	g	f	0	1
g	a	f	0	1

States "e" and "g" — equivalent states

(ii) Then, we identify the equivalent states.

Two states are said to be equivalent if, for each member of the set of inputs, they give exactly the same output and send the circuit either to the same state or an equivalent state. We apply this principle to the state table of Table 8.26. Look at the state table for two present states which go the same next states and have same output (Y) for both the input conditions (X = 0 and X = 1). The states e and g in Table 8.26 are found to be equivalent states, because both go to the next states a and f for X = 0 and 1 respectively with Y = 0 and 1 for X = 0 and 1 respectively.

(iii) Next, we eliminate one of the equivalent states and write a new state table.

As the state e and g are equivalent, we can remove one of them. Let us remove state g, i.e., the last row containing the present state g and then replace g by e each time it appears in the next state column shown in the new state table of Table 8.27.

Table 8.27. State table

Present state	Next state		Output Y	
	X = 0	X = 1	X = 0	X = 1
a	a	b	0	0
b	c	d	0	0
c	a	d	0	0
d	e	f	0	1
e	a	f	0	1
f	e	f	0	1

After replacing g by e the states d and *f* have become equivalent

Here, state g has been eliminated, g is replaced by e. Note that in the last row, we have replaced g by e. After doing so, the states d and f become equivalent states. Hence, we eliminate f.

(iv) Next, we replace f by d and prepare a new state table.

The reduced state table is as follows. Note that each f in the next state column is replaced by d and the last row for present state f has been removed completely.

Table 8.28. Reduced state table

Present state	Next state		Output Y	
	X = 0	X = 1	X = 0	X = 1
a	a	b	0	0
b	c	d	0	0
c	a	d	0	0
d	e	d	0	1
e	a	d	0	1

f is replaced by d

Here, state *f* has been eliminated. In Table 8.28, all the 5 states are different from each other. Hence, the process of state reduction is complete.

(v) Finally, we draw the state diagram.

The state diagram for reduced state table is shown in figure 8.69.

How does reduction in number of FFs take place due to state reduction ?

The reduction of number of flip-flops in a sequential circuit is called as *state reduction problem.* State reduction algorithms will reduce the number of states in a state table while keeping the external input-output requirement unchanged. As m flip-flops produce 2^m states, a reduction in number of states may (or may not) result in a reduction in number of flip-flops.

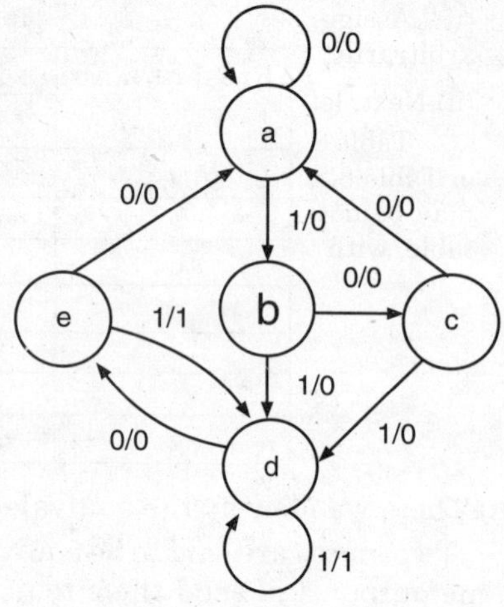

Fig. 8.69. Reduced state diagram

8.15 STATE ASSIGNMENT

In last article, we have discussed the state reduction technique to minimize the number of flip-flops. Now, let us discuss another technique to reduce the cost of the combinational circuit. Such a technique is called as state assignment technique. It plays an important role in minimizing the combinational gates. The state assignment procedure is based on the methods which assign binary values to the states in such a way that it reduces the cost of the combinational circuit that drives the flip-flops. The concept of state assignment will be obvious after solving next example.

EXAMPLE 8.15. For the given state table, obtain the reduced state table using state assignment procedure.

Table 8.29. Given state table

Present state	Next state		Output Y	
	X = 0	X = 1	X = 0	X = 1
a	a	b	0	0
b	c	d	0	0
c	a	d	0	0
d	e	d	0	1
e	a	d	0	1

(i) First, let us assign various possible binary values to different states.

The assignment of various possible binary values to different states is demonstrated in Table 8.30. Here, it may be noted that the binary values of the state are not important as long as their sequence maintains the proper input output relationship. Hence, assignment of any binary number to any state is satisfactory if we assign a unique number to each state. Table 8.30 shows three examples of possible binary assignment.

Table 8.30. Three possible binary state assignments

State	Assignment 1	Assignment 2	Assignment 3
a	001	000	000
b	010	010	100
c	011	011	010
d	100	101	101
e	101	111	011

Assignment-1 in Table 8.30 is a straight forward binary assignment, whereas the remaining two are done arbitrarily.

(ii) Next, let us obtain the reduced state table.

Table 8.31 represents the reduced state table with binary assignment 1 substituted in place of letter symbols of Table 8.29. Such a state table with binary assignment in sometimes called as a **transition table.** Here, it may be noted that if we substitute binary assignment 2 or 3, then we shall get a completely new transition table, with the same input output relationship.

Table 8.31. Reduced state table or transition table

Present state	Next state		Output	
	X = 0	X = 1	X = 0	X = 1
001 a	001	010	0	0
010 b	011	100	0	0
011 c	001	100	0	0
100 d	101	100	0	1
101 e	001	100	0	1

This type of state table is used to derive the combinational part of the sequential circuit. The complexity of the combinational circuit obtained depends on the binary state assignment chosen.

Therefore various procedures have been tried to select a particular binary assignment from so many possible ones. The most important criterion is that the selected binary assignment should result in a simpler combinational circuit. But, it is surprising to know that there is no state assignment procedure that will guarantee a minimal-cost combinational circuit.

8.16 DESIGN PROCEDURE FOR CLOCKED SEQUENTIAL CIRCUITS

Design of a clocked sequential circuits will start from the set of given specifications and it will end with drawing a logic diagram. The stepwise design procedure is as under :

(i) A state diagram or timing diagram or some other information is given, which describes the behaviour of the circuit that is to be designed.

(ii) Then, we obtain the state table.

(iii) The number of states can be reduced by state reduction methods if the sequential circuit can be characterized by input-output relationships, independent of number of states.

(iv) Then, we assign binary values to each state if the state table obtained in step 2 or 3 contain letter symbols.

(v) Next, we determine the number of flip-flops required and assign a letter symbol to each one.

(vi) Next, we decide the type of FF to be used.

(vii) Then, we derive the circuit excitation table and output table from the state table.

(viii) Next, we obtain the expressions for the circuit output and flip-flop inputs.

(ix) After that, we draw the logic diagram.

The design procedure of clocked sequential circuits will be obvious by solving the following example.

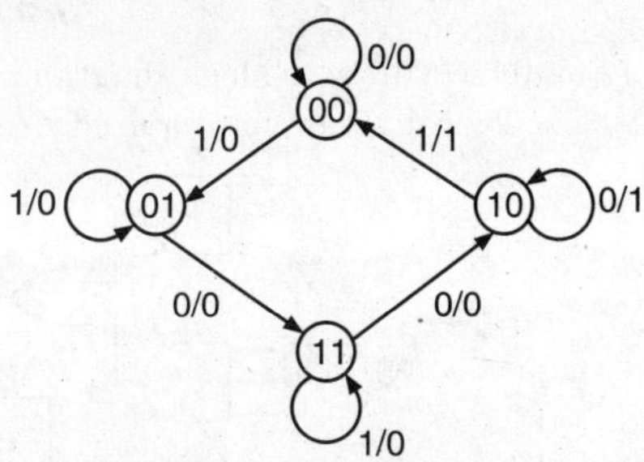

Fig. 8.70. Given state diagram

EXAMPLE 8.16. For the state diagram given in figure 8.70, draw the clocked sequential circuit using T flip-flops.

Solution. (*i*) First, let us write the state table.

The state table is written from the given state diagram as shown in Table 8.32.

Table 8.32. State table

Present state		Next state		Output	
		X = 0	X = 1	X = 0	X = 1
A	B	A_{n+1} B_{n+1}	A_{n+1} B_{n+1}		
0	0	0 0	0 1	0	0
0	1	1 1	0 1	0	0
1	0	1 0	0 0	1	1
1	1	1 0	1 1	0	0

(ii) Next, let us find the number of flip-flops.

As observed from the state table, there are no equivalent states. Hence, there won't be any reduction in the state diagram. The circuit goes through four states. Therefore, we need to use 2 flip-flops.

(iii) Next, we write the circuit excitation table.

The circuit excitation table is shown in Table 8.33. The type of FF used is T type FF.

Table 8.33. Circuit excitation table

Present state		Input	Next state		FF inputs		Output
A	B	X	A_{n+1}	B_{n+1}	T_A	T_B	Y
0	0	0	0	0	0	0	0
0	0	1	0	1	0	1	0
0	1	0	1	1	1	0	0
0	1	1	0	1	0	0	0
1	0	0	1	1	0	0	1
1	0	1	0	1	1	0	1
1	1	0	1	1	0	1	0
1	1	1	1	1	0	0	0

(iv) Next, we make K-maps and carry out simplification.

Figure 8.71 shows the K-maps and corresponding simplified expressions for T_A T_B, i.e., input of the two T flip-flops and the Y output.

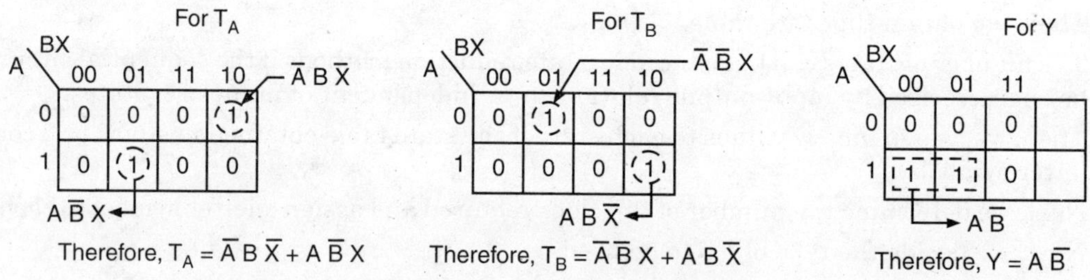

Fig. 8.71. K-maps and simplification

(v) Finally, we draw the logic diagram.

The logic diagram for required clocked sequential circuit has been shown in figure 8.72.

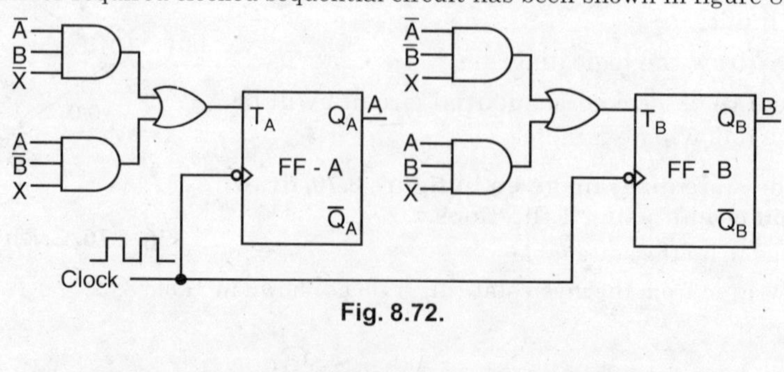

Fig. 8.72.

8.17 DESIGN USING UNUSED STATES

Sometimes a sequential circuit may use less than the available maximum number of states. While designing such a sequential circuit, we may consider the unused states as don't care conditions. This is called as design with unused state. Let us solve the following example to get the idea about designing with unused states.

EXAMPLE 8.17. Implement the following state diagram using D flip-flops.

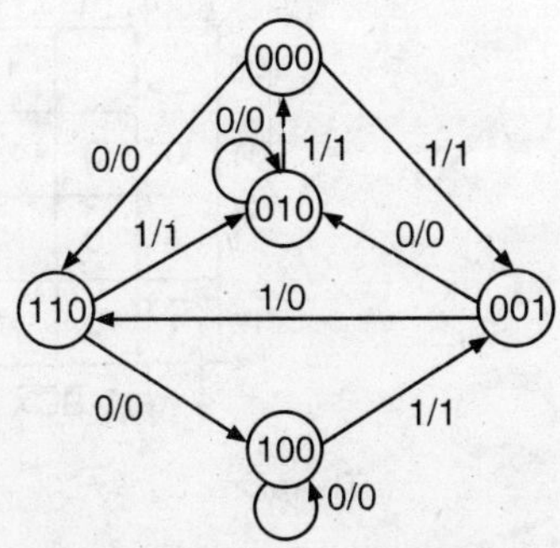

Fig. 8.73. Given state diagram

Solution: From the given state table, it is evident that 3 flip-flops are required to be used. The unused states are 011, 101 and 111.

(i) First, we write the circuit excitation table.

The required circuit excitation table is shown in Table 8.34.

This state table has been derived from the given state diagram.

Table 8.34. Circuit excitation table

Present state			Input	Next state			Output	Flip-flop inputs		
A	B	C	X	A_{n+1}	B_{n+1}	C_{n+1}	Y	D_A	D_B	D_C
0	0	0	0	1	1	0	0	1	1	0
0	0	0	1	0	0	1	1	0	0	1
0	0	1	0	0	1	0	0	0	1	0
0	0	1	1	1	1	0	0	1	1	0
0	1	0	0	0	1	0	0	0	1	0
0	1	0	1	0	0	0	1	0	0	0
1	0	0	0	1	0	0	0	1	0	0
1	0	0	1	0	0	1	1	0	0	1
1	1	0	0	1	0	0	0	1	0	0
1	1	0	1	0	1	0	1	0	1	0

Here, it may be noted that the entries in the column D_A have been made by taking into consideration the entries in the columns A and A_{n+1}. Similarly, the entries in the columns of D_B and D_C have been made by considering B, B_{n+1} and C, C_{n+1} respectively. The excitation table for a general D FF has been shown in Table 8.35.

Table 8.35. Excitation table for a D FF

Present state	Next state	D
0	0	0
0	1	1
1	0	0
1	1	1

(ii) Next, let us make K-maps and carryout simplifications.

The K-maps for D_A, D_B and D_C and output Y are shown in figure 8.74.

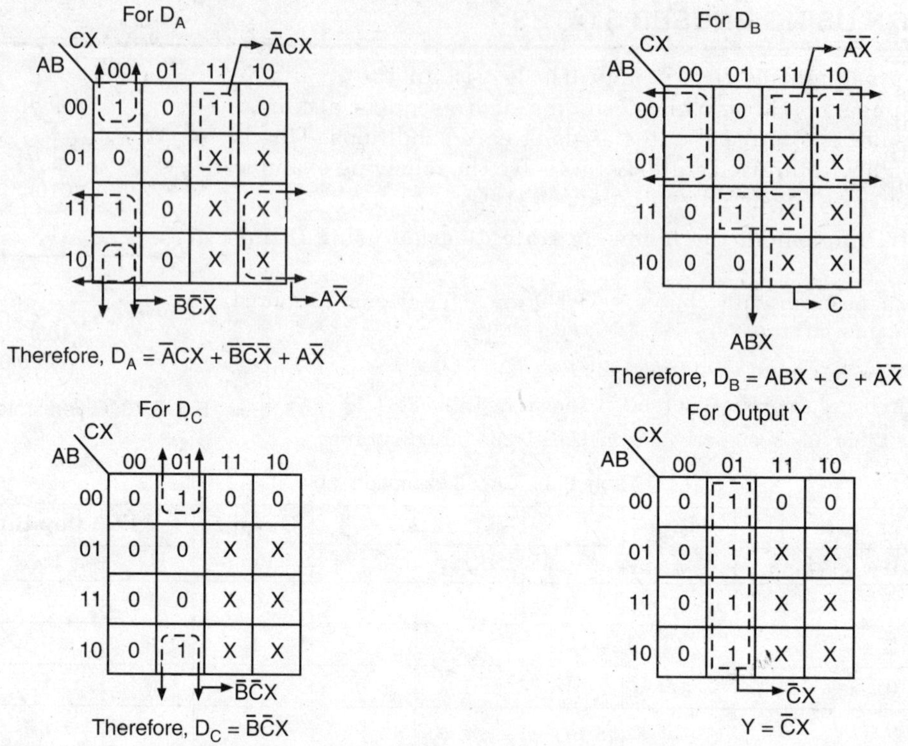

Fig. 8.74. K-maps and simplification

The simplified expression are also derived. Note that these K-maps have been written from the circuit excitation table shown in Table 8.35. The inputs are A, B, C and X. But the excitation Table 8.35 does not contain all the possible combinations of A, B, C and X. Hence, the missing combinations are treated as don't care conditions and denoted by X in the K-maps. They correspond to 011, 101 and 111.

(iii) Finally, let us make the logic diagram.

The logic diagram of the required sequential circuit is shown in figure 8.75.

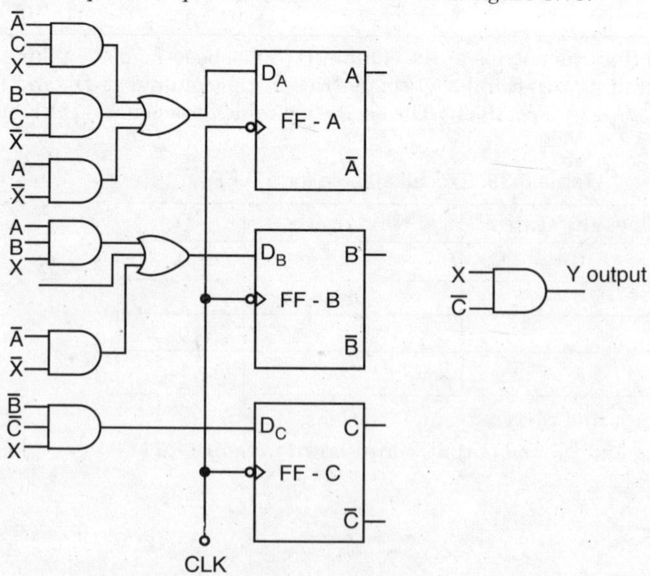

Fig. 8.75. Logic diagram of the required clocked sequential circuit

8.18 LOCKOUT CONDITION

(PTU, Jalandhar, Sem. Exam., 2008–09)

A counter is supposed to follow the sequence of only the desired states as shown in figure 8.76 (a). If it enters into an unused or unwanted state, then it is expected to return back to a desired state. Instead, if the next state of an unwanted state is again an unwanted state as shown in figure 8.76, then the counter is said to be in the lockout conditions.

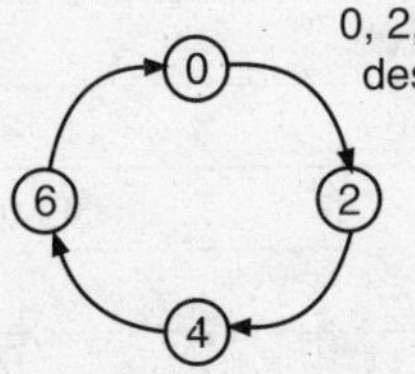

(a) State diagram showing the desired sequence

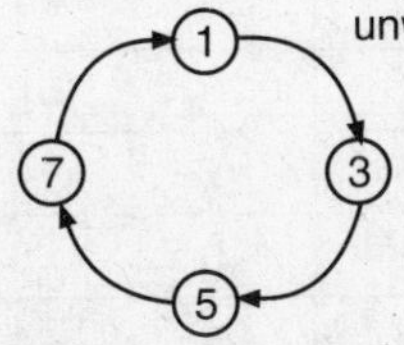

(b) State diagram showing the lockout condition

Fig. 8.76.

8.18.1 Bushless Circuit

The sequential circuit which enters into the lockout condition is called as the bushless circuit.

How to Avoid Lockout Situation ?

In order to avoid the lockout condition, we have to use some additional logic circuit to force next state of the counter to be the initial state, if it enters into an undesired state.

Hence, the next state of each unwanted state must be the initial state as shown in figure 8.77 (a).

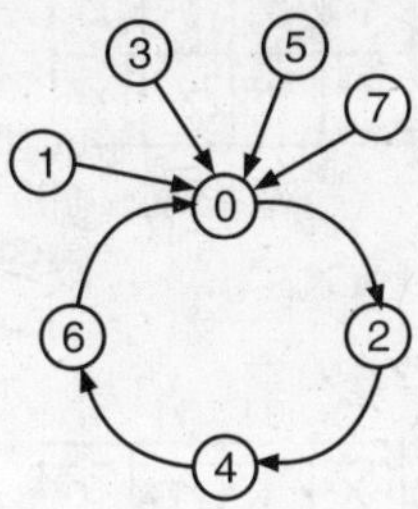

(a)

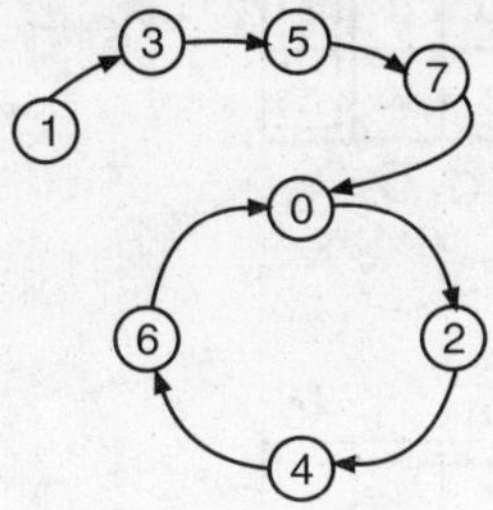

(b)

Fig. 8.77. Ways to avoid lockout condition

However it is not essential to terminate all the undesired states into the initial state as shown in figure 8.77 (a). It will be sufficient to terminate only one unused state into the initial state shown in figure 8.77 (b). If the circuits enter into say state 3 then its next states will be 5 and 8. As 7 has n state of 0 which is the desired state, the lockout condition is avoided.

EXAMPLE 8.18. By using suitable flip-flops, design a counter to go through the following states 0 – 2 – 4 – 6 – 7 – 0 ... Avoid the lockout condition.

Solution: (i) First, let us draw the state diagram.

The state diagram has been shown in figure 8.78. It shows that the next state of all the unused state (1, 3 and 5) is forced to be the initial state (0). This is essential in order to avoid the lockout condition.

(ii) Next, let us determine the number of flip-flops and type of flip-flop.

Since the highest state is 7, i.e., 111, we require to use three flip-flops. Let us use J-K flip-flops.

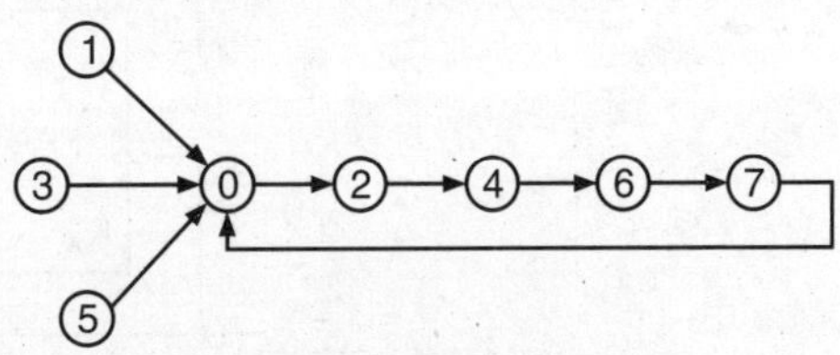

Fig. 8.78. State diagram of desired counter

(iii) Then, we write the state table.

Table 8.36 shows the state table for the required counter.

Table 8.36.

Present state			Next state			Flip-flop inputs					
Q_C	Q_B	Q_A	Q_{C+1}	Q_{B+1}	Q_{A+1}	J_C	K_C	J_B	K_B	J_A	K_A
0	0	0	0	1	0	0	×	1	×	0	×
0	0	1	0	0	0	0	×	0	×	×	1
0	1	0	1	0	0	1	×	×	1	0	×
0	1	1	0	0	0	0	×	×	1	×	1
1	0	0	1	1	0	×	0	1	×	0	×
1	0	1	0	0	0	×	1	0	×	×	1
1	1	0	1	1	1	×	0	×	0	1	×
1	1	1	0	0	0	×	1	×	1	×	1

(iv) Next, we make K-maps and and carryout simplification.

Figure 8.79 illustrates the K-maps for all the FF inputs and the corresponding simplified expressions.

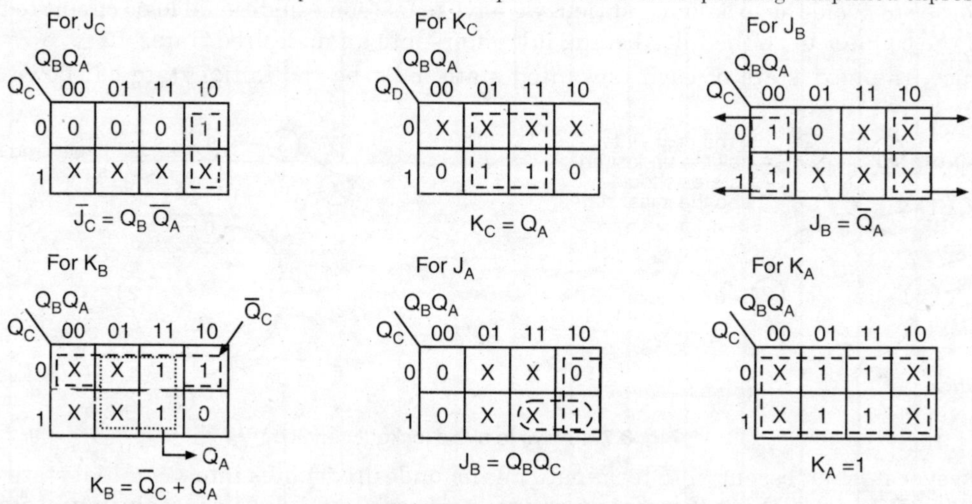

Fig. 8.79. K-maps and simplification

(v) Finally, we draw the logic diagram.

The logic diagram for the required counter has been shown in figure 8.80.

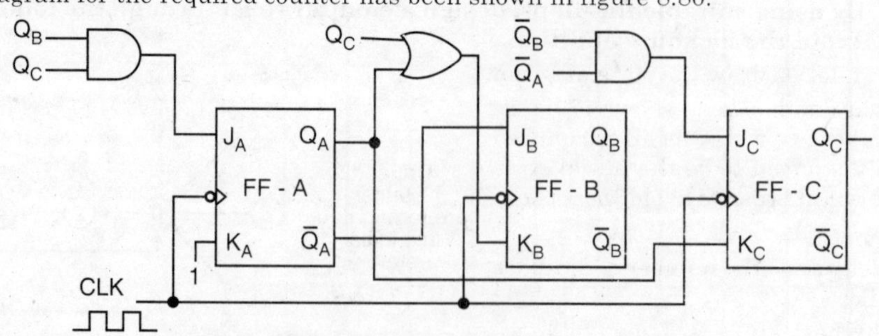

Fig. 8.80. Logic circuit of the required counter

8.19 SEQUENCE GENERATOR

A sequence generator is a sequential circuit which generates a prescribed sequence at its output. The output sequence is in synchronization with the clock input. It is possible to design a sequence generator using counters or using the shift registers.

8.19.1 Sequence Generator Using Counters

Figure 8.81 shows the general block diagram of a sequence generator using counter.

The next state decoder is a combinational circuit. The input to it are obtained from the flip-flop outputs and its outputs are applied to the inputs of the flip-flops.

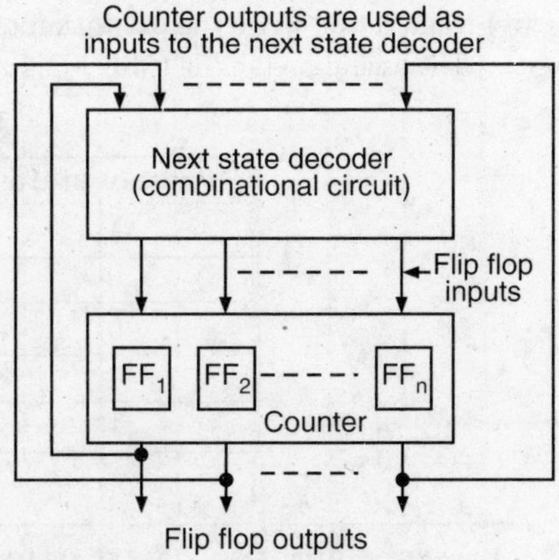

Fig. 8.81. A general block diagram of a sequence generator

Deciding the Number of Flip-flops

We will be given the required sequence of 1's and 0's that is to be generated such as 1011011....

Based on the given sequence we have to decide the number of flip-flops as under :

(i) First, we count the number of 1's and 0's in the given sequence.

(ii) Then, we choose the higher number of the two. Let this number be N.

(iii) Then, the number of FFs (n) may be calculated as under :

$$N \leq 2^{n-1}$$

Let us apply this to the given sequence, i.e., 1011011..........

Number of 1's = 5, number of 0's = 2. Hence, we select higher one of them, i.e., 5. Therefore, N = 5.

But, $N = 2^{n-1}$. Hence, $5 \leq 2^{n-1}$

or $n = 4$

Therefore, four FFs will be required

EXAMPLE 8.19. For the following sequence, design a sequence generator using JK flip-flops

1 0 0 1 0 0 1

Solution: (i) First, we decide the number of flip-flops.

Number of 1's = 3 and number of 0's = 4. Hence, we select higher one of them, i.e., 4. Therefore, N = 4. Hence $N = 2^{n-1}$, so $4 \leq 2^{n-1}$. Therefore, number of flip-flops. n = 3.

(ii) Next, we write the state table.

We assume that the required sequence is obtained at the output of flip-flop C. The state assignment for the other outputs is shown in Table 8.37.

Table 8.37.

Flip flop outputs			State
C	B	A	
1	0	0	4
0	0	1	1
0	1	0	2
1	1	1	7
0	0	0	0
0	1	1	3
1	1	0	6

These entries (B, A) have been made in an arbitrary manner

This (C) is the given sequence

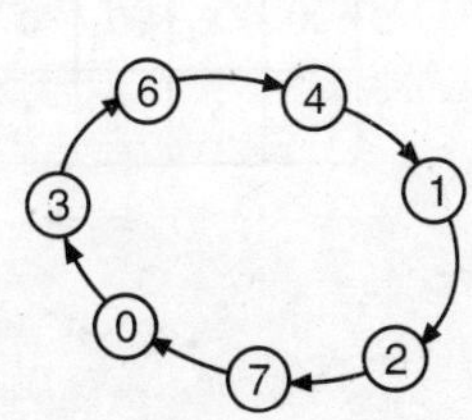

Fig. 8.82. State diagram

There is only one unused state, i.e., 5. We assume that the next state of 5 is don't care (xxx).

(iii) Next, let us draw the state diagram.

From the state table, we draw the state diagram as shown in figure 8.82.

(iv) Next, let us write the excitation table.

The excitation table of a J-K FF is given in Table 8.38 and the circuit excitation table is shown in Table 8.39.

Table 8.38. Excitation table of J-K FF

Present state	Next state	FF inputs	
Q_n	Q_{n+1}	J	K
0	0	0	×
0	1	1	×
1	0	×	1
1	1	×	0

Table 8.39. Circuit excitation table

Present state			Next state			Flip-flop inputs					
C	B	A	C_{n+1}	B_{n+1}	A_{n+1}	J_C	K_C	J_B	K_B	J_A	K_A
0	0	0	0	1	1	0	×	1	×	1	×
0	0	1	0	1	0	0	×	1	×	×	1
0	1	0	1	1	1	1	×	×	0	1	×
0	1	1	1	1	0	1	×	×	0	×	1
1	0	0	0	0	1	×	1	0	×	1	×
1	0	1	×	×	×	×	×	×	×	×	×
1	1	0	1	0	0	×	0	×	1	0	×
1	1	1	0	0	0	×	1	×	1	×	1

← The next state and FF inputs are don't care for the unused state

(v) Then, we obtain K-maps and simplified Boolean expressions.

Figure 8.83 shows the K-maps and corresponding simplified boolean expressions for all the FF inputs.

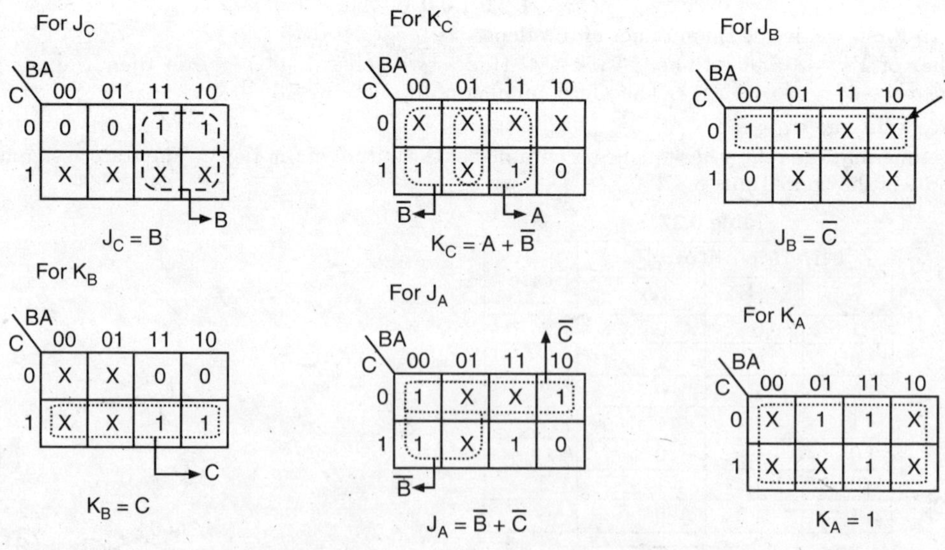

Fig. 8.83. K-map and simplification

(vi) Finally, let us draw the logic diagram.

The logic diagram has been shown in figure 8.84.

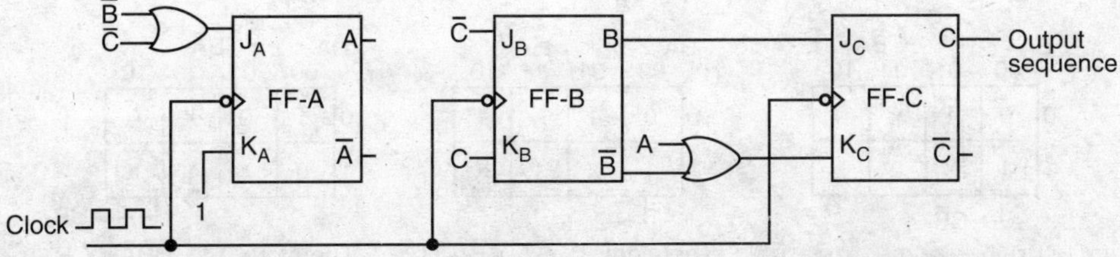

Fig. 8.84. Logic diagram of the required sequence generator

Note. Here, it may be noted that the required sequence is obtained at the C output. All the flip-flops are clocked simultaneously.

EXAMPLE 8.20. Design a sequence generator using T FFs.

0 → 1 → 7 → 4 → 2 (→ 0)

Fig. 8.85. Given state diagram

Draw the bush diagram to check for the lockout conditions.

Solution: We shall assume that the next state of each unused state is a don't care condition, while writing the circuit excitation table.

(i) First, let us draw the state diagram.

The state diagram has been shown in figure 8.86.

(ii) Next, let us determine number of FFs.

Since the highest state is 7, i.e., 111, we have to use three T flip-flops.

(iii) Next, we write the excitation table.

Table 8.40 shows the excitation table for a T FF.

Table 8.41 shows the circuit excitation table.

Table 8.40. Excitation table for a T FF

Present state	Next state	T input
0	0	0
0	1	1
1	0	1
1	1	0

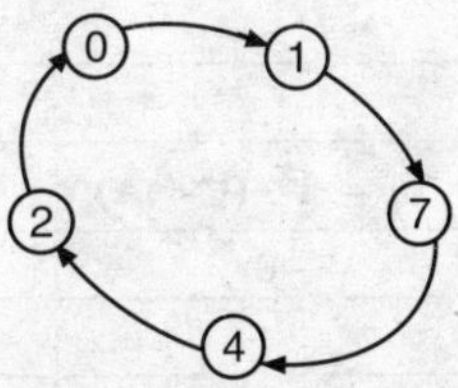

Fig. 8.86. State diagram

Table 8.41. Circuit excitation table

Present state			Next state			Flip-flop inputs		
C	B	A	C_{n+1}	B_{n+1}	A_{n+1}	T_C	T_B	T_A
0	0	0	0	0	1	0	0	1
0	0	1	1	1	1	1	1	0
0	1	0	0	0	0	0	1	0
→ 0	1	1	×	×	×	×	×	× ←
1	0	0	0	1	0	1	1	0
→ 1	0	1	×	×	×	×	×	×
→ 1	1	0	×	×	×	×	×	× ←
1	1	1	1	0	0	0	1	1

FF inputs are don't care for the unused inputs

→ Next state of each unused state is don't care. We can assume don't care conditions to be logic 1's.

(iv) Now, let us find K-maps and carry out simplifications.

Let us consider figure 8.87 for the K-maps and simplified boolean expressions.

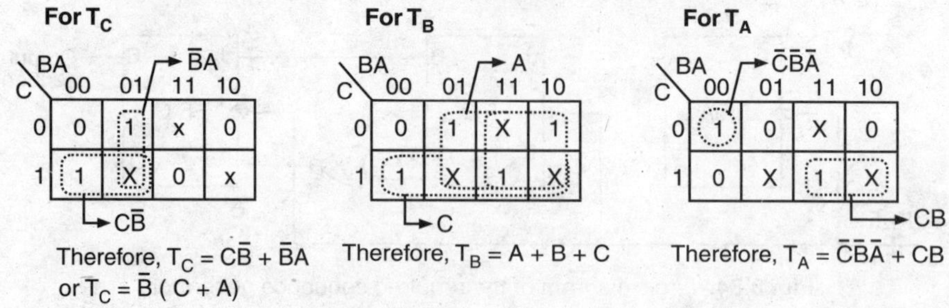

Fig. 8.87. K-maps and simplifications

(v) Next, let us make logic diagram.

Figure 8.88 shows the required logic diagram.

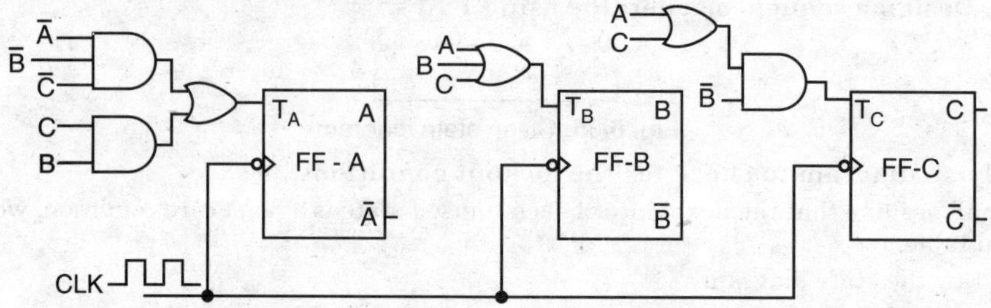

Fig. 8.88. Logic diagram of the required counter

(vi) Next, we examine the counter action for the unused states.

The unused states are 3, 5 and 6.

For state 3

C = 0, B = 1, A = 1

Table 8.42.

Present state			FF inputs			Next state		
C	B	A	$T_C = \bar{B}\,(C + A)$	$T_B = A + B + C$	$T_A = \bar{C}\,\bar{B}\,\bar{A} + CB$	C_{n+1}	B_{n+1}	A_{n+1}
0	1	1	0	1	0	0	0	1
1	0	1	1	1	0	0	1	1
1	1	0	0	1	1	1	0	1

→ Valid state

Thus, if counter goes to state 3 it returns to state 1 which is a valid state.

For state 5

Let us consider Table 8.42. The next state of 5 is 3 which altimately terminates into 1.

For state 6

Let us consider Table 8.42. The next state of 6 is 5 and that of 5 is 3 which ultimately gets terminated into 1 which is a valid state.

(vii) Finally, we draw the bush diagram.

The bush diagram has been shown in figure 8.89.

From the bush diagram, it is evident that lockout condition shall be avoided.

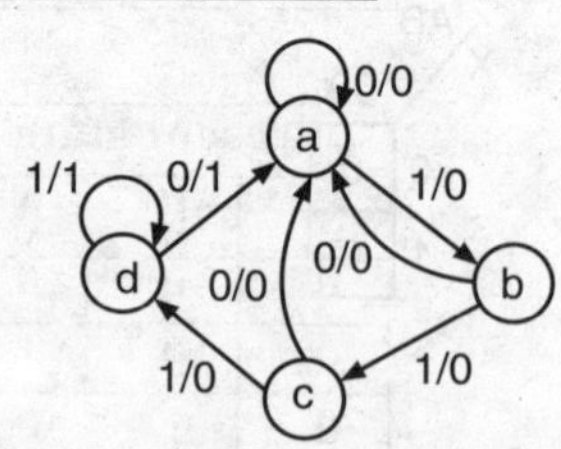

Fig. 8.89. Complete state diagram

FEW SOLVED EXAMPLES

EXAMPLE 8.21. For the state diagram shown in figure 8.90, obtain the state table and design the circuit using minimum number of J-K flip-flops.

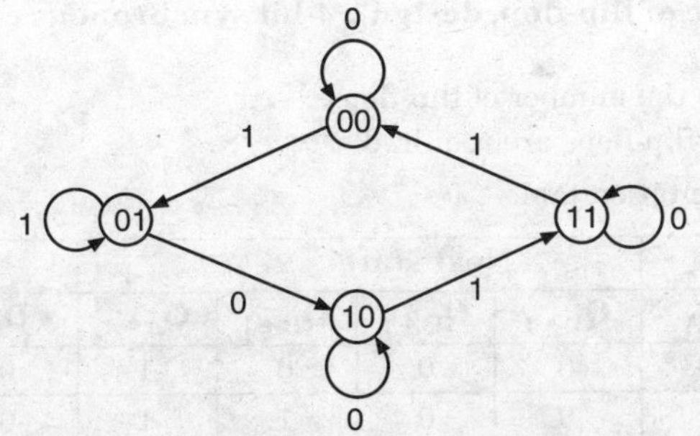

Fig. 8.90.

Solution : (i) First, we write the state table.

Present state		Next state			
		X = 0		X = 1	
A	B	A_{+1}	B_{+1}	A_{+1}	B_{+1}
0	0	0	0	0	1
0	1	1	0	0	1
1	0	1	0	1	1
1	1	1	1	0	0

(ii) Next, we decide the number of flip-flops.

The circuit goes through four state. Hence, number of flip-flops used are 2.

(iii) Next, we write the circuit excitation table

Present state		Input	Next state		Flip-flop inputs			
A	B	X	A_{+1}	B_{+1}	J_B	K_B	J_A	K_A
0	0	0	0	0	0	X	0	X
0	1	0	1	0	1	X	X	1
1	0	0	1	0	X	0	0	X
1	1	0	1	1	X	0	X	0
0	0	1	0	1	0	X	1	X
0	1	1	0	1	0	X	X	0
1	0	1	1	1	X	0	1	X
1	1	1	1	0	X	0	X	1

(iv) Then, we obtain K-maps and carry out simplification

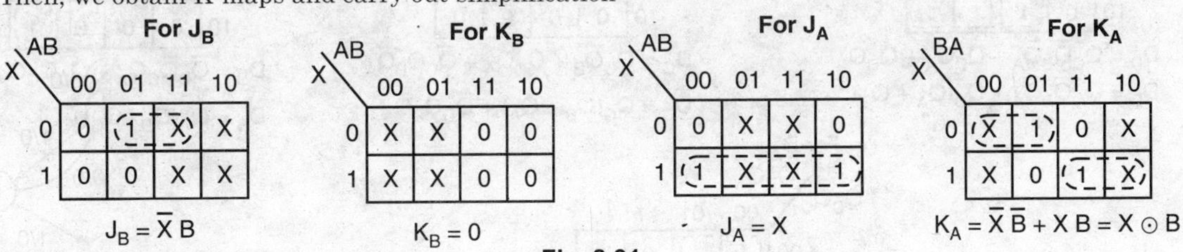

Fig. 8.91.

(v) Finally, we make the logic diagram.

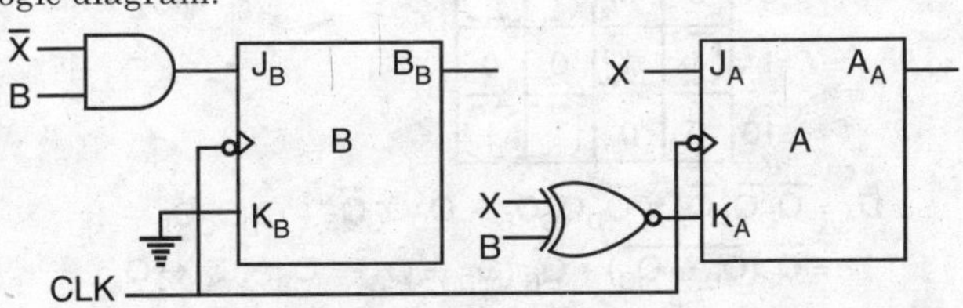

Fig. 8.92. Logic diagram

EXAMPLE 8.22. Using any type of flip-flop, design a 4-bit synchronous counter that upcounts in gray code.

Solution: (i) First, we determine the number of flip-flops.

To design a 4-bit counter, 4 flip-flops are required.

(ii) Next, we write the circuit excitation table.

Present state				Next state				Flip-flop inputs			
Q_D	Q_C	Q_B	Q_A	Q_{D+1}	Q_{C+1}	Q_{B+1}	Q_{A+1}	D_D	D_C	D_B	D_A
0	0	0	0	0	0	0	1	0	0	0	1
0	0	0	1	0	0	1	1	0	0	1	1
0	0	1	1	0	0	1	0	0	0	1	0
0	0	1	0	0	1	1	0	0	1	1	0
0	1	1	0	0	1	1	1	0	1	1	1
0	1	1	1	0	1	0	1	0	1	0	1
0	1	0	1	0	1	0	0	0	1	0	0
0	1	0	0	1	1	0	0	1	1	0	0
1	1	0	0	1	1	0	1	1	1	0	1
1	1	0	1	1	1	1	1	1	1	1	1
1	1	1	1	1	1	1	0	1	1	1	0
1	1	1	0	1	0	1	0	1	0	1	0
1	0	1	0	1	0	1	1	1	0	1	1
1	0	1	1	1	0	0	1	1	0	0	1
1	0	0	1	1	0	0	0	1	0	0	0
1	0	0	0	0	0	0	0	0	0	0	0

(iii) Next, we make K-maps and carry out simplification.

For D_D

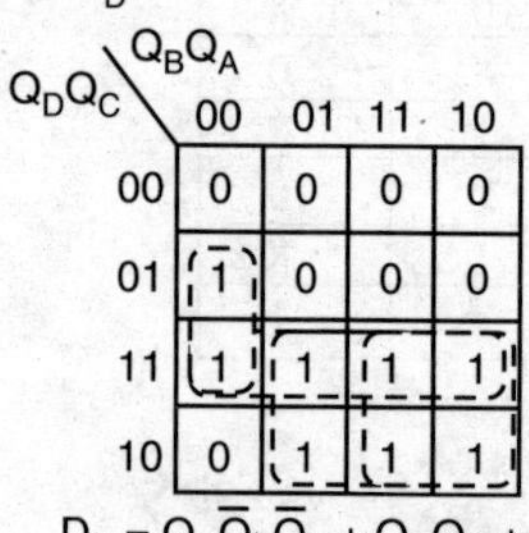

$D_D = Q_C\overline{Q}_B\overline{Q}_A + Q_DQ_A + Q_DQ_B$

$D_D = Q_C\overline{Q}_B\overline{Q}_A + Q_D(Q_A + Q_B)$

For D_C

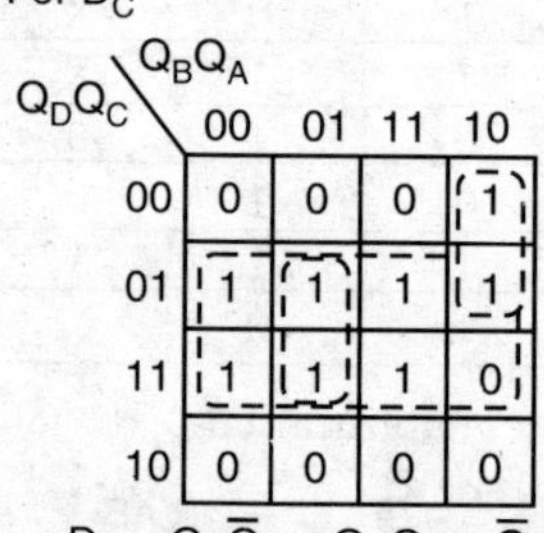

$D_C = Q_C\overline{Q}_B + Q_CQ_A + \overline{Q}_DQ_B\overline{Q}_A$

$D_C = Q_C(\overline{Q}_B + Q_A) + \overline{Q}_DQ_B\overline{Q}_A$

For D_B

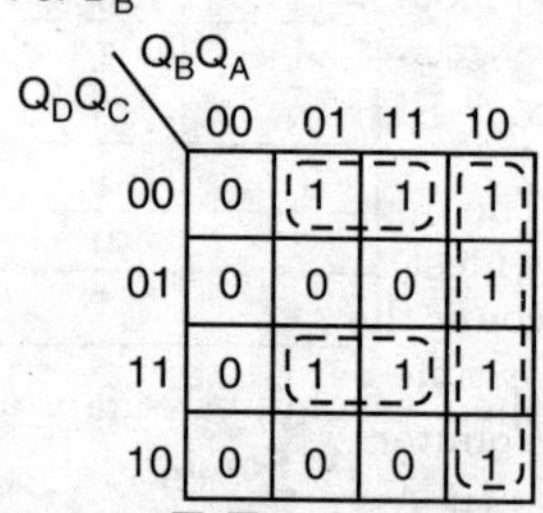

$D_B = \overline{Q}_D\overline{Q}_CQ_A + Q_DQ_CQ_A$

$D_B = Q_A(Q_D \odot Q_C)$

For D_A

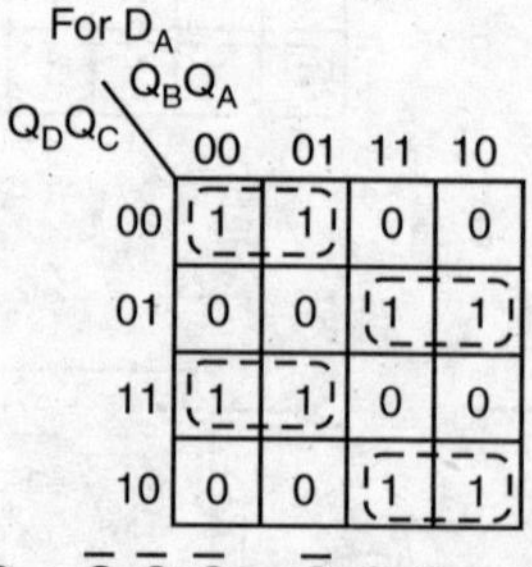

$$D_A = \overline{Q}_D\overline{Q}_C\overline{Q}_B + \overline{Q}_DQ_CQ_B + Q_DQ_C\overline{Q}_B + Q_D\overline{Q}_CQ_B$$

$$= \overline{Q}_B(\overline{Q_D \oplus Q_C}) + Q_B\,(Q_D \oplus Q_C) = Q_B\odot[Q_D \oplus Q_C]$$

Fig. 8.93.

(iv) Finally, let us draw the logic diagram.

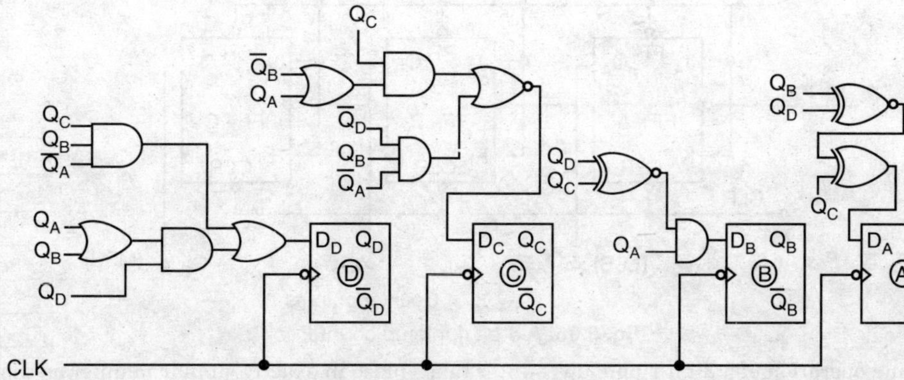

Fig. 8.94. Logic diagram

GLOSSARY

1. **BCD counter :** A 4-bit counter that commonly counts from 0000 to 1001 and resets to 0000.
2. **Ripple counter :** Simple binary counter where the changing state of the LSB flip-flop triggers the clock input of the next, etc. A time delay results from the rippling of the count from LSB to MSB.
3. **Ring counter :** A recirculating shift register which is loaded with a pattern of 1s (such as a single 1) which continue to circulate around in the circle or repeated clock pulses.
4. **Sequential logic :** A logic circuit whose logic states depend on asynchronous and synchronous inputs. Exhibit memory characteristics.
5. **Serial data :** The transmission of data one bit at a time.
6. **Signal :** The information transmitted within, to, and from electronics circuits.
7. **Three-state output :** Conditions of outputs on certain digital ICs, or high impedance. Also commonly referred to as Tristate (trademark of National Semiconductor).
8. **Toggle :** To change the opposite logic state. A pulse that changes logic circuits state to opposite condition. A mode of operation in a flip-flop where the output goes to opposite state on each successive clock pulse.

SUMMARY

- The digital circuit used for counting pulses is known as counter. It is a sequential circuit. Counter is the widest application of flip-flops. It is a group of flip-flops with a clock signal applied. Basically, counters count the number of clock pulses. Hence, with some modifications, it can be used for measuring frequency or time period.
- Counters are basically of following two types :
  (i) Asynchronous or ripple counters.
  (ii) Synchronous counters.
- For these counters, the external clock signal is applied to one flip-flop and then the output of preceding flip-flop is connected to the clock of next flip-flop.
- In synchronous counters, all the flip-flops receive the external clock pulse simultaneously. Ring counter and Johnson counter are the examples of synchronous counters.
- All the counters discussed so far have counted upwards from zero. Hence, they can be called as **up counters.** However, the counters which can count in the **downward** direction, i.e., from the maximum count to zero are called **down counters.**
- We have designed the up counters and the down counters separately. However, in practice, both these modes are generally combined together to obtain an UP/DOWN counter. A mode control (M) input is also provided to select either up or down mode. A combinational circuit is required to be designed and used between each pair of flip-flops in order to achieve the up/down operation.
- We can apply all the basic concepts which were introduced for the 2-bit ripple up counter to the 3-bit ripple up counter. Figure 8.95 shows the logic diagram of a 3-bit ripple up counter. Since it is a 3-bit counter, we require to use 3-flip-flops.

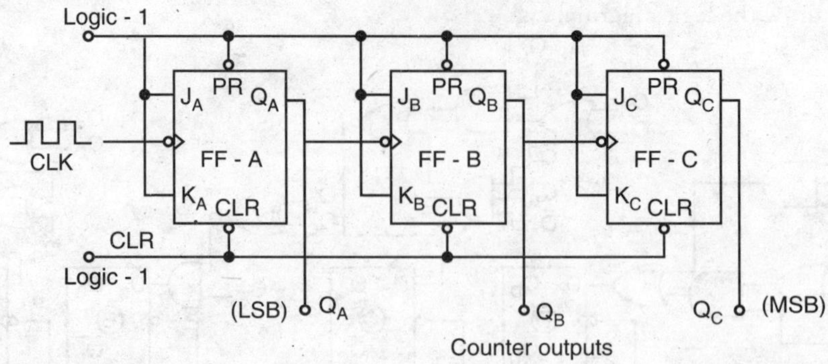

Fig. 8.95. A 3-bit ripple up counter

The working operation of a 3-bit ripple up counter takes place in exactly similar manner as that of a 2-bit counter.

- The two major problems associated with the ripple counters may be listed as under :
  (i) Generation of unwanted short duration pulses called glitch.
  (ii) Propagation delay.
- If the clock pulses are applied to all the flip-flops in a counter simultaneously, then such a counter is called as **synchronous counter**.
- Some of the applications of counters may be listed as under :
  1. In digital clock
  2. In the frequency counters
  3. In time measurement
  4. In digital voltmeters
  5. In the counter type A to D converter
  6. In the digital triangular wave generator
  7. In the frequency divider circuits.
- The operating principle of a synchronous up/down counter is same as that of the up/down ripple counters. However, the design procedure and realization are different. The mode control (M) is used for selecting the mode of operation. The procedural steps to be followed for the design may be listed as under.
- The present state is stored in the memory elements. These present states are applied to the combinational circuit alongwith the external inputs. Therefore, the outputs and the next state of the sequential circuit are decided by the external inputs and the present state.
- The sequential circuits may be classified into following two categories :
  (i) Synchronous sequential circuits
  (ii) Asynchronous sequential circuits
- In the synchronous sequential circuits, the contents of memory elements can be changed only at the rising or falling edges of clock signal.
- In the asynchronous sequential circuit, the contents of memory elements can be changed at any instant of time.
  (i) First, we write the circuit excitation table.
  (ii) Then, we write K-maps and obtain the simplified expressions.
  (iii) Next, we draw the logic diagram.
- There are two models developed for representing the synchronous sequential circuits as under :
  (i) Moore circuit
  (ii) Mealy circuit.

  In the synchronous sequential circuits, the combinational circuit consists of a combination of logic gates whereas the flip flops are used as memory elements. The synchronous sequential circuit is called as a **Moore circuit** if the output depends only on the present state of flip-flops. The circuit is called as a **Mealy circuit** if the output is dependent on the present state of flip-flops and the external inputs.
- The reduction of number of flip-flops in a sequential circuit is called as state reduction problem. State reduction algorithms will reduce the number of states in a state table while keeping the external input-output requirement unchanged. As m flip-flops produce 2^m states, a reduction in number of states may (or may not) result in a reduction in number of flip-flops.

SHORT QUESTIONS WITH ANSWER

Q.1. How can you classify counters.

Ans. Counters are basically of following two types :

(i) Asynchronous or ripple counters.

(ii) Synchronous counters.

(i) Asynchronous or Ripple Counters

For these counters, the external clock signal is applied to one flip-flop and then the output of preceding flip-flop is connected to the clock of next flip-flop.

(ii) Synchronous Counters

In synchronous counters, all the flip-flops receive the external clock pulse simultaneously. Ring counter and Johnson counter are the examples of synchronous counters.

Q.2. Define asynchronous down counter.

Ans. All the counters discussed so far have counted upwards from zero. Hence, they can be called as **up counters.** However, the counters which can count in the **downward** direction, i.e., from the maximum count to zero are called **down counters.**

Q.3. Write the drawbacks of ripple counters.

Ans. (i) Every flip-flop has its own propagation delay. In ripple counter, the output of the previous FF is used as clock for the next FF.

(ii) Hence the propagation delay goes on accumulating. For a 3-bit ripple counter, the propagation delay of the first FF gets added to that of the second FF to decide the transition time for the third stage.

(iii) This accumulated time delay is the main problem with the ripple counters, because the propagation delay goes on increasing with increase in number of flip-flops.

(iv) This will put a limitation on the maximum clock frequency.

(v) The frequency f of a clock pulse for reliable working operation of the counter is given by,

$$f \leq \frac{1}{n(t_d) + T_s}$$

where n = Number of flip-flops

T_s = Width of strobe pulse (if used)

and t_d = Propagation delay of one flip-flop.

Q.4. Define introduction to up/down counters.

Ans. We have designed the up counters and the down counters separately. However, in practice, both these modes are generally combined together to obtain an UP/DOWN counter. A mode control (M) input is also provided to select either up or down mode. A combinational circuit is required to be designed and used between each pair of flip-flops in order to achieve the up/down operation.

Q.5. Write the problems associated with ripple counters.

Ans. The two major problems associated with the ripple counters may be listed as under :

(i) Generation of unwanted short duration pulses called glitch.

(ii) Propagation delay.

Q.6. Define synchronous counters.

Ans. If the clock pulses are applied to all the flip-flops in a counter simultaneously, then such a counter is called as **synchronous counter**.

Q.7. Write the salient Features of a Synchronous Counter.

Ans. (i) As all the FFs are clocked at the same instant, the propagation delay problem is reduced to a great extent.

(ii) The total propagation delay between the instant of application of clock edge and the instant at which the MSB output changes is equal to sum of propagation delay of one FF and that of one AND gate. Therefore, total propagation delay = t_d of one FF + t_d of one gate.

(iii) This is much less than the propagation delay of an asynchronous counter.

Q.8. Carryout Performance Comparison of Synchronous and Asynchronous

Ans. Performance comparison of synchronous and asynchronous counters is given in following table.

Table : Performance comparison of synchronous and asynchronous counters

S. No.	Parameter of Comparison	Asynchronous Counter	Synchronous Counter
1.	Circuit complexity	Logic circuit is simple.	With increase in number of states, the logic circuit becomes complicated.
2.	Connection pattern	Output of the preceding FF is connected to clock of the next FF.	There is no connection between output of preceding FF and CLK of next one.
3.	Clock input	All the FFs are not clocked simultaneously.	All FFs receive clock signal simultaneously.
4.	Propagation delay	P.D. = n × (t_d) where n is number of FFs and t_d is p.d. per FF.	P.D. = $(t_d)_{FF} + (t_d)_{gate}$. It is much shorter than that of asynchronous counter.
5.	Maximum frequency of operation.	Low because of the long propagation delay.	High due to shorter propagation delay.

Q.9. Write Other Applications of Counters

Ans. Some of the applications of counters may be listed as under :

1. In digital clock
2. In the frequency counters
3. In time measurement
4. In digital voltmeters
5. In the counter type A to D converter
6. In the digital triangular wave generator
7. In the frequency divider circuits.

Q.10. Explain about Up/Down Synchronous Counter.

Ans. The operating principle of a synchronous up/down counter is same as that of the up/down ripple counters. However, the design procedure and realization are different. The mode control (M) is used for selecting the mode of operation. The procedural steps to be followed for the design may be listed as under:

(i) First, we write the circuit excitation table.
(ii) Then, we write K-maps and obtain the simplified expressions.
(iii) Next, we draw the logic diagram.

Q.11. Explain Classification of Sequential Circuits.

Ans. The sequential circuits may be classified into following two categories :

(i) Synchronous sequential circuits
(ii) Asynchronous sequential circuits

(i) Synchronous Sequential Circuits

In the synchronous sequential circuits, the contents of memory elements can be changed only at the rising or falling edges of clock signal.

(ii) Asynchronous Sequential Circuits

In the asynchronous sequential circuit, the contents of memory elements can be changed at any instant of time.

Q.12. Explain Models for Synchronous Sequential Circuits.

Ans. There are two models developed for representing the synchronous sequential circuits as under :

(i) Moore circuit, (ii) Mealy circuit.

In the synchronous sequential circuits, the combinational circuit consists of a combination of logic gates whereas the flip-flops are used as memory elements. The synchronous sequential circuit is called as a **Moore circuit** if the output depends only on the present state of flip-flops. The circuit is called as a **Mealy circuit** if the output is dependent on the present state of flip-flops and the external inputs.

Q.13. What is meant by a counter?

Ans. Counter is a digital circuit which is used to count pulses and give the output in the binary form.

Q.14. What is difference between synchronous and asynchronous counters?

Ans. In a synchronous counter all the flip-flops get clocked simultaneously where as in an asynchronous counter the output of one flip-flop drives another flip-flop.

Q.15. Write about Moore circuit

Ans. Figure 8.96 shows the general form of Moore circuit.

Here, it may be noted that the most important thing for a Moore circuit that its output depends only on the present state of the memory elements. The external inputs do not influence the outputs. The combinational circuit C_1 on the input side is sometimes called as the next state decoder. The external inputs and the feedback signals from the memory elements are applied to this combinational circuit. The outputs of the memory elements are applied to another combinational circuit C_2 which is also called as output decoder.

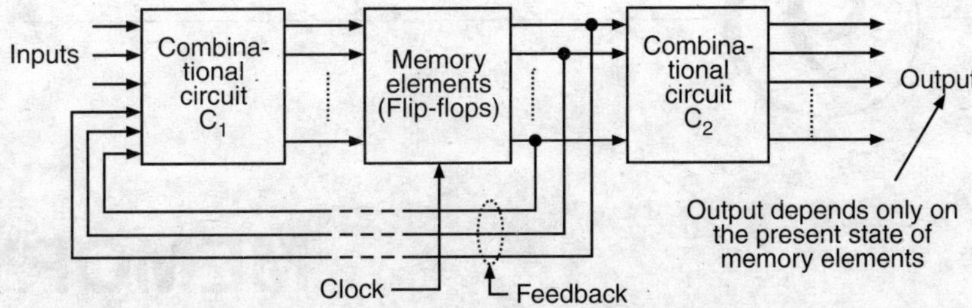

Fig. 8.96. A general form of Moore circuit

Q.16. What is a ripple counter ?

Ans. When the output of one flip-flop drives another flip-flop, the counter is called a ripple counter.

Q.17. Why is a parallel counter faster than ripple counter?

Ans. In a ripple counter the output of one flip-flop drives another flip-flop while in a parallel counter all the flip-flops get clocked simultaneously. So parallel counter is faster than ripple counter.

Q.18. What is a ring counter?

Ans. In a ring counter, the Q output of the flip-flop and the D input of the successive flip-flop are connected in such a way that it forms a ring.

REVIEW QUESTIONS

1. What is counter ? What are its types ?
2. Explain the terms synchronous and asynchronous.
3. What is the difference between synchronous and asynchronous counter ?
4. What are the merits and demerits of the synchronous counter over the asynchronous counter ?
5. Explain the working of 4-bit ripple counter (asynchronous counter) using T flip-flop with suitable circuit diagram and timing diagram.
6. Write a note on MOD-8 synchronous down counter using T flip-flop.
7. Write a note on MOD-8 synchronous up- counter using T flip-flop.
8. What do you mean by 'ring counter'? Explain it with truth table and timing diagram.
9. Explain with truth table and waveforms a 4-bit Johnson counter.
10. State the procedure for designing MOD counter from N-bit ripple counter.
11. Draw the circuit for MOD-12 counter. Explain the same with neat waveforms.
12. Sketch pin-out of IC 7490.
13. What is the similarity between a decade counter and a binary counter ? Explain in brief.
14. What are the applications of counter ?
15. How is a counter different from a register ?
16. Write the count sequence of three bit binary down counter.
17. Explain with diagram 3-bit ring counter using D Flip-flop.
18. Draw the divide by 7 asynchronous up counter using T flip-flop. Write truth table Draw timing diagram.
19. How does a counter work as frequency divider ?
20. What is synchronous counter ?
21. Define modulus of counter. State the state of MOD 5 counter.
22. Design a MOD-5 ripple counter using J-K FF.
23. Explain with neat diagram a 3-bit ripple counter using T flip-flops.

❑❑❑

9 MEMORY DEVICES

Syllabus

Memory Devices: Classification of memories, RAM organization, Write operation, Read operation, Memory cycle, Static RAM Cell-Bipolar, RAM cell, MOSFET RAM cell, Dynamic RAM cell, ROM organization, PROM, EPROM, EEPROM, Field Programmable Gate Arrays (FPGA).

9.1 INTRODUCTION

As a matter of fact, a system which processes various kinds digital data requires a facility to store the unprocessed, partially processed and completely processed data. A sub-system of such a digital processing system which can store all the above mentioned data is known as **memory.** In earlier times, the memory was of magnetic type. But now-a-days, **semiconductor memories** are available of various types and size.

9.1.1 Salient Features of Semiconductor Memory

The salient features of semiconductor memories may be listed as under :

(i) Low cost
(ii) Small size
(iii) Better reliability
(iv) High speed
(v) It is quite easy to expand memory size.

In present chapter, let us discuss the principle of operation and also limitations of various semiconductor memory.

9.2 MEMORY ORGANIZATION AND OPERATION

We already know that a flip-flop is equivalent to a single bit memory cell. Because of this fact, the basic element of a semiconductor memory is a flip-flop.

Chapter Outline

- Introduction
- Memory Organization and Operation
- Classification and Characteristics of Memories
- Classification based on Principle of Operation
- Classification based on Physical Characteristics
- Classification based on Mode of Access
- Classification based on Fabrication Technology
- Internal Organization of RAM
- Random access memory (RAM)
- Static RAM (SRAM)
- Dynamic RAM (DRAM) or Dynamic Memory Element
- Memory Expansion
- Read only memory (ROM)
- EEPROM (E^2 PROM)
- NVRAM
- Flash Memory
- Introduction to plds
- Generic Array Logic Devices
- Programming of PLDs
- Field Programmable Gate Array
- Sequential Programmable Devices

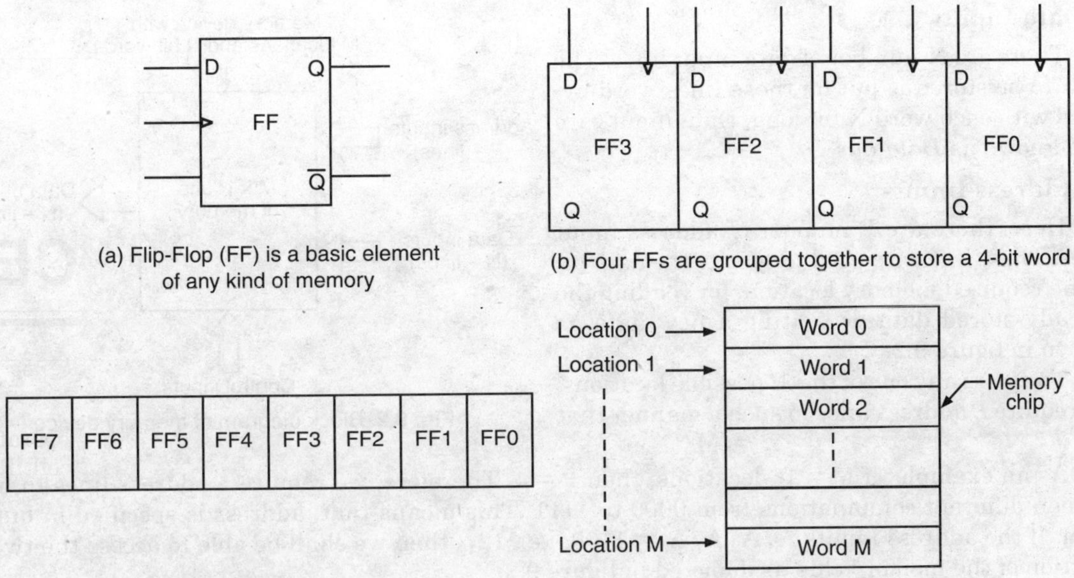

(a) Flip-Flop (FF) is a basic element of any kind of memory

(b) Four FFs are grouped together to store a 4-bit word

(c) Eight FFs are grouped together to store an 8-bit word

(d) Memory consists of many locations. Each location stores one word

Fig. 9.1.

Further, each flip-flop is capable of storing one binary bit, i.e., 0 or 1 only as shown in figure 9.1 (a). Therefore, to store a 4-bit word, we shall require four flip-flops. To store an 8-bit word, eight flip-flops are required to be used and so on as shown in figure 9.1 (b) and 9.1 (c). In addition, since, we have to store a number of 4 bit, 8 bit or 16 bit words in a memory, therefore, there are a number of locations in a memory with each location storing a word of required length as shown in figure 9.1 (d).

> **DO YOU KNOW?**
>
> All memory devices store binary logic levels (1s and 0s) in an array structure. The size of each binary word (number of bits) that is stored varies depending on the memory device. These binary values are referred to as data.

9.2.1 Memory Size

The number of locations and number of bits per word vary from memory to memory. As an example, if a particular memory chip is capable of storing M words with each word having N bits in it, then the size of the memory shall be M × N. Hence, if the size of a memory chip is specified as 16 × 4 then it means that this memory consists of 16 locations and each location can store one 4-bit word. The commonly used values of number of words per chip are 64, 128, 256, 512, 1024, 2048, 4096 etc. Also, the common word size are 1, 4 and 8. Now, the question arise what to be done if we require to store 16 bit words ? The answer is that memories with higher number of words and/or large word size can be formed by cascading a number of chips of smaller size.

9.2.2 Block Diagram of a Memory Device **(PTU, Jalandhar, Sem. Exam. 2007-08)(10 marks)**

1. Types of input

Figure 9.2 shows the block diagram of a memory device. There are following three types of inputs to an M × N memory device :

(i) Address input lines

(ii) Data input lines

(iii) Control inputs.

Also, there are N number of data output lines.

2. Data Inputs

There are N-number of data input lines. The data to be stored is put on these lines word-by-word with each word N-bit long. Data inputs are also known as *Data bus*.

3. Address Inputs

Here, there are P number of address input lines. These lines are used to specify the *address* of the required memory location, for reading the already stored data or writing a new data as shown in figure 9.2.

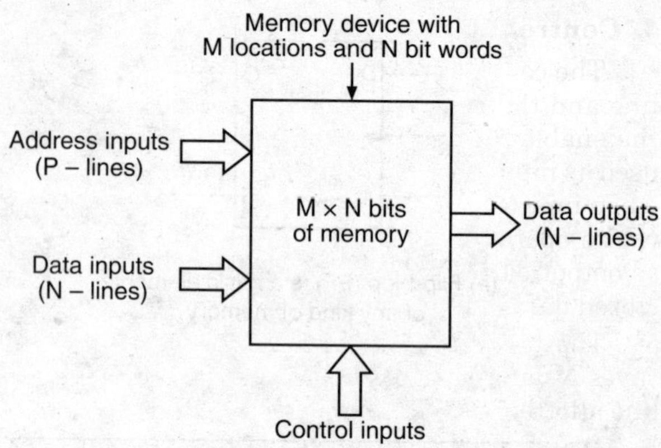

Fig. 9.2. Block diagram of memory device

To access any one of the *M* possible locations, we require *P* address lines in such a manner that

$$2^P = M$$

As an example, if M = 16 locations, then P = 4. Therefore, we require 4 address lines to have sixteen different combinations from 0000 to 1111. This means that, address is specified in binary form. If the address inputs $A_3\ A_2\ A_1\ A_0 = 1100$, i.e., 12_{10} then we shall be able to access the twelth location of the memory chip as depicted in figure 9.3.

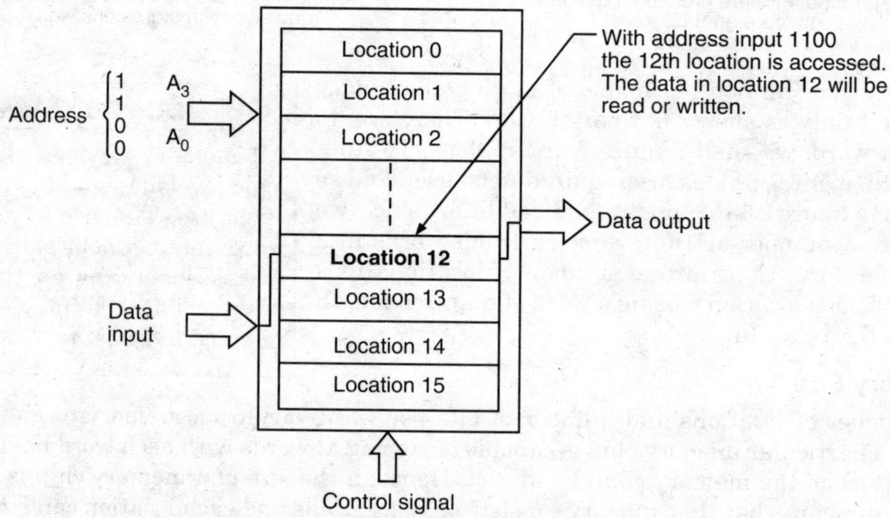

Fig. 9.3. Selection of a location using the address lines.

It is quite interesting to note that the P address lines are known as address inputs or address bus. In practice, the address input is applied to a P to M decoder. Depending upon the binary number adjusted on the P lines, the decoder outputs select the desired memory location.

4. Data Output Lines

As a matter of fact, the data available in the selected memory location can be read on the data output lines. The number of data lines is N, i.e., equal to the number of bits per word. The output data lines are also known as output data bus. Input and output data buses are unidirectional. This means that the data flows only in one direction. In most of memory chips, the same set of data lines is used for data input as well as data output. Such a data bus is known as bi-directional data bus and it is represented with arrows on both sides as shown in figure 9.4.

5. Control Lines

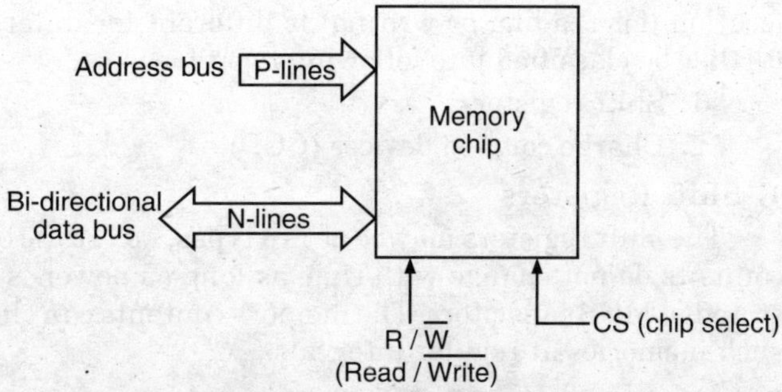

Fig. 9.4. Memory chip with bi-directional data bus.

The control lines include the read/write line and the chip select line (which acts as the enable input). The bi-directional bus is used as input data bus for some specific time when input data is to be loaded into memory (write operation). Also, it finds application as output data bus for specific time when the stored data is to be read (read operation).

The bi-directional data bus and hence saves *N* data lines. There is only one control line allotted to the read and write operation. It is denoted by $R/\overline{W}$ line in figure 9.4. If this line is at logic 1 level, then *reading* operation takes place whereas for logic 0 on this line, the *write* operation takes place. With $R/\overline{W} = 1$, the data bus acts as an output data bus and with $R/\overline{W} = 0$, it acts as the input data bus.

6. Input Chip Enable

The chip select input is an active high input. When CS = 1, then only the chip is enabled and reading or writing will take place.

7. Power Supply Lines

In addition to the address, control and data lines, there are two more pins required for the power supply and ground.

DO YOU KNOW?

The place (location) in the memory device where any data value is stored is identified by another binary number referred to as an address. Each memory location has a unique address.

9.3 CLASSIFICATION AND CHARACTERISTICS OF MEMORIES

The memory devices may be classified on the basis of various parameters. The parameters used for basis of classification may be listed as under :

(i) Operating principle
(ii) Physical characteristics
(iii) Mode of access
(iv) Technology used for fabrication process.

9.4 CLASSIFICATION BASED ON OPERATING PRINCIPLE

The classification of memories based on the principle of operation is shown in figure 9.5.

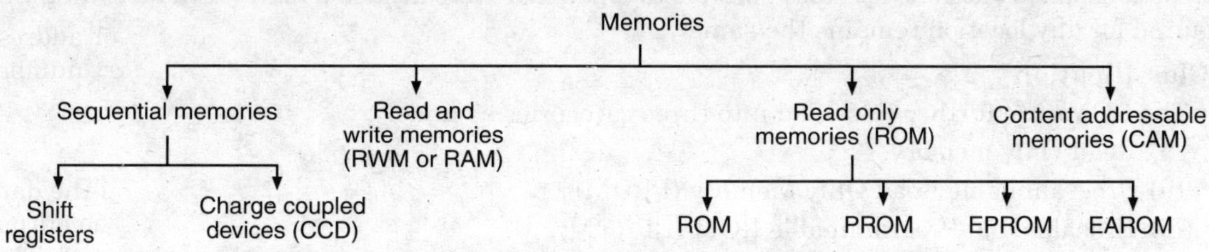

Fig. 9.5. Memory classification based on principle of operation.

9.4.1 Sequential Memories (Sequential Access)

The examples of sequential memory are magnetic tape audio/video cassette. In the sequential memories, the memory locations are organized in a sequence (one after the other). The reading/writing from such memories is a sequential process. Hence, the time required to access a memory location (for reading or writing) is different for different locations. The sequential memories may further be classified into following two types :

1. Shift registers
2. Charge coupled devices (CCD)

1. Shift Registers

The shift registers may be of two types, i.e., static or dynamic. In the static memory, the memory contents do not change with time as long as power is ON. In the dynamic memory, information is stored in MOS capacitors. The memory contents can change with time. Then, it is essential to refresh such memories at regular intervals.

Salient Features of Dynamic Memories

Few salient features of dynamic memories may be described as under :

(i) They are simpler than the static memories
(ii) They are less expensive
(iii) Require less power for operation
(iv) They have high packaging densities.

Due to all these salient features, the dynamic memories are widely used in digital systems. The only drawback of dynamic memories is the additional circuitry for refreshing.

2. Charge Coupled Devices (CCD)

The CCD are manufactured using MOS technology. The salient features of CCD are high density and low cost.

9.4.2 Random Access Memory (RWM or RAM)

Random access memory is also known as read write memory (RWM). The memory locations in this type of memory are organized in such a manner that the access time required for accessing any location remains the same. Actually, this is an advantage of using RAM over the sequential memories. RAMs also may be further classified into two types namely the static RAM and dynamic RAM. RAMs may be fabricated using either bipolar technology or unipolar technology.

9.4.3 Read Only Memories (ROM)

1. Basic Concepts

These memories are designed only for reading the information which is already stored on them. The user cannot write any new information on them. These are similar to the prerecorder cassettes. A manufacturer or someone else can write ROMs, but the writing process is much more complicated as compared to that of a RAM. Also, ROM is used to store fixed informations such as look up tables, instructions, fixed data etc. It is also possible to organize ROMs in such a manner that reading time required for any location remains the same.

2. Classification

The ROMs are further classified into three categories as under :

(i) Read-Only memory.
(ii) Programmable Read Only Memory (PROM).
(iii) Erasable and Programmable ROM (EPROM).

(i) Read-Only Memory

These memories in true sense are the read only memories, because they are programmed at the time of manufacturing. The programming is formulated as per the requirements of the user and it is the last process in manufacturing of ROMs. Such ROMs are known as *custom programmed* or *mask programmed.* We cannot change the programmed data after fabrication. These ROMs are quite costly and hence are suitable only for bulk use (number of chips in millions).

(ii) Programmable Read Only Memories (PROM)

A user can program these ROMs using a PROM programmer. A PROM can be programmed only once after its fabrication. After programming, the contents will become permanently fixed in the ROM. The PROMs are manufactured without any data stored into them (i.e., they are blank).

(iii) Erasable and Programmable ROM

As the name suggests, these ROMs can be erased and programmed again and again. The commonly used techniques for erasing may be described as under :

(a) Erasing using ultraviolet radiation

(b) Erasing using electricity.

The erasable programmable ROM using ultraviolet erasing is known as **EPROM.** Also, the ROM that makes use of the electrical voltage for erasing is known as **electrically alterable ROM (EAROM).**

9.4.4 Content Accessible Memories (CAM)

It is a special type of RAM. It can perform association operation in addition to the read/write operations performed by the conventional memories.

9.5 CLASSIFICATION BASED ON PHYSICAL CHARACTERISTICS

Figure 9.6 shows the classification based upon the physical characteristics.

9.5.1 Erasable or Non-erasable Memories

An **erasable** memory is that kind of memory in which the stored information can be erased and new information can be stored. The information stored in the non-erasable memory cannot be erased. As an example, ROM is a non-erasable memory. The erasable memories can further be categorised in two classes as under :

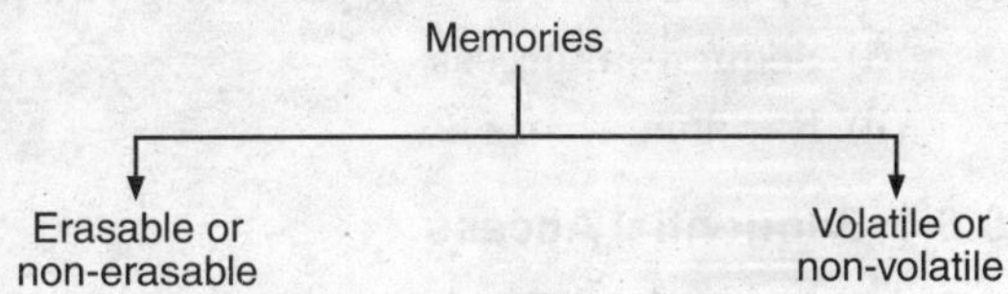

Fig. 9.6. Classification based on physical characteristics.

(i) Location-by-location erasable memories.

(ii) All locations simultaneously erasable.

(i) Location by Location Erasable Memories

In this type, the desired memory locations can be erased one by one and then the new information is stored. EAROM, RAM and CAM are the examples of location by location erasable memory. In EAROM, the contents of the location must be erased first before entering the new information. However, in RAM and CAM, erasing operation is performed automatically when a new information is to be entered.

(ii) All Locations Simultaneously Erasable Memories

These are the memories in which, contents of all the locations of a memory are erased simultaneously. As an example, when we expose an EPROM to the ultraviolet radiation, the contents of all the locations get erased simultaneously.

9.5.2 Ultraviolet Erasable Programmable ROM

The EPROM which can be erased using the ultra violet rays is shown in figure 9.7. In this EPROM is a quartz lid or window on the package. We can erase this PROM by exposing it to the ultraviolet (UV) light through the quartz window. The exposure period should be 10 to 15 minutes. Note that the UV rays will erase all the cells at time and all the locations in the erased EPROM will store a 1.

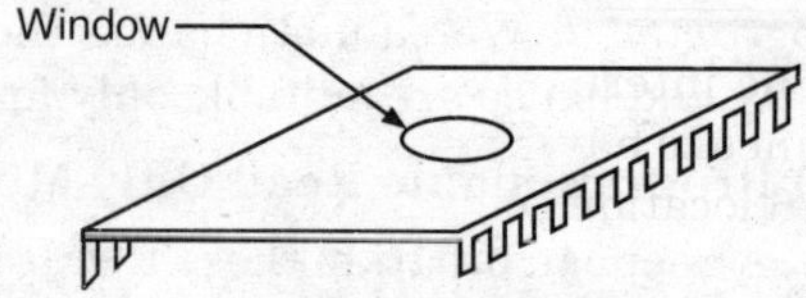

Fig. 9.7. UV EPROM

Darwbacks

(i) Selective erasing is not possible to achieve. All the locations get erased.

(ii) The EPROM has to be removed from the socket and put in the eraser for erasing.

> **DO YOU KNOW?**
>
> All memory device operate in the same general way. To write data in memory, the address to be accessed is placed on the address input, the data value to be stored is applied to the data inputs, and the control signals are manipulated to store the data.

9.5.3 Volatile or Non-Volatile Memories

(*i*) Volatile Memory

If the information stored in a memory chip is lost when the electrical power is switched OFF, then the memory is called as volatile memory. RAM is a volatile memory.

(ii) Non-Volatile Memory

If the information once stored in memory chip does not change unless altered deliberately are known as non-volatile memory. Such memories can hold the information even after switching off the power supply. Infact, read only memories (ROMs) of all types are examples of non-volatile memory.

9.6 CLASSIFICATION BASED ON MODE OF ACCESS

Mode of access means the manner in which a memory location is accessed while performing the read or write operation. There are following two possible modes of access :

(i) Sequential access

(ii) Random access

9.6.1 Sequential Access

The sequential memories have a sequential access to various memory locations. Hence, the time required to access various locations shall be different.

9.6.2 Random Access

In this type, we can access any memory location without going sequentially. Accessing any memory location for read or write requires same amount of time. Examples of random access are RAM, ROM and CAM.

9.7 CLASSIFICATION BASED ON FABRICATION TECHNOLOGY PROCESS

There are following two basic fabrication technologies used as under :

(i) Bipolar technology

(ii) MOS technology (unipolar technology).

The examples of bipolar technologies are TTL, ECL and RAM, ROM, EPROMs can be fabricated using either bipolar or MOS technology. The dynamic RAM, EPROM and EAROM are fabricated using MOS technology only using the MOSFETs.

> Four steps to achievement : plan purposefully, prepare prayerfully, proceed positively, pursue persistently.
>
> — **William Arthur Ward**

9.8 INTERNAL ORGANIZATION OF RAM (SELECTION ARRANGEMENT)

The internal organization of a 16 × 4 memory chip is shown in figure 9.8. It consists of 4-address lines to have 16 addresses. For 4 to 16 line conversion, we have a 4 to 16 line decoder used. There are 16 locations each one capable of storing a **4-bit word.**

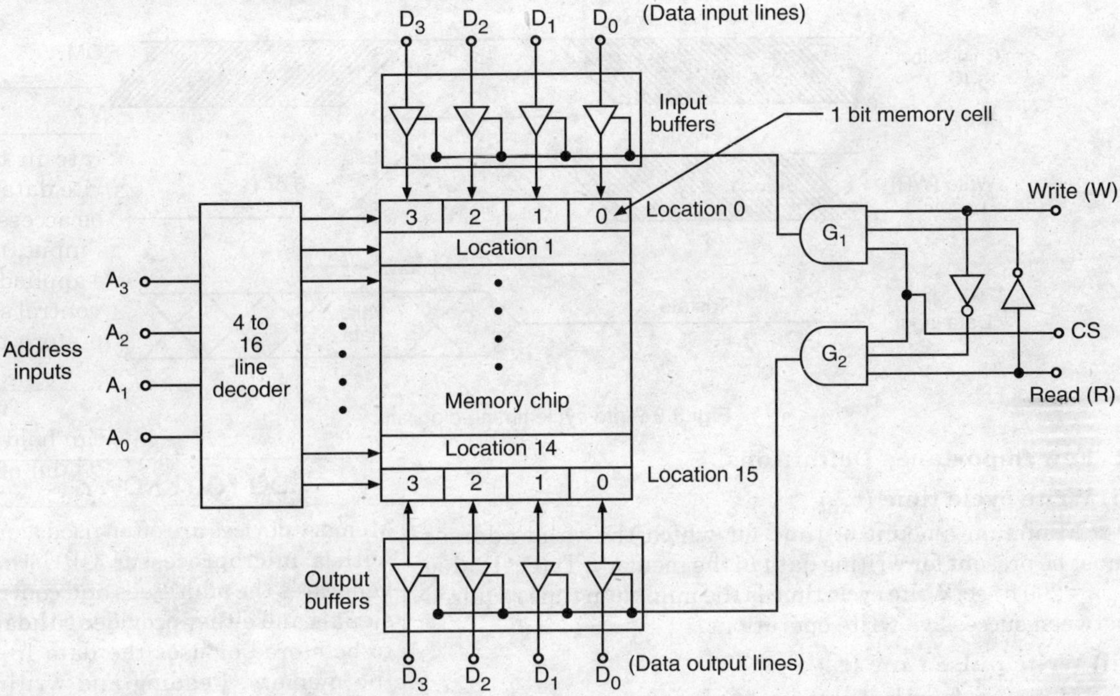

Fig. 9.8.

Actually, input and output buffers are connected between the data input and memory or memory and data output. Here, it may be noted that separate input and output lines are being used. Control logic circuit consists two AND gates and two NOT gates. The inputs to this circuit are read, write and chip select CS.

9.8.1 Write Operation

The sequence of events taking place for writing the desired data (1001) at the desired memory location (0110) is as under :

We adjust $A_3\ A_2\ A_1\ A_0 = 0110$ so as to select the desired memory location. Then, we adjust $D_3\ D_2\ D_1\ D_0 = 1001$ at the data inputs. We make CS = 1, W = 1 and R = 0. This will enable gate G_1 and disable gate G_2. Hence, the input buffers are enabled and output buffers are disabled. This will load the word 1001 into the memory location 0110. The output of gate G_2 is 0, hence the output buffers go into tristate (high impedance state). Therefore, the data is not available on the output lines. Here, it may be noted that the data which is originally present at the selected memory location is cleared off first and the data present at the data inputs replace it.

9.8.2 Timing Characteristics of the Write Cycle

1. Waveforms

Figure 9.9 shows the waveforms for the write cycle. Let us define few important terms with reference to these waveforms.

One who thinks in terms of silver, cannot act in terms of gold. — **Henry G. Weaver**

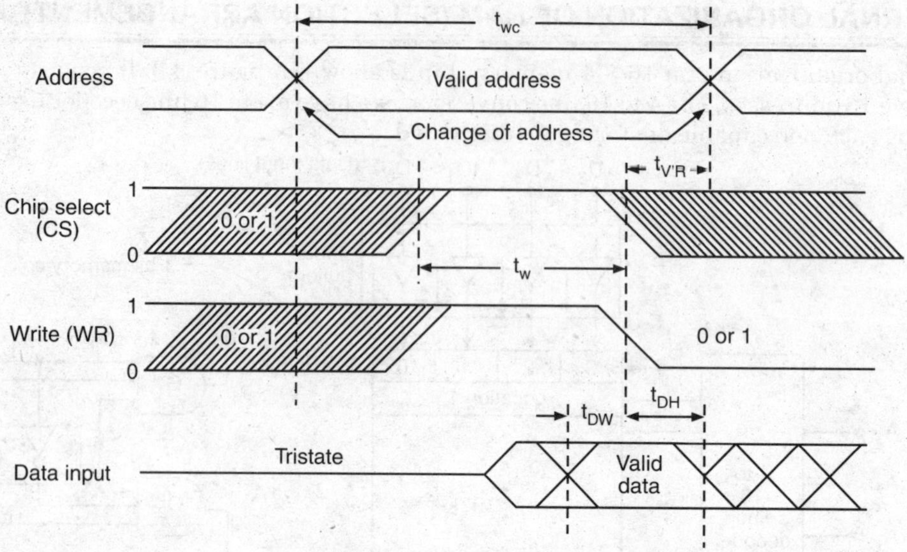

Fig. 9.9 Write cycle timing diagram.

2. Few Importance Definitions

(i) Write cycle time (t_{WC})

Minimum amount of time for which the valid address must be present for writing data in the memory. Typical value t_{WC} = 200 n sec. Write cycle time is the minimum time required between successive write operations.

(ii) Write pulse time (t_W)

Minimum length of write pulse. Infact, typical value t_W = 120 n sec.

> **DO YOU KNOW?**
>
> Memory devices are often used along with a microprocessor CPU that generates the addresses and control signals and either provides the data to be stored or uses the data from the memory. Reading and writing are always done from the CPU's perspective.

(iii) Write release time (t_{WR})

Minimum amount of time for which address must be valid after write pulse ends.

(iv) Data hold up time (t_{DH})

Minimum amount of time for which data must be valid after write pulse ends.

(v) Data set up time (t_{DW})

Minimum amount of time for which data must be valid before write pulse ends.

9.8.3 Read Operation

1. Definition

Basically, read operation is essential for reading the contents of a selected memory location. For read operation, the CS and read (R) inputs must be adjusted to logic 1 while the write input must be at logic 0. This enables gate G_2 and disables gate G_1. Therefore, the input buffers are disabled and output buffers are enabled. The input buffers go into the tristate (high impedance state) and does not allow the connection with the input data. The enabled output buffer makes the data of selected location available on the output lines. This is all about read operation.

2. Procedure steps to be followed for Read Operation

To read data from the memory chip, we have to follow the following procedure :

(i) First, we apply chip select signal (CS) to the chip. This enables the chip.

(ii) Then, we apply address of the memory location to the address lines.

(iii) Now, we apply Read (R) = 1 and write (W) = 0.

9.8.4 Timing Characteristics of Read Cycle

1. Waveforms

Figure 9.10 illustrates the waveforms for the read operation.

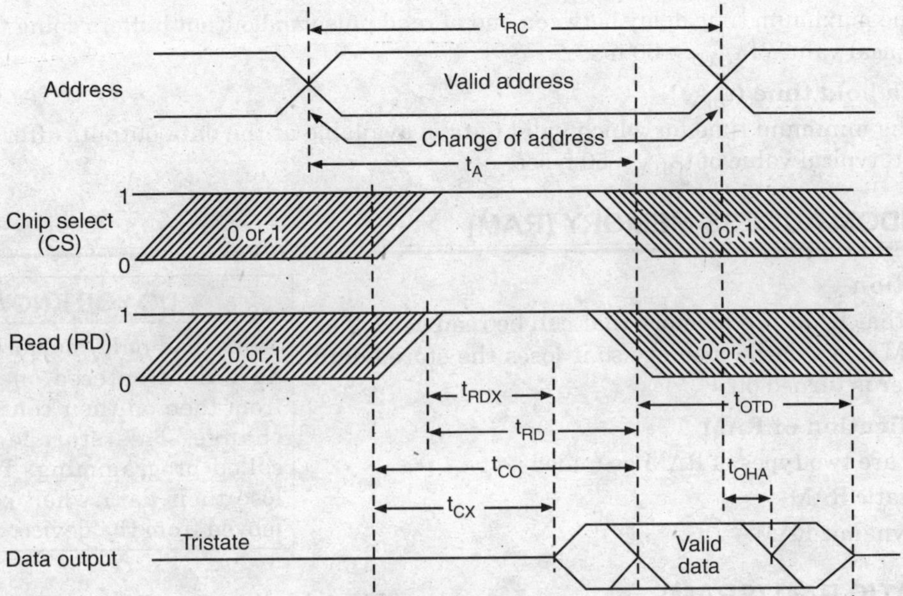

Fig. 9.10. Read cycle timing diagram.

Let us define some important terms related with the read operation, with respect to the read cycle timing diagrams shown in figure 9.10. Following above steps, one can read the data from the chip. You are aware that basic element in any IC is transistor, transistor has its own way of functioning, *i.e.*, it requires, hold up, setup, propagation delay time etc.

2. Timing specifications

Hence, for above read operation IC manufacturers provide following timing specifications for Read Operation :

(i) Read cycle time (t_{RC})

It is the minimum amount of valid time for which address must be present for reading data from memory. Infact, typical value of t_{RC} = 200 n sec. In other words, it is the minimum time required between the successive read operations.

(ii) Access time (t_A)

It is maximum time gap between valid address present on address line and valid data available at an output. Usually, $t_A \leq t_{RC}$. Infact, typical value of t_A = 20 n sec. This means that the data outputs must be ready before the memory is ready for next read operation.

(iii) Read to output valid time (t_{RD})

This is the maximum time-delay between the beginning of read pulse and availability of valid data bus. Infact, typical value of t_{RD} = 70 n sec.

(iv) Read to output active time (t_{RDX})

This is the maximum time-delay between beginning of read pulse and the output buffers coming to active state (from Hi-z condition). Infact, its typical value = 20 n sec.

(v) Chip select to output valid time (t_{CO})

It is the maximum time delay between beginning of chip select pulse and the availability of valid data at the data output. Infact, its typical value of t_{CO} = 70 n sec.

(vi) Chip select to output active time (t_{CX})

It is the minimum time delay time between beginning of chip select pulse and output buffer coming to active state. Infact, typical value of t_{CX} = 20 n sec.

(vii) Output tristate from read (t_{OTD})

It is the maximum time delay between end of read pulse and output buffers going to Hi-z state. In fact, typical value of t_{OTD} = 60 n sec.

(viii) Data hold time (t_{OHA})

It is the minimum time for which valid data is available at the data outputs after the address end. Infact, typical value of t_{OHA} = 50 n sec.

9.9 RANDOM ACCESS MEMORY (RAM)

1. Definition

We know that RAM is a memory that can be read as well as write. RAM is a volatile memory so it loses the stored data when power is turned off.

2. Classification of RAM

There are two types of RAM as under :

(i) Static RAM

(ii) Dynamic RAM.

DO YOU KNOW?

Most read-only memories (ROMs) have data entered one time, and from then on their contents do not change. This storage process is called programming. They do not lose their data when power is removed from the device.

9.10 STATIC RAM (SRAM)

(PTU, Jalandhar, Sem. Exam. 2006-07) (10 marks)

It is possible to implement the static RAM using the bipolar as well as MOS technology. It is possible to store data as long as power is applied to the chip. Static RAM cells are basically flip-flops which can stay in a given state (i.e., store a bit) as long as power to the circuit is not interrupted.

9.10.1 TTL RAM Cell

1. Circuit Diagram

Figure 9.11 illustrates the circuit diagram of a simplified TTL RAM cell. Here, multiple emitter transistors are used to form a flip-flop with the set and reset facility and also the facility for select line connections. Actually, this cell is capable of storing only 1 bit.

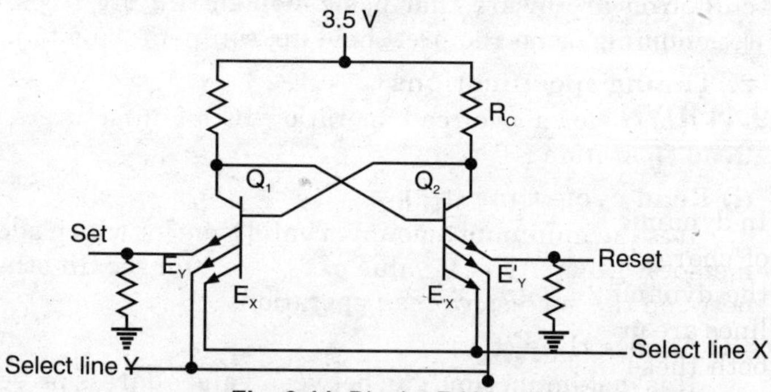

Fig. 9.11. Bipolar RAM cell.

2. Working Operation

Here, Q_1 and Q_2 are cross coupled, i.e., collector of one is applied to the base of other. Because of such connection, only one transistor is ON at any given instant. The X and Y are the select lines that are useful in connecting the cell in a matrix. A 1 is said to be stored in the cell if Q_2 is OFF and Q_1 is ON. Whereas a 0 is said to be stored if Q_2 is ON and Q_1 is OFF. Note that the transistors act as switches. If the set input is pulsed high, then Q_1 is turned OFF. This raises the collector voltage of Q_1 to V_{CC}. Therefore, base voltage of Q_2 goes high and it goes into saturation. Thus, a 0 is stored. In a similar manner, when reset input is pulsed high, Q_2 is turned OFF, its collector voltage goes to V_{CC} and a 1 is stored in the cell. A large number of such cells are connected in the matrix form (rows and columns) to form a memory chip.

9.10.2 MOS Static RAM Cell (Static Memory Element)

1. Definition

Figure 9.12 shows a six MOSFET static RAM cell. This circuit consists of two cross coupled inverters T_1 and T_3. T_2 and T_4 are working as resistances. They act as loads for T_1 and T_3. X and Y lines are used for addressing the cell. When X = Y = 1 (High), the cell is selected. When X = 1, the MOSFET T_5 and T_6 are turned ON. This connects the memory cell to the data and $\overline{\text{data}}$ lines. When Y = 1, the MOSFETs T_7 and T_8 are turned ON. This makes read, or write operation possible.

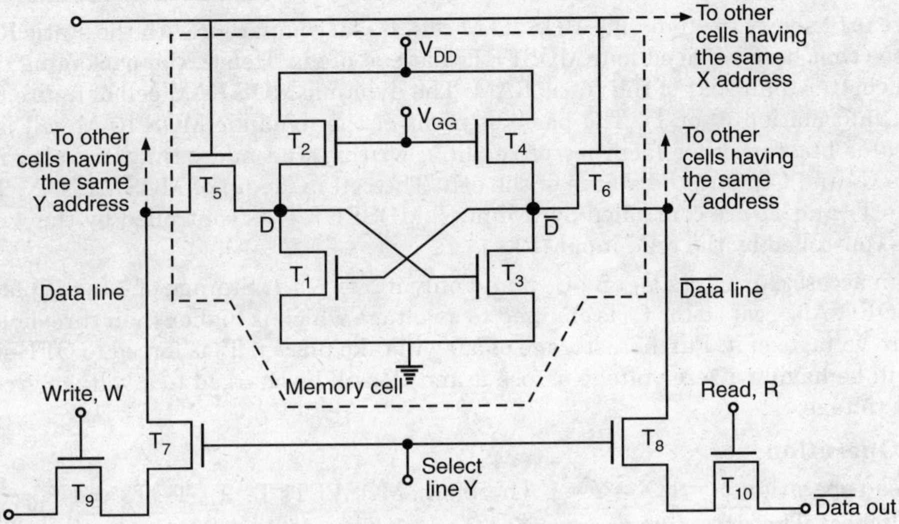

Fig. 9.12. MOS static RAM cell

3. Read Operation

For read operation, we have to apply a logic 1 to the read (R) input. This turns on MOSFET T_{10}. This will connect the $\overline{\text{data}}$ line to data output. Thus, the complement of the stored data is read.

9.11 DYNAMIC RAM (DRAM) OR DYNAMIC MEMORY ELEMENT

(PTU, Jalandhar, Sem. Exam, 2006-07) (10 marks)

In dynamic RAM, the data is stored in the form of charge on the capacitor. Figure 9.13 shows the dynamic RAM cell. The sense and control lines are used as column and row signals. When both these lines are high, the MOSFET works as a closed switch and charges the capacitor. When the control and sense both are low, the MOSFET turns OFF and the capacitor retains its charge. In this manner, a logic 1 bit is stored.

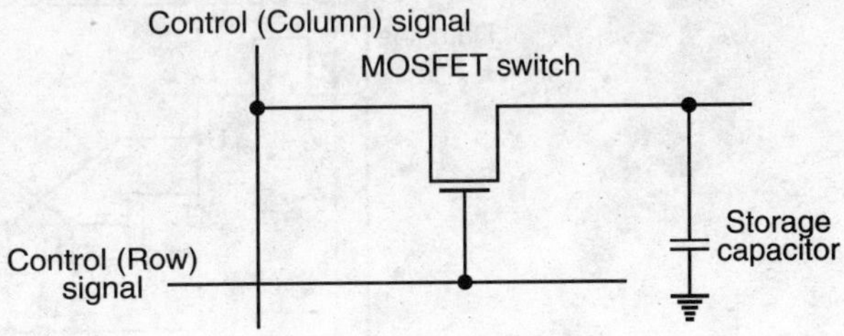

Fig. 9.13. Dynamic RAM

9.11.1 Salient Features of Dynamic RAM

(i) The single bit dynamic RAM is formed using only two components, one MOSFET and a capacitor.

(ii) Hence, it requires very small area as compared to static RAM. This means that, the DRAM consists of more memory cell as compared to static RAM per unit area.

9.11.2 Drawback of Dynamic RAM

The charge hold by the capacitor requires to be refreshed after every few milliseconds. Therefore, alongwith the DRAM cell, we have to use an additional circuitry for refreshing the charge. This is the main drawback of a DRAM.

DO YOU KNOW?

Random access memory (RAM) is a generic term given to devices that can have data easily stored and retrieved. Data are retained in a RAM device only as long as power is applied.

9.11.3 Dynamic MOS RAM Cell

1. Definition and Circuit Diagram

Figure 9.14 shows the dynamic MOS RAM cell. If you compare it with the static RAM cell, then you will see that here we need four MOSFETs instead of six. Hence, the packaging density of this memory is better than that of the static RAM. The dynamic MOS RAM cell of figure 9.14 can store one bit of information (0 or 1). The basic element of the dynamic MOS RAM cell is the bistable multivibrator but the other facilities of reading, writing and addressing are also included. The capacitors C_1 and C_2 store the status of the cell. The cell made up of MOSFETs T_1, T_2, T_5 and T_6. MOSFETs T_7 and T_8 are controlled by Y inputs. MOSFET T_9 is controlled by the write input (W) and T_{10} is controlled by the read input (R).

We can access capacitors C_1 and C_2 if and only if X = Y = 1. Storage of 1 takes place if T_2 is ON and T_1 is OFF. Also, capacitor C_2 is charged to a voltage which is higher than threshold voltage and C_1 has zero voltage on it. Further, storage of a 0 will take place if T_2 is forced to OFF state and T_1 is ON. C_2 will be having a zero voltage across it and C_1 will be charged to a voltage greater than the threshold voltage.

2. Read Operation

For read operation, R = 1, X = Y = 1. Therefore, MOSFETs T_5, T_6, T_7, T_8 and T_{10} conduct (act as closed switches). The $\overline{\text{data}}$ line is connected through T_8 and T_{10} to the $\overline{\text{data}}$ output line. Thus, the read operation will take place.

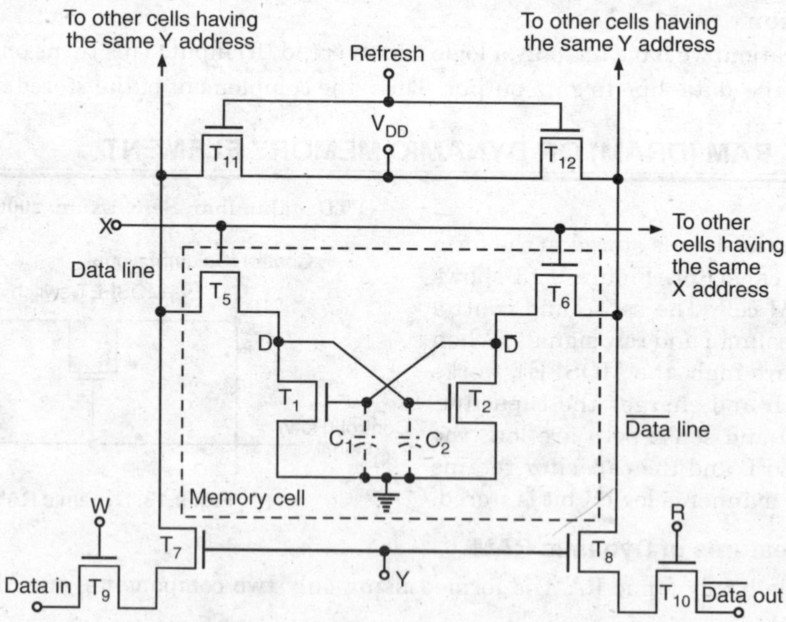

Fig. 9.14. Four transistors MOS dynamic cell.

3. Write Operation

For write operation, W = 1, X = Y = 1. Therefore, the MOSFETs, T_6, T_7, T_8 and T_9 will conduct (act as closed switches). If Data in = 1, then capacitor C_2 is charged through T_9, T_7, T_5 and turns on

T_2. This turns off T_1 and force D = 1. If Data in = 0, then C_2 discharges through T_5, T_7 and T_9. This turns off T_2. Hence, T_1 is turned on and D = 0. In this manner, a zero is stored.

4. Refresh Operation

The stored charge on the capacitors requires to be refreshed after sometime otherwise it leaks off and the stored information shall be lost. For refreshing the charged capacitors, we have to make X = 1 and the refresh input = 1. Hence, MOSFETs T_{11}, T_{12}, T_5, T_6 will get turned ON. If C_1 is fully charged and C_2 does not carry any charge then the lost charge on C_1 is compensated by charging it through T_{12} and T_6 from V_{DD}.

9.11.4 Performance Comparison of SRAM and DRAM

Table 9.1 presents the performance comparison of DRAM and SRAM.

Table 9.1. Comparison of SRAM and DRAM

S.No.	Parameter of comparison	SRAM	DRAM
1.	Circuit configuration	Each SRAM cell is a flip-flop.	Each DRAM unit consists of one MOSFET and a capacitor.
2.	Number of components per cell	More	Less (only two)
3.	Memory cells/unit area	Less than DRAM	More than SRAM
4.	Refreshing	Not required	Required
5.	Cost	More	Less
6.	Access time	Less. So these are faster memories.	More. Hence, these are slower memories.

9.12 MEMORY EXPANSION (i.e., MEMORY BANKS)

(PTU, Jalandhar, Sem. Exam, 2006-07)(10 marks)

In several applications, the capacity of the single available memory chip is not sufficient. Because of this, we have to increase it by using more than one chips. This is known as expansion of memory. The memory expansion can be of following two types :

(i) Expanding the word size (increase in N).

(ii) Expanding the word capacity (increase in M).

> **DO YOU KNOW?**
>
> Static RAM (SRAM) uses storage elements that are basically latch circuits. Once the data are stored, they will remain unchanged as long as power is applied to the chip.

9.12.1 Expanding the Word Size

Here, we want a word size n which is greater than the word size N of the available memory chip. As an example, conversion of 16 × 4 memory into a 16 × 8 memory requires expansion of word size from 4 to 8. It is interesting to know that number of words M = 16 remains constant. The next example explains the word size expansion process.

EXAMPLE 9.1. A 16 × 4 size memory is available. Expand its word size so as to obtain a 16 × 8 memory.
(PTU, Jalandhar, Sem. Exam, 2007-08)(10 Marks)

Solution : (i) First, we decide number of memory chips.

If the required word size is denoted by n and the available word size is denoted by N, then the number of ICs required is an integer given by,

Number of ICs, $X \geq n/N$.

Here, n = 8 and N = 4, hence X = 2. We have to use 2 ICs, each with a capacity of 16 × 4.

(ii) Then, we make connections.

Since the number of locations does not change (16), the number of address lines will not change. It will be equal to 4 and they will be common to both the memory ICs. As the word size is to be expanded, we will use the data lines of IC-2 as the MSB lines (D_7, D_6, D_5, D_4) whereas the data lines of IC-1 will be used as LSB lines (D_3, D_2, D_1, D_0). Connect the read (RD) input of both ICs together. It will now act as read (RD) input of the overall memory. Same thing is true for the write (WR) input, as well as the chip select (CS) input. The connections are as shown in figure 9.15.

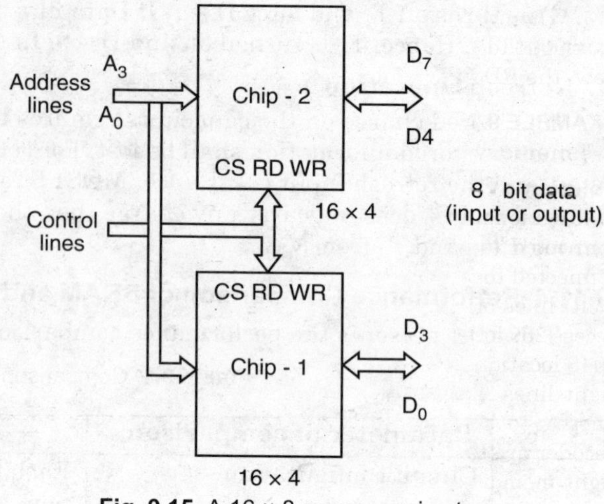

Fig. 9.15. A 16 × 8 memory using two 16 × 4 memory chips

9.12.2 Expanding The Word Capacity

Expanding word capacity means increasing the value of *M*, keeping the word size (*N*) unchanged. For example, if 16 × 4 chips are available and we want to have a 32 × 4 memory, then we have to perform the word capacity expansion. The following example illustrates the word capacity expansion.

EXAMPLE 9.2. Obtain a 32 × 4 memory using 16 × 4 memory chips.

Solution : (i) Number of Memory Chips

To obtain a memory of capacity m words using memory chips with M words each, the number of chips required is given by

$$X \geq m/M$$

where X is an integer.

In this example, m = 32 and M = 16

Therefore, X = 2, i.e., 2 chips of capacity 16 × 4 are required.

(ii) Connections

The two 16 × 4 chips are connected in the following manner :

The address lines are connected to the address lines of two chips individually. This means that out of the 5 address lines (we need 5 address lines), the 4 LSB lines A_3 A_2 A_1 A_0 will be same for each chip and the MSB line A_4 should be used to select chip-1 or chip-2.

with $A_4 = 0$, chip-1 should be selected

with $A_4 = 1$, chip-2 should be selected.

As the word size is same (4 bit), the data input/output lines of both the chips should be connected to each other.

This means that D_0 of chip-1 is connected to D_0 of chip-2 and the common D_0 line now works as the D_0 line of the overall memory. In a similar manner, the connections are made for D_1, D_2 and D_3 lines.

Then, we connect the read input of each chip together and write inputs also together.

Figure 9.16 shows the connection diagram.

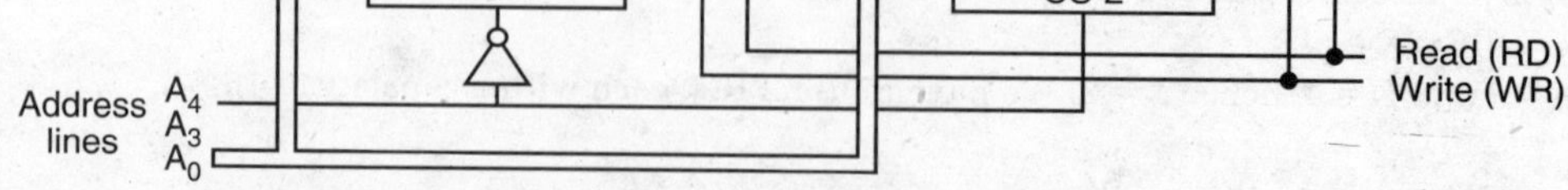

Fig. 9.16. A 32 × 4 size memory obtained using two 16 × 4 memory chips

When $A_4 = 0$, the first chip is selected. Therefore, the first 16 locations can be accessed for $A_3 A_2 A_1 A_0$ = 0000 to 1111. With $A_4 = 1$, first chip is disabled and second chip is selected. For $A_3 A_2 A_1 A_0$ = 0000 to 1111, now, the RD or WR operation can be performed on the 16 locations of chip-2.

EXAMPLE 9.3. How many 256 × 8 memory chips are required to obtain a 2048 × 8 memory ? Also show the memory addresses associated with each memory chip.

Solution : To obtain 2048 × 8 memory using 256 × 8 chips, we require to use 2048/256 = 8 chips. The bi-directional data lines of all the chips are connected to each other. This means that D_0 lines of all chips are connected together, D_1 lines are connected together ... and so on. The read (RD) lines of all the chips are connected together and this line works as the read input of the combined memory. In a similar manner, write (WR) lines of all the chips are connected together to form the write (WR) input of the combined memory. To access 256 locations, we require 8 address lines A_0 to A_7 and to access 2048 locations, we need 11 address lines A_0 to A_{10}. Hence, the LSB eight lines A_0 to A_7 are common to all the chips. The MSB three lines A_8 to A_{10} are applied to a 3 : 8 decoder and the 8 outputs of the decoder are then applied one by one-to-the chip select inputs of the eight memory chips. Figure 9.17 shows the connection diagram of 2048 × 8 memory using eight 256 × 8 memory chips.

DO YOU KNOW?

Dynamic RAM (DRAM) uses capacitors to store data by charging or discharging them.

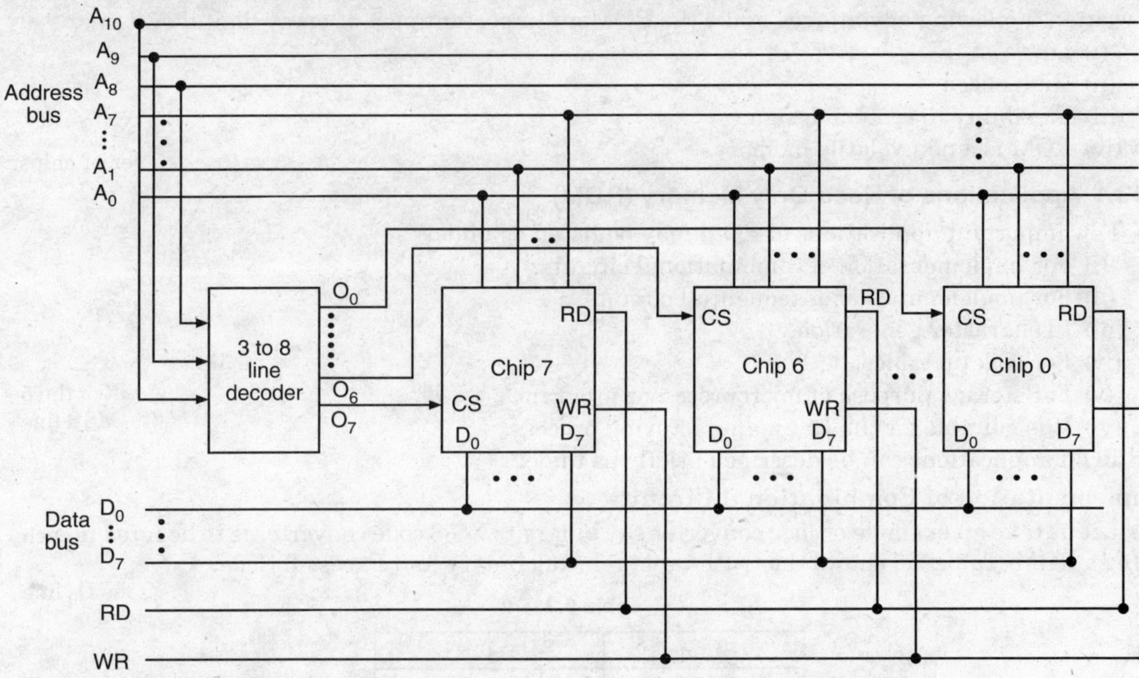

Fig. 9.17. 2048 × 8 size memory using eight 256 × 8 memory chips

Memory Addresses

Table 9.2 describes the addresses of memory locations for eight memory chips

Table 9.2. Memory addresses for various chips

S.No.	Memory chip number	Address (Hex)
1.	Chip – 0	000 – 0FF (256 locations)
2.	Chip – 1	100 – 1FF (Next 256 locations)
3.	Chip – 2	200 – 2FF (Next 256 locations)
4.	Chip – 3	300 – 3FF (Next 256 locations)
5.	Chip – 4	400 – 4FF (Next 256 locations)
6.	Chip – 5	500 – 5FF (Next 256 locations)
7.	Chip – 6	600 – 6FF (Next 256 locations)
8.	Chip – 7	700 – 7FF (Next 256 locations)

9.13 READ ONLY MEMORY (ROM)

(PTU, Jalandhar, Sem. Exam., 2006-07)

We know that the information stored in a ROM is permanent nature. It can be only read and cannot be written. Following advantages make the ROM an important part of many digital systems :

(i) Low cost
(ii) High speed
(iii) Flexibility in system design
(iv) ROM is a non-volatile memory.

9.13.1 Applications of Read Only Memory (ROM)

Few important applications of ROM may be listed as under :

(i) For implementation of combinational circuits.
(ii) For implementation of sequential circuits.
(iii) In character generation.
(iv) For look up tables.
(v) For storage purpose of microprocessor program.
(vi) It is suitable for the LSI manufacturing process.

ROM applications can be described briefly as under :

Implementation of Combinational Circuits

Let us take an example of code converter say binary to XS-3 code converter is to be implemented using a ROM. Table 9.3 shows the truth table relating binary and Excess-3 Code.

Table 9.3.

Binary	XS - 3
0 0 0 0	0 0 1 1
0 0 0 1	0 1 0 0
0 0 1 0	0 1 0 1
0 0 1 1	0 1 1 0
0 1 0 0	0 1 1 1
0 1 0 1	1 0 0 0
0 1 1 0	1 0 0 1
0 1 1 1	1 0 1 0

Nothing is useless to the man of sense, he turns everything to account. — **La Fountaine**

Now, we store the XS-3 numbers serially from location 0000 to 0111 into the ROM. We give the binary number which is to be converted into XS-3, at the address input. As an example, if we apply $A_3\,A_2\,A_1\,A_0 = 0\;1\;0\;1$, then, we set the ROM into read mode. Therefore, on the data output lines, we obtain the data of location 0101 which would be 1000. Thus, we got the Excess-3 equivalent of 0101.

9.13.2 ROM Manufacturing

Various technologies used for manufacturing of ROM may be listed as under :

(i) Bipolar Technology, and

(ii) MOS Technology.

Bipolar ROMs are faster and they have higher driving capability whereas MOS ROMs require less area and consume less power. The improved MOS ROMs have improved speeds (typically equivalent to those of bipolar devices).

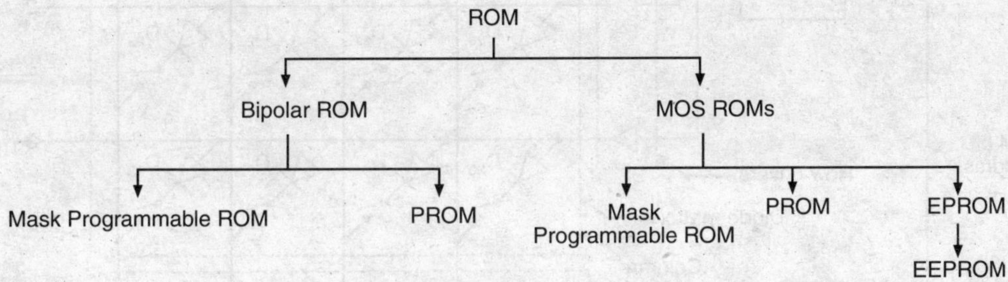

Fig. 9.18. Classification of ROM

9.13.3 ROM Programming

The process of entering information into a ROM is known as ROM programming. The ROMs may be classified based on the programming process as described under :

(i) Mask programmable ROMs.

(ii) Programmable ROM (PROM).

(iii) Erasable programmable ROMs (EPROMs).

The important information regarding the ROMs is as given below :

ROM

Mask programmable ROM	Programmable ROM	Erasable programmable ROM
(i) In this, data pattern must be programmed as a part of manufacturing process.	(i) This device is electrically programmable. Here user can program the chip. Normally PROM are OTP (one time programmable).	(i) In this, memory data can be written any number of times.
(ii) Not reprogrammed.	(ii) Not reprogrammable.	(ii) Chip can be erased by exposing it to UV rays called EPROM OR it can be electrically erased referred as EEPROM (E^2 PROM).
(iii) Normally shifted for high volume usage due to low cost.	(iii) Normally used for production purpose.	(iii) This is used when one wants to develop digital computer system.

Self-confidence is the first requisite to great undertakings. — **Samuel Johnson**

9.13.4 ROM Organization

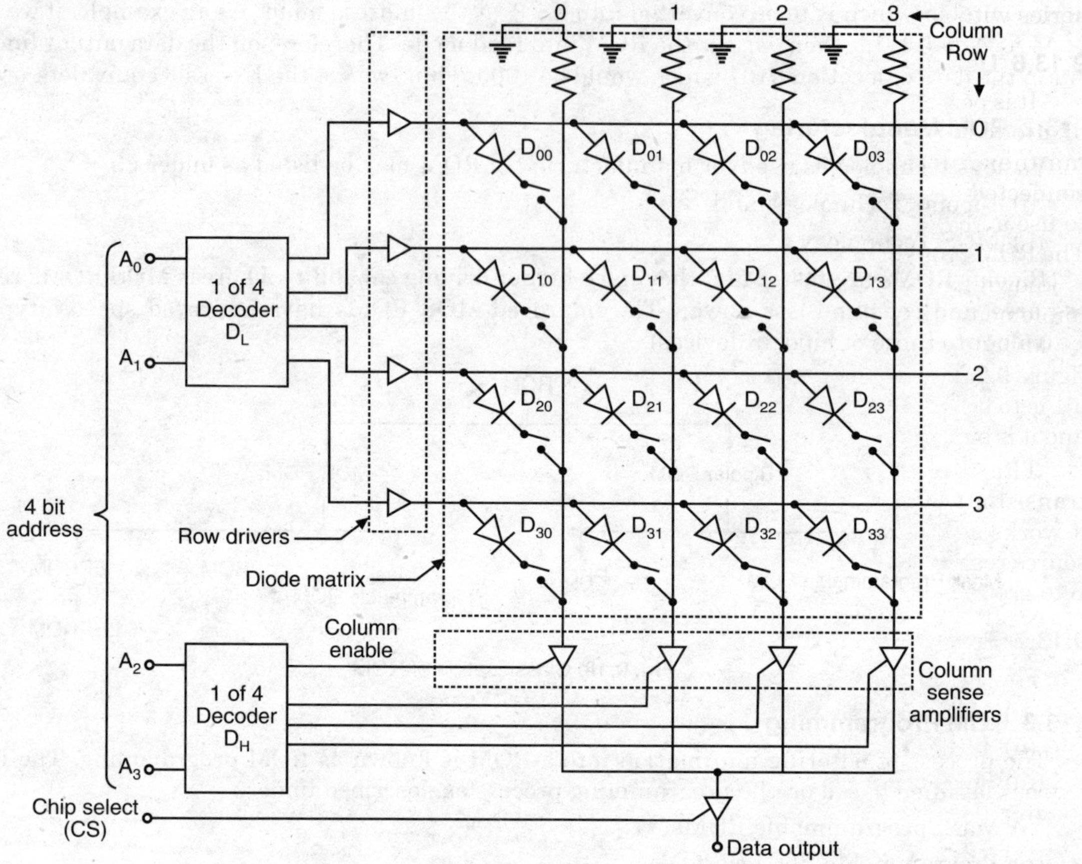

Fig. 9.19. Illustration of a 16-bit ROM array

Basically, a ROM is an array of selectively unidirectional contacts. The contacts can be open or closed selectively to program the required information into a ROM. Figure 9.19 shows a 16-bit ROM array. It makes use of two 2-4 address, 16 locations and 1-data output. Therefore, this is a 16 × 1 ROM. The method of addressing is known as two dimensional, X – Y or coincident selection, addressing. A unidirectional switch is included at the junction of every row and column. The lower two address bits (A_1 A_0) are decoded by the decoder D_L. The outputs of this decoder are used to select one out of the four rows. The high two address bits (A_3 A_2) are decoded by the decoder D_H. The four outputs of this decoder are used to select one out of the four column sense amplifiers.

Diode Matrix

Diode matrix is the heart of the ROM array shown in figure 9.20. It is formed by connecting a diode and a switch between each row and column. As an example, let us consider figure 9.20 which illustrates diode D_{03} connected between row 0 and column 3.

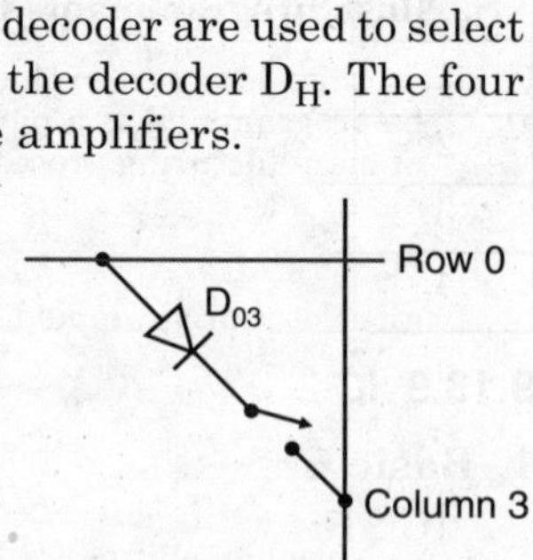

Fig. 9.20.

The chip select (CS) input is utilized to enable the output. A ROM is programmed by selectively opening and closing the switches connected in series with the diodes. This has been explained below.

9.13.5 ROM Programming

If switch of D_{03} is in closed position and the address input is 1100, then row 0 is activated through D_L and column 3 through D_H. The sense amplifier of column 3 is enabled. This yields output

= Logic 1 if CS = 1. Thus, a logic 1 is stored at the address 1100. On the other hand, if the switch in series with D_{03} is open, then a logic 0 is stored at the address 1100.

9.13.6 Use of Other Devices in Place of Diodes

It is possible to implement ROMs by using the bipolar junction transistors or MOSFETs in place of diodes. Figure 9.21(a) shows the manner in which a BJT must be connected and figure 9.21(b) shows how to use a MOSFET to implement a ROM. The ROM implemented using transistors is known as a bipolar memory and that implemented using MOSFETs is called as MOS memory. The fuse shown in figure 9.21(a) is for programming. When a 1 is to be stored, this fuse is kept intact and it is blown off when a 0 is to be stored.

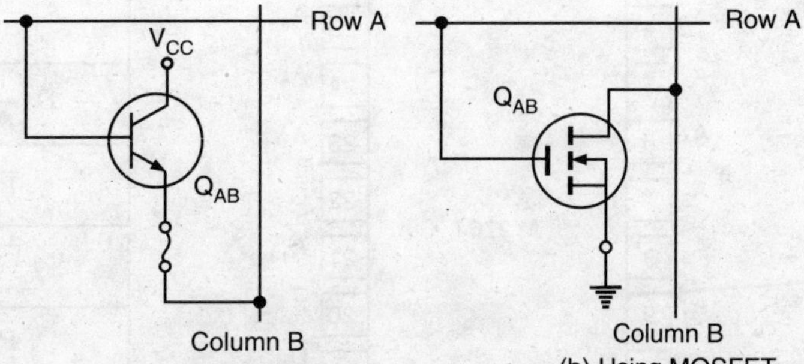

(a) Using BJT (b) Using MOSFET

Fig. 9.21. Unidirectional switches

The base emitter junction of transistor Q_{AB} is forward biased, hence, it works as a diode. In figure 9.21(b), a MOSFET is used as a switch. When it is ON, the drain to source resistance $R_{DS(ON)}$ is very small and the MOSFET works as a closed switch connecting column B to ground.

9.13.7 Few Important Characteristics of Memory

Few important characteristics of memory may be described as under :

(i) Memory capacity and its organization
(ii) Physical dimensions
(iii) Type of packaging
(iv) Power consumption
(v) Timing characteristics
(vi) Reliability
(vii) Cost.

9.13.8 Performance Comparison of RAM and ROM

Table 9.4

S. No.	Parameter of comparison	RAM	ROM
1.	Operations involved	Reading and writing	Reading only
2.	Type of storage	Temporary	Permanent
3.	Types	SRAM, DRAM	PROM, EPROM, EEPROM
4.	Application	Calculators, computer	Computers, microprocessor

9.13.9 IC 2764 (EPROM)

1. Basic Concepts

IC 2764 is 8192 × 8 EPROM. This means that it can store 8192 words of size 8 bit each. So, the capacity of this EPROM is 65536 bits or 64 K. The number of locations = 8192. To access them, we need to use 12 address lines since $2^{12} = 8192$. Hence, A_0 to A_{11} address lines are being used. Since, the word size = 8, eight bi-directional data input/output lines (D_0 to D_7) are used. Figure 9.22 shows the pin configuration of IC 2764. It is a 28 pin window ceramic frit dual in line package. Since it is an UV erasable and electrically programmable EPROM, a transparent lid is provided in the package.

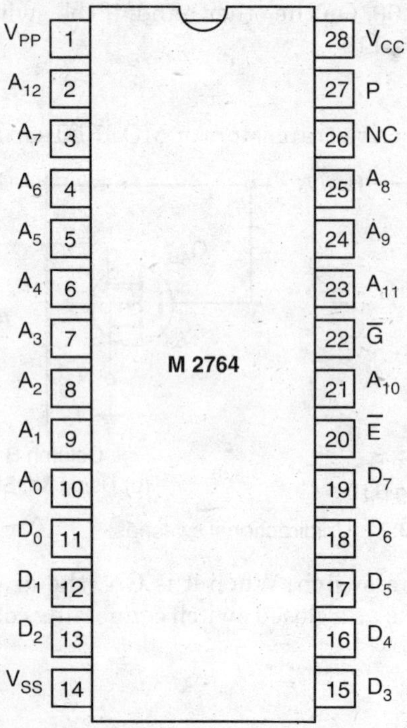

Fig. 9.22. Pin configuration of IC 2764

Pin No.	Description
$A_0 - A_{12}$	Address lines
$D_0 - D_7$	Data I/O lines
$\overline{E}$	Chip enable
$\overline{G}$	Output enable
$\overline{P}$	Program
V_{PP}	Power supply
V_{CC}	Supply
V_{SS}	Ground

Fig. 9.23. Functions of various pins of IC 2764

2. Read Operation

Just to read the stored data, the chip enable $\overline{E}$ must be connected to ground. In addition to this, the output enable $\overline{G}$ also should be connected to ground. Then, we apply the address of desired memory location on the address lines and achieve the corresponding data on the data lines.

3. Programming Operation

Before programming (writing) we have to erase the EPROM by exposing it to the UV radiation for a sufficiently long duration of time. After erasure, all the bits in 2764 will be in the logic 1 state. In fact, same will be the condition of a new EPROM purchased from the market. The EPROM is programmed by selective programming of only 0s. In the programming mode, we have to connect the chip enable ($\overline{E}$) and program pin ($\overline{P}$) to TTL low and apply a voltage of 12.5 V to the V_{PP} pin. Then, apply the 8 bit data to be stored to the data line D_0 to D_7. Apply the desired address to program this data at the required location.

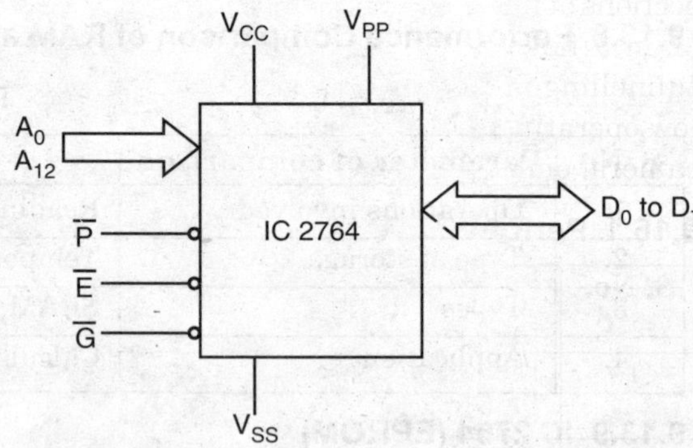

Fig. 9.24. Logic diagram of IC 2764

4. Logic Diagram

Figure 9.24 shows the logic diagram of IC 2764.

9.14 EEPROM (E^2 PROM)

The long form of EEPROM is electrically erasable PROM. These are very similar to EPROMs and

The good life is one inspired by love and guided by knowledge. — **Bertrand Russell**

use the MOS circuitry. In EEPROM, data is stored in the form of charge. Actually, presence of charge represents 1 and its absence represents a 0. A voltage as low as 20 to 25 volts can be used for erasing or programming. It is possible to erase and program a particular memory location. This is known of selective erasing. It is not essential to erase all the memory locations. It is however possible to erase all the locations within about 10 ms by operating the EEPROM in a special mode. This period is very short as compared to that required for the ultraviolet radiation erasing used for EPROM. Another advantage of EEPROM is that it can be erased and reprogrammed when the memory chip is connected in the circuit. However, the drawback of EEPROM is very expensive and has very low density.

9.15 NVRAM

The long form of NVRAM is non-volatile RAM. It is a special type of RAM because the conventional RAM is a volatile memory. NVRAM are formed by a combination of a static RAM and EEPROM. When electrically powered, NVRAM works like a RAM, whereas when power fails, it will save the data in EEPROM. As soon as power fails, all the contents of RAM are parallely transferred to EEPROM. The transfer time is typically 4 ms. As soon as the power is reapplied all the saved contents can be transferred back to RAM. Thus, the RAM is used for normal operation and EEPROM is used for temporary storage in the event of power failure. A $\overline{\text{store}}$ signal is used for storing the RAM data into EEPROM whereas the $\overline{\text{Recall}}$ signal is used to transfer the stored data into RAM. NVRAM does not need any battery backup, in order to save the working data in the event of power failure. This is the greatest advantage of NVRAM. NVRAMs are usually used to store the information such as data, time, monitor setting etc.

9.16 FLASH MEMORY

It is a special type of RAM. It is a non-volatile memory which is powered continuously. The erasing and programming of this memory takes place block-by-block. Because of this process, the flash memories are faster than EEPROMs that erase and write new data at the byte level. Flash memory cannot be used as a random access memory because RAM is to be addressed at byte level. The sections of flash memory chip are organized in such a manner that they can be erased instantaneously or in a flash. Hence, its name is **Flash Memory**. A special technique called Fowler-Nordheim tunnelling is utilized to erase the contents. Important features of a flash memory are high speed, low operating voltages, low power consumption and durability. Typical application areas are digital camera, embedded controllers, cellular phones, etc.

9.16.1 Performance Comparison of EPROM and EEPROM

S. No.	Parameter of comparison	EPROM	EEPROM
1.	Technique use for erasing	Exposure to ultraviolet light	A voltage of 20 to 25 volts is applied.
2.	Selective erasing	Not possible. All the locations get erased.	Possible-A particular location only can be erased.
3.	Time required for erasing	Long 10 to 15 min.	Short 10 ms.
4.	Requirement to remove PROM from circuit	It is necessary to remove the PROM	Not essential to remove PROM.
5.	Cost	Less expensive	Very expensive.

9.17 INTRODUCTION TO PLDs

In earlier chapters, we have discussed the combinational and sequential circuits. We have also discussed different ICs such as multiplexers, comparators, adders, code converters, demultiplexers etc. Such ICs are called as *fixed function* ICs. The advantages of designing logic circuits using the fixed function ICs are low development cost and easy testing. But the disadvantages are requirement of large board space, large power requirements no security, additional costs are required if the existing circuit is to be modified. The remedy to this problem is to use the *application specific integrated circuits* (ASICs). These ICs are designed to meet specific requirements of a user or application. But, the design of ASICs is too complex. The advantages of ASICs are reduction in space requirements, reduction in power requirements, more security because designs implemented using ASICs are impossible to copy, cost reduction if produced in bulk. But the disadvantages are enormous initial development cost, new testing methods will add to the cost and efforts.

Use of PLDs

The third approach to the problem of digital circuit design is to use the programmable logic devices (PLDs). This method has the advantages of both the approaches discussed earlier.

What is a PLD ?

PLD are special type of ICs which can be programmed by the users as per their requirements. Therefore it is possible to implement a combinational or sequential circuit using the PLD ICs.

GEORGE BOOLE

George Boole was born in Lincoin, England on November 2, 1815. He was a self-taught mathematician who invented modern symbolic logic and pioneered the calculus of operators. Around 1850, George Boole created Boolean algebra which underlies the theory of logic.

9.17.1 Salient Features of PLDs

Following are some of the salient features or advantages of PLDs.

1. Reduced space requirement
2. Reduction in power requirement
3. Increased speed of switching
4. Better security (copying is less likely to take place)
5. Higher density
6. Low production cost as compared to ASICs
7. Low development cost
8. Short design cycle, i.e., time required to implement the design is short (as compared to ASICs)
9. More flexibility to the designer
10. Reprogramming is possible to be done within few seconds
11. Modifications can be carried out within a short span of time
12. An important application of PLDs is that we can prepare prototype ASIC design. This will reduce the time required to design ASICs.

DO YOU KNOW?

A programmable logic device (PLD) is an IC that contains a large number of logic gates whose interconnections can be programmed by the user to generate the desired logic relationship between inputs and outputs.

9.17.2 Types of PLDs

Various types of PLDs are as under :

(i) Read Only Memories (ROMs)
(ii) Programmable Logic Array (PLAs)
(iii) Programmable Array Logic (PALs)

(iv) Simple Programmable Logic Devices (SPLDs)
(v) Complex Programmable Logic Devices (CPLDs)
(vi) Field Programmable Gate Arrays (FPGAs).

9.18 GENERIC ARRAY LOGIC DEVICES (GAL)

This is another type of configurable PAL. We can replace a wide variety of PAL devices using a GAL device. GAL 16 V8 can replace almost all the 20 pin PAL devices and GAL 20 V8 is capable of replacing most of the 24 pin PAL devices. GAL 16 V8 uses the macrocells discussed earlier on the output side. We can configure its macrocells in one of the three basic configurations. The speciality of GAL devices is that the OR gate is treated as a part of each macrocell. This will help us to obtain various types of I/O configurations obtainable in those PAL devices that are to be replaced by GAL.

9.18.1. EX-OR PALs

If the output polarity is to be made configurable then EX-OR gates should be used as discussed earlier. It is possible to use the EX-OR gates in the AND-OR-EX-OR configuration shown in figure 9.25 such devices are called as the EX-OR PAL devices. The EX-OR outputs are applied to the flip-flop inputs.] The inputs to the EX-OR gates are obtained from two SOP a4rrays consisting of gates 1 through 6. Each sum of product array produces two product terms each. This configuration is used for reducing the logic circuit required for many application. PALs can also be used for the asynchronous state machines.

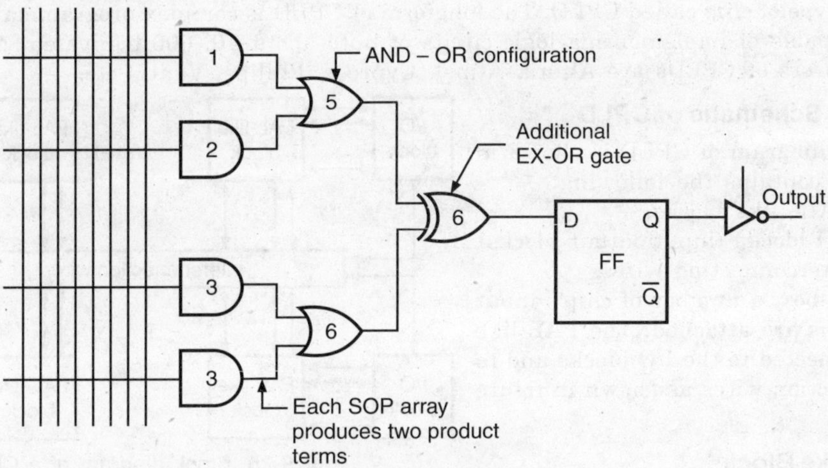

Fig. 9.25. EX-OR PAL

9.18.2. Commercially Available PALs

Some of the PALs available in market and their characteristics are given in Table 9.5.

Table. 9.5. PAL devices available in market

Device	Inputs			Outputs			
	External	Feedback	Total	Bidirectional Combinational	Registered	Combinational	
PAL 10L8	10	0	10	0	0	8	8
PAL 16L8	10	6	16	6	0	2	8
PAL 20L8	14	6	20	6	0	2	8
PAL 16R8	8	8	16	0	8	0	8
PAL 20R8	12	8	20	2	6	0	8

9.18.3. Simple PLDs (SPLDs)

The programmable logic devices discussed till now such as PLAs, PALs etc. are called as simple PLDs or SPLDs.

9.18.4. Comparison of PROM, PLA and PAL

Prom	PLA	PAL
It is cheap and easy to use.	It is expensive than PAL and PROM and complicated to use.	Moderately expensive and moderately complicated.
There are fixed AND arrays and programmable OR arrays.	AND and OR arrays are programmable	Only the AND array is programmable. Or array is fixed.
SOP functions in the standard from only can be implemented.	Any SOP function can be implemented	Any SOP function can be implemented.
It is possible to decode any minterm	We can get any desired minterm by programming the AND matrix	We can get any desired minterm by programming the ADC matrix.

9.18.5. Complex Programmable Logic Devices (CPLDs)

The major disadvantage of the PLDs such as PLA and PALs is that they have limited number of inputs, product terms and outputs. Therefore we can implement logic circuit with number of inputs and output upto 32 only, using such PLDs. If the number of inputs and outputs is higher than 32, than we can use more than one SPLD devices. An alternative to this is to use a jore sophisticated type of chip called CPLD. The longform of CPLD is complex programmale logic device. CPLDs are capable of implementing logic circits of upto about 10, 000 equivalent gates. Some of the manufacturers of CPLDs are Altera, Atmel, Cypress, Philips, Vantis etc.

9.18.6. Block Schematic of CPLD

The block diagram of CPLD is shown in figure 9.26. It contains the following:

(a) PAL– like blocks.

(b) I/O blocks (input/output blocks)

(c) Interconnecting wires.

To the I/O box, a number of chip's input and output pins are attached. The PAL like blocks are connected to the I/O blocks and to the interconnection wires as shown in figure 9.26.

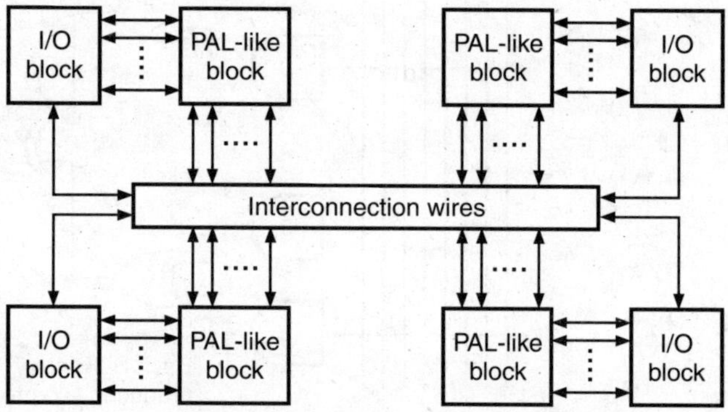

Fig. 9.26. Block diagram of a CPLD.

9.19.7. PAL like Block

The PAL like block contains 16 macrocells. The simplified internal diagram of a microcell is shown in figure 9.27, which shows that each macrocell consists of an AND-or combination followed by an EX-OR gate, flip-flop multiplexers and tristate buffers.

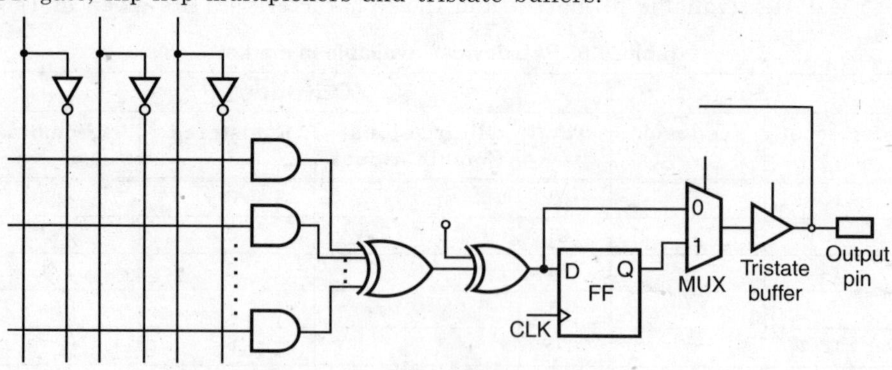

Fig. 9.27.

Every AND-OR combination contains about 5 to 20 AND gates and 1 or gate. So OR gate has 5 to 20 inputs. The OR gate output is connected to the input of a EX-OR gate. The EX-OR gate output to the tristate buffer depending on the stte of its select input (0 to 1). The tri-state buffer acts as a switch. The corresponding pin of the chip acts as output if the buffer is enable and the same pin acts as input pin if the buffer is disabled. But if the pin is used as input pin, then the macrocell is disabled.

9.19 PROGRAMMING OF PLDS

The SPLDs and CPLDs are programmed using the electrically erasable programmable rear only memory (EEPROM) technology. That means SPLDs and CPLDs are programmed in exactly the identical manner as that of EEPROM. The SPLD pins are taken out and put into the programming unit to carry out the programming. This is possible because the number of pins for SPLDs is limited. However for CPLDs the in system programming (ISP) is used. In ISP the CPLD chip being programmed on the cirucit board itself and we don't rely on the programmer unit for programjking of chip.

9.19.1. Reasons for Using ISP

The reasons for using ISP may be described as under:

(i) CPLDs have a large number of pins, some times even greater than 200. In addition to large number, these pins are fragile and can bend easily.

(ii) If programming unit requires a socket for holding the chip. For the CPLDs of large size, such sockets are sometimes more expensive than the CPLDs themselves.

(iii) To programme CPLDs and SPLDs we have to use alrge number of programmable switches, that can be configured as per requirements.

(iv) Due to large number of switches, it is not possible to configure them manually. Henf ecomputer aided design (CAD) systems are employed for this purpose.

(v) After configuring the switches using CAD, the user has to prepare a programming file or fuse map. This map specifies the state of each programmable switch.

(vi) A computer which runs the CAD is connected to the programming unit. For IsP technique as small connector is included on the printed circuit board, that has CPLD on it. And the computer is connected to CPLD using a cable.

(vii) In programming, the programming file is transferred from computer into the CPLD through the cable.

(viii) The CPLD circuitry that allows the ISP is called as JTAG port as standardized by IEEE.

(ix) JTAG stands for Joint Test Action Group. The JTAG programming of two CPLDs is shown in figure 9.28.

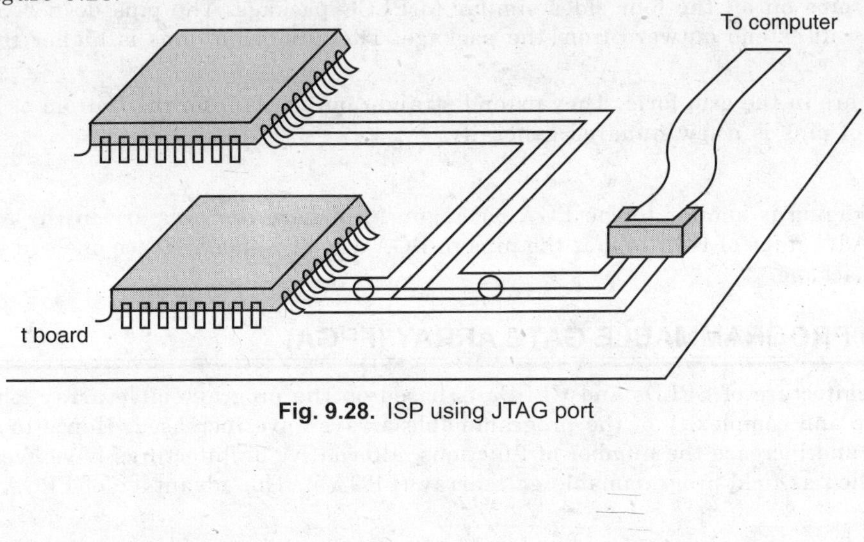

Fig. 9.28. ISP using JTAG port

9.19.2. CPLD Packaging

The dual in line package (DIP) is not convenient for the CPLD ICs due to large number of pins involved. Hence, the following special type of packages are used.

(i) Plastic – leaded Chip Carrier (PLCC)

(ii) Quad Flas Pack (QFP)

(iii) Ceramic Pin Grid Array (PGA)

(iv) Ball Grid Array (BGA)

(i) PLCC

The PLCC package has been shown in figure 9.29. It has pins on all the four sides and they wrap around the chip edges. A special socket for PLCC is soldered to the PCB.

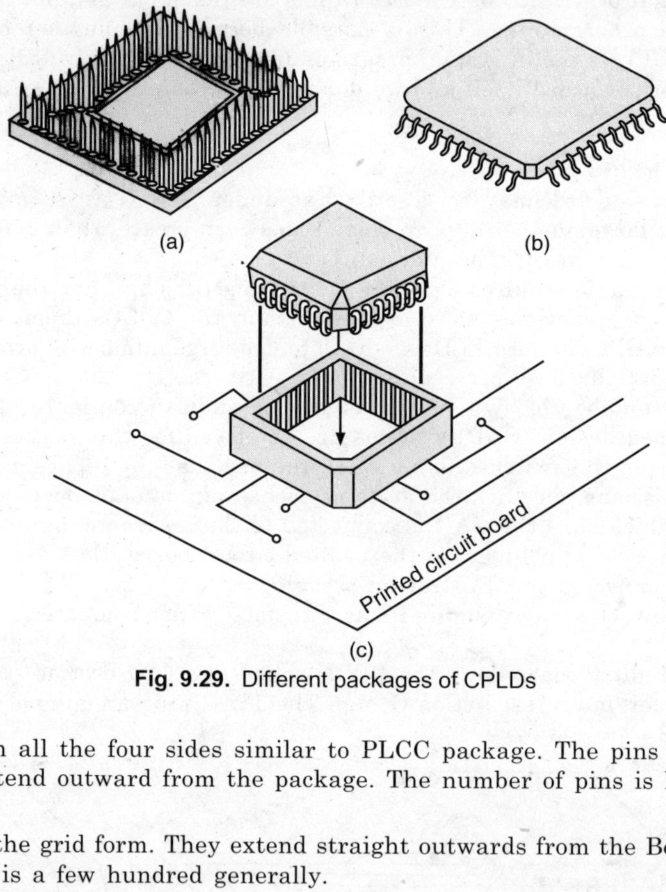

Fig. 9.29. Different packages of CPLDs

(ii) QFP

QFP has pins on all the four sides similar to PLCC package. The pins do not wrap around. Instead they will extend outward from the package. The number of pins is higher than 200.

(iii) PGA

The pins are in the grid form. They extend straight outwards from the Bottom of the package. The number of pins is a few hundred generally.

(iv) BGA

This packaging is similar to the PGA packaging. But here the pins are in the form of small round balls. Advantage of BGA is that the pins in BGA are very small. Hence one can provide more pins on the package.

9.20 FIELD PROGRAMMABLE GATE ARRAY (FPGA)

The basic architecture of SPLDs and CPLDs is based on the programmable array logic. Over the years the size and complexity of the programmable arrays have increased. Hence to increase the effective size and increase the number of functions, alternative architectures have been developed, which are called as field programmable gate array (FPGA's). The advantage of FPGA is that they have very high logic densities, typically from 10,000 to few hundred thousand equivalent gates. Therefore FPGA are used for implementation of relatively large logic circuits.

9.20.1. What is difference between FPGA and SPLDs or CPLDs?

(i) The FPGA does not contain the AND or OR matrices. Instread of that logic blocks are provided for implementing the required Boolean function.
(ii) FPGA consists of independent logic modules. I/O cells and routing channels.
(iii) The logic modules and I/O cells are fonfigurable and the routing channels are nothing but programmable interconnecting paths.
(iv) For the best possible utilization of an FPGA, it is necessary to optimize the size of logic modules and number of interconnecting paths.
(v) Note that a given FPGA device can be connected in many different ways depending on the requirement of an application.

General architecture of FPGA chip is shown in figure 9.30.

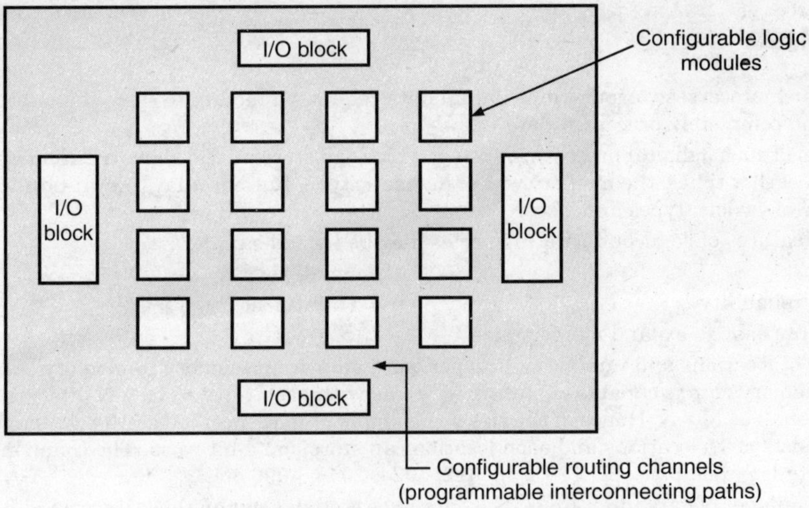

Fig. 9.30. General architecture of FPGA chip

Different Types of FPGAS

A number of FPGAs are available with different types of architectures, packaging and programming technologies.

9.21 SEQUENTIAL PROGRAMMABLE DEVICES

The PLDs designed for combinational circuits consists of only gates. But the sequential programmable devices include gates as well as flip-flops. The registered PLA, PAL, CPLD and FPGA are the examples of sequential programmable devices. The macrocells which contain AND-OR gates, flip-flops and multiplexers are used, in all the sequential programmable devices. The implementation of sequential circuits using the programmable devices is a two level implementation as shown in figure 9.31.

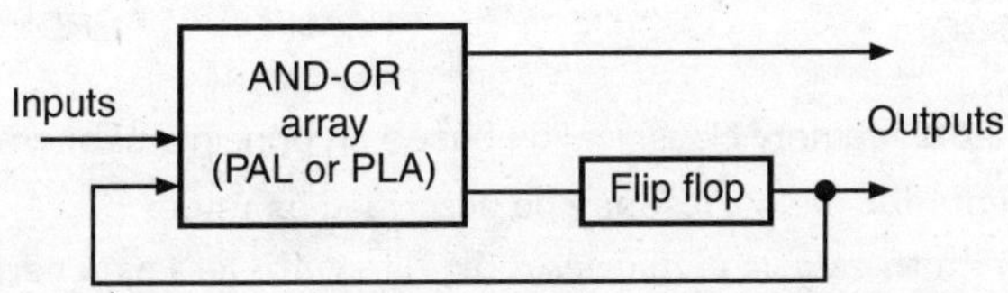

Fig. 9.31. Block diagram of sequential programmable logic device.

GLOSSARY

1. **Cell :** In memories, a single storage element.
2. **Non-volatile memory :** Memory which retains data even if the power is turned off.
3. **Non-volatile RAM :** Read/write memory that will hold its data even when the power is turned off.
4. **Phase-change technology :** Used in DVD-RW and DVD + RW optical discs. A phase-change alloy is employed for reading, writing, and erasing information. The tiny "pit" and "no pit" area of optical disc are either dark/non-reflective if the alloy is in its amorphous state or reflective if the phase-change alloy is in its crystalline state. These discs are rewritable.
5. **RDRAM :** In computer technology, an extremely fast dynamic RAM. Acronym for Rambus dynamic RAM. Compare with DRAM, SDRAM.
6. **Read-only memory :** Non-volatile memory which is not usually changed once it is programmed. ROM commonly used to refer to mask-programmable read-only memory.
7. **Write-once read-many :** An optical CD recordable disc can be recorded on once using your PC and it then is permanent like a CD-ROM.

SUMMARY

- A system which processes various kinds digital data requires a facility to store the unprocessed, partially processed and completely processed data.
- A sub-system of such a digital processing system which can store all the above mentioned data is known as **memory.** In earlier times, the memory was of magnetic type. But nowadays, **semi-conductor memories** are available of various types and size.
- The salient features of semiconductor memories may be listed as under :
  (i) Low cost (ii) Small size
  (iii) Better reliability (iv) High speed
  (v) It is quite easy to expand memory size.
- The number of locations and number of bits per word vary from memory to memory. As an example, if a particular memory chip is capable of storing M words with each word having N bits in it, then the size of the memory shall be M × N. Hence, if the size of a memory chip is specified as 16 × 4 then it means that this memory consists of 16 locations and each location can store one 4-bit word. The commonly used values of number of words per chip are 64, 128, 256, 512, 1024, 2048, 4096 etc.
- There are N-number of data input lines. The data to be stored is put on these lines word-by-word with each word N-bit long. Data inputs are also known as Data bus.
- The memory devices may be classified on the basis of various parameters. The parameters used for basis of classification may be listed as under :
  (i) Operating principle (ii) Physical characteristics
  (iii) Mode of access (iv) Technology used for fabrication process.
- The classification of memories based on the principle of operation is shown in figure 9.32.

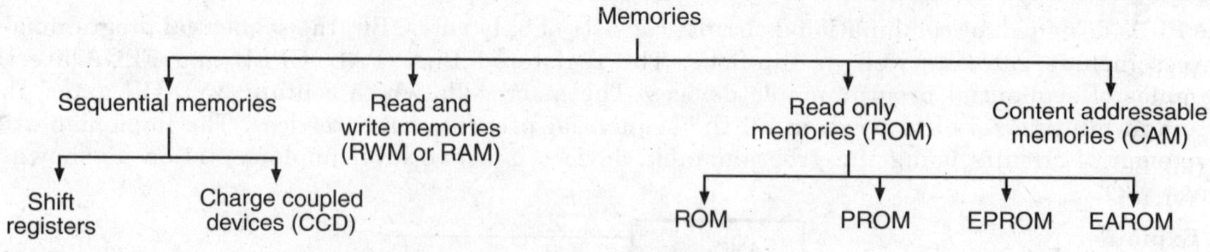

Fig. 9.32. Memory classification based on principle of operation

- Few salient features of dynamic memories may be described as under :
  (i) They are simpler than the static memories (ii) They are less expensive
  (iii) Require less power for operation (iv) They have high packaging densities.

Due to all these salient features, the dynamic memories are widely used in digital systems. The only drawback of dynamic memories is the additional circuitry for refreshing.

- The ROMs are further classified into three categories as under :
  - (i) Read-Only memory.
  - (ii) Programmable Read Only Memory (PROM).
  - (iii) Erasable and Programmable ROM (EPROM).
- As the name suggests, these ROMs can be erased and programmed again and again. The commonly used techniques for erasing may be described as under :
  - (a) Erasing using ultraviolet radiation
  - (b) Erasing using electricity.

  The erasable programmable ROM using ultraviolet erasing is known as **EPROM.** Also, the ROM that makes use of the electrical voltage for erasing is known as **electrically alterable ROM (EAROM).**
- An **erasable** memory is that kind of memory in which the stored information can be erased and new information can be stored. The information stored in the non-erasable memory cannot be erased. As an for following example, ROM is a non-erasable memory. The erasable memories can further be categorised in two classes as under :
  - (i) Location-by-location erasable memories.
  - (ii) All locations simultaneously erasable.
- In this type, the desired memory locations can be erased one by one and then the new information is stored. EAROM, RAM and CAM are the examples of location by location erasable memory. In EAROM, the contents of the location must be erased first before entering the new information. However, in RAM and CAM, erasing operation is performed automatically when a new information is to be entered.
- It is possible to implement the static RAM using the bipolar as well as MOS technology. It is possible to store data as long as power is applied to the chip. Static RAM cells are basically flip-flops which can stay in a given state (i.e., store a bit) as long as power to the circuit is not interrupted.
- The charge held by the capacitor requires to be refreshed after every few milliseconds. Therefore, alongwith the DRAM cell, we have to use an additional circuitry for refreshing the charge. This is the main drawback of a DRAM.
- We know that the information stored in a ROM is of permanent nature. It can be only read and cannot be write. Following advantages make the ROM an important part of many digital systems :
  - (i) Low cost
  - (ii) High speed
  - (iii) Flexibility in system design
  - (iv) ROM is a non-volatile memory.
- Few important applications of ROM may be listed as under :
  - (i) For implementation of combinational circuits.
  - (ii) For implementation of sequential circuits.
  - (iii) In character generation.
  - (iv) For look up tables.
  - (v) For storage purpose of microprocessor program.
  - (vi) It is suitable for the LSI manufacturing process.

SHORT QUESTIONS WITH ANSWER

Q.1. Write the salient features of semiconductor memory.

Ans. The salient features of semiconductor memories may be listed as under :

(i) Low cost
(ii) Small size
(iii) Better reliability
(iv) High speed
(v) It is quite easy to expand memory size.

Q.2. Explain about Memory size?

Ans. The number of locations and number of bits per word vary from memory-to-memory. As an example, if a particular memory chip is capable of storing M words with each word having N bits in it, then the size of the memory shall be M × N. Hence, if the size of a memory chip is specified as 16 × 4 then it means that this memory consists of 16 locations and each location can store one 4-bit word. The commonly used values of number of words per chip are 64, 256, 512, 1024, 2048, 4096 etc.

Q.3. Write the classification of ROMs?

Ans. The ROMs are further classified into three categories as under :

(i) Read-Only memory. (ii) Programmable Read Only Memory (PROM).

(iii) Erasable and Programmable ROM (EPROM).

Q.4. Write salient features of Dynamic RAM.

Ans. (i)The single bit dynamic RAM is formed using only two components, one MOSFET and a capacitor.

(ii) Hence, it requires very small area as compared to static RAM. This means that, the DRAM consists of more memory cell as compared to static RAM per unit area.

Q.5. Write the Drawback of Dynamic RAM?

Ans. The charge held by the capacitor requires to be refreshed after every few milliseconds. Therefore, alongwith the DRAM cell, we have to use an additional circuitry for refreshing the charge. This is the main drawback of a DRAM.

REVIEW QUESTIONS

1. Explain the classification of memories based on their principle of operation.
2. Explain the classification of memories based on their physical characteristics.
3. Explain the concept of RAM. Mention its types. Explain with diagram working of any one type of RAM.
4. Differentiate between ROM and RAM.
5. Explain with circuit diagram the dynamic MOS memory.
6. How can you differentiate between sequential access memory and RAM ?
7. Explain it with a neat block diagram IC 2716. Sketch its Pin-out.
8. Explain with logic diagram IC 7481. Sketch its Pin-out.
9. Explain the meaning of static and dynamic memories. State their applications.
10. Draw a circuit of basic static MOS RAM cell. Explain how information is stored in it.
11. What is meant by ROM ? What are the types of ROM ? Compare between them.
12. Explain RAM, ROM and their types.
13. Differentiate between static and dynamic RAM on the basis of 4 parameters.
14. Differentiate between RAM and ROM on the basis of 4 parameters.
15. Draw the circuit diagram of static RAM cell. Explain its operation.
16. Explain four points of differentiation between EPROM and EEPROM.
17. Write a short-note on semiconductor memories.
18. Discuss various types of read-only memories (ROMs).
19. With neat and clear circuit diagram, explain the working operation of bipolar static RAM cell.
20. Discuss various types of ROMs and their applications in brief.
21. Write short notes on following :
    (i) Static and Dynamic MOS RAM (ii) ROM, EPROM, EEPROM.
22. Distinguish between volatile and non-volatile memories.
23. Explain the differences between SRAM and DRAM.
24. Write a short note on static bipolar RAM cell.
25. With the help of neat and clean circuit diagram, explain the working of static MOS RAM cell.
26. Write a short note on dynamic RAM.
27. Determine how many address bits are needed for a memory which has 1024 and 16,384 number of bits.
28. Explain the most appropriate memory type to use for each of the following in a small computer :
    (i) The working memory. (ii) The memory used to store permanent programs.
    (iii) A memory used to store development programs.
29. Explain in brief EPROM and EEPROM.
30. Differentiate between SRAM and DRAM.
31. State two salient features of using a PROM instead of a SRAM in the memory logic.
32. State and explain in brief, different types of read only memories.

❑❑❑

10

SIGNAL CONVERSIONS

Syllabus

Signal Conversions: Analog and Digital signals, A/D and D/A conversion techniques (Weighted type, R-2R Ladder type, Counter Type, Dual Slope type, Successive Approximation type).

10.1 INTRODUCTION

In previous chapter, we have discussed the advantages of using digital systems. Due to the advantages offered by the digital systems, they are widely used in many fields such as instrumentation, computers, communication and control.

In many such applications, the signal are not available in the digital form. Most of the physical quantities such as temperature, pressure, displacement, vibrations etc. are available in analog form. These quantities are represented accurately in analog form but it is difficult to process, store or transmit the analog signal because error gets introduced easily, due to noise.

Hence, to reduce these errors it is always better to express these physical quantities in the digital form. The digital representation of a signal makes storage possible, processing simpler and transmission easier. Therefore A to D conversion is necessary. Now, once the processing, transmission etc. is done the signal should be brought back to its analog form, for which the D to A conversion is essential. Both ADC and DAC circuits are known as **data converters** and they are available in the IC form as well.

10.2 INTRODUCTION TO DIGITAL SYSTEMS

In modern world of Electronics, the term 'digital' is usually associated with a computer. This is due

Chapter Outline

- Introduction to Digital Systems
- Introduction to Analog and Digital Signals and Systems
- Analog Systems
- Salient Features of Digital Systems
- Detailed Comparison of Analog and Digital Systems
- Process of A to D Conversion
- DAC Fundamentals
- Digital to Analog (D to A) Conversion
- Binary Weighted Resistor DAC
- R-2R Ladder DAC
- Inverted R-2R Ladder DAC (Current Mode R-2R Ladder)
- Specifications for DAC
- Source of Errors in DACs
- Monolithic IC Dac 0808
- Sampling Theorem and Choice of Sampling Frequency
- Sample And Hold Amplifiers (SHA)
- Analog to Digital Converters (ADC)
- Types of A To D Converters
- ADC Circuits
- ADC Characteristics
- Monolithic IC ADC 0809
- Applications of ADC

We make a living by what we get, we make a life by what we give. – **Sir Winston Churchill.**

to the fact that the term 'digital' is derived from the way the computers perform operations, by counting digits. Now a days, digital concepts are usually applied to various problems that could only be solved by analog methods several years ago. Fast and reliable solutions using digital techniques proved the tremendous power and usefulness of digital systems. In modern times, digital systems find applications in a wide variety of industrial and consumer products such as automated industrial machinery, pocket calculators, microprocessors, digital computers, digital watches, TV games, signal processing and so on. Infact, the areas of application of digital systems have been increasing every day.

Also, the digital computer can be programmed to perform different tasks. The users are allowed to specify and change the program or the data according to the specific requirements. Because of this flexibility, general purpose digital computers can perform a variety of information processing tasks that range over a very wide spectrum of applications.

For many years, the main application of digital electronics was only in the computer systems. But today, the digital electronics is used in many other application such as:

1. Television
2. Communication systems
3. Radar, navigation and guidance systems
4. Military systems
5. Medical equipments
6. Industrial process control
7. Consumer electronics

10.3. INTRODUCTION TO ANALOG AND DIGITAL SIGNALS AND SYSTEMS*

In day-to-day life, we come across various types of signals and systems.

The signals may be broadly classified into following two categories :

(i) Analog signals

(ii) Digital signals

(i) Analog Signals

Analog signals are the signals which may have infinite number of different magnitudes or values. They vary continuously with time. Sine wave, triangular wave etc. are few examples of analog signals.

(ii) Digital Signals

A signal is known as a digital signal if it has only a finite number of predetermined distinct magnitudes. Depending upon the number of distinct magnitudes, the digital signals may be classified as under :

S.No.	Number of distinct magnitudes	Types of digital signal
1.	2	Binary
2.	8	Octal
3.	16	Hexadecimal

Actually, the digital signals are discrete time signals, i.e., they are not continuous with time as shown in figure 10.1.

* Analog signals vary gradually and continuously, while digital signals produce discrete voltage levels commonly referred to as HIGH and LOW.

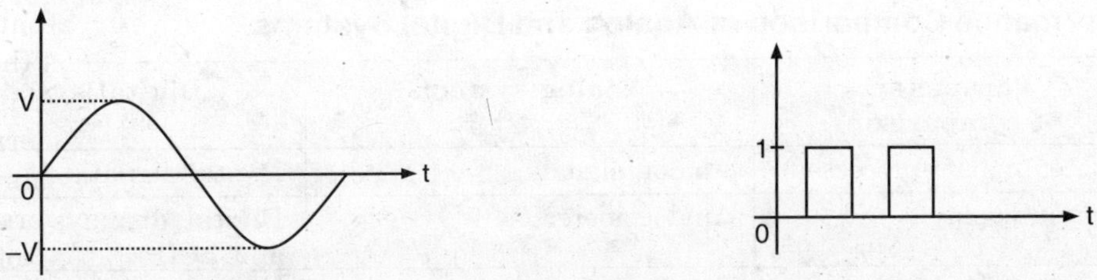

(a) Analog signal (b) Digital signal

Fig. 10.1.

10.3.1. Performance Comparison of Digital and Analog Signals

S.No.	Parameter of comparison	Analog signals	Digital signals
1.	Number of values	Infinite	Finite (2, 8, 16 etc.)
2.	Nature of signals	Continuous-time	Discrete-time
3.	Sources of signals	Signals generators, transducers etc.	Computers, A to D converters etc.
4.	Examples	Sine wave, triangular wave	Binary signal

10.4 ANALOG SYSTEMS

The systems which process the analog signals are known as the analog systems. The analog systems, and analog measurement systems have been used for a long time. There are some certain advantages of these systems, but more than the advantages, there are some drawbacks.

Examples

Few examples of analog systems may be listed as under:

(i) Filters circuits
(ii) Amplifier circuits
(iii) Signal generators
(iv) Motor speed controllers

> **DO YOU KNOW?**
>
> The two basic ways of representing the numerical value of physical quantities are analog (continuous) and digital (discrete).

It is important to note that the displays used for the analog systems are the analog meters (having moving pointers).

10.4.1. Drawbacks of Analog Systems

(i) Analog systems are less accurate.
(ii) Analysis of analog systems require time-domain analysis which requires very strong mathematical background.
(iii) Analog systems are usually affected more by the undesired electrical disturbance called noise.
(iv) Once the noise contaminates an analog signal, it is impossible to separate the signal from noise. The signal contaminated by noise leads to degradation of quality of output.
(v) The performance of analog systems degrades because of component ageing.
(vi) Their performance changes with variation in temperature.
(vii) Analog systems have less reliability.
(viii) Analog systems have less versatility since no software is included.
(ix) Because of the use of analog meters, it is quite difficult to take accurate reading.
(x) Storage and processing of data is not possible since they do not have memory.
(xi) In analog systems, computerized control cannot be exercised.

10.4.2 Performance Comparison of Analog and Digital Systems

S. No.	Parameter of comparison	Analog systems	Digital systems
1.	Type of signals processed	Analog signals	Digital signals
2.	Type of display	Analog meters	Digital displays using LED and LCD
3.	Accuracy of system	Small	High
4.	Design complexity	Difficult to design	Easier to design
5.	Memory attached	No memory	They have memory
6.	Storage of information	Not possible	Possible
7.	Noise effect	High	Small
8.	Versatility	Small	High
9.	Distortion	High	Small
10.	Effect of temperature and ageing on performance	High	Small
11.	Communication between systems	Not easy	Easy
12.	Few Examples	Filters, amplifiers, power supplies, signal generators	Counters, registers, micro-processors, computers

10.5 SALIENT FEATURES OF DIGITAL SYSTEMS

As a matter of fact, digital systems have following salient features (advantages) over analog systems :

(i) Digital systems manipulate discrete elements of information.

(ii) Discrete elements are nothing but the *digits* such as 10 decimal digits, 26 letters of alphabets and so on.

(iii) Digital systems make use of physical quantities called *signals* to represent discrete elements.

(iv) In digital systems, the signals have two discrete values and are therefore said to be binary.

(v) A signal in digital system represents one binary digit, called a *bit*. The bit has a value either 0 or 1.

(vi) Discrete elements of information are represented with group of bits called *binary codes*. As an example, the decimal digits 0 through 9 are represented in a digital system with 4-bit code called *BCD (Binary Coded Decimal) code*.

(vii) The digital systems like a digital computer is a programmable device which may be programmed to perform a variety of tasks.

KARL FRIEDRICH GAUSS

In 1799 the *German mathematician Karl Friedrich Gauss*, at the ripe age of 22, proved the fundamental theorem of algebra, namely that every algebraic equation in one unknown has a root in the form of a complex number. He showed that every equation of the *n*th order has exactly n solutions (roots), no more and no less. Gauss was also one of the first to give a coherent account of complex numbers and to interpret them as points in a complex

(viii) The digital systems like a digital computer is an interconnection of digital modules. The major units/modules of digital computer are central processing unit, memory unit and input-output unit. The memory unit stores programs and data. The central processing unit (CPU) performs arithmetic and logical operations on data stored in the memory specified by the program. Further, input device such as keyboard is used to store program and data into the memory and display device such as printer, monitor is used to print/display the result of operation.

Thus, based upon discussion, we may define digital system as under :

Definition : A digital system may be defined as an interconnection of digital modules and it is a system that manipulates discrete elements of information that is represented internally in the binary form.

10.6 DETAILED COMPARISON OF ANALOG AND DIGITAL SYSTEMS

Analog systems process information that varies continuously, i.e., they process time-varying signals that may take on any value across a continuous range of voltage, current or any physical parameter. On the other hand, the digital systems use discrete quantities to represent information. Discrete means distinct or separated as opposed to continuous or connected. Figure 11.2 shows the difference between two signals : continuous and discrete.

The digital systems represent levels of discrete signals, usually, in a binary number system. The binary number system has only two characters : 0 and 1. The numbers formed by these two characters are used to represent the discrete levels of digital signal.

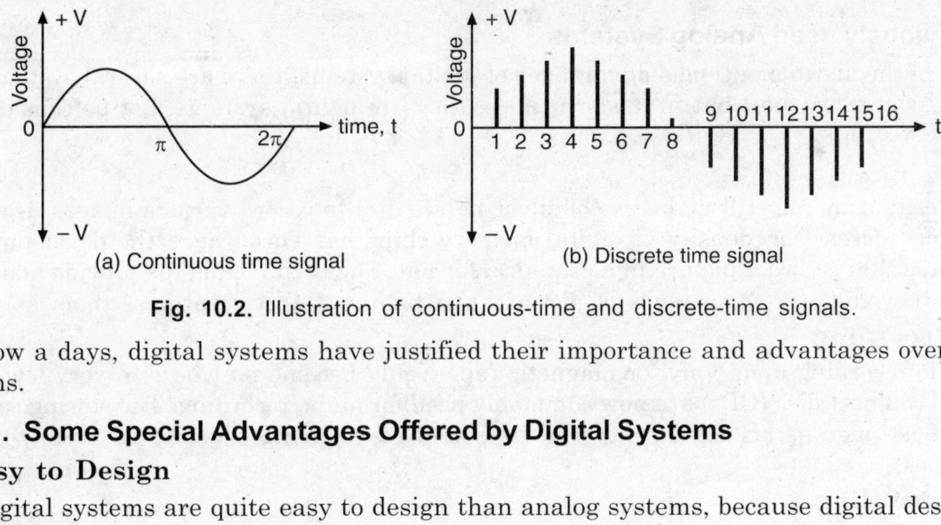

Fig. 10.2. Illustration of continuous-time and discrete-time signals.

Now a days, digital systems have justified their importance and advantages over the analog systems.

10.6.1. Some Special Advantages Offered by Digital Systems

1. Easy to Design

Digital systems are quite easy to design than analog systems, because digital design involves logic design which does not require special math skills and its behaviour may be visualized part-by-part.

To visualize the behaviour of analog systems is quite difficult. Actually, it requires the special insights about the analog components such as capacitors, inductors and transistors used in the analog systems.

2. Reliability and Reproducibility of Results

The output of analog systems varies with the variation in temperature, power-supply voltage, component-aging, and several other factors. Therefore, it is difficult to reproduce the same result each time with the same sets of inputs and circuit components. However, this is not the case with digital systems. Digital systems always reproduce exactly same results with same set of inputs and circuit components.

3. Flexibility

Digital systems are more flexible to design since their design involves a set of logical steps and have various discrete and integrated options available to the user.

4. Functionality

It is quite easy to provide additional functionality in digital systems than analog systems. As an example, in digital systems, it is quite easy to add password facility. However, in analog systems, it is a difficult job.

5. Programmability

Now a days, digital design is carried out by writing programs, in *hardware description language* (HDL). These languages allow us to stimulate and test the performance of digital circuits. This feature is quite useful in designing critical digital systems for various applications.

6. Device Speed

Today's digital devices are very fast and they can produce 500 millions or more results since their operating speed is smaller than 2 nanoseconds.

7. Economy in Manufacturing

Digital circuits can easily be integrated into a single chip and can be produced in large quantity to have smaller cost compared to discrete circuits.

8. Upgrading Technology

As digital technology is becoming more and more popular, more research is going on in this field. Therefore, more technological upgrade is expected in the digital world and we know that after about every six months, there is a launching of a new processor, new memory technology etc.

10.6.2. Previously used Analog Systems

Because of the above mentioned advantages of digital systems, there are several systems that were analog systems earlier, but in modern times, they are digital systems. Let us consider few of them.

1. Cameras

The majority cameras still make use of silver-halide film to record various images. However, the continuous increase in density of digital memory chips has encouraged the development of digital cameras that record a picture in a graphical format. The digital cameras provide additional facility that they can store thousands of pictures and has an ability to process them as well.

2. Audio Recording

The audio recording upon vinyl or magnetic tape would become no where in very few years. Since, digital compact disc (CDs) are now commonly used for audio recording. The storing capacity of one CD is as huge as six billion bits and its sound reproducing capacity is far better than magnetic tapes.

3. Video Recording

Like audio recording, now a days, the digital recording is achieved using digital versatile disc (DVD). Actually, a DVD can store video information in a highly compressed digital format known as MPEG-2. Infact, the capacity of a single layer, single-side DVD is about 35 billion bits, which is sufficient for about 2 hours of high-quality video, and a two-layer, double-sided disc has capacity four times that of a single-side.

4. Telephone System

Most of the telephone exchanges now make use of digital signals in place of analog signals.

5. Special Effects in Advertisement and Movies

The digital processing and creation have allowed to add special effects in advertisement and movies for various applications.

10.6.3. Examples of Digital Systems*

Some of the examples of digital systems used in day-to-day life may be listed as under :

1. In communication systems
2. Business transactions
3. Traffic control
4. Medical treatment
5. Industrial applications
6. Digital telephones
7. Digital TV
8. Digital cameras
9. Weather monitoring
10. Digital computers

10.7 PROCESS OF A TO D CONVERSION

The analog to digital (A/D) conversion involves the following steps:

(i) Sampling
(ii) Holding
(iii) Quantizing
(iv) Encoding

A sample and hold circuit performs the sampling and holding operations simultaneously. Quantizing and encoding operations are performed simultaneously by the A to D converter or ADC. The A to D and D to A converters are used in the microprocessor based control systems, they are referred as peripherals or I/O (input/output) devices.

10.8 DAC FUNDAMENTALS

Since D to A converters are used as a part of some A to D converters, let us first discuss the DAC.

10.8.1. Circuit Symbol and Transfer Characteristics

The circuit symbol and input-output characteristics of a 4-bit DAC have been shown in figure 10.3. There are four digital inputs and one analog output. d_0 is the most significant bit MSB and d_3 is the least significant bit LSB. The 4-bit digital word will have sixteen different possible combinations from 0000 to 1111. Figure 10.4 shows the ideal transfer characteristics for a four bit DAC.

In figure 10.4, we have plotted the 16 different combinations of the four bit digital input (0000 to 1111) on the X-axis and the corresponding output voltages on the Y-axis.

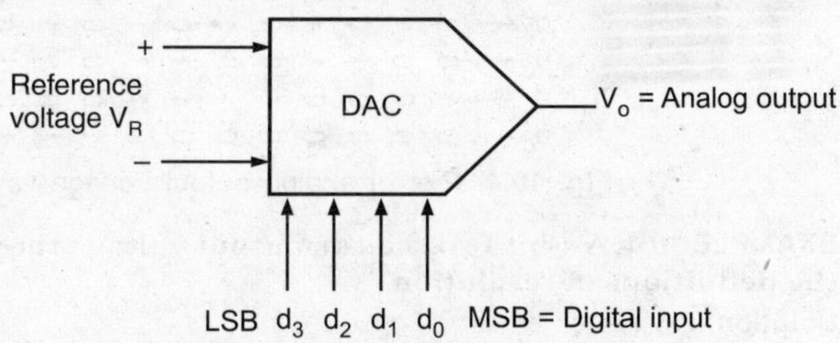

Fig. 10.3. Circuit symbol of DAC

Definition of Resolution

Resolution of a DAC can be defined in following two different ways:

(i) Resolution is the number of different analog output voltage values that can be provided by a DAC. For an n-bit DAC,

$$\text{Resolution} = 2^n \qquad ...(10.1)$$

Hence, the resolution of a 4-bit DAC is $2^4 = 16$ and that of a 3-bit DAC is $2^3 = 8$. Therefore, resolution increases with the increase in number of bits.

(ii) Resolution is defined as the ratio of change in analog output voltage resulting from a change of 1 LSB at the digital input.

* Most modern electronics equipment contains both analog and digital circuitry.

To calculate the resolution with this definition, we need to know the full scale analog output voltage V_{FS} and the number of digital inputs n. The full scale analog output voltage V_{FS} is defined as the output voltage corresponding to the digital input with all digits 1. Hence, the resolution is given by,

$$\text{Resolution} = \frac{V_{FS}}{2^n - 1} \quad ...(10.2)$$

where, n = Number of digital inputs.

Resolution can be treated as the smallest change in the analog output voltage. Resolution will improve with reduction in the smallest change in the output voltage.

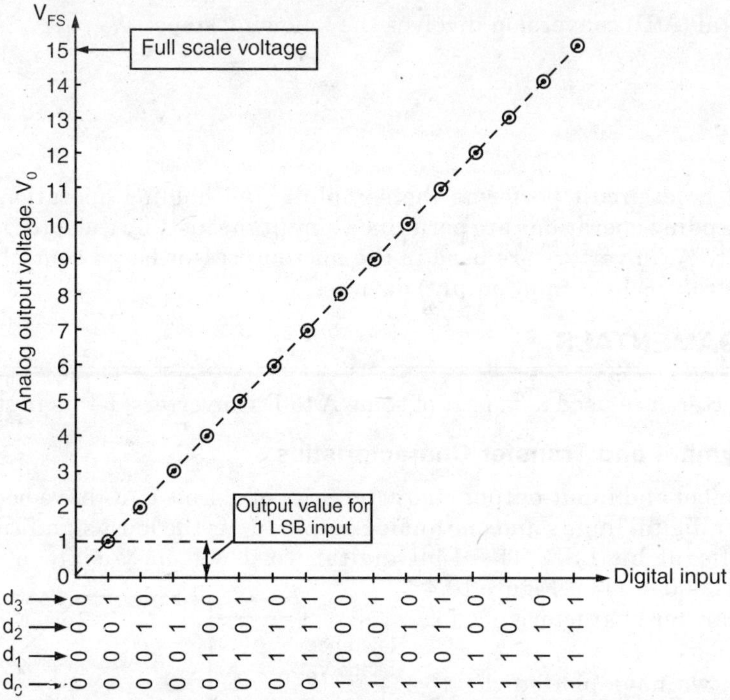

Fig. 10.4. Plot of analog output voltage vs digital input code for a 4-bit DAC

EXAMPLE 10.1. A 4-bit DAC has an output voltage range 0 – 5 V. Calculate its resolution using both the definitions of resolution.

Solution:

(*i*) We know that the resolution is given by,

$$\text{Resolution} = 2^n = 2^4 = 16 \quad \textbf{Ans.}$$

(*ii*) The resolution is also given by,

$$\text{Resolution} = \frac{V_{FS}}{2^n - 1} = \frac{5V}{2^4 - 1} = 0.333 \text{ V} \quad \textbf{Ans.}$$

Thus, an input change of 1 LSB causes the output to change by 0.333 Volts.

10.8.2. Input-Output Equation

(i) The input output equation of a DAC can be used to calculate the analog output voltage corresponding to the given digital input word. If the input to a DAC block is an n bit digital word and analog output voltage is V_o, then the expression for the analog output voltage V_o is given by,

$$V_o = V_{FS}\,[d_1\,2^{-1} + d_2\,2^{-2} + ... + d_n 2^{-n}] \qquad ...(10.3)$$

where, V_{FS} = Full scale output voltage.

$d_1, d_2 .....d_n$ = n-bit binary fractional word with the decimal point at extreme left

d_1 = MSB (most significant bit) with a weight of $\frac{V_{FS}}{2}$.

d_n = LSB (least significant bit) with a weight of $\frac{V_{FS}}{2^n}$.

(*ii*) Let us substitute V_{FS} by V_R i.e. the reference voltage to write

$$V_o = K\,V_R\,[d_1\,2^{-1} + d_2\,2^{-2} + ... + d_n 2^{-n}] \qquad ...(10.4)$$

Where, K = scaling factor

(*iii*) If we substitute $D = [d_1\,2^{-1} + d_2\,2^{-2} + ... + d_n 2^{-n}]$,

Then $V_o = K\,V_R\,D$ ...(10.5)

where, D is called as the fractional binary value.

Multiplying DAC (MDAC)

From equation (10.5), it is evident that the output of a DAC is obtained by multiplying the reference voltage (V_R) and the digital variable (D). Hence, the DAC is also called as a multiplying DAC or MDAC.

10.8.3. Types of DAC

Basically, the DACs are of two types as shown in figure 10.5.

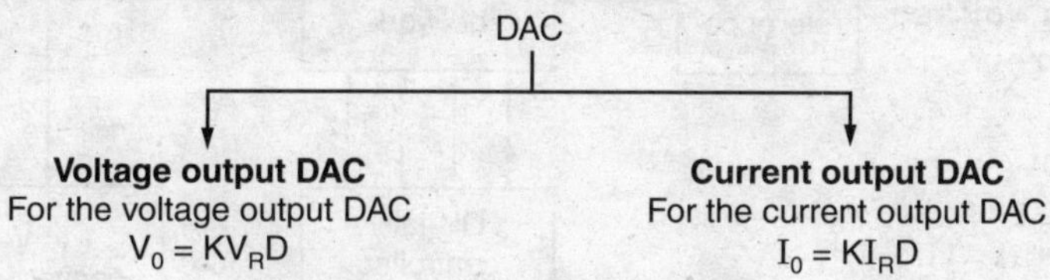

Fig. 10.5. Classification of DACs

EXAMPLE 10.2. Calculate the analog output of a 4-bit DAC if the digital inputs is 1011. Assume V_{FS} = 5V.

Solution: The 4-bit digital word is $d_1\,d_2\,d_3\,d_4$ - 1011 with V_{FS} = 5 Volts and K = 1, then, the corresponding analog output voltage V_o can be obtained as under:

$$V_o = 5\left[\left(1\times2^{-1}\right)+0+\left(1\times2^{-3}\right)+\left(1\times2^{-4}\right)\right]$$

$$V_o = 5\,[0.5 + 0 + 0.125 + 0.0625\,]$$

$$V_o = 3.4375 \text{ Volts}$$

Similarly, we can convert the digital input words $d_1\,d_2\,d_3\,d_4$ = 0000 and 1111 into analog voltages of V_o = 0 Volts and 4.6875 Volts respectively. An alternative method to obtain the analog output voltage for a given digital input as under:

The output voltage V_o can be obtained by multiplying the resolution by the change in LSB.

This means that, V_o = Resolution × D ...(10.6)

where D = Decimal value of the digital input.

EXAMPLE 10.3. The digital input word for a 4-bit DAC changes from 0000 to 0101. Determine the DAC output.

Solution: Let us assume V_{FS} = 15V.

(i) The decimal equivalent of 0101 i.e. D is 5. It has been given that n = 4.

Therefore, $$\text{Resolution} = \frac{V_{FS}}{2^n - 1} = \frac{15\text{ V}}{2^4 - 1} = 1\text{ V/LSB} \quad \textbf{Ans.}$$

(*ii*) From equation (10.6), we have

$$V_o = \text{Resolution} \times D = \frac{1\text{V}}{\text{LSB}} \times 5 \text{ LSB} = 5 \text{ Volts} \quad \textbf{Ans.}$$

EXAMPLE 10.4. An 8-bit DAC has a resolution of 15 mV/LSB. Determine:

(i) Full scale output voltage V_{FS}.

(ii) Output voltage when input digital word is 0001 0010.

Solution: (i) Full scale output voltage V_{FS}

The full scale output voltage corresponds to the digital input with all digits equal to 1. i.e. when the input is equal to 1111 1111. The decimal equivalent of D is = 255.

Therefore,
$$V_{FS} = \text{Resolution} \times D = \frac{15 \text{ mV}}{\text{LSB}} \times 255 \text{ LSB} = 3.825 \text{ Volts} \quad \textbf{Ans.}$$

(ii) Output voltage

We have
$$V_o = \text{Resolution} \times D$$
$$D = 18 \text{ corresponding to } 0001\ 0010$$

or,
$$V_o = \frac{15 \text{ mV}}{\text{LSB}} \times 18 \text{ LSB} = 270 \text{ mV} \quad \textbf{Ans.}$$

10.9 DIGITAL TO ANALOG (D TO A) CONVERSION

The block diagram of a resistive type DAC has been shown in figure 10.6. The basic building blocks of a DAC are a resistive network, digitally controlled electronic switches, a voltage reference and current to voltage converter.

(i) A digital input code is applied to the resistive network via the digitally controlled switches.

(ii) The digitally controlled switches are turned on or off by the digital inputs bits (0's and 1's).

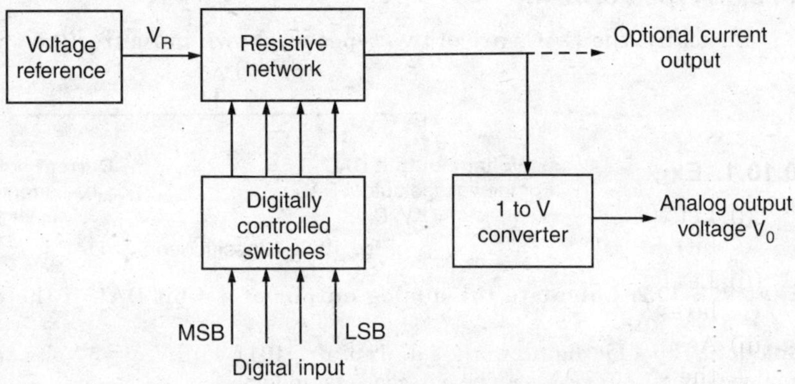

Fig. 10.6. Block diagram of a DAC

(iii) The output of the resistive network is in the form of current. It can be converted into a proportional voltage with the help of a current to voltage converter. Thus we obtain an analog output voltage proportional to the digital input code.

(iv) The actual digital to analog conversion takes place within the resistive network.

Depending on the type of resistive network used, we have different types of D to A converter circuits. Some of them are as under:

(a) Binary weighted resistor DAC

(b) R – 2 R ladder type DAC

(c) Inverted R – 2R ladder type DAC.

10.10 BINARY WEIGHTED RESISTOR DAC

The circuit diagram of binary weighted resistor type DAC has been shown in figure 10.7. This circuit uses a network of binary weighted resistors and a summing amplifier. The resistors 2^1R, 2^2R ...., 2^nR are from the network of binary weighted resistors.

There are n number of electronic switches used, one per digital bit. They are single pole double throw (SPDT) type switches. The position of the moving arm of every switch is controlled by the binary input word. For example, if the binary bit $d_1 = 1$, then, the first SPDT switch will connect resistor 2^1R to a negative reference voltage $(-V_R)$ and when the binary input is 0, then this switch will connect this resistor to ground. Thus, an n-bit digital word will decide the positions of all the SPDT switches and connect their corresponding binary weighted resistor to either $-V_R$ or ground. Depending upon the positions of various switches, the currents I_1 to I_n will start flowing through the resistors 2^1R to 2^nR respectively as shown in figure 10.7.

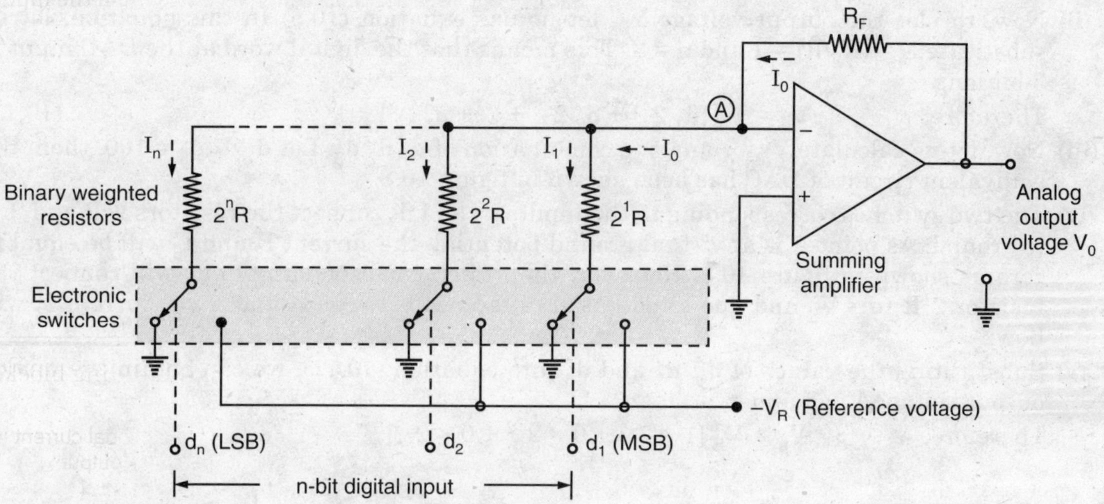

Fig. 10.7. Weighted resistor DAC

10.10.1. Expression for Analog Output Voltage

(i) Let the n-bit digital input word to the DAC be $d_1\, d_2 \ldots d_n$ with d_1 as MSB (most significant bit) and d_n as LSB (least significant bit).

(ii) Let R_F be the feedback resistor and I_0 be the output current. Let the op-amp in figure 10.7 be an ideal op-amp so that the current flowing between its input terminals is zero.

(iii) Assuming the op-amp of figure 10.7, to be ideal, the output current I_0 can be expressed as the sum of individual currents through the weighted resistors.

Thus,
$$I_0 = I_1 + I_2 + \ldots + I_n = \frac{V_R}{2R} d_1 + \frac{V_R}{2^2 R} d_2 + \ldots + \frac{V_R}{2^n R} d_n \qquad \ldots(10.7)$$

Here, it may be noted that $d_1, d_2 \ldots\ldots d_n$ can have a value of either 0 or 1.

Thus,
$$I_0 = \frac{V_R}{R}\left[d_1 2^{-1} + d_2 2^{-2} + \ldots + d_n 2^{-n}\right] \qquad \ldots(10.8)$$

(iv) The output voltage V_0 is given by,

$$V_0 = I_o R_F = V_R \frac{R_F}{R}\left[d_1 2^{-1} + d_2 2^{-2} + \ldots + d_n 2^{-n}\right] \qquad \ldots(10.9)$$

This is the required expression for output voltage.

(v) Now, comparing equations (10.8) and (10.9), we conclude that.

$$K = \frac{R_F}{R} \text{ and } V_{FS} = V_R.$$

Hence, if $R_F = R$, then $K = 1$

i.e., $$V_0 = V_R\,[d_1\, 2^{-1} + d_2\, 2^{-2} + ... + d_n 2^{-n}] \quad ...(10.10)$$

By substituting the values of d_1, d_2.....d_n in equation (10.9), we can obtain the value of corresponding analog output voltage V_o.

10.10.2. Output Voltage Waveform of a 3-bit DAC

(i) The circuit of DAC shown in figure 10.7 uses an inverting summing amplifier. Hence, the use of a negative reference voltage ($-V_R$) gives a positive analog output voltage.

(ii) Now, to plot the output voltage V_o, let us use equation (10.9) In this equation, let us substitute $K = (R_F/R) = 1$ and $n = 3$. This means that the digital word at the DAC input is 3-bit long.

Therefore, $$V_o = V_R\,[d_1\, 2^{-1} + d_2\, 2^{-2} + ... + d_3 2^{-3}] \quad ...(10.11)$$

(iii) Now, let us calculate V_o for any one combination of $d_1\,d_2\,d_3$. Let $d_1\,d_2\,d_3 = 100$, then, the equivalent circuit of DAC has been shown in figure 10.8.

(iv) The two switches corresponding to the inputs d_2 and d_3 connect the resistors 2^2R and 2^3R to ground. As point A is at virtual ground potential, the current I_2 and I_3 will be equal to zero as shown in figure 10.8. However, the switch corresponding to d_1 will connect the resistor 2^1R to $-V_R$ and due to potential difference between A and $-V_R$, current I_1 will flow.

(v) Substituting the values of d_1, d_2 and d_3 into equation (10.11), we can obtain the analog output voltage V_o as under:

Therefore, $$V_o = V_R\,[1 \times 2^{-1} + 0 \times 2^{-2} + 0 \times 2^{-3}]$$

or $$V_o = \frac{V_R}{2}$$

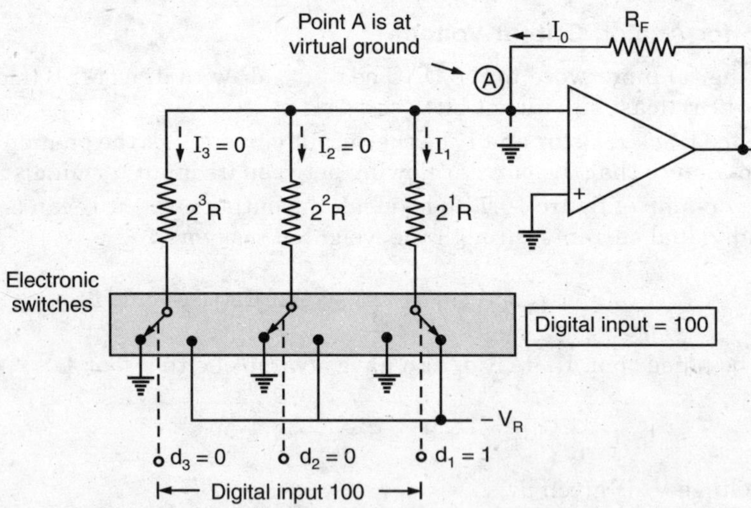

Fig. 10.8. Equivalent circuit of DAC for digital input $d_1\,d_2\,d_3 = 100$

Similarly, we can obtain the output voltage for all the possible input digital words as shown in table 10.1. The graph of digital input code versus output voltage V_o is plotted in figure 10.9. Thus, the DAC output is a staircase waveform.

Table 10.1.

Digital input			Analog output voltage V_0
d_1	d_2	d_3	
0	0	0	0
0	0	1	$V_R/8$
0	1	0	$2\,V_R/8$
0	1	1	$3\,V_R/8$
1	0	0	$4\,V_R/8$
1	0	1	$5\,V_R/8$
1	1	0	$6\,V_R/8$
1	1	1	$7\,V_R/8$

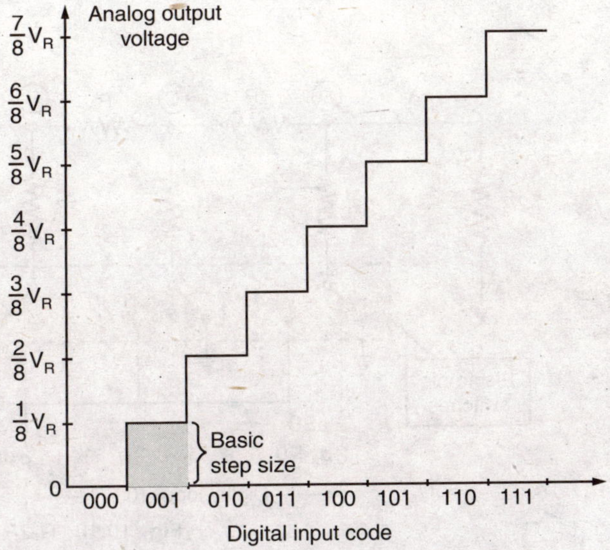

Fig. 10.9. Staircase output for DAC

Important Points about Weighted Resistor DAC

Some other important points about the weighted resistor DAC may be listed as under:

(i) It is possible to connect the op-amp in the non-inverting summing amplifier configuration instead of using the inverting configuration.

(ii) In this circuit (figure 10.7), the op-amp operates as a voltage to current converter.

(iii) The polarity of reference voltage depends also on the type of switches used. For example, if we use the TTL switches, then, the reference voltage $V_R = +5V$ and with the DAC circuit of figure 10.7, the output voltage will be negative.

Drawbacks of Weighted Resistor DAC

(i) The accuracy and stability of this type of DAC depends on the accuracy of the resistors used.

(ii) This type of DAC requires a wide range of resistor values. If the number of digits n per binary word is 8, then, the smallest resistor is $2^0 \times R = \Omega$, while the largest one is 2^7 R i.e. 128 RΩ. Thus, the largest resistor is 128 times, the smallest one. This proportion will be worse for still higher values of n (say n = 12).

(iii) For n = 12, if the smallest resistance is 2.5 KΩ, then, the largest one will be $2^{11} \times 2.5$ KΩ = 5.12 MΩ. The fabrication of such a large resistance in an IC form is not practically possible. The smallest resistance cannot be smaller than 2.5 KΩ. This is necessary to avoid the loading effect.

(iv) The difficulty to achieve and maintain accurate ratios over such a wide range of resistor values restricts the use of weighted resistor DAC for the values of n less than 8-bits, therefore, the resolution of such a DAC is poor.

(v) The finite resistance of the switches will disturb the currents particularly in the most significant bit (MSB) position, where the current setting resistors are small in value.

10.11 R-2R LADDER DAC

The problem of using a wide range of resistor values can be solved by using the R-2R ladder type DAC.

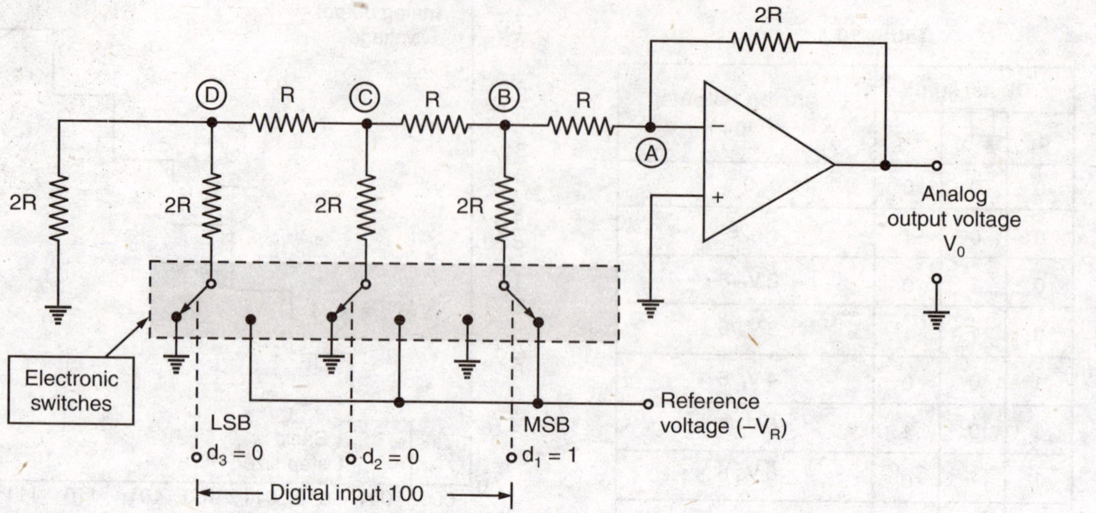

Fig. 10.10. R-2R ladder DAC

The circuit diagram for R – 2R ladder type DAC has been shown in figure 10.10. It shows that only two values of resistors are required. This method is therefore suitable for the integrated circuit realization. The value of R in figure 10.10 can be anywhere between 2.5 KΩ to 10 KΩ.

10.11.1. Working Operation of 2-2R ladder DAC

(i) Let us consider figure 10.10, where the number of digits per binary word is assumed to be 3 (i.e. n = 3). The switch positions indicate that the binary words is $d_1\, d_2\, d_3 = 100$.

(ii) The original circuit can now be simplified as shown in figures 10.11 and 10.12 respectively.

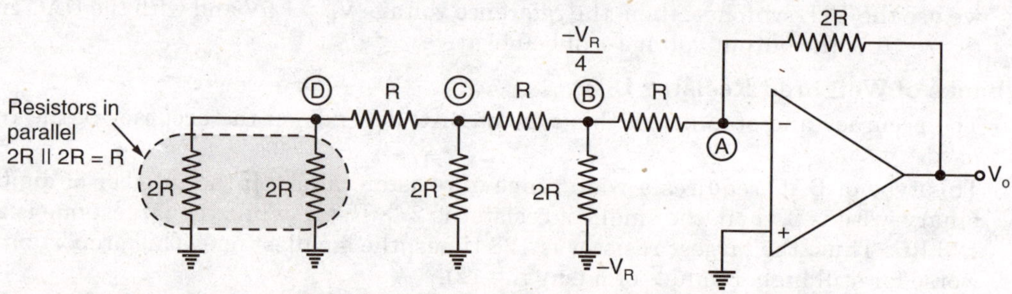

Fig. 10.11. Simplified R-2R ladder DAC

(iii) The simplified circuit shown in figure 10.11 gets further reduced to the simplified equivalent circuit shown in figure 10.12. The equivalent resistance to the left of node B in figure 10.10 is only 2R and node A is at virtual ground potential.

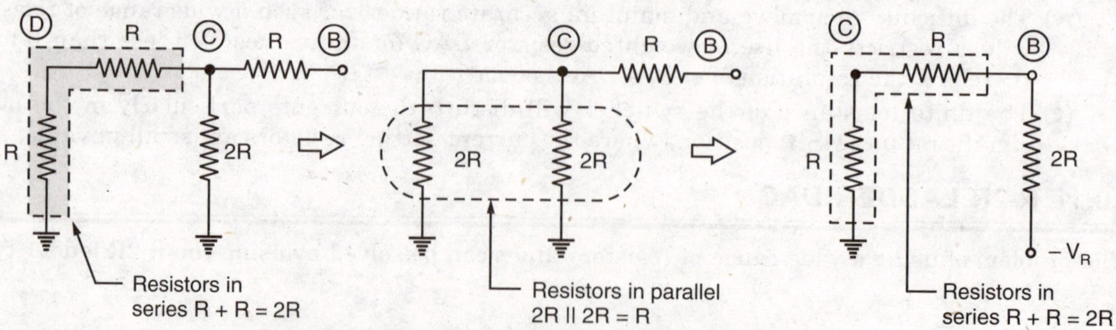

Fig. 10.12. Simplified equivalent circuits to the left of B

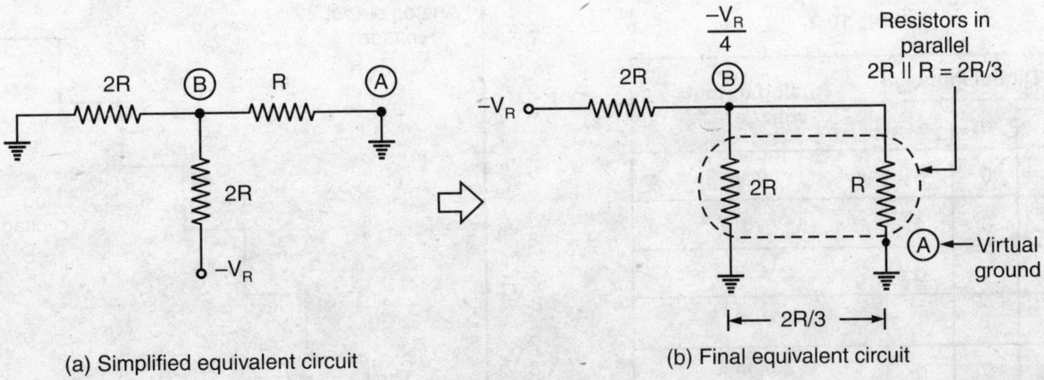

Fig. 10.13.

(iv) The simplified equivalent circuit figure 10.13(a) is redrawn as shown in figure 10.13(b). As the two resistors R and 2R are in parallel with each other, their parallel combination results in a resistance of (2R/3)Ω.

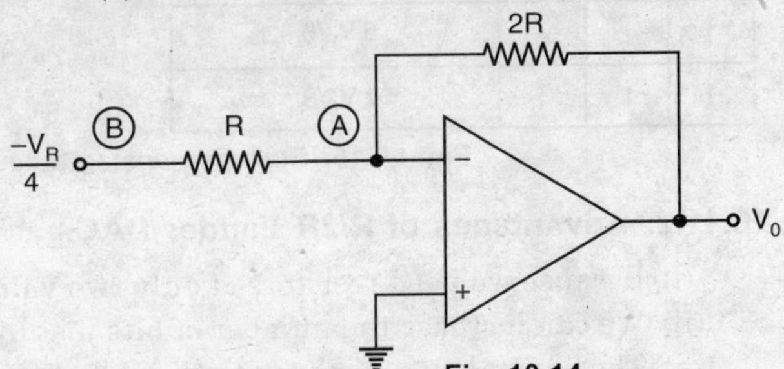

Fig. 10.14.

(v) Voltage at node B is given by,

$$V_B = \frac{(2R/3)}{2R + (2R/3)} \times -V_R$$

$$\text{or } V_B = \frac{(2R/3)}{\left[\dfrac{6R + 2R}{3}\right]} \times -V_R = \frac{-V_R}{4} \qquad \text{...(10.12)}$$

(vi) Considering the op-amp to be an inverting amplifier, the equivalent circuit of the DAC has been shown in figure 10.14.

Hence, the output voltage is given by,

$$V_o = \left[\frac{-R_F}{R_i}\right] V_{in} = \frac{-2R}{R} \times \frac{-V_R}{4} = \frac{V_R}{2} = \frac{V_{FS}}{2} \qquad \text{...(10.13)}$$

Thus, for a binary input of 100, the analog output produced is $\frac{V_R}{2}$.

(vii) Similarly, it is possible to obtain the analog output voltages for other digital input words. Table 10.2 shows the digital inputs and corresponding analog output voltages and figure 10.15 shows the staircase output voltage waveform obtained for R-2R ladder DAC.

Table 10.2

Digital input			Analog output voltage V_0
d_1	d_2	d_3	
0	0	0	0
0	0	1	$V_R/8$
0	1	0	$2\ V_R/8$
0	1	1	$3\ V_R/8$
1	0	0	$4\ V_R/8$
1	0	1	$5\ V_R/8$
1	1	0	$6\ V_R/8$
1	1	1	$7\ V_R/8$

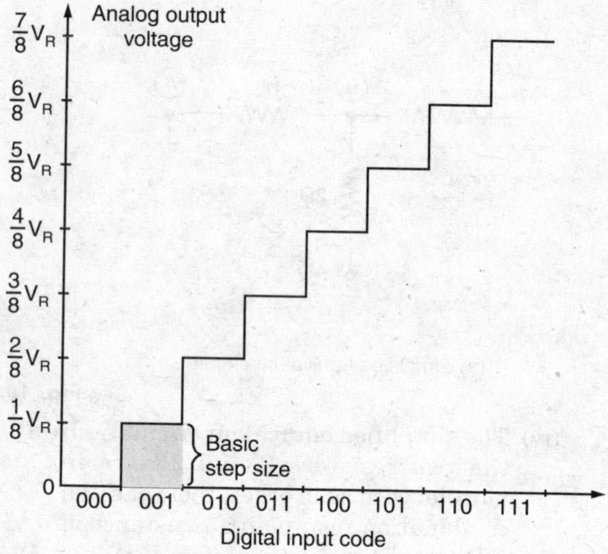

Fig. 10.15. Staircase analog output voltage for R-2R ladder DAC

10.11.2. Advantages of R/2R Ladder DACs

(i) Because we need resistors of only two values, it is easier to build this circuit accurately.

(ii) We can increase the number of bits just by adding more sections of same R/2R values.

(iii) The equivalent resistance to the right of each labelled node (A, B, C..) will be equal to 2R. Hence, current flowing downwards, away from each node is equal to the current flowing towards right.

(iv) Due to the small resistance spread (only 2 to 1), the R-2R ladder can be fabricated monolithically, with a high accuracy and stability.

EXAMPLE 10.5. For the 2-2R ladder DAC shown in figure 10.10 prove that the output voltage $V_o = V_R/8$ Volts for a three bit digital input $d_1\ d_2\ d_3 = 001$. Draw the necessary equivalent circuits to support your answer.

Solution: With the digital input to be $d_1\ d_2\ d_3 = 001$, switches d_1 and d_2 will connect their corresponding resistors to ground and d_3 will connect it to $-V_R$. The equivalent circuit will then be as shown in figure 10.16.

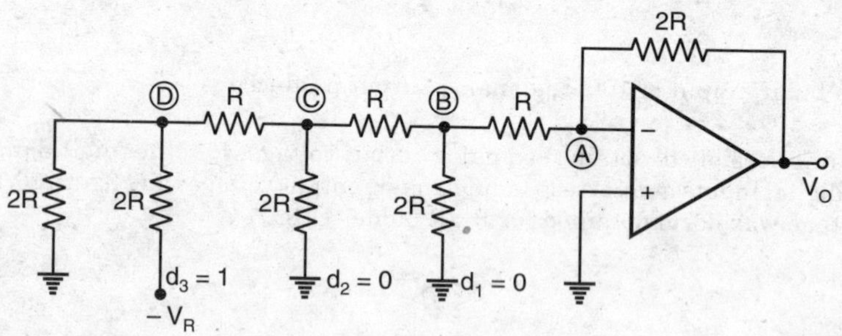

Fig. 10.16.

This circuit can be further simplified as shown in figures 10.17(a) and (b).

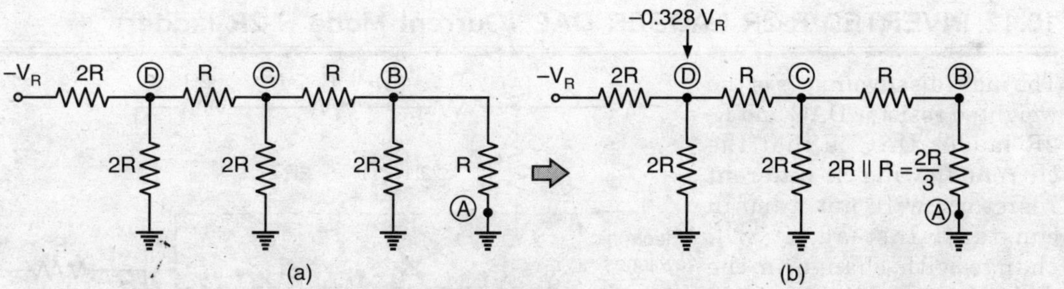

Fig. 10.17

The Potential at node D is given by,

$$V_D = \frac{R_{eq}}{2R + R_{eq}} \times -V_R \qquad \text{...(i)}$$

where
$$R_{eq} = \left\{\left[\left(\frac{2R}{3} + R\right) \| 2R\right] + R\right\} \| 2R$$

$$R_{eq} = \left\{\left[\left(\frac{5R}{3} \| 2R\right)\right] + R\right\} \| 2R + \left\{\left[\frac{10R}{11} + R\right]\right\} \| 2R = \frac{21R}{11} \| 2R$$

or
$$R_{eq} = \frac{\frac{21R}{11} \times 2R}{\frac{21R}{11} + 2R} = \frac{42R}{43} = 0.976\ R$$

Substituting in equation (i), we get,

$$V_D = \frac{0.976\ R}{2R + 0.976\ R} \times -V_R = -0.328\ V_R \qquad \text{...(ii)}$$

Now, let us consider the resistive network to the right of node C i.e. C to A.

The equivalent resistance between C and A is given by,

$$R'_{eq} = \left[\left(\frac{2R}{3} + R\right)\right] \| 2R = \frac{5R}{3} \| 2R = \frac{\left(\frac{5R}{3}\right) \times 2R}{\left(\frac{5R}{3}\right) + 2R} = \frac{10R}{11} = 0.909\ R$$

Therefore, voltage at node C is $V_C = \frac{R'_{eq}}{R'_{eq} + R} \times V_D = \frac{0.909\ R}{0.909\ R + R} \times -0.328\ V_R$

or $\qquad V_C = -0.156\ V_R \qquad \text{...(iii)}$

Similarly, we can obtain the voltage at node B as under:

$$V_B = \frac{\left(\frac{2R}{3}\right)}{R + \left(\frac{2R}{3}\right)} \times -0.156\ V_R$$

or $\qquad V_R = -0.0625\ V_R \qquad \text{...(iv)}$

The output voltage is given by

$V_o =$ (Gain of inverting amplifier)

$$\times V_B = \frac{-2R}{R} \times V_B$$

or $\qquad V_o = -2 \times -0.0625\ V_R$

or $\qquad V_o = 0.125\ V_R$ Volts **Ans.**

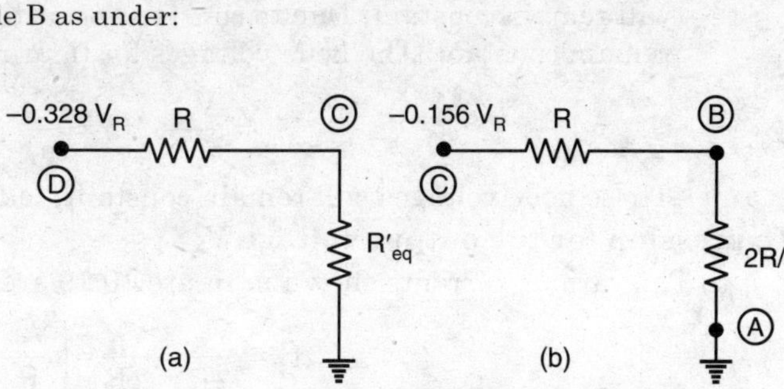

Fig. 10.18.

Here, $0.125\ V_R$ is same as $\frac{V_R}{8}$, hence, we have proved that for a digital input of 001, the analog output is $\frac{V_R}{8}$ Volts.

10.12 INVERTED R-2R LADDER DAC (Current Mode R-2R ladder)

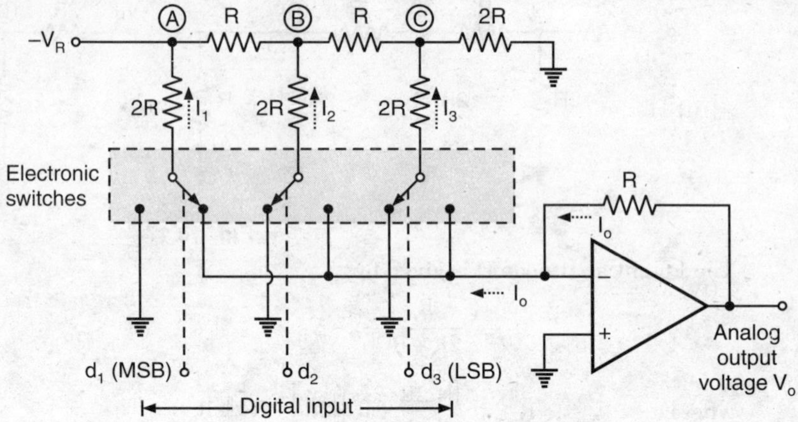

Fig. 10.19. Inverted R-2R ladder DAC (current mode R-2R DAC)

The main disadvantages of the weighted resistor DAC and R-2R ladder DAC is that the current flowing in different resistors will not remain constant. Instead, it will change with change in the digital input to DAC. This will cause more power dissipation in the resistors carrying more current which in turn causes heating of the resistors. The resistance value changes due to heating which will introduce non-linearity in DAC. This problem can be solved by using the *Inverted R-2R ladder* type DAC. The circuit diagram of a 3-bit inverted R-2R ladder type DAC has been shown in figure 10.19 This is called as *inverted* R-2R ladder DAC because the positions of MSB and LSB in the original R-2R ladder DAC have been interchanged. The circuit shown in figure 10.19 is also called as the **current mode R-2R ladder DAC.** This is because this circuit operates on the ladder currents.

Working Operation of the circuit

(i) Figure 10.19 shows a 3-bit inverted R-2R ladder type DAC.

(ii) In this circuit, each input binary word connects the corresponding switch to either ground or to the inverting terminal of the op-amp. Here, it may be noted that the inverting (–) terminal of the op-amp is at virtual ground potential.

(iii) Thus, both the terminals of each switch are at ground potential. Hence, the currents flowing through the resistors will remain constant irrespective of the switch position i.e. independent of the input binary word.

(iv) The another important property of this circuit is that the total current gets divided equally at each node. This is due to the fact that the resistance to the left and to the right of each node is same i.e., equal to 2R.

(v) Due to equal current division at each node, the current in each branch of the R-2R ladder will remain constant. Due to constant branch currents, the ladder node voltages also will remain constant. The node voltages for figure 10.19 will be,

$$V_A = \frac{V_R}{2^0},\ V_B = \frac{V_R}{2^1},\ V_C = \frac{V_R}{2^2}$$

These node voltages will remain constant, independent of the binary inputs.

Expression for the output voltage V_o

(i) The various currents shown in figure 10.19 are given by

$$I_1 = \frac{V_A}{2R} = \frac{V_R}{2R} = \left(\frac{V_R}{R}\right)2^{-1}$$

$$I_2 = \frac{V_B}{2R} = \frac{V_R}{2R \times 2} = \left(\frac{V_R}{R}\right)2^{-2}$$

and

$$I_3 = \frac{V_C}{2R} = \frac{V_R}{2R \times 2^2} = \left(\frac{V_R}{R}\right)2^{-3}$$

These currents will either come from the ground point or from the virtual ground point.

(ii) Hence $I_o = d_1 I_1 + d_2 I_2 + d_3 I_3$

or $$I_o = \frac{V_R}{R}\left[2^{-1} b_1 + 2^{-2} b_2 + 2^{-3} b_3\right]$$

(iii) Hence, the output voltage is given by,

$$V_o = -R_F \times I_o$$

$$V_o = -\frac{R_F}{R} \times V_R\left[b_1 2^{-1} + b_2 2^{-2} + b_3 2^{-3}\right] \quad ...(10.15)$$

(iv) We can substitute $\left(-\frac{R_F}{R}\right) = K$ to get

$$V_o = K V_R\left[b_1 2^{-1} + b_2 2^{-2} + b_3 2^{-3}\right] \quad ...(10.16)$$

(v) Due to the constant node voltages, the stray capacitances cannot produce the *slow down* effect on the performance of the DAC circuit. This is the biggest advantage of the inverted R-2R ladder DAC.

10.12.1. Advantages of Inverted R-2R Ladder DAC

(i) Current flowing through resistors remain constant irrespective of the switch position.

(ii) Total current gets divided equally at each node.

(iii) Current in each branch of R-2R ladder will remain constant and the node voltages are also constant.

(iv) The *slow down* effect due to stray capacitance is not present.

EXAMPLE 10.6. The basic step size of a 4-bit DAC is 312.5 mV. If 0000 represents 0V, calculate the analog output voltage produced for the digital input of 1101.

Solution: Let us consider following expression:

$$V_o = V_{FS}\left[d_1 2^{-1} + d_2 2^{-2} + d_3 2^{-3} \;....\; + d_n 2^{-n}\right] \quad ...(i)$$

Assuming n = 4, we get $$V_o = V_{FS}\left[d_1 2^{-1} + d_2 2^{-2} + d_3 2^{-3} + d_4 2^{-4}\right] \quad ...(ii)$$

The basic step size is the output voltage which corresponds to the digital input of $d_1\, d_2\, d_3\, d_4$ = 0001.

Substituting these values into equation (ii), we get,

$$V_o = V_{FS}\left[0 \times 2^{-1} + 0 \times 2^{-2} + 0 \times 2^{-3} + 1 \times 2^{-4}\right]$$

Therefore, basic step size $$V_o = \frac{V_{FS}}{2^4} = \frac{V_{FS}}{16} \quad ...(iii)$$

It has been given that basic step size is 312.5 mV.

Therefore, $$312.5 \times 10^{-3} = \frac{V_{FS}}{16}$$

or $$V_{FS} = 312.5 \times 10^{-3} \times 16 = 5 \text{ Volts} \quad ...(iv)$$

New, let us calculate the analog output for the digital input of 1101.

Thus, $$V_o = 5\left[1 \times 2^{-1} + 1 \times 2^{-2} + 0 \times 2^{-3} + 1 \times 2^{-4}\right]$$

$$V_o = 5\left[\frac{1}{2} + \frac{1}{4} + \frac{1}{16}\right]$$

$$V_o = 4.0625 \text{ Volts}$$ **Ans.**

EXAMPLE 10.7. For an 8-bit DAC, calculate the values of LSB, MSB and full scale output voltage for the range of 0 to 10V.

Solution: Given that n = 8 and V_R = 10V

(i) The value of LSB

Let us use following expression to calculate the value of LSB :

$$V_o = V_R\left[d_1 2^{-1} + d_2\, 2^{-2} + ... + d_n 2^{-n}\right] \quad ...(i)$$

To calculate the value of LSB, only the LSB should be high and all other bits should be low i.e. the digital input should be 00000001. Substituting this in the expression above, we get,

$$V_o = 10\ [0 + 0 + 0 + 0 + 0 + 0 + 0 + 1 \times 2^{-8}]$$

$$V_o = \frac{10}{2^8} = \frac{10}{256} = 39.06 \text{ mV} \quad \textbf{Ans.}$$

Thus, the value of LSB is 39.6 mV.

(ii) The value of MSB

To calculate the value of MSB, the digital input should be 1000 0000. Substituting this in the equation for output voltage, we get,

$$V_o = 10\ [1 \times 2^{-1} + 0 + 0 + 0 + 0 + 0 + 0]$$

Therefore, value of MSB, $V_o = \frac{10}{2^1} = 5 \text{ Volts.}$ **Ans.**

(iii) The full scale output voltage

To calculate the full scale output voltage, the digital input should be 1111 1111. Substituting this in equation (i), we get,

Full scale output voltage $= 10\left[1\times 2^{-1} + 1\times 2^{-2} + 1\times 2^{-3} + ... + 1\times 2^{-8}\right]$

$$= 10\left[\frac{1}{2} + \frac{1}{4} + \frac{1}{8} + \frac{1}{16} + \frac{1}{32} + \frac{1}{64} + \frac{1}{128} + \frac{1}{256}\right] = 9.961 \text{ Volts} \quad \textbf{Ans.}$$

An Alternative method

Full scale output = Full scale voltage – 1 LSB

= 10 – 39.06 mV.

Therefore, full scale output = 9.961 V. **Ans.**

EXAMPLE 10.8. Calculate the analog output voltage of a DAC whose output voltage range is 0 to 5V, for the following digital inputs:

(i) 10 (for a 2-bit DAC)

(ii) 1010 (for a 4-bit DAC)

Solution: (i) For a 2-bit DAC with input to be 10, the output voltage is given by,

$$V_o = 5\left[1\times 2^{-1} + 0\times 2^{-2}\right] = 5/2 = 2.5 \text{ Volts} \quad \textbf{Ans.}$$

(ii) For a 4-bit DAC with input to be 1010, the output voltage is

$$V_o = 5\ \left[1\times 2^{-1} + 0\times 2^{-2} + 1\times 2^{-3} + 0\times 2^{-4}\right]$$

or $$V_o = 5\ \left[\frac{1}{2} + \frac{1}{8}\right] = 3.125 \text{ Volts} \quad \textbf{Ans.}$$

10.13 SPECIFICATIONS FOR DAC

The designers of D/A converter specify the following important characteristics (specifications of a D/A converter:

(i) Resolution
(ii) Accuracy
(iii) Linearity
(iv) Temperature sensitivity
(v) Settling time
(vi) Speed
(vii) Long term drift
(viii) Supply rejection.

Let us discuss them one by one.

(i) Resolution

Resolution describes the smallest possible change in the analog output voltage. Resolution should be as high as possible. Its value depends on the number of bits in the digital input applied to DAC. Higher the number of bits, higher is the resolution.

(ii) Accuracy

Accuracy indicates how close the analog output voltage is to its theoretical value. In short, it indicates the deviation of actual output from the theoretical value. Accuracy depends on the accuracy of the resistors used in the ladder and the precision of the reference voltage used. Accuracy is always specified in terms of percentage of the full scale output that means maximum output voltage e.g. if the full scale output is 15V and accuracy is + 0.1 percent, then the maximum error is given by $0.001 \times 15 = 0.015$ or 15 mV.

(iii) Linearity

The relation between the digital input and analog output should be linear. However, practically, it is not so due to the error in the values of resistors used for the resistive networks.

(iv) Temperature Sensitivity

The analog output voltage of D to A converter should not change due to changes in temperature. But, practically, the analog output is a function of temperature. It is so because the resistance values and op-amp parameters change with changes in temperature.

(v) Settling Time

Theoretically, the analog output voltage should change instantaneously in response to the change in its digital input. Practically, the analog output of a D to A converter does not change instantaneously. Due to the resistors and op-amp in the circuit, oscillations are observed at the output. The time required to settle the analog output within 1/2 LSB of the final value, after the change in digital input is called as **settling time**. The settling time should be as short as possible.

(vi) Speed

It is defined as the time needed to perform a conversion from digital to analog. It is also defined as the number of conversions that can be performed per second. The speed of DAC should be as high as possible.

(vii) Long Term Drift

This results mainly due to resistor and semiconductor ageing and can affect all the characteristics. Characteristics mainly affected are linearity, speed etc.

(viii) Supply Rejection

It indicates the ability of DAC to maintain scale, linearity and other important characteristics when the supply voltage is varied. Supply rejection is usually specified as percentage of full scale change at or near full scale at 25°C.

10.14 SOURCE OF ERRORS IN DACs

There are three type of errors present in the transfer characteristics of DACs. They are linearity, offset and gain errors. Let us discuss them one by one.

10.14.1. Linearity Error

This is defined as the amount by which the actual output of a DAC differs from the ideal straight line transfer characteristics. This error is introduced due to the error in the current source resistance values. Figure 10.20(a) shows the linearity error in the transfer characteristics of a DAC.

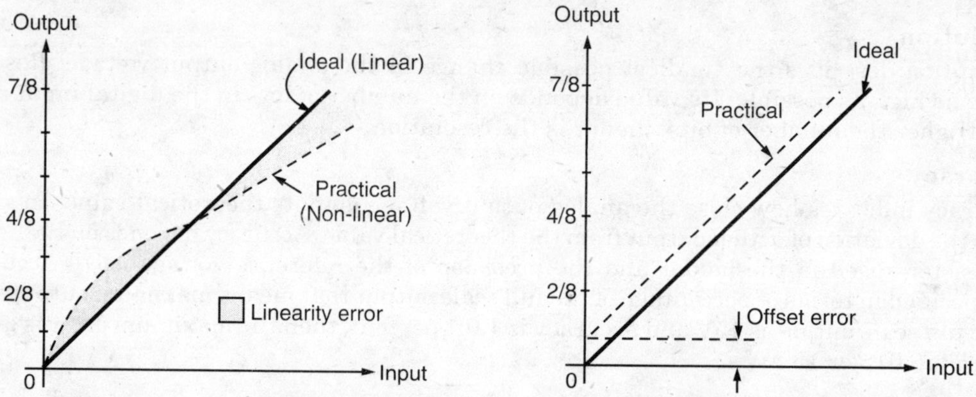

(a) Linearity error in transfer characteristics of D/A converter

(b) Offset error in transfer characteristics of a D/A converter

Fig. 10.20.

10.14.2. Offset Error

When all the digital inputs are 0, the analog output also is expected to be 0. But , in practice, it does not happen. As shown in figure 10.20(b), some non-zero analog output voltage is present even for a zero digital input. This is called as the **offset error**. Thus, the offset error is defined as the non-zero level of analog output when all the digital inputs are 0. The offset error is due to the offset voltages of op-amps and leakage currents in the switches.

10.14.3. Gain Error

In a DAC, we use a current to voltage converter. The gain of this converter determines the analog output voltage of the DAC. The gain error is defined as the difference between the calculated gain and practically obtained gain of the I to V converter. This error exists due to the error in the feedback resistor value. The gain error has been shown in figure 10.21.

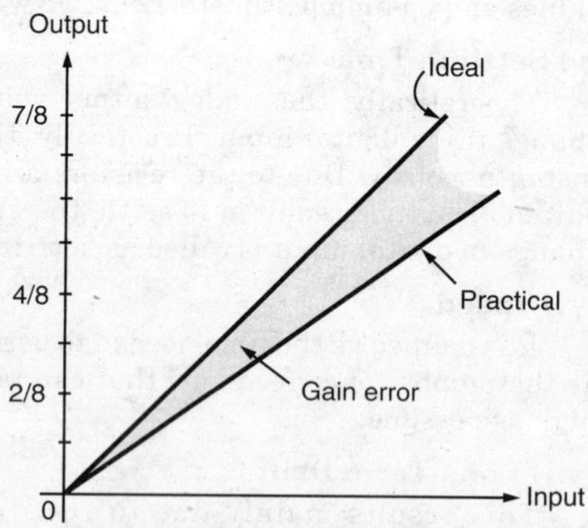

Fig. 10.21. Gain error in the transfer characteristics of a D/A converter

10.15 MONOLITHIC IC DAC 0808

The DAC 0808 is an 8-bit current output monolithic DAC manufactured by the National semiconductor corporation. It is a 16-pin IC available in dual in Line DIP plastic package. The analog output is available in the form of current I_o. This means that I_o is proportional to the 8-bit digital input. The important features of DAC 0808 are as under:

(i) Fast settling time 150 nsec. typically
(ii) Power supply voltage range + 4.5 mW at + 5V.
(iii) Low power consumption 33 mW at + 5V.
(iv) High speed multiplying input slew rate 8 mA/μsec.
(v) Interfaces directly with TTL, DTL and CMOS logic levels.

10.15.1. Pin Configuration and Functional Block Diagram

The pin configuration and functional block diagram of DAC 0808 have been shown in figures 10.22 and 10.23 respectively.

The internal block diagram shows that DAC 0808 consists of R-2R ladder along with current switches and reference current amplifier. Here, A_1 to A_8 are the 8-digital input lines with A_1 as the most significant bit and A_8 as the least significant bit. The analog output is available in the form of current I_o. Therefore, we need to use an external current to voltage converter if the analog output in the form of voltage is required. DAC 0808 requires a dual polarity (+) supply voltage, typically + 15 V, for its operation. The reference voltage can be either positive or negative. An external reference voltage should be applied to either V_{REF} (+) or V_{REF} (–) depending on the polarity of reference voltage.

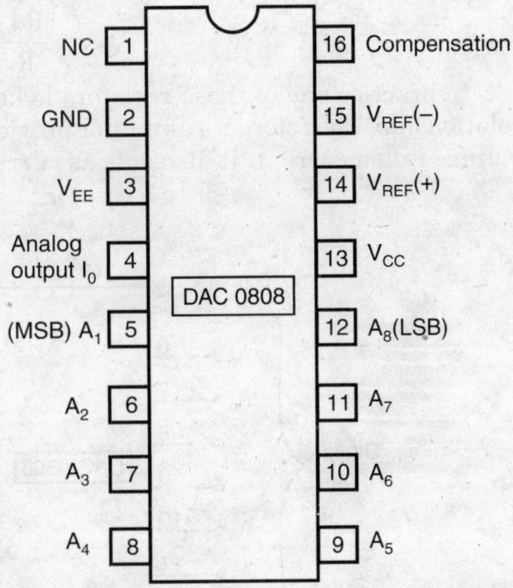

Fig. 10.22. Pin configuration of DAC 0808

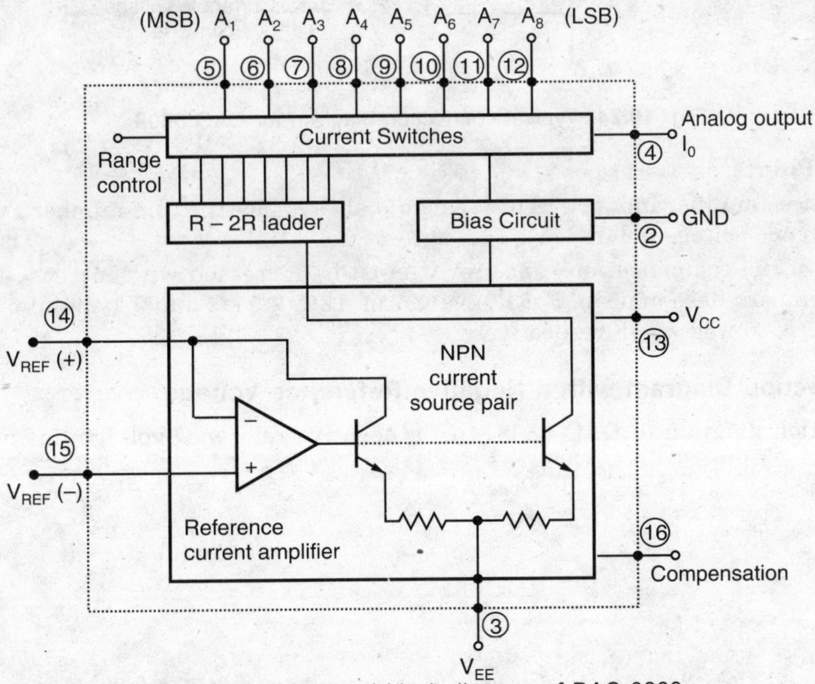

Fig. 10.23. Functional block diagram of DAC 0808

10.15.2. Typical Connection Diagram

The typical connection diagram for DAC 0808 with a positive reference voltage is shown in figure 10.24. This circuit provides a unipolar positive output voltage due to the use of an external I to V converter. As the reference voltage has been decided to be positive, an external positive voltage has been applied at pin number 14 (V_{REF} +) of the IC through R_{14}, connecting V_{REF} (–) i.e. pin number 15 to ground through R_{15}. Assuming that $R_{14} = R_{15} = R$, the full scale output voltage is given by,

$$V_{FS} = \frac{R_F}{R} \times V_{REF} \qquad ...(10.17)$$

In practice one of these resistors is kept adjustable in order to provide more accurate scaling relationship. The reference amplifier provides a voltage at pin number 14 for converting the reference voltage into a current. It also acts as current mirror for feeding the internal R-2R ladder network.

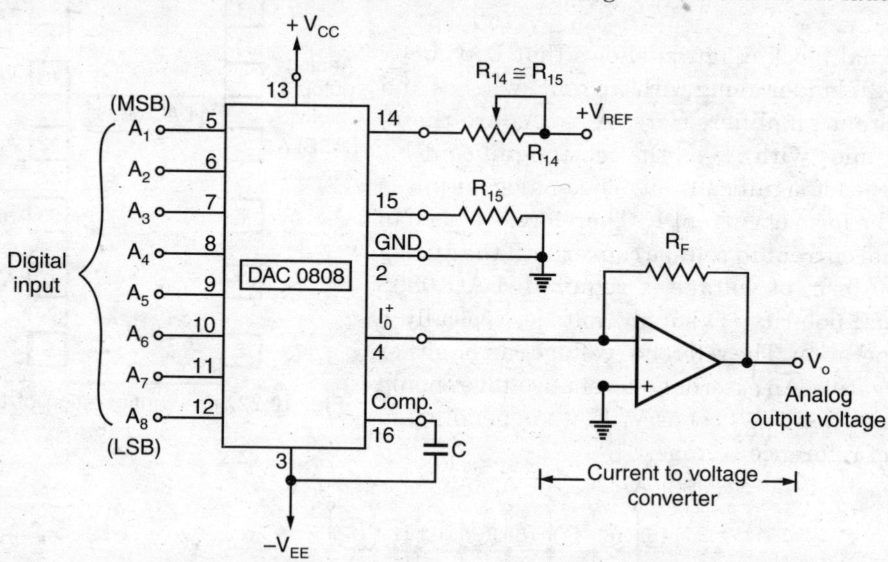

Fig. 10.24. Typical connection diagram for DAC 0808

Important Points:

(i) The reference amplifier input current I_{14} should always flow into pin number 14 irrespective of the reference voltage polarities.

(ii) The value of the compensation capacitor C must be increased with increase in the R_{14} to maintain proper phase margin. For R_{14} values of 1kΩ, 2.5 kΩ and 5 kΩ minimum capacitor values are 15, 37 and 75 pF respectively.

10.15.3. Connection Diagram with a Negative Reference Voltage

The connection diagram of DAC 0808 with a negative reference voltage has been shown in figure 10.25.

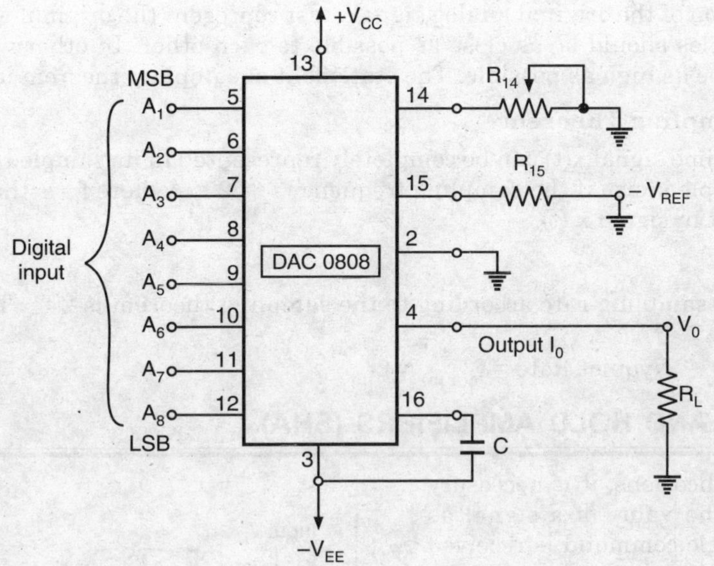

Fig. 10.25. Connection diagram for DAC 0808 with a negative reference voltage

10.16 SAMPLING THEOREM AND CHOICE OF SAMPLING FREQUENCY

In A to D converters, it is necessary that the analog input voltage V_A remains constant during the conversion time of ADC. Otherwise, the digital output produced by it will not be corresponding to the analog input signal. In order to maintain the analog voltage to be converted, constant during the conversion period, a technique called 'sampling' is used. Instead of using only sampler, we have to use a **sample and hold** circuit. The sampler samples the analog input voltage and the hold circuit will hold it constant for a sufficiently long period as shown in figure 10.26(b).

(i) In figure 10.26(a) , the analog signal V_A and a clock signal are applied to a sample and hold circuit. The analog signal is shown by a dotted line in figure 10.26(b). The clock signal is a train of narrow pulses at frequency f_s called sampling frequency.

(ii) Corresponding to the rising edges of the clock pulses i.e. at instants t_1, t_2, t_3, t_4 ......as shown in figure 10.26(b), the analog signal is sampled. The sample values are shown by small circles in figure 10.26(b).

(iii) The hold circuit will then maintain its output voltage constant equal to the sample value. The time for which a constant output maintained is τ sec., as shown in figure 10.26 (b).

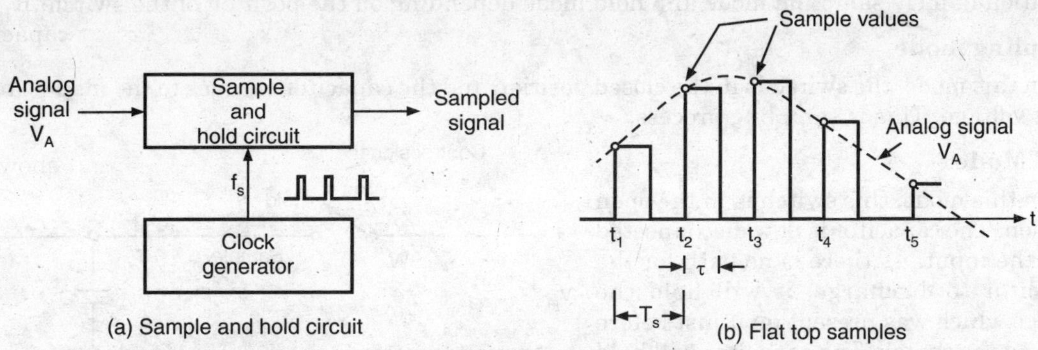

Fig. 10.26.

Thus, the sampled signal is produced. Here, it may be noted that the time duration between the successive sample is T_s sec. where $T_s = (1/f_s)$. Therefore, T_s is also called as **sampling period**.

The sampled version of the original analog signal must represent the original waveform faithfully. For this, the samples should be as close as possible to each other. In other words, the sampling frequency should be as high as possible. The statement of sampling theorem is as under:

Statement of Sampling Theorem

A continuous time signal x(t) can be completely represented in its sampled form and recovered back from the sampled form if the sampling frequency $f_s > 2\ f_m$, where f_m is the highest frequency component present in signal x (t).

Nyquist Rate

The minimum sampling rate according to the sampling theorem is $2\ f_m$. This is called as the Nyquist rate.

Therefore, Nyquist Rate $= f_{s(min)} = 2f_m$

10.17 SAMPLE AND HOLD AMPLIFIERS (SHA)

(i) In some applications, it is necessary to capture the value of a signal as soon as a logic command is received and hold the captured value until the new capture command is received.

(ii) The sample and hold amplifier (SHA) is a circuit which captures the value of input signal instantaneously, as shown in figure 10.27, and holds it in between the sampling instants.

(iii) The instantaneous capture (instantaneous sampling) shown in figure 10.27 is not practically feasible since the capture time τ tends to zero. Hence, the capture time is increased and the system is called as **track and hold amplifier (THA).**

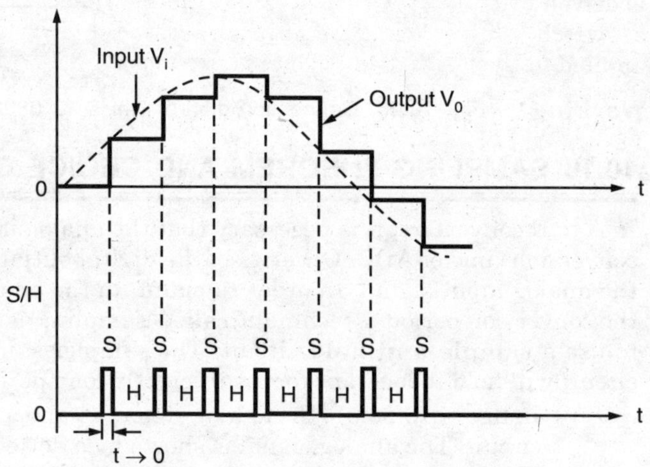

Fig. 10.27. Response of a sample and hold amplifier (SHA)

10.17.1 Basic Sample and Hold Circuit

The basic sample and hold circuit shown in figure 10.28, has one input, one output and an electronic switch which can be controlled by a control signal. There are two possible modes of operation namely sampling mode and hold mode depending on the position of the switch.

Sampling mode

In this mode, the switch is in the closed position and the capacitor charges to the instantaneous input voltage. This is sampling process.

Hold Mode

In this mode, this switch is in the open position. The capacitor is now disconnected from the input. As there is no path for the capacitor to discharge, it will hold the voltage which was present on it just before opening the switch. The capacitor will hold this voltage till the next sampling instant.

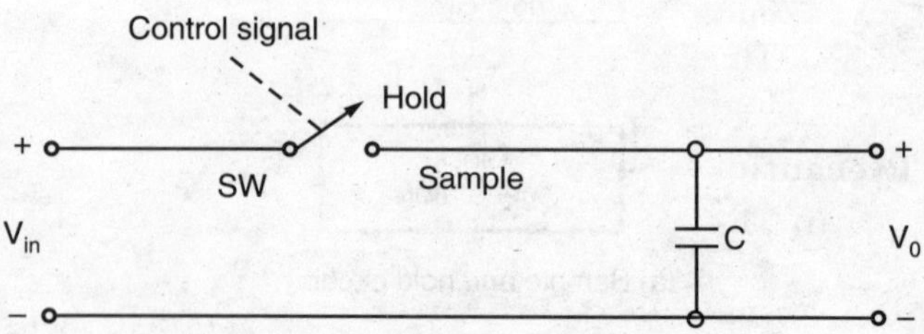

Fig. 10.28. Basic sample and hold circuit

10.17.2. Track and Hold Amplifier (THA)

Track and hold amplifier is actually an improved version of sample and hold amplifier. Instead of capturing the input for a very short time, the capture time is increased. The output will track the input for a finite time and the final value captured at the end of the tracking interval is held as shown in figure 10.29.

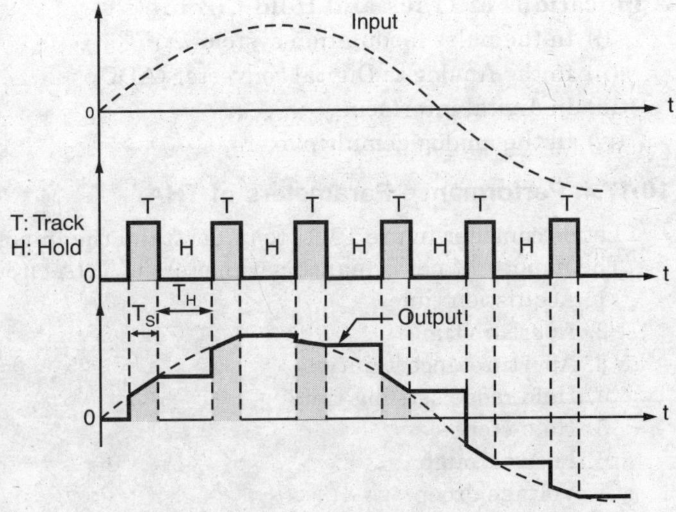

Fig. 10.29. Waveforms for track and hold amplifier.

10.17.3. Track and Hold Circuit Using op-amp

The track and hold circuit to obtain the waveforms of figure 10.29 is shown in figure 10.30. The N-channel MOSFET is driven by a control voltage V_C acts as a switch. The control voltage V_C is applied to the gate of the MOSFET.

Working Operation of the Circuit

(i) The circuit diagram of figure 10.30 can be split into three stages: First stage is the voltage follower, second one is the switch and capacitor and the third one is again the voltage follower.

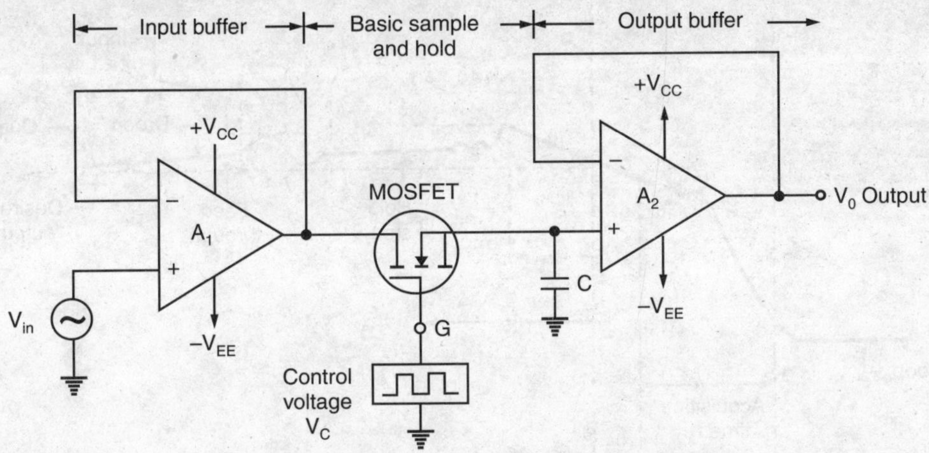

Fig. 10.30. Track and hold circuit using op-amp.

(ii) When V_C is high, the MOSFET turns on and acts like a closed switch. This is sampling mode. The capacitor charges through the MOSFET to the instantaneous input voltage.

(iii) As soon as $V_C = 0$, the MOSFET turns off and the capacitor is disconnected from op-amp 1 output. Capacitor cannot discharge through amplifier A_2 due to its high input impedance. Thus, this is the hold mode in which the capacitor holds the latest sample value. Waveforms for both these modes have been shown in figure 10.29.

Precautions

(i) The total period $T = T_H + T_S$ as shown in figure 10.29, where, T_H = holding time and T_S = sampling time. This total period should be at least half of one cycle period of the input signal. In other words frequency of V_C should be at least twice the frequency of V_{in}.

(ii) The capacitor C should be leak proof. We can use the polyester or Teflon capacitors.

Applications of Track and Hold Circuit

(i) In the pulse modulation systems
(ii) In the Analog to Digital converter (ADC)
(iii) In digital interfacing
(iv) In the analog demultiplexers.

10.17.4. Performance Parameters of THA

Let us consider figure 10.31 to understand the important parameters of THA.

The important performance parameters of THA circuit may/be described as under:

1. Acquistion time
2. Aperture time
3. Aperture uncertainty
4. Hold mode settling time
5. Hold step
6. Feed through
7. Voltage droop.

Let us understand these parameters one by one.

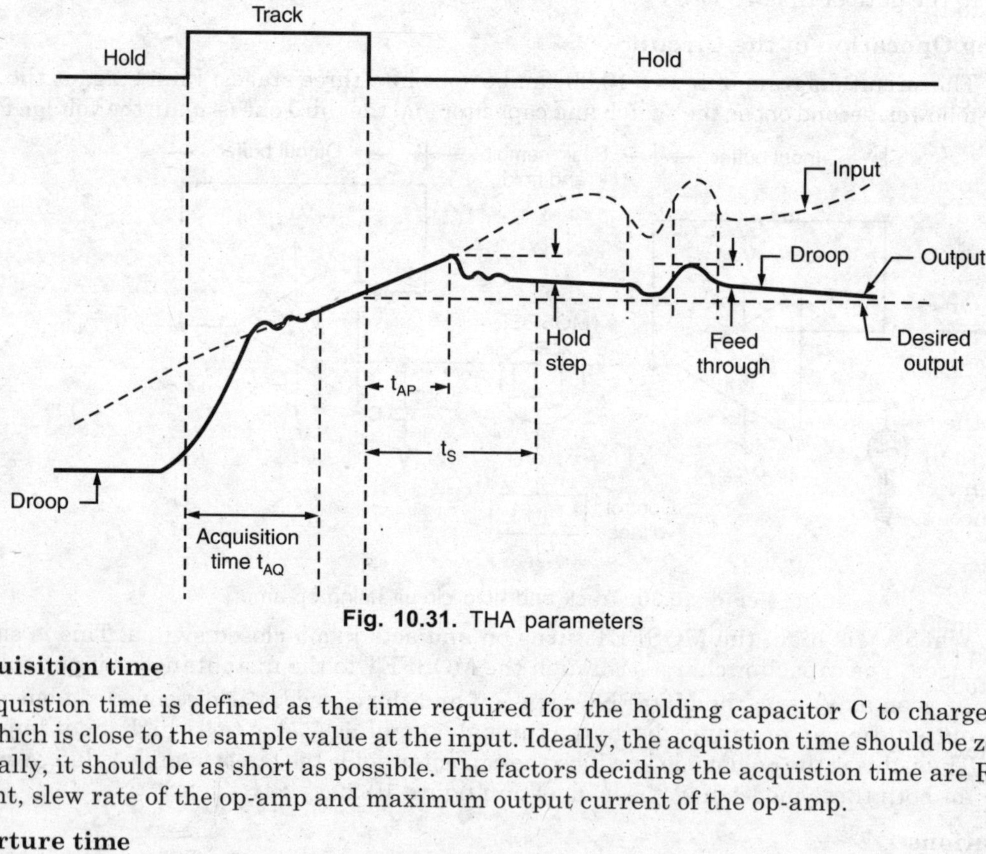

Fig. 10.31. THA parameters

1. Acquisition time

Acquistion time is defined as the time required for the holding capacitor C to charge upto a level which is close to the sample value at the input. Ideally, the acquistion time should be zero and practically, it should be as short as possible. The factors deciding the acquistion time are RC time constant, slew rate of the op-amp and maximum output current of the op-amp.

2. Aperture time

Ideally, as soon as the hold command is given to the S and H circuit, the circuit should stop following any changes taking place in the input and hold the latest sample value. But, practically, the S and H circuit will follow the changes in input voltage for a short time, even after receiving the hold command. This period is called as **aperture time**. It comes into existence due to the propagation delays of driver and the switch.

3. Aperture Uncertainty

The aperture time is not constant. It keeps varying depending on the level of sample. This is called as **aperture uncertainty**. Due to the aperture uncertainty, it becomes very difficult to compensate for the aperture time.

4. Hold mode settling time

It is the time taken by the output voltage V_o to settle within a specific error band, after the application of hold command to the S/H circuit. The error band can be of a specific value such as 1%, 0.1% or 0.01%

5. Hold step

At the time of switching from sample to hold mode, there is an unwanted charge transfer from the switch driver to the capacitor. This charge transfer takes place due to the parasitic capacitance associated with the switch. Due to this unwanted charge transfer the output voltage also changes. These unwanted changes in the output voltage are called as **hold step**. The other names for hold step are pedestal error and sample and hold offset.

6. Feed through

Due to the stray capacitance associated with the switch, a small amount of ac coupling takes place between the input and output during the hold mode. Thus, there are small variations in the output voltage during the hold mode which is called as **feed through**. Feed through is generally given as feed through rejection ratio (FRR) and it is expressed in dB.

7. Voltage droop

Due to non-ideal capacitor used, the voltage on the capacitor does not remain exactly constant in the hold mode. Instead, it reduces slightly due to the leakage current. This is called as voltage droop.

10.18 ANALOG TO DIGITAL CONVERTERS (ADC)

(i) The simple block diagram of an ADC has been shown in figure 10.32. ADC performs a function which is exactly opposite to that of a DAC. The input to an ADC is the analog voltage V_A and at the output we get an n bit digital word.

(ii) In the n-bit digital output, d_1 represents the most significant bit (MSB) and d_n is the least significant bit (LSB). The analog input voltage V_A produces an output digital word, having a functional value D given by,

$$D = \left[d_1 2^{-1} + d_2\, 2^{-2} + \ldots\ldots + d_n 2^{-n}\right] \qquad \ldots(10.18)$$

(iii) In addition to the analog input voltage V_A, the ADC block has a reference voltage V_R input and two control lines SOC and EOC. The start of conversion (SOC) input is used to start the A to D conversion whereas the end of conversion (EOC) output goes high to indicate that the conversion is complete.

(iv) The relation between the analog input voltage V_i and the digital output D is given by,

$$D = d_1 2^{-1} + d_2\, 2^{-2} + \ldots\ldots + d_n 2^{-n} = \frac{V_i}{K\, V_R} = \frac{V_i}{V_{FS}} \qquad \ldots(10.19)$$

Where, K = Constant

V_R = Reference voltage

V_{FS} = Full scale input voltage

V_i = Analog input voltage

Conversion Time

Let us consider figure 10.33. At $t = t_0$, a SOC signal is given to the ADC and at $t = t_1$, we get EOC output. The time difference between these two instants is called as **conversion time**. The conversion time should be as small as possible. Practically, it can have values from a few hundreds of μs to a few msec.

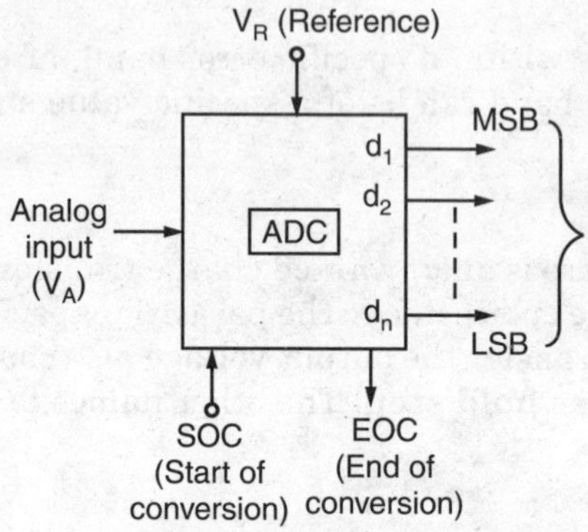

Fig. 10.32. Basic block diagram

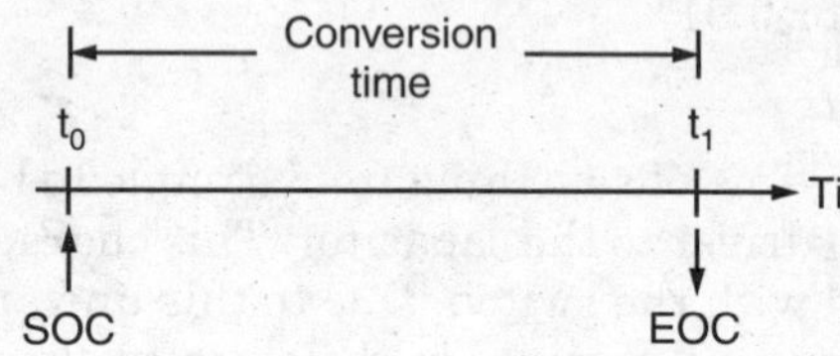

Fig. 10.33. Graphical representation of conversion time

10.19 TYPES OF A TO D CONVERTERS

There are various circuits available for the A to D conversions. They use different principles of operation for conversion. Some of them are as under :

(i) Flash type or simultaneous ADC
(ii) Counter type ADC
(iii) Tracking or servo type
(iv) Successive approximation A to D converter
(v) Dual slope A to D converter (Integrator type)

Out of them, the most commonly used ADCs are successive approximation type and integrator type ADC. All the circuits mentioned above have their advantages and demerits and they are preferred in specific applications. The A to D converters are broadly categorized into two types as **direct type** and **indirect type** ADCs.

10.20 ADC CIRCUITS

(PTU, Jalandhar, Sem. Exam., May, 2006)

In this section, let us discuss the following direct type ADCs:

(i) The parallel comparator ADC (Flash type ADC)
(ii) Digital ramp ADC
(iii) Successive approximation ADC.

10.20.1. The Parallel Comparator or Flash ADC

The parallel comparator ADC uses a number of op-amp comparators. Therefore, let us first revise the principle of a comparator. The principle of a comparator has been summarized in figure 10.34.

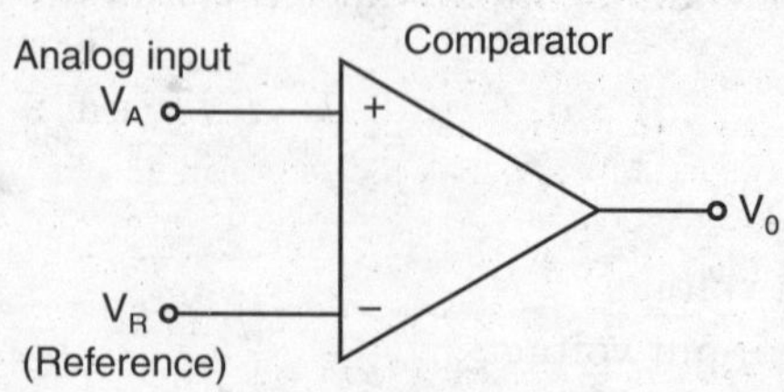

Condition of inputs	Output voltages
$V_A > V_R$	$V_0 = 1$
$V_A < V_R$	$V_0 = 0$
$V_A = V_R$	Previous value

Fig. 10.34. Principle of operation of a comparator

A Two Bit Parallel Comparator

The circuit diagram for a two bit parallel comparator ADC has been shown in figure 10.35. As observed from the circuit diagram, three comparators are being used. The non-inverting (+) terminal of each comparator has been connected to the analog input voltage V_A. The inverting (–) terminal of each comparator is connected to a different reference voltage. Here, V_1, V_2 and V_3 are the reference voltages for the three comparators respectively as shown in figure 10.35. These reference voltages have been derived from the voltages source V_R by using a resistive divider. The reference voltages are as under:

$$V_1 = \frac{V_R}{4},\ V_2 = \frac{V_R}{2}$$

$$V_3 = \frac{3\,V_R}{4}$$

The comparator outputs C_1, C_2 and C_3 are connected to an encoding network which will produce 2-bits equivalent to the analog input voltage. The encoder used is **priority encoder**.

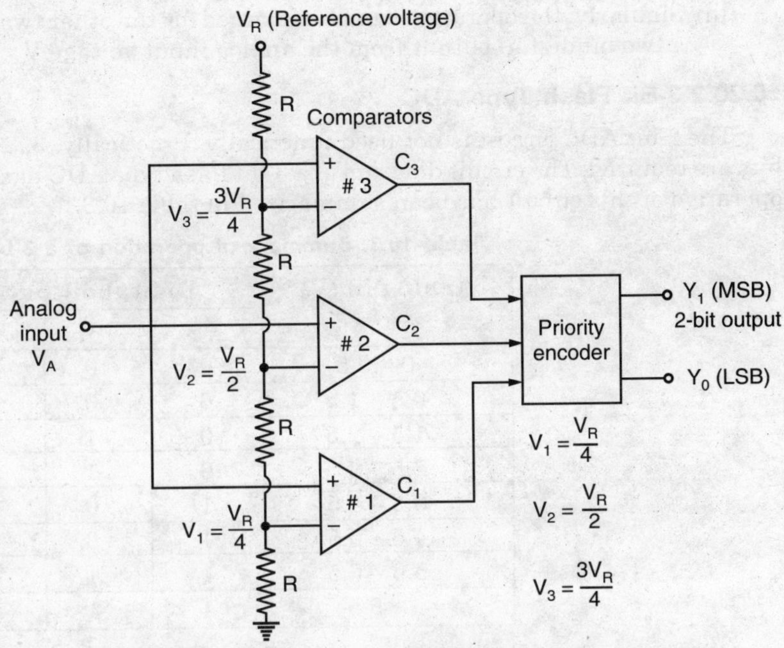

Fig. 10.35. 2 bit flash type (parallel comparator) ADC

The comparator outputs for various windows, and the priority encoder output are given in table 10.2.

Table 10.2. Summary of operation of 2-bit flash type ADC

Range Number	Analog input Voltage V_A Volts	Comparator outputs			Priority encoder outputs	
		C_3	C_2	C_1	Y_1	Y_0
1.	$0 < V_A < \frac{V_R}{4}$	0	0	0	0	0
2.	$\frac{V_R}{4} < V_A < \frac{V_R}{2}$	0	0	1	0	1
3.	$\frac{V_R}{2} < V_A < \frac{3V_R}{4}$	0	1	1	1	0
4.	$\frac{3V_R}{4} < V_A < V_R$	1	1	1	1	1

Working operation of two bit flash type ADC

(i) Let us consider the first range of input voltage, i.e., $0 < V_A < \frac{V_R}{4}$. In this range, the reference voltages V_1, V_2 and V_3 are all higher than the input voltage. Therefore, the output voltages of all the comparators are low, i.e., $C_1\ C_2\ C_3 = 000$ and the corresponding output of the priority encoder is 00 as shown in table 10.2.

(ii) In the second of input voltage, i.e., $\frac{V_R}{4} < V_A < \frac{V_R}{2}$, the reference voltage $V_1 = \frac{V_R}{4}$ is less than V_A. Therefore, the output of comparator 1 only is high. ($C_1 = 1$). The other two comparators produce a 0 output voltage ($C_2 = C_3 = 0$). The output of the priority encoder is 01 as shown in table 10.2.

(iii) Similarly, the operation can be explained for the other two voltage ranges. Thus, we obtain a two bit digital output from the analog input voltage V_A.

10.20.2 3-Bit Flash Type ADC

The 2-bit ADC circuit is not used practically. Practically, 3, 4 or even more number of output bits are required. The circuit diagram of a 3-bit flash type ADC has been shown in figure 10.36. The operation of this circuit has been summarized in table 10.3.

Table 10.3. Summary of operation of a 3-bit ADC

Analog in (V)	Digital output		
	d_2	d_1	d_0
0 – 0.5	0	0	0
0.5 – 1.5	0	0	1
1.5 – 2.5	0	1	0
2.5 – 3.5	0	1	1
3.5 – 4.5	1	0	0
4.5 – 5.5	1	0	1
5.5 – 6.5	1	1	0
> 6.5	1	1	1

(i) The resistance ladder shown in figure 10.36. is used to generate different reference voltages ranging from 0.5 V to 6.5 V.

(ii) These reference voltages are applied to the inverting terminals of comparators. The analog input V_A is applied to the non-inverting terminals of all the comparators.

(iii) The comparator outputs are applied to a priority encoder which produces a 3-bit digital output as shown in table 10.3.

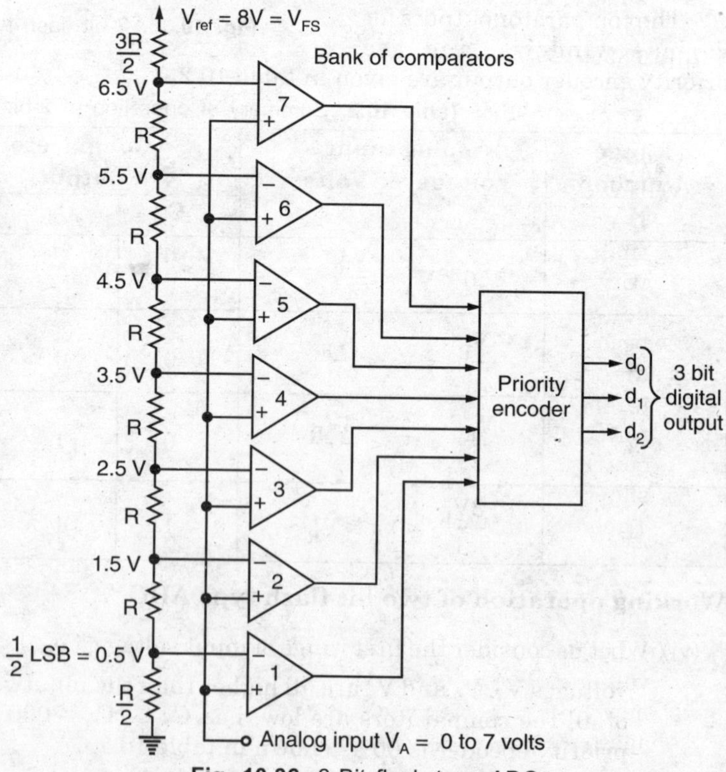

Fig. 10.36. 3-Bit flash type ADC

Advantages of Flash Type ADC

This circuit has the advantage of high speed of conversion as the A to D conversion takes place simultaneously and not sequentially. Typically, the conversion time of this type of ADC is 100 nanosec or less. Thus, this is the fastest ADC.

Disadvantages of Flash ADC

(i) The biggest disadvantage of this type of ADC is that a large number of comparators are required to be used. For a 2-bit ADC, three comparators are used, for a 3-bit ADC, we need seven comparators and for a 4-bit ADC, the number of comparators required to be used is fifteen. Thus, with increase in the number of bits by 1, the number of comparators will approximately be doubled.

(ii) With increased number of bits "n" the priority encoder becomes increasingly complex.

DAC based A-D conversion

It is possible to obtain the A-D conversion by using DAC. Some of the techniques that use the DAC are as under :

(i) Counter type ADC

(ii) Successive approximation (SAR) type ADC.

10.20.3. Counter Type ADC

For this type of A to D converter, we use a D to A converter i.e. DAC. Here, the digital input to this DAC is adjusted until the DAC's output comes within + (1/2) LSB of the analog output of DAC, i.e., V_A. (This is to be converted into a digital form by the ADC).

The circuit diagram of counter type ADC has been shown in figure 10.37.

Working Operation

(i) Figure 10.37 shows an n-bit counter type ADC. The counter is Reset to zero first, by applying a reset pulse. Then, after releasing the Reset pulse, the Clock pulses are applied to the counter through an AND gate.

(ii) Initially, the DAC output is zero. Therefore, the analog input voltage V_A is higher than the DAC output V_d. i.e. ($V_A > V_d$). The comparator output is high and the AND gate is enabled. Thus, the clock pulses are allowed to pass through the AND gate to the counter.

(iii) The counter starts counting these clock pulses, its output goes on increasing. As the counter output acts as input to DAC, the DAC output (which is a staircase waveform) also increases.

(iv) As long as DAC output $V_d < V_A$, this process will continue, as the comparator output remains high enabling the AND gate.

(v) However, as soon as the DAC output is higher than the input analog voltage ($V_d > V_A$), the comparator output becomes low. This will disable the AND gate and stop the clock pulses. Thus, the counting stops as shown in figure 10.38. The digital output of the counter then represents the analog input voltage V_A.

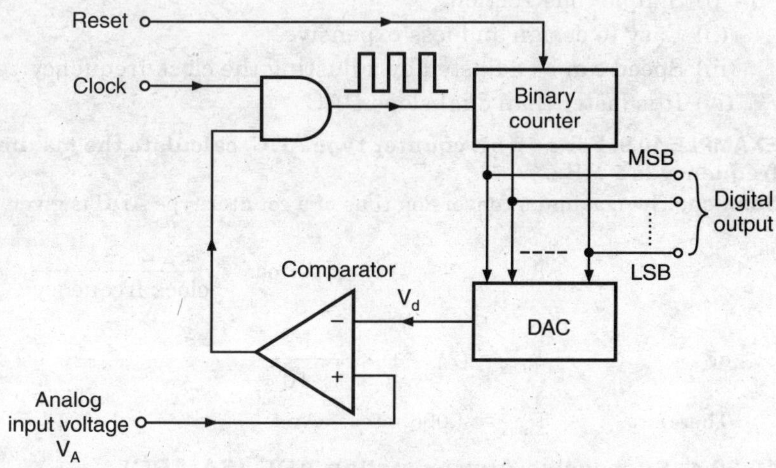

Fig. 10.37. Counter type ADC

(vi) When the analog input voltage is changed to a new value, a second reset pulse is applied to the counter to clear it and the counting begins again as shown in figure 10.38.

Disadvantages

The clock frequency should be low enough so as to give a sufficient time for the comparator to respond and for the DAC to settle down. However, due to the low clock frequency, the speed of conversion decreases and the conversion time increases. This will make the ADC slow. For the Digital ramp ADC, low speed is the most serious drawback. The conversion time does not remain constant. It is dependent on the analog input voltage V_A. This is another drawback. The conversion time can be as long as $(2^n - 1)$ clock cycle periods for high input voltage magnitudes.

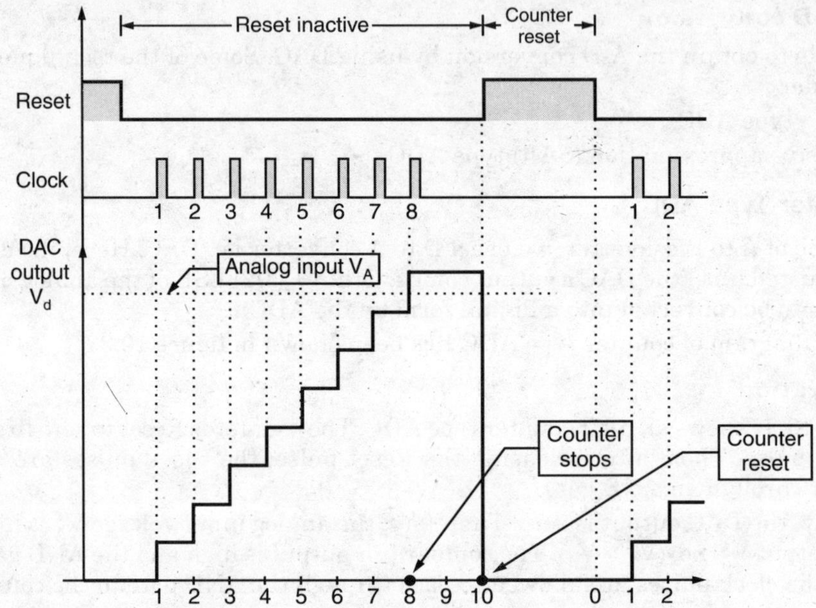

Fig. 10.38. Waveforms for counter type ADC.

Advantages

(i) Simple construction.
(ii) Easy to design and less expensive.
(iii) Speed can be adjusted by adjusting the clock frequency.
(iv) It is faster than dual slope DAC.

EXAMPLE 10.9. For a 12-bit counter type ADC, calculate the maximum conversion time if the clock frequency is 1 MHz.

Solution: The maximum conversion time of a counter type ADC is given by,

$$T_{max} = (2^n - 1) \text{ clock periods} = \frac{2^n - 1}{\text{clock frequency}}$$

or $$T_{max} = (2^{12} - 1) \times \frac{1}{1 \times 10^6}$$

Therefore, $$T_{max} = 4.095 \text{ msec.} \quad \textbf{Ans.}$$

10.20.4. Successive Approximation ADC (SA-ADC)

Successive approximation ADC is one of the most widely and popularly used ADC technique. This technique uses an efficient **code search** strategy to complete the n-bit conversion in just n-block periods. Thus, it takes much shorter conversion time as compared to the counter type ADC.

Block Diagram

The block diagram/circuit diagram of the successive approximation ADC has been shown in figure 10.39.

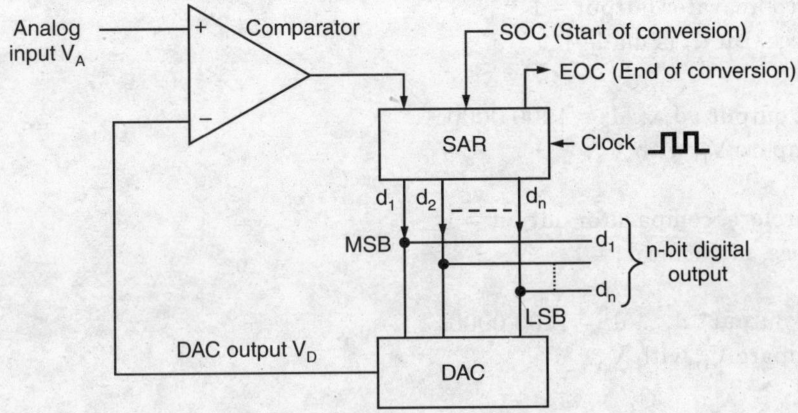

Fig. 10.39. Successive approximation ADC

10.20.5 Working Operation

(i) The successive approximation register SAR receives the comparator output, clock and start conversion signals and produces an n-bit digital output along with the end of conversion, i.e., EOC signal.

(ii) As soon as we initiate the SOC input, the SAR will set the MSB $d_1 = 1$ with all other bits to zero. Thus, the trial code at the SAR output is 1000 0000 for an 8-bit ADC. This trial code is then applied at the input of a DAC. The corresponding output of DAC i.e. V_D is applied to the comparator. If ($V_D < V_A$) i.e. if trial code is less than the correct digital representation then comparator output goes high which is applied to SAR.

(iii) In response to high comparator output, the MSB d_1 is maintained at 1 and the next lower significant bit d_2 is made 1. The trial code at the SAR output now becomes 1000 0000. The corresponding DAC output is compared with V_A and the process continues as explained in (ii).

(iv) However, for the first trial code of 1000 0000 only if $V_A < V_D$, then, the comparator output will go low i.e. 0. The SAR will respond to it by resetting its MSB bit d_1 to 0 and making the next bit $d_2 = 1$ so that the new trial code is 0100 0000.

(v) This procedure is repeated for all the subsequent bits one at a time, until all bit positions are tested.

(vi) As soon as the DAC output V_D crosses V_A (i.e. $V_D > V_A$), the comparator changes state and this is taken as end of conversion (EOC) command.

Let us suppose that for the given analog input voltage, the correct digital output is 1101 0010. Then, the SAC ADC reaches this word in the way shown on next page

Sequence of operation in the successive approximation ADC technique

1. Initially : SAR output : $d_1 .... d_8$ = 0000 0000
   But $V_D = 0$
   So, comparator output = 1
   Hence, Bit d_1 is made 1

↓

2. New SAR output : $d_1 ... d_8$ = 1000 0000.
   Compare V_D with V_A.
   $V_D < V_A$
   Therefore, comparator output = 1
   Hence, make bit $d_2 = 1$

↓

3. New SAR output : $d_1 ... d_8$ = 1100 0000.
   Compare V_D with V_A.
   $V_D > V_A$
   Comparator output = 1
   So, Make bit $d_3 = 1$.

↓

4. New SAR output : $d_1 ... d_8$ = 1110 0000.
   Compare V_D with V_A.
   $V_D > V_A$
   Comparator output = 0
   So, make $d_3 = 0$ and $d_4 = 1$

↓

5. New SAR output : $d_1 ... d_8$ = 1101 0000.
   Compare V_D with V_A.
   $V_D < V_A$
   Comparator output = 0
   So, make bit $d_5 = 1$.

↓

6. New SAR output : $d_1 ... d_8$ = 1101 1000.
   Continue the same process as under :

SAR output $d_1\ d_2\ d_3\ d_4\ \ d_5\ d_6\ d_7\ d_8$		Comparator output
1 1 0 1 1 0 0 0	$V_D > V_A$ →	0
	Make $d_5 = 0$ and $d_6 = 1$	
1 1 0 1 0 1 0 0	$V_D > V_A$ →	0
	Make $d_6 = 0$ and $d_7 = 1$	
1 1 0 1 0 0 1 0	→ Stop as $V_D = V_A$	

This shows that for $V_A > V_D$, the comparator output becomes high and for $V_A < V_D$, the comparator output is low.

Advantages

(i) The conversion time is equal to the n clock cycle period for an n-bit ADC. Thus, conversion time is very short. For example, for a 10-bit ADC with a clock frequency of 1 MHz, the conversion time will be 10×10^{-6}, i.e., 10 μsec only.

(ii) Conversion time is constant and independent of the amplitude of analog signal V_A.

Disadvantages

(i) The circuit is complex

(ii) The conversion time is more as compared to flash type ADC.

Applications

Due to the advantages mentioned earlier, the successive approximation ADC is most widely used in the microprocessor based data acquistion systems. IC ADC 0809 uses the concept of successive approximation.

10.20.6. Integrating Type ADCs

The integrating type ADCs use an integrator circuit. Therefore, if the analog input changes during the conversion, then, the ADC digital output will be proportional to the analog input averaged over the integration period. This is because integration is a process of averaging. There are two types of integrating ADCs:

(i) Charge balancing ADC

(ii) Dual slope ADC

Out of these, the dual slope ADC is very popularly used in all types of practical applications. Hence, let us study only the dual slope integrator type ADC.

10.20.7. Single Slope Integrating ADC

These converters are called as integrating ADCs because in these converters, either a reference voltage or the signal is integrated during the conversion process. The ADC output code is proportional to the value of the input averaged over the integration period.

Basic Principle

Basically, this technique involves comparison of the unknown input voltage with a reference voltage that begins at zero and increases linearly with time. The time required for the reference voltage to increase to the value of the unknown voltage is directly proportional to the magnitude of the unknown voltage, and this time period is measured with a digital counter. This is referred to as a **single ramp method**, since the reference voltage is sloped like a ramp.

Circuit/block diagram

Figure 10.40 shows the block diagram of single slope ADC.

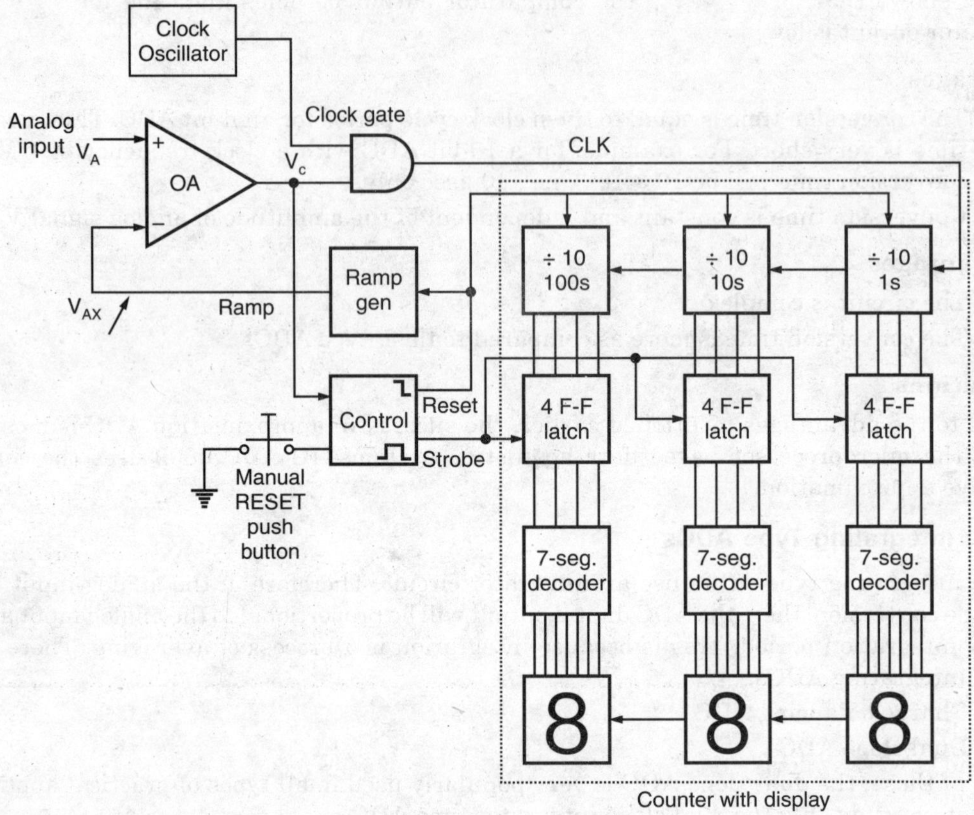

Fig. 10.40. Single slope A/D converter

Description

Dotted box shown is nothing but counter circuit with display. There are also control circuit, RESETS ramp generator as well as counter. The output of ramp generator is fed to comparator, which compares the same with analog input voltage. Here, V_C (output of comparator) controls gating to the clock and also informs control circuit about completion of the conversion.

Working Operation

First, let us observe the waveforms shown in figure 10.41(*a*). The working operation may be described as under :

Manual RESET, will reset ramp generator as well as counter. V_A has to be positive. RAMP begins at 0 V. As $V_{AX} < V_A$, $V_C = 1$ (HIGH). This will enable CLOCK gate allowing the CLK input, to be applied to the counter. Here it may noted that ramp generator may be of two types :

(a) Using DAC: This will resemble with counter-ramp ADC.

(b) Using a simple integrator as shown in figure 10.41(b).

For integrator, if V_i is constant, the output volage is given by equation $V_o = (V_i/RC)$ t. Since V_i, R and C are all constants, this is the equation of a straight line that has a slope (V_i/RC). As counter receives clock pulses, it will count up; and the RAMP continues upward. RAMP voltages rises till it reaches to V_A input voltage. At this point, time t_1, output $V_C = 0$ (LOW) and it will disable clock gate and counter cease to advance. The negative transition of V_C, simultaneously generates a strobe signal in the CONTROL box that shifts the contents of the three decade counters into the three 4FF latch circuit.

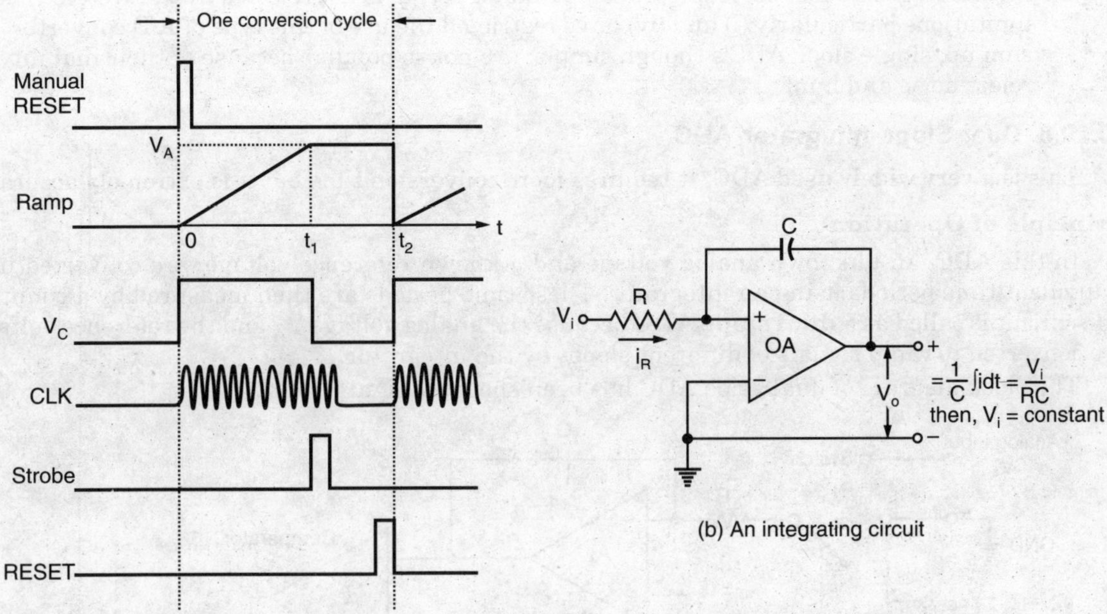

(a) Single slope A/D converter

(b) An integrating circuit

Fig. 10.41.

Shortly after that a reset pulse is generated (time t_2) by the CONTROL box that reset the RAMP and clears the decade counter to 0s and another conversion cycle begins. In the meantime, the contents of the previous conversion are contained in the latches and are displayed on the seven segment display.

Advantages of Single Slope ADC

(i) Linear conversion from voltage to time or frequency.

(ii) The output is unaffected by any sudden change in input.

Limitations of Single Slope ADC

Some of the limitations of this converter may be classified as under:

(i) The ramp generator is perhaps the most critical circuit in this type of converter. As it can be observed, the ramp waveform is a function of the value of the step voltage applied and the value of resistance and capacitance and their temperature coefficients. Even though precise resistors of negligible temperature coefficients are available, it is very difficult to find a precise and temperature stable capacitance. Also V, the amplitude of the step, in general, is a function of supply voltages. Hence, the ramp cannot be held very stable over a wide range. An amplitude stability of the order of 0.1% is achieved with difficulty.

(ii) The comparators are assumed to be prefect, i.e., their gain is infinite and offset zero. It is practically impossible to realize this.

In practice, to get good accuracy, comparator of less than 1 mV sensitivity is required. The response time of the comparator should be less than that of the period of the system clock. It is possible to compensate for any offsets by special methods.

(iii) This A/D converter cannot take bipolar signals. This difficulty can be overcome by some modifications.

(iv) Any noise of zero means, riding the input analog signal is not rejected by the system. These limitations particularly (i) and (iv) have restricted the use of this type of A/D converter. To sum up, single slope ADC's though simple, are not so popular because of their inability to reject noise and hum.

10.20.8. Dual Slope Integrator ADC

This is a very widely used ADC. It requires more conversion time but it is extremely accurate.

Principle of Operation

In this ADC, an unknown analog voltage and a known reference voltage, are converted into equivalent time periods using an integrator. These time periods are then measured by a counter. This circuit is called as a **dual slope ADC** because the analog voltage V_A and the reference voltage are converted to ramp signals of different slopes by the integrator.

The block diagram of dual slope ADC has been shown in figure 10.42.

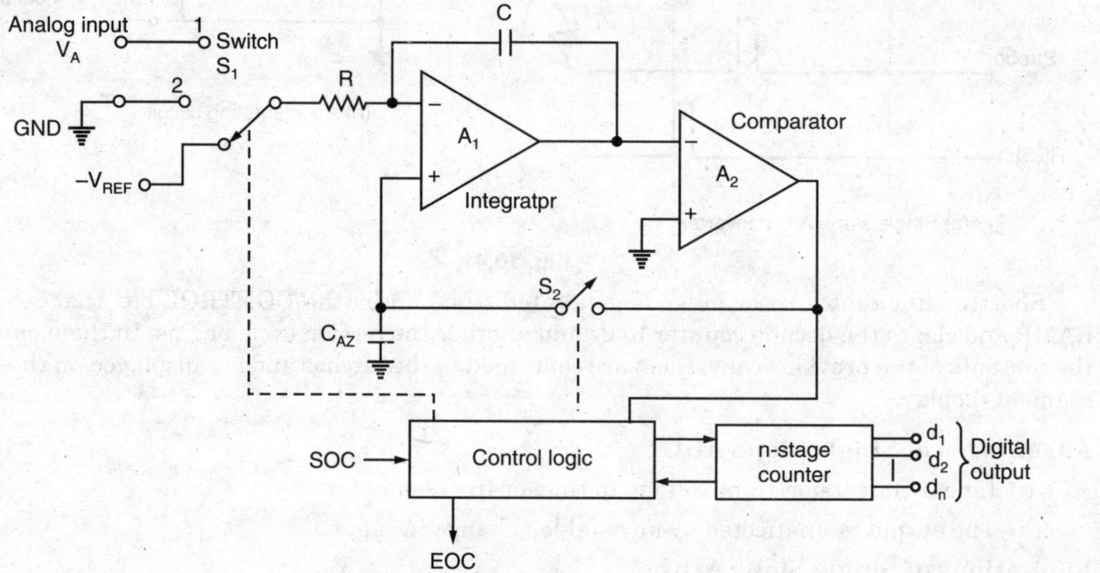

Fig. 10.42. A dual slope ADC

Circuit Description

The control logic accepts the SOC signal and generates EOC signal when the conversion is over. It also controls the two switches S_1 and S_2, out of which S_1 is a single pole three way switch whose one terminal to connected to analog voltage V_A, second one is connected to ground and the third one is connected to a negative reference voltage $-V_{REF}$.

Here, op-amp A_1 is a simple integrator and A_2 is a comparator. At the output of the counter, we get the n-bit digital output.

Working Operation

(i) Initially, let us assume that the integrator output voltage $V_o = 0$ and the counter is in RESET condition i.e. counter output is 00.

(ii) Let us consider figure 10.43. At $t = t_0$, switch S_1 is connected to ground and switch S_2 is closed. The capacitor C_{AZ} gets connected across the comparator output. Any offset voltage present in the op-amps will appear across the capacitor C_{AZ}. This will provide an automatic compensation for the input offset voltage of all the amplifiers. Therefore, integrator output voltage is zero for the interval t_0 to t_1 in figure 10.43.

(iii) At instant t_1, the SOC command is given to the control logic. Switch S_1 is connected to V_A and S_2 is open circuited. C_{AZ} then acts as a memory to hold the voltage required to keep the offset zero. Hence, C_{AZ} is known as the **autozero capacitor**.

(iv) From t_1 to t_2, this ADC will integrate the analog input V_A, for a fixed duration of 2^n number of clock cycles. This time interval is required for the counter to advance through all its possible output states, because for an n-bit counter, there will be 2^n possible output states. The counter output then reduces to zero. The time duration t_1 to t_2 is represented by T_1 in figure 10.43. The integrator output during this period is given by,

$$V_o = -\frac{1}{RC}\int_{t_1}^{t_2} V_A \, dt = \frac{-V_A T_1}{RC} \quad ...(10.20)$$

This expression represents a straight line with a slope of $-V_A/RC$. Thus, we get a decreasing ramp as shown in figure 10.43. The time period T_1 is thus represented by 2^n clock cycles. Therefore, $T_1 = 2^n \times T$

where, T = One clock cycle period.

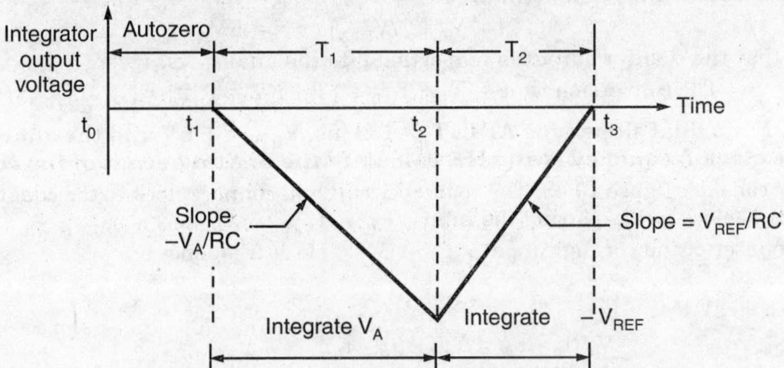

Fig. 10.43. Waveforms of a dual slope ADC

(v) At the end of interval T_1, the integrator input is connected to a fixed negative reference voltage $-V_{REF}$ via switch S_1. The integrator output now starts increasing towards zero with a positive slope as shown in figure 10.43. The slope is V_{REF}/RC for the duration t_2 to t_3. The counter starts counting from 0. The integration will continue till the integrator output is non-zero. At instant t_3, the integrator output reduces to zero, the comparator output goes from HIGH to LOW and the clock pulses given to the counter are stopped.

(vi) At t_3, the counter output shows a number corresponding to N clock cycles it has counted during period t_2 to t_3. Thus, this number N represents the time taken for integrator output to reduce from $-V_A$ to 0. Hence, N represents the desired digital output code proportional to the analog input V_A.

(vii) If V_A increases, then the integrator capacitor will charge to a higher negative voltage during the time interval T_1. Therefore, the time T_2 required to reduce the integrator output to zero increases. Therefore, the counter output count (N) will be higher. Thus, N is proportional to V_A.

EXAMPLE 10.10. Prove that the reading or output of the counter in the dual slope ADC is proportional to the analog input V_A and obtain the expression for conversion time.

Solution: Let us consider figure 10.43.

(i) $T_1 = t_2 - t_1 = 2^n$ counts $\times$ T, where T = One clock cycle period, and

$$T_2 = t_3 - t_2 = \text{Digital count } N \times T.$$

(ii) For an integrator, the output voltage V_o will be equal to V_1 at instant t_2.

Therefore,
$$V_o = V_1 = -\frac{1}{RC}.V_A T_1$$

or $$V_o = -\frac{1}{RC} V_A (t_2 - t_1) \quad ...(i)$$

V_1 can also be expressed as under :

$$-V_1 = -\frac{1}{RC} (-V_{REF}) (t_3 - t_2)$$

or $$V_1 = \frac{1}{RC} V_{REF} (t_2 - t_3) \quad ...(ii)$$

(iii) Equating equations (i) and (ii), we get,

$$-\frac{1}{RC} V_A (t_2 - t_1) = \frac{1}{RC} V_{REF} (t_2 - t_3)$$

or $$V_A (t_2 - t_1) = V_{REF} (t_3 - t_2) \quad ...(iii)$$

(iv) But $$(t_2 - t_1) = 2^n \times T \text{ and } (t_3 - t_2) = N \times T$$

or $$V_A = V_{REF} [N/2^n] \quad ...(iv)$$

In this expression, N is the count which appears at the counter output at instant $t = t_3$. As V_{REF} and 2^n are constant terms, equation (iv) shows that.

$$N = V_A [2^n/V_{REF}] \text{ or } N = KV_A \quad ...(v)$$

This means that the counter output is proportional to the analog voltage V_A.

$$\text{The conversion time} = T_1 + T_2 = 2^n T + NT = T [2^n + N] \quad ...(10.8)$$

EXAMPLE 10.11. For a dual slope type ADC, $T_1 = 100$ ms, $V_{REF} = +1$ V and the analog input voltage is $V_A = +1$V. If the clock frequency is 10 kHz, calculate the digital output of the counter.

Solution: Let uc consider figure 10.43. The control circuit will connect clock to the counter at instant t_2. The counter will count N clock pulses during the interval t_2 to t_3.

Therefore, counter output or digiral output = $(t_3 - t_2)$ clock frequency.

or $$N = T_2 f \quad ...(i)$$

Now, we have

$$V_A (t_2 - t_1) = V_{REF} (t_3 - t_2)$$

or $$V_A T_1 = V_{REF} T_2$$

or $$T_2 = \frac{V_A}{V_{REF}} T_1$$

Substituting this expression into equation (i), we get,

$$N = \frac{V_A}{V_{REF}} T_1 . f \quad ...(ii)$$

or $$N = 1 \times 100 \text{ ms} \times 10 \text{ kHz}$$

or $$N = 1 \times 0.1 \times 10 \times 10^3 = 1000 \text{ counts}$$

Thus, the digital count at the counter output is 1000 for $V_A = 1$V.

Advantages of Dual Slope ADC

(i) If the expression obtained for V_A is observed, then, it is observed that since V_{REF} and n are constant, V_A is proportional to the count reading N. It is independent of R, C and T. Thus, drifts in any of the components affects T_1 and T_2 in the same proportion and ADC output is unaffected.

(ii) If T_1 is chosen so that it is related to power supply period, the noise and hum on the input line get largely averaged out. Therefore, dual slope ADC is capable of rejecting noise and hum.

(iii) Offset correction can be introduced by a relatively simple circuit, facilitating auto-zeroing.

(iv) Low cost.

(v) Accuracy of the dual slope ADC can be of the order of 0.05%, which is adequate for most applications.

Disadvantage of Dual Slope ADC

The only major drawback of a dual slope type ADC is its long conversion time as compared to other ADCs.

10.20.9. Servo Tracking Type ADC

Tracking type ADC is actually the improved version of counter ADC. The block diagram of a tracking ADC is shown in figure 10.44 and the waveforms are shown in figure 10.45.

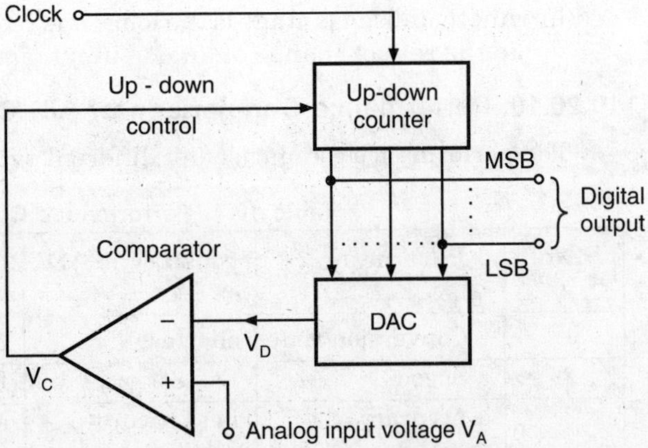

Fig. 10.44. Tracking type ADC

Working Operation

(i) Instead of a simple counter, this ADC consists of an up/down counter. The direction of counting (up/down) is controlled by the status of comparator output V_C.

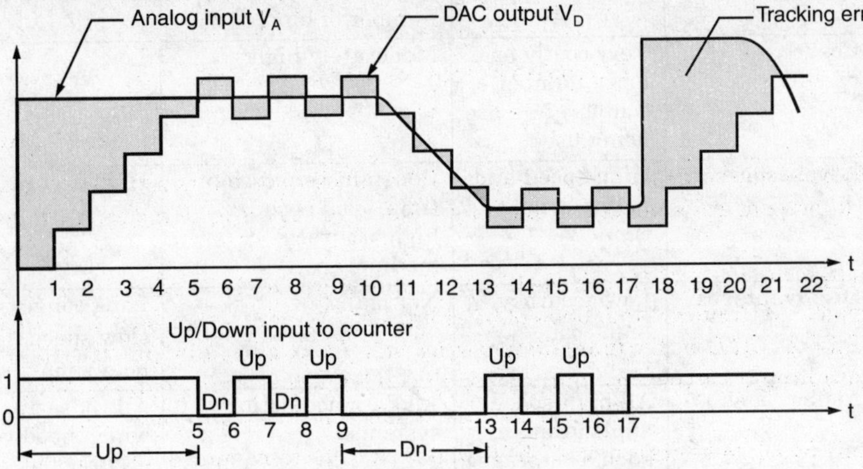

Fig. 10.45. Waveforms for tracking type ADC

(ii) The analog input V_A and DAC output V_D are compared by a comparator. If $V_A > V_D$, then comparator output goes high and the counter will count up.

(iii) The counter output is applied to the DAC input. Therefore, as the counter counts up, the DAC output will increase with each incoming clock pulse.

(iv) As soon as the DAC output becomes higher than the analog input ($V_D > V_A$), the comparator output goes low and the counter goes in Down counting mode.

(v) In this way, the staircase type DAC output V_D keeps tracking the variations in the analog voltage V_A. The corresponding counter output is the digital output of the tracking type ADC.

(vi) The process of switching between up and down counting modes continues and the digital output changes back and forth by + 1 LSB around the correct value.

Advantages and Disadvantages

(i) The only advantage of the tracking ADC is its simplicity.

(ii) The disadvantage is that as long as the analog input changes rapidly, the DAC output can not track it correctly within ± 1 LSB. But, when the analog input changes rapidly, the DAC output cannot track it giving rise to the tracking error as shown in figure 10.45.

(iii) Another disadvantage is the long time required to stabilize because it is directly proportional to the rate of change of analog input signal V_A.

10.20.10. Performance Comparison of A/D Converters

The performance comparison of different types of A to D converters is given in Table 10.4.

Table 10.4. Performance Comparison of A/D converters

S.No.	Parameter	Flash type ADC	Successive Approximation type ADC	Dual slope type ADC
1.	Conversion time	shortest	Moderately long	Longest
2.	Speed	Fastest	Faster than dual slope	slowest
3.	Accuracy of conversion	Less accurate	Moderately accurate	Very accurate
4.	Input hold time	Almost zero	Higher than that of flash type ADC. It depends on the number of bits.	Longest. Therefore, a sample and hold circuit is necessary.
5.	Cost	Very costly as a large number of comparators are being used.	Moderately high	Low
6.	Advantages	High speed and no input hold time	Constant conversion time, good speed, high accuracy, medium cost.	High accuracy, low cost.
7.	Disadvantages	Low accuracy, high cost	Not many.	Long conversion time, slow speed, long input hold time.
8.	Applications	All high speed applications such as storage CRO, fiber optic communication etc.	All data acquistion systems.	All those applications which need very high accuracy.

10.20.11. ADC Selection

In this article, let us discuss the general guidelines about selecting an ADC. While selecting the ADC, the following points should be considered.

1. The number of bits

Depending on requirement of resolution, we can select an 8 bit, 12 bit or 16 bit ADC. Higher the number of bits, better is the resolution

2. The required accuracy

The accuracy required for different applications is going to be different, so, selection of ADC should be done accordingly.

3. Speed or conversion time

The speed of an ADC should be large enough or conversion time should be small in order to convert the fast changing analog signal successfully into digital form.

4. Range of the input signal

The ADC full scale input range should match with the complete range of the analog input signal.

5. Cost budget

Cheaper ADCs do not provide high performance and high performance ADCs are costly. Hence, selection of an ADC is a compromise between performance and cost.

EXAMPLE 10.12. A dual-slope ADC has, for the integrator unit, the following components :

$R = 1 M\Omega$, and $C = 0.01\ \mu F$.

The reference voltage (V_R) is 10 V, and the fixed integration time is 5 ms. Find the conversion time for a 7.2 Volts input.

Solution: We have the output of the integrator after integration time $T_1 = 5$ ms, as under :

$$V_{o1} = \frac{1}{RC} V_A T_1$$

Hence, with input voltage $V_A = 7.2$ V, we get,

$$V_{o1} = \frac{1}{(1\times10^6)(0.01\times10^{-6})}\times7.2\times(5\times10^{-3}) \text{ Volts} = 3.6 \text{ Volts}$$

The time required to integrate $V_{o1} = 3.6$ V, to zero is given by,

$$T_2 = \frac{V_A}{V_R}\times T_1 = \frac{7.2}{10}\times(5\times10^{-3}) \text{ sec} = 3.6 \text{ ms}$$

Hence, total conversion time will be,

$$T = T_1 + T_2$$
$$= 5 \text{ ms} + 3.6 \text{ ms} = 8.6 \text{ ms.} \quad \textbf{Ans.}$$

EXAMPLE 10.13. A dual slope ADC uses a 16-bit counter and a 4 MHz clock rate. The maximum input voltage is + 10V. The maximum integrator output voltage should be – 8V when the counter has cycled through 2^n counts. Capacitor used in the integrator is 0.1 μF. Find the value of resistor of the integrator.

Solution:

We have, time period $T_1 = t_2 - t_1 = \dfrac{2^{16}}{4 \text{ MHz}}$

Also, $$T_1 = \frac{65536}{4\times10^6} = 16.38 \text{ msec.}$$

For the integrator, $\Delta V_o = -\dfrac{1}{RC} V_A \Delta t = -\dfrac{1}{RC} V_A (t_2 - t_1)$

Also, $$RC = -\frac{10}{(-8)}(16.38\times10^{-3}) = 20.47 \text{ msec.}$$

$$R = \frac{20.47\times10^{-3}}{0.1\times10^{-6}} = 204.7\,k\Omega = 205\ k\Omega \quad \textbf{Ans.}$$

EXAMPLE 10.14. The output of an 8-bit ADC is 1111 1111 when analog input voltage V_A = 2.55 Volts.

Calculate: (i) Resolution and (ii) Digital output for V_A = 0.64 V.

Solution: (i) Resolution

The resolution of an ADC is given by,

$$\text{Resolution} = 2^n = 2^8 = 256 \quad \textbf{Ans.}$$

We also know that

$$\text{Resolution} = \frac{V_{FS}}{2^n - 1} = \frac{2.55}{2^8 - 1}$$

$$\text{Resolution} = \frac{2.55}{255} = 10 \text{ mV/LSB.}$$

(ii) Digital output

$$\text{Digital output code} = \frac{V_A}{\text{Resolution}} = \frac{0.64\text{V}}{10 \text{ mV/ LSB}}$$

or $$\text{Digital output code} = \frac{0.64 \times 10^3}{10} = 64 \text{ LSB}$$

Digital input code = binary equivalent of 64

= 0100 0000 **Ans.**

10.21 ADC CHARACTERISTICS

Some of the important characteristics of ADC are as under :

(i) Resolution
(ii) Conversion time
(iii) Quantization error.

10.21.1. Resolution

Similar to DAC, the resolution of an ADC is defined in following two ways :

(i) Resolution may be defined as the maximum number of digital output codes. This is same as that of a DAC.

Therefore, Resolution = 2^n ...(10.21)

(ii) Alternatively, resolution may be defined as the ratio of the change in the value of the input analog voltage V_A, required to change the digital output by 1 LSB.

Thus, $$\text{Resolution} = \frac{V_{FS}}{2^n - 1}$$ ...(10.22)

10.21.2. Conversion Time

It is the total time required to convert the analog input signal into a corresponding digital output. As we know that the conversion time depends on the conversion technique used for an ADC. The conversion time is also dependent on the propagation delays introduced by the circuit components. Conversion time should ideally be zero and practically be as small as possible.

10.21.3. Quantization Error

As shown in figure 10.46, the digital output is not always the accurate representation of the analog input. For example, any input voltage between 1/8 to 2/8 of full scale will be converted to a digital word of 001.

This approximation process is called as quantization and the error due to the quantization process is called as **quantization error**. The maximum value of quantization error is + 1/2 LSB. The quantization error should be as small as possible. It can be reduced by increasing the number of bits. The increase in number of bits will also improve the resolution.

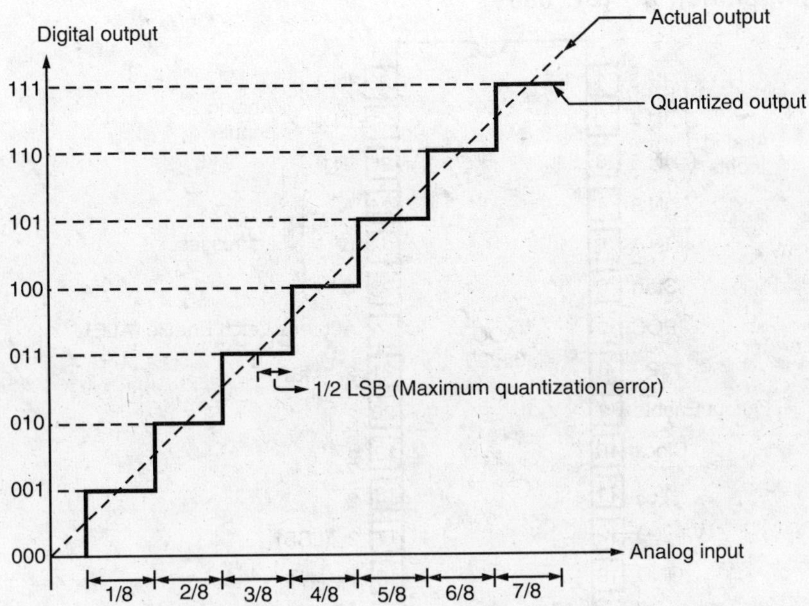

Fig. 10.46. A graph of input and output voltages for an ADC

10.22 MONOLITHIC IC ADC 0809

A large number of ADC ICs have been produced by the manufacturing companies like National semiconductors, Motorola, Intersil etc. to meet various demands such as speed of response resolution, compatibility and ease of interfacing with microprocessors etc. The National semiconductor produces ADC 0809 which is an 8-bit ADC whereas Intersil produces IC, ICL 7109 which is a 12-bit ADC. Let us discuss the ADC 0809 in details.

Principle of A to D Conversion in ADC 0809

The ADC 0809 operates on the successive approximation technique of A to D conversion. It is a CMOS device with 8-analog inputs, an 8 channel multiplexer and microprocessor compatible control logic. As the number of bits n = 8, it includes a 256 resistor voltage divider, a group of analog switches and a successive approximation register (SAR). As there are 8-analog channels, we can connect upto 8 analog inputs to this IC. However, due to the use of a multiplexer, at a time only one analog input will be converted into an equivalent 8-bit digital output. The analog input channels can be selected using the three address lines A, B and C.

Salient Features of ADC 0809

(i) Inbuilt 8 analog channels with multiplexer.
(ii) Zero or full scale adjustment is not required.
(iii) 0 to 5 V input voltage range with a single polarity 5 V supply.
(iv) Output is TTL compatible.
(v) High speed
(vi) Low conversion time (100μs).
(vii) High accuracy.
(viii) 8-bit resolution.
(ix) Low power consumption (less than 15 mW).
(x) Easy to interface with all microprocessors.
(xi) Minimum temperature dependence.

10.22.1. Pin Configuration of ADC 0809

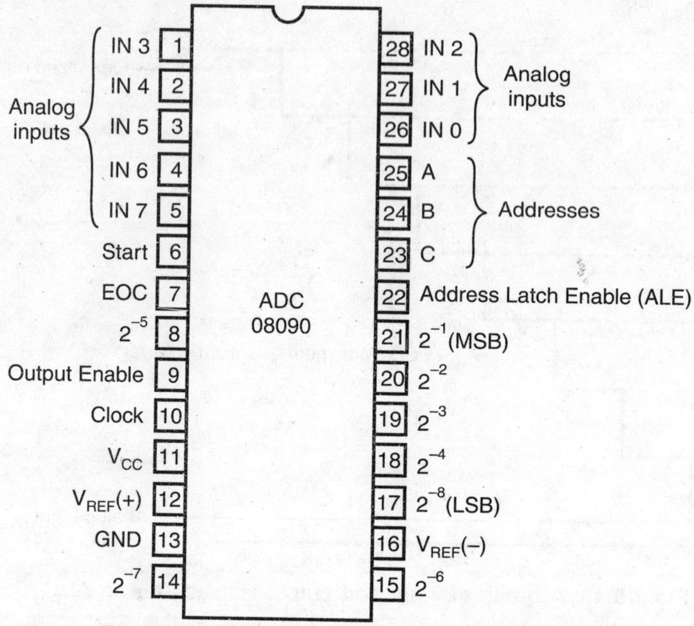

Fig. 10.47. Pin configuration of IC ADC 0809

Table 10.5. Selection of one of the analog inputs using the address lines A, B and C

Selection analog Channel	Address		
	C	B	A
IN 0	0	0	0
IN 1	0	0	1
IN 2	0	1	0
IN 3	0	1	1
IN 4	1	0	0
IN 5	1	0	1
IN 6	1	1	0
IN 7	1	1	1

Description

(i) Analog Inputs (IN 0 to IN 7) : Pin number 1 to 5 and 26 to 28, designated as IN 0 to IN 7 are the eight analog inputs of this IC. We can connect signals coming from eight different transducers to these inputs. Each one of these inputs will be converted to an 8-bit equivalent digital (binary word). However, these inputs are converted into digital form one by one and not all at a time. Hence, one of these eight inputs should be selected for conversion. This selection is done by means of the address pins A, B and C.

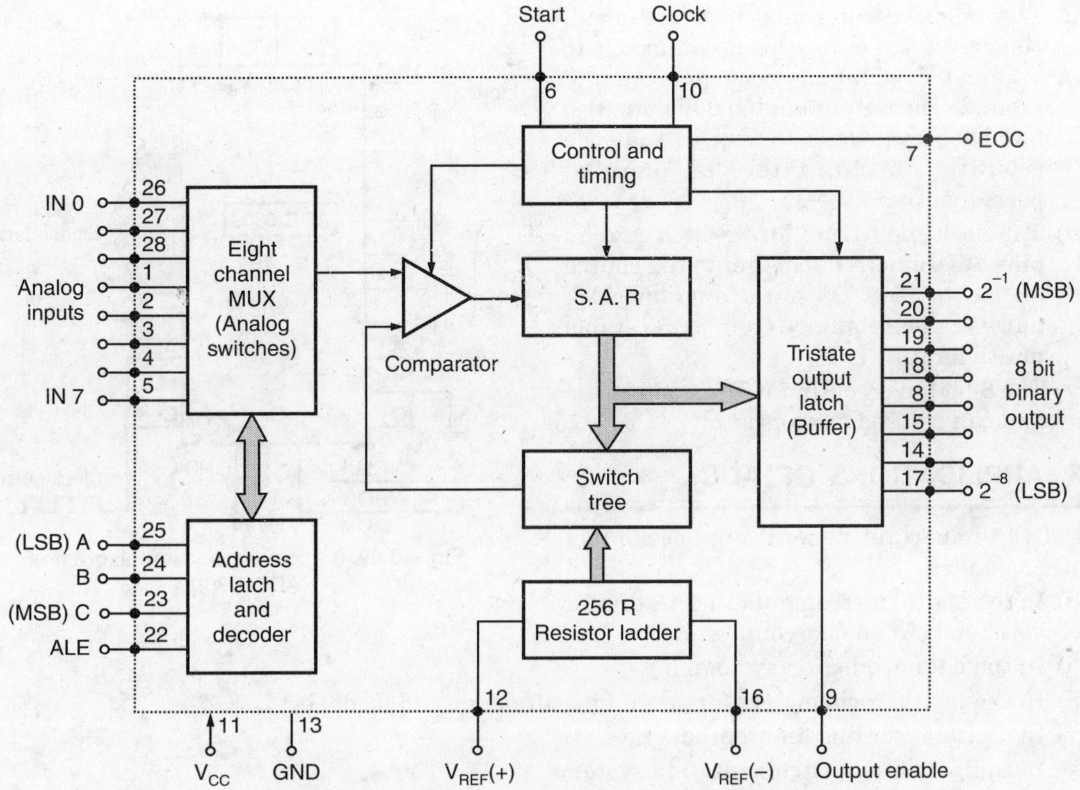

Fig. 10.48. Functional block diagram of IC ADC 0809

(ii) Address Pins A, B, C (Pin 23, 24, 25): These pins will decide or select one out of the eight analog inputs, for conversion into digital form. For example, if CBA = 010, then the IN 2 is selected and the analog signal at this input is converted to equivalent digital form.

(iii) Reference Voltage [V_{REF} (+) and V_{REF} (–)] : Depending on the desired polarity of the reference voltage, we can connect a positive or negative reference voltage externally to these pins.

(iv) ALE and Output Enable : As shown in the functional block diagram of ADC 0809 (figure 10.48), the address latch enable (ALE) input is useful in enabling the address latch which stores the address on lines A, B and C. The output enable pin, when activated will make the digital output available on the output pins (2^{-1} to 2^{-8}).

(v) Start and EOC : As explained earlier, we have to enable the start input to begin the A to D conversion. The end of conversion is indicated by EOC output.

(vi) Digital Outputs [2^{-1} to 2^{-8}] : The digital output is available to these pins 2^{-1} represents the MSB and 2^{-8} represents the LSB digital output. The functional block diagram of ADC 0809 is shown in figure 10.48 and the typical connection diagram has been shown in figure 10.49.

10.22.2 Typical Connection Diagram of ADC 0809

(i) Figure 10.49 shows the connection diagram for converting the analog input connected to IN 0. All the other analog inputs are left open. In order to select IN 0, the status of address lines ABC should be 000. Therefore, these pins have been connected to ground.

(ii) The start signal is generated either by using a switch or from a microprocessor. The EOC output can be connected to an LED or it can be returned back to the microprocessor.

(iii) The positive reference is being used. Hence, V_{REF} (+) has been connected to V_{CC} and V_{REF} (–) has been connected to ground. The output enable (OE) pin also has been connected to V_{CC} so that the tristate output latch is enabled permanently.

(iv) The clock and ALE (address latch enable) pins are connected together to the source of clock pulses. These rectangular clock pulses can be obtained from a 555 stable multivibrator.

(v) The 8-bit binary output is TTL compatible and can be used appropriately.

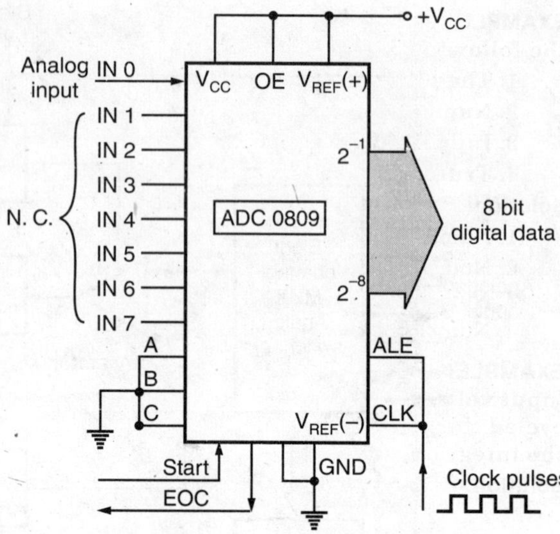

Fig. 10.49. Typical connection diagram of ADC 0809

10.23. APPLICATIONS OF ADC

Some of the important general applications of ADC are as under :

(i) In the digital instruments such as digital voltmeter, frequency counter etc.
(ii) In the data acquistion system.
(iii) In the digital tachometers for speed measurement and feedback.
(iv) In digital recording and reproduction.
(v) In computerised instrumentation systems.
(vi) NC and CNC machines.

FEW IMPORTANT SOLVED EXAMPLES

EXAMPLE 10.15. A 6 bit DAC has an input of 100101 and 10 V reference. Find
(i) Output voltage produced
(ii) Conversion resolution

Solution: Given that number of bits = 6,

V_{ref} = 10 V

Digital input = 100101

(i) Output voltage produced

We have $V_o = V_{ref}\,[d_1 2^{-1} + d_2 2^{-2} + d_3 2^{-3} + d_4 2^{-4} + d_5 2^{-5} + d_6 2^{-6}]$

where, d_1 = MSB = 1, $d_4 = 1$

$d_2 = d_3 = 0,\ d_5 = 0$

and d_6 = LSB = 1

Therefore, $V_o = 10\,[(1 \times 2^{-1}) + (1 \times 2^{-4}) + (1 \times 2^{-6})]$

or V_o = 5.78 Volts **Ans.**

We also have $V_{FS} = [(1 \times 2^{-1}) + (1 \times 2^{-2}) + (1 \times 2^{-3}) + (1 \times 2^{-4}) + (1 \times 2^{-5}) + (1 \times 2^{-6})]$

or V_{FS} = 9.8443 Volts

(ii) Conversion Resolution

Resolution can be obtained as under :

Resolution = $2^n = 2^6 = 64$

or Resolution = $\dfrac{V_{FS}}{2^n - 1} = \dfrac{9.8443}{2^6 - 1} = 0.1562$ Volts. **Ans.**

EXAMPLE 10.16. A 3-bit comparator type ADC is required for the conversion application. Find out the following:

1. **The number of comparators required.**
2. **Number of resistors in potential divider to generate reference voltage.**
3. **Input and output lines required for the priority encoder.**
4. **Truth table required to explain the working of ADC.**

Solution:

1. Number of comparators used = $2^N - 1 = 2^3 - 1 = 7$
2. Number of resistors in the potential divider = $2^N = 2^3 = 8$
3. Number of input lines for the priority encoder = 7
4. Number of output lines for the priority encoder = 3.

EXAMPLE 10.17. A dual slope ADC uses an 8 bit counter and a 3 MHz clock rate. The maximum input voltage is + 10V. The maximum integrator output voltage should be – 8 V, when the counter cycled through 2's counts. The capacitor used for the integrator is 0.01μF. Find the value of R of the integrator. How the RC product limit the speed of ADC.

Solution: Given : $n = 8$, $f = 3$ MHz, $V_{A\,(max)} = 10$ V

(i) Once clock cycle period = $T = \frac{1}{f} = \frac{1}{3 \times 10^6} = 0.33\ \mu s$

(ii) Time period $T_1 = 2^n \times T = 2^8 \times 0.33 \times 10^{-6} = 84.48\ \mu s$

(iii) The output voltage is given by,

$$V_o = \frac{-1}{RC} \int_{t_1}^{t_2} V_A\, dt$$

or

$$-8 = \frac{-1}{R \times 0.01 \times 10^{-6}} \int_{t_1}^{t_2} 10\, dt = \frac{10 \times 84.48 \times 10^{-6}}{R \times 0.01 \times 10^{-6}}$$

or

$$R = 10.56\ k\Omega \quad \text{Ans}$$

EXAMPLE 10.18. For the same ADC of previous example, if V_A = 4.129 V and V_R = 8V, obtain the digital count N and also obtain the digital output in 2's complement form.

Solution: We know that the digital count N is given by

$$N = V_A \cdot \left[\frac{2^n}{V_{REF}}\right] = 4.129 \left[\frac{2^8}{8}\right]$$

$$N = 132.128 \equiv 132 \quad \textbf{Ans.}$$

(ii) The digital output in the normal form = 1000 0100

Therefore, output in the 1's complement form = 0111 1011

Therefore, output in the 2's complement form = 0111 1100 **Ans.**

EXAMPLE 10.19. Derive an expression for the R– 2R D/A converter output voltage. An 8 bit DAC has an input of 1001 1011 and 10 V reference, find:

(i) Corresponding output voltage and

(ii) Conversion resolution

Solution: Given that : $n = 8$, $V_{REF} = 10$ V and input = 1001 1011

(i) Output voltage

$$V_o = KV_{ref}\,[d_1\, 2^{-1} + d_2\, 2^{-2} + d_3\, 2^{-3} + d_4\, 2^{-4} + d_5\, 2^{-5} + d_6\, 2^{-6} + d_7\, 2^{-7} + d_8\, 2^{-8}]$$

Assuming K = 1, we get

$$V_o = 10\,[(1 \times 2^{-1}) + 0 + 0 + (1 \times 2^{-4}) + (1 \times 2^{-5}) + 0 + (1 \times 2^{-7}) + (1 \times 2^{-8})] = 6.0546 \text{ Volts} \quad \textbf{Ans.}$$

(ii) The full scale output voltage V_{FS}

$$V_{FS} = V_R\,[(1 \times 2^{-1}) + (1 \times 2^{-2}) + ...+ (1 \times 2^{-8})]$$
$$= 10 \times 0.9961 = 9.961 \text{ Volts} \quad \textbf{Ans.}$$

(iii) Conversion Resolution

The conversion resolution can be obtained by one of the following methods :

$$\text{Resolution} = 2^n = 2^8 = 256$$

or
$$\text{Resolution} = \frac{V_{FS}}{2^n - 1} = \frac{9.961}{2^8 - 1} = 39.06 \text{ mV} \quad \textbf{Ans.}$$

EXAMPLE 10.20. Derive an expression for R – 2R D/A converter output voltage. An 8-bit DAC has an input of 1010 1011 and 10V reference. Find:

(i) Corresponding output voltage, and

(ii) Conversion resolution

Solution: First part of this example is identical to Example 10.5.

(i)s Output voltage

We have
$$V_o = KV_{ref}\,[d_1 2^{-1} + d_2 2^{-2} + d_3 2^{-3} + d_4 2^{-4} + d_5 2^{-5} + d_6 2^{-6} + d_7 2^{-7} + d_8 2^{-8}]$$

Assuming K = 1, we get,

$$V_o = 10\,[(1 \times 2^{-1}) + 0 + (1 \times 2^{-3}) + 0 + (1 \times 2^{-5}) + 0 + (1 \times 2^{-7}) + (1 \times 2^{-8})]$$
$$= 6.6796 \text{ Volts} \quad \textbf{Ans.}$$

(iii) Conversion resolution

The conversion resolution can be obtained as under :

(a) Resolution = $2^n = 2^8 = 256$ **Ans.**

(b) Resolution = $\frac{V_{FS}}{2^n - 1} \cong \frac{9.961}{2^8 - 1} = 39.06$ mV **Ans.**

SUMMARY

- In many such applications, the signal are not available in the digital form. Most of the physical quantities such as temperature, pressure, displacement, vibrations etc. are available in analog form.
- The digital representation of a signal makes storage possible, processing simpler and transmission easier. Therefore A to D conversion is necessary.
- In modern world of Electronics, the term 'digital' is usually associated with a computer. This is due to the fact that the term 'digital' is derived from the way the computers perform operations, by counting digits.
- For many years, the main application of digital electronics was only in the computer systems. But today, the digital electronics is used in many other application such as:
  1. Television
  2. Communication systems
  3. Radar, navigation and guidance systems
  4. Military systems
  5. Medical equipments
  6. Industrial process control
  7. Consumer electronics
- The main disadvantages of the weighted resistor DAC and R-2R ladder DAC is that the current flowing in different resistors will not remain constant. Instead, it will change with change in the digital input to DAC. This will cause more power dissipation in the resistors carrying more current which in turn causes heating of the resistors.
- In A to D converters, it is necessary that the analog input voltage V_A remains constant during the conversion time of ADC. Otherwise, the digital output produced by it will not be corresponding to the analog input signal. In order to maintain the analog voltage to be converted, constant during the conversion period, a technique called 'sampling' is used. Instead of using only sampler, we have to use a sample and hold circuit.

- A continuous time signal x(t) can be completely represented in its sampled form and recovered back from the sampled form if the sampling frequency $f_s > 2\ f_m$, where f_m is the highest frequency component present in signal x (t).
- The minimum sampling rate according to the sampling theorem is $2\ f_m$. This is called as the Nyquist rate. Therefore, Nyquist Rate = $f_{s(min)} = 2f_m$
- There are various circuits available for the A to D conversions. They use different principles of operation for conversion. Some of them are as under :
  (i) Flash type or simultaneous ADC
  (ii) Counter type ADC
  (iii) Tracking or servo type
  (iv) Successive approximation A to D converter
  (v) Dual slope A to D converter (Integrator type)
- Advantages of Single Slope ADC are as under :
  (i) Linear conversion from voltage to time or frequency.
  (ii) The output is unaffected by any sudden change in input.
- Some of the important general applications of ADC are as under :
  (i) In the digital instruments such as digital voltmeter, frequency counter etc.
  (ii) In the data acquistion system.
  (iii) In the digital tachometers for speed measurement and feedback.
  (iv) In digital recording and reproduction.
  (v) In computerised instrumentation systems.
  (vi) NC and CNC machines.

SHORT QUESTIONS WITH ANSWER

Q.1. Write the Advantages of R/2R Ladder DACs.

Ans. (i) Because we need resistors of only two values, it is easier to build this circuit accurately.

(ii) We can increase the number of bits just by adding more sections of same R/2R values.

(iii) The equivalent resistance to the right of each labelled node (A, B, C..) will be equal to 2R. Hence, current flowing downwards, away from each node is equal to the current flowing towards right.

(iv) Due to the small resistance spread (only 2 to 1), the R-2R ladder can be fabricated monolithically, with a high accuracy and stability.

Q.2. Write the Advantages of Inverted R-2R Ladder DAC.

Ans. (i) Current flowing through resistors remain constant irrespective of the switch position.

(ii) Total current gets divided equally at each node.

(iii) Current in each branch of R-2R ladder will remain constant and the node voltages are also constant.

(iv) The slow down effect due to stray capacitance is not present.

Q.3. Write the Statement of Sampling theorem.

Ans. A continuous time signal x(t) can be completely represented in its sampled form and recovered back from the sampled form if the sampling frequency $f_s > 2\ f_m$, where f_m is the highest frequency component present in signal x (t).

Q.4. What is Nyquist Rate?

Ans. The minimum sampling rate according to the sampling theorem is $2\ f_m$. This is called as the Nyquist rate.

Therefore, Nyquist Rate = $f_{s(min)} = 2f_m$

Q.5. Write the Types of A to D Converters.

Ans. There are various circuits available for the A to D conversions. They use different principles of operation for conversion. Some of them are as under :

(i) Flash type or simultaneous ADC
(ii) Counter type ADC
(iii) Tracking or servo type
(iv) Successive approximation A to D converter
(v) Dual slope A to D converter (Integrator type) .

Q.6. Write the advantages of Single Slop ADC.

Ans. Advantages of Single Slope ADC

(i) Linear conversion from voltage to time or frequency.

(ii) The output is unaffected by any sudden change in input.

Q.7. Write Applications of ADC.

Ans. Some of the important general applications of ADC are as under :

(i) In the digital instruments such as digital voltmeter, frequency counter etc.

(ii) In the data acquistion system.

(iii) In the digital tachometers for speed measurement and feedback.

(iv) In digital recording and reproduction.

(v) In computerised instrumentation systems.

(vi) NC and CNC machines.

(ii) Small size.

REVIEW QUESTIONS

1. What is meant by the term binary weighted resistors?
2. Explain the necessity of using the data converters.
3. Explain about the basic principle of a D to A converter.
4. Define and explain the term Resolution for D to A converters.
5. Define and explain the input-output equation of D to A converters.
6. What are the advantages of inverted R-2R ladder DAC?
7. Name various techniques used for DAC implementation.
8. With the help of neat circuit diagram, explain the working operation of binary weighted resistor DAC.
9. How can you compare various DAC techniques.
10. Derive the expression for the analog output voltage of a binary weighted resistor DAC and plot the output voltage for a 4-bit digital input word.
11. What are the drawbacks of binary weighted resistor DAC?
12. With the help of neat circuit diagram, explain the operation of an R-2R ladder DAC.
13. State and explain the important specifications of ADC.
14. R-2R ladder DAC is superior in performance to weighted resistance DAC-justify.
15. State and explain the advantages of R-2R ladder DAC.
16. With the help of circuit diagram, explain the operation of the inverted R-2R ladder type DAC.
17. State various important specifications of DAC and explain any four of them.
18. Explain in brief different types of ADC techniques
19. Explain the operation of dual slope ADC and state its advantages.
20. Compare the flash type, successive approximation and dual slope A/D converters.
21. Give the significance of conversion time and accuracy with respect to ADC.

□□□

MODEL QUESTION PAPERS FOR PRACTICE

MODEL QUESTION PAPER–I

DIGITAL CIRCUITS AND LOGIC DESIGN

Time : 3 Hours | *Maximum marks : 60*

Notes: Section-A is compulsory. Attempts any Four questions from section-B. Attempt any two questions from Section-C.

SECTION-A **(2×10=20)**

1. (a) Choose the correct or best alternative in the following :

(i) 8 is equal to signed binary number

(a) 10001000 (b) 10000000

(c) 00001000 (d) 11000000

(ii) De-Morgan's first theorem shows the equivalence of

(a) OR gate and Exclusive OR gate (b) NOR gate and Bubbled AND gate

(c) NOR gate and NAND gate (d) NAND gate and NOT gate

(iii) The digital logic family which has the lowest propagation delay time is

(a) ECL (b) CMOS

(c) TTL (d) PMOS

(iv) The device which changes from serial data to parallel data is

(a) Counter (b) Demultiplexer

(c) Multiplexer (d) Flip-Flop

(v) The device which changes BCD to Seven Segment is called

(a) Encoder (b) Multiplexer

(c) Decoder (d) Demultiplexer

(vi) In successive-approximation A/D converter, offset voltage equal to ½ LSB is added to the D/A converter's output. This is done to

(a) Improve the speed of operation (b) Reduce the maximum quantity error

(c) Increase the number of bits at the output

(d) Increase the range of input voltage that can be converted

(b) (i) Convert 2222 in Hexadecimal number.

(ii) Subtract-27 from 68 using 2's complements.

(iii) Divide $(101110)_2$ by $(101)_2$

SECTION-B **(4×5=20)**

2. State and prove De-Morgan's theorems.

3. Prove the following identities using Boolean algebra :

(*a*) $(A + B)(A + \bar{A}B)C + A(B + C) + \bar{A}B + ABC = C(A + B)\bar{A}(B + C)$

4. A combinational circuit has 3 inputs A, B, C and output F.F is true for following input combinations.

A is False, B is True

A is False, C is True

A, B, C are False

A, B, C are True

(a) Write Truth table for F. Use the convention True = 1 and False = 0.

(b) Write the simplified expression for F in SOP form

(c) Write the simplified expression for F in POS form

(d) Draw logic circuit using minimum number of 2-input NAND gates.

5. Minimise the logic function

$$F(A, B, C, D) = \Pi M(1, 2, 3, 8, 9, 10, 11, 14).D(7, 15)$$

Use karnaugh map, draw the logic circuit for the simplified function using NOR gates only.

6. What is necessity of Interfacing in digital ICs and what are the points to be kept in view, while interfacing between TTL gate and CMOS gate?

SECTION-C **(2×10=20)**

7. (a) What is a Multiplexer Tree? Why is it needed? Draw the block diagram of 32:1 Multiplexer Tree and explain how input is directed to the output in this sytem.
   (b) What is Decoder? Compare a decoder and a demultiplexer with suitable block diagrams.

8. (a) Draw the logic diagram of 4-bit Twisted Ring counter and explain its operation with the help of timing diagram
   (b) Design a MOD-3 synchronous counter using J-K Flip-Flops

9. (a) Differentiate between static MOS and Dynamic MOS RAM. Draw the circuit of a static MOS RAM cell and explain its working.
   (b) The capacity of 2K × 16 PROM is to be expanded to 16 K × 16. Find the number of PROM chips required and the number of address lines in the expanded memory.

SOLUTIONS

Section-A

1. **(a)**
   (i) (B) 00001000
   (ii) (B) NOR gate and Bubbled AND gate
   (iii) (A) ECL
   (iv) (A) COUNTER
   (v) (B) Decoder
   (vi) (A) Improive the speed of operation

   (b)
   (i) $(8AE)_{16}$
   (ii) 68 – (–27) = 68 + 27

	Binary Rep.	2's Comp.
68	1000100	
+ 27	0011011	

(iii)
```
        1001
101 ) 101110
      101
      ------
      000110
         101
      ------
         001
      ------
   remainder
```

MODEL QUESTION PAPER–II

DIGITAL CIRCUITS AND LOGIC DESIGN

Time : 3 Hours **Maximum marks : 60**

Notes: Section-A is **compulsory**. Attempts any **Four** questions from section-B. Attempt any **two** questions from Section-C.

SECTION-A (10×2=20)

1. (a) The decimal equivalent of Binary number 11010 is

(a) $(26)_{10}$ (b) 36
(c) 16 (d) 23

(b) 1's complement representation of decimal number of –17 by using 8 bit representation is

(a) 11101110 (b) 11011101
(c) 11001100 (d) 00010001

(c) The excess-3 code of decimal number 26 is

(a) 01001001 (b) 01011001
(c) 10001001 (d) 01001101

(d) How many AND gates are required to realize Y = CD + EF + G

(a) 4 (b) 5 (c) 3 (d) 2

(e) How mamy select lines will a 16 to 1 multiplexer will have

(a) 4 (b) 3 (c) 5 (d) 1

(f) How many flip flops are required to construct a decade counter

(a) 10 (b) 3 (c) 4 (d) 2

(g) Which TTL logic gate is used for wired ANDing

(a) Open collector output (b) Totem Pole
(c) Tristate output (d) ECL gates

(h) CMOS circuits consume power

(a) Equal to TTL (b) Less than TTL
(c) Twice of TTL (d) Thrice of TTL

(i) IC7490 contains flip flops

(a) 4 (b) 3 (c) 2 (d) 10

(j) In a RAM, information can be stored

(a) By the user, number of times (b) By the user, only once
(c) By the manufacturer, a number of times
(d) By the manufacturer only once

SECTION-B (4×5=20)

2. (a) Convert decimal 177.25 to octal number.

(b) Perform following subtraction.

(i) 11001-10110 using 1's complement
(ii) 11011-11001 using 2's complement

3. (a) Reduce the following equation using K-map :

$Y = \overline{ABC} + \overline{ACD} + \overline{AB} + ABC\overline{D} + \overline{A}BC$

(b) Write the expression for Boolean function :

F(A, B, C) = Σm(1, 4, 5, 6, 7) in standard POS form.

4. Explain working of three states TTL.

5. What do you mean by interfacing? Explain its need. How will you interface TTL to CMOS?

6. Implement the following function using a 3 line to 8 line decoder

S(A, B, C) = Σm(1, 2, 4, 7)
C(A, B, C) = Σm(3, 5, 6, 7)

SECTION-C **(2×10=20)**

7. (a) Explain the operation of Octal to Binary Encoder.
(b) Explain the working of master slave J-K flip flop.

8. (a) Explain how parallel In Serial Out (PISO) shift registers works.
(b) Design a mod-6 up counter.

9. (a) Explain how EPROM memory cell works.
(b) Explain the working of dual shape A/D converter.

SOLUTION

Section-A

1. **(a)** The correct answer is (a)
(Decimal equivalent of 11011 $= 1 \times 2^4 + 1 \times 2^3 + 0 \times 2^2 + 1 \times 2^1 + 0 \times 2^0$
$= 16 + 8 + 0 + 2 + 0 = (26)_{10}$

(b) The correct answer is (a)
1's complement representation of $(-17)_{10}$
Sign Magnitude Representation (–ve sign) take 1.
7-bit magnitude = 0010001
Hence, $(-17)_{10}$ = 10010001
In one's complement form $(-17)_{10}$ = 11101110

(c) The correct answer is (b).
26
33
59 – (01011001)

(d) The correct answer is (a)

(e) In multiplexer, we have n-data inputs, are output and m select inputs with $2^m = n$
Also, $n = 16$, $2^4 = 16$
∴ 4 select lines

(f) The correct answer is (c)
(Decade counter or MOD 10 Counter)
In general, m no. of flip-lops are required to construct Mod n counter where,
$n \leq 2^m$, m(no. of flip-flops) = 4 flip-flops required)

(g) The correct answer is (a)
(Wire ANDing means typing the outputs of gates together to obtain AND function. It is possible to connect the outputs of two or more gates together). Open Collector TTL gates is used for Wired ANDing.

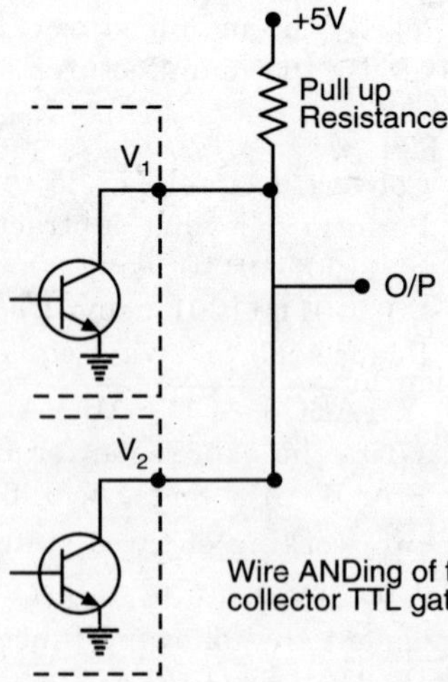

Wire ANDing of two open collector TTL gate

(h) CMOS circuits consume power less than TTL. In case of CMOS, power dissipation P_D = 0.1 mW. In case of TTL, power dissipation = P_D = 10 mW

(i) IC 7490 is a MOD-10 or decade counter. No. of flip-flops required to construct MOD-10 counter : $n \leq 2^m$, 4 flip-flops are required.

(j) RAM is a volatile memory, so, it loses the stored data when power is turned off.
In a RAM, information can be stored by the user, numbers of times.

PREVIOUS YEAR QUESTION PAPERS

THIRD SEMESTER EXAMINATION

DIGITAL CIRCUITS AND LOGIC DESIGN, May 2011

Time : 3 Hours *Maximum marks : 60*

Instruction to Candidates: Section-A is compulsory. Attempt any Four questions from section-B. Attempt any Two questions from Section-C.

SECTION-A **(10×2=20)**

1. (a) How many select lines are required for a 10 to 1 MUX?
   (b) List the various types of A/D converters.
   (c) How ROM is different from RAM?
   (d) Divide $(10101011)_2$ by $(101)_2$.
   (e) Construct the truth table for $Z = xy + \overline{xy}$
   (f) If A = 1010 and B = 1001, fin A – B using 2's complement method.
   (g) How sequential circuits are different from combinational circuits?
   (h) Compare TTL and ECL.
   (i) Compare synchronous counters with asynchronous counter.
   (j) Minimize the following expressions:
      (i) $A + \overline{(\overline{B} + \overline{C})}$
      (ii) $\overline{(A\overline{B}(C + \overline{D}) + \overline{C})}$

SECTION-B **(4×5=20)**

2. Minimize the following expressions using K-map
   (a) $Y = (A + B)(A + \overline{B})(A + \overline{C})$
   (b) $Y = \overline{A}B + A\overline{B}C + AB$
3. Design 8 : 1 MUX by using two 4 : 1 MUX.
4. What is race around condition? How it is avoided in Master Slave Flip-Flop?
5. Explain the working of successive approximation A/D converter.
6. Write a short note on the following :
   (a) CMOS
   (b) RTL

SECTION-C **(4×5=20)**

7. (a) Design full adder using logic gates.
   (b) Design EX-OR gate using NAND gates only.
8. Design mode-8 synchronous counter using T Flip Flops.
9. Explain the different types of ROMs. Discuss their advantages and disadvantages.

❑❑❑

THIRD SEMESTER EXAMINATION

DIGITAL CIRCUITS AND LOGIC DESIGN, Dec. 2012

Time : 3 Hours **Maximum marks : 60**

Instruction to Candidates:

1. Section-A is compulsory consisting of TEN questions carrying TWO marks each.
2. Section-B contains FIVE questions carrying FIVE marks each and students has to attempt any FOUR questions.
3. Section-C contains THREE questions carrying TEN marks each and students has to attempt any TWO questions.

SECTION-A (10×2=20)

1. Write briefly:
   (a) Represent the function (x,y,z) = y using K-map.
   (b) What is the advantage of R-2R ladder?
   (c) Explain clearly how a Flip-flop is used as a memory unit.
   (d) Rewrite the following expression in a form that required as few inversions as possible : $b'c + acd' + a'c + c(a + c)(a' + d')$
   (e) Given that $(79)_{10} - (142)_b$ determine the value of b.
   (f) What is edge-triggering?
   (g) What is the major advantage of D-flip flop over S-R?
   (h) What is the purpose of shift register?
   (i) What are De-Morgan's laws?
   (j) What is DAC?

SECTION-B (4×5=20)

2. What is the difference between 1's and 2's complement? Which is better of the two for the representation of the negative numbers and why?
3. Using 2's complement notation perform the following arithmetic operations using 8 bit register(s):
   (i) 25 + (–12) (ii) 17 – 6
   (iii) –18 – 16 (iv) – 8 × (18)
   (v) 12 – (–19)
4. What is Ring Counter? Explain the working of a 4-bit ring counter.
5. Mention all the Boolean laws and explain any two of them.
6. Explain the structure of a typical RAM cell.

SECTION-C (4×5=20)

7. (a) Explain the following terms used in a DAC:
   (i) Resolution (ii) Accuracy
   (iii) Settling time (iv) Mon tonicity
   (b) How are R-2R ladder used in DACs?
8. Given the function $T(w,x,y,z) = \Sigma(1,3,4,5,7,8,9,11,14,15)$
   (a) Use the K-map to determine the set of all prime implicants. Indicate specifically the essential ones. Find three distinct minimal expressions for T.
   (b) Assume that only unprimed variables are available. Construct a circuit which realizes T.
9. Design a decade counter using J-K Flip-flops.

INDEX

Win Prizes

ATTENTION READERS

Win Prizes

We request for your frank assessment regarding various aspects of the book as given below :

DIGITAL CIRCUITS & LOGIC DESIGN

By: Dr. SANJAY SHARMA

Please Fill up the Sheet in Neat and Clean Writing

(i) What topics of your syllabus, important for your examination point of view, are not covered in this book ?

(ii) What are the chapters wherein the treatment of the subject matter is not systematic or organised or updated ?

(iii) Have you come across misprints/mistakes/factual inaccuracies in the book? Please specify the chapters and the page numbers.

(iv) What other books on the same subject have you found/heard better than the present book? Please specify in terms of price and quality

(v) Further suggestions and comments for the improvement of this book.

(PS: If need be please attach a separate sheet to write your views)

OTHER DETAILS

(i) Who recommended you this book ? (Please tick (✓) in the box near the option relevant to you.

☐ Teacher ☐ Friends ☐ Bookseller

....

....

....

....

(ii) Name of the recommending teacher, his designation and address :

....

....

....

....

(iii) Name and address of the bookseller you purchased the book from :

....

....

....

....

(iv) Name and address of your institution (Please mention the University or Board as the case may be)

....

....

....

....

(v) Your name and complete postal address :

....

....

....

....

(vi) Course you are enrolled in :

....

....

....

....

The best assessment will be awarded each month: The award will be in the form of our publications as decided by the Editorial Board.

Please mail the filled up coupon at your earliest to :

S.K. KATARIA & SONS®
4760-61,23, Ansari Road,
Daryaganj, New Delhi-110002
☎: +91-11-23243489, 23269384
E-mail : katariabooks@yahoo.com
katariabook@gmail.com